Geophysical Monograph 277

Compressional Tectonics
Plate Convergence to Mountain Building

Tectonic Processes: A Global View, Volume 1

Editors

Elizabeth J. Catlos
İbrahim Çemen

This Work is a co-publication of the American Geophysical Union and John Wiley and Sons, Inc.

AGU
ADVANCING EARTH AND SPACE SCIENCE

WILEY

This edition first published 2023

Published under the aegis of the AGU Publications Committee

Matthew Giampoala, Vice President, Publications
Carol Frost, Chair, Publications Committee

For details about the American Geophysical Union visit us at www.agu.org.

The right of Elizabeth J. Catlos and İbrahim Çemen to be identified as the editors of this work has been asserted in accordance with law.

Registered Office
John Wiley & Sons, Inc., 111 River Street, Hoboken, NJ 07030, USA

Editorial Office
111 River Street, Hoboken, NJ 07030, USA

For details of our global editorial offices, customer services, and more information about Wiley products visit us at www.wiley.com.

Wiley also publishes its books in a variety of electronic formats and by print-on-demand. Some content that appears in standard print versions of this book may not be available in other formats.

Library of Congress Cataloging-in-Publication Data
Names: Catlos, Elizabeth J., 1971– editor. | Çemen, İbrahim, 1951– editor. | American Geophysical Union, publisher.
Title: Compressional tectonics : plate convergence to mountain building / Elizabeth J. Catlos, İbrahim Çemen.
Other titles: Geophysical monograph
Description: Hoboken, NJ : American Geophysical Union, 2023. | Series: Geophysical monograph series | Includes index.
Identifiers: LCCN 2022054054 (print) | LCCN 2022054055 (ebook) | ISBN 9781119773849 (hardback) | ISBN 9781119773887 (adobe pdf) | ISBN 9781119773863 (epub)
Subjects: LCSH: Plate tectonics. | Convergent margins. | Subduction zones. | Orogeny.
Classification: LCC QE511.4 .C64 2023 (print) | LCC QE511.4 (ebook) | DDC 551.1/36–dc23/eng20230302
LC record available at https://lccn.loc.gov/2022054054
LC ebook record available at https://lccn.loc.gov/2022054055

Cover Design: Wiley
Cover Image: © Inigo Cia/Getty Images

Set in 10/12pt Times New Roman by Straive, Pondicherry, India

CONTENTS

LIST OF CONTRIBUTORS

Philippe Agard
Sorbonne Université
CNRS-INSU
Institut des Sciences de la Terre Paris
Paris, France

Songjian Ao
State Key Laboratory of Lithospheric Evolution
Institute of Geology and Geophysics
Chinese Academy of Sciences
Beijing, China

Guillaume Bonnet
Sorbonne Université
CNRS-INSU
Institut des Sciences de la Terre Paris
Paris, France

Elizabeth J. Catlos
Jackson School of Geosciences
Department of Geological Sciences
The University of Texas at Austin
Austin, Texas, USA

İbrahim Çemen
Department of Geological Sciences
The University of Alabama
Tuscaloosa, Alabama, USA

Ling Chen
State Key Laboratory of Lithospheric Evolution
Institute of Geology and Geophysics
Chinese Academy of Sciences
Beijing, China

Yang Chu
State Key Laboratory of Lithospheric Evolution
Institute of Geology and Geophysics
Chinese Academy of Sciences
Beijing, China

David A. Foster
Department of Geological Sciences
University of Florida
Gainesville, Florida, USA

Bibek Giri
Department of Earth Sciences
Montana State University
Bozeman, Montana, USA

Ben D. Goscombe
Department of Geological Sciences
University of Florida
Gainesville, Florida, USA
and
Integrated Terrane Analysis Research
Adelaide, Australia

Mary Hubbard
Department of Earth Sciences
Montana State University
Bozeman, Montana, USA

Chong Ma
Mineral Exploration Research Centre
Harquail School of Earth Sciences
Laurentian University
Sudbury, Ontario, Canada

Aaron J. Martin
Division de Geociencias Aplicadas
Instituto Potosino de Investigación Científica y Tecnológica
San Luis Potosí, S.L.P., Mexico

Paul A. Mueller
Department of Geological Sciences
University of Florida
Gainesville, Florida, USA

Dia Ninkabou
Sorbonne Université
CNRS-INSU
Institut des Sciences de la Terre Paris
Paris, France

Jack C. Pashin
Boone Pickens School of Geology
Oklahoma State University
Stillwater, Oklahoma, USA

Alexis Plunder
BRGM (French Geological Survey)
Université d'Orléans
Orléans, France

Cécile Prigent
Insitut de Physique du Globe de Paris
Sorbonne Paris Cité
Université Paris Diderot
Paris, France

Delores M. Robinson
Department of Geological Sciences, and
Center for Sedimentary Basin Studies
The University of Alabama
Tuscaloosa, Alabama, USA

Mathieu Soret
Sorbonne Université
CNRS-INSU
Institut des Sciences de la Terre Paris
Paris, France
and
Institut des Sciences de la Terre d'Orléans
Université d'Orléans
Orléans, France

Morteza Talebian
Research Institute for Earth Sciences
Geological Survey of Iran
Tehran, Iran

Bo Wan
State Key Laboratory of Lithospheric Evolution
Institute of Geology and Geophysics
Chinese Academy of Sciences
Beijing, China

Philippe Yamato
Géosciences Rennes
Université de Rennes 1
Rennes, France

Donald J. Yezerski
Chevron Corporation
Houston, Texas, USA

Erdinç Yiğitbaş
Department of Geology
Çanakkale Onsekiz Mart Üniversity
Çanakkale, Turkey

Yücel Yılmaz
Department of Geology
Istanbul Technical University
Istanbul, Turkey

Zhiyong Zhang
State Key Laboratory of Lithospheric Evolution
Institute of Geology and Geophysics
Chinese Academy of Sciences
Beijing, China

PREFACE

TECTONIC PROCESSES: A GLOBAL VIEW

Tectonic processes control the shape and structure of the Earth and these processes affect the Earth's climate, geomorphology, magmatism, geochemistry, sedimentary environments, and economic resources. The evolution of these features through geologic times can be explained within the framework of 'plate tectonics' that is the overwhelmingly accepted unified theory of Earth sciences. This makes plate tectonics the central core discipline in geoscience research.

The Earth's lithosphere is divided into major plates and microplates that interact with each other along divergent (extensional), convergent (compressional), and transform (strike-slip) plate boundaries. Within the past few decades, Earth sciences have made tremendous advances in our understanding of plate tectonics processes along these three types of plate boundaries.

The overarching objective of the three-volume "Tectonic Processes: A Global View" is to present an up-to-date compendium and valuable reference for students of Earth sciences at all levels, from advanced undergraduate and graduate to doctoral and postdoctoral researchers, as well as for educators, policymakers, and research professionals in academia and industry. The collection contains three volumes:

• Volume 1: *Compressional Tectonics: Plate Convergence to Mountain Building*

• Volume 2: *Extensional Tectonics: Continental Breakup to Formation of Oceanic Basins*

• Volume 3: *Strike-Slip Tectonics: Oceanic Transform Faults to Continental Plate Boundaries*

COMPRESSIONAL TECTONICS

Major mountain belts on Earth, such as the Alps, Himalayas, Cordillera, and Appalachians, have been built by compressional tectonics processes during the continent-continent and arc-continent collisions. They are made of two major parts: a collisional fold-thrust belt and a peripheral foreland basin. Ever since the early field-oriented geological studies in the Alps and Appalachians, geologists have been working on providing a better understanding of collisional mountains and associated basins. This process accelerated after the development of the Plate Tectonics Theory in the mid-1970s.

This volume, *Compressional Tectonics: Plate Convergence to Mountain Building*, reviews present-day knowledge of the tectonic evolution of the Alpine-Himalayan and Appalachian belts. The papers in this volume are important for our scientific curiosity to understand our planet and for industrial development because collisional mountain belts contain many important economic resources, such as precious metals, rare earth elements, oil and gas, and coal. This volume will provide an essential reference for researchers and graduate students working on compressional tectonics. This volume contains three parts and 12 chapters.

Part I: Plate Convergence

Chapter 1 by Catlos and Çemen introduces compressional tectonics, including plate tectonics processes. Chapter 2 by Agard et al. look at subduction and obduction processes, which are vital for compressional tectonics along convergent plate margins. The chapter describes these processes with a close look into blueschists, eclogites, and ophiolites. Chapter 3 by Giri and Hubbard summarizes lateral heterogeneity in convergent mountain belt settings with an example along the Himalayan fold-thrust belt. Chapter 4 by Catlos and Çemen reviews the dynamics of subduction zone initiation in the Aegean Region.

Part II: Alpine-Himalayan Collision

Chapter 5 by Robinson and Martin summarizes the genesis of Himalayan stratigraphy and tectonic development. Chapter 6 by Catlos presents a thorough and well-developed explanation of the recent findings on metamorphism and compressional tectonics in the central Himalayas. Chapter 7 by Yılmaz et al. provides a detailed tectonic history of the Southeast Anatolian Orogenic Belt. Chapter 8 by Yılmaz et al. focuses on the tectonics of the Eastern Anatolian Plateau. Chapter 9 by Wan et al. presents a case study from Iran to explore when and why the Neo-Tethyan Ocean began to subduct along the Eurasian margin.

Part III: North America Mountain Building

Chapter 10 by Pashin looks at the stratigraphic and thermal maturity evidence for a break-backward thrust sequence in the Southern Appalachian Thrust Belt in Alabama. Chapter 11 by Çemen and Yezerski examines the subsurface geology of strain partitioning along the Ouachita fold-thrust belt, Arkoma Basin Transition Zone in southeastern Oklahoma and western Arkansas. Finally, in Chapter 12, Foster et al. review the extensional collapse of orogens with an example from the Southern Appalachian Orogen. This chapter deals with the last stages in the life of compressional orogeny, when a collisional mountain belt starts to collapse under its weight through gravitational collapse.

ACKNOWLEDGMENTS

We appreciate the contributions of all authors to this volume, and acknowledge the time, effort, and diverse perspectives of a large number of insightful reviewers. We thank the AGU and Wiley for allowing us to work on this multivolume project to present a global view of tectonic processes. We appreciate the unwavering support provided by Noel McGlinchey, Keerthana Govindarajan and Lesley Fenske from Wiley, Jenny Lunn from AGU's Publications Department. Rituparna Bose from Wiley initiated these volumes and provided continuous support in various stages of this project.

Elizabeth J. Catlos
Jackson School of Geosciences
The University of Texas at Austin, USA
İbrahim Çemen
Department of Geological Sciences
The University of Alabama, USA

Part I
Plate Convergence

1

When Plates Collide

Elizabeth J. Catlos[1] **and İbrahim Çemen**[2]

ABSTRACT

Compressional and contractional tectonics are of interest to various researchers, from rock mechanics and engineering to those studying the hazards, dynamics, and evolution of plate boundaries. We summarize here the terminology regarding deformation associated with compressional and contractional tectonics. We describe the now largely discarded geosyncline theory, which has its roots in contraction. Today, plate tectonics is the primary theory for explaining the processes shaping the Earth, including earthquakes, volcanoes, and mountain ranges. We emphasize the importance of subduction zones, the most extensive recycling system on the planet, and suture zones, complex boundaries marking the collision zone between two plates. The effects and hazards associated with convergent and collisional plate boundaries are felt far afield and for long distances.

1.1. INTRODUCTORY NOTES ABOUT TERMINOLOGY

Compressional tectonics is associated with terminology that will be defined here and in other sections. Rock *deformation* is divided into basic components: *translation* (change position), *rotation* (change orientation), *dilation* (change size passively), *dilatation* (change size in response to an active force), and *distortion* (change shape). In basic terms, compressive forces are directed toward each other (→←) and work to squeeze and shorten rock volumes (Fig. 1.1a). A rock responds to stress (σ), including compressional stress, by changing volume or form. Stress has units of force per area (N/m^2 or lb/in^2 or Pa, pascals) and is characterized by both a magnitude and an orientation on the surface in which it acts (Fig. 1.1). Deformed rocks result from total (finite) deformation over time, from which the forces and mechanisms that created rock textures or structures are interpreted.

[1] *Jackson School of Geosciences, Department of Geological Sciences, The University of Texas at Austin, Austin, Texas, USA*

[2] *Department of Geological Sciences, The University of Alabama, Tuscaloosa, Alabama, USA*

Stress can be *normal* (perpendicular to the surface) or *shear* (parallel). Anderson (1905, 1951) linked the orientation of the causative stress tensor relative to the Earth's surface relation to fault types in the upper, shallower levels of the crust (see reviews in Simpson, 1997; Sorkhabi, 2013). The magnitude of stress may not be the same in all directions and thus is defined as maximum $\sigma1 >$ intermediate $\sigma2 >$ minimum $\sigma3$.

A rock experiences *uniaxial* or *unconfined compression* when stress is directed toward the center of a rock mass, but more force is applied in one direction, and lateral component forces are zero ($\sigma1 > 0, \sigma2 = \sigma3 = 0$) (Fig. 1.1a). *Shortening strain* is the change in rock volume due to compressive stress. *Compressional stress* results in shortening features in rocks from the microscale to mesoscale, depending on the pressure-temperature (P-T) environment and the nature of the materials composing the rock.

Rock composition and temperature are critical factors in evaluating how rocks respond to compressional stress. The initial deformation rock experiences during gradually increasing stress is elastic. During this time, changes in stress induce an instantaneous change in sample dimensions as measured by strain. With *elastic* deformation, the strain completely disappears when the stress is

Compressional Tectonics: Plate Convergence to Mountain Building, Geophysical Monograph 277, First Edition.
Edited by Elizabeth J. Catlos and İbrahim Çemen.

DOI:10.1002/9781119773856.ch01

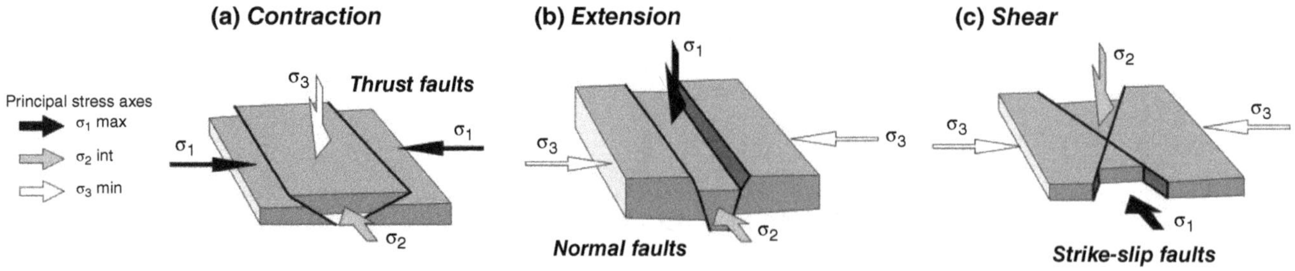

Figure 1.1 Relationship between stress axes and fault types (after Butler, 2021). (a) Rocks are displaced by contraction, (b) extension, and (c) shear. The principal stress axes are identified.

removed, and strain is recoverable (Twiss & Moores, 1992). Brittle materials fracture under compressive stress to release stored energy, whereas *ductile* materials deform and compress without failure. Rock layers may fold, or objects change shape, as evidenced by distributed strain. *Plastic* materials flow readily without fracture when the applied stress reaches conditions at or above specific yield stress (Twiss & Moores, 1992).

This book focuses on the processes that occur when the maximum compressive stress is in a horizontal orientation (*contraction*) (Fig. 1.1a). In this case, thrust faulting or folding occurs, shortening and thickening a rock or rock layers. Contraction is also observed as rocks lose volume through crushing, consolidation, or shear. In rock mechanics, *contraction* is a term that results in a reversible reduction in size, whereas *compression* results in a density increase. Contraction is exposed in the rock record as the shortening of rock layers, thrust or reverse faults, and folds. Thrust faults occur when rocks break along low angles and result in large earthquakes due to the large surface area affected by the process. In this volume, the dynamics of thrust faulting are described by Pashin (Chapter 10, Stratigraphic and Thermal Maturity Evidence for a Break-Back Thrust Sequence in the Southern Appalachian Thrust Belt, Alabama, USA) and Çemen and Yezerski (Chapter 11, Strain Partitioning in Foreland Basins: An Example from the Ouachita Fold-Thrust Belt Arkoma Basin Transition Zone in Southeastern Oklahoma and Western Arkansas). Reverse faults result from the rock breaking at high angles in response to compression (Fig. 1.1a). Normal faults occur when the maximum compressive stress is vertical, horizontally extending, and vertically thinning rock (Fig. 1.1b). We cover extensional tectonics in the second volume and strike-slip tectonics (Fig. 1.1c) in the third volume of this series.

1.2. SETTING THE STAGE: GEOSYNCLINE THEORY

The origin of mountains on the Earth has always been debated among philosophers, geographers, and Earth scientists. Since the late 1960s, plate tectonics has been a unifying theory of mountain building (see section 1.3). Although many theories before plate tectonics were proposed regarding the formation of mountains, one that received wide recognition is the *geosynclinal or geosyncline theory*, commonly attributed to James Hall and his coworkers (Hall, 1859; Dana, 1873; see Fisher, 1978; Frankel, 1982; Friedman, 2012; De Graciansky et al., 2011; Kay, 2014). James Hall and coworkers based their theory on field observations in the Appalachian Mountains of New York and Pennsylvania, where they observed features characteristic of shallow-water sedimentation, such as ripple marks, mud cracks, and shallow-water fossils in sedimentary units that were over 10,000 m in thickness. But they knew these sediments were deposited in basins where water was only about 100 m deep. Consequently, Hall proposed that these thick Paleozoic shallow-water sediments must have been deposited in a slowly subsiding basin, receiving a thick succession of shallow-water sediments as it subsided. They coined the term *geosyncline* for this subsiding basin (Fig. 1.2) (Glaessner & Teichert, 1947; De Graciansky et al., 2011). The formation can be further divided into miogeosynclines, eugeosynclines, and orthogeosynclines, depending on the rock strata, location, and nature of the mountain system.

To explain the deformation that they observed in the Appalachian Mountains, Hall and his coworkers proposed that after thick sediments accumulated, horizontal compressional forces directed from the seaward side of the geosyncline squeezed the sediments, shortened, and thickened the crust, and produced a high-standing mountain chain while pushing much of sediments into the crust. In the 1873, Dana proposed that the deeply buried sediments melted in high temperature and pressure conditions and generated magma that intruded into the sediments. During the 1890s and early 1990s, geosynclinal theory was widely recognized for explaining the formation of mountain chains, like the Appalachians, Ouachitas, Cordillerans, Urals, Alps, and Himalayas (see Mark, 1992; Şengör, 2021). However, Schaer and Şengör (2008) indicate that the geosyncline theory is not a "made in America" concept. For example, geologists in the Alps had noted the behavior of sediments in deep-water basins

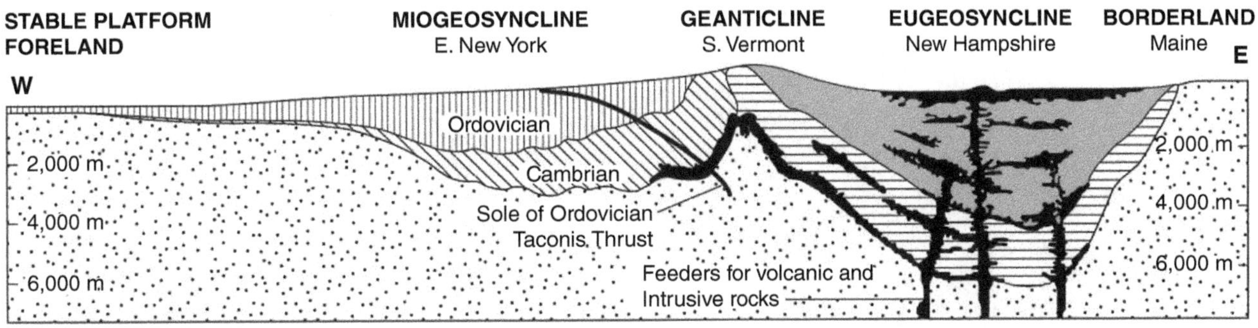

Figure 1.2 A diagram showing an imagined cross section of the northern Appalachians before the Appalachian Orogeny (after Kay, 1948). A geanticline is a ridge that separates two belts of sedimentary rocks. A eugeosyncline is a deep-water trough with abundant volcanic rocks and deep-water sediments. A miogeosyncline is a basin of mainly shallow-water sediments (De Graciansky et al., 2011).

and ascribed their formation to synclines (e.g., 1828, Elie de Beaumont) (Schaer, 2010).

In 1912, Alfred Wegener published a paradigm-changing hypothesis in his book *The Origin of Continents and Oceans*. His hypothesis, called *continental drift*, suggested that the Earth's ocean basins and continents changed their positions throughout geological time. Wegener also suggested that all of the continents were together at one time. He called this supercontinent Pangea. Most scientists did not accept Wegener's idea of continental drift in the early part of the first half of the 20th century because his lines of evidence were thought to be mostly coincidental. The acceptance of his idea had to wait until the late 1960s, when the data collected from the ocean floor provided evidence that the oceans were indeed temporary: they were opening and closing, and continents were drifting.

Vine and Matthews (1963) worked on magnetic lineations obtained on either side of the mid-Atlantic ridge south of Iceland. They proposed that new oceanic crust is created by the solidification of magma injected and extruded at the crest of a Mid Ocean Ridge (MOR). When this magma cools below the Curie point, ferromagnetic behavior becomes possible, and magnetite in the basalt gets magnetized. The solidified magma (basaltic rocks) acquires a magnetization with the same orientation as the geomagnetic field. They based their hypothesis on the presence of stripes of magnetic anomalies on either side of the MOR. Their findings and those of others who studied the aspects of the geophysical dynamics of MOR gave birth to a unifying theory of Earth sciences called plate tectonics (see review by Marvin, 2005).

Although geosyncline theory for the evolution of the Earth is today largely discarded, the term is still retained by geologists describing specific basins (e.g., Arabian Gulf geosyncline, Elobaid et al., 2020; Adelaide Geosyncline of South Australia, Preiss, 2000; West Siberian geosyncline, Yolkin et al., 2007). Today, the term is a historical, practical, descriptive, and nongenetic term not meant to be associated with interpretations of a specific tectonic environment (e.g., Preiss, 2000).

1.3. PLATE TECTONICS AND COMPRESSIONAL MOTION

1.3.1. What Are Plates?

Plate tectonic theory divides the Earth into rigid layers of crust and upper mantle (*lithosphere*) above the Earth's *asthenosphere*, which can flow at much lower stress levels (Fig. 1.3) (e.g., Anderson, 1995). By their original definition, *tectonic plates* are rigid and include ocean or continental crust or a combination. However, plates do not always correspond with continental margins (e.g., Gordon, 1998). Identifying tectonic plates requires examining geological, geophysical, and geodetic data at multiple sources and scales. These include detailed field mapping and structural analysis, earthquake fault plane solutions, estimates of average rates of plate and fault motion, transform fault azimuths, very long baseline interferometry, satellite laser ranging, Doppler Orbitography and Radiopositioning Integrated by Satellite, and Global Positioning System data (DeMets et al., 2010; Harrison, 2016). Information from these sources helps identify how many plates exist, which has dramatically increased with the technology used to identify them (e.g., n = 52, Bird, 2003; n = 159, Harrison, 2016). Only 25 tectonic plates occupy 97% of Earth's surface (DeMets et al., 2010). The other 3% are *microplates*, defined as relatively small-scale, rigid, geological blocks with a consistent motion or behavior in present-day space with boundaries that behave as plate boundaries (Li et al., 2018). Microplates are located at the major plate boundaries but rotate and behave independently (Hey, 2021). These features may grow into larger plates

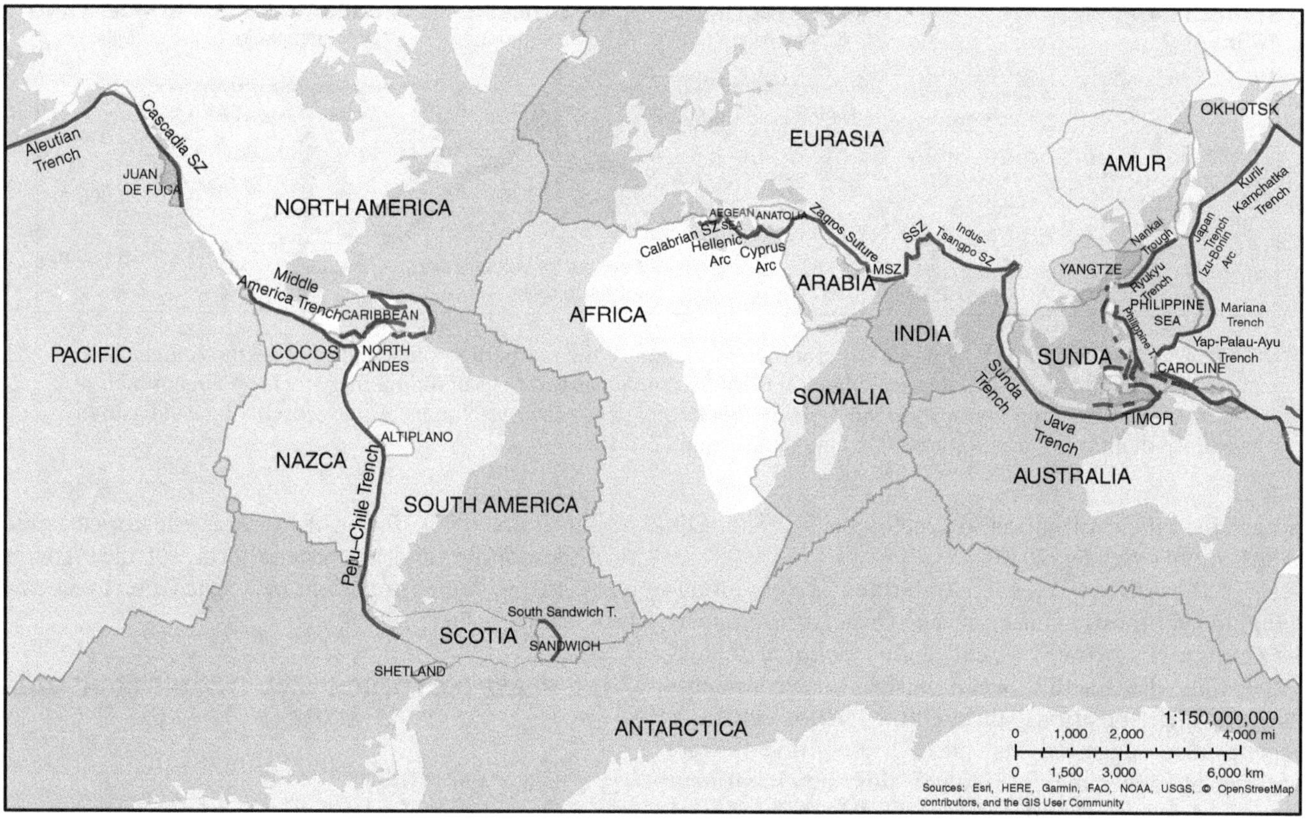

Figure 1.3 Map of the Earth showing present-day plate configurations and convergent and collisional plate boundaries. Labels are included for some plates and plate boundaries. The map was created using ArcGIS (ESRI) with data from Bird (2003). Convergent and collisional plate boundaries are identified (Coffin et al., 1998). Abbreviations: SZ = suture zone; SSZ = Shyok suture zone; MSZ = Makran suture zone; Philippine T = Philippine Trench.

over time (Seton et al., 2012; Boschman & van Hinsbergen, 2016) or are transient (Hey, 2021).

Plates are composed of *oceanic lithosphere* and/or *continental lithosphere*. The *lithosphere* is the Earth's strong, solid outer shell (Anderson, 1995). The oceanic lithosphere is produced at ocean ridges by decompression melting of upwelling mantle, which cools, thickens, and increases in age as it moves away from ridges (e.g., Condie, 2022). The process creates *midocean ridge basalt (MORB)*. This most abundant magma type can be recognized and classified geochemically by source and degree with interaction material recycled in the mantle, spreading rate, and even ocean basin (e.g., Anderson, 1995; Perfit, 2001; Wallace, 2021). The oceanic lithosphere covers ~60% of the Earth's surface (Minshull, 2002; Fowler, 2012), with ocean crust on average 6–8 km thick. Oceanic crust averages 7.1 ± 0.8 km thick away from fracture zones and hot spots and ranges from 5.0–8.5 km (White et al., 1992).

The continental lithosphere is the part of the continental crust and upper mantle that can support long-term geological loads (Anderson, 1995). This layer covers ~40% of the Earth and has a granitic upper portion (32–56 km thick) underlain by mantle peridotite (96–130 km thick) (DiPietro, 2013). The origin of continental lithosphere differs significantly from mantle lithosphere in that the modification of existing rock creates it through thinning or replacement (Condie, 2005; Sleep, 2005; Eagles, 2020; Şengör et al., 2021). On average, continents are thought mainly to be intermediate (andesitic) in composition with a felsic upper crust and mafic lower crust (Palin et al., 2021). However, based on seismic refraction data, the lower crust may be more felsic in some locations (49–62 wt% SiO_2; Gao et al., 1998; Hacker et al., 2015). This portion of the Earth experiences complex and dynamic interactions that can significantly change its nature, including metamorphism, mixing with mantle-derived melts or other reservoirs, and delamination (e.g., Kay & Mahlburg-Kay, 1991).

Craton lithosphere or *continental platforms* are thick (~200 km) portions of continental thicknesses but differ in age and the mantle dynamics beneath them. Cratons formed during the Archean and platforms are younger features, not underlain by a buoyant mantle that drives convection (Sleep, 2005). Continental lithosphere can thin through extension, orogenic collapse, or underlying mantle processes (e.g., Dewey, 1988; Ruppel, 1995; Lee et al., 2000; Rey et al., 2001; Lavier & Manatschal, 2006). The subcontinental lithospheric mantle (SCLM) can also be sheared away by cold, shallowly subducting crust, which has an impact on plate buoyancy (e.g., Hernández-Uribe & Palin, 2019) and magmatism (e.g., Wei et al., 2017). Although the oceanic lithosphere assumes the plates are located underwater, some continental lithospheric plates are underwater (e.g., Aegean microplate).

1.3.2. What Are Plate Boundaries?

Plate boundaries are edges that mark the contact between two plates. Plate boundaries are classified into *divergent* (*extensional*, plates move apart), *conservative* (*strike-slip* if plates slide past each other, and *transform* if they also connect divergent plate boundaries), *convergent* (plates move together and a plate is consumed in a subduction zone), or *collisional* (plates move together and plates are joined at a *suture zone*) (see reviews in Cox & Hart, 2009; Le Pichon et al., 2013). Convergent and collisional plate boundaries are classified into a single group (convergent) by most introductory textbooks. These textbooks will also discuss conservative plate boundaries as transform only, with faults classified as strike-slip. Figure 1.3 highlights the locations of convergent and collisional plate boundaries on Earth as bolder lines, many of which are in the Northern Hemisphere. Most of Earth's tectonic plates, including many smaller microplates, have a portion in compression (Harrison, 2016).

Although plate boundaries are classified into end-member types, convergent and collisional plate boundaries may also be affected by strike-slip or normal deformation, especially when the plates interact obliquely (Fitch, 1972; Haq & Davis, 1997; Burbidge & Braun, 1998; Bevis & Martel, 2001; Gaidzik & Więsek, 2021). It has long been known that a significant number of plate boundaries have relative velocity vectors that are oblique from normal (>22°, n = 59%) and parallel to the boundary (n = 14%) (e.g., Woodcock, 1986). *Composite transform convergent (CTC)* plate boundaries define convergent margin plate boundaries that are affected by regional strike-slip faulting along trends that parallel or subparallel the boundary (Ryan & Coleman, 1992). Examples of CTC boundaries may be primarily at *subduction zones* (Fig. 1.4). Subduction zones occur when two lithospheric plates converge, and one plate abruptly descends beneath the other (e.g., Stern & Gerya, 2018; Crameri et al., 2020). CTC boundaries have been identified near *volcanic island arcs* at the Aleutian Ridge and the Philippines (Ryan & Coleman, 1992). Volcanic island arcs are an arcuate continuation of islands with present-day prominent volcanic and seismic activity (Sugimura & Uyeda, 1973). CTCs are present if strike-slip faults develop in the overriding plate (Fig. 1.4) (Beck et al., 1983; McCaffrey, 1993; Bevis & Martel, 2001). The rate of strike-slip faulting in subduction zones is governed by both convergence obliquity and rate (Jarrard, 1986). Normal motion and strike-slip fault motion in oblique subduction zones have been observed to generate large earthquakes and significantly contribute to seismic hazards (e.g., Fitch, 1972; McCaffrey, 1996; McCaffrey et al., 2000; Moreno et al., 2008; Melnick et al., 2009; Gaidzik & Więsek, 2021).

Convergent and collisional plate boundaries are characterized by distinct topographical or bathymetric features (Fig. 1.5). Those associated with the oceanic lithosphere will show deep ocean trenches, shallower troughs, ridges of sediment accretion, volcanoes, including seamounts and island arcs, fault lines, and ridges. The U.S. Board on Geographic Names (BGN) Advisory Committee on Undersea Features (ACUF) recommends names of undersea features and official standard names for use in the field or on hydrographic and bathymetric charts. Plate boundaries are often named based on those adopted by the ACUF or by their location, followed by the topographical features they generate (trough, trench, ridge), shape (arc), or nature of deformation (suture, subduction).

However, based on the researcher's focus, the same convergent plate boundary may have several names. For example, the Hellenic subduction zone extends ~1,200 km from approximately 37.5°N, 20.0°E offshore of the island of Zakynthos to 36.0°N, 29.0°E offshore of the island of Rhodes (Ganas & Parsons, 2009; Le Pichon et al., 2019). The same feature is sometimes referred to as the Aegean subduction zone (Wortel et al., 1990; Berk Biryol et al., 2011; Crameri et al., 2020), Hellenic arc (Ganas & Parsons, 2009; Royden & Papanikolaou, 2011), or Hellenic arc and trench system (Le Pichon & Angelier, 1979; Papadopoulos et al., 2007). The ACUF assigns the same feature to the Hellenic Trough, Hellenic Trench, or Ionia Basin.

Trenches, troughs, and *arcs* are often associated with ocean-continent or ocean-ocean subduction zones. Trenches are deeper water regions and exist on the oceanic side of an island arc, whereas a shallow sea exists on the continental side (Figs. 1.4 and 1.5). Trenches have steep sides like river gorges (e.g., Bellaiche, 1980). Troughs are asymmetrical shallow depressions at the foot of a slope. For example, the Nankai Trough near Japan (Figs. 1.3 and 1.5) has a maximum water depth that does not exceed 5,000 m (Yamano et al., 1984). In contrast, the Izu-Bonin Trench reaches 9,780 m (e.g., Bellaiche, 1980). Arcs are curved subduction zones with the curvature associated with the negative buoyancy and steep dip of the down-going slab (Turcotte & Schubert, 2002), rates

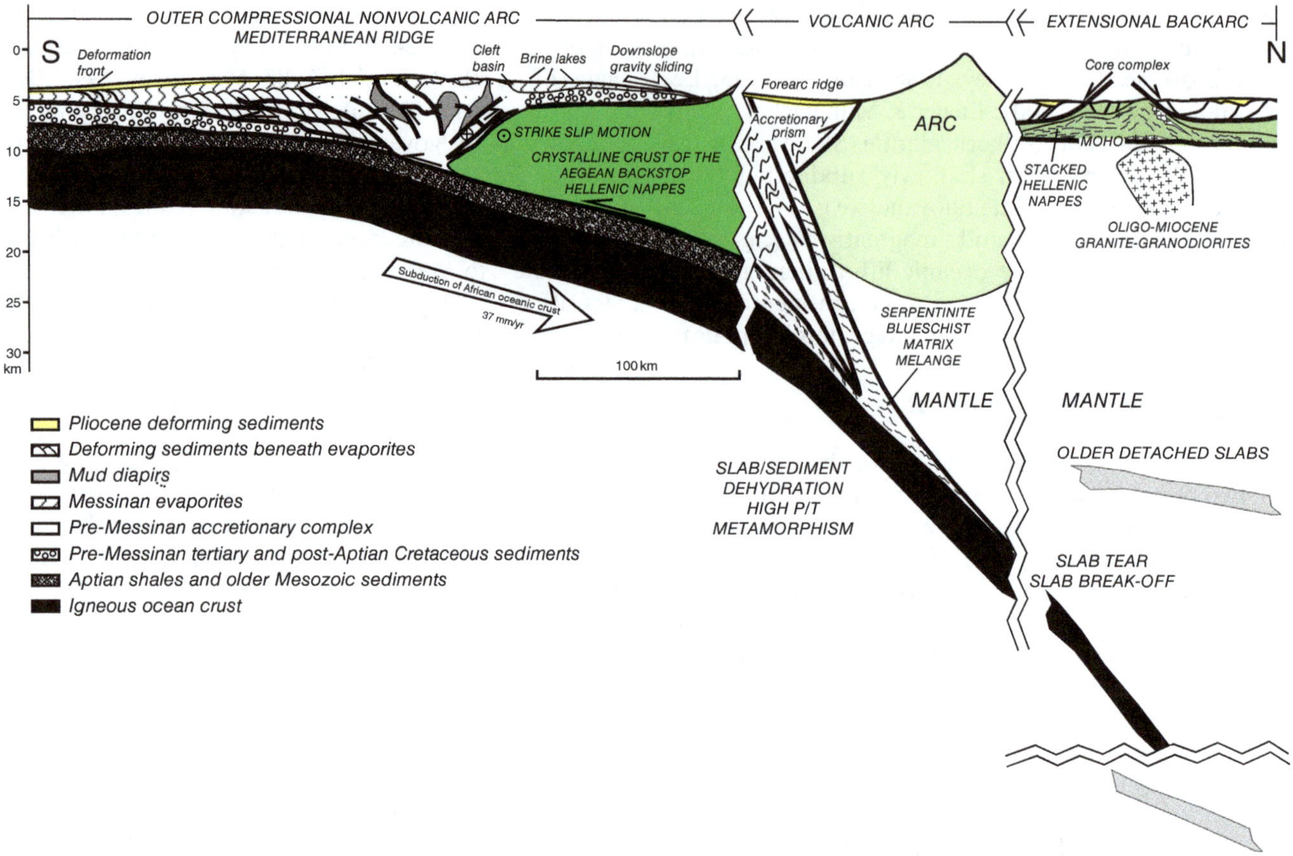

Figure 1.4 North-south generalized cross section through the accretionary Hellenic subduction zone showing the structural elements map of the Mediterranean Ridge after Westbrook and Reston (2002).

of the plate motion, or specific mechanical conditions that govern their geometry (Mahadevan et al., 2010).

1.3.3. Subduction and Suture Zones

Subduction zones are considered the most extensive recycling system on the planet and play a key role in Earth's geodynamics and crustal evolution (e.g., Li et al., 2013). The majority of the driving force of plate motion today is generally thought to be slab pull caused by the densification of subducted ocean crust (Forsyth & Uyeda, 1975; Chen et al., 2020; Palin & Santosh, 2021). Subduction zones also form large-scale metal ore deposits (e.g., Sawkins, 1972; Glasby, 1996; Rosenbaum et al., 2005; Kerrich et al., 2005; Li et al., 2013). Igneous activity within these zones forms most of the world's ore deposits (Stern, 2002). These include porphyry copper ± molybdenum ± gold deposits (PCDs), considered the most representative and valuable magmatic-hydrothermal metallogenic systems (Sillitoe, 2010; Rosenbaum et al., 2005; Chen & Wu, 2020). PCDs are located in magmatic-hydrothermal systems in the crust above subduction zones (Sillitoe, 2010; Chen & Wu, 2020; Xue et al., 2021). Here, ore-forming elements are enriched in the mantle wedge due to metasomatism driven by subducting slab-derived fluids (e.g., Zheng, 2019).

Subduction zones are classified based on the fate of ocean basin sediment and detritus accumulated through the erosion of continent and volcanoes that accumulate in the trench or trough (von Huene & Scholl, 1991). A thorough discussion of subduction zone dynamics is provided in this volume by Agard and coauthors (Chapter 2, Subduction and Obduction Processes: The Fate of Oceanic Lithosphere Revealed by Blueschists, Eclogites, and Ophiolites). *Erosive* subduction zones have crustal sedimentary material removed through subduction, whereas *accretionary* subduction zones show upper plate growth due to frontal accretion or underplating (e.g., von Huene & Scholl, 1991; Clift & Vannuchhi, 2004; Straub et al., 2020). Subduction erosion can still occur beneath accretionary margins and contribute to the geochemistry of arc volcanoes (Clift & Vannuchhi, 2004; Straub et al., 2020).

Convergent plate boundaries are often evident on bathymetry maps based on the subduction of one plate as it is consumed (Fig. 1.5). However, Dewey (1977) noted that *suture*

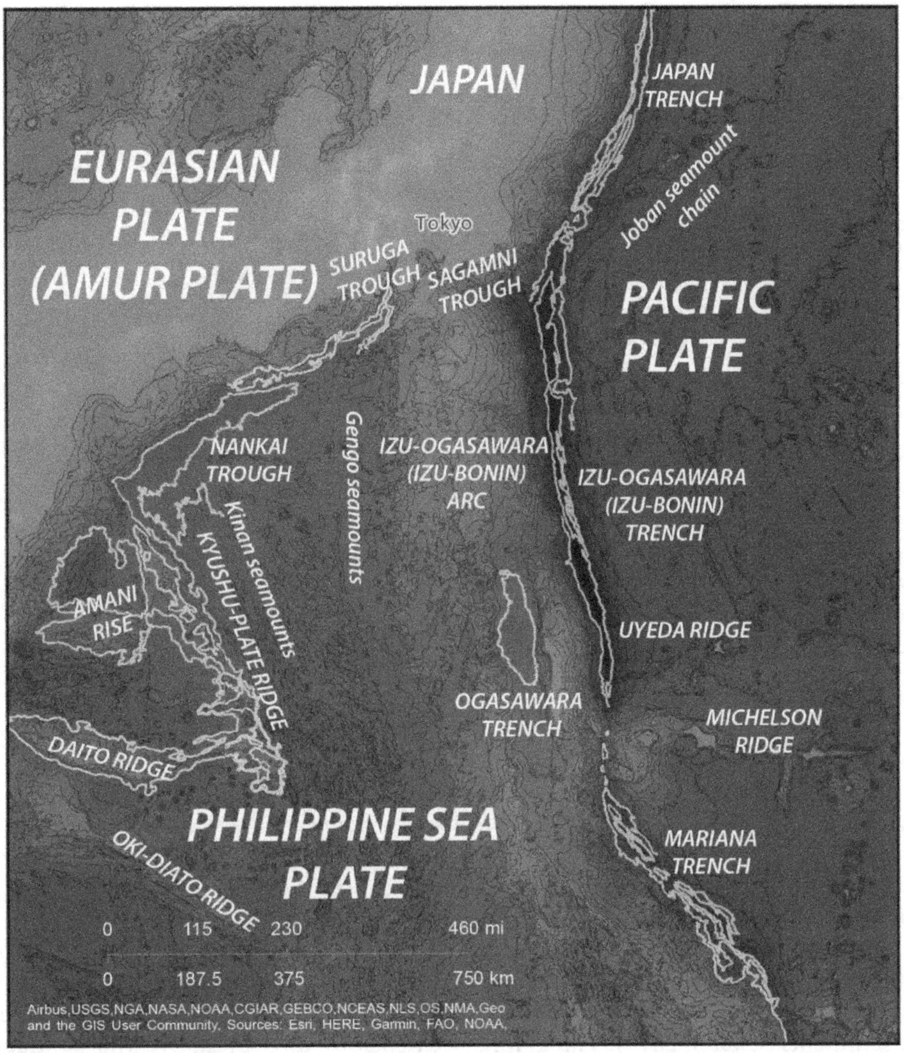

Figure 1.5 Bathymetry map of subduction zones located near Japan. Some contour lines are highlighted to emphasize particular boundaries and features. The names are after the U.S. Board on Geographic Names (BGN) Advisory Committee on Undersea Features (ACUF).

zones that delineate the zones of collision between two continents are rarely simple and rarely create easily recognizable lines (Fig. 1.6). These zones are locations where oceans and back-arc basins are closed (Burke et al., 1977). Their complexity is attributed to the irregular margins of colliding continental plates that generate broad and complex deformation zones (e.g., Chetty, 2017). These locations can involve multiple fault structures, with many experiencing high-strain, intense, and sometimes multistage deformation (Abdelsalam & Stern, 1996). Paleolocation of crusts on either side of the zone helps identify such zones, often facilitated by paleomagnetism studies.

As seen in Figure 1.6, suture zones incorporate a wide range of rock materials. They are critical locations for developing orogenic gold deposits where hydrothermal fluids are localized near and along convergent margins and in the middle and upper crusts (e.g., Goldfarb et al., 2001; Pour et al., 2016). Goldfarb et al. (2001) document numerous goldfields worldwide associated with suture zones over Earth's history. Collision granitoids within suture zones can concentrate economically critical minerals, such as tungsten (scheelite) and gold, rare-metal granites and pegmatite, and colored gemstones (e.g., Koroteev et al., 2009). Although these mountain-building events occur with lower thermal gradients than subduction zone settings and thus are not favorable for the hydrothermal mobilization of ore-forming elements, they are sometimes preceded by subduction zone convergence, which provides ample preliminary enrichment before collision (Zheng et al., 2019).

Sedimentary rocks in suture zones have recorded multiple facies types attributed to the deep-water ocean's

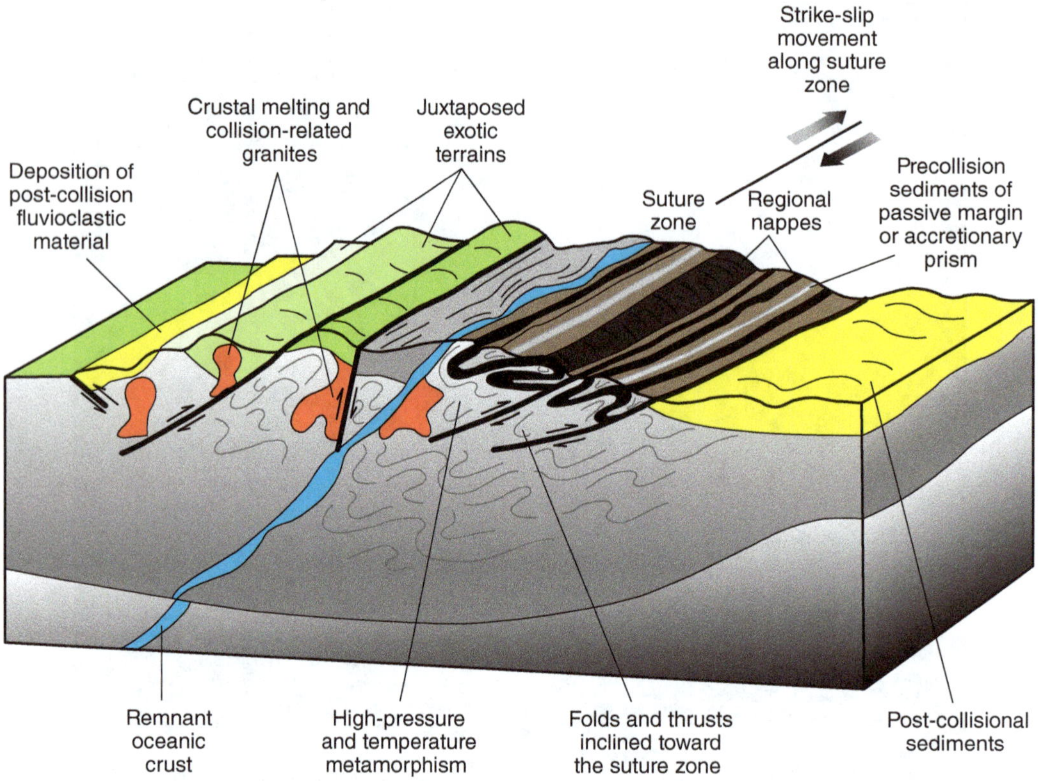

Figure 1.6 A schematic example of a suture zone. The picture is from the Open University (Geological processes in the British Isles).

nature to erosion from the overriding continental plate. Shales, turbidites, and deep-water radiolarian chert are recorded in suture zones (e.g., Chakrabarti, 2016). Suture zones can contain chemically and mineralogically matured multicycle sediments (Chetty, 2017). Thick units of sedimentary rocks can be partially subducted under the overriding lithosphere, creating metamorphic assemblages that record the collisional process. Depending on protolith and collision conditions, these metamorphic assemblages can be high-pressure eclogites and Barrovian-grade metapelites. Suture zones are often characterized by high-pressure blueschist–eclogite belts to even ultrahigh-pressure metamorphic (UHPM) complexes, remnants of the subduction zone that existed between two continents (Chetty, 2017).

Various igneous rocks may be present within suture zones, including mafic (ophiolites, serpentinized gabbro, sheared volcanic, blueschists) and felsic assemblages (syntectonic high Si, peraluminous granites). Deformed alkaline rocks and carbonatites (DARCS) delineate the boundaries of major Proterozoic suture zones (e.g., Burke et al., 2003; Leelanandam et al., 2006; Catlos et al., 2008). Perhaps the most recognizable feature of suture zones is stratigraphically intact *ophiolites*, remnants of the crust and upper mantle portions of ocean lithosphere or back-arc basins that disappeared between the two continents (e.g., Steinmann, 1906; Hess, 1955; Hawkins, 2003). *Supra-subduction zone (SSZ) ophiolites* are obducted oceanic crust with island arc geochemical characteristics that formed via seafloor spreading (synmagmatic extension) directly above the subducted oceanic lithosphere (Miyashiro, 1973; Pearce et al., 1984; Shervais & Kimbrough, 1985; Hawkins, 2003; Pearce, 2003). Ophiolites in suture zones provide a critical record of deep oceanic crust and ancient seafloor processes (Chetty, 2017).

The timing of collision and convergence of particular subduction and suture zones can be challenging and is often disputed. See a discussion about this topic as it relates to the development in the Himalayas by Robinson and Martin (Chapter 5, Genesis of Himalayan Stratigraphy and the Tectonic Development of the Thrust Belt) and Catlos (Chapter 6, Records of Himalayan Metamorphism and Contractional Tectonics in the Central Himalayas: Darondi Khola, Nepal). For example, although the Himalayan collision is often cited as during the Paleocene (Patriat & Achache, 1984; Klootwijk et al., 1992; Rowley, 1996; Yin & Harrison, 2000; Najman et al., 2001; Ding et al., 2005), much younger constraints are also suggested (e.g., Eocene/Oligocene boundary; Aitchison et al., 2007). Collision may have been a two-stage process, with events occurring in the Paleocene (soft) and Miocene (hard) collision (van Hinsbergen et al., 2012; see review in Parsons et al., 2020). Each component in the suture zone environment has the potential to provide evidence for its history, including the onset of sediment deposition, timing of metamorphism and recrystallization, and paleomagnetic evidence for the locations of the continental block before the collision. Suture zones are often at sites of high topography, but the development of large mountain belts associated with plate convergence occurs significantly after initial contact. In this volume, Giri and Hubbard (Chapter 3, Lateral Heterogeneity in Compressional Mountain Belt Settings) discuss how orogenic belts worldwide record deformation along strike.

Subduction zone initiation (SZI) is the onset of downward plate motion forming a new slab, which later evolves into a self-sustaining subduction zone (Crameri et al., 2020). In this volume, SZI is discussed as relevant to the Eurasian margin by Bo et al. (Chapter 9, When and Why the Neo-Tethyan Subduction Initiated Along the Eurasian Margin: A Case Study From a Jurassic Eclogite in Southern Iran) and along the Hellenic arc by Catlos and Çemen (Chapter 4, A Review of the Dynamics of Subduction Zone Initiation in the Aegean Region). The Hellenic arc (Fig. 1.4) has perhaps the most significant discrepancy between the onset subduction of the African (Nubian) slab beneath the Aegean microplate. Some studies suggest a Cenozoic SZI age, although estimates are from the Eocene-Pliocene (e.g., Meulenkamp et al., 1988; Spakman et al., 1988; Papadopoulos, 1997; Brun & Sokoutis, 2010; Le Pichon et al., 2019) to Mesozoic (Late Cretaceous-Jurassic) (Faccenna et al., 2003; van Hinsbergen et al., 2005; Royden & Papanikolaou, 2011; Jolivet et al., 2013; Crameri et al., 2020; van Hinsbergen et al., 2021). Tools used to time SZI are similar to those at suture zones. They include sediment deposition in the accretionary prism (Fig. 1.4), paleomagnetism, the analysis of topography combined with estimates of slab age and depth, reconstructions of subducted slabs using tomography, and the timing of metamorphism and volcanic activity that parallels the subduction zone (e.g., Crameri et al., 2020).

1.3.4. Hazards Associated with Compressional Plate Boundaries

The theory of plate tectonics suggests that plate interaction occurs primarily at the plate boundaries (see review by Gordon, 1998). Plate boundaries are often shown as thin lines and narrow zones (e.g., Figs. 1.3 and 1.5). However, the effects of convergent and collisional plate boundaries are felt far afield. Figure 1.7 shows the compressional fault systems associated with convergent and collisional plate boundaries in parts of Europe, the Middle East, and Asia. The effects of these plate boundaries extend far beyond their contact zones. The figure also outlines several *orogenic belts,* which are deformation zones due to horizontal compression, gravity, heat, and climate-driven erosion (DiPietro, 2018). Orogenic belts are explicitly discussed in this volume by Yilmaz et al. (Chapter 7 Tectonics of the Southeast Anatolian Orogenic Belt and Chapter 8 Tectonics of Eastern Anatolian Plateau; Final Stages of Collisional Orogeny in Anatolia). Orogens imply not only collisional dynamics and the nature of the kinematics in that region, but also a culturally relative statement that the velocity field in that region has more degrees of freedom than present data constrain (Bird, 2003). Orogenic belts form due to a collage of processes, including magmatism, metamorphism, sedimentation, and deformation (Chetty, 2017). The end stages of orogenic belts are described in this volume by Foster et al. (Chapter 12, Extensional Collapse of Orogens: A Review and Example From the Southern Appalachian Orogen).

Figure 1.7 shows the relationship between some of Earth's largest earthquakes and destructive volcanoes and convergent and collisional plate boundaries. According to the USGS, all of the Earth's most destructive and largest magnitude earthquakes occurred at convergent or collisional plate boundaries (Table 1.1). According to Table 1.1, subduction zones around the Pacific plate account for most of these events, including the Aleutian arc, Japan Trench, Peru-Chile, Columbia-Ecuador, and Kurile-Kamchatka subduction zones. Subduction zones host Earth's most destructive megathrust earthquakes, which are also associated with devastating *tsunamis* (e.g., Plafker, 1969; Cisternas et al., 2005; McCaffrey, 2008; Melnick et al., 2009; Toda & Tsutsumi, 2013; Bletery et al., 2016). Tsunamis are catastrophic wave motions generated by shock waves that cover large parts of the sea and behave intricately in coastal zones (Sugawara et al., 2008). All events in Table 1.1, except for the 1950 Assam-Tibet earthquake, are *tsunamigenic earthquakes*. Tsunamis triggered by earthquakes are partially generated due to a shallow focus coupled with large rupture areas associated with lower-angle megathrust faulting at subduction zones (e.g., Sugawara et al., 2008; Bilek & Lay, 2018). The largest earthquakes in Table 1.1 were associated with significant rupture areas: the 1960 Great Chilean earthquake (Valdivia) at the Peru-Chile trench had a rupture length of 920 ± 100 km (e.g., Cifuentes, 1989), whereas the 1964 Aleutian-Alaska megathrust fault ruptured a length of 600–800 km (Ichinose et al., 2007). The 2004 Sumatra-Andaman Islands earthquake resulted in a rupture length of 1,500 km (e.g., Gahalaut et al., 2006).

The 1950 Assam-Tibet earthquake (Fig. 1.7, Table 1.1) influenced rivers in India, Burma, East Pakistan, Tibet, and China. Many flooded and changed their courses permanently (Ben-Menahem et al., 1974; Mrinalinee Devi &

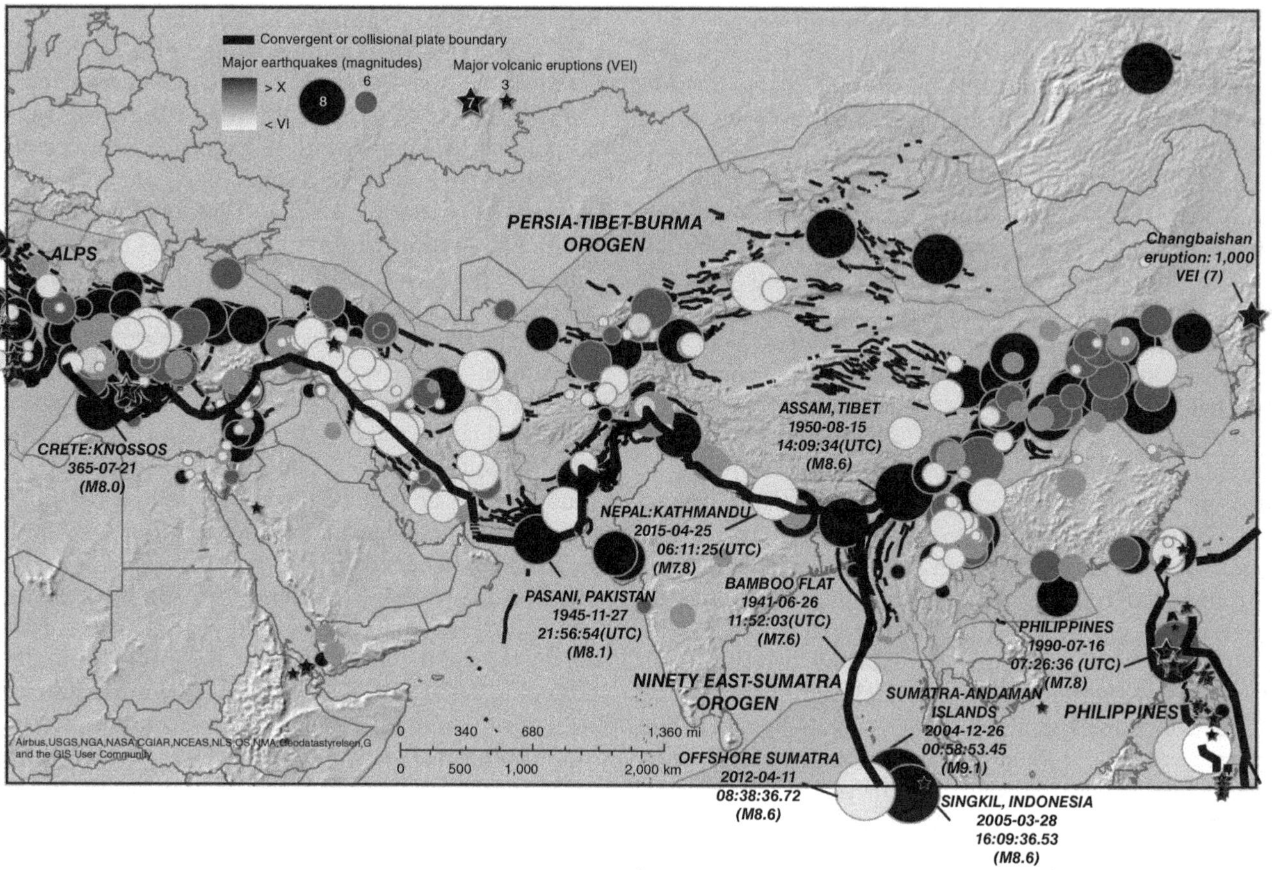

Figure 1.7 Map (ArcGIS) showing the major collisional and convergent plate boundaries with significant earthquakes and volcanic eruptions overlain. Also included are the boundaries of orogenic belts (Bird, 2002) and fault systems with an element of compression only. Convergent and collisional plate boundaries are identified by Coffin et al. (1998). Global active fault lines from information collected by the Global Earthquake Model Foundation.

Bora, 2016). Sharma and Zaman (2019) describe the ecological impact of the Assam-Tibet earthquake on the Brahmaputra River as it was affected by liquefaction and contamination by sulfur emanating from underground coal beds and oil seepages. In addition, *seismic seiches* related to the earthquake were recorded in several fjords and lakes over 7,000 km away in Norway (Kvale, 1955; McGarr, 2011). Seismic seiches are standing waves in closed or partially closed bodies of water due to the passage of seismic waves from an earthquake (McGarr, 2020). Based on a historical assessment, earthquakes in the Himalayan region may not be expected to be as large as those in subduction zones (Srivastava et al., 2013). However, the variations in seismicity of collisional mountain belts are related to a complex interplay between rheology, fault style, kinematics, and tectonic stress regime, but the parameters that control earthquake behavior in orogenic mountain belts remain unclear (e.g., Dal Zilio et al., 2018).

Ground shaking due to earthquakes at convergent and collisional boundaries often triggers significant mass wasting events, including landslides, rockfalls, and liquefaction. Evidence for giant terrestrial landslides is present along several convergent and collisional plate boundaries worldwide (Mather et al., 2014; Roberts et al., 2014). Landslides develop over steepened slopes and are triggered by large earthquakes or volcanic eruptions. If these events are located near coastal areas, tsunamis can develop. Significant triggers for tsunamis are subaqueous earthquakes and slides (Sugawara et al., 2008). Submarine landslides generated by earthquakes have triggered devastating tsunamis in the Aegean region (e.g., Dominey-Howes, 2002; Okal et al., 2009; Ebeling et al., 2012). The sloping bottom of the Hellenic arc, coupled with thick accumulations and high rates of recent sedimentation, closely spaced active faults, active earthquakes, and *magmatic diapirism* (where less dense rock rises through buoyant forces;

Table 1.1 Earth's 20 largest earthquakes

Location	Day and time	Lat.	Long.	Mag*	Depth	Location
Great Chilean earthquake (Valdivia)	1960-05-22 19:11:20.00	−38.143	−73.407	9.5	25	Peru-Chile Trench
Prince William Sound (Great Alaska)	1964-03-28 03:36:16.00	60.908	−147.339	9.2	25	Aleutian subduction zone
Sumatra - Andaman Islands	2004-12-26 00:58:53.45	3.295	95.982	9.1	30	Sumatra-Andaman subduction zone
Great Tohoku Japan	2011-03-11 05:46:24.12	38.297	142.373	9.1	29	Japan Trench
Kamchatka, Russia	1952-11-04 16:58:30.00	52.623	159.779	9	21.6	Kuril-Kamchatka subduction zone
Ecuador-Colombia	1906-01-31 15:36:10.00	0.955	−79.369	8.8	20	Subduction zone
Quirihue, Chile	2010-02-27 06:34:11.53	−36.122	−72.898	8.8	22.9	Peru-Chile Trench
Rat Islands, Aleutian Islands, Alaska	1965-02-04 05:01:22.00	51.251	178.715	8.7	30.3	Aleutian subduction zone
Unimak Island, Aleutian Islands Alaska	1946-04-01 12:29:01.00	53.492	−162.832	8.6	15	Aleutian subduction zone
Assam-Tibet	1950-08-15 14:09:34.00	28.363	96.445	8.6	15	Indo-Asia Collision (Mishmi Thrust)
Offshore Sumatra	2012-04-11 08:38:36.72	2.327	93.063	8.6	20	Sumatra-Andaman subduction zone
Singkil, Indonesia	2005-03-28 16:09:36.53	2.085	97.108	8.6	30	Sumatra-Andaman subduction zone
Adak, Alaska	1957-03-09 14:22:33.00	51.499	−175.626	8.6	25	Aleutian subduction zone
Vallenar, Chile	1922-11-11 04:32:51.00	−28.293	−69.852	8.5	70	Peru-Chile Trench
Tual, Indonesia	1938-02-01 19:04:22.00	−5.045	131.614	8.5	25	Banda Sea Arc
Kuril'sk, Russia	1963-10-13 05:17:59.00	44.872	149.483	8.5	35	Kurile-Kamchatka subduction zone
Mil'kovo, Russia	1923-02-03 16:01:50.00	54.486	160.472	8.4	15	Kurile-Kamchatka subduction zone
Atico, Peru	2001-06-23 20:33:14.13	−16.265	−73.641	8.4	33	Peru-Chile Trench
Sanriku-oki, Japan	1933-03-02 17:31:00.00	39.209	144.59	8.4	15	Japan Trench
Bengkulu, Indonesia	2007-09-12 11:10:26.83	−4.438	101.367	8.4	34	Sumatra–Andaman subduction zone

Source: USGS Earthquakes Hazards (2019).
* Magnitude estimated using the moment magnitude scale (Mw) or Moment W-phase. Some locations are seen in Figure 1.7.

Rajput & Thakur, 2016) contribute to its high hazards of tsunamis in the region (e.g., Ferentinos, 1990; Hooft et al., 2017). The eruption of Santorini in 1610 BCE generated a tsunami that affected civilizations throughout the eastern Mediterranean (Dominey-Howes, 2004; Friedrich, 2006; Marinatos, 1939; Hooft et al., 2017). Detailed bathymetry across the Mediterranean is critical in understanding tsunami propagation and mitigating its impacts (e.g., CIESM, 2011).

Figure 1.7 shows the relationship between convergent plate boundaries and significant volcanic eruptions. The Earth's most extensive volcanic fields in terms of basaltic and silicic eruptions are not found at convergent plate boundaries but are over *large igneous provinces* (LIPS) (e.g., Coffin & Eldholm, 1994; Bryan & Ernst, 2008; Bryan et al., 2010). However, the origin of LIPS may lie in the subduction process that perturbs mantle dynamics, forces extension in the back-arc region, thins the lithosphere, and triggers large-scale and voluminous basalt eruption (Zhu et al., 2019). The return flow of *slab avalanches* from the mantle transition zone can also generate LIPS (Gurnis, 1988; Coltice et al., 2007; Condie et al., 2021). Slab avalanches develop when large-volume subducted slabs temporarily stagnate within the transition zone and periodically penetrate the lower mantle (e.g., Solheim & Peltier, 1994; Deschamps & Tackley, 2009; Yang et al., 2018). Slab avalanches are controlled by mantle thermal instabilities and accelerate as slab sinking rates increase with time (e.g., Solheim & Peltier, 1994; Yang et al., 2018).

Subduction zones also produce eruptions that are most commonly observed and most dangerous to human populations (Siebert et al., 2015). Subduction zone volcanism propels volcanic gases (e.g., SO_2, CO_2, H_2S) and ash into the stratosphere or troposphere and has affected short-term climate (Bryan et al., 2010; Cooper et al., 2018) and the carbon cycle (Zhu et al., 2021). Some sulfur gases convert to sulfate aerosols in the stratosphere and scatter radiation (e.g., Robock, 2000). The dust veil index (DVI/Emax) measures an eruption's release of dust and aerosols over the years following the event, especially the impact on the Earth's energy balance (Lamb, 1985). For example, the AD 1835 eruption of Volcan Cosiguina, Nicaragua, which is located on a convergent margin where the oceanic crust of the Cocos plate subducts beneath the western edge of the Caribbean plate, is recorded as a DVI/Emax of 4,000, with ashfall recorded as far away as 1,900 km (Scott et al., 2006).

Climate change is intrinsically related to collisional plate boundaries, as topographic barriers interact with the Earth's atmosphere (e.g., Burbank, 1992; Cronin, 2009; Ruddiman, 2013; Song et al., 2021) and subducting slabs at collisional boundaries eliminate megatons of carbon (e.g., Clift, 2017; Plank & Manning, 2019). Controls on the subduction process may be related to climate change (Lamb & Davis, 2003; Iaffaldano et al., 2006). The onset of the Himalayan monsoon is related to India-Asia convergence and is widely studied for understanding the timing of mountain building (e.g., Clift et al., 2008; Allen & Armstrong, 2012; Webb et al., 2017). Mountain ranges are barriers to atmospheric circulation, and exposures of rocks in the mountainous regions can also drive the drawdown of atmospheric gasses through weathering processes that may be directly related to climate change (e.g., Stern & Miller, 2018).

1.4. OBJECTIVES AND ORGANIZATION OF THE BOOK

This volume was written to create an up-to-date and relevant compendium and valuable reference for Earth sciences students, including advanced undergraduate and graduate students, postdocs, educators, research professionals, and policy makers in academia and industry. These papers aim to synthesize current knowledge of complex geological topics surrounding global collisional and convergent plate boundaries with an accessible approach and transparent organization. The papers are meant to be readable for a range of consumers. Several reviewers helped to identify topical oversights and assure that citations fairly represent the body of existing information. The topics are mentioned in the preface and in the text of this introduction, and are highlighted in the volume's table of contents.

ACKNOWLEDGMENTS

No real or perceived financial conflicts of interests exist for any author. We appreciate the time and effort by the authors of this volume and the reviewers of these papers who provided constructive comments. We appreciate discussions regarding the book title with John Waldron (University of Alberta), who suggested an alternative volume title could be contractional or convergent tectonics. We appreciate constructive comments from Richard Palin (University of Oxford) and two anonymous reviewers. Finally, we appreciate drafting assistance from Jeffrey S. Horowitz.

REFERENCES

Abdelsalam, M. G., & Stern, R. J. (1996). Sutures and shear zones in the Arabian-Nubian Shield. *Journal of African Earth Sciences*, *23*(3), 289–310.

Aitchison, J. C., Ali, J. R., & Davis, A. M. (2007). When and where did India and Asia collide?. *Journal of Geophysical Research: Solid Earth*, *112*(B5). https://doi.org/10.1029/2006JB004706

Allen, M. B., & Armstrong, H. A. (2012). Reconciling the intertropical convergence zone, Himalayan/Tibetan tectonics, and the onset of the Asian monsoon system. *Journal of Asian Earth Sciences*, *44*, 36–47.

Anderson, D. L. (1995). Lithosphere, asthenosphere, and perisphere. *Reviews of Geophysics*, *33*(1), 125–149.

Anderson, E. M. (1905). The dynamics of faulting. *Transactions of the Edinburgh Geological Society*, *8*(3), 387–402.

Anderson, T. W. (1951). Estimating linear restrictions on regression coefficients for multivariate normal distributions. *The Annals of Mathematical Statistics*, 327–351.

Beck, M. E., Jr. (1983). On the mechanism of tectonic transport in zones of oblique subduction. *Tectonophysics*, *93*(1–2), 1–11.

Bellaiche, G. (1980). Sedimentation and structure of the Izu-Ogasawara (Bonin) Trench off Tokyo: New lights on the results of a diving campaign with the Bathyscape "Archimede." *Earth and Planetary Science Letters*, *47*(1), 124–130.

Ben-Menahem, A., Aboodi, E., & Schild, R. (1974). The source of the great Assam earthquake: An interplate wedge motion. *Physics of the Earth and Planetary Interiors*, *9*(4), 265–289.

Berk Biryol, C., Beck, S. L., Zandt, G., & Özacar, A. A. (2011). Segmented African lithosphere beneath the Anatolian region inferred from teleseismic P-wave tomography. *Geophysical Journal International*, *184*(3), 1037–1057.

Bevis, M., & Martel, S. J. (2001). Oblique plate convergence and interseismic strain accumulation. *Geochemistry, Geophysics, Geosystems*, *2*(8). https://doi.org/10.1029/2000GC000125

Bilek, S. L., & Lay, T. (2018). *Subduction zone megathrust earthquakes. Geosphere*, *14*(4), 1468–1500.

Bird, P. (2003). An updated digital model of plate boundaries. *Geochemistry, Geophysics, Geosystems*, *4*(3).

Bletery, Q., Thomas, A. M., Rempel, A. W., Karlstrom, L., Sladen, A., & De Barros, L. (2016). Mega-earthquakes rupture flat megathrusts. *Science*, *354*(6315), 1027–1031.

Boschman, L. M., & Van Hinsbergen, D. J. (2016). On the enigmatic birth of the Pacific Plate within the Panthalassa Ocean. *Science Advances*, *2*(7), e1600022. https://doi.org/10.1126/sciadv.1600022

Brun, J. P., & Sokoutis, D. (2010). 45 my of Aegean crust and mantle flow driven by trench retreat. *Geology*, *38*(9), 815–818.

Bryan, S. E., & Ernst, R. E. (2008). Revised definition of large igneous provinces (LIPs). *Earth-Science Reviews*, *86*(1–4), 175–202.

Bryan, S. E., Peate, I. U., Peate, D. W., Self, S., Jerram, D. A., Mawby, M. R., et al. (2010). The largest volcanic eruptions on Earth. *Earth-Science Reviews*, *102*(3–4), 207–229.

Burbank, D. W. (1992). Causes of recent Himalayan uplift deduced from deposited patterns in the Ganges basin. *Nature*, *357*(6380), 680–683.

Burbidge, D. R., & Braun, J. (1998). Analogue models of obliquely convergent continental plate boundaries. *Journal of Geophysical Research: Solid Earth*, *103*(B7), 15221–15237.

Burke, K., Ashwal, L. D., & Webb, S. J. (2003). New way to map old sutures using deformed alkaline rocks and carbonatites. *Geology*, *31*(5), 391–394.

Burke, K., Dewey, J. F., & Kidd, W. S. F. (1977). World distribution of sutures: The sites of former oceans. *Tectonophysics*, *40*(1–2), 69–99.

Butler, R. (2021). Faults and stress. https://www.youtube.com/watch?v=cQ6zgeM4PN8.

Catlos, E. J., Dubey, C. S., & Sivasubramanian, P. (2008). Monazite ages from carbonatites and high-grade assemblages along the Kambam Fault (Southern Granulite Terrane, South India). *American Mineralogist*, *93*(8–9), 1230–1244.

Chakrabarti, B. K. (2016). Lithotectonic subdivisions of the Himalaya. In B. K. Chakrabarti (Ed.), *Geology of the Himalayan Belt* (pp. 1–9). Elsevier. https://doi.org/10.1016/B978-0-12-802021-0.00001-2

Chen, H., & Wu, C. (2020). Metallogenesis and major challenges of porphyry copper systems above subduction zones. *Science China Earth Sciences*, *63*(7), 899–918.

Chen, L., Wang, X., Liang, X., Wan, B. & Liu, L. (2020). Subduction tectonics vs. plume tectonics: Discussion on driving forces for plate motion. *Science China Earth Sciences*, *63*(3), 315–328.

Chetty, T. R. K. (2017). *Proterozoic orogens of India: A critical window to Gondwana*. Elsevier.

CIESM (2011). Marine geohazards in the Mediterranean. N°42. In F. Briand (Ed.). *CIESM workshop monographs: Marine geohazards in the Mediterranean*. The Mediterranean Science Commission Workshop in Nicosia, 2–5 February 2011.

Cifuentes, I. L. (1989). The 1960 Chilean earthquakes. *Journal of Geophysical Research: Solid Earth*, *94*(B1), 665–680.

Cisternas, M., Atwater, B. F., Torrejón, F., Sawai, Y., Machuca, G., Lagos, M., et al. (2005). Predecessors of the giant 1960 Chile earthquake. *Nature*, *437*(7057), 404–407.

Clift, P., & Vannucchi, P. (2004). Controls on tectonic accretion versus erosion in subduction zones: Implications for the origin and recycling of the continental crust. *Reviews of Geophysics*, *42*(2).

Clift, P. D. (2017). A revised budget for Cenozoic sedimentary carbon subduction. *Reviews of Geophysics*, *55*(1), 97–125.

Clift, P. D., Hodges, K. V., Heslop, D., Hannigan, R., Van Long, H., & Calves, G. (2008). Correlation of Himalayan exhumation rates and Asian monsoon intensity. *Nature Geoscience*, *i(12)*, 875–880.

Coffin, M. F., & Eldholm, O. (1994). Large igneous provinces: Crustal structure, dimensions, and external consequences. *Reviews of Geophysics*, *32*(1), 1–36.

Coffin, M. F., Gahagan, L. M., & Lawver, L. A. (1998). *Present-day plate boundary digital data compilation*. University of Texas Institute for Geophysics Technical Report *No. 174*, 5.

Coltice, N., Phillips, B. R., Bertrand, H., Ricard, Y., & Rey, P. (2007). Global warming of the mantle at the origin of flood basalts over supercontinents. *Geology*, *35*(5), 391–394.

Condie, K. C. (2005). High field strength element ratios in Archean basalts: A window to evolving sources of mantle plumes? *Lithos*, *79*(3–4), 491–504.

Condie, K. C. (2022). The mantle. In K. C. Condie (Ed.), *Earth as an evolving planetary system, 4th ed.* (pp. 81–125). Academic Press. https://doi.org/10.1016/B978-0-12-819914-5.00010-X

Condie, K. C., Pisarevsky, S. A., & Puetz, S. J. (2021). LIPs, orogens and supercontinents: The ongoing saga. *Gondwana Research*, *96*, 105–121.

Cooper, C. L., Swindles, G. T., Savov, I. P., Schmidt, A., & Bacon, K. L. (2018). Evaluating the relationship between climate change and volcanism. *Earth-Science Reviews*, *177*, 238–247. https://doi.org/10.1016/j.earscirev.2017.11.009

Cox, A., & Hart, R. B. (2009). *Plate tectonics: How it works*. New Jersey: John Wiley & Sons.

Crameri, F., Magni, V., Domeier, M., Shephard, G. E., Chotalia, K., Cooper, G., et al. (2020). A transdisciplinary and community-driven database to unravel subduction zone initiation. *Nature Communications*, *11*(1), 1–14.

Cronin, T. M. (2009). *Paleoclimates: Understanding climate change past and present*. Columbia University Press.

Dal Zilio, L., Faccenda, M., & Capitanio, F. (2018). The role of deep subduction in supercontinent breakup. *Tectonophysics*, *746*, 312–324.

Dana, J. D. (1873). ART. XLVI: On some results of the Earth's contraction from cooling, including a discussion of the origin of mountains, and the nature of the Earth's interior. *American Journal of Science and Arts (1820–1879)*, *5*(30), 423.

De Graciansky, P. C., Roberts, D. G., & Tricart, P. (2011). The birth of the western and central Alps: Subduction, obduction, collision. In *Developments in Earth surface processes* (vol. *14*, pp. 289–315). Elsevier.

DeMets, C., Gordon, R. G., & Argus, D. F. (2010). Geologically current plate motions. *Geophysical Journal International*, *181*(1), 1–80.

Deschamps, F., & Tackley, P. J. (2009). Searching for models of thermo-chemical convection that explain probabilistic tomography. II: Influence of physical and compositional parameters. *Physics of the Earth and Planetary Interiors*, *176*(1–2), 1–18.

Dewey, J. F. (1977). Suture zone complexities: A review. *Tectonophysics*, *40*(1–2), 53–67.

Dewey, J. F. (1988). Extensional collapse of orogens. *Tectonics*, *7*(6), 1123–1139.

DiPietro, J. A. (2013). Formation, collapse, and erosional decay of mountain systems. In J. A. DiPietro (Ed.), *Landscape evolution in the United States* (pp. 365–373). Elsevier. https://doi.org/10.1016/B978-0-12-397799-1.00022-1

DiPietro, J. A. (2018). Forcing agent: The tectonic system. In J. A. DiPietro (Ed.), *Geology and landscape evolution, 2d ed.* (pp. 59–77). Elsevier. https://doi.org/10.1016/B978-0-12-811191-8.00005-1

Dominey-Howes, D. (2002). Documentary and geological records of tsunamis in the Aegean Sea region of Greece and their potential value to risk assessment and disaster management. *Natural Hazards*, *25*(3), 195–224.

Dominey-Howes, D. (2004). A re-analysis of the Late Bronze Age eruption and tsunami of Santorini, Greece, and the implications for the volcano-tsunami hazard. *Journal of Volcanology and Geothermal Research*, *130*(1–2), 107–132.

Eagles, G. (2020). Plate boundaries and driving mechanisms. In N. Scarselli, J. Adam, D. Chiarella, D. G. Roberts, & A. W. Bally (Eds), *Regional geology and tectonics, 2d ed.* (pp. 41–59). Elsevier. https://doi.org/10.1016/B978-0-444-64134-2.00004-3

Elobaid, E. A., Sadooni, F., & Al Saad, H. (2020). *Tectonic and geologic settings of Halul and Al-Alyia offshore islands, examples of different evolution models, within the emergence of the Arabian Gulf geosyncline: A review*. Qatar University Annual Research Forum and Exhibition (QUARFE 2020), Doha, 2020. https://doi.org/10.29117/quarfe.2020.0044

Faccenna, C., Jolivet, L., Piromallo, C., & Morelli, A. (2003). Subduction and the depth of convection in the Mediterranean mantle. *Journal of Geophysical Research: Solid Earth*, *108*(B2).

Ferentinos, G. (1990). Offshore geological hazards in the Hellenic Arc. *Marine Georesources and Geotechnology*, *9*(4), 261–277.

Fisher, D. W. (1978). James Hall: Patriarch of American paleontology, geological organizations, and state geological surveys. *Journal of Geological Education*, *26*(4), 146–152. https://doi.org/10.5408/0022-1368-26.4.146.

Fitch, T. J. (1972). Plate convergence, transcurrent faults, and internal deformation adjacent to southeast Asia and the western Pacific. *Journal of Geophysical Research*, *77*(23), 4432–4460.

Forsyth, D., & Uyeda, S. (1975). On the relative importance of the driving forces of plate motion. *Geophysical Journal International*, *43*(1), 163–200.

Fowler, C. M. R. (2012). Ocean floor tectonics. In D. G. Roberts & A. W. Bally (Eds.), *Regional geology and tectonics: Principles of geologic analysis* (pp. 732–818). Elsevier. https://doi.org/10.1016/B978-0-444-53042-4.00026-1

Frankel, H. (1982). The development, reception, and acceptance of the Vine-Matthews-Morley hypothesis. *Historical Studies in the Physical Sciences*, *13*(1), 1–39.

Friedman, G. M. (2012). The great American carbonate bank in the northern Appalachians: Cambrian-Ordovician (Sauk), Albany Basin, New York. In J. R. Derby, R. D. Fritz, S. A. Longacre, W. A. Morgan, & C. A. Sternbach (Eds.), *The great American carbonate bank: The geology and economic resources of the Cambrian-Ordovician Sauk megasequence of Laurentia* (pp. 493–497). American Association of Petroleum Geologists. https://doi.org/10.1306/M981333

Gahalaut, V. K., Nagarajan, B., Catherine, J. K., & Kumar, S. (2006). Constraints on 2004 Sumatra-Andaman earthquake rupture from GPS measurements in Andaman-Nicobar Islands. *Earth and Planetary Science Letters*, *242*(3–4), 365–374.

Gaidzik, K., & Więsek, M. (2021). Seismo-lineaments and potentially seismogenic faults in the overriding plate of the Nazca–South American subduction zone (S. Peru). *Journal of South American Earth Sciences*, *109*, 103303.

Ganas, A., & Parsons, T. (2009). Three-dimensional model of Hellenic arc deformation and origin of the Cretan uplift. *Journal of Geophysical Research: Solid Earth*, *114*(B6).

Gao, S., Zhang, B. R., Jin, Z. M., Kern, H., Luo, T. C., & Zhao, Z. D. (1998). How mafic is the lower continental crust? *Earth and Planetary Science Letters*, *161*(1–4), 101–117.

Glaessner, M. F., & Teichert, C. (1947). Geosynclines, a fundamental concept in geology. *American Journal of Science*, *245*(8), 465–482.

Goldfarb, R. J., Groves, D. I., & Gardoll, S. (2001). Orogenic gold and geologic time: A global synthesis. *Ore Geology Reviews*, *18*(1–2), 1–75.

Gordon, R. G. (1998). The plate tectonic approximation: Plate nonrigidity, diffuse plate boundaries, and global plate reconstructions. *Annual Review of Earth and Planetary Sciences*, *26*(1), 615–642.

Gurnis, M. (1988). Large-scale mantle convection and the aggregation and dispersal of supercontinents. *Nature*, *332*(6166), 695–699.

Hacker, B. R., Kelemen, P. B., & Behn, M. D. (2015). Continental lower crust. *Annual Review of Earth and Planetary Sciences*, *43*, 167–205. https://doi.org/10.1016/j.epsl.2011.05.024.

Hall, J. (1859). Introduction, geological survey of New York palaeontology, *3*, 1–90.

Haq, S. S., & Davis, D. M. (1997). Oblique convergence and the lobate mountain belts of western Pakistan. *Geology*, *25*(1), 23–26.

Harrison, C. G. A. (2016). The present-day number of tectonic plates. *Earth, Planets, and Space*, *68*, 37. https://doi.org/10.1186/s40623-016-0400-x

Hawkins, J. W. (2003). *Geology of supra-subduction zones: Implications for the origin of ophiolites*. Special Papers, Geological Society of America, 227–268.

Hernández-Uribe, D., & Palin, R. M. (2019). Catastrophic shear-removal of subcontinental lithospheric mantle beneath the Colorado Plateau by the subducted Farallon slab.

Scientific Reports, *9*, 8153. https://doi.org/10.1038/s41598-019-44628-y

Hess, H. H. (1955). *Serpentines, orogeny, and epeirogeny*. GSA Special Papers, Crust of the Earth: A symposium. https://doi.org/10.1130/SPE62-p391

Hey, R. (2021). Propagating rifts and microplates at mid-ocean ridges. In D. Alderton & S. A. Elias (Eds.), *Encyclopedia of geology, 2d ed.* (pp. 855–867). Academic Press. https://doi.org/10.1016/B978-0-12-409548-9.03027-X

Hooft, E. E., Nomikou, P., Toomey, D. R., Lampridou, D., Getz, C., Christopoulou, M. E., et al. (2017). Backarc tectonism, volcanism, and mass wasting shape seafloor morphology in the Santorini-Christiana-Amorgos region of the Hellenic Volcanic Arc. *Tectonophysics*, *712*, 396–414.

Iaffaldano, G., Bunge, H-P., Dixon, T. H. (2006). Feedback between mountain belt growth and plate convergence. *Geology*, *34* (10), 893–896. https://doi.org/10.1130/G22661.1

Ichinose, G., Somerville, P., Thio, H. K., Graves, R., & O'Connell, D. (2007). Rupture process of the 1964 Prince William Sound, Alaska, earthquake from the combined inversion of seismic, tsunami, and geodetic data. *Journal of Geophysical Research: Solid Earth*, *112*(B7).

Jarrard, R. D. (1986). Relations among subduction parameters. *Reviews of Geophysics*, *24*(2), 217–284.

Jolivet, L., Faccenna, C., Huet, B., Labrousse, L., Le Pourhiet, L., Lacombe, O., et al. (2013). Aegean tectonics: Strain localization, slab tearing and trench retreat. *Tectonophysics*, *597*, 1–33.

Kay, M. (1948). Summary of Middle Ordovician bordering Allegheny synclinorium. *AAPG Bulletin*, *32*(8), 1397–1416.

Kay, R.W., & Mahlburg-Kay, S. (1991). Creation and destruction of lower continental crust. *Geologische Rundschau*, *80*(2), 259–278.

Kay, S. M. (2014). 125th anniversary of the Geological Society of America: Looking at the past and into the future of science at GSA. *GSA Today*, *24*(3), 4–11.

Kerrich, R., Goldfarb, R. J., & Richards, J. P. (2005). Metallogenic provinces in an evolving geodynamic framework. *Economic Geology: One Hundredth Anniversary Volume*. https://doi.org/10.5382/AV100.33

Klootwijk, C. T., Gee, J. S., Peirce, J. W., Smith, G. M., & McFadden, P. L. (1992). An early India-Asia contact: Paleomagnetic constraints from Ninetyeast ridge, ODP Leg 121. *Geology*, *20*(5), 395–398.

Koroteev, V. A., Sazonov, V. N., Ogorodnikov, V. N., & Polenov, Y. A. (2009). Suture zones of the Urals as integral prospective ore-bearing tectonic structures. *Geology of Ore Deposits*, *51*(2), 93–108.

Kvale, A. (1955). Seismic seiches in Norway and England during the Assam earthquake of August 15, 1950. *Bulletin of the Seismological Society of America*, *45*(2), 93–113.

Lamb, H. H. (1985). Volcanic loading: The dust veil index (No. NDP-013). Oak Ridge National Lab.(ORNL), Oak Ridge, Tennessee.

Lamb, S., & Davis, P. (2003). Cenozoic climate change as a possible cause for the rise of the Andes. *Nature*, *425*, 792–797. https://doi.org/10.1038/nature02049

Lavier, L. L., & Manatschal, G. (2006). A mechanism to thin the continental lithosphere at magma-poor margins. *Nature*, *440*(7082), 324–328.

Lee, C. T., Yin, Q., Rudnick, R. L., Chesley, J. T., & Jacobsen, S. B. (2000). Osmium isotopic evidence for Mesozoic removal of lithospheric mantle beneath the Sierra Nevada, *California. Science*, *289*(5486), 1912–1916.

Leelanandam, C., Burke, K., Ashwal, L. D., & Webb, S. J. (2006). Proterozoic mountain building in Peninsular India: An analysis based primarily on alkaline rock distribution. *Geological Magazine*, *143*(2), 195–212.

Le Pichon, X., & Angelier, J. (1979). The Hellenic arc and trench system: A key to the neotectonic evolution of the eastern Mediterranean area. *Tectonophysics*, *60*(1–2), 1–42.

Le Pichon, X., Francheteau, J., & Bonnin, J. (2013). *Plate tectonics* (Vol. *6*). Elsevier.

Le Pichon, X., Şengör, A. C., & Imren, C. (2019). A new approach to the opening of the eastern Mediterranean Sea and the origin of the Hellenic subduction zone. Part 2: The Hellenic subduction zone. *Canadian Journal of Earth Sciences*, *56*(11), 1144–1162.

Li, J. L., Gao, J., John, T., Klemd, R., & Su, W. (2013). Fluid-mediated metal transport in subduction zones and its link to arc-related giant ore deposits: Constraints from a sulfide-bearing HP vein in lawsonite eclogite (Tianshan, China). *Geochimica et Cosmochimica Acta*, *120*, 326–362.

Li, S., Suo, Y., Li, X., Liu, B., Dai, L., Wang, G., et al. (2018). Microplate tectonics: New insights from micro-blocks in the global oceans, continental margins and deep mantle. *Earth-Science Reviews*, *185*, 1029–1064.

Mahadevan, L., Bendick, R., & Liang, H. (2010). Why subduction zones are curved. *Tectonics*, *29*(6).

Marinatos, S. (1939). The volcanic destruction of Minoan Crete. *Antiquity*, *13*(52), 425–439.

Mark, K. (1992). From geosynclinal to geosyncline. *Earth Sciences History*, *11*(2), 68–69. https://doi.org/10.17704/eshi.11.2.48j84852842rg203

Marvin, U. B. (2005). *History of geology since 1962*. Encyclopedia of Geology. Elsevier. https://doi.org/10.1016/B0-12-369396-9/00371-3

Mather, A. E., Hartley, A. J., & Griffiths, J. S. (2014). The giant coastal landslides of northern Chile: Tectonic and climate interactions on a classic convergent plate margin. *Earth and Planetary Science Letters*, *388*, 249–256.

McCaffrey, R. (1993). On the role of the upper plate in great subduction zone earthquakes. *Journal of Geophysical Research: Solid Earth*, *98*(B7), 11953–11966.

McCaffrey, R. (1996). Estimates of modern arc-parallel strain rates in fore arcs. *Geology*, *24*(1), 27–30.

McCaffrey, R. (2008). Global frequency of magnitude 9 earthquakes. *Geology*, *36*(3), 263–266.

McCaffrey, R., Long, M. D., Goldfinger, C., Zwick, P. C., Nabelek, J. L., Johnson, C. K., & Smith, C. (2000). Rotation and plate locking at the southern Cascadia subduction zone. *Geophysical Research Letters*, *27*(19), 3117–3120.

McGarr A. (2011). Seismic seiches. In H. K. Gupta (Ed.), *Encyclopedia of solid Earth geophysics*. Encyclopedia of Earth Sciences Series. Springer, Dordrecht. https://doi.org/10.1007/978-90-481-8702-7_186

McGarr, A. (2020). Seismic seiches. In H. Gupta (Ed.) *Encyclopedia of Solid Earth Geophysics*. Encyclopedia of

Earth Sciences Series. Springer, Cham. https://doi.org/10.1007/978-3-030-10475-7_186-1

Melnick, D., Bookhagen, B., Strecker, M. R., & Echtler, H. P. (2009). Segmentation of megathrust rupture zones from fore-arc deformation patterns over hundreds to millions of years, Arauco peninsula, Chile. *Journal of Geophysical Research: Solid Earth*, *114*(B1).

Meulenkamp, J. E., Wortel, M. J. R., Van Wamel, W. A., Spakman, W., & Strating, E. H. (1988). On the Hellenic subduction zone and the geodynamic evolution of Crete since the late Middle Miocene. *Tectonophysics*, *146*(1–4), 203–215.

Minshull, T. A. (2002). The breakup of continents and the formation of new ocean basins. *Philosophical Transactions of the Royal Society of London. Series A: Mathematical, Physical and Engineering Sciences*, *360*(1801), 2839–2852.

Miyashiro, A. (1973). The Troodos ophiolitic complex was probably formed in an island arc. *Earth and Planetary Science Letters*, *19*(2), 218–224.

Moreno, M. S., Klotz, J., Melnick, D., Echtler, H., & Bataille, K. (2008). Active faulting and heterogeneous deformation across a megathrust segment boundary from GPS data, south central Chile (36-39 S). *Geochemistry, Geophysics, Geosystems*, *9*(12).

Mrinalinee Devi, R. K., & Bora, P. K. (2016). The impact of the Great 1950 Assam Earthquake on the frontal regions of the northeast Himalaya. In S. D'Amico (Ed.), *Earthquakes and their impact on society*. Springer Natural Hazards. Springer, Cham. https://doi.org/10.1007/978-3-319-21753-6_19

Najman, Y., Pringle, M., Godin, L., & Oliver, G. (2001). Dating of the oldest continental sediments from the Himalayan foreland basin. *Nature*, *410*(6825), 194–197.

Okal, E. A., Synolakis, C. E., Uslu, B., Kalligeris, N., & Voukouvalas, E. (2009). The 1956 earthquake and tsunami in Amorgos, *Greece. Geophysical Journal International*, *178*(3), 1533–1554.

Palin, R. M., & Santosh, M. (2021). Plate tectonics: What, where, why, and when? *Gondwana Research*, *100*, 3–24.

Palin, R. M., Moore, J. D. P., Zhang, Z., Huang, G., Wade, J., & Dyck, B. (2021). Mafic Archean continental crust prohibited exhumation of orogenic UHP eclogite. *Geoscience Frontiers*, *12*(5), 101225. https://doi.org/10.1016/j.gsf.2021.101225

Papadopoulos, G. A. (1997). On the interpretation of large-scale seismic tomography images in the Aegean sea area. *Annals of Geophysics*, *40*(1).

Papadopoulos, G. A., Daskalaki, E., Fokaefs, A., & Giraleas, N. (2007). Tsunami hazards in the Eastern Mediterranean: Strong earthquakes and tsunamis in the east Hellenic arc and trench system. *Natural Hazards and Earth System Sciences*, *7*(1), 57–64.

Parsons, A. J., Hosseini, K., Palin, R. M., & Sigloch, K. (2020). Geological, geophysical and plate kinematic constraints for models of the India-Asia collision and the post-Triassic central Tethys oceans. *Earth-Science Reviews*, *208*, 103084.

Patriat, P., & Achache, J. (1984). India-Eurasia collision chronology has implications for crustal shortening and driving mechanism of plates. *Nature*, *311*(5987), 615–621.

Pearce, J. A. (2003). Supra-subduction zone ophiolites: The search for modern analogues. *Special Papers, Geological Society of America*, 269–294.

Pearce, J. A., Harris, N. B., & Tindle, A. G. (1984). Trace element discrimination diagrams for the tectonic interpretation of granitic rocks. *Journal of Petrology*, *25*(4), 956–983.

Perfit, M. R. (2001). Mid-ocean ridge geochemistry and petrology. *Encyclopedia of Ocean Sciences*, *3*, 1778–1788.

Plafker, G. (1969). *Tectonics of the March 27, 1964 Alaska earthquake*. U.S. Geological Survey Professional Paper 543-I.

Plank, T., & Manning, C. E. (2019). Subducting carbon. *Nature*, *574*(7778), 343–352.

Pour, A. B., Hashim, M., Makoundi, C., & Zaw, K. (2016). Structural mapping of the Bentong-Raub suture zone using PALSAR remote sensing data, Peninsular Malaysia: Implications for sediment-hosted/orogenic gold mineral systems exploration. *Resource Geology*, *66*(4), 368–385.

Preiss, W. V. (2000). The Adelaide geosyncline of South Australia and its significance in Neoproterozoic continental reconstruction. *Precambrian Research*, *100* (1–3), 21–63.

Rajput, S., & Thakur, N. K. (2016). *Geological controls for gas hydrates and unconventionals*. Elsevier.

Rey, P., Vanderhaeghe, O., & Teyssier, C. (2001). Gravitational collapse of the continental crust: definition, regimes and modes. *Tectonophysics*, *342*(3–4), 435–449.

Roberts, N. J., McKillop, R., Hermanns, R. L., Clague, J. J., & Oppikofer, T. (2014). Preliminary global catalogue of displacement waves from subaerial landslides. In *Landslide science for a safer geoenvironment* (pp. 687–692). Springer, Cham.

Robock, A. (2000). Volcanic eruptions and climate. *Reviews of Geophysics*, *38*(2), 191–219.

Rosenbaum, G., Giles, D., Saxon, M., Betts, P. G., Weinberg, R. F., & Duboz, C. (2005). Subduction of the Nazca Ridge and the Inca Plateau: Insights into the formation of ore deposits in Peru. *Earth and Planetary Science Letters*, *239*(1–2), 18–32.

Rowley, D. B. (1996). Age of initiation of collision between India and Asia: A review of stratigraphic data. *Earth and Planetary Science Letters*, *145*(1–4), 1–13.

Royden, L. H., & Papanikolaou, D. J. (2011). Slab segmentation and late Cenozoic disruption of the Hellenic arc. *Geochemistry, Geophysics, Geosystems*, *12*(3).

Ruddiman, W. F. (Ed.) (2013). *Tectonic uplift and climate change*. Springer Science & Business Media.

Ruppel, C. (1995). Extensional processes in continental lithosphere. *Journal of Geophysical Research: Solid Earth*, *100*(B12), 24187–24215.

Ryan, H. F., & Coleman, P. J. (1992). Composite transform-convergent plate boundaries: description and discussion. *Marine and Petroleum Geology*, *9*(1), 89–97.

Sawkins, F. J. (1972). Sulfide ore deposits in relation to plate tectonics. *The Journal of Geology*, *80*(4), 377–397.

Schaer, J. P. (2010). Swiss and Alpine geologists between two tectonic revolutions. Part 1: From the discovery of nappes to the hypothesis of continental drift. *Swiss Journal of Geoscience*, *103*, 503–522. https://doi.org/10.1007/s00015-010-0037-x

Schaer, J. P., & Şengör, A. M. C. (2008). *Alpine geology and geosynclines: Birth and death of the concept in a small mountain range*. 2009 International Annual Meetings ASA-CSSA-SSSA.

Paper 245-2. https://a-c-s.confex.com/crops/2008am/webprogram/Paper48181.html

Scott, W., Gardner, C., Devoli, G., & Alvarez, A. (2006). The AD 1835 eruption of Volcán Cosigüina, Nicaragua: A guide for assessing local volcanic hazards. *Volcanic Hazards in Central America*, *412*, 167.

Şengör, A. C., Lom, N., & Polat, A. (2021). The nature and origin of cratons constrained by their surface geology. *GSA Bulletin 2021*. https://doi.org/10.1130/B36079.1

Şengör, A. M. C. (2021). History of geology. In D. Alderton & S. A. Elias (Eds.), *Encyclopedia of geology*, *2d ed.* Academic Press. https://doi.org/10.1016/B978-0-08-102908-4.00084-9

Seton, M., Müller, R. D., Zahirovic, S., Gaina, C., Torsvik, T., Shephard, G., et al. (2012). Global continental and ocean basin reconstructions since 200 Ma. *Earth-Science Reviews*, *113*(3–4), 212–270.

Sharma, A., & Zaman, F. (2019). The Great Assam Earthquake of 1950: A historical review. *Senhri Journal of Multidisciplinary Studies*, *4*, 1–10. https://senhrijournal.ac.in/wp-content/uploads/2020/12/The-Great-Assam-Earthquake-of-1950-A-Historical-Review.pdf

Shervais, J. W., & Kimbrough, D. L. (1985). Geochemical evidence for the tectonic setting of the Coast Range ophiolite: A composite island arc-oceanic crust terrain in western California. *Geology*, *13*(1), 35–38.

Siebert, L., Cottrell, E., Venzke, E., & Andrews, B. (2015). Earth's volcanoes and their eruptions: An overview. *The Encyclopedia of Volcanoes*, 239–255.

Sillitoe, R. H. (2010). Porphyry copper systems. *Economic Geology*, *105*(1), 3–41.

Simpson, R. W. (1997). Quantifying Anderson's fault types. *Journal of Geophysical Research: Solid Earth*, *102*(B8), 17909–17919.

Sleep, N. H. (2005). Evolution of the continental lithosphere. *Annual Review of Earth and Planetary Sciences*, *33*, 369–393.

Solheim, L. P., & Peltier, W. R. (1994). Avalanche effects in phase transition modulated thermal convection: A model of Earth's mantle. *Journal of Geophysical Research: Solid Earth*, *99*(B4), 6997–7018.

Song, Z., Wan, S., Colin, C., Yu, Z., Révillon, S., Jin, H., Zhang, J., et al. (2021). Paleoenvironmental evolution of South Asia and its link to Himalayan uplift and climatic change since the late Eocene. *Global and Planetary Change*, *200*, 103459. https://doi.org/10.1016/j.gloplacha.2021.103459

Sorkhabi, R. (2013). Know your faults! Part I. *Geoeducation World Wide*, *9*(5), 64–68, https://www.geoexpro.com/articles/2013/03/know-your-faults-part-i

Spakman, W., Wortel, M. J. R., & Vlaar, N. J. (1988). The Hellenic subduction zone: A tomographic image and its geodynamic implications. *Geophysical Research Letters*, *15*(1), 60–63.

Srivastava, H. N., Bansal, B. K., & Verma, M. (2013). Largest earthquake in Himalaya: An appraisal. *Journal of the Geological Society of India*, *82*, 15–22. https://doi.org/10.1007/s12594-013-0117-4

Steinmann, G. (1906). *Geologische probleme des alpengebirges: Eine einführung in das verständnis des gebirgsbaues der Alpen. Deutscher und Österreichischer Alpenverein.*

Stern, R. J. (2002). Subduction zones. *Reviews of Geophysics*, *40*(4), 3–1.

Stern, R. J., & Gerya, T. (2018). Subduction initiation in nature and models: A review. *Tectonophysics*, *746*, 173–198.

Stern, R. J., & Miller, N. R. (2018). Did the transition to plate tectonics cause Neoproterozoic Snowball Earth? *Terra Nova*, *30*(2), 87–94.

Straub, S. M., Gómez-Tuena, A., & Vannucchi, P. (2020). Subduction erosion and arc volcanism. *Nature Reviews Earth & Environment*, *1*(11), 574–589.

Sugawara, D., Minoura, K., & Imamura, F. (2008). Tsunamis and tsunami sedimentology. In *Tsunamiites* (pp. 9–49). Elsevier.

Sugimura, A., & Uyeda, S. (1973). *Island arcs: Japan and its environs. Developments in Geotectonics*. Elsevier Scientific Publishing Company.

Toda, S., & Tsutsumi, H. (2013). Simultaneous reactivation of two, subparallel, inland normal faults during the M w 6.6, 11 April 2011 Iwaki earthquake triggered by the M w 9.0 Tohoku-oki, Japan, earthquake. *Bulletin of the Seismological Society of America*, *103*(2B), 1584–1602.

Turcotte, D. L., & Schubert, G. (2002). *Geodynamics*. New York: Cambridge University Press.

Twiss, R. J., & Moores, E. M. (1992). *Structural geology*. New York: Macmillan.

USGS Earthquakes Hazards (2019). *20 largest earthquakes in the world*. https://www.usgs.gov/programs/earthquake-hazards/science/20-largest-earthquakes-world

Van Hinsbergen, D. J., Lippert, P. C., Dupont-Nivet, G., McQuarrie, N., Doubrovine, P. V., Spakman, W., & Torsvik, T. H. (2012). Greater India Basin hypothesis and a two-stage Cenozoic collision between India and Asia. *Proceedings of the National Academy of Sciences*, *109*(20), 7659–7664.

Van Hinsbergen, D. J., Steinberger, B., Guilmette, C., Maffione, M., Gürer, D., Peters, K., et al. (2021). A record of plume-induced plate rotation triggering subduction initiation. *Nature Geoscience*, *14*(8), 626–630.

Van Hinsbergen, D. J. J., Hafkenscheid, E., Spakman, W., Meulenkamp, J. E., & Wortel, R. (2005). Nappe stacking resulting from subduction of oceanic and continental lithosphere below Greece. *Geology*, *33*(4), 325–328.

Vine, F. J., & Matthews, D. H. (1963). Magnetic anomalies over oceanic ridges. *A century. Nature*, *4897*, 947–949.

Von Huene, R., & Scholl, D. W. (1991). Observations at convergent margins concerning sediment subduction, subduction erosion, and the growth of continental crust. *Reviews of Geophysics*, *29*(3), 279–316.

Wallace, P. J. (2021). Magmatic volatiles. *In Encyclopedia of geology, 2d ed.* Academic Press. https://doi.org/10.1016/B978-0-08-102908-4.00097-7.

Webb, A. A. G., Guo, H., Clift, P. D., Husson, L., Müller, T., Costantino, D., et al. (2017). The Himalaya in 3D: Slab dynamics controlled mountain building and monsoon intensification. *Lithosphere*, *9*(4), 637–651.

Wegener, A. (1912). Die Entstehung der Kontinente. *Geologische Rundschau*, **3**, 276–292. https://doi.org/10.1007/BF02202896

Wei, F., Prytulak, J., Xu, J., Wei, W., Hammond, J. O., & Zhao, B. (2017). The cause and source of melting for the most recent volcanism in Tibet: A combined geochemical and geophysical perspective. *Lithos*, *288*, 175–190.

Westbrook, G. K., & Reston, T. J. (2002). The accretionary complex of the Mediterranean Ridge: Tectonics, fluid flow and the formation of brine lakes, an introduction to the special issue of marine geology. *Marine Geology*, *1*(186), 1–8.

White, R. S., McKenzie, D., & O'Nions, R. K. (1992). Oceanic crustal thickness from seismic measurements and rare earth element inversions. *Journal of Geophysical Research*, *97*(B13), 19683–19715. https://doi.org/10.1029/92JB01749

Woodcock, N. H. (1986). The role of strike-slip fault systems at plate boundaries. *Philosophical Transactions of the Royal Society of London. Series A, Mathematical and Physical Sciences*, *317*(1539), 13–29.

Wortel, M. J. R., Goes, S. D. B., & Spakman, W. (1990). Structure and seismicity of the Aegean subduction zone. *Terra Nova*, *2*(6), 554–562.

Xue, S., Deng, J., Wang, Q., Xie, W., & Wang, Y. (2021). The redox conditions and C isotopes of magmatic Ni-Cu sulfide deposits in convergent tectonic settings: The role of reduction process in ore genesis. *Geochimica et Cosmochimica Acta*, *306*, 210–225.

Yamano, M., Honda, S., & Uyeda, S. (1984). Nankai Trough: A hot trench? *Marine Geophysical Researches*, *6*(2), 187–203.

Yang, T., Gurnis, M., & Zahirovic, S. (2018). Slab avalanche-induced tectonics in self-consistent dynamic models. *Tectonophysics*, *746*, 251–265.

Yin, A., & Harrison, T. M. (2000). Geologic evolution of the Himalayan-Tibetan orogen. *Annual Review of Earth and Planetary Sciences*, *28*(1), 211–280.

Yolkin, E. A., Kontorovich, A. E., Bakharev, N. K., Belyaev, S. Y., Varlamov, A. I., Izokh, N. G., et al. (2007). Paleozoic facies megazones in the basement of the West Siberian geosyncline. *Russian Geology and Geophysics*, *48*(6), 491–504.

Zheng, Y. F. (2019). Subduction zone geochemistry. *Geoscience Frontiers*, *10*(4), 1223–1254.

Zhu, B., Guo, Z., Zhang, S., Ukstins, I., Du, W., & Liu, R. (2019). What triggered the early-stage eruption of the Emeishan large igneous province? *GSA Bulletin*, *131*(11–12), 1837–1856.

Zhu, J., Zhang, Z., Santosh, M., Tan, S., Deng, Y., & Xie, Q. (2021). Recycled carbon degassed from the Emeishan plume as the potential driver for the major end-Guadalupian carbon cycle perturbations. *Geoscience Frontiers*, *12*(4). https://doi.org/10.1016/j.gsf.2021.101140

2

Subduction and Obduction Processes: The Fate of Oceanic Lithosphere Revealed by Blueschists, Eclogites, and Ophiolites

Philippe Agard[1], Mathieu Soret[1,2], Guillaume Bonnet[1], Dia Ninkabou[1], Alexis Plunder[3], Cécile Prigent[4], and Philippe Yamato[5]

ABSTRACT

Fragments of ancient oceanic lithosphere preserved in mountain belts, though volumetrically subordinate, provide essential insights into past geodynamics and formation and destruction of oceanic lithosphere. This contribution shows how the two types of oceanic fragments, blueschists and eclogites, on one hand, and ophiolites on the other, preserve crucial information on the dynamics of oceanic convergence, that is, subduction and obduction. Their mutual relationships, as well as the similarities and differences in the mechanisms leading to their preservation, allow tracking the evolution of the subduction process through time, from the onset of intraoceanic subduction to the cessation of continental subduction, and, in some cases, to the obduction of ophiolites. Fragments located at the base and immediately below unmetamorphosed (true) ophiolites represent witnesses of intraoceanic subduction initiation and reveal, in particular, initial mechanical resistance to subduction, subsequent cooling, and gradual strain localization. Subducted fragments of oceanic lithosphere metamorphosed as blueschists and eclogites, scraped off the downgoing slab episodically, at shallow or great depths, provide direct access to the composition, structure, and rheology of rocks at the plate interface. Both types reflect the mechanical behavior and "hiccups" of the subduction plate boundary, during subduction initiation and mature subduction, respectively.

2.1. INTRODUCTION

Blueschists and eclogites have now been collected and studied for more than two centuries (de Saussure, 1804; Ernst, 1971). Ophiolites, at least coined as such, have been studied for about 50 yr (Coleman, 1971; Dewey, 1976; Moores, 1982). Both represent important milestones for the theory of plate tectonics. More often than not, however, these rock associations are studied independently. The present chapter provides a short odyssey through these two types of oceanic fragments to look at their intimate links, as well as to show the advantage of studying them jointly for understanding the dynamics of oceanic convergence.

Oceanic lithosphere, formed by partial melting along ocean ridges or by extreme thinning of the mantle, makes up more than 60% of the surface of our planet. Despite this extreme prevalence, ocean floors are younger than ~200 Ma around most of the globe (Fig. 2.1a; except possibly in the eastern Mediterranean). Oceanic lithosphere

[1] *Sorbonne Université, CNRS-INSU, Institut des Sciences de la Terre Paris, Paris, France*

[2] *Institut des Sciences de la Terre d'Orléans, Université d'Orléans, Orléans, France*

[3] *BRGM (French Geological Survey), Université d'Orléans, Orléans, France*

[4] *Institut de Physique du Globe de Paris, Sorbonne Paris Cité, Université Paris Diderot, Paris, France*

[5] *Géosciences Rennes, Université de Rennes 1, Rennes, France*

Compressional Tectonics: Plate Convergence to Mountain Building, Geophysical Monograph 277, First Edition.
Edited by Elizabeth J. Catlos and İbrahim Çemen.

DOI:10.1002/9781119773856.ch02

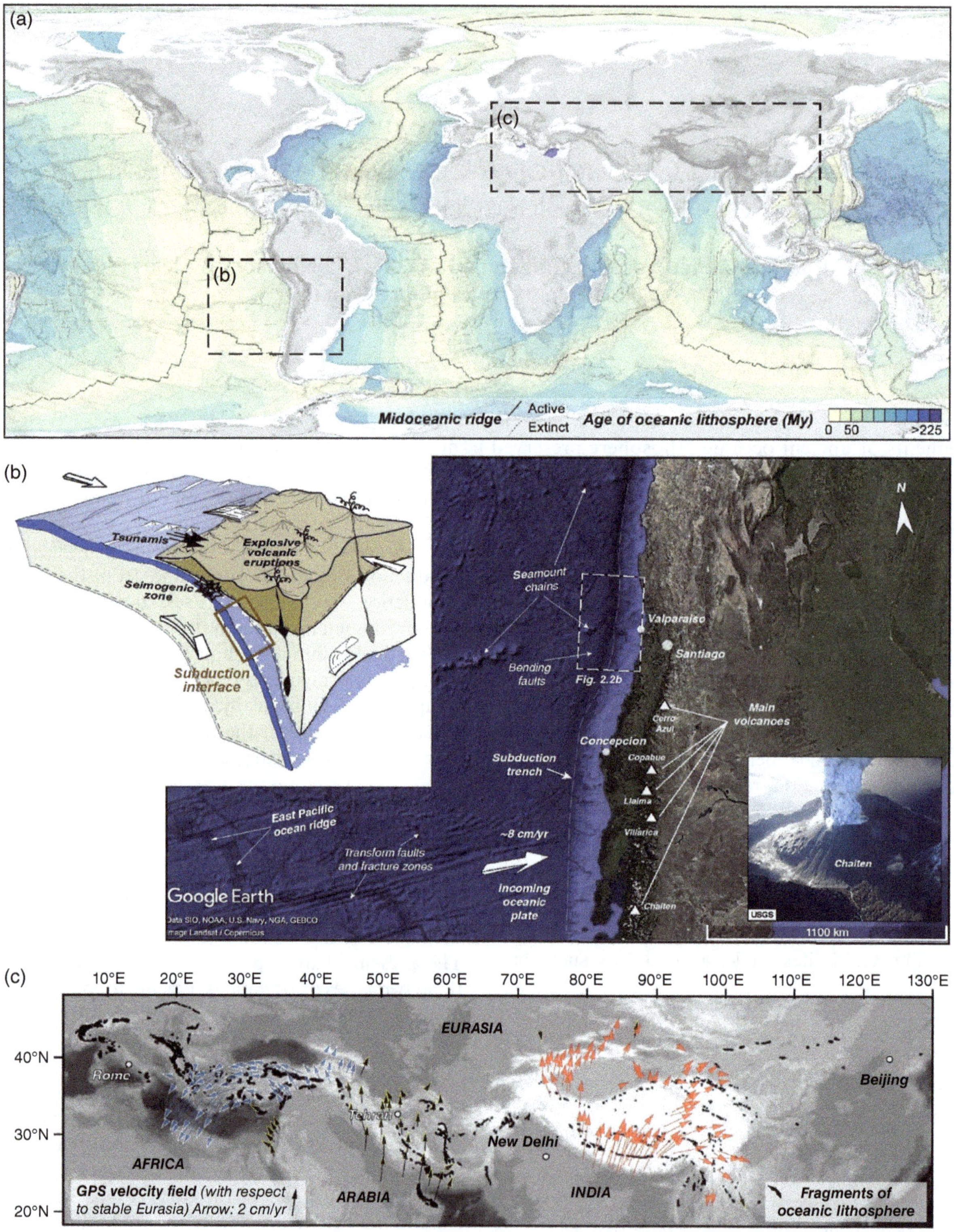

Figure 2.1 (a) Age map of present-day oceanic lithosphere, after Seton et al. (2020). Almost all of it is younger than 200 Ma (save, perhaps, for part of the eastern Mediterranean domain whose age is debated). (b) The subduction process, where oceanic lithosphere goes down the "escalator" with devastating earthquakes triggered along the subduction interface, tsunamis, and explosive volcanic eruptions. About half of the entire seismic energy of the globe is being released in the Chilean subduction zone! (c) The fragments of oceanic lithosphere disseminated along the suture zones of the Alpine-Himalayan mountain belts mark the location of former Tethyan oceans. Arrows indicate present-day displacements, with respect to stable Eurasia, deduced from satellite data (after Agard et al., 2011).

is young because it is doomed: Most of it, save for a few fragments, irreversibly disappears in convergent zones through the subduction process.

The subduction machine drives ocean recycling and triggers the largest known earthquakes, as well as devastating volcanic eruptions (Stern, 2002; Fig. 2.1b). Subduction, together with mantle convection (Forsyth & Uyeda, 1975; Coltice et al., 2019), not only governs solid Earth dynamics but also fundamentally connects the atmosphere and the biosphere with the deep Earth: Water, carbon, or any other element of the Mendeleev table travels down through subduction zones.

Some fragments of oceanic lithosphere nevertheless escape their tragic fate and are preserved as slivers in recent as well as ancient mountain belts. They form oceanic sutures outlining fossil plate boundaries where former oceans have disappeared (Fig. 2.1c) and constitute major lithospheric scars. As such, they commonly play an important role in later deformation of the Earth's crust.

The preserved fragments of oceanic lithosphere (mantle, crust, sediments), recognizable through their petrological and structural features, are of two major types: (1) ophiolites, that is, hundreds of kilometer-long fragments with a diagnostic lithological sequence that have largely escaped later metamorphic transformations, thanks to a process called obduction (Coleman, 1971; Moores, 1982); (2) blueschists or eclogites that have experienced high-pressure low-temperature (HP-LT) metamorphic conditions during subduction to variable depths (Agard et al., 2009). While the latter are sometimes referred to as ophiolites (or ophiolitic) to underline their oceanic origin, we shall restrict the use of the term *ophiolite* to the former type for clarity. We will see that this distinction echoes a more fundamental one related to their mode of emplacement.

The first type of fragment informs us about the detailed constitution of pristine oceanic lithosphere. While surveyed and partly sampled at sea, oceanic lithosphere is conveniently investigated on foot, on land: The Semail, Josephine, or Newfoundland ophiolites, for example, reveal important details on the genesis of oceanic lithosphere (cf. section 2.2; Table 2.1). We will see that ophiolites also preserve crucial information on the process of subduction initiation. The second type provides an essential glimpse onto the mechanisms of its destruction, that is on the subduction process itself: Metamorphosed fragments keep traces of the transformations experienced

Table 2.1 Selected list of the most characteristic and well-studied ophiolites in the world

Ophiolite	Formation age (Ma)	Onset of obduction (Ma)	Type	Approx. size (km)	References*
Bay of Islands, Newfoundland	485	460	MOR (back-arc?)	>>100 × 30	Suhr and Cawood, 1993; Dewey and Casey, 2013
Coast Range, USA	170–165	165–160	SSZ	400 × 40	McLaughlin et al., 1988; Choi et al., 2008
Josephine, USA	165–160	155–150	SSZ	100 × 15	Harper et al., 1994; Harper, 2003
Lizard, England	>375	370–365	MOR	20 × 15	Jones, 1997; Strachan et al., 2014
Lycian nappes, Turkey	100–95	95–90	SSZ	200 × 50	Celik et al., 2011; Plunder et al., 2016
Mirdita, Albania	175–165	165–160	SSZ	200 × 25	Nicolas et al., 1999; Dilek et al., 2007
Nappe des péridotites, New Caledonia	100, 80?	55	SSZ (forearc) + MOR	300 × 50	Ulrich et al., 2010; Cluzel et al., 2012
PUB, Papoua-New Guinea	70–65	60–55	MOR	400 × 40	Davies and Jacques, 1984; Lus et al., 2004
Semail, Oman-UAE	96–95	95–90	SSZ	500 × 100	Nicolas et al., 2000; Rioux et al., 2012
Sevan, Armenia	165	95–90	MOR	100 × 30	Galoyan et al., 2009; Hassig et al., 2016
Sistan, Iran	110–100	70	MOR	300 × 25	Zarrinkoub et al., 2012; Jentzer, 2022
Tanimbar, Timor	10–5	5–2	SSZ (forearc)	150 × 20	Linthout et al., 1997; Kaneko et al., 2007
Troodos, Cyprus	100–95	95–90	SSZ	50 × 15	Moores and Vine, 1971; Pearce and Robinson, 2010

* For a larger compilation see Furnes et al. (2014).

at depth, during their transient burial and exhumation along the subduction plate boundary.

After briefly recalling some of the basic petrological features of the oceanic lithosphere (section 2.2), this short odyssey starts by considering the relics left over by subduction and the precious insights they give us. Obduction is comparatively less frequent and in fact corresponds, as shown below, to one of subduction dead ends.

2.2. DIVERSITY OF OCEANIC LITHOSPHERES

Two end-member types of oceanic lithosphere are recognized, depending on whether expansion rates and magmatic accretion are fast and profuse, or instead (ultra-)slow and limited (Fig. 2.2a). Pacific-type lithospheres belong to the first type, whereas the Atlantic, Southwest Indian, and Tethyan lithospheres exemplify the second one.

Fast-spreading oceans are characterized by a ~5–7 km thick and continuous crust made of gabbros and basalts (mafic, or basic rocks), onto which a veneer of sediments is deposited, typically ~100 m thick far from subduction trenches (Clift & Vannucchi, 2004). This lithological structure is also referred to as the "ocean plate stratigraphy" (Isozaki et al., 1990; Kusky et al., 2014; Wakabayashi et al., 2015). Magmatic production can increase significantly and the oceanic crust may reach up to ~20 km in oceanic plateaus (e.g., Ontong-Java, Aleutians).

At the other end of the spectrum, slow- and ultra-slow spreading oceans are heterogeneous, characterized by a partly serpentinized mantle lithosphere intruded by gabbroic bodies exhumed to the seafloor through detachment faults (Cannat et al., 2006) and sparse magmatic segments or centers made of basalts and gabbros (Fig. 2.2a).

Oceanic lithosphere can thus be quite heterogeneous vertically or laterally, through variations of morphology, structure, lithology, and crustal thickness. Another source of heterogeneity comes from the considerable variations of surface features and rugosity (Lallemand et al., 2018): Seamounts and ridges, magmatic or not, as well as major fracture zones, transform faults or bending faults formed near subduction trenches (Ranero et al., 2003). These features can be very unevenly distributed, as observed along strike Chile for example (Fig. 2.2b).

2.3. BLUESCHISTS AND ECLOGITES: FRAGMENTS THAT HAVE ESCAPED IRREVERSIBLE BURIAL

Oceans vanish on geological timescales (Fig. 2.1). Fragments recovered from subduction depths, albeit precious to probe Earth's interiors, are very subordinate compared to the vast amounts of oceanic lithosphere metamorphosed and irreversibly subducted. The fate of oceanic lithosphere in subduction zones is illustrated in the following through two main examples: (1) a former seamount recently discovered in southwest Iran, exposed in the Neo-Tethyan suture within the Zagros orogen; and (2) the extensively documented domain of the European Alps, with many well-preserved blueschists and eclogites.

2.3.1. Siah Kuh (Zagros, Iran): A Seamount Subducted at Shallow Depths and Later Exhumed

The oceanic origin of this massif, largely volcanic, is reflected in its architecture and petrology (Bonnet et al., 2019). From bottom to top, its succession comprises peridotites, gabbros, with typical oceanic textures, a >2 km thick sequence of basalts capped by reef limestones and other sediments. Younger lava flows were emplaced on top of the sediments, revealing the existence of a prolonged magmatic activity (Fig. 2.3a–c). The sedimentary sequence reveals a progressive subsidence of this massif on the seafloor: shallow-water coral-bearing limestones, then slope deposits with limestone fragments and débris flows, and finally deep-sea radiolarites. The dimensions of Siah Kuh, ~15 km*20 km and about 2 km high (its base is not exposed), appear comparable to seamounts observed on the seafloor (Fig. 2.2b; Cloos, 1992).

This massif, though largely intact, is made up of two distinct units (A and B) separated by a several kilometer-long thrust that exposes the mantle and gabbros of unit B (Fig. 2.3d). The massif is not only deformed but also shows mineral transformations that attest to its partial burial during subduction. Transformations are more pervasive in the northeast quarter of the edifice (Fig. 2.3c). The presence of metamorphic aragonite and lawsonite or recrystallization of blue amphibole reveal subduction of the B unit to a pressure of 0.8 GPa (Bonnet et al., 2019; Fig. 2.3c,e). The Siah Kuh massif therefore consists of two adjacent portions of oceanic lithosphere subducted down to a maximum ~25–30 km depth, considering pressure estimates as lithostatic pressure, then scraped off the downgoing plate (i.e., the slab). They were finally embedded in the Eurasian margin as a result of the collision between the Arabian and Eurasian plates during the Tertiary.

This nicely preserved seamount has therefore evolved and recrystallized at the seismogenic depths of subduction zones, where the largest seismic ruptures reach moment magnitudes Mw greater than 8 (i.e., hundred-kilometer-long ruptures, such as the 2010 Maule earthquake in Chile or the 2011 Tohoku earthquake in Japan; Figs. 2.1b, 2.3g). Evaluating seismic risk and predicting

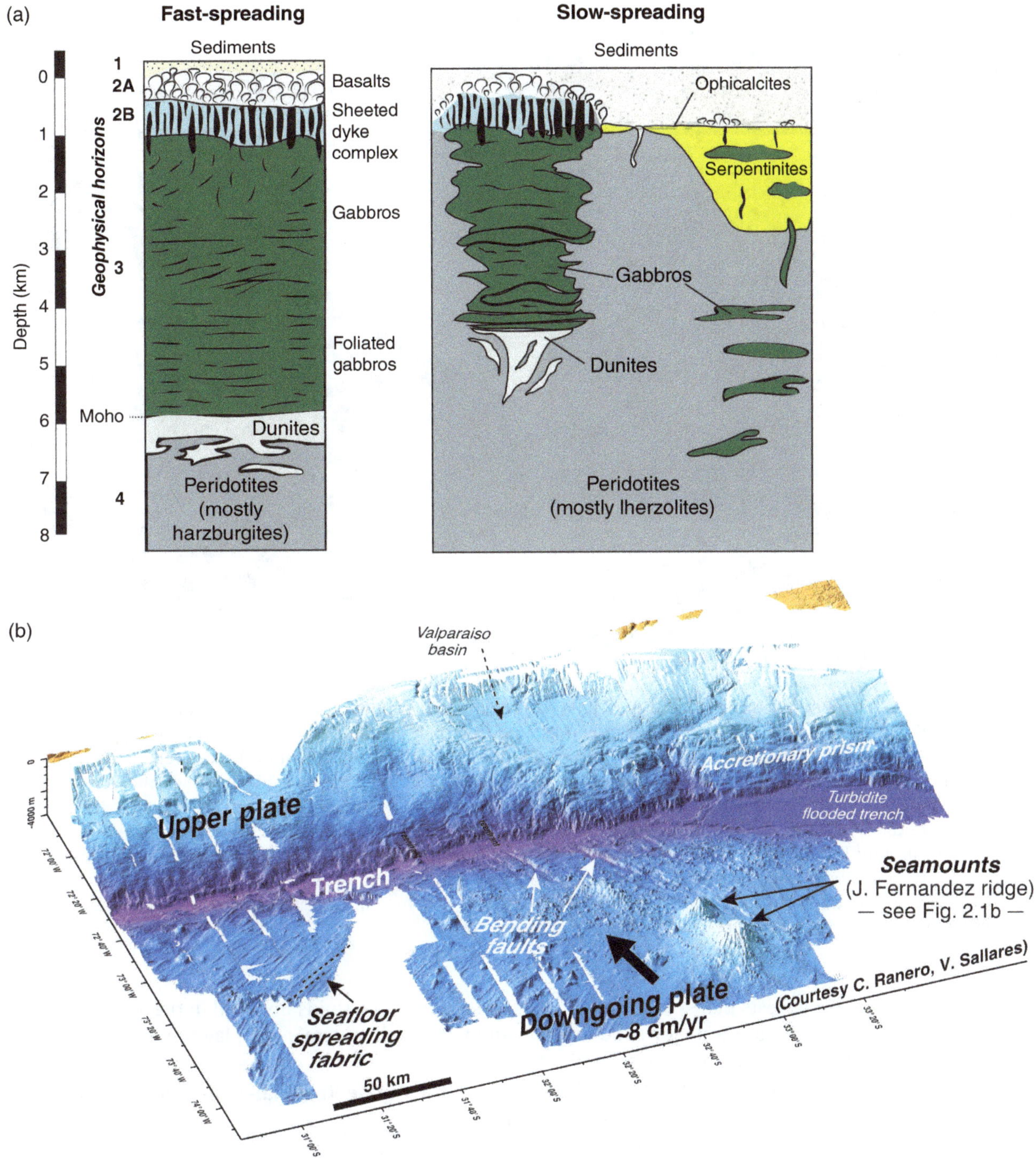

Figure 2.2 (a) First-order characteristics of oceanic lithosphere formed at fast-spreading (left) or slow-spreading oceans (right). (b) Seafloor heterogeneities are remarkable: fracture zones, bending and transform faults, ocean ridges and seamounts, as shown here off Santiago, Chile (location in Fig. 2.1b).

the magnitude of (future) earthquakes requires assessing the mechanical behavior and along-strike segmentation of the subduction zone (Fig. 2.1b). Therefore, whether morphological and lithological heterogeneities entering the subduction trench (such as seamounts or seamount chains; Figs. 2.1b, 2.2b) constitute barriers to the propagation of earthquakes, and, therefore, control earthquake size, or instead form asperities likely to trigger them, is currently much debated (Wang & Bilek, 2011; Lallemand et al., 2018).

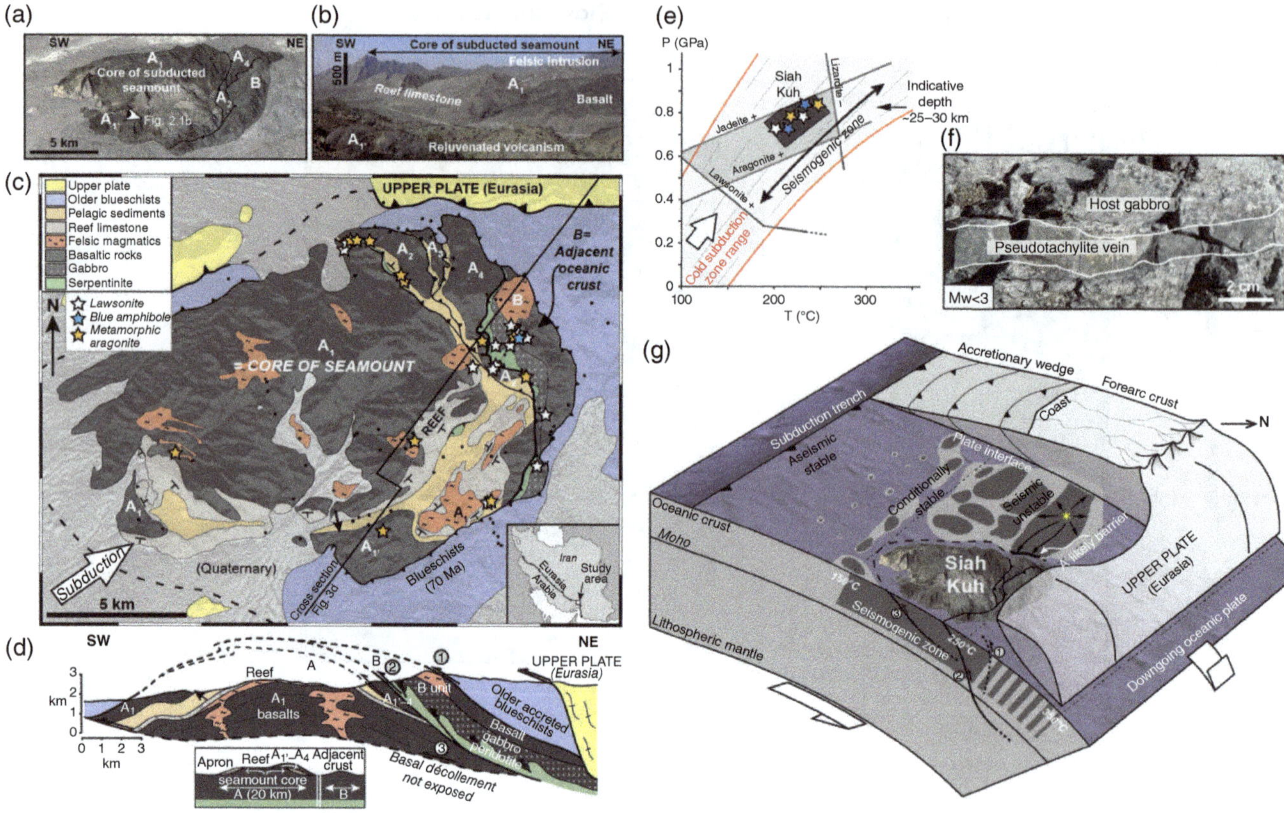

Figure 2.3 (a) A former seamount stands out of the Quaternary infill, southwest Iran: the Siah Kuh massif ("black mountain"). (b) Reef limestones form like a crown around the main volcanic edifice. (c) Simplified geological map of the massif outlining the main edifice (unit A) and the one to the northeast (unit B), as well as the many mineral metamorphic recrystallizations, particularly in the vicinity of the large thrust separating units A and B. (d) Cross-section view showing the juxtaposition of the two portions of oceanic lithosphere and tentative restoration before offscraping along the subduction interface. (e) The stability of metamorphic minerals yields pressure estimates around 0.8 GPa for unit B. The greater extent of mineral transformations in the northeast suggests a greater burial for the B unit, consistent with the former subduction direction to the northeast, beneath Eurasia. (f) Only a few pseudotachylite veins are observed in the Siah Kuh massif. (g) Idealized view of the Siah Kuh seamount at peak burial conditions, within the seismogenic zone.

The Siah Kuh seamount suggests that stresses accumulated on this massif did not induce a major subduction earthquake. Pseudotachylites, which are considered as the most reliable evidence for the local and brief melting of rocks by the frictional heat released during an earthquake, are rare and their size would indicate magnitudes Mw less than 3 (Fig. 2.3g; Bonnet et al., 2019). If there were once major earthquakes, they carefully avoided the interior of Siah Kuh (Fig. 2.3g).

Subducted seamounts, although rarely preserved intact, are found as fragments offscraped and later accreted to shallow depths in former accretionary prisms in Japan, California, Turkey, or New Zealand. The Iranian example, however, compared with others, commonly hm to km scale at best, provides access to the entire structure in 3D. This example also provides an opportunity to study the deformation processes and mineral transformations at these depths, as well as the nature and pathways of fluids involved in the formation of veins and water-rich minerals (e.g., lawsonite).

2.3.2. The Alpine Regional-Scale Record of Subduction Processes

In the heart of the Alps, the internal domains preserve the trace of two extinct oceanic branches of the Alpine Ocean: the Liguro-Piemont and Valais domains (Figs. 2.4a,b; Lemoine et al., 1986; Vissers et al., 2013). The small ~700 km wide Alpine Ocean can be compared with a rather narrow zone of today's North Atlantic Ocean, such as between Norway and Greenland. Like the Atlantic, its slow-spreading nature is attested by the type of peridotite, commonly partly refertilized and showing the full range of compositions between inherited subcontinental and asthenospheric mantle (Manatschal & Müntener, 2009; Picazo et al., 2016), the abundance of

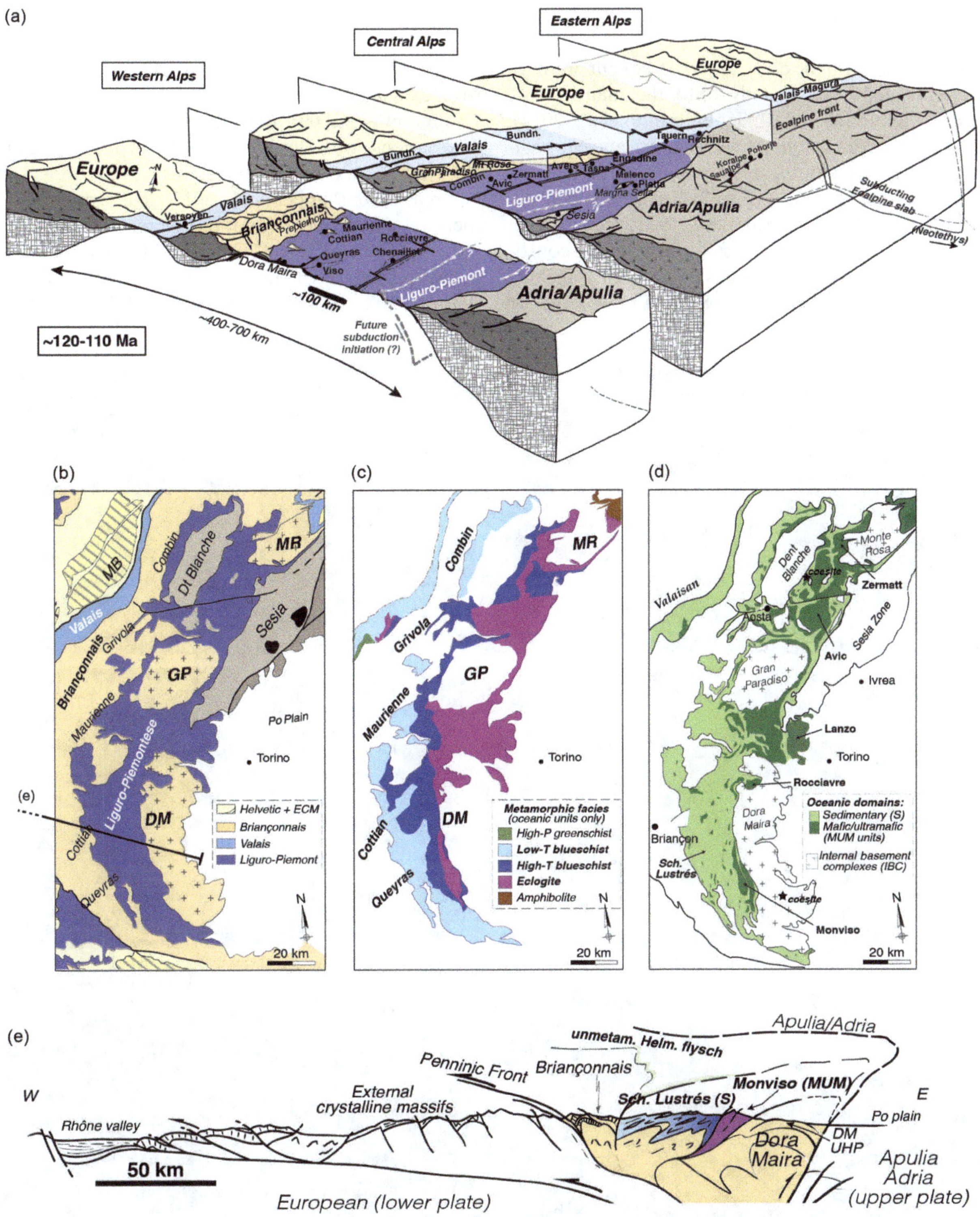

Figure 2.4 (a) Schematic paleogeography of the Alpine ocean around 120 Ma, approximately at the time of its greatest extension (Agard, 2021). Some localities are indicated for reference. (b) Structural sketch map of the western Alps. Color coding is the same as for Figure 2.4a and highlights, for example, the presence of an intervening continental block (the Briançonnais) between and along most of the Liguro-Piemont and Valais branches of the ocean. (c) Distribution of metamorphic facies along the western Alps, showing a marked increase toward the east. (d) Distribution of lithologies along the western Alps: note the spatial contrast, somewhat coincident with that of metamorphic facies, between the S (dominantly sedimentary) and MUM (dominantly mafic and ultramafic) units. (e) Crustal-scale section across the western Alps, at the latitude of Monviso and Dora Maira (Fig. 2.4b). Remnants of the oceanic lithosphere include (i) the unmetamorphosed Helminthoid Flysch deposits (Helm. flysch) representing the former trench infill (see Fig. 2.7a), the dominantly sedimentary Schistes Lustrés, and the mafic and ultramafic bodies like Monviso.

mantle compared with the crust, and the frequent presence of sediments deposited directly onto the mantle (e.g., ophicalcites; Lemoine et al., 1986; Beltrando et al., 2014).

The various oceanic relics (pillow lavas, ancient sediments, gabbros, and mantle) are now transformed to blueschist and eclogite facies rocks (Figs. 2.4c–e, 2.5), evidencing subduction of the Alpine Ocean (Berger & Bousquet, 2008; Agard et al., 2009). In the metabasalts, the minerals constituting the characteristic associations of blueschists include blue amphibole associated with lawsonite or epidote, or garnet, and for the eclogites, garnet and sodium clinopyroxene of jadeitic nature (Fig. 2.5; Angiboust et al., 2009; Groppo et al., 2009). Metasediments show mineralogical associations with ferro-magnesio carpholite, lawsonite, chloritoid or garnet, from blueschists to eclogites (Goffé & Chopin, 1986; Plunder et al., 2012). These minerals reflect thermodynamic reequilibration at pressures ranging from 1 to 1.3 GPa at 300–400°C to 2.8–3.0 GPa at about 550–570°C, respectively (Fig. 2.6a; Agard, 2021). Considering that pressure estimates correspond to lithostatic pressure,

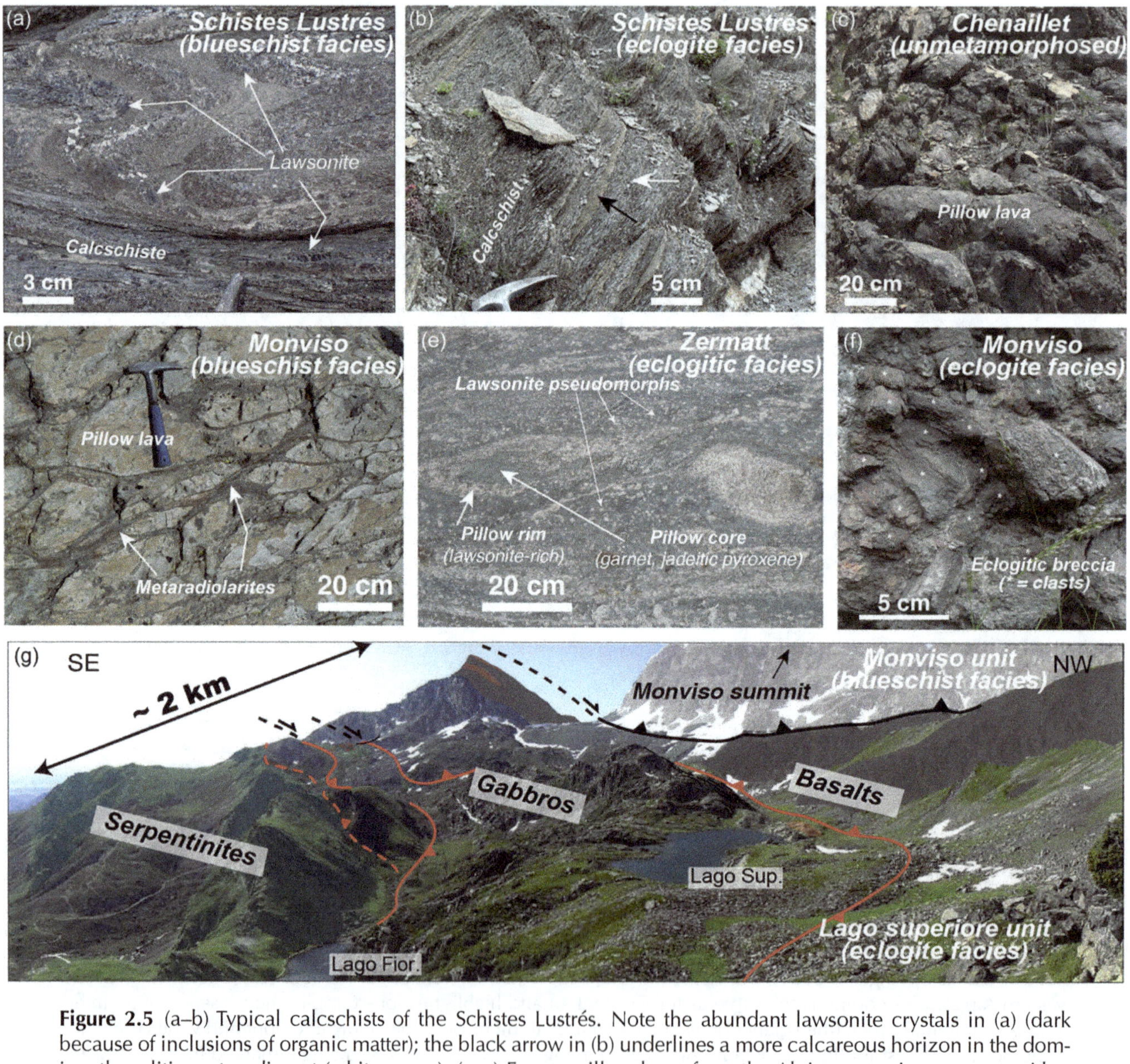

Figure 2.5 (a–b) Typical calcschists of the Schistes Lustrés. Note the abundant lawsonite crystals in (a) (dark because of inclusions of organic matter); the black arrow in (b) underlines a more calcareous horizon in the dominantly pelitic metasediment (white arrow). (c–e) Former pillow lavas from the Alpine oceanic crust now either unmetamorphosed (c), blueschist facies (d), or eclogite facies (e). (f) Metagabbros brecciated at eclogite facies conditions: the same eclogitic mineralogy is found in the clasts (underlined by stars) and (g) in the cement. One of the best-preserved pieces of slab in the world, and one of the deepest known so far: (g) the Lago Superiore eclogitic tectonic slice (Italian side of Monviso). See Angiboust et al. (2012) and Locatelli et al. (2018) for details.

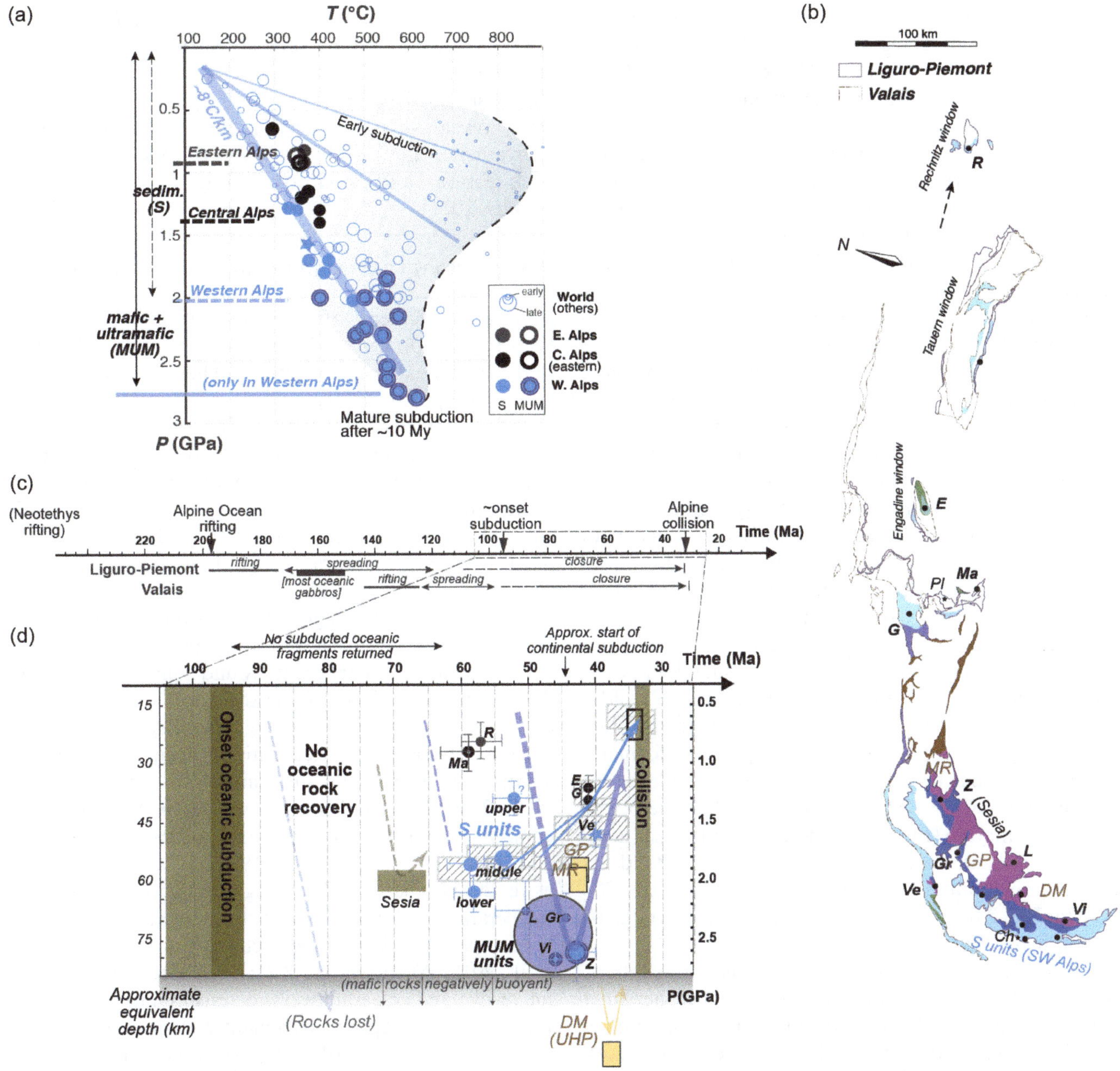

Figure 2.6 (a) Maximum pressure (and burial) reached by representative subducted fragments from across the entire Alps (see Agard, 2021). Note the decrease of peak pressure toward the east, and the alignment of all remnants on the P-T gradient typical of mature subduction (hence suggesting ~8°C/km considering pressure as lithostatic). These estimates are set back against worldwide literature data (Agard et al., 2018). (b) Map of the exposed remnants of oceanic lithosphere in the Alps. Metamorphosed oceanic fragments are color coded as in Figure 2.4c. Abbreviations: E: Engadine, G: Grisons, Gr: Grivola, L: Lanzo, Ma: Malenco, R: Rechnitz, Ve: Versoyen, Vi: Monviso. Some largely unmetamorphosed oceanic fragments are also found, like Chenaillet (Ch) or Platta (Pl). Large, eclogitized subducted continental fragments, in addition to Sesia, include Dora Maira (DM), Gran Paradiso (GP), and Monte Rosa (MR). (c) Opening and closing of the two branches of the Alpine ocean. Oceanic subduction lasted approximately 50–60 Myr. (d) Pressure-time paths of subducted fragments of oceanic lithosphere.

these rocks indicate equivalent burial depths from approximately ~30 km down to 80–100 km.

Fragments of the upper few kilometers of the slab are found almost intact in places. In Monviso, the mantle/gabbro/basalt/sediment sequence is overall preserved (Angiboust et al., 2011), although these rocks were likely buried at a depth of 80 km (Fig. 2.5g). Fracturing at peak burial is even found in some eclogites (e.g., eclogitic breccias;

Fig. 2.5f), otherwise deformed in an intimate, ductile manner; they likely represent the scars of past subduction earthquakes (Angiboust et al., 2012; Locatelli et al., 2018).

The distribution of these alpine fragments reveals first-order contrasts:

1. Across the French-Italian Alps, the most intensely metamorphosed rocks are found in the eastern part of the Liguro-Piemont domain (Figs. 2.4c, 2.6b), indicating that the ocean subducted to the east, under Apulia (i.e., the present-day Po Plain). These rocks furthermore formed along a gradient of ~8°C–10°C/km typical of mature subduction zones, considering pressure as lithostatic (Agard et al., 2018; Fig. 2.6a).

2. A much larger fraction of sediments (than mafic crust) has been exhumed, mostly from pressures between 1.3 and 2.1 GPa (Agard, 2021). They form the S (sedimentary) units, commonly referred to as "Schistes Lustrés" or "Bündnerschiefer" (Fig. 2.4d).

3. Mafic eclogites (metabasalts, metagabbros), mainly present in the western Alps, are closely associated with serpentinites in MUM (mafic and ultramafic) units structurally distinct from their former seafloor sedimentary cover (Fig. 2.4d). They reached higher pressures, mostly between 2.3 and 2.8 GPa (hence equivalent depths up to 80 km), late in the history of oceanic subduction, at the time continental subduction began (Fig. 2.6c): No older crustal fragments have survived. These remnants of oceanic lithosphere correspond to fragments formerly located at the edge of the continent: Their burial ages are only slightly older (~5 Ma) than those at the edge of the continental margin (Fig. 2.6c), and the age of initial magmatic crystallization of these gabbros (and basalts) dates from the beginning of the history of expansion (~170–155 Ma; Fig. 2.6d), which places them well at the edge of the continent (Fig. 2.7a).

4. Nonmetamorphic fragments of oceanic lithosphere, sometimes called "ophiolites," are found in the Chenaillet and Platta massifs (Figs. 2.5d, 2.6b). However, unlike ophiolites presented in the following (see section 2.4), their spatial extension is on the 1–10 km scale and they have no metamorphic sole. They were likely sliced off the slab at shallow depth and emplaced in the subduction prism, like the Marine Headlands sliver in California, which was sliced off at a maximum 8–10 km depth (and is now exposed at the northern end of the Golden Gate; Meneghini & Moore, 2007).

5. While subduction lasted for about 60 Myr through slow, cm/year closure between ~100 and 40 Ma (Fig. 2.6c), no fragment buried during the first half of this period has come back to us (Fig. 2.6d): that part of the ocean entirely disappeared. During the second half of the subduction period, only a small fraction of metasediments was returned, and finally a few rare mafic eclogites at the end of oceanic subduction (Fig. 2.6d).

These fragments of oceanic lithosphere allow one to reconstruct the history of subduction from a chronological and mechanical point of view (Figs. 2.7a,b). In detail, the spatial contrasts in type, abundance, and burial of these fragments allow recognizing several sectors with distinct ocean subduction dynamics (sectors A to D; Fig. 2.7c). They reveal, in addition, that the imprint of oceanic convergence has deeply influenced today's collisional architecture and contrasts (Fig. 2.7c; Agard & Handy, 2021).

2.3.3. Lessons Learned from Blueschists and Eclogites

Oceanic fragments offscraped from the slab provide us with key samples to study their mineralogical, chemical, and mechanical transformations. They give direct access to the composition, structure, and rheology of rocks at the plate interface down to great depths, with a high spatial resolution inaccessible (yet) to geophysical observations. Since the subduction plate contact must transiently jump from the base of the upper plate to within the slab to detach slices from it (Figs. 2.3g, 2.7a), they shed light on mechanical coupling across the plate boundary.

Subducted oceanic fragments worldwide reveal that variations in interplate mechanical coupling (Agard et al., 2018) are statistically more frequent around equivalent depths of 30–40 km (at the base of the seismogenic zone; Menant et al., 2019) or 70–80 km (Fig. 2.8a). Beyond ~2.8 GPa, nothing is recovered: The eclogitized oceanic crust is significantly denser than the surrounding dry mantle (Agard et al., 2009, 2018) and is moreover dragged by the strong coupling established between the two plates beyond ~80 km (see section 2.4; Wada & Wang, 2009; van Keken et al., 2011). The size of the fragments and their pressure-temperature evolution through time suggest that the plate boundary is a few hundred meters thick at most (Agard et al., 2018; Wakabayashi, 2021; Fig. 2.8b,c), at odds with the results and/or assumptions of most numerical models. Fragments also appear dominantly derived from slow-spreading oceans (Agard, 2021), whose heterogeneous lithosphere and former detachment faults (Fig. 2.2a) make it easier to slice off fragments during subduction (Ruh et al., 2015). This is, for example, the case for the Alps and for the spectacular blueschists and eclogites of the Cyclades, a similarly restricted and once slow-spreading oceanic domain (Schmid et al., 2020; Gyomlai et al., 2021).

2.4. FRAGMENTS OF OCEANIC LITHOSPHERE SPARED FROM SUBDUCTION: OPHIOLITES

2.4.1. The Semail Ophiolite and Beyond

Let's now turn to almost fully preserved fragments of oceanic lithosphere: ophiolites. The iconic Semail ophiolite straddles the northern part of Oman and the United Arab Emirates (Fig. 2.9a). Perfectly exposed in an arid environment, ~500 km long, ~100 km wide, and locally

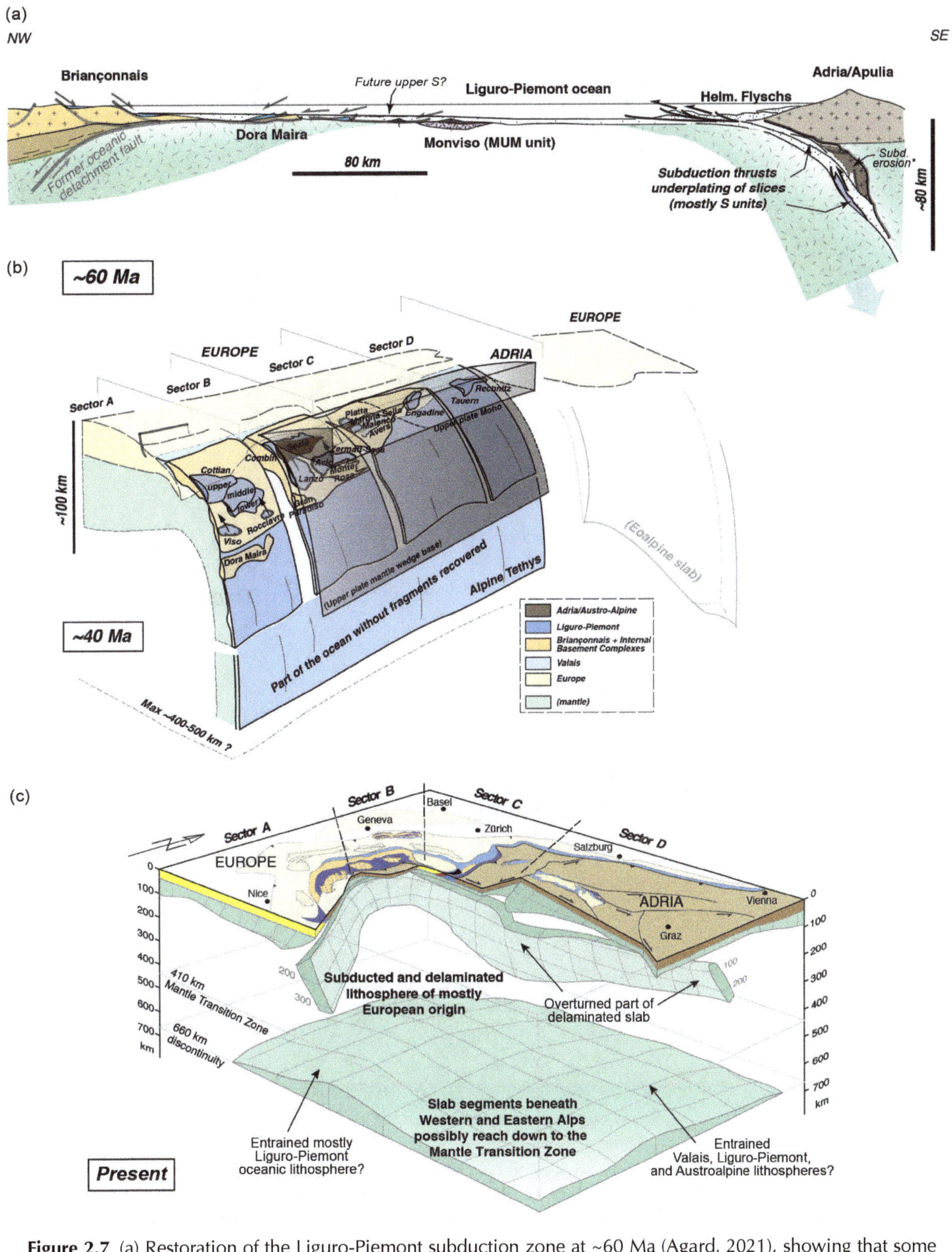

Figure 2.7 (a) Restoration of the Liguro-Piemont subduction zone at ~60 Ma (Agard, 2021), showing that some metasedimentary units from the slab were offscraped during subduction. Recovered MUM units, like in Monviso, were subducted later and saved from irreversible burial into the mantle, partly thanks to continental subduction (Fig. 2.6c). (b) Schematic 3D view, from the western to the eastern Alps of all oceanic units recovered from subduction (Agard, 2021): petrological and lithological contrasts reveal the existence of distinct sectors with distinct subduction dynamics. This lateral segmentation was likely controlled, at least in part, by structures inherited from the former Variscan and/or rifting events. (c) Tomographic imaging outlining slab fragments recognized beneath the Alps (Agard & Handy, 2021; courtesy M. Handy and collaborators). Lateral contrasts seem to match those deduced from contrasts in oceanic subduction dynamics recognized independently. The entire oceanic lithosphere now lies in the Mantle Transition Zone, between 440 and 670 km depth.

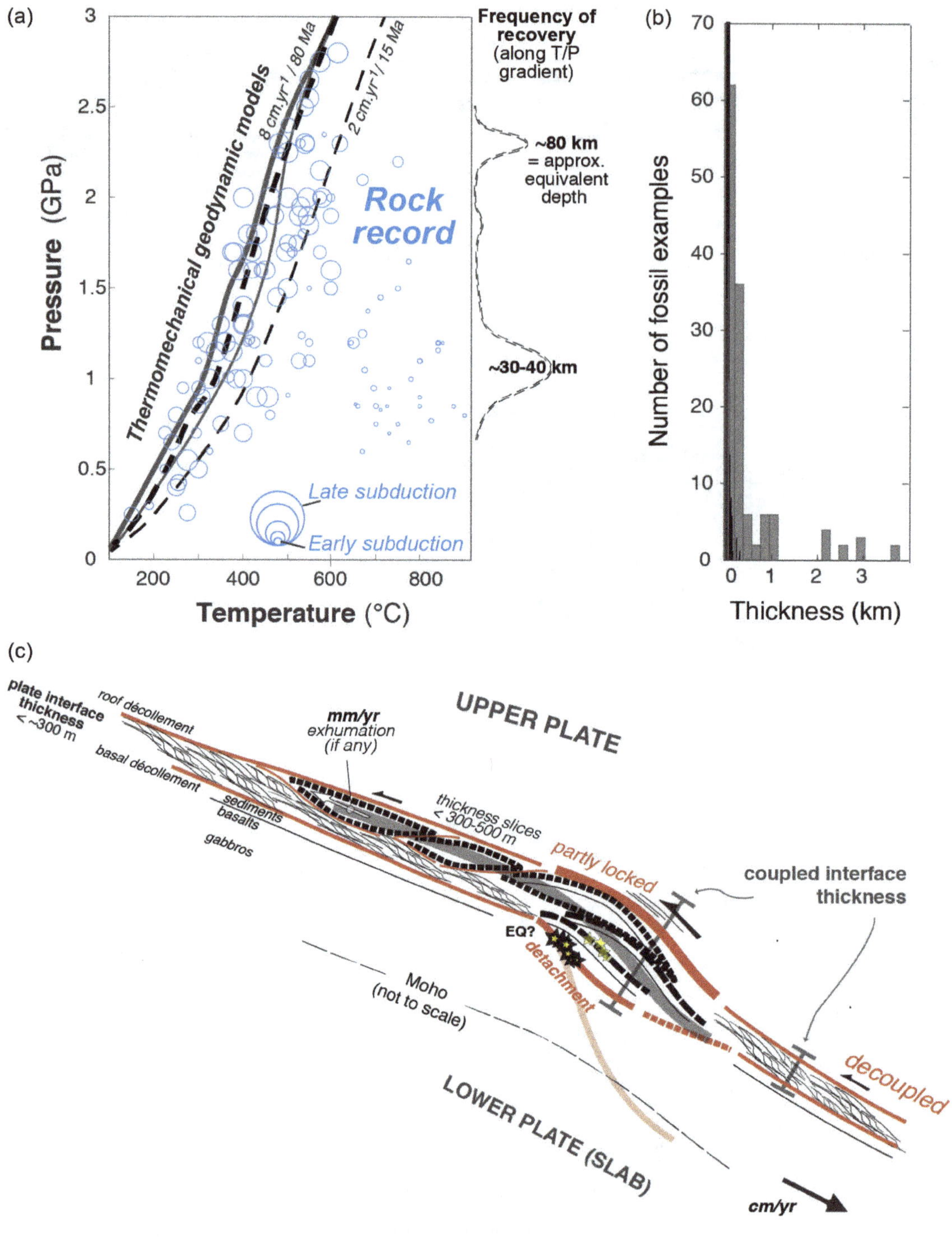

Figure 2.8 Geological and dynamic constraints on the plate interface (see Agard et al., 2018, for details). (a) The subduction rock record is shown as a function of its relative timing of recovery in a given subduction history (e.g., early or late). It can be compared with the results of thermomechanical models for mature subduction. Rock recovery, though transient and relatively rare, is favored at depths of 30–40 and 70–80 km (assuming P as lithostatic pressure). (b) The thickness of the various tectonic slices allows placing constraints on the thickness of the plate interface and variations in mechanical coupling. (c) Integrated view of the subduction interface: thickness, composition, structure, and mechanical behavior.

12–15 km thick, unaffected by any subsequent collision, it is by far the most studied in the world. It is regarded, since the Penrose conference of 1972, as representative of the Pacific-type oceanic lithosphere. A mafic crust, composed from top to bottom of kilometer-scale basaltic flows (Fig. 2.9c), a kilometer-thick sheeted dyke complex, and a few kilometers of gabbros (Fig. 2.9d), overly mantle rocks that are dominantly depleted harzburgites (Fig. 2.9e). This succession is similar to that recognized in marine geophysics (2A, 2B, 3, and 4, respectively, in

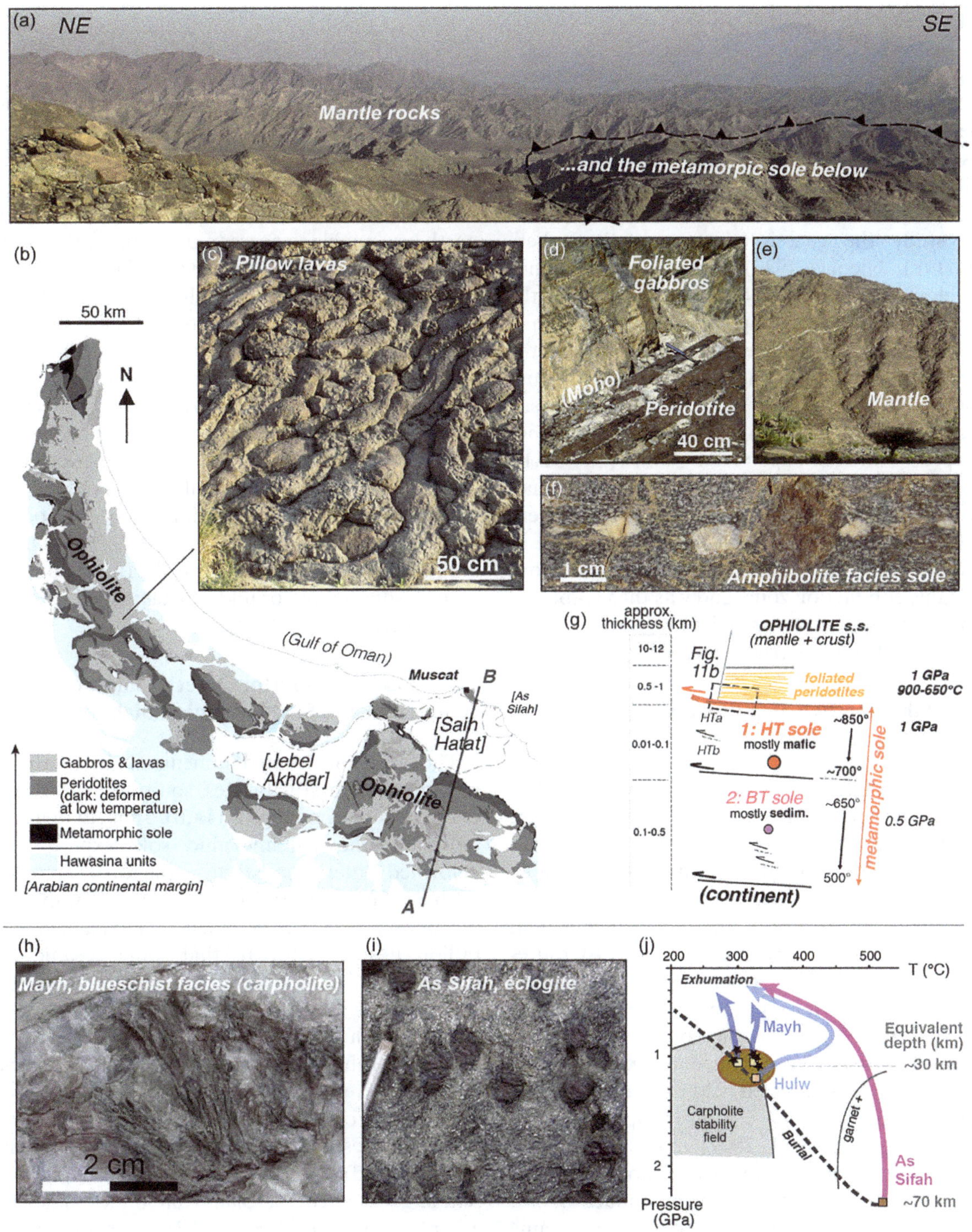

Figure 2.9 (a) Mantle rocks peaking above the continental margin of Arabia over vast distances (across several hundreds of kilometers!) with a systematic though thin metamorphic sole plastered beneath it. (b) Simplified tectonic map of northeastern Oman and the UAE. The Semail ophiolite (grey color) is overlying the Hawasina units (light blue), that is, pelagic sediments deposited on the ocean floor and on the transitional domain between the continent and the Neo-Tethys ocean. Structurally below is the continental lithosphere, with basement and recent sedimentary rocks nicely exposed in the Saih Hatat and Jabal Akhdar tectonic windows. Blueschists and eclogites crop out in the former, whereas the latter shows more modest mineral transformations. (c–e) Typical horizons within the ophiolite: the famous pillow lavas from Wadi Jizi (c), the Moho transition zone (d; courtesy C. Nicollet), and the variably deformed peridotites (e). (f) Close-up view on the metamorphic sole: here a former basal transformed to a (amphibolite- to granulite-facies) garnet-amphibole-plagioclase ± clinopyroxene rock. (g) Structural organization at the base of the ophiolite, showing the genetic association between the strongly foliated peridotites and the metamorphic sole (particularly the HT type). (h–i) Carpholite-bearing blueschist and eclogite with garnet and clinopyroxene relics (in a matrix of epidote, blue amphibole, and white mica) from the Saih Hatat window. (j) P-T trajectories followed by the HP-LT metamorphic rocks of the Saih Hatat units (h, i) during exhumation.

Fig. 2.2a). Below the ophiolite, a distinctive hundred-meter-thick metamorphic sole is present, which comprises deformed and metamorphosed basalts and sediments (Fig. 2.9f,g). Further below, the continental crust exhibits blueschists and eclogites (Fig. 2.9h–j).

The Semail ophiolite has been scrutinized over the last 50 yr to study the genesis of oceanic lithosphere (Coleman, 1981; Nicolas, 1989), revealing the existence of (mappable) mantle diapirs (Nicolas et al., 2000) or the interactions between ascending magmas and mantle rocks (Kelemen et al., 1995). The lessons learned from the Semail ophiolite are of general relevance and pertain, with some variations, to other large ophiolitic massifs worldwide such as in New Caledonia, Newfoundland, Turkey, Cyprus, Timor, or Papua New Guinea (Table 2.1). While the exact geodynamic context in which the Semail ophiolite formed is still partly debated, most recent studies suggest that it formed around 95 Ma in a suprasubduction context (Rioux et al., 2012), that is, right above an intraoceanic subduction zone, and was then emplaced (obducted) when it was a few million years old only (Agard et al., 2020, and references therein).

2.4.2. The Obduction Two-Step Process: Triggering and Ophiolite Emplacement

Ophiolites attest that large, several hundred kilometer-long pieces of oceanic lithosphere can be emplaced on top of continental lithosphere (Table 2.1; Fig. 2.10a). How can a dense, rigid, and thin oceanic lithosphere be emplaced on top of a continental lithosphere with the exact opposite characteristics: light, deformable, and thick (Fig. 2.10b)? Superimposing lithospheres requires lithospheric-scale shortening, which hints to subduction initiation. This provides a simple answer: If intraoceanic subduction is initiated close to the continental margin, yet dipping away from the continent (Fig. 2.10c), the margin may in turn be subducted below the adjacent oceanic lithosphere, that is, below the future ophiolite (e.g., Searle et al., 2004; Agard et al., 2007). In this case, ophiolites are the witnesses of an intraoceanic subduction that has turned to a dead end, stopped by the subduction of a continental margin (or of an island arc: e.g., Josephine ophiolite). Docking of the Cordilleran ophiolites (Moores, 1982) is more enigmatic, but also relates to intraoceanic subduction dynamics (Wakabayashi & Dilek, 2000; Agard et al., 2020, their Fig. 3a).

In the case of the Semail ophiolite, why was a new subduction zone initiated near the Arabian margin since one already existed to the north, beneath Eurasia (Fig. 2.10c,d)? The Semail obduction followed an episode of major plate kinematics reorganization around 110 ± 5 Ma, marked by a sharp increase in convergence velocity between Arabia and Eurasia (Agard et al., 2007), the opening of the Atlantic Ocean, and the formation of large mantle plumes in the eastern Pacific (Vaughan & Scarrow, 2003) and in the Indian Ocean (Rodriguez et al., 2021). Onset of subduction to the south of the Neotethys (Fig. 2.10c; at ~104 Ma: Guilmette et al., 2018) indicates that the preexisting subduction zone to the north was unable to accommodate the entire convergence between Arabia and Eurasia (Agard et al., 2007). Similar perturbations also predated the obduction of most ophiolites whose geodynamic context can still be reconstructed (e.g., Turkey, Timor, New Caledonia).

The scenario, from the abyssal plain to the exposure of the ophiolite, can be reconstructed quite precisely in the case of Semail, thanks to the metamorphic and sedimentary record (Fig. 2.9g,j). Two crucial stages of the obduction process are preferentially preserved: (1) its start, which shows the (difficult) initiation of intraoceanic subduction (section 2.4.3); (2) its termination, which results in the effective emplacement of the (rather passive) ophiolite by subduction, or underthrusting, of the continental lithosphere (section 2.4.4).

2.4.3. Obduction Birth: Onset of Intraoceanic Subduction and Slab Dynamics (Slabitization)

The fragments located immediately below the ophiolite represent our best witnesses, so far, of intraoceanic subduction initiation (Fig. 2.11a,b). Immediately below the ophiolite, the metamorphic sole exposes extremely deformed material constituting the top of an oceanic lithosphere (Gnos, 1998; Fig. 2.9f,g): metabasalts and metamorphosed pelagic sediments, in particular former radiolarites. This reveals that some oceanic crust was underplated below oceanic lithosphere, thereby demonstrating the existence of intraoceanic subduction (Wakabayashi & Dilek, 2000; Figs. 2.9c, 2.11a).

Metamorphic soles furthermore testify to the initial mechanical resistance to subduction (Fig. 2.11b,c). The sequence of rocks encountered in the sole (Fig. 2.9g) attests to the progressive offscraping of the newly forming slab (Fig. 2.11a–d). This is reminiscent of the accidental offscraping observed for Siah Kuh or for blueschists and eclogites worldwide (section 2.3), yet happens during the birth of subduction. The internal organization of the metamorphic sole allows drawing major inferences: (1) the stripping of the slab first affects the entire crust (basalts, and sometimes the sheeted dyke complex) then is confined to the most superficial levels, the sediments (Fig. 2.11d); (2) the formation and progressive accretion of the sole involves materials that are less and less metamorphic with time (HT then LT: Figs. 2.9g, 2.11d): first granulite- then amphibolite- and finally greenschist-facies rocks, revealing a general cooling of the system at the same time as a more and more pellicular offscraping of the slab (Soret et al., 2017, 2019).

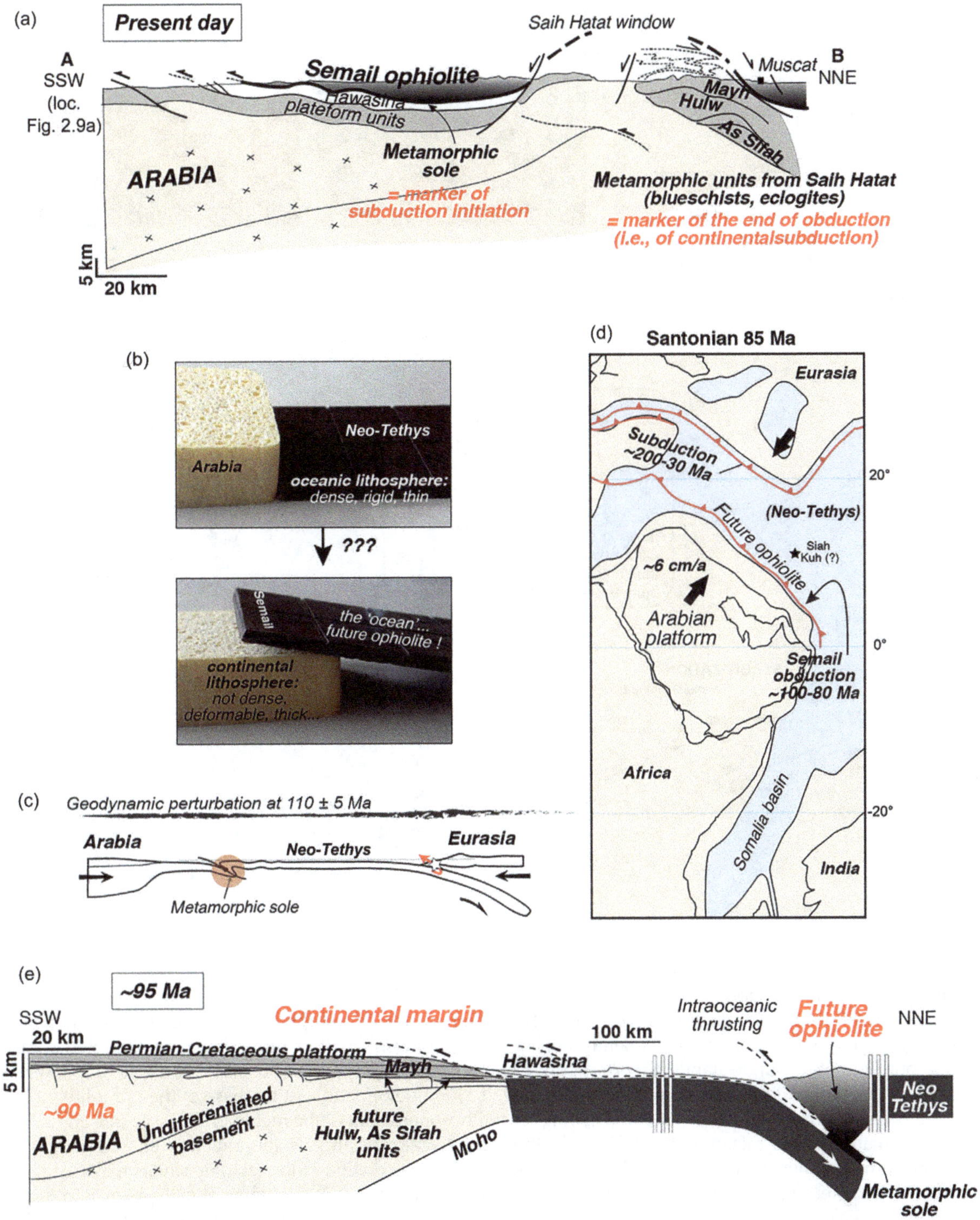

Figure 2.10 (a) Present-day section across northeastern Oman, with emphasis on the metamorphic landmarks of the start and end of the obduction history. (b) How can oceanic lithosphere be emplaced on top of continental lithosphere? (c) Onset of intraoceanic subduction at the southern end of the Neo-Tethys following Cretaceous kinematic reorganization (Agard et al., 2007; Matthews et al., 2012). (d) Late Cretaceous (85 Ma) paleogeography of the Arabia-Eurasia region (after Fournier et al., 2006) showing the disappearance of the Neo-Tethys beneath Eurasia. This ocean had been subducting since ~200 Ma and was therefore much larger than the Alpine one (Fig. 2.4). The approximate location of the Siah Kuh seamount is indicated here (see section 2.3.1; Bonnet et al., 2020). (e) Restoration of the northern margin of Arabia at 100–95 Ma (simplified after Béchennec et al., 1989 ; Searle et al., 2004). Comparing figures (a) and (e) allows capturing the emplacement of the Semail ophiolite: first through intraoceanic subduction and then throughunderthrusting of the continental margin beneath the ophiolite.

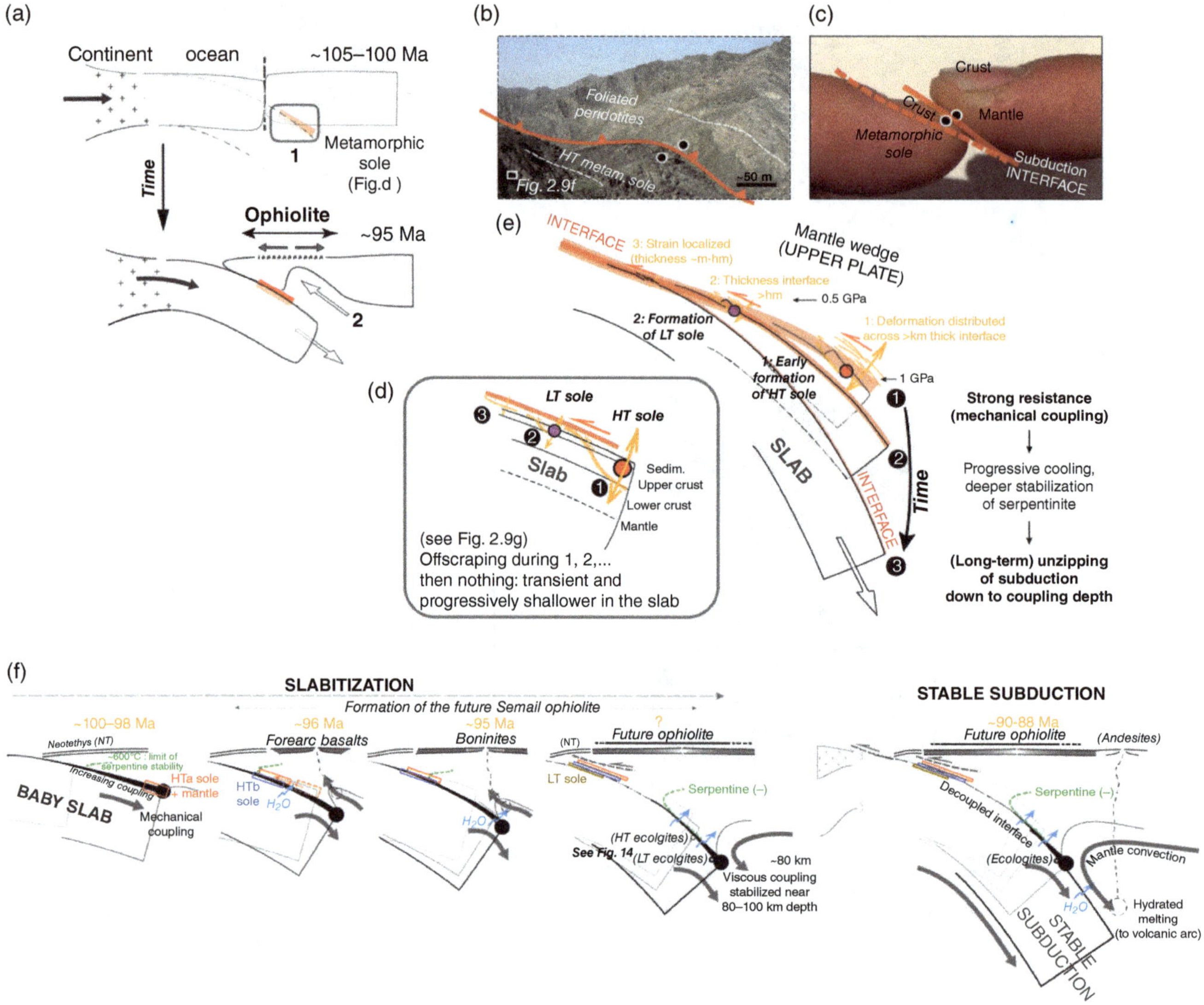

Figure 2.11 (a) Simple tectonic evolution outlining the genetic link between intraoceanic subduction and obduction. The start of oceanic subduction is systematically accompanied by the stripping of fragments of oceanic crust (the future metamorphic sole) from incipiently sinking oceanic lithosphere (the future slab). The obducted ophiolite, in our present-day understanding, corresponds to a portion of newly formed oceanic lithosphere as a result of mantle upwelling above the young subduction zone. (b) Joint deformation, at the base of the ophiolite, of the banded peridotites and of the top of the downgoing crust/lithosphere, that is, the metamorphic sole (see Fig. 2.9g). (c) The formation of the metamorphic sole marks the mechanical resistance to the initiation of the subduction process, compared here with stripping the nail. (d) The successive accretion of the tectonic slivers making up the metamorphic sole, progressively colder and with a larger sedimentary content (schematically from 1 to 3; see Fig. 2.9g), reveals a progressively shallower stripping or offscraping of the slab along with a cooling subduction regime. (e) This evolution of accretion accompanies the gradual lubrication and unzipping of the subduction plate boundary, enabling the downward progress of the slab. See text for details. (f) Slabitization, from subduction nucleation to stable subduction: This tectonic scenario relies on observations from sole-peridotite pairs across the world, and uses observations from subduction initiation across the Izu-Bonin/Marianas forearc (e.g., Stern et al., 2012, and references in text). It links early slab dynamics with the onset and progressive downward migration of viscous coupling as the subduction zone cools, and with the onset of mantle upwelling.

This process ceases with the gradual cooling of the subduction and the lubrication of the new interface between the plates, in part due to the serpentinization of the mantle of the upper plate (Agard et al., 2016; Fig. 2.11e).

The base of the mantle is made up of "low temperature" foliated peridotites (Fig. 2.9e,g; Ceuleneer et al., 1988; Prigent et al., 2018a). These portions of the ophiolite are characterized by a very intense deformation, distributed over a thickness of a kilometer or so, which is similar to that of the underlying metamorphic sole (Fig. 2.11b): synchronous, in the same direction and in the same temperature range decreasing with time (from ~900 to 650°C, around 1 GPa). With time, deformation gets more and more localized near the contact with the metamorphic sole. Geochemical data show that this deformation is accompanied by the percolation of slab-derived fluids into the ophiolite base (travelling at meters per year; Prigent et al., 2018b), which represents an essential witness of element transfer during subduction (Fig. 2.11f; in comparison, no intact interplate interface is preserved in the mature subduction environments probed by blueschists and eclogites).

The mechanical coupling between the metamorphic sole and the foliated peridotites records the very early stages of the disappearance of the oceanic lithosphere (Fig. 2.11e), while subduction is still in its infancy (Agard et al., 2016). Deformation is first distributed over a characteristic thickness exceeding one kilometer before deformation gets localized along a mature, 1–100 m thick plate contact (stages 1–3, Fig. 2.11e).

As the subduction thermal structure cools and the new slab progresses downward, the zone of strong mechanical resistance/coupling between the plates deepens (black dot, Fig. 2.11f): This process, coined as "slabitization" (Agard et al., 2020), unzips the subduction plate interface down to a depth referred to as the common depth of viscous coupling (CDVC, ~80 km depth; Wada & Wang, 2009). The onset of mechanical coupling gradually initiates a mantle counterflow and leads, through decompression and fluid ingression, to mantle melting and embryonic arc magmatism (with forearc basalts, boninites; Stern & Bloomer, 1992; Ishizuka et al., 2020; Fig. 2.11f). This explains the formation of the Semail ophiolite in a supra-subduction context (and of many others: Table 2.1). After a few million years, once mechanical coupling has stabilized at the CDVC (Fig. 2.11f), hydrated melting appears and generates the classical subduction-related andesitic magmas (Fig. 2.1; Stern, 2002; Bonnet et al., 2020a&b).

As a result of this evolution, the subduction interface has become mechanically decoupled in the long term down to a depth of about 100 km (Fig. 2.11e–f): Slab fragments will not be recovered except during mechanical and/or geodynamic perturbations able to trigger the offscraping of blueschists and eclogites (§ 3; Agard et al., 2018). This process also explains the close relationship between subduction initiation and the genesis (and future emplacement) of the ophiolite, and their close ages. In the case of Semail, the whole process lasts about 10 Myr (Fig. 2.11a,f): Transformation and burial of the metamorphic soles range from 105–100 to 95 Ma (Rioux et al., 2016; Guilmette et al., 2018), while the ophiolite has yielded a narrow age range, around 96–95 Ma (Rioux et al., 2012). More fundamentally, the processes of subduction initiation and "slabitization" accompany the birth of a new slab, before it becomes connected to and starts interacting with the convecting asthenospheric mantle.

In most cases, preserved ophiolites correspond to "fresh" oceanic lithosphere newly formed in a supra-subduction setting (Table 2.1). There are, however, less common examples of ophiolites that correspond to tracts of oceanic lithosphere formed well before any convergence started (e.g., Armenia, Sistan; Table 2.1; Agard et al., 2020, their Fig. 2.3b). This indicates that intraoceanic subduction was unable to drive mantle upwelling, probably as a result of shorter-lived subduction (those examples also show cooler metamorphic soles; Agard et al., 2020).

2.4.4. Obduction Death: Ophiolites Preserved Through Continental Subduction

Unmetamorphosed pelagic sediments are found below the base of the Semail ophiolite and its metamorphic sole (Hawasina units; Figs. 2.9b, 2.10a), as for the Bay of Island or Lycian ophiolites (Tab. 1). These were scraped off during the underthrusting of the distal part of the continental margin. Even lower, in the Saih Hatat and Jabal Akhdar tectonic windows, blueschist and eclogite facies metamorphic rocks crop out (Figs.2.9b,h–j, 2.10a). These fragments are again witnesses of a subduction process, but this time that of continent subduction (Goffé et al., 1984; as for the Briançonnais domain in the Alps; section 2.3, Figs. 2.5, 2.7). These portions of the Arabian continental margin are easily recognizable through their Permian to Late Cretaceous sedimentary successions and their Proterozoic metasediments already metamorphosed during the ~600–550 Ma Panafrican orogeny (Béchennec et al., 1989; Cozzi et al., 2012; Fig. 2.10a). The rocks from the Saih Hatat window were dragged into subduction, reaching variable pressures from 1 to 2.3 GPa (Yamato et al., 2007; Massonne et al., 2013).

Peak burial of these continental rocks occurred around 80 Ma. Their exhumation is marked by spectacular ductile deformation, including kilometer-scale sheath folds (Searle & Alsop, 2007; Scharf et al., 2021). Here again, preservation was selective: While the burial history is partially fossilized in minerals (Yamato et al., 2007), the macroscopic structures associated with burial have been largely erased, and only the exhumation dynamics can be restored (Fig. 2.12a; Yamato et al., 2007; Agard et al., 2010). The large folds and intense shearing accompanying exhumation of these rocks attest that they must have moved upwards, likely along the subduction interface, during or following the choking of the subduction zone (Searle et al., 2004). Continental subduction was transient (Agard & Vitale-Brovarone, 2013), the low density and greater thickness of the continental margin preventing long-lived subduction (Fig. 2.12b; like sponge below chocolate in Fig. 2.10b). All convergence had disappeared by 75–70 Ma.

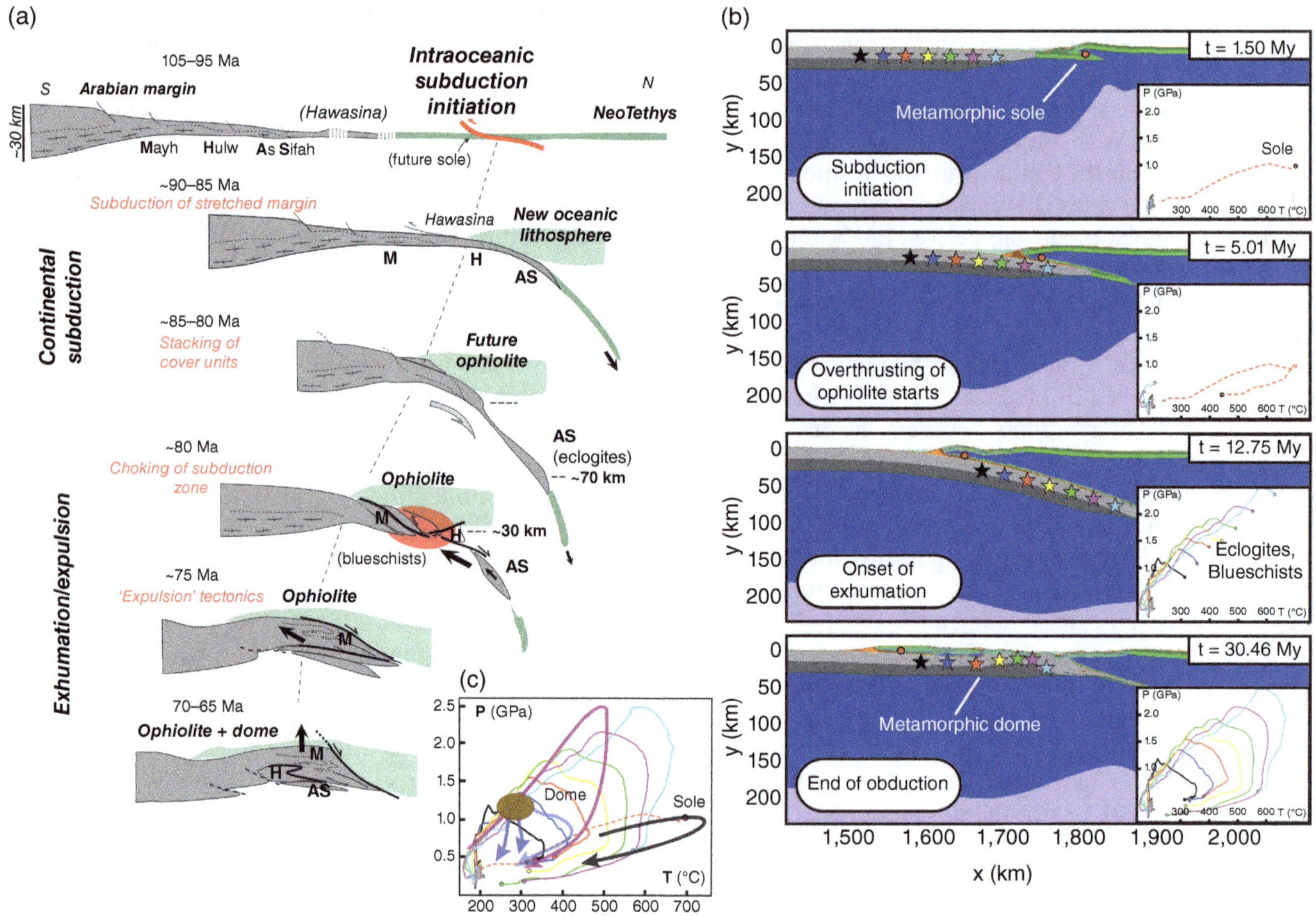

Figure 2.12 (a) Geodynamic reconstruction based on the P-T-t paths of the metamorphic sole and of the continental blueschists and eclogites found in the Saih Hatat metamorphic dome, as well as on the sedimentary, stratigraphic, and kinematic record (after Agard et al., 2010). (b) Fully coupled thermomechanical models of obduction (Duretz et al., 2016): Stars outline the evolution and P-T-t paths of some model markers at depth. (c) The comparison between predicted P-T-t paths and those estimated from mineral transformations (see Fig. 2.9j) allows placing constraints on convergence rates, lithosphere ages, or rheologies.

The entire obduction crisis therefore lasted about 25 Myr (Fig. 2.12a), as for obduction events elsewhere (cf. New Caledonia: Cluzel et al., 2001; Soret et al., 2016; Vitale-Brovarone et al., 2018; Timor: Linthout et al., 1997). In essence, what ultimately allows for the effective emplacement of the ophiolite and its good preservation (thanks also to the absence of later collision, as opposed to the ophiolite fragments scattered in the Himalayan/Tibet sutures; Fig. 2.1c), is continental subduction.

Obduction is therefore both an accident and a subduction dead end. At the plate tectonics scale, it can be seen as the result of a change in the partitioning of deformation between the different convergence zones (Fig. 2.10c,d; Agard et al., 2014). Thermomechanical geodynamic models allow testing the obduction scenario quantitatively by estimating the theoretical P-T-t paths of the rocks and characteristic duration of the process (Fig. 2.12b; Duretz et al., 2016; Porkolab et al., 2021). To a first approximation, the agreement between predictions and nature is satisfactory (Fig. 2.12c), notwithstanding the fact that extension is somewhat imposed in the model (i.e., subduction "choking" is not produced self-consistently by the model), the "spontaneous" formation of supra-subduction lithosphere is not accounted for and the role of 3D variations is not considered.

In fact, the case of the Semail ophiolite reveals the important impact of crustal heterogeneities and lateral contrasts within the continental margin on the final structure of the ophiolite (as also observed in New Caledonia; Cluzel et al., 2001). The two tectonic windows below the ophiolite, the Saih Hatat (to the east) and Jebel Akhdar windows (to the west), evidence first-order differences (Fig. 2.13a). In the Saih Hatat, rocks were subducted down much higher pressure and exhumation is characterized by major shearing and deformation; this sector of the continental margin was also cut across by mafic

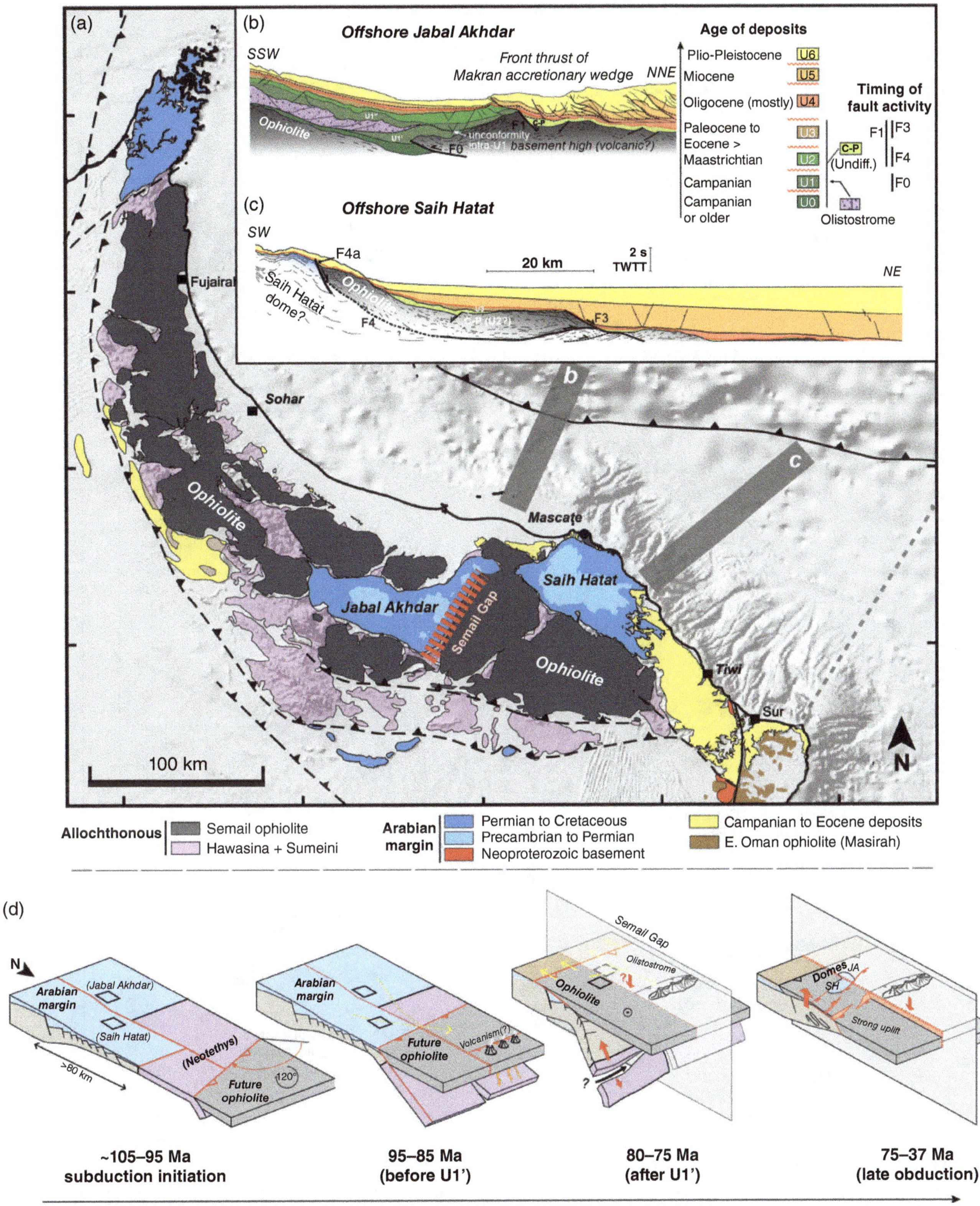

Figure 2.13 (a) Simplified geological map of northeastern Oman and United Arab Emirates. (b–c) Selection of interpreted seismic profiles across the offshore northern Oman margin (Ninkabou et al., 2021) highlighting the sharp contrasts in sedimentary and tectonic evolution across the Semail Gap, a major crustal-scale divide inherited from the Panafrican orogeny. (d) Obduction and continental subduction of the Oman margin through time. The metamorphic, tectonic, and sedimentary records reveal important lateral contrasts across the Semail Gap.

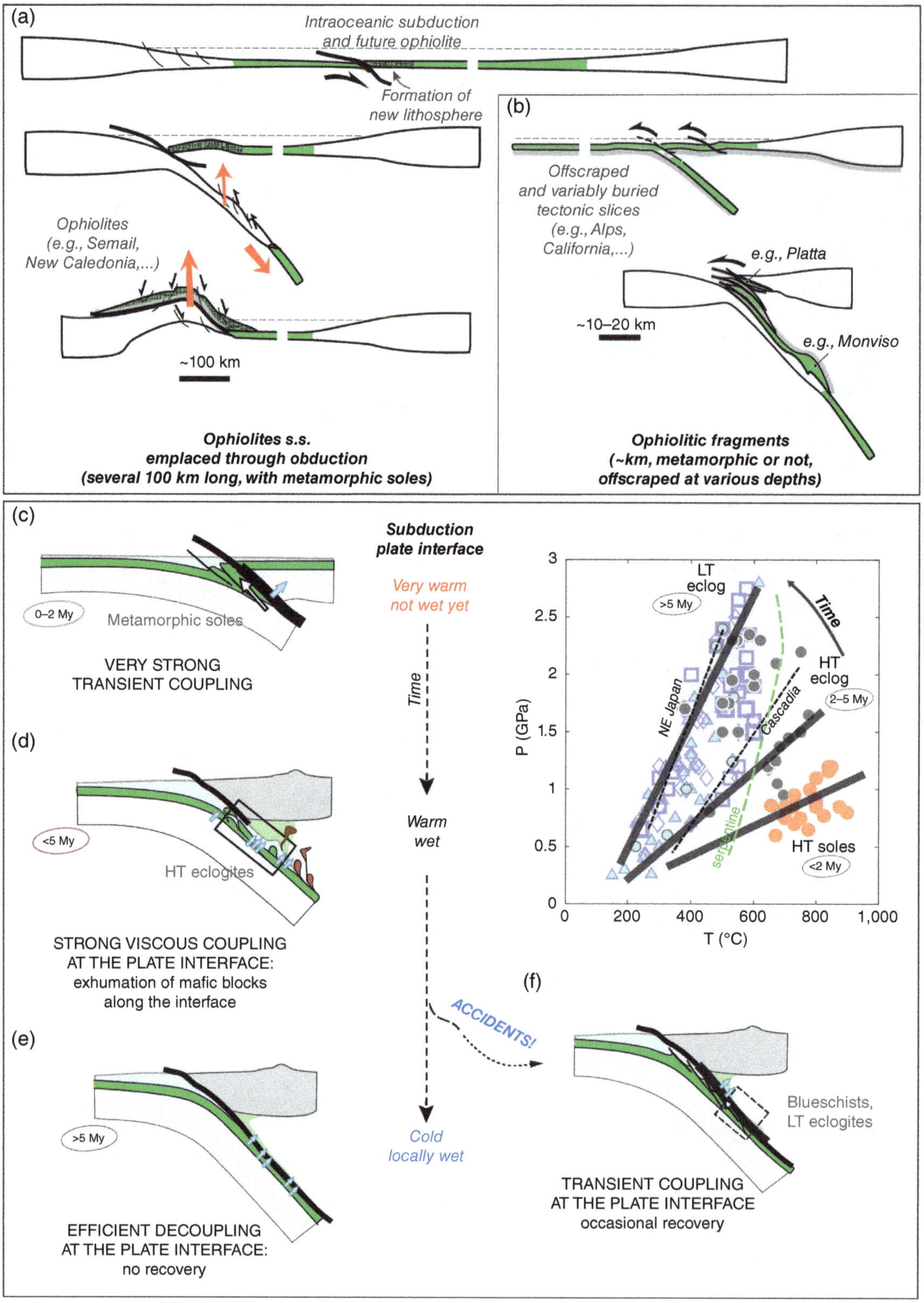

Figure 2.14 (a–b) Distinguishing real ophiolites from fragments/slivers/slices of oceanic lithosphere (sometimes referred to as ophiolitic) is a matter of scale and of quite distinct geodynamic context! (c–f) Long-term trend in mechanical coupling and rock recovery associated with subduction cooling illustrating the fundamental link between metamorphic soles and blueschists and eclogites: The progressive change in mechanical coupling, from strong to weak is governed by an increase in the effective viscosity contrasts across the subduction interface. The combined evolution of temperature and fluid migration, and fluid storage in the upper plate notably as serpentinite, plays a key role. When young and still relatively dry, (c) the plate interface experiences strong viscous coupling, resulting in the offscraping of metamorphic soles. During subsequent subduction and cooling, (d) high-temperature eclogite-facies rocks are recovered, mostly in serpentinite mélanges. The steady-state, locally fluid-rich, colder, older, and weaker interface leads to a decoupled plate interface (e), similar to most subduction zones today, except during occasional accidents (f), as for the Alps (section 2.3; see also Agard et al., 2018, for details).

intrusions during the Permian rifting of the Neotethys (Chauvet et al., 2009). On the contrary, the Jebel Akhdar, which is devoid of mafic intrusions of Permian age, only reached a peak pressure of ~0.7 GPa (Breton, 2004; hence, in terms of equivalent depths, not much more than the 10–15 km-thick ophiolite above) and experienced only moderate obduction-related deformation.

Large contrasts are also observed offshore, in the Gulf of Oman (Fig. 2.13b–c). The Jebel Akhdar transect reveals the existence of a major ~80 Ma thrust affecting the offshore part of the Semail ophiolite and of a strongly subsiding sedimentary basin, with profuse infill and debris-flows (olistostrome in Fig. 2.13b) dated from the Campanian (i.e., 80–75 Ma; Ninkabou et al., 2021). The Saih Hatat transect, on the contrary, is marked by pronounced Maastrichtian uplift and Maastrichtian to lower Tertiary extension and exposes continental material below the ophiolite (Fig. 2.13c). These onshore and offshore contrasts are observed across the Semail Gap (Fig. 2.13a), a major fault zone considered to be inherited from the Pan-African orogeny (Cozzi et al., 2012; Scharf et al., 2019). They are comparable in scale to those observed in the subducted continental margin of the Alps (Fig. 2.7; Agard, 2021).

Worthy of note, the structure of the ophiolite shows comparatively limited lateral contrasts (whether petrological or paleomagnetic; Nicolas et al., 2000; van Hinsbergen et al., 2019), except for the larger suprasubduction imprint west of the Semail Gap (Python et al., 2008). This shows that the late exhumation of the Saih Hatat blueschists and eclogites occurred without deforming much the ophiolite above, suggesting that the tectonic contact in between them was then rheologically rather weak. Importantly, the very limited syn-convergence sedimentary record of obduction (restricted to the west of the Semail Gap) suggests that most of the obduction-related shortening proceeded without significant erosion, hence presumably underwater.

2.5. THE FATE OF OCEANIC LITHOSPHERE: TRAGIC YET INSIGHTFUL

Blueschists, eclogites and ophiolites are all that remain of ancient oceanic lithosphere. They are precious relics allowing to probe depths inaccessible to direct observation and to study the onset of plate tectonics on Earth (Moores, 2002). These fragments of oceanic lithosphere provide us with unique information on the mechanical behavior of rocks, which is essential to understand seismic risks, or on the transfer of fluids and elements, which is crucial to understand volcanic processes or elemental recycling within the globe.

This overview shows the intimate links and the differences between subduction and obduction processes. It allows distinguishing true ophiolites (constitutive of the upper plate) from fragments of oceanic lithosphere off-scraped from the downgoing slab (Fig. 2.14a–b), whether these were deeply subducted or only shallowly accreted to the upper plate. It also underlines the importance of the evolution of the subduction regime through time (Figs. 2.11e, 2.14c–f).

What blueschists and eclogites record are the dynamics of mature subduction zones, the mechanics of the subduction plate interface and its hiccups. Ophiolites and their metamorphic soles as well, but at the time of subduction initiation and with complementary insights. Both types of relics even give us a kind of wink: The initiation of subduction is enlightened by obduction, while subduction proves to be essential to obduction. Yet, as for any historical landmark, they are primarily accidents: They reveal transient episodes in the otherwise rather smooth, well-established subduction process, which dooms oceans to disappear.

ACKNOWLEDGMENTS

Immense thanks are due to S. Angiboust, M. Locatelli, B. Dubacq, S. Guillot, M. Handy, J. Wakabayashi, A. Vitale-Brovarone, T. Duretz, M. Jentzer and many others, for fruitful and friendly discussions over the years.

REFERENCES

Agard, P. (2021). Subduction of oceanic lithosphere in the Alps: Selective and archetypal from (slow-spreading) oceans. *Earth-Science Reviews, 103517.*

Agard, P., & Handy, M. R. (2021). Ocean subduction dynamics in the Alps. *Elements: An International Magazine of Mineralogy, Geochemistry, and Petrology, 17*(1), 9–16.

Agard, P., & Vitale-Brovarone, A. (2013). Thermal regime of continental subduction: the record from exhumed HP–LT terranes (New Caledonia, Oman, Corsica). *Tectonophysics, 601*, 206–215.

Agard, P., Jolivet, L., Vrielynck, B., Burov, E., & Monie, P. (2007). Plate acceleration: The obduction trigger? *Earth and Planetary Science Letters, 258*(3–4), 428–441. https://doi.org/10.1016/j.epsl.2007.04.002

Agard, P., Plunder, A., Angiboust, S., Bonnet, G., & Ruh, J. (2018). The subduction plate interface: Rock record and mechanical coupling (from long to short timescales). *Lithos, 320–321*, 537–566. https://doi.org/10.1016/j.lithos.2018.09.029

Agard, P., Prigent, C., Soret, M., Dubacq, B., Guillot, S., & Deldicque, D. (2020). Slabitization: Mechanisms controlling subduction development and viscous coupling. *Earth-Science Reviews, 208*, 103259.

Agard, P., Searle, M. P., Alsop, G. I., & Dubacq, B. (2010). Crustal stacking and expulsion tectonics during continental subduction: P-T deformation constraints from Oman. *Tectonics, 29*(5).

Agard, P., Yamato, P., Jolivet, L., & Burov, E. (2009). Exhumation of oceanic blueschists and eclogites in subduction zones: Timing and mechanisms. *Earth-Science Reviews*, *92*(1–2), 53–79. https://doi.org/10.1016/j.earscirev.2008.11.002

Agard, P., Yamato, P., Soret, M., Prigent, C., Guillot, S., Plunder, A., et al. (2016). Plate interface rheological switches during subduction infancy: Control on slab penetration and metamorphic sole formation. *Earth and Planetary Science Letters*, *451*, 208–220. https://doi.org/10.1016/j.epsl.2016.06.054

Agard, P., Zuo, X., Funiciello, F., Bellahsen, N., Faccenna, C., & Savva, D. (2014). Obduction: Why, how and where. Clues from analog models. *Earth and Planetary Science Letters*, *393*, 132–145.

Angiboust, S., Agard, P., Jolivet, L., & Beyssac, O. (2009). The Zermatt-Saas ophiolite: The largest (60 km wide) and deepest (c. 70–80 km) continuous slice of oceanic lithosphere detached from a subduction zone? *Terra Nova*, *21*(3), 171–180. https://doi.org/10.1111/j.1365-3121.2009.00870.x

Angiboust, S., Agard, P., Raimbourg, H., Yamato, P., & Huet, B. (2011). Subduction interface processes recorded by eclogite-facies shear zones (Monviso, W. Alps). *Lithos*, *127*(1–2), 222–238. https://doi.org/10.1016/j.lithos.2011.09.004

Angiboust, S., Agard, P., Yamato, P., & Raimbourg, H. (2012). Eclogite breccias in a subducted ophiolite: A record of intermediate-depth earthquakes? *Geology*, *40*(8), 707–710. https://doi.org/10.1130/G32925.1

Béchennec, F., Le Metour, J., Rabu, D., Beurrier, M., Bourdillon-Jeudy-de-Grissac, C., De Wever, P., et al. (1989). Geologie d'une chaine issue de la Tethys; les montagnes d'Oman. *Bulletin de la Société Géologique de France*, *2*, 167–188.

Beltrando, M., Manatschal, G., Mohn, G., Dal Piaz, G. V., Brovarone, A. V., & Masini, E. (2014). Recognizing remnants of magma-poor rifted margins in high-pressure orogenic belts: The Alpine case study. *Earth-Science Reviews*, *131*, 88–115.

Berger, A., & Bousquet, R. (2008). Subduction-related metamorphism in the Alps: Review of isotopic ages based on petrology and their geodynamic consequences. *Geological Society, London, Special Publications*, *298*(1), 117–144.

Bonnet, G., Agard, P., Angiboust, S., Fournier, M., & Omrani, J. (2019). No large earthquakes in fully exposed subducted seamount. *Geology*, *47*(5), 407–410.

Bonnet, G., Agard, P., Whitechurch, H., Fournier, M., Angiboust, S., Caron, B., et al. (2020). Fossil seamount in southeast Zagros records intraoceanic arc to back-arc transition: New constraints for the evolution of the Neotethys. *Gondwana Research*, *81*, 423–444.

Cannat, M., Sauter, D., Mendel, V., Ruellan, E., Okino, K., Escartin, J., et al. (2006). Modes of seafloor generation at a melt-poor ultraslow-spreading ridge. *Geology*, *34*(7), 605–608.

Celik, Ö. F., Marzoli, A., Marschik, R., Chiaradia, M., Neubauer, F., & Öz, İ. (2011). Early-middle Jurassic intra-oceanic subduction in the İzmir-Ankara-Erzincan Ocean, northern Turkey. *Tectonophysics*, *509*(1–2), 120–134.

Ceuleneer, G., Nicolas, A., & Boudier, F. (1988). Mantle flow patterns at an oceanic spreading centre: The Oman peridotites record. *Tectonophysics*, *151*(1–4), 1–26.

Chauvet, F., Dumont, T., & Basile, C. (2009). Structures and timing of Permian rifting in the central Oman Mountains (Saih Hatat). *Tectonophysics*, *475*(3–4), 563–574.

Choi, S. H., Shervais, J. W., & Mukasa, S. B. (2008). Supra-subduction and abyssal mantle peridotites of the Coast Range ophiolite, *California. Contributions to Mineralogy and Petrology*, *156*(5), 551–576.

Clift, P., & Vannucchi, P. (2004). Controls on tectonic accretion versus erosion in subduction zones: Implications for the origin and recycling of the continental crust. *Reviews of Geophysics*, *42*(2). https://doi.org/10.1029/2003RG000127

Cloos, M. (1992). Thrust-type subduction-zone earthquakes and seamount asperities: A physical model for seismic rupture. *Geology*, *20*(7), 601–604. https://doi.org/10.1130/0091-7613(1992)020<0601:TTSZEA>2.3.CO;2

Cluzel, D., Aitchison, J. C., & Picard, C. (2001). Tectonic accretion and underplating of mafic terranes in the Late Eocene intraoceanic fore-arc of New Caledonia (Southwest Pacific): geodynamic implications. *Tectonophysics*, *340*(1–2), 23–59.

Cluzel, D., Jourdan, F., Meffre, S., Maurizot, P., & Lesimple, S. (2012). The metamorphic sole of New Caledonia ophiolite: 40Ar/39Ar, U-Pb, and geochemical evidence for subduction inception at a spreading ridge. *Tectonics*, *31*(3).

Coleman, R. G. (1971). Plate tectonic emplacement of upper mantle peridotites along continental edges. *Journal of Geophysical Research*, *76*(5), 1212–1222.

Coleman, R. G. (1981). Tectonic setting for ophiolite obduction in Oman. *Journal of Geophysical Research: Solid Earth*, *86*(B4), 2497–2508.

Coltice, N., Gérault, M., & Ulvrová, M. (2019). A mantle convection perspective on global tectonics. *Earth-Science Reviews*, *165*, 120–150.

Cozzi, A., Rea, G., & Craig, J. (2012). From global geology to hydrocarbon exploration: Ediacaran-Early Cambrian petroleum plays of India, Pakistan and Oman. *Geological Society, London, Special Publications*, *366*(1), 131–162.

Davies, H., & Jaques, A. (1984). Emplacement of ophiolite in Papua New Guinea. *Geological Society, London, Special Publications*, *13*(1), 341–349.

De Saussure, H. B. (1804). *Voyages dans les Alpes: précédés d'un essai sur l'histoire naturelle des environs de Genève (vol. 2)*. L. Fauche-Borel, imprimeur.

Dewey, J. F. (1976). Ophiolite obduction. *Tectonophysics*, *31*(1–2), 93–120.

Dewey, J. F., & Casey, J. F. (2013). The sole of an ophiolite: The Ordovician Bay of Islands Complex, *Newfoundland. Journal of the Geological Society*, *170*(5), 715–722.

Dilek, Y., & Furnes, H. (2014). Ophiolites and their origins. *Elements*, *10*(2), 93–100.

Dilek, Y., Furnes, H., & Shallo, M. (2007). Suprasubduction zone ophiolite formation along the periphery of Mesozoic Gondwana. *Gondwana Research*, *11*(4), 453–475.

Duretz, T., Agard, P., Yamato, P., Ducassou, C., Burov, E. B., & Gerya, T. V. (2016). Thermo-mechanical modeling of the obduction process based on the Oman ophiolite case. *Gondwana Research*, *32*, 1–10.

Ernst, W. (1971). Metamorphic zonations on presumably subducted lithospheric plates from Japan, California and the Alps. *Contributions to Mineralogy and Petrology*, *34*(1), 43–59.

Forsyth, D., & Uyeda, S. (1975). On the relative importance of the driving forces of plate motion. *Geophysical Journal International*, *43*(1), 163–200.

Fournier, M., Lepvrier, C., Razin, P., & Jolivet, L. (2006). Late Cretaceous to Paleogene post-obduction extension and subsequent Neogene compression in the Oman Mountains. *GeoArabia*, *11*(4), 17–40.

Galoyan, G., Rolland, Y., Sosson, M., Corsini, M., Billo, S., Verati, C., et al. (2009). Geology, geochemistry and $^{40}Ar/^{39}Ar$ dating of Sevan ophiolites (Lesser Caucasus, Armenia): evidence for Jurassic Back-arc opening and hot spot event between the South Armenian Block and Eurasia. *Journal of Asian Earth Sciences*, *34*(2), 135–153.

Gnos, E. (1998). Peak metamorphic conditions of garnet amphibolites beneath the Semail Ophiolite: Implications for an inverted pressure gradient. *International Geology Review*, *40*(4), 281–304. https://doi.org/10.1080/00206819809465210

Goffé, B., & Chopin, C. (1986). High-pressure metamorphism in the Western Alps: Zoneography of metapelites, chronology and consequences. *Schweizerische Mineralogische Und Petrographische Mitteilungen*, *66*(1–2), 41–52.

Groppo, C., Beltrando, M., & Compagnoni, R. (2009). The P–T path of the ultra-high pressure Lago di Cignana and adjoining high-pressure meta-ophiolitic units: Insights into the evolution of the subducting Tethyan slab. *Journal of Metamorphic Geology*, *27*(3), 207–231.

Guilmette, C., Smit, M. A., van Hinsbergen, D. J., Gürer, D., Corfu, F., Charette, B., et al. (2018). Forced subduction initiation recorded in the sole and crust of the Semail Ophiolite of Oman. *Nature Geoscience*, *11*(9), 688–695.

Gyomlai, T., Agard, P., Marschall, H. R., Jolivet, L., & Gerdes, A. (2021). Metasomatism and deformation of block-in-matrix structures in Syros: The role of inheritance and fluid-rock interactions along the subduction interface. *Lithos*, *386*, 105996.

Haessig, M., Duretz, T., Rolland, Y., & Sosson, M. (2016). Obduction of old oceanic lithosphere due to reheating and plate reorganization: insights from numerical modelling and the NE Anatolia-Lesser Caucasus case example. *Journal of Geodynamics*, *96*, 35–49.

Harper, G. D. (2003). Tectonic implications of boninite, arc tholeiite, and MORB magma types in the Josephine Ophiolite, California-Oregon. *Geological Society, London, Special Publications*, *218*(1), 207–230.

Harper, G. D., Saleeby, J. B., & Heizler, M. (1994). Formation and emplacement of the Josephine ophiolite and the Nevadan orogeny in the Klamath Mountains, California-Oregon: U/Pb zircon and $^{40}Ar/^{39}Ar$ geochronology. *Journal of Geophysical Research: Solid Earth*, *99*(B3), 4293–4321.

Isozaki, Y., Maruyama, S., & Furuoka, F. (1990). Accreted oceanic materials in Japan. *Tectonophysics*, *181*(1–4), 179–205.

Jentzer, M., Agard, P., Bonnet, G., Monié, P., Fournier, M., Whitechurch, H., et al. (2022). The North Sistan orogen (Eastern Iran): Tectono-metamorphic evolution and significance within the Tethyan realm. *Gondwana Research*.

Jones, K. (1997). Deformation and emplacement of the Lizard Ophiolite Complex, SW England, based on evidence from the Basal Unit. *Journal of the Geological Society*, *154*(5), 871–885.

Kaneko, Y., Maruyama, S., Kadarusman, A., Ota, T., Ishikawa, M., Tsujimori, T., et al. (2007). On-going orogeny in the outer-arc of the Timor-Tanimbar region, eastern Indonesia. *Gondwana Research*, *11*(1–2), 218–233.

Kelemen, P. B., Shimizu, N., & Salters, V. J. (1995). Extraction of mid-ocean-ridge basalt from the upwelling mantle by focused flow of melt in dunite channels. *Nature*, *375*(6534), 747–753.

Kusky, T. M., Windley, B. F., Safonova, I., Wakita, K., Wakabayashi, J., Polat, A., et al. (2014). Recognition of ocean plate stratigraphy in accretionary orogens through Earth history: A record of 3.8 billion years of sea floor spreading, subduction, and accretion. *Gondwana Research*, *24*(2), 501–547. https://doi.org/10.1016/j.gr.2013.01.004

Lallemand, S., Peyret, M., van Rijsingen, E., Arcay, D., & Heuret, A. (2018). Roughness characteristics of oceanic seafloor prior to subduction in relation to the seismogenic potential of subduction zones. *Geochemistry, Geophysics, Geosystems*, *19*(7), 2121–2146.

Lemoine, M., Bas, T., Arnaud-Vanneau, A., Arnaud, H., Dumont, T., Gidon, M., et al. (1986). The continental margin of the Mesozoic Tethys in the Western Alps. *Marine and Petroleum Geology*, *3*(3), 179–199.

Linthout, K., Helmers, H., & Sopaheluwakan, J. (1997). Late Miocene obduction and microplate migration around the southern Banda Sea and the closure of the Indonesian Seaway. *Tectonophysics*, *281*(1–2), 17–30.

Locatelli, M., Verlaguet, A., Agard, P., Federico, L., & Angiboust, S. (2018). Intermediate-depth brecciation along the subduction plate interface (Monviso eclogite, W. Alps). *Lithos*, *320*, 378–402.

Lus, W. Y., McDougall, I., & Davies, H. L. (2004). Age of the metamorphic sole of the Papuan Ultramafic Belt ophiolite, *Papua New Guinea. Tectonophysics*, *392*(1–4), 85–101.

Manatschal, G., & Müntener, O. (2009). A type sequence across an ancient magma-poor ocean–continent transition: the example of the western Alpine Tethys ophiolites. *Tectonophysics*, *473*(1–2), 4–19.

Massonne, H.-J., Opitz, J., Theye, T., & Nasir, S. (2013). Evolution of a very deeply subducted metasediment from As Sifah, northeastern coast of Oman. *Lithos*, *156*, 171–185.

Matthews, K. J., Seton, M., & Müller, R. D. (2012). A global-scale plate reorganization event at 105–100 Ma. *Earth and Planetary Science Letters*, *355*, 283–298.

McLaughlin, R., Blake, M., Jr., Griscom, A., Blome, C., & Murchey, B. (1988). Tectonics of formation, translation, and dispersal of the Coast Range ophiolite of California. *Tectonics*, *7*(5), 1033–1056.

Menant, A., Angiboust, S., & Gerya, T. (2019). Stress-driven fluid flow controls long-term megathrust strength and deep accretionary dynamics. *Scientific Reports*, *9*(1), 1–11.

Meneghini, F., & Moore, J. C. (2007). Deformation and hydrofracture in a subduction thrust at seismogenic depths: The Rodeo Cove thrust zone, Marin Headlands, *California. Geological Society of America Bulletin*, *119*(1–2), 174–183. https://doi.org/10.1130/B25807.1

Moores, E. M. (1982). Origin and emplacement of ophiolites. *Reviews of Geophysics*, *20*(4), 735–760.

Moores, E. M. (2002). Pre-1 Ga (pre-Rodinian) ophiolites: Their tectonic and environmental implications. *Geological Society of America Bulletin*, *114*(1), 80–95.

Moores, E., & Vine, F. J. (1971). The Troodos Massif, Cyprus and other ophiolites as oceanic crust: Evaluation and implications. *Philosophical Transactions of the Royal Society of London. Series A, Mathematical and Physical Sciences*, *268*(1192), 443–467.

Nicolas, A. (1989). Structures of ophiolites and dynamics of oceanic lithosphere. *Structures of Ophiolites and Dynamics of Oceanic Lithosphere*. https://www.scopus.com/inward/record.uri?eid=2-s2.0-85040886915&partnerID=40&md5=17b50b83cc6eca46fb5331ec22375966

Nicolas, A., Boudier, F., & Meshi, A. (1999). Slow spreading accretion and mantle denudation in the Mirdita ophiolite (Albania). *Journal of Geophysical Research: Solid Earth*, *104*(B7), 15155–15167.

Nicolas, A., Boudier, F., Ildefonse, B., & Ball, E. (2000). Accretion of Oman and United Arab Emirates ophiolite-discussion of a new structural map. *Marine Geophysical Researches*, *21*(3–4), 147–180.

Ninkabou, D., Agard, P., Nielsen, C., Smit, J., Gorini, C., Rodriguez, M., et al. (2021). Structure of the offshore obducted Oman margin: Emplacement of Semail ophiolite and role of tectonic inheritance. *Journal of Geophysical Research: Solid Earth*, *126*(2), 2020JB020187.

Pearce, J. A., & Robinson, P. (2010). The Troodos ophiolitic complex probably formed in a subduction initiation, slab edge setting. *Gondwana Research*, *18*(1), 60–81.

Picazo, S., Müntener, O., Manatschal, G., Bauville, A., Karner, G., & Johnson, C. (2016). Mapping the nature of mantle domains in Western and Central Europe based on clinopyroxene and spinel chemistry: Evidence for mantle modification during an extensional cycle. *Lithos*, *266*, 233–263.

Plunder, A., Agard, P., Chopin, C., Soret, M., Okay, A. I., & Whitechurch, H. (2016). Metamorphic sole formation, emplacement and blueschist facies overprint: early subduction dynamics witnessed by western Turkey ophiolites. *Terra Nova*, *28*(5), 329–339.

Plunder, A., Agard, P., Dubacq, B., Chopin, C., & Bellanger, M. (2012). How continuous and precise is the record of P-T paths? Insights from combined thermobarometry and thermodynamic modelling into subduction dynamics (Schistes Lustres, W. Alps). *Journal of Metamorphic Geology*, *30*(3), 323–346. https://doi.org/10.1111/j.1525-1314.2011.00969.x

Porkolab, K., Duretz, T., Yamato, P., Auzemery, A., & Willingshofer, E. (2021). Extrusion of subducted crust explains the emplacement of far-travelled ophiolites. *Nature Communications*, *12*(1), 1–11.

Prigent, C., Agard, P., Guillot, S., Godard, M., & Dubacq, B. (2018a). Mantle wedge (De) formation during subduction infancy: Evidence from the base of the Semail ophiolitic mantle. *Journal of Petrology*, *59*(11), 2061–2092.

Prigent, C., Guillot, S., Agard, P., Lemarchand, D., Soret, M., & Ulrich, M. (2018b). Transfer of subduction fluids into the deforming mantle wedge during nascent subduction: Evidence from trace elements and boron isotopes (Semail ophiolite, Oman). *Earth and Planetary Science Letters*, *484*, 213–228. https://doi.org/10.1016/j.epsl.2017.12.008

Python, M., Ceuleneer, G., & Arai, S. (2008). Chromian spinels in mafic-ultramafic mantle dykes: Evidence for a two-stage melt production during the evolution of the Oman ophiolite. *Lithos*, *106*(1–2), 137–154.

Ranero, C. R., Morgan, J. P., McIntosh, K., & Reichert, C. (2003). Bending-related faulting and mantle serpentinization at the Middle America trench. *Nature*, *425*(6956), 367–373. https://doi.org/10.1038/nature01961

Rioux, M., Bowring, S., Kelemen, P., Gordon, S., Dudás, F., & Miller, R. (2012). Rapid crustal accretion and magma assimilation in the Oman-UAE ophiolite: High precision U-Pb zircon geochronology of the gabbroic crust. *Journal of Geophysical Research: Solid Earth*, *117*(B7).

Rioux, M., Garber, J., Bauer, A., Bowring, S., Searle, M., Kelemen, P., et al. (2016). Synchronous formation of the metamorphic sole and igneous crust of the Semail ophiolite: New constraints on the tectonic evolution during ophiolite formation from high-precision U-Pb zircon geochronology. *Earth and Planetary Science Letters*, *451*, 185–195.

Rodriguez, M., Arnould, M., Coltice, N., & Soret, M. (2021). Long-term evolution of a plume-induced subduction in the Neotethys realm. *Earth and Planetary Science Letters*, *561*, 116798.

Ruh, J. B., Le Pourhiet, L., Agard, P., Burov, E., & Gerya, T. (2015). Tectonic slicing of subducting oceanic crust along plate interfaces: Numerical modeling. *Geochemistry, Geophysics, Geosystems*, *16*(10), 3505–3531.

Scharf, A., Mattern, F., Al-Wardi, M., Frijia, G., Moraetis, D., Pracejus, B., et al. (2021). Tectonic evolution of the Oman Mountains. *Geological Society, London, Memoirs*, *54*(1), 67–103.

Scharf, A, Mattern, F., Moraetis, D., Callegari, I., & Weidle, C. (2019). Postobductional kinematic evolution and geomorphology of a major regional structure: The Semail Gap Fault Zone (Oman Mountains). *Tectonics*, *38*(8), 2756–2778.

Schmid, S. M., Fügenschuh, B., Kounov, A., Maţenco, L., Nievergelt, P., Oberhänsli, R., et al. (2020). Tectonic units of the Alpine collision zone between Eastern Alps and western Turkey. *Gondwana Research*, *78*, 308–374.

Searle, M., & Alsop, G. (2007). Eye-to-eye with a mega-sheath fold: A case study from Wadi Mayh, northern Oman Mountains. *Geology*, *35*(11), 1043–1046.

Searle, M., Warren, C., Waters, D., & Parrish, R. (2004). Structural evolution, metamorphism and restoration of the Arabian continental margin, Saih Hatat region, Oman Mountains. *Journal of Structural Geology*, *26*(3), 451–473.

Seton, M., Müller, R. D., Zahirovic, S., Williams, S., Wright, N. M., Cannon, J., et al. (2020). A global data set of present-day oceanic crustal age and seafloor spreading parameters. *Geochemistry, Geophysics, Geosystems*, *21*(10), e2020GC009214

Soret, M., Agard, P., Dubacq, B., Plunder, A., & Yamato, P. (2017). Petrological evidence for stepwise accretion of metamorphic soles during subduction infancy (Semail ophiolite, Oman and UAE). *Journal of Metamorphic Geology*, *35*(9), 1051–1080. https://doi.org/10.1111/jmg.12267

Soret, M., Agard, P., Dubacq, B., Vitale-Brovarone, A., Monie, P., Chauvet, A., et al. (2016). Strain localization and fluid infiltration in the mantle wedge during subduction initiation: Evidence from the base of the New Caledonia ophiolite. *Lithos*, *244*, 1–19.

Soret, M., Agard, P., Ildefonse, B., Dubacq, B., Prigent, C., & Rosenberg, C. (2019). Deformation mechanisms in mafic amphibolites and granulites: Record from the Semail metamorphic sole during subduction infancy. *Solid Earth Discussions*, *1–36*. https://doi.org/10.5194/se-2019-28

Stern, R. J. (2002). Subduction zones. *Reviews of Geophysics*, *40*(4), 3-1-3-38.

Stern, R. J., & Bloomer, S. H. (1992). Subduction zone infancy: Examples from the Eocene Izu-Bonin-Mariana and Jurassic California arcs. *Geological Society of America Bulletin*, *104*(12), 1621–1636.

Stern, R. J., Reagan, M., Ishizuka, O., Ohara, Y., & Whattam, S. (2012). To understand subduction initiation, study forearc crust: To understand forearc crust, study ophiolites. *Lithosphere*, *4*(6), 469–483.

Strachan, R. A., Linnemann, U., Jeffries, T., Drost, K., & Ulrich, J. (2014). Armorican provenance for the mélange deposits below the Lizard ophiolite (Cornwall, UK): evidence for Devonian obduction of Cadomian and Lower Palaeozoic crust onto the southern margin of Avalonia. *International Journal of Earth Sciences*, *103*(5), 1359–1383.

SUHR, G., & Cawood, P. A. (1993). Structural history of ophiolite obduction, Bay of Islands, Newfoundland. *Geological Society of America Bulletin*, *105*(3), 399–410.

Ulrich, M., Picard, C., Guillot, S., Chauvel, C., Cluzel, D., & Meffre, S. (2010). Multiple melting stages and refertilization as indicators for ridge to subduction formation: The New Caledonia ophiolite. *Lithos*, *115*(1–4), 223–236.

Van Hinsbergen, D. J., Maffione, M., Koornneef, L. M., & Guilmette, C. (2019). Kinematic and paleomagnetic restoration of the Semail ophiolite (Oman) reveals subduction initiation along an ancient Neotethyan fracture zone. *Earth and Planetary Science Letters*, *518*, 183–196.

Van Keken, P. E., Hacker, B. R., Syracuse, E. M., & Abers, G. A. (2011). Subduction factory: 4. Depth-dependent flux of H_2O from subducting slabs worldwide. *Journal of Geophysical Research: Solid Earth*, *116*(B1).

Vaughan, A. P., & Scarrow, J. H. (2003). Ophiolite obduction pulses as a proxy indicator of superplume events? *Earth and Planetary Science Letters*, *213*(3–4), 407–416.

Vissers, R. L., van Hinsbergen, D. J., Meijer, P. T., & Piccardo, G. B. (2013). Kinematics of Jurassic ultra-slow spreading in the Piemonte Ligurian ocean. *Earth and Planetary Science Letters*, *380*, 138–150.

Vitale-Brovarone, A. V., Agard, P., Monie, P., Chauvet, A., & Rabaute, A. (2018). Tectonic and metamorphic architecture of the HP belt of New Caledonia. *Earth-Science Reviews*, *178*, 48–67.

Wada, I., & Wang, K. (2009). Common depth of slab-mantle decoupling: Reconciling diversity and uniformity of subduction zones. *Geochemistry, Geophysics, Geosystems*, *10*(10). https://doi.org/10.1029/2009GC002570

Wakabayashi, J. (2015). Anatomy of a subduction complex: Architecture of the Franciscan Complex, California, at multiple length and time scales. *International Geology Review*, *57*(5–8), 669–746. https://doi.org/10.1080/00206814.2014.998728

Wakabayashi, J. (2021). Subduction and exhumation slip accommodation at depths of 10–80 km inferred from field geology of exhumed rocks: Evidence for temporal-spatial localization of slip. *Plate Tectonics, Ophiolites, and Societal Significance of Geology: A Celebration of the Career of Eldridge Moores*, *552*, 257.

Wakabayashi, J., & Dilek, Y. (2000). Spatial and temporal relationships between ophiolites and their metamorphic soles: A test of models of forearc ophiolite genesis. In Y. Dilek, E. M. Moores, D. Elthon, & A. Nicolas (Eds.), *Ophiolites and oceanic crust: New insights from field studies and ocean drilling program* (pp. 53–64). Boulder: Geological Society of America, Inc.

Wang, K., & Bilek, S. L. (2011). Do subducting seamounts generate or stop large earthquakes? *Geology*, *39*(9), 819–822.

Yamato, P., Agard, P., Goffé, B., De Andrade, V., Vidal, O., & Jolivet, L. (2007). New, high-precision P-T estimates for Oman blueschists: Implications for obduction, nappe stacking and exhumation processes. *Journal of Metamorphic Geology*, *25*(6), 657–682.

Zarrinkoub, M. H., Pang, K. N., Chung, S. L., Khatib, M. M., Mohammadi, S. S., Chiu, H.-Y., et al. (2012). Zircon U–Pb age and geochemical constraints on the origin of the Birjand ophiolite, Sistan suture zone, eastern Iran. *Lithos*, *154*, 392–405.

3

Lateral Heterogeneity in Compressional Mountain Belt Settings

Bibek Giri and Mary Hubbard

ABSTRACT

Analysis of convergent orogens reveals important orthogonal/cross structures and laterally heterogenous deformation style, igneous activity, metamorphic grade geomorphology and seismic activity. To assess the occurrence, causal mechanisms, and implications of these lateral heterogeneities, we review a selection of convergent orogens with different tectonic settings and history. We find that heterogeneities are primarily driven by the preexisting deformational history of the cratonic blocks involved, lateral change in lithology of crustal rocks, variations in rheologic properties, or changes in plate kinematics. The Appalachian orogenic front mimics the Iapetan rift margin. As seen in the Zagros and the Cordillera, preexisting basement structures have control on preorogen and synorogenic sedimentation, which subsequently impact how the orogenic wedge evolves. Several cross faults in the Himalaya spatially coincide with basement structures on the downgoing plate. Similarly, physiography of the subducting oceanic plate influences deformation in the overriding plate of Andean-type margins. Orogenic extension in the Alps has been accompanied by a system of orogen-parallel strike slip faults and extensional cross faults. It is evident that lateral heterogeneities can form crucial control on the evolution of orogenic belts and can influence seismic rupture patterns, resource occurrence, and landslide hazards.

3.1. INTRODUCTION

Many of the world's major mountain belts are found along modern or former convergent plate boundaries and they are typically dominated by major thrust structures (Elliott, 1976; Chapple, 1978; Boyer & Elliott, 1982; Dahlen, 1990). The terms *continental collision* or *terrain accretion* have been used for convergent boundaries when both the involved plates are continental, though mountain building also occurs along convergent plate boundaries where an oceanic plate is subducting beneath continental crust (e.g., Dewey & Bird, 1970). In both tectonic settings, compressive stress is accommodated through imbrication and shortening of upper crustal material as thrust sheets and/or nappes, within a system of fault surfaces, which generally strike perpendicular to the direction of compression (or parallel to the orogen). In general, these faults and related structures are laterally continuous both temporally and spatially. Detailed geological and geophysical research, however, has found that while these convergent mountain belts exhibit overall lateral continuity, there are also lateral heterogeneities that have made an important contribution to the development of the mountain belt. Driven by various mechanisms, these lateral changes serve to induce important structural, geological, and geomorphological lateral heterogeneities along mountain belts that may (or may have) influence(d) igneous or seismic processes or even the occurrence of economic resources. Commonly, significant along-strike changes are known to occur across structures, which strike orthogonal to the overall trend of the orogen. In this article, the term *cross structure* has been utilized to refer to all generic, syn, and postorogenic structural

Department of Earth Sciences, Montana State University, Bozeman, Montana, USA

Compressional Tectonics: Plate Convergence to Mountain Building, Geophysical Monograph 277, First Edition.
Edited by Elizabeth J. Catlos and İbrahim Çemen.

DOI:10.1002/9781119773856.ch03

features that bound lateral changes along a mountain belt, though the terms *cross-strike structural discontinuities (CSD)*, *transverse zones*, and *cross faults* have all been utilized to describe large-scale structures that strike across convergent mountain belts (Drahovzal & Thomas, 1976; Wheeler, 1980; Hubbard et al., 2018; Hubbard et al., 2021). Thomas (1990) suggested that transverse zones (TZ) along fold and thrust belts are domains of several, structurally aligned lateral connectors, which encompass transverse/tear faults, lateral ramps, and displacement transfer faults/zones. In general, the term *transverse zone* has been used in the literature to denote large-scale, inherited basement features with a high-angled orientation to the evolving orogen. Lateral connecters like tear faults, lateral ramps, and transfer faults/zones form during thrust propagation and are rather localized, only spanning across a few thrust sheets. Formed during the propagation of thrust belts, these lateral connectors could be spatially controlled by the inherited structures in the crystalline basement as well as the overlying stratigraphy of the evolving thrust sheet (Thomas, 1990). During orogenesis, inherited basement structures are commonly reactivated or inverted and impart a great control on the tectonic and structural evolution of mountain belts (e.g., Tavarnelli et al., 2004; Butler et al., 2006). Evidence has also been presented suggesting that oceanic transform faults or seamounts in a subducting oceanic plate can influence the structural development of the mountain belt in the overriding continental plate, thus causing lateral heterogeneity or the presence of cross structures (e.g., Gutscher et al., 2000; Chapin, 2012).

Mountain belts formed in convergent plate tectonic settings at different time periods in Earth's history all exhibit some degree of lateral heterogeneity. The following review explores a selection of mountain belts from convergent boundaries, examining the proposed driving mechanisms and implications of lateral heterogeneities and the presence of cross structures during and after orogenesis. The Appalachians, the North American Cordillera, the Alps, the Himalaya, the Zagros, and the Andes have been looked at in detail, and several other mountain belts are briefly discussed (Fig. 3.1). This selection represents some of the most well studied and prominent mountain belts around the world and presents a wide array of tectonic settings. Within each mountain belt, presented examples of lateral heterogeneities and cross structures are representative of the lateral variability for that orogen. For each mountain belt, we present a brief introduction of the tectonic setting, observed lateral heterogeneities, and proposed causal factors for those heterogeneities. It is noted that lateral heterogeneities can be expressed in several ways including (1)

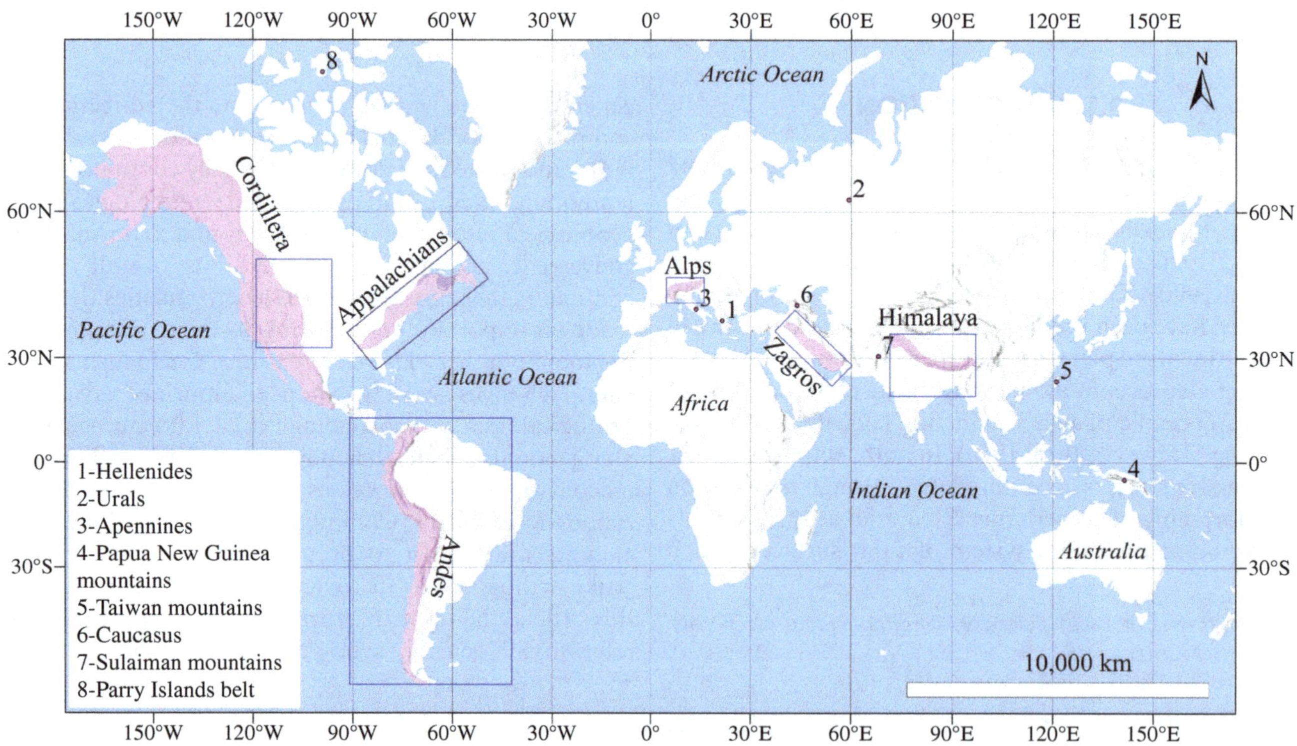

Figure 3.1 Map of the world showing locations of mountain belts described in this chapter. The pink shades are spatial extent of respective orogenic belts. Rectangular boxes indicate the extent of tectonic maps presented in the text. Numbered areas are other convergent orogens mentioned in the text.

along-strike changes in deformation style; (2) variation in igneous activity or metamorphic grade; (3) variation in seismic activity; (4) changes in geomorphology; or (5) abrupt lateral stratigraphic change. Drivers of these along-strike changes can include irregularities along continental margin, inherited basement structures of an overriding plate, features from a subducting plate (oceanic) or footwall block (continental), lateral variation in lithospheric or crustal properties, lateral changes in thickness and facies of the deforming sedimentary sequence, lateral variation in dynamic plate tectonic setting, or a combination of these factors. Apart from their influence in the structural/tectonic evolution of the mountain belt, the lateral heterogeneity may also have important seismic, geomorphic, and economic implications.

3.2. APPALACHIANS

3.2.1. Tectonic Setting and Lateral Heterogeneities

The Appalachian orogenic belt, along the eastern margin of the North American continent, is the structural culmination of a prolonged tectonic history consisting of multiple episodes of subduction, terrain accretion, continental collision, and rifting (Fig. 3.2). The prominent Grenville orogeny (1.3–0.9 Ga; Li et al., 2008; Hynes & Rivers, 2010; Rivers, 2012) was the coalescence of the Laurentian (modern-day North American continent) and Amazonian cratons, which formed the Rodinia supercontinent (Li et al., 2008; Rivers, 2008, 2015). Following the Grenville orogeny, the Laurentian craton remained tectonically inactive for about 250 Ma before being detached from the Rodinia craton in various episodes from 759 ± 12 Ma to 530 Ma (Thomas, 1991; Aleinikoff et al., 1995; Hogan & Gilbert, 1998). The Iapetan rifting, during which the Iapetus Ocean opened between the Laurentian and Amazonian (Gondwana) cratons (Domeier, 2016), marked the end of this series of tectonic events (Rankin, 1976; Thomas, 1977; O'Brien & van der Pluijm, 2012). The Middle Ordovician to Late Silurian Taconian orogeny represents the aggregate contractional tectonic activity along the boundary between the Laurentian and Iapetan plates (Hatcher, 1972; Drake et al., 1989; van Staal et al., 2007; Hatcher, 2010). The Late Silurian to Devonian Acadian orogeny has been interpreted to have formed along an Andean-type margin

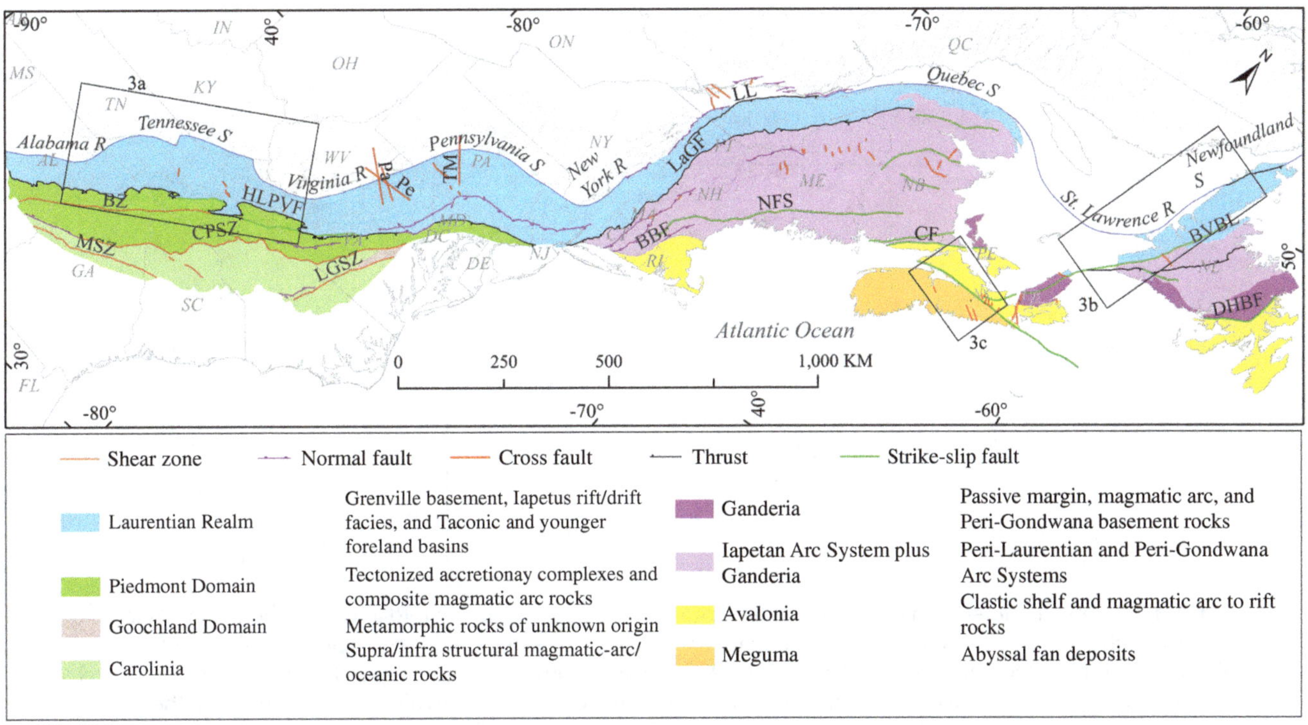

Figure 3.2 Simplified tectonic map of the Appalachians belt showing the major faults, the salient (S) and recesses (R), and the extent of the Appalachian deformation front. Insets show the locations of Figure 3.3a,b, and c (simplified from Southworth, 1986; Hibbard et al., 2006; Waldron et al., 2015). Note: BBF = Bloody Bluff Fault; BVBL = Baie Verte-Brompton Line; BZ = Brevard Zone; CF = Caledonia fault; CPSZ = Central Piedmont Shear Zone; DHBF = Dover Hermitage Bay Fault; HLPVF = Hollins Line-Pleasant Valley Fault System; LaGF = La Gaudaloupe Fault; LGSZ = Lake Gordon Shear Zone; LL = Logan's Line; MSZ = Modoc Shear Zone; NFS = Norumbega Fault System; Pa = Parson's Lineament; Pe = Petersburg Lineament; TM = Tyrone-Mount Union lineament.

(flat slab subduction) following the accretion of Avalonia to the Laurentia margin (Keppie & Krogh, 1999; Murphy & Keppie, 2005). The youngest event in the building of the Appalachians was the Late Mississippian to Late Permian Alleghenian orogeny (Laurentia and Gondwana) (Hatcher et al., 1989), which also marked the formation of the Pangean supercontinent. In the Mesozoic, the modern-day North American plate rifted westward from the African continent. Dominant structures from all of these convergent and divergent tectonic events are range-parallel faults and fold axes. Close inspection, however, reveals lateral heterogeneity that includes changes in the mountain belt orientation, structural style, and metamorphic grade, and the presence of cross structures.

One of the obvious heterogeneities in topography is the fact that the northeast-trending Appalachian mountain front curves around a series of recesses and salients along the length of its western margin (Fig. 3.2). These recesses and salients are proposed to be bound by broad transverse structures that offset or truncate thrust structures (Thomas, 1990). In the southern Appalachians, the large-scale thrust faults and related structures are predominantly striking northeast and verging toward the northwest (Bayona et al., 2003). Abrupt along-strike terminations of these structures are known to spatially coincide with these transverse zones (TZ). In the Alabama recess (Fig. 3.3a), major TZs include the Bessemer TZ, the Happersville TZ, the Anniston TZ, and the Rising Fawn TZ from southwest

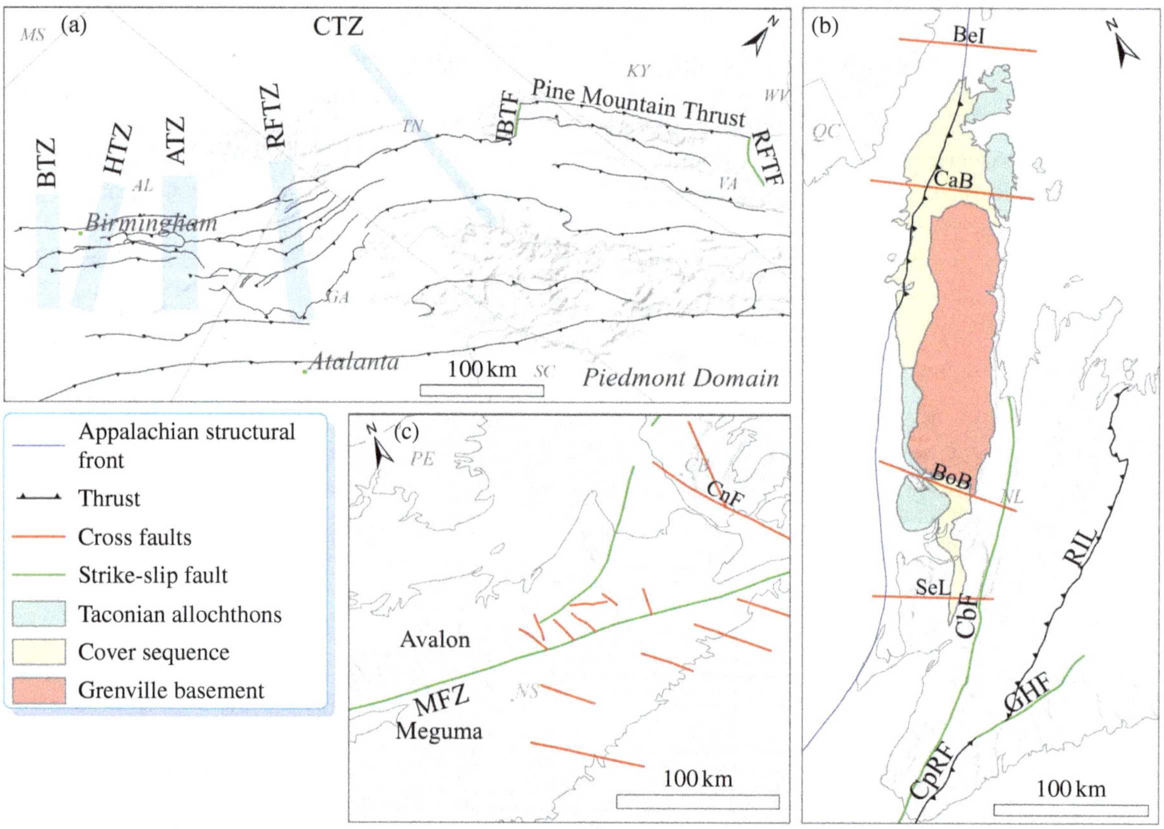

Figure 3.3 Cross structures along the Appalachians. (a) Map of the Alleghenian fold and thrust belt in the Alabama recess and the Tennessee salient showing the major transverse zones (light blue rectangles) and the tear faults on either side of Pine Mountain (Mitra, 1988; Cook & Thomas, 2009). (b) Tectonic-sketch map of Newfoundland Island showing the major cross faults. Note the lateral termination of the Grenville basement inliers at the Bonne Bay cross-strike discontinuity (BoB) and the Canada Bay (CaB) cross-strike discontinuity (Cawood & Botsford, 1991; Hibbard et al., 2006; Waldron et al., 2015). (c) The Minas Fault Zone (MFZ) between the Avalon and Meguma terrains and its conjugate faults including the Canso Fault (CnF) (Hibbard et al., 2006; Waldron et al., 2015). Note: ATZ = Anniston Transverse Zone; Bel = Belle Isle cross-strike discontinuity; BoB = Bonne Bay cross-strike discontinuity; BTZ = Bessemer Transverse Zone; CaB = Canada Bay cross-strike discontinuity; CbF = Cabot Fault; CnF = Canso Fault; CpRF = Cape Ray Fault; CTZ = Cartersville Transverse Zone, GHF = Gunflap Hills Fault; HTZ = Happersville Transverse Zone; JBTF = Jacksboro Tear Fault; MFZ = Minas Fault Zone; RFTF = Russel Fork Tear Fault; RFTZ = Rising Fawn Transverse Zone; RIL = Red Indian Line; SeL = Serpentine Lake cross-strike discontinuity.

to northeast, respectively (Thomas, 1990; Bayona et al., 2003; Brewer, 2004). In general, these transfer zones are tens of kilometers wide and hundreds of kilometers long, and are cut into the Proterozoic basement rocks (Thomas & Bayona, 2002; Brewer, 2004). Thrust-related structures plunge into these TZs, abruptly terminate at lateral or oblique ramps, show offsets along transverse faults (tear faults), or break into enechelon segments of transfer zones (Thomas & Bayona, 2002; Thomas, 2007; Cook & Thomas, 2009). Structural strike deviates within these TZs, while propagating over oblique/lateral ramp and irregular basement and at junctions between lateral/oblique and frontal ramps (Bayona et al., 2003). Above a lateral/oblique ramp, hanging wall rocks are generally folded into upright, isoclinal folds driven by local-scale, ramp perpendicular contraction in the hanging wall due to crowding of materials along that trend (Cook & Thomas, 2009). Across the Cartersville TZ, deformation style, metamorphic grade, and thrust sheet stratigraphy show sharp contrast between the Alabama recess and the Tennessee salient (Tull & Holm, 2005). In the salient, passive margin and foreland sedimentary sequences are much thicker and have been deformed and metamorphosed (biotite grade of the Barrovian metamorphic series) more strongly than in the recess (Tull & Holm, 2005). Toward the adjoining Virginia recess, a section of the Pine Mountain belt, bounded by the left-lateral Jacksboro tear fault and the right- lateral Russel Fork tear fault (Fig. 3.3a), demonstrates a double ramp geometry in the middle section, while a single ramp geometry around the edges (Mitra, 1988). Presence of the orogen-parallel, large-scaled (about 35 km wide; Brewer, 2004) Birmingham graben beneath the southern Alleghenian thrust belt also had major controls on the foreland sedimentation and subsequent propagation of the thrust sheet (Thomas, 2007). Transverse zones in the southern Appalachians have been interpreted as inherited transform fault related to the Iapetan rifting along the continental margin of Laurentia (e.g., Fig. 3.4; Rankin, 1976; Thomas, 1977, 2014).

Cross structures in the Appalachians of Pennsylvania and West Virginia (central Appalachians) such as the Parsons, Petersburg, and Tyrone-Mount Union lineaments (Fig. 3.2) also have similar effects on the orogen-parallel structures as is seen in the southern Appalachians (Wheeler, 1980; Southworth, 1986). In the northern Appalachians, however, genesis of cross structures presents much more diversity. The Serpentine Lake cross-strike discontinuity (CSD), the Bonne Bay CSD, the Canada Bay CSD, and Belle Isle CSD in Newfoundland mark the boundaries of second-order salients and recesses and had controls on the Taconic and younger orogenic events (Fig. 3.3b) (Cawood & Botsford, 1991). Across these structures, the depositional facies and structural style of the rock units overlying the Grenville

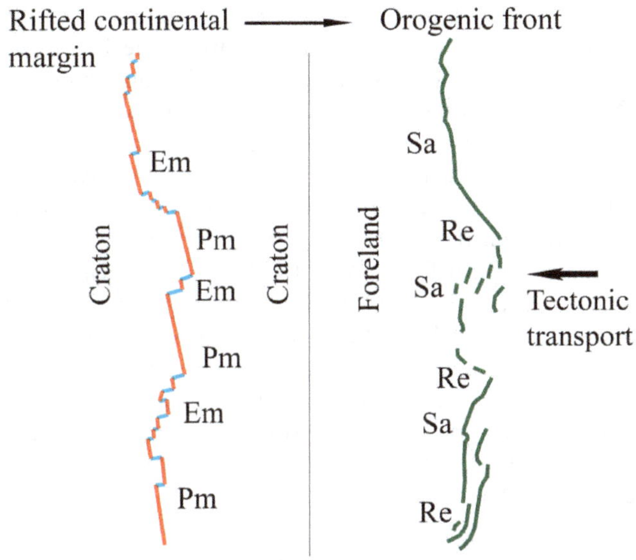

Figure 3.4 Impact of the continental margin geometry on the geometry of an orogenic front. Embayments (Em) evolve into salients (Sa), and promontories (Pm) evolve into recesses (Re).

basement change abruptly suggesting these faults have Precambrian roots. The Grenville basement inlier terminates southward at the Bonne Bay CSD (Fig. 3.3b), which also delineates the southern limit of the Acadian uplift of the Grenville inlier (Cawood & Botsford, 1991). North of the Bonne Bay CSD, thrusts and associated structures are west verging, while toward the south the vergence reverses (Williams & Cawood, 1986). In southern Newfoundland and the Cape Breton Islands, the Silurian tectonic wedging of the Avalon craton into the Laurentian craton has resulted in a very narrow Appalachian central mobile belt characterized by stronger syncollisional deformation, high-grade Barrovian metamorphism (Kyanite and Sillimanite grade, 700°C–750°C at 6–10 Kbar (Plint & Jamieson, 1989), 5 Kbar elsewhere), and a greater volume of Silurian magmatism than elsewhere (Lin et al., 1994). This intensely deformed and metamorphosed narrow band is bound by the northwest-striking Canso fault (Fig. 3.3c) to the southwest and the east-striking Gunflap Hills fault to the northwest, interpreted as transverse, right-lateral wrench faults that accommodated the differential movement on either side of the wedging (Lin et al., 1994). In its western end, the Gunflap Hills fault (Colman-Sadd et al., 1992) merges with the northwest-striking Cape Ray fault zone (Fig. 3.3b). In eastern Nova Scotia, several northwest-striking, transverse, left-lateral strike-slip faults (e.g., the Country Harbour fault) have been described as conjugates of the Minas Fault Zone (Fig. 3.3c; Waldron et al., 2015), which is an east-west striking, 300 km long, continental transform fault in New Brunswick and Nova Scotia that marks the boundary between the Avalon and Meguma terrains (Murphy et al., 2011). Toward the west, the Paleozoic Norumbega Fault System (Fig. 3.2) demonstrates the effect of the subducting plate physiography on the upper plate structures. This northeast-striking, 300 km long, transpressive, dextral fault system (West & Hubbard, 1997; Hubbard, 1999) has recently been interpreted to be a result of subduction of an oceanic ridge and related transform fault (Kuiper, 2016; Kuiper & Wakabayashi, 2018). Around the orogenic front in Quebec and northern Vermont, cross structures were formed due to reactivation of earlier-formed transverse, normal faults with an oblique thrust-sense or served as an oblique ramp during the westward thrusting of the Lower to Middle Ordovician shelf-carbonates (Séjourné & Malo, 2007).

3.2.2. Proposed Factors Controlling Lateral Heterogeneities

Lateral heterogeneity along the Appalachians has largely been controlled by the irregularities along precollisional continental margins and by inherited basement structures. The eastern continental margin of Laurentia went through a series of rifting events including the Iapetan rifting, all of which acted in concert to develop a series of embayments (concave oceanward) and promontories (convex oceanward) separated by transform faults (Rankin, 1976; Thomas, 1977, 2014). During subsequent orogenic events (Taconic, Acadian, and Alleghenian orogenies), embayments evolved into salients (convex toward the foreland), promontories evolved into recesses (concave toward the foreland), and the transform fault boundaries evolved into transverse zones (Fig. 3.4). As Tull and Holm (2005) noted, foreland sedimentary sections along salients are much thicker than along recesses, which has a great control on the geometry and deformation style of the evolving thrust sheets (discussed in detail in section 3.3.1, the North American Cordillera section of this chapter). Irregularity along continental margins also has another important control, which results in a sharp lateral heterogeneity along an orogen. A collisional event between two promontories (of two different landmasses) results in a narrower but stronger orogenic deformation and a higher grade of metamorphism that a promontory-embayment collision (Lin et al., 1994). The evolution of rift-related transform faults into transverse zones also had a profound impact on the evolution of the Appalachian mountains by creating barriers or zones that truncated or offset major thrust faults. Along the range front, transverse zones have served as nucleation sites for cross structures in the form of lateral ramps or tear faults.

3.3. CORDILLERA

3.3.1. Tectonic Setting and Lateral Heterogeneities

The North American Cordilleran orogenic system records the Jurassic (and possibly earlier) to Paleogene history of terrain accretion and the eastward convergence of the Farallon and Kula plates beneath the western margin of Laurentia (North American continent) (Oldow et al., 1989; Burchfiel et al., 1992; Saleeby et al., 1992; DeCelles, 2004; Dickinson, 2004). This convergence resulted in a number of different ranges including the Sierra Nevada, the Rocky Mountains, and the various mountain systems of Alaska (Fig. 3.5a). Changing plate dynamics has resulted in varied deformation style and distribution throughout the Cordilleran evolution. The Late Cretaceous (Beck et al., 1988; DeCelles et al., 1991) transition of deformation style from the thin-skinned, taper-wedge in the U.S. Sevier fold-thrust belt to predominantly thick-skinned, basement-involved Laramide orogenesis is one of the remarkable changes. Initiation of the Laramide system uplifts has generally been correlated to the subduction of a flat-slab segment (e.g., Coney & Reynolds, 1977; Dickinson & Snyder, 1978; Saleeby, 2003; Chapin, 2012). North of the Canadian border, the Laramide-style deformation is absent. Orogenic collapse of the Cordilleran system, focused on the Sevier belt, initiated during Middle to Late Eocene due to the change in relative plate kinematics and has resulted in the active basin and range province of extensional deformation (e.g., Constenius, 1996; Yonkee & Weil, 2015). The contractional phase of deformation resulted in spectacular range-parallel fold and thrust belts (Yonkee & Weil, 2015), yet there are important lateral heterogeneities and a number of cross structures along the Cordillera.

Basement provinces and their suture zones and major faults along the western margin of the North American continent include the Snowbird Tectonic Zone (Ross et al., 1991; Hope & Eaton, 2002), the Archean Hearne Province (Hope & Eaton, 2002), the Archean Medicine Hat Block (Ross et al., 1991; Lemieux et al., 2000), the Archean Wyoming Province (Wooden & Mueller, 1988; Mogk et al., 1992), and the Paleoproterozoic Yavapai and Mazatzal Provinces (Foster et al., 2006; Whitmeyer & Karlstrom, 2007), from north to south (Fig. 3.5b). The Trans-Hudson Orogen and the Superior Craton lie toward the east, around the central region of North America. Toward the west, the basement block includes the Archean Grouse Creek Block and the Selway terrain (Foster et al., 2006).

Transverse crustal boundaries between these provinces and the other Archean/Paleoproterozoic transverse basement structures are one of the primary causes of lateral heterogeneity in the Cordillera due to their influence in

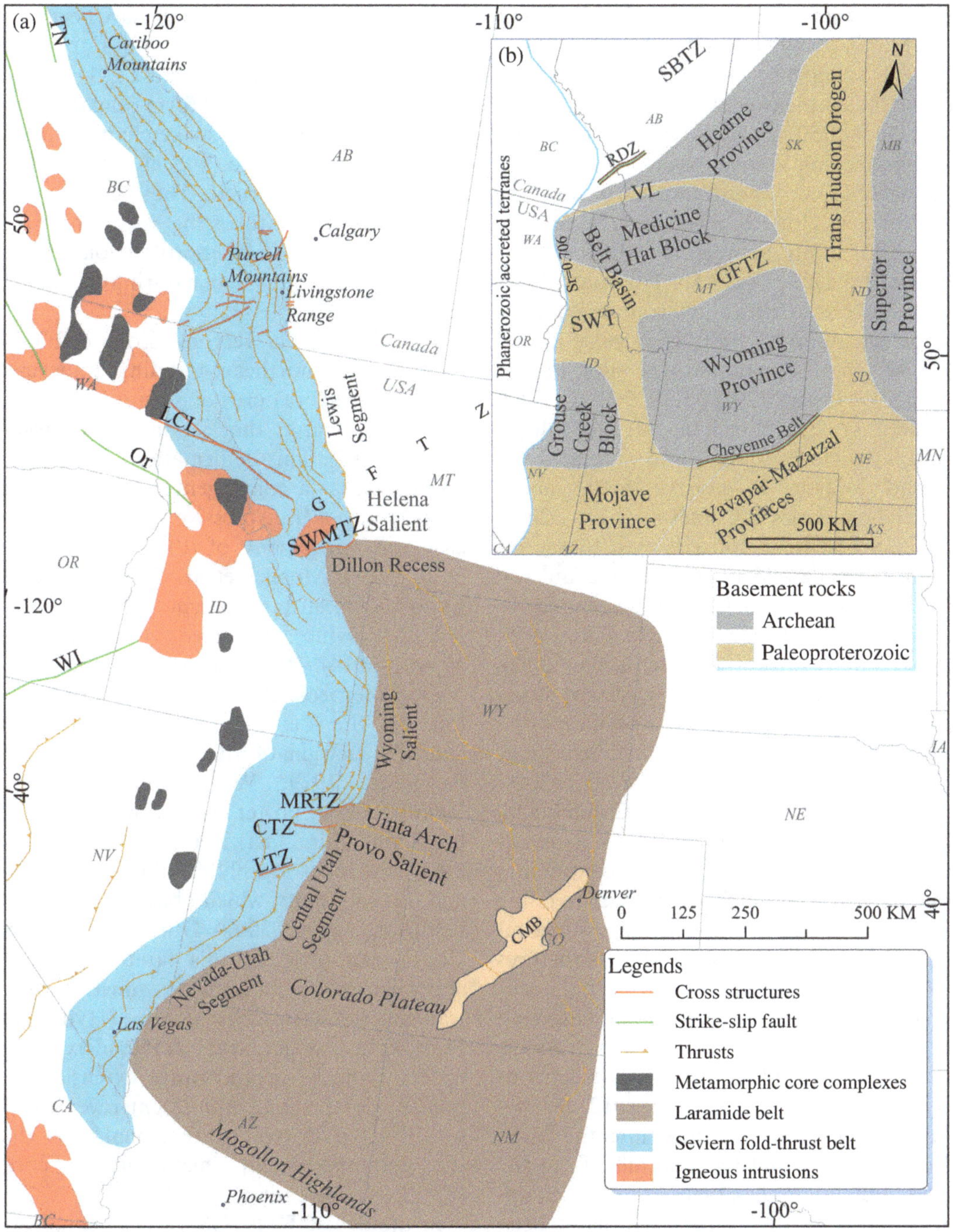

Figure 3.5 Simplified tectonic and basement map around the western North American Cordilleran belt. (a) Tectonic map of the United States and southern Canadian Cordillera showing the major orogenic curvature and the major cross structures (modified from Moffat & Spang, 1984; McGugan, 1987; Kwon & Mitra, 2006; Cooley et al., 2011; McMechan, 2012; Whisner et al., 2014; Yonkee & Weil, 2015). (b) Generalized basement map of western North America. The Sr = 0.706 line represents the eastern boundary of the accreted terrains (Foster et al., 2006; Whitmeyer & Karlstrom, 2007; McMechan, 2012; Yonkee & Weil, 2015; Ronemus et al., 2020). Note: CMB = Colorado Mineral Belt; CTZ = Charleston Transverse Zone; GFTZ = Great Falls Tectonic Zone; LCL = Lewis and Clark Line; LTZ = Leamington Transverse Zone; MRTZ = Mount Raymond Transverse Zone; Or = Orfino Shear Zone; RDZ = Red Deer Zone; SBTZ = Snowbird tectonic zone (the Thorsby Low is located within the SBTZ); SWMTZ = Southwestern Montana Transverse Zone; SWT = Selway terrance; TN = Tintina-Rocky Mountain Shear Zone; VL = Vulcan Low; WI = West-Idaho Fault.

the subsequent sedimentary and deformational history (e.g., Paulsen & Marshak, 1999; Sears & Hendrix, 2004; McMechan, 2012). Significant crustal-scale, transverse boundaries around the southeastern Canadian Cordilleran front (Alberta) include the Thorsby Low, a crustal suture within the Snowbird tectonic zone (Ross et al., 1991; Hope & Eaton, 2002), the Red Deer zone, the northern boundary of the Hearne Province (Hope & Eaton, 2002), and the Vulcan Low, a crustal suture between the Medicine Hat Block and Hearne Province (Fig. 3.5b; Hoffman, 1988; Ross et al., 1991). These structures and terrain boundaries served as loci for activation of transverse zones during the deposition of the Mesoproterozoic Belt-Purcell sequence (Benvenuto & Price, 1979; Foo, 1979; Root, 1987; Anderson & Davis, 1995; McMechan, 2012). Multiple other transverse structures were also initiated in Mississippian and Triassic as normal faults above these transverse boundaries (Cooley et al., 2011; McMechan, 2012). Segments of some of these transverse zones were reactivated as dextral, oblique reverse faults or as tear faults during Jurassic to Eocene in response to compressional stresses (Bielenstein, 1969; Foo, 1979; Price, 1981; McMechan & Price, 1982; McMechan, 2000; Cooley et al., 2011). Similar impacts of basement structures on the evolution of thrust belts have been reported from other domains in the Canadian Cordillera (e.g., Pride et al., 1986; Thompson, 1989; Pilkington et al., 2000; Berger et al., 2008). In northern British Columbia, a geometric inversion of the northern end of a northwest-trending basement uplift during the Sevier contraction resulted in the formation of northeast to east-trending (transverse) contractional structures, which were superimposed on the regional northwest-striking structures (McMechan, 2007). Aside from their influence in the sedimentary and deformational history, these basement structures and terrain boundaries also formed suitable conduits for melt transport and served as loci for igneous intrusions and mineralization (McMechan, 2012).

The Sevier FTB is generally classified into the Main Ranges, the Front Ranges, and the Foothills, from the hinterland to the foreland, in the northern United States and southern Canada. Cross structures have been observed in all these subdivisions at different scales. In the Main Ranges of British Columbia (Purcell Mountains), Mesoproterozoic and Eocambrian synsedimentary faults have been reactivated as dextral, east-striking, transverse, thrust faults (Fig. 3.5a) during Mesozoic and Cenozoic (Höy & Heyden, 1988; Larson et al., 2006; McMechan, 2012). Several transverse faults have also been identified in the Front Ranges (e.g., Moffat & Spang, 1984; McGugan, 1987; Fermor, 1999; McMechan, 2001; Norris, 2001). In the foothills, the north to northwest-trending Livingstone Range anticlinorium consists of enechelon, chevron-style fault-propagation folds, which have been compartmentalized by multiple tear faults and transverse zones of intense fracturing (Cooley et al., 2011). Anticlines plunge into these tear faults and form domal culminations. Tear faults also mark along-strike changes in fold style, from chevron to concentric, and changes in the presence or absence of back thrusts. These tears are thought to be the results of synthrusting reactivation of preexisting northeast-striking Paleoproterozoic basement faults, some of which were active during Mississippian (Cooley et al., 2011).

The Great Falls Tectonic Zone (GFTZ; Fig. 3.5) in Montana is a northeast-trending, transverse (in relation to the Cordillera) crustal suture between the Medicine Hat Block and the Wyoming Craton (O'Neill & Lopez, 1985). This deformation zone was initiated prior to ~1.9 Ga as a convergent boundary and subsequently evolved into a transpressive boundary in response to the eastward movement of the Wyoming Craton (Dahl et al., 1999; Mueller et al., 2002; Gifford et al., 2014). Cenozoic alkali magmatism, belonging to the Montana Alkali Province, occurred within this zone (Gifford et al., 2014). In central Montana and Idaho, the Mesoproterozoic Belt rifting most likely led to the initiation of the Lewis and Clark fault zone from the southern margin of the GFTZ (Harrison et al., 1974; Reynolds, 1979; Wallace et al., 1990; Sears & Hendrix, 2004; Foster et al., 2006). The Lewis and Clark fault zone (Fig. 3.5a) is a northwest-trending, >800 km long, system of steeply dipping strike-slip, oblique-slip, and dip-slip faults, which obliquely cuts across the Sevier belt (Foster et al., 2007). During the Mesozoic-Cenozoic Cordilleran orogeny, this fault zone deformed as an approximately 40 km wide, sinistral shear zone with transpressional flower structures (Hyndman et al., 1988; Sears et al., 2000; Sears & Hendrix, 2004). With the onset of the basin and range extension in the Eocene, polarity of the fault movement reversed, and this zone served as a dextral, extensional, TZ that facilitated the exhumation of metamorphic core complexes (Fig. 3.5a; Foster et al., 2007).

In the Sevier FTB (USA), the geometry of basement structures had first-order control on the formation of orogenic curvatures and on the evolution of their transverse boundaries (e.g., Paulsen & Marshak, 1999). A deeper basin generally corresponds to a thicker sedimentary column and more material available to be incorporated into the deforming taper, which results in a wider wedge (salient) (e.g., Marshak & Wilkerson, 1992; Boyer, 1995). The Helena salient likely formed over an east-trending, asymmetrical depositional trough of the Middle Proterozoic belt basin called the Helena embayment (Fig. 3.5b; Harrison et al., 1974). A north-dipping Middle Proterozoic normal fault on the southern edge of the embayment probably evolved into the Southwest Montana Transverse Zone (SWMTZ), which forms the southern boundary of the salient with the Dillon recess (Whisner et al., 2014). Thrusts and related structures in the southern domain of this salient are strongly converging into the right-lateral, reverse faults within the SWMTZ, due to a gradual clockwise rotation of the shortening direction during their evolution (Whisner et al., 2014).

Further south, controls of the basin boundary geometry on the evolution of transverse zones have been exemplified by the Mount Raymond Transverse Zone (MRTZ) and the Charleston TZ. The MRTZ and the Charleston TZ form the northern and the southern boundaries of the Uinta/Cottonwood arch (recess) with the Wyoming Salient and the Provo Salient respectively (Fig. 3.5a; Paulsen & Marshak, 1997, 1998). Paulsen and Marshak (1999) noted contrasting structural styles between these two zones and explained this contrast in light of corresponding basement structures. An east-west trending asymmetric basement high, with a gentle northern flank and a steep southern flank, existed just north of the present Uinta/Cottonwood arch, along the boundary between the Archean Wyoming province (north) and Proterozoic terrain (south) (Paulsen & Marshak, 1999). A gentler northern flank meant that the sedimentary thickness gradually increased northward from the Uinta recess into the Wyoming Salient. The MRTZ initiated above this flank as northeast-trending thrusts along the southern margin of the Wyoming salient, which were later tilted northward creating an east-west strike during the uplift of the Uinta/Cottonwood arch (Paulsen & Marshak, 1997). The steeper southern flank, however, marked an abrupt increase in sedimentary thickness toward the south and thereby formed the boundary between two contrasting taper wedges. The Charleston TZ (Fig. 3.5a) served as zone of lateral ramp between the two contrasting tapers and gradually evolved into a left-lateral strike-slip zone, which accommodated the differential motion between the Uinta recess and the Provo salient (Paulsen & Marshak, 1998, and references therein). The southern boundary of the Provo salient with the central Utah segment is the Leamington TZ, which is an east-northeast trending, >50 m long, complex, cross structure (Lawton et al., 1997; Kwon & Mitra, 2006). In the salient, an initial east-directed vergence over the TZ rotated clockwise during subsequent deformational phases, which likely reflects the interaction between a deforming wedge and an oblique ramp (Lawton et al., 1997; Paulsen & Marshak, 1999; Kwon & Mitra, 2006).

In southern Wyoming, the east- to northeast-trending Cheyenne belt (Fig. 3.5b) represents the transverse, crustal suture/ transpressional shear zone between the Wyoming and Yavapai-Mazatzal provinces, across which the Precambrian geology, metamorphism, and metallogenesis abruptly change between adjacent blocks (Karlstrom & Houston, 1984). During the Laramide orogeny, this weak crustal zone was reactivated as a left-lateral transpressional structure and subsequently as a right-lateral transtensional zone during the Tertiary extension (Bader, 2008). Just to the south, east-west trending Precambrian basement fault zones, genetically linked to the Cheyenne belt (Sims et al., 2001; Whitmeyer & Karlstrom, 2007), have been identified to influence Laramide uplift resulting in accumulation of oil and gas resources (Bader, 2009).

A peculiar feature in the Laramide belt of Colorado is a 500 km long, 25–50 km wide, linear zone of numerous magmatic intrusions, known as the Colorado Mineral Belt (Fig. 3.5a). This zone marks an abrupt along-strike change in (1) the structural trend of the Laramide uplifts (north-trending in the southern region versus northwest-trending northward); (2) the thickness of the Late Cretaceous and Paleogene sedimentary sequences; and (3) the composition of the Laramide plutons (Chapin, 2012). This economically valuable belt has been interpreted to have formed over an extensional boundary between two adjacent segments of the underlying Farallon flat slab (Chapin, 2012), which serves to demonstrate the effects of the downgoing plate features and processes on the upper plate structures. In the Laramide belt, preexisting basement structures and lateral heterogeneities are suggested to have a control on the stress field and thereby led to the development of structures with an orientation different from the regional trend (Weil et al., 2016).

3.3.2. Proposed Factors Controlling Lateral Heterogeneities

Along the Cordillera (primarily along the Sevier FTB), preexisting basement features/structures likely had the greatest impact on creating lateral heterogeneity, which has generally manifested as curvatures along the orogenic front and as variation in the geometry of cross structures. Prior to the Cordilleran orogeny, transverse crustal boundaries served as loci for activation of transverse zones during various rift-related extensional features (McMechan, 2012). Subsequently, these transverse zones were reactivated as cross structures during the orogenic contraction/extension and served to partition or distort deformation of the evolving crustal wedge. Both transverse structures and irregular basement topography further added lateral heterogeneity through their profound control on the lateral continuity of facies and thickness of the pre and syn orogenic sedimentary succession. Abrupt lateral changes in the facies and thickness of the sedimentary column caused the deforming wedge to partition into segments, which are separated by cross structures of various geometries and genesis. As Paulsen and Marshak (1999) explained, a thicker sedimentary column corresponds to a further propagation of the deforming wedge (salients) than a thinner column (recesses). Further, the geometry of the basement irregularities and transverse structures dictate the geometry of cross structures. A near vertical transverse structure is more likely to evolve into a tear fault, whereas an inclined structure evolves into a lateral ramp (Paulsen & Marshak, 1999). And finally, features of the subducting plate (Farallon) may have influenced the development of cross structures and lateral heterogeneity along the range (e.g., Chapin, 2012).

3.4. ALPS

3.4.1. Tectonic Setting and Lateral Heterogeneities

The European Alps formed during a Late Cretaceous collision between the European and African plates following the closure of the Alpine Tethys, which consisted of the northern Valais ocean and the southern Piemont-Liguria ocean, separated by continental crust in the middle known as the Brianconnais (Tricart, 1984; Stampfli & Borel, 2002; Schmid et al., 2004; Handy et al., 2010). The European side of the collision was the lower plate with the Apulian/Adriatic blocks of the African plate forming the upper plate (Dewey et al., 1998; Handy et al., 2010). The Apulian plate refers to all continental domains located south of the Alpine Tethys. The Adriatic microplate or Adriatic indenter, a part of the Apulian plate, is situated south of the modern-day Periadriatic Fault System (PA; Stampfli & Borel, 2002; Schmid et al., 2004; Handy et al., 2010). Today this orogen trends approximately east-west in the eastern and central Alps and has a northeast-southwest orientation in the western Alps (Fig. 3.6). Major range-parallel lithotectonic units of the Alpine orogen are from north to south, the European foreland, Jura Mountains, Molasse basin (Oligocene-Miocene Foreland), Helvetics, Penninic zone, Austroalpine, Southern Alpine, and Po Basin (retroarc basin for the Alps and foreland basin for the Apennines) (Pfiffner, 2014). The Helvetic domains are derived from the Mesozoic and Cenozoic shelf and upper slope deposits along the southern European plate margin, and in places include the pre-Triassic basement (Zerlauth et al., 2014). For simplicity, the classic term *Penninic nappes/zone* has been largely used in the literature to denote the tectonic units derived from the subducted European margin, the Valais ocean, the Piemont-Liguria ocean, and the Brianconnais continental crust (e.g., Schmid et al., 2004; Pfiffner, 2014). Parts of the Apulian plate to the north and south of the PA are represented by the Austoalpine and southern Alpine units, respectively (Polinski & Eisbacher, 1992; Schmid et al., 2004). While the Austroalpine unit dominates the eastern Alps, this unit has fully eroded away in the western Alps, exposing up to blueschist to eclogite facies rocks of the Penninic unit and subgreenschist facies rocks of the Helvetics (Pfiffner, 2014). The rheologically strong Dolomites, the Adriatic indenter, lie within the southern Alpine unit.

Following the initial collision, the eastern Alps underwent east-west directed orogen parallel extension in the Miocene (Oligocene; Ring, 1994; Steck, 2008). This extension has been referred to as "lateral extrusion" (Ratschbacher et al., 1991). The lateral extrusion has been interpreted as (1) coupling of compression and gravitational collapse (Ratschbacher et al., 1991), (2) upper plate extension due to roll back of a subduction zone beneath the Carpathian orogen (Royden et al., 1983; Royden, 1993; Horváth & Cloetingh, 1996; Sperner et al., 2002), (3) northward indentation of the southern Dolomites (Rosenberg et al., 2004; Rosenberg & Garcia, 2011; Reiter et al., 2018), and (4) an extension related to the roll back of the Mediterranean plate in the west (Ring & Gerdes, 2016). This Miocene extension has been accommodated along a series of orogen-parallel, strike-slip faults and orogen-perpendicular, normal faults. Modern-day topographic evolution was largely controlled by these faults as much of the lateral heterogeneity we see along the range (Bartosch et al., 2017). Major orogen-parallel strike-slip faults in the Eastern and Central Alps are the Periadriatic Fault System (PA), the Defreggen-Antholz-Vals Fault (DAV), Salzach-Ennstal-Mariazell-Puchberg Fault (SEPM), Inntal Fault (IN), and Mur-Murz Valley Fault (MM) (Fig. 3.6). The PA trends east-west for ~700 km and forms a rheological boundary between the weaker Eastern Alps and the relatively stronger Southern Alps (Robl & Stüwe, 2005). From west to east, the PA consists of the Tonale (or Insubric) Line, Giudicarie Fault System, Mauls, Puresetral, and Gailtal segments (Fig. 3.6). Different segments of the PA were active during 32–29 Ma and 22–16 Ma (Müller et al., 2001). North of the PA (Pustertal and Mauls segments), the sinistral, normal DAV runs east-west, for ~80 km and forms the southern boundary of alpine metamorphism (Müller et al., 2000; Bartosch et al., 2017). Within the Austroalpine domain, the East-west-striking, ~400 km long, SEPM has a cumulative left-lateral slip of about 60 km (Urbanek et al., 2002) and separates the Mesozoic Northern Calcareous Alps (NCA) from the Middle Austroalpine basement rocks (Bartosch et al., 2017). The northeast-striking, sinistral MM was active during 17–13 Ma (Dunkl et al., 2005) as a conjugate of the northwest-striking, dextral Pols-Lavanttal Fault System (PL) (Bartosch et al., 2017).

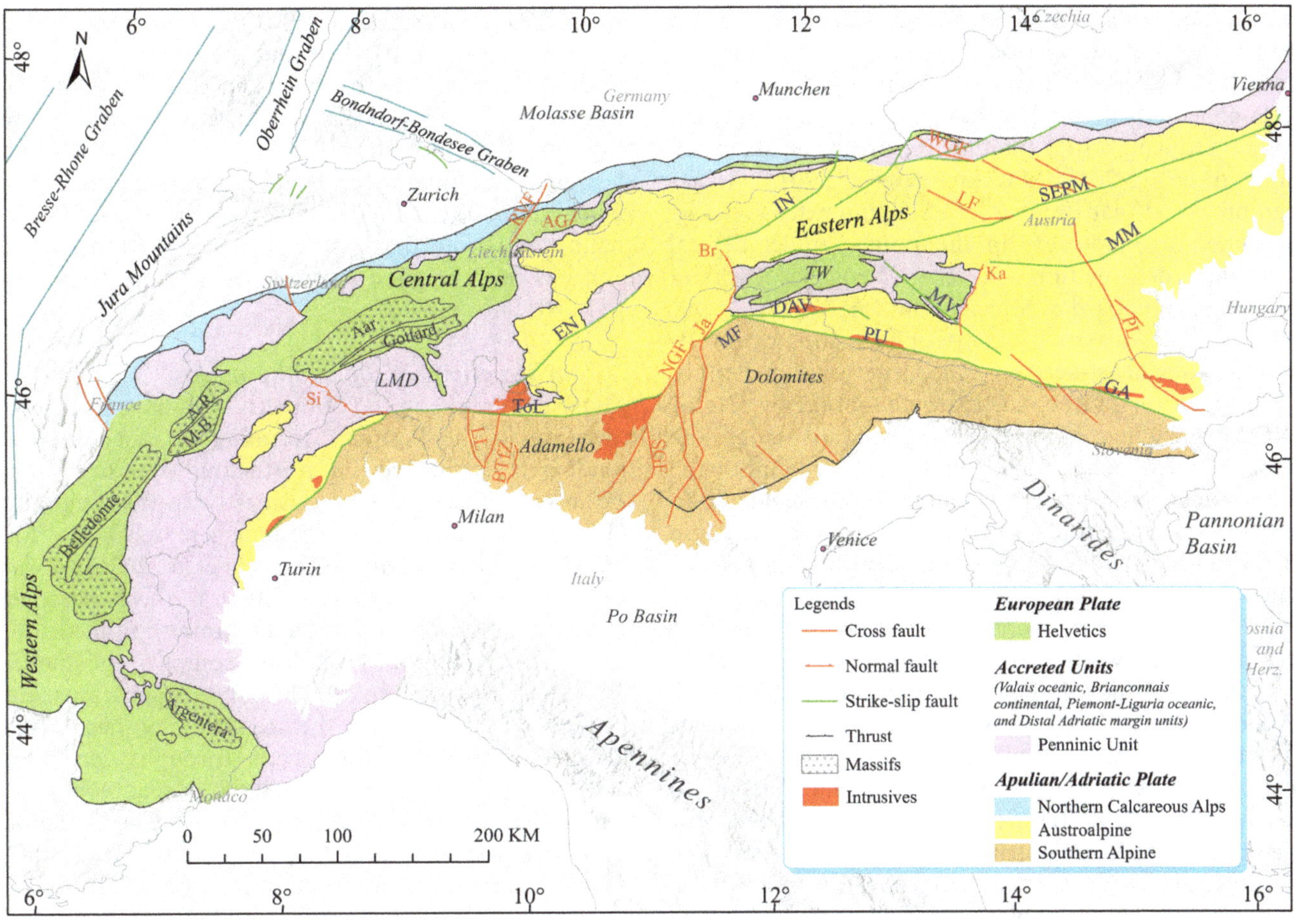

Figure 3.6 Geological map of the Alps showing the major lithotectonic units, major strike-slip faults (golden lines), and major cross faults (red lines) (after Laubscher, 1985; Polinski & Eisbacher, 1992; Schönborn, 1992; Linzer et al., 1995; Castellarin et al., 2006; Pfiffner, 2014; Zerlauth et al., 2014; Ring & Gerdes, 2016; Bartosch et al., 2017). Note: AG = Alpenrhein Graben; A-R = Aiguilles Rouges Massif; Br = Brenner Fault; BTfZ = Ballabio-Barzio Transfer Zone; EN = Enagdin Fault; GA = Gailtal Fault; IN = Inntal Fault; Ja = Jaufen Fault; Ka = Katschberg Faults; LF = Lammertal Fault; LL = Lecco Line; LMD = Lepontine Metamorphic Dome; M-B = Mont-Blanc Massif; MF = Meran-Mauls Fault; MM = Murz-Valley Fault; MV = Moll Valley Fault; NGF = North Giudicarie Fault System; PL = Pols-Lavanttal fault system; PU = Purestal Fault; RVF = Rhine Valley Fault; SEPM = Salzach-Ennstal-Mariazell-Puchberg Fault; SGF = South Giudicarie Fault System; Si = Simplon Fault Zone; ToL = Tonale (Insurbic) Line; TW = Tauern Window; WGF = Wolfgangsee Fault.

Cross Structures in the Eastern and Central Alps

Major transverse, normal faults in the eastern and central Alps are the Simplon Fault Zone (Si), the Pols-Lavanttal Fault System (PL), the Brenner Fault (Br), the Moll Valley Fault (MV), and the Katschberg Faults (Ka) (Fig. 3.6). The Brenner and Katschberg Faults mark the western and eastern edges of the Tauern Window, respectively (Behrmann, 1988; Selverstone, 1988; Fiigenschuh et al., 1997), where blueschist and eclogitic facies rocks of the Penninic zone have been exposed (Pfiffner, 2014). Exhumation of the Tauern Window has been widely linked to Miocene extension. Rosenberg and Garcia (2011) argue that localized intensive folding deformation due to an irregular geometry of the Dolomite indenter coupled with erosion can also exhume deep-seated rocks without a significant crustal extension. The southern edge of the window has been cut by the northwest-striking dextral, normal MV fault. Gently west-dipping mylonitic fabric, with top-to-the-west shear along the Brenner Fault (Behrmann, 1988; Selverstone, 1988) has been overprinted by steeply west-dipping cataclastic zones (Prey, 1989). Much like the Brenner fault, the Simplon fault zone is a low-angle, southwest-dipping, extensional fault that exhumes the Lepontine Metamorphic dome in its footwall. Mylonitic shearing that formed the Simplon Fault Zone initiated at 30 Ma (Campani et al., 2010) and was overprinted by brittle detachment faults of the Simplon line from

14.5–10 Ma until 3–5 Ma (Hubbard & Mancktelow, 1992; Mancktelow, 1992; Grosjean et al., 2004; Campani et al., 2010). Both the Brenner and Simplon faults expose a telescoped crustal section via an initial ductile shearing and subsequent brittle faulting (Grosjean et al., 2004; Campani et al., 2010; Mancktelow et al., 2015). It has been proposed that Brenner and Simplon extension may be related to step-overs in the range-parallel dextral strike-slip deformation in the Neogene (Schmid et al., 1989; Hubbard & Mancktelow, 1992).

Around the eastern Alpine periphery, the northwest-trending, northeast-verging, Paleogene folds in Permo-Mesozoic cover of the Austroalpine are cut by sets of northeast-striking, high-angle, cross faults (Fig. 3.6; Polinski & Eisbacher, 1992). In the Miocene, both the Austroalpine and southern Alpine units were folded into northeast-trending folds, which are dissected by a system of northwest-trending high angle, dextral cross faults, which includes the 150 km long PL. These Miocene cross faults either offset or merge into the PA and partition deformation between the contractional Alpine domain and the extensional Pannonian basin (Polinski & Eisbacher, 1992). In the western Alps, northeast to east-striking transverse faults are represented by the Neogene normal faults with a few faults showing minor left-lateral reactivation (Sue & Tricart, 2003) and the late to post Oligocene right-lateral, strike-slip faults (Malusà, 2004; Perello et al., 2004; Perrone et al., 2011). Anticlockwise rotation of the Apulian plate has been attributed as the cause of the orogen-parallel dextral displacement and the northeast to southwest extension (Hubbard & Mancktelow, 1992; Calais et al., 2002).

The Giudicarie Fault System (GFS) is a system of west-northwest dipping fault-links between the Tonale and Pustertal lines of the PA (Castellarin & Cantelli, 2000; Mancktelow et al., 2001; Müller et al., 2001; Viola et al., 2001). The Meran-Mauls (MF) and the Northern Giudicarie Fault (NGF) within the GFS form the trace of the PA, while the Southern Giudicarie Fault (SGS) is entirely located in the Southern Alps. The Jaufen Fault, as a secondary fault within the GFS, has been considered as a potential continuity of the Brenner Fault fault (Rosenberg & Garcia, 2011). An explanation of the genesis of the GFS assumes evolution of an initially curved section of the PA, which underwent Neogene sinistral transpression over an inherited northeast-trending horst and graben structure (Castellarin & Cantelli, 2000; Müller et al., 2001; Viola et al., 2001). A second explanation, however, considers an initially straight PA bent by the Late Oligocene to Early Miocene, northward movement of the Dolomite indenter. The MF represents a section of the bend, which was subsequently cut by the NGS (Laubscher, 1971; Schmid et al., 1996; Frisch et al., 1998; Pomella et al., 2012). The SGF has been interpreted as a Serravallian-Tortonian transfer fault between the Giudicarie belt and the pre-Adamello belt of the southern Alps (Castellarin et al., 2006). The Adamello batholith is an Upper Eocene and Lower Oligocene batholith, the northern rim of which has been sheared by the Tonale Line (Brack, 1981; Laubscher, 1985; Zanchetta et al., 2011). Just to the east of the NGS, a shallow, north-northeast-striking, sinistral transpressive fault has been mapped in the southern Alps (Fondriest et al., 2015).

Cross Structures in the Southern Alps

Basement rocks in the Southern Alps have undergone extension in the Permian, resulting in east-northeast-trending faults in Triassic, north-south-trending faults, and in Jurassic and Cretaceous east-west-trending faults (Gaetani et al., 1986; Schumacher, 1990; Bertotti et al., 1993; Picotti et al., 1997; Festa et al., 2020). During the south-verging tectonic transport, these basement structures acted as lateral/oblique ramps, initiated more lateral ramps, and produced complex geometries like enechelon ramp folds and back thrusting (Schönborn, 1992). In the central Southern Alps, inherited transverse zones cut through decoupling surfaces, partition thrust sheets into discrete blocks, and thereby serve to generate a laterally heterogenous deformation style (Laubscher, 1985; Schönborn, 1992). Several transverse structures that have been identified in this area include the Lecco Line and Ballabio-Barzio TfZ (Fig. 3.6; Laubscher, 1985; Schönborn, 1992; Zanchi et al., 2012). The central southern Alpine thrust belt has been compartmentalized by these transverse zones (Schönborn, 1992). Moreover, many upper Triassic to Jurassic (Bernoulli, 2007), orogen-perpendicular (e.g., Berra & Carminati, 2010), normal faults were reactivated during the Neo-Alpine deformation (Oligocene-Miocene; Castellarin & Cantelli, 2000) along with nucleation of several transverse, sinistral, strike-slip faults (Prosser, 1998; Zanchi et al., 2012).

Cross Structures in the Northern Calcareous Alps (A Subdivision of the Austroalpine Unit)

The Permo-Mesozoic sedimentary succession (3–5 km thick) of the Northern Calcareous Alps were deformed into roughly northeast-trending contractional structures during Early Cretaceous to Late Eocence (Gaupp & Batten, 1983; Kralik et al., 1987). The synchronous, enechelon, west-northwest-striking dextral, tear/transfer faults cut these contractional structures at all scales (Linzer et al., 1995). Some examples include the Lammertal and Wolfgangsee fault systems in the northeastern Alps (Fig. 3.6). Linzer et al. (1995) observed a peculiar deformation decoupling between the sedimentary cover and basement during the ongoing oblique convergence. The sedimentary cover accommodated the convergence obliquity via deformation partitioning between the contractional and strike-slip structures, while the basement deformed through crystal-plastic flow (Linzer et al., 1995). Together these displacement transfer faults and deformation decoupling caused a 30° clockwise rotation of the entire NCA belt about a vertical axis (Linzer et al., 1995). Moreover, tear faults related to the synorogenic inversion of a Jurassic rift-related graben shoulder have also been identified in the NCA (Oswald et al., 2019). The Miocene, extensional, northeast-trending, sinistral, strike-slip faults, such as the SEPM and IN, cut all the earlier structures and divide the NCA into a number of rhombohedral crustal blocks (Linzer et al., 1995).

Cross Structures in the Helvetics

In the Helvetic units, northwest to north-striking tear faults were formed due to lateral variation in shortening of the nappe stack, primarily during 35–30 Ma (Hunziker et al., 1986). Nappe imbrication in the Helvetics changes laterally due to the absence or presence of a decoupling layer (Zerlauth et al., 2014). The Permo-Carboniferous and Jurassic extensional structures along the European margin resulted in these along-strike facies changes, which subsequently controlled the deformation style. During the nappe formation, synsedimentary normal faults were also reactivated as lateral ramps (e.g., the Rhine Valley Fault) and tear faults (Zerlauth et al., 2014). In the Oligocene, the northwest-trending, sinistral, transtensional Alpenrhein graben was formed in the Helvetics (Ring & Gerdes, 2016). Its conjugate, the Bonndorf-Bodensee in the Jura Mountains, is a northeast-trending graben formed due to dextral transtension prior to 18 Ma (Hofmann et al., 2000; Ring & Gerdes, 2016). The Bresse-Rhone and Oberrhien grabens in the Alpine foreland are related to the European Cenozoic Rift System (Ring & Gerdes, 2016).

3.4.2. Proposed Factors Controlling Lateral Heterogeneities

The Eastern and Western Alps show remarkable heterogeneity in timing of major orogenic events (Late Cretaceous in the Eastern Alps Cenozoic in the Western Alps), timing of the regional metamorphism (older in the Eastern Alps and younger in the Western Alps), and overall direction of tectonic transport (northwest- to west-directed in the Eastern Alps versus north to northwest directed transport in the western Alps) (Handy et al., 2010, p. 123, and references therein). Within each lateral subdivision of the Alps, several factors have contributed to the development of cross structures and lateral heterogeneity. Extension of the upper plate has given rise to several cross faults such as the Simplon and Brenner faults, which serve as a large-scale, displacement transfer fault between major, range-parallel, strike-slip faults (e.g., Selverstone, 1988; Hubbard & Mancktelow, 1992). Lateral connecters (tear faults, lateral ramps, and displacement transfer faults) are common in most of the tectonic units of the Alps. As observed in the North American Cordillera and the Appalachians, extension-related basement structures, in both the European and Apulian plates, had controls on the lateral continuity of facies and thicknesses of sedimentary rocks and thereby on subsequent deformation, primarily in the Southern Alps and Helvetics (e.g., Laubscher, 1985; Schönborn, 1992; Zanchi et al., 2012; Zerlauth et al., 2014). Lateral variability in the composition of the sedimentary section has influenced deformation style based on the presence or absence of decollement horizons. Oblique convergence, coupled with partitioning of deformation between the basement and sedimentary cover, has resulted in multiple cross faults in the northern Calcareous Alps (Linzer et al., 1995).

3.5. HIMALAYA

3.5.1. Tectonic Setting and Lateral Heterogeneities

The Himalaya are our planet's most prominent example of a collisional mountain belt. While the geology of this orogen is generally presented in the context of range-parallel thrust faults that separate lithotectonic units of differing metamorphic grade (Fig. 3.7; Hodges, 2000), there is also lateral heterogeneity in various aspects of the geology along the range and in some areas cross structures have played a role in the segmentation of the Himalaya (Mukul, 2010; Godin & Harris, 2014). Early geological and geophysical studies on the Indo-Gangetic plain presented evidence for lateral variations in sediment thickness and geophysical properties along the Himalayan foreland (Burrard, 1915; Oldham, 1917). In the 1970s, new data from the oil and gas industry connected these lateral variations to a series of northeast-trending basement ridges (Sastri et al., 1971; Rao, 1973; Raiverman, 1983). It was also proposed that these transverse basement structures may influence Himalayan deformation (Valdiya, 1976). In recent years, evidence from field surveys and laboratory analyses confirms the lateral heterogeneity and locates structural transfer zones and cross structures (Mugnier et al., 1999; Mukul, 2010; Godin & Harris, 2014; Hubbard et al., 2018; DeCelles et al., 2020; Duvall et al., 2020).

The Himalaya are characterized by range-parallel zones of differing geologic characteristics separated by north-dipping thrust faults (Fig. 3.7). From south to north, the Indo-Gangetic plain is separated from the

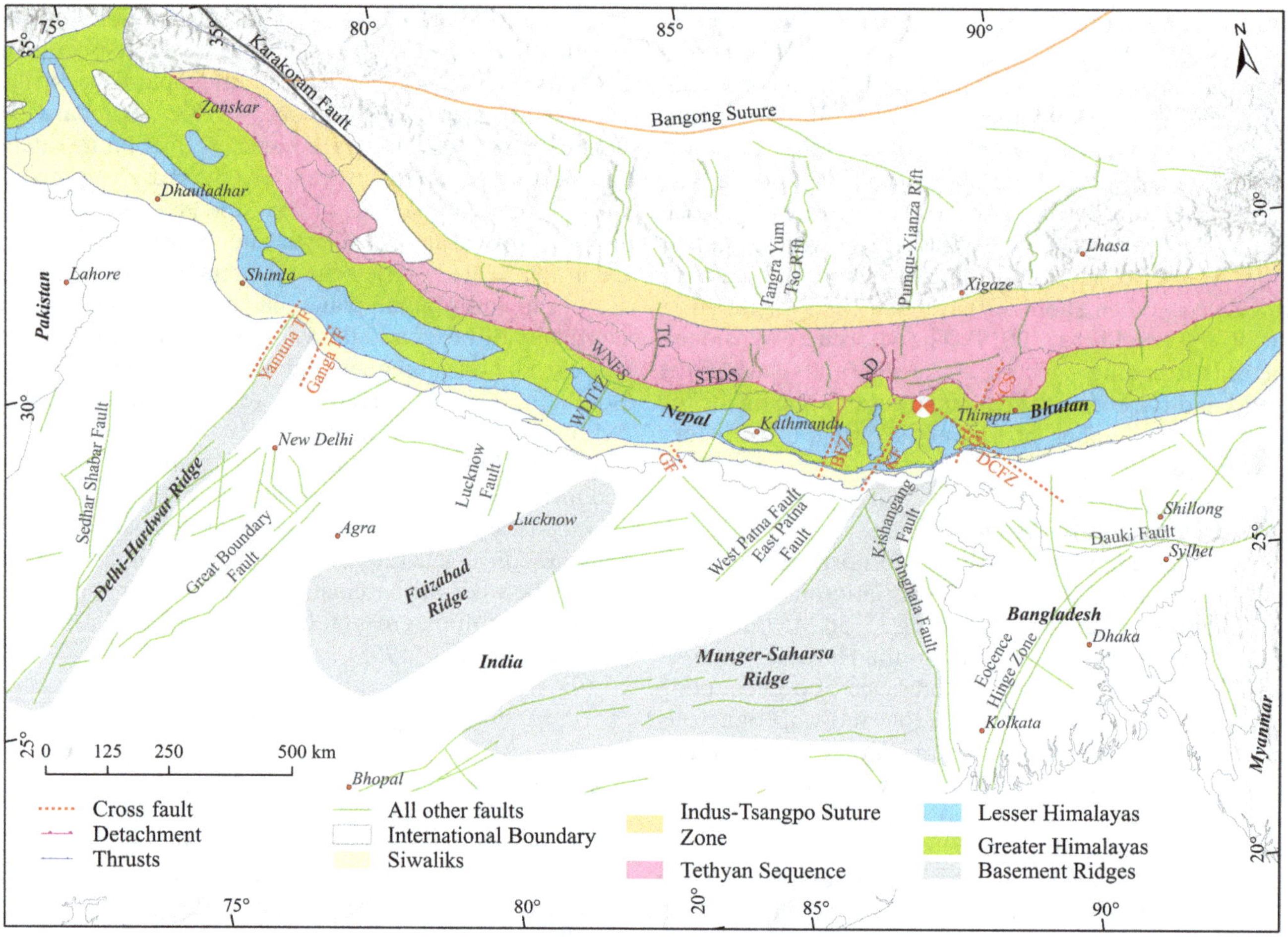

Figure 3.7 Simplified tectonic map of the Himalaya showing the major cross structures. The beach ball shows focal mechanism of the 2011 M_w 6.9 Sikkim earthquake (Department of Mines and Geology, Nepal; the Geological Survey of India; Gansser, 1964; Sastri et al., 1971; Mugnier et al., 1999; Sahoo et al., 2000; Searle et al., 2003; Guillot et al., 2008; Jessup et al., 2008; Godin & Harris, 2014; Silver et al., 2015; Diehl et al., 2017; Mukul et al., 2018; Divyadarshini & Singh, 2019; Seifert, 2019). Note: AD = Ama Drime Detachment; BFZ = Benkar Fault Zone; DCFZ = Dhurbi-Chungthan Fault Zone; GF = Gardi Tear Fault; Gi = Gish Fault; KF = Kosi Fault; MBT = Main Boundary Thrust; MCT = Main Central Thrust,; MFT = Main Frontal Thrust; MMT = Main Mantle Thrust; MZB = Main Zanskar Back Thrust; STDS = South Tibetan Detachment System; TF = Tear Fault; WDTfZ = Western Dang Transfer Zone; WNFS = Western Nepal fault system; YCS = Yadong Cross Structure.

Sub-Himalaya by the active Main Frontal Thrust (MFT). The Sub-Himalaya are bound to the north by the Main Boundary Thrust (MBT). The hanging wall to the MBT is the Lesser Himalayan zone, which is bound to the north by the Main Central Thrust (MCT). The unit that includes the highest peaks of the range is the Greater Himalayan Sequence (GHS). Geophysical data support the merging of these thrusts into a master fault known as the main Himalayan thrust (Zhao et al., 1993; Avouac, 2003; Nabelek, 2009). There is an extensional fault system bounding the northern GHS, the South Tibetan Detachment System (STDS; Hodges, 2000). The hanging wall of the STDS is the Tibetan or Tethyan Sedimentary Sequence (Gansser, 1964; DeCelles et al., 2020). Each of these zones from the Sub-Himalaya to the Tibetan Sedimentary Sequence and their bounding structures exhibits some degree of lateral heterogeneity along the length of the range in either the expression of structural style, depositional history, geomorphology, or seismicity.

In the Sub-Himalayan there is notable variability in the morphology of the range front including the local presence of dun structures and a series of recesses and salients (Yeats & Lillie, 1991; Mukul, 2010). In some areas the irregular mountain front has been linked to differences in shortening (Dubey, 2001) and in other areas there is a connection with cross faults that mark the transitions from salients to recesses. Examples of cross faults in the Sub-Himalaya include the Yamuna and Ganga Tear Faults in the northwest Himalaya and the Kosi and Gish faults in eastern Nepal and Sikkim (India), respectively (Fig. 3.7; Sahoo et al., 2000; Srivastava et al., 2018). In the Lesser Himalaya, lateral heterogeneity is seen in topographic data, seismic data, and cooling history data (Harvey et al., 2015; van der Beek et al., 2016; Soucy La Roche & Godin, 2019). In western Nepal, there is a zone of change in a variety of parameters that has led researchers to conclude that the MHT may have a ramp at that location, possibly coupled with a change in strike (Harvey et al., 2015; van der Beek et al., 2016; Soucy La Roche & Godin, 2019). This change also aligns with the proposed West Dang Transfer Zone (Fig. 3.7; Mugnier et al., 1999). Soucy La Roche and Godin (2019) propose a connection between this structural change and the northeast-striking Lucknow fault in the Indian basement. In several areas, along-strike changes in the geometry of duplex structures in the Lesser Himalaya have also been noted (Hauck et al., 1998; DeCelles et al., 2001; Grujic et al., 2002; Long, 2011).

In the Greater Himalaya there is lateral heterogeneity in exhumation rates (Eugster, 2018), topographic profiles (Duncan et al., 2003), the presence of leucogranitic intrusions (Weinberg, 2016), and the presence or absence of discontinuities (Carosi, 2010; Larson & Cottle, 2014). Hubbard et al. (2018) recently recognized a cross structure, the Benkar fault zone, in the Greater Himalaya of eastern Nepal (Fig. 3.7). This structure has dextral normal displacement and was active in the past ~12 Ma. The Benkar fault zone may continue into the Lesser Himalaya, however it is yet to be mapped to the south.

Microseismicity in the Himalaya has occurred along cross-strike trends, has terminated at cross-strike zones, and has made other changes along cross-strike boundaries (Rajaure et al., 2013; Mugnier et al., 2017; Bilham, 2019; Mendoza, 2019). The aftershock seismicity from the 2015 Gorkha earthquake terminates along a sharp northeast-striking boundary east of Kathmandu in Nepal (Hubbard et al., 2016; Mendoza, 2019). In the eastern Himalaya, several seismic events have been consistent with strike-slip displacement on transverse structures (Drukpa et al., 2006; Paul, 2015; Diehl, 2017). While a number of these events are minor, there also have been major earthquakes such as the 2011 M_w 6.9 Sikkim event that was interpreted to have occurred along a northwest-striking (cross-strike) plane with dextral kinematics (Fig. 3.7; Paul, 2015). The Sikkim event and a number of the smaller events have originated at depths of ~50–60 km suggesting that rupture initiation was below the MHT (Drukpa et al., 2006; Paul, 2015) but possibly penetrating the hanging wall. Major historic thrust-related earthquake events are known to have had a finite rupture area (Bilham et al., 2001) and tracking of these areas has facilitated the recognition of seismic gaps that may represent regions with higher risk (Bilham, 2019). It has been proposed that cross structures may limit the seismic rupture area, thus contributing to the lateral heterogeneity in seismic activity (Mugnier et al., 2017; Hubbard et al., 2021).

3.5.2. Proposed Factors Controlling Lateral Heterogeneities

The combination of preexisting transverse basement structures and the lateral variation in sedimentary history may both contribute to the lateral heterogeneity and cross structures that we see today in the Himalaya. Godin and Harris (2014) proposed a connection between variations in gravity data across a broad region in the Himalaya and Tibet to ridges in the Indian basement detected in the foreland. Soucy La Roche and Godin (2019) demonstrated lateral variations in cooling histories in western Nepal and proposed a cross structure in the form of a lateral ramp or tear fault separating the regions with differing cooling histories. This cross structure aligns with the Lucknow basement fault in the Indian foreland and these authors propose that the Himalayan cross structure is linked to the basement fault. Boundaries between salient and recesses in range front geomorphology has also been linked to basement faults (Hubbard et al., 2021). These basement faults in the foreland may also separate down-dropped blocks with thicker sedimentary sequences from higher blocks and these differences in sedimentary thickness may continue into the Himalaya, thus influencing lateral variations in structural style (DeCelles et al., 2020). Understanding the nature of segmentation and segment boundaries may shed light on details of the mountain building process in collisional orogens but may also help us understand how convergence is accommodated and factors that control seismic energy propagation.

3.6. ZAGROS

3.6.1. Tectonic Setting and Lateral Heterogeneities

The Zagros orogenic belt runs for about 2,000 km along the northeastern margin of the Arabian plate (Fig. 3.8) and is the product of Miocene collision between the Arabian and Iranian continental plates and subsequent convergence (Allen & Armstrong, 2008; McQuarrie & van Hinsbergen, 2013). The eastern boundary is the dextral Zagros-Makran Transfer Zone (Regard et al., 2005), and its western boundary is the sinistral East Anatolian Fault (Falcon, 1974; Haynes & McQuillan, 1974). During Permian to lower Cretaceous, the Neo-Tethys Ocean opened between the Arabian and Iranian plates. Closure of this short-lived ocean began in Late Cretaceous along

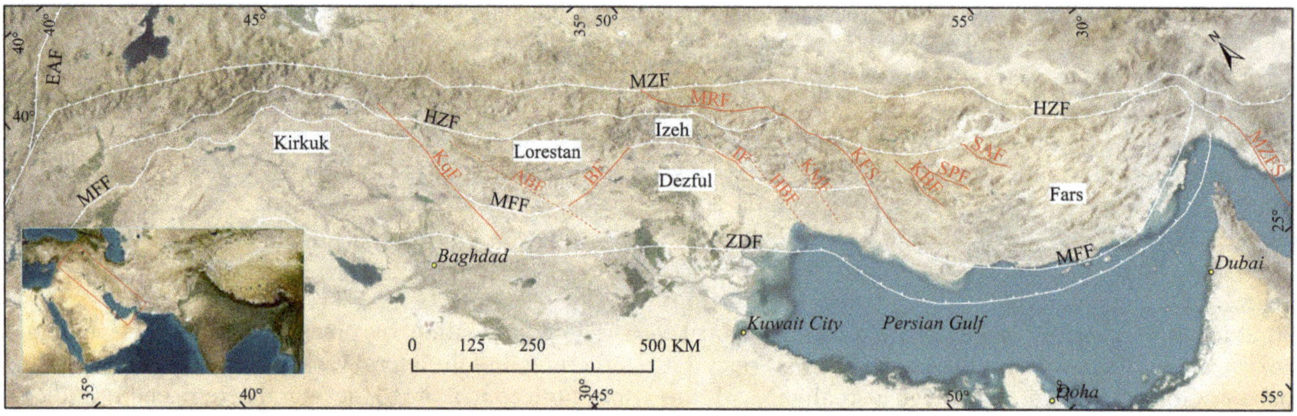

Figure 3.8 Simplified tectonic map of the Zagros mountains showing the major faults and the recesses and salients. The inset on the bottom-left corner shows the location of the larger map. The Minab-Zendan fault system (MZFS) is a major fault of the Zagros-Makran Transfer Zone (modified from Authemayou et al., 2006; Farzipour-Saein et al., 2013; Joudaki et al., 2016; Le Garzic et al., 2019). The white lines are range parallel faults and red lines indicate cross structures. Background imagery adapted from the Environmental Systems Research Institute Online Resources. Note: ABF = Anaran Basement Fault; BF = Belarud Fault; EAF = East Anatolian Fault; HBF = Hendijan-Izeh Basement Fault; HZF = High Zagros Fault; IF = Izeh Fault; KBF = Kareh-Bas Fault; KFS = Kazerun Fault System; KMF = Khark-Mish Basement Fault; KqF = Khanequin Fault; MRF = Main Recent Fault; MZFS = Minab-Zendan Fault System; MZF = Main Zagros Reverse Fault; SAF = Saverstan Fault; SPF = Sabz Pushan Fault; ZDF = Zagros Deformation Front (also known as the Zagros Foredeep Fault).

with the tectonic emplacement of ophiolites and trench sediments on the Arabian plate (Haynes & Reynolds, 1980; Berberian & King, 1981). While there are dominant range-parallel thrust faults, there are also several important cross structures and other geologic features that vary along the length of the range.

The major, range-parallel tectonic elements in the Zagros have a northwest-southeast trend (Fig. 3.8) and include the Main Zagros Reverse Fault (the continental suture), the Zagros Imbricate Zone, the High Zagros Faults, the Zagros Folded Belt, Mountain Front Fault, the Zagros Foredeep, and the Zagros Foredeep Fault, from northeast to southwest (Stocklin, 1968; Berberian, 1995). A peculiar feature in the Zagros is that crustal shortening has been accommodated by deep-seated, high-angled reverse faults, which have been interpreted as reactivated, basement normal-faults initiated during the Neo-Tethyan rifting and spreading (Falcon, 1974; Jackson, 1980; Chauvet et al., 2004; Mouthereau et al., 2012). Generally, these faults are segmented with about 85–100 km gaps in between segments and are confined in the Precambrian basement and the lower stratigraphic levels of the overlying sedimentary cover (Berberian, 1995; Bigi et al., 2018). The Cambrian to Pliocene sedimentary cover (~5–13 km thick) overlies the Proterozoic to early Cambrian Hormoz Salt (Stocklin, 1968; Falcon, 1974; Colman-Sadd, 1978). This incompetent salt unit acted as a decollement horizon between the thick-skinned deformation in the basement and the thin-skinned deformation in the sedimentary cover (e.g., Berberian, 1995; Lacombe et al., 2006). Deformation initiated in the Zagros Imbricate Zone at about 20 Ma or earlier (Fakhari et al., 2008), propagated toward the southwest and reached the Zagros Folded belt at 14 Ma (Khadivi et al., 2010).

Since the Neogene, a system of strike-slip faults has also been active in the Zagros (Berberian, 1995). The Main Recent Fault (MRF) was initiated in late Pliocene, which is an active, orogen-parallel, strike-slip fault that follows the trace of the Main Zagros Reverse Fault (Tchalenko & Braud, 1974; Berberian, 1995; Authemayou et al., 2006). Oblique convergence between the Iranian and Arabian plate is partially accommodated along this hinterland fault (Authemayou et al., 2006). At its southeastern end, the MRF diffuses into a series of north to north-northwest-striking right-lateral, basement-inherited faults, grouped as the Karezun Fault System (Authemayou et al., 2006). Major faults belonging to this zone are the Saverstan, Sabz Pushan, and Kareh-Bas faults (Fig. 3.8) with an offset of about 100–150 km along each one (Berberian, 1995; Authemayou et al., 2006). During rifting of the Proto-Tethys and initial stages of the Neo-Tethys, these faults were activated as right-lateral transform faults in the basement rocks (Talbot & Alavi, 1996). Neogene reactivation of these transverse faults has produced a dragging effect on the earlier-formed, orogen-parallel folds. These faults serve to transmit and distribute the slip along the MRF toward the southeast into the Zagros Folded Belt and the Foredeep (Berberian, 1995; Authemayou et al., 2006).

Salt diapirs have been intruded along these faults at multiple locations, which suggests that these faults cut into the top of the basement (e.g., Kent, 1979; McQuillan, 1991; Talbot & Alavi, 1996). A certain degree of seismic hazard is associated with these transverse faults (Berberian, 1995; Authemayou et al., 2006). West of the Karezun Fault System, important transverse faults are the Izeh, Balarud Fault (east-west, 130 km left-lateral), the Anaran Basement Fault, and the Khanequin Fault (Fig. 3.8; Hessami et al., 2001; Joudaki et al., 2016; Sadeghi & Yassaghi, 2016). All these transverse faults were active basin-bounding, normal faults, reactivated as strike-slip faults, and had controls on pre- and synorogenic sedimentation and subsequent deformation in the Zagros (Sepehr & Cosgrove, 2004).

Along its strike, the Zagros curves into a series of salients and recesses bounded by the transverse faults (Fig. 3.8). Generally, these curvatures are grouped into domains and include the Fars salient, the Dezful embayment, the Izeh zone (juxtaposed with the Dezful embayment in the north), the Lorestan salient, and the Kirkuk recess from sooutheast to northwest (e.g., Stocklin, 1968; Falcon, 1974; McQuarrie, 2004). The deformation zone is widest in the Fars salient and narrows toward the northwest. The style and distribution of deformation changes notably along these curvatures, mostly driven by the presence or absence of decollement layers (e.g., Bahroudi & Koyi, 2003).

Lateral heterogeneity in the Zagros is expressed through the transverse faults, the presence of salt diapirs, and the salient and recesses on the mountain front as discussed above. Further heterogeneity is also expressed in changes in fold geometry and the amount of shortening in adjacent regions. The Fars salient is characterized by a very low taper angle, several salt intrusions, and concentric folds with large amplitude (Talbot & Alavi, 1996; Sepehr et al., 2006; Mukherjee et al., 2010). Toward the west, box folds and large concentric folds are dominant in the Izeh zone. In the Dezful embayment, however, exposed folds are concentric folds with small wavelengths and probably overlie large concentric folds beneath a detachment surface (Sepehr et al., 2006). The presence of a paleo-basement high, bounded by the north-striking, Hendijan-Izeh and Khark-Mish basement faults, has been suggested beneath the Izeh and Dezful domains (Farzipour-Saein et al., 2013, and references therein). In the Lorestan salient, folds are rounded but with much smaller wavelength than in the Fars salient (Sepehr et al., 2006). Several small-scale transverse faults have also been identified around the inflection zone between the Lorestan salient and the Kirkuk recess (e.g., Sadeghi & Yassaghi, 2016). Amounts of shortening in the Fars, Dezful, and Lorestan segments are 67 km, 85 km, and 57 km, respectively (McQuarrie, 2004).

Spatial changes in folding style have been linked to the mechanical anisotropy within the deforming sedimentary column and the depth at which those anisotropies occur (e.g., Sepehr et al., 2006). Mechanical anisotropy (presence of an incompetent layer) has a major control on the shape of the folds (lower anisotropy equals more rounded folds) and the depth of anisotropy controls the fold wavelength (deeper anisotropy equals large folds) (Sepehr et al., 2006). The deep-seated Cambrian Hormuz Salt is about 1–2 km thick beneath the Fars salient, and possibly present in the Izeh zone but absent beneath the sedimentary columns in the Lorestan and Dezful domains (Bahroudi & Koyi, 2003). In contrast, the Mesozoic Kazhumdi shales or the Dashtak Evaporites in the Lorestan, and the Miocene Gachsaran Formation in the Dezful Embayment act as higher level decoupling surfaces, while the overlying sedimentary cover has a uniform mechanical stratigraphy (Sepehr & Cosgrove, 2004; Sepehr et al., 2006). The presence of decollement horizons at different structural levels in the Dezful and Izeh domains has generated a peculiar ramp and flat geometry in the sedimentary cover in the central Zagros (McQuarrie, 2004; Sepehr et al., 2006).

Stratigraphy and deformation style are also generally different from the recesses to the salients (Sepehr et al., 2006). It has been suggested that the Zagros foreland in the salient deform by both thin-skinned and thick-skinned deformation, whereas, the foreland deformation in recesses is accommodated only by thin-skinned tectonics (Malekzade et al., 2016). Around the central Zagros foreland, both the intensity of deformation and the amount of shortening increases toward the west and the deformation has been partitioned by left-lateral, blind-tear faults reactivated on preexisting basement structures (Sarkarinejad et al., 2018; Pash et al., 2020).

3.6.2. Proposed Factors Controlling Lateral Heterogeneities

Several ideas have been proposed to explain the sinuosity and lateral heterogeneity of deformation style in the Zagros. These ideas include (1) rotations of crustal blocks (Hessami et al., 2001; Edey et al., 2020), (2) along strike variations in a viscous decollement horizon between the basement and the sedimentary cover (Bahroudi & Koyi, 2003; McQuarrie, 2004), (3) the presence of a lateral buttress that serves to partition deformation (Cotton & Koyi, 2000; Bahroudi & Koyi, 2003), (4) lateral variation in the degree of oblique convergence (McQuarrie et al., 2003; Vernant et al., 2004; Vernant & Chéry, 2006), and (5) heterogenous rigidity along the plate margin (Malekzade et al., 2016). The block rotation model suggests that adjacent crustal blocks rotate with a reverse polarity along a vertical axis, driven by movements along strike-slip faults and the presence of a rigid backstop (e.g., Hessami et al., 2001; Edey et al., 2020). These strike-slip faults were activated along north-south trending, inherited basement structures attributed to the Pan-African tectonics (Koop & Stoneley, 1982; Husseini, 1988; Hessami et al., 2001) and subsequent Tethyan rifting (Talbot & Alavi, 1996).

Along-strike variation in the distribution of the Cambrian salt unit has a major control in the deformation along the Zagros Fold Belt. The presence of this incompetent unit served as a viscous detachment and deformation propagated much farther where the salt was present than in the domains where it is missing (Bahroudi & Koyi, 2003; McQuarrie, 2004; Farzipour-Saein et al., 2013). The patchy arrangement of salt deposition has been attributed to the aforementioned inherited basement structures, which could have also created tear fault-type transverse structures in the hanging walls of the thrust sheets. The style of deformation is also dependent on the nature of detachment horizons. Above a viscous decollement, the overlying units generally deform by ductile thickening, whereas above a frictional detachment, imbrication and folding deformation dominates (Bahroudi & Koyi, 2003). During deformation, the presence of transverse basement faults or a lateral change in facies can act as a lateral buttress and reorient local kinematics (Cotton & Koyi, 2000; Bahroudi & Koyi, 2003). Obliquity of plate convergence and the frictional strength along the MRF (a hinterland fault) together dictate if strike-slip partitioning initiates or not and thereby can have controls on internal deformation in the Zagros belt (McQuarrie et al., 2003; Vernant et al., 2004; Vernant & Chéry, 2006). Moreover, heterogenous plate-margin rigidity along with the presence of embayments and indenters can cause along-strike changes in deformation patterns by affecting the local obliquity angle and favoring the escape of the upper crustal material into adjacent reentrants (Malekzade et al., 2016).

It is likely that a number of factors have contributed to lateral heterogeneity in the Zagros. The stratigraphy has played a role in terms of thickness changes causing differences in folding patterns and fault geometry. The presence of localized salt layers and salt diapirs further contribute to differences in thrust displacement and local geology. The geometry of the extensional or rifts structures that predate collision may have impacted the geometry of postcollisional faults.

3.7. ANDES

3.7.1. Tectonic Setting and Lateral Heterogeneities

The Andean mountain belt is a classic example of an active subduction margin and likely represents processes that were active in the world's collisional mountain belts prior to collision. Modern-day Andean topography is largely the manifestation of crustal shortening and thickening in response to the ongoing oblique subduction of the Nazca oceanic plate beneath the South American plate (e.g., Mpodozis & Ramos, 1989). The two plates are currently converging along a N78°E vector at a rate of about 66 mm/yr (Angermann et al., 1999; Kendrick et al., 2003). The Andean orogen follows a north-south trend along the western margin of the South American plate, except in the Central Andes where the orogen makes a curve known as the Arica bend (Fig. 3.9a). Along the length of the mountain belt there is lateral heterogeneity exhibited at several scales. Lateral heterogeneity along the range has been observed in terms of the upper and lower plate dynamics, topography, and deformation style of the retroarc belt, nature of the foreland basin, and igneous activity.

At a coarse scale, the Andes exhibit lateral heterogeneity in topographic changes from north to south that has led to the characterization of the Northern Andes, the Central Andes, and the Southern Andes (Fig. 3.9a). The major visible difference is in the width of the mountain belt where the Northern and Southern Andes are narrow while the Central Andes is much wider and includes the Altiplano and Puna plateau regions. These general topographic differences are the product of differences in their terrain accretion history and subduction zone dynamics. Other broad differences seen along the range include the presence or absence of active volcanism, which is likely related to changes in the dip angle of the subduction slab (Ramos & Folguera, 2009). During early Paleozoic, the southwestern margin of South America had a history of exotic terrain accretion and subduction (e.g., Ramos, 1988). The modern-day subduction margin initiated in late Paleozoic (Giambiagi et al., 2012). The Late Permian to Early Jurassic tectonic history was characterized by crustal extension and the associated volcanism (Llambías et al., 1993). In the southern Andean front, there was backarc extension from late Triassic to Early Jurassic (Vergani et al., 1995; Giambiagi et al., 2012) while the Central Andes experienced extension from Late Jurassic to Early Cretaceous (Galliski & Viramonte, 1988; Salfity & Marquillas, 1994). The resulting orogen-parallel, rift-related, normal faults and/or transfer faults were reactivated as reverse faults with strike-slip components in Cretaceous to Paleogene (Kley et al., 2005; Mescua & Giambiagi, 2012). In the Southern Andes, this Neogene inversion of the extensional basins led to lateral variations in the thickness and facies of the sedimentary sequences, which were subsequently reflected in heterogeneities of the deforming fold and thrust belt (e.g., Ghiglione et al., 2009; Likerman et al., 2013).

The style of subduction of the downgoing Nazca Plate in the Central Andean margin differs greatly from the

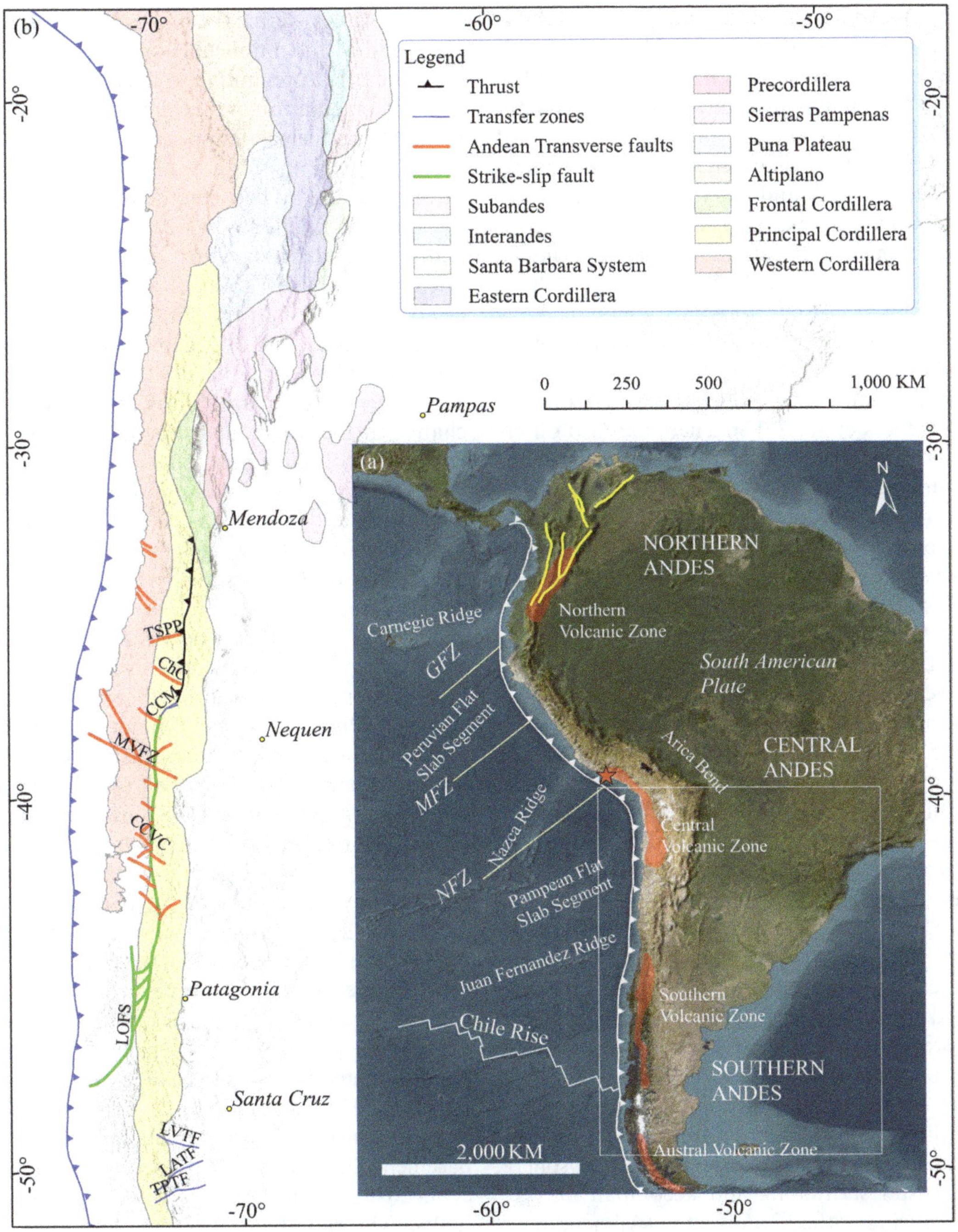

Figure 3.9 Tectonic map of the Andes. (a) Map of the South American plate and the Nazca oceanic plate (west) showing major transverse features on the subducting Nazca oceanic plate. The red star shows the epicenter of the M_w 8.4 2001 Peru earthquake. Red shades indicate volcanic zones. Note the large-scale strike-slip faults, indicated by yellow lines, in the northern Andes. The inset shows the extent of Figure 3.9b (modified from Gutscher et al., 2000; Robinson et al., 2006; Egbue & Kellogg, 2010; Schepers et al., 2017). (b) Geological map of the central and southern Andes showing the location of major cross structures. The deformation style and distribution are quite different from the central to the southern Andes. (modified from Ghiglione et al., 2009; Stanton-Yonge et al., 2016; Schepers et al., 2017). The red lines in Figure 3.9b represent the Andean transverse faults. Note: CCM = Calliqui-Copahue-Mandolegue Transfer Zone; CCVC = Cordon Caulle Volcanic Complex; ChC = Chillan-Cortaderas Lineament; GFZ = Grijalva Fracture Zone; LATF = Lago Argentino Transfer Fault; LOFS = Liquine-Ofqui Fault System; LVTF = Lago Videma Transfer Fault; MFZ = Mendana Fracture Zone; MVFZ = Mocha-Villarrica Fault Zone; NFZ = Nazca Fracture Zone; TPTF = Torres del Paine Transfer Fault; TSPP = Tarta-San Pedro-Pellado Volcanic Alignment.

adjacent margins to the north and south. The Central segment has had a steep subducting plate while to the north and south the oceanic plate has been subducting at a low angle since 12 Ma in the respective trench segments, namely, the northern Peruvian and southern Pampean flat slabs (Schepers et al., 2017). The trench retreat has greatly outpaced the slab roll back and this difference has generated the 200–300 km long flat slab segments. The slab is thought to have been anchored at the 660 km discontinuity, which is preventing the roll back (Schepers et al., 2017; Chen et al., 2019). The crustal shortening along the Arica bend in the Central Andes is 420 km, which is much greater than shortening values of 160 km and 150 km in the southern and northern, respectively (Arriagada et al., 2008; Schepers et al., 2017). There are also differences in the style of deformation related to flat slabs including the inland development of thick-skinned deformation east of the crest of the Andes. Another lateral variation related to the flat slab zones is the absence of active volcanism. Several causes for the flattening of the slab have included subduction of buoyant oceanic crust around the Nazca and Juan Fernandez ridges or possible changes in the thermal thickness of the overriding lithosphere (Pilger, 1981; Manea et al., 2012; Schepers et al., 2017).

At a finer scale, there are lateral differences in the timing and nature of deformation along the Andes. In the Northern Andes, crustal blocks are escaping northeastward toward the free Caribbean-North Andes boundary and away from the rigid South American Plate along the system of large-scaled, strike-slip faults (Fig. 3.9a; Audemard et al., 2005; Audemard, 2009; Egbue & Kellogg, 2010; Monod et al., 2010). The trench-parallel component of the oblique plate convergence has been driving this escape for the last 1.8 Ma. This escape was likely triggered by an increase in coupling between the upper and lower plate when the Carnegie ridge entered the subduction zone (Egbue & Kellogg, 2010).

The Central Andes segment is dominated by anomalous topography of the Altiplano-Puna plateau, which was uplifted due to thermal softening of the lithosphere followed by crustal shortening and accompanying arc magmatism during the orogeny (Allmendinger et al., 1997; Coutand et al., 2001). Within the plateau region, researchers have documented significant differences in the timing of the surface uplift and crustal structure of adjacent blocks (e.g., Bianchi et al., 2013; Leier et al., 2013; Canavan et al., 2014). The fault boundaries of these blocks are thought to be related to inherited basement faults, possibly bounding older accreted terrains (Jordan et al., 1983). Toward the foreland, spatial location of the Cenozoic fault systems were likely controlled by the presence of Paleozoic and Mesozoic basement structures (Coutand et al., 2001; Gillis et al., 2006). The plateaus and the adjacent ranges in the Central Andes, define the highest topographic features in the world within a noncollisional setting (Isacks, 1988).

Another feature unique to the southern Central Andes is the fragmentation and exhumation of the retroarc foreland basin (Sierras-Pampeanas) along inherited structures (Fig. 3.9b; Jordan et al., 1983; Japas et al., 2016). Stronger coupling between the upper and lower plates in the flat-slab segments has been described as the driver behind inland propagation of deformation and subsequently the basin uplift (Ramos et al., 2002; Oriolo et al., 2014). Just to the south of the flat-slab segment, Giambiagi et al. (2012) documented an abrupt southward decrease in topographic uplift and crustal shortening. This sharp, lateral variation is driven by a strong lateral change in the upper plate rheology, which controls the degree of coupling between the upper and lower crust, such that a thick and more felsic crust has strongly coupled upper and lower slabs (Giambiagi et al., 2012).

Thick-skin deformation in the Central Andes terminates toward the south at the northern segment of the Southern Andes, which is characterized by trench-parallel, thin-skinned, fold and thrust deformation. The southern segment, however, is dominated by a system of strike-slip fault systems. The east-northeast-striking, transverse, dextral Callaqui-Copahue Mandolegue (CCM) fault is thought to decouple the contrasting deformation styles between the retroarc thrust belt (north) and the Liquine-Ofqui Fault System (LOFS) in the south (Folguera et al., 2004).The LOFS (Fig. 3.9b) is a ~1,200 km long, north-northeast-striking, right-lateral, reverse, intra-arc fault in the southern segment (Cembrano et al., 1996; Thomson, 2002; Vargas et al., 2013). Here, a series of northeast-striking, enechelon, dextral, normal faults splay off from the LOFS and form a strike-slip duplex between two subparallel fault branches of the LOFS (Cembrano et al., 1996). Another set of strike-slip faults cut the LOFS and associated faults and is called the Andean Transverse Faults (ATF). The ATF (Fig. 3.9b) are northwest-striking, reverse, sinistral faults, primarily formed on inherited basement structures (e.g., Roquer et al., 2017; Sielfeld et al., 2019). Deformation partitioning occurs within these faults, whereby the trench-parallel component of the oblique convergence is accommodated by the LOFS and its splay faults and the ATFs accommodate the trench-perpendicular component (e.g., Stanton-Yonge et al., 2016). In the northern termination of the LOFS, the strike-slip dominated domain sharply transitions into margin-parallel fold and thrust deformation (Fig. 3.9b) (Stanton-Yonge et al., 2016). These strike-slip fault systems also form excellent conduits as well as reservoirs for magma and hydrothermal fluids and thereby control the locus of volcanic complexes (e.g., Petrinovic et al., 2006; Pérez-Flores et al., 2016; Roquer et al., 2017; Sielfeld et al., 2019; Lupi et al., 2020; Piquer et al., 2020).

Effects of the subduction of oceanic fracture zones on seismic activity have also been observed in the central Andes. In the subducted Peruvian slab, generally deeper earthquakes are generated around a subducted segment boundary, the Mendana Fracture Zone (Fig. 3.9a) (Gutscher et al., 2000). Toward the south, a fracture zone in the downgoing Nazca Plate (most likely the Nazca Fracture Zone) is thought to have induced a fracture in the overriding plate that acted as a temporary lateral barrier during the initial seismic rupture propagation but subsequently allowed energy to pass through this vertical plane releasing the energy of the M_w 8.4 2001 Peru earthquake (Robinson et al., 2006). These examples indicate that transverse discontinuities in the lower plate (or subducting plate) can influence structures in the upper plate and can have great seismic implications.

3.7.2. Proposed Factors Controlling Lateral Heterogeneities

Along the Andean belt, lateral heterogeneities have been primarily expressed as variation in crustal shortening and topographic uplift (e.g., Allmendinger et al., 1997; Arriagada et al., 2008; Giambiagi et al., 2012; Schepers et al., 2017), along-strike changes in the styles and timing of deformation (e.g., Ramos et al., 2002; Bianchi et al., 2013; Leier et al., 2013; Canavan et al., 2014; Oriolo et al., 2014; Stanton-Yonge et al., 2016), laterally contrasting seismicity (Gutscher et al., 2000; Robinson et al., 2006), and changes in volcanic activity (e.g., Pérez-Flores et al., 2016; Roquer et al., 2017; Sielfeld et al., 2019; Lupi et al., 2020; Piquer et al., 2020). Researchers interpret that these heterogeneities are governed by the angle of lower plate subduction (flat versus steep slab), physiography of the lower plate, obliquity of the subduction vector, preexisting basement structures in the upper plate, and the rheology of the upper plate. The angle of the lower plate subduction controls the degree of coupling between the lower and upper plates (lower angle equals stronger coupling), which in turn impacts the nature and distribution of the upper plate deformation (e.g., Ramos et al., 2002; Oriolo et al., 2014). Subduction of physiographical features in the lower plate also can change the amount of plate coupling and generate respective topographic and seismic signatures in the upper plate or it can influence the development of cross structures (Gutscher et al., 2000; Robinson et al., 2006). Oblique plate convergence in the Southern Andes has been accommodated by a system of strike-slip and transverse faults (e.g., Stanton-Yonge et al., 2016). In the northern Andes, however, the dynamic plate setting has favored crustal escape (e.g., Egbue & Kellogg, 2010). Rheology of the upper plate dictates the amount of plate coupling and, therefore, has control on the rate and amount of crustal shortening and the overall topographic uplift (Giambiagi et al., 2012).

3.8. OTHER OROGENS

Lateral heterogeneities have been documented in several other mountain belts around the world (Fig. 3.1). While there are groups of along-strike variations common across multiple mountain belts, some of the heterogeneities are the result of the broad tectonic setting and the nature of the sedimentary cover, and therefore can be unique to only a few mountain belts. Comparable to the Appalachians, the Mesozoic rift structures in the external Hellenides thrust belt in Greece were reactivated as transverse zones (e.g., the Corinth Gulf, Ierapetra, and Omalos transverse zones), which partitioned the belt into a series of salients, recesses, and linear segments (Skourlis & Doutsos, 2003; Kokkalas & Doutsos, 2004; Chatzaras et al., 2013). These inherited structures also disrupted the lateral continuity of the foreland sedimentary sequences and served as crustal-scale lateral/oblique ramps during thrust propagation and thereby had a significant control on the nature and deformation style of the evolving taper wedge (Robertson et al., 1991; Doutsos et al., 2006; Chatzaras et al., 2013). Similarly, in the Urals, Precambrian aulacogens (failed rift arms) served as transverse basement structures, above which tear faults and lateral ramps were developed during fold-thrust propagation (Rodgers, 1990; Brown et al., 1997; Perez-Estaun et al., 1997). Cross faults are common in the Apennines and are generally called anti-Apennine faults (e.g., Coltorti et al., 1996; Sorgi et al., 1998; Butler et al., 2006; Elter et al., 2011). The north-northwest-striking Apennine belt is underlain by the north-northeast-trending crustal tectonic lineaments (Valnerina Line, Ancona-Anzio Line, Ortona-Rocca Monfina Line). These large-scale lineaments act as structural barriers (zones of abrupt lateral changes in tectonic and structural style, wedge stratigraphy, and topography) during tectonic transport and laterally limit the propagation of seismic fault rupture (Pizzi & Galadini, 2009; Satolli et al., 2014). There are also smaller-scaled transverse basement structures that may either serve as seismic segment boundaries/seismic loci or get reactivated as transfer zones (Valensise & Pantosti, 2001; Pizzi & Galadini, 2009). Furthermore, transfer zones in the Apennines are known to form suitable loci for magma emplacement (Dini et al., 2008). On a much larger scale, the differential retreat along the adjacent segments of the Adriatic plate during the last 5 Ma has been manifested as a lithospheric tear/transfer zone across the Apennines (Scrocca, 2006), which may share a similar genesis to the Colorado Mineral Belt in the Cordilleran belt.

Cross structures have also been interpreted to have great economic and seismic impacts besides their influence in tectonic, stratigraphic, and structural evolution (e.g., Mahoney et al., 2017). In the Papua New Guinea fold-thrust belt, the Jurassic, extensional, transverse, crustal structures evolved as zones of economically significant copper-gold mineralization during their Late Miocene-Pliocene inversion (Davies, 1991; Corbett, 1994; Hill et al., 2002). Like in the Apennines and the Himalaya, active cross faults in the Taiwan mountains are known to serve as earthquake nucleation sites and as rupture segment boundaries (Deffontaines et al., 1997; Lacombe et al., 2001; Ching et al., 2011). These cross faults have genetically been linked to changing deformation style (thick-skinned versus thin-skinned) and to a lateral change in thickness of the deforming sedimentary sequences (Lacombe et al., 2001; Mouthereau et al., 2002; Mouthereau & Lacombe, 2006; Ching et al., 2007). Much similar to the Zagros belt, controls of the presence or absence of a decoupling layer on the style of the evolving taper wedge (narrow wedge with a high-taper angle, when decoupling layer is absent and vice-versa) have been noted in the Caucasus belt (Upper Jurassic salt and Middle-Lower Jurassic shales) in Russia (Sobornov, 1996), the Sulaiman belt (Paleozoic, Lower Cretaceous, and Eocene strata) in Pakistan (Jadoon et al., 1994), and the Parry Island belts (Ordovician salt) in the Canadian Arctic (Harrison & Bally, 1988).

3.9. DISCUSSION

Despite the general temporal and spatial continuity of crustal deformation along convergent mountain belts, significant lateral heterogeneities have been observed in several orogenic belts from around the world. There are certain lateral variations, which are unique to only a few orogens such as the lithospheric tear in the Apennines and the asynchronous orogenic events in the Alps. More frequently, however, similar groups of lateral variations have been observed across multiple mountain belts. Generally, lateral heterogeneity in convergent mountain belt settings have been expressed as (1) an along-strike change in deformation style (thick-skinned versus thin-skinned, imbrication versus duplexing/changes in ramp geometries); (2) variation in igneous activity and metamorphic grade; (3) variation in seismic activity; (4) differential topographic uplift/features along the strike; and/or (5) abrupt changes in thickness and facies of sedimentary sequences in both foreland basins and within the fold-thrust belt. Oftentimes such lateral variations are abrupt rather than gradual and are marked by geological structures, mostly faults, that are nearly orthogonal to the strike of the orogen, generically referred to as cross structures in this review. The causal factors/mechanisms behind the observed lateral heterogeneities are discussed in this section.

3.9.1. Irregular Continental Margins

The geometry of the continental margin(s), prior to the collision/convergence, has a profound effect on the geometry of the evolving orogenic belt. The orogenic front may mimic the continental margin geometry (Fig. 3.4) such that any irregularities along the margin are reflected along the deformational font. This phenomenon has been proposed in the Appalachians. During the Iapetan rifting (plus the other rifting events) along the Laurentian margin, a series of embayments (concave oceanward) and promontories (convex oceanward) separated by transform faults were produced (Thomas, 2014). Along such an irregular margin, the depositional environment is bound to vary laterally. During subsequent orogenic events, promontories evolved into recesses (concave toward the foreland), embayments evolved into salients (convex toward the foreland), and the transform fault boundaries evolved into transverse zones or cross structures (Thomas, 2014). Kwon and Mitra (2004) have compiled five end-member models for the kinematic development of salients as a function of principal transport direction, amount of shortening, degree of vertical axis rotation, thrust displacement, and presence or absence of tear faults or lateral/oblique ramp boundaries. These models, however, can be viewed as second-order, structural controls on formation of orogenic curvatures.

Lin et al. (1994) further noted that a collisional event between two promontories (of two different landmasses) results in a narrower but stronger orogenic deformation and a higher grade of metamorphism than a promontory-embayment collision. It can be concluded that an irregularity along a continental margin may serve as a nucleation site for lateral heterogeneity.

3.9.2. Inherited Basement Structures

Inherited basement structures form a first-order control on formation of cross structures. Basement structures are primarily related to continental rifting, hyperextension of rifted margins (e.g., Ribes et al., 2019) , crustal sutures, or backarc extension (Fig. 3.10). Furthermore, it has been noted that a failed rift-arm (aulacogen) can also serve as an inherited basement structure and control structural evolution (Rodgers, 1990; Brown et al., 1997; Perez-Estaun et al., 1997). Crustal sutures mark boundaries between two continental blocks and form as a result of collisional tectonics or terrain accretion. Generally, suture zones consist of highly deformed rock units and

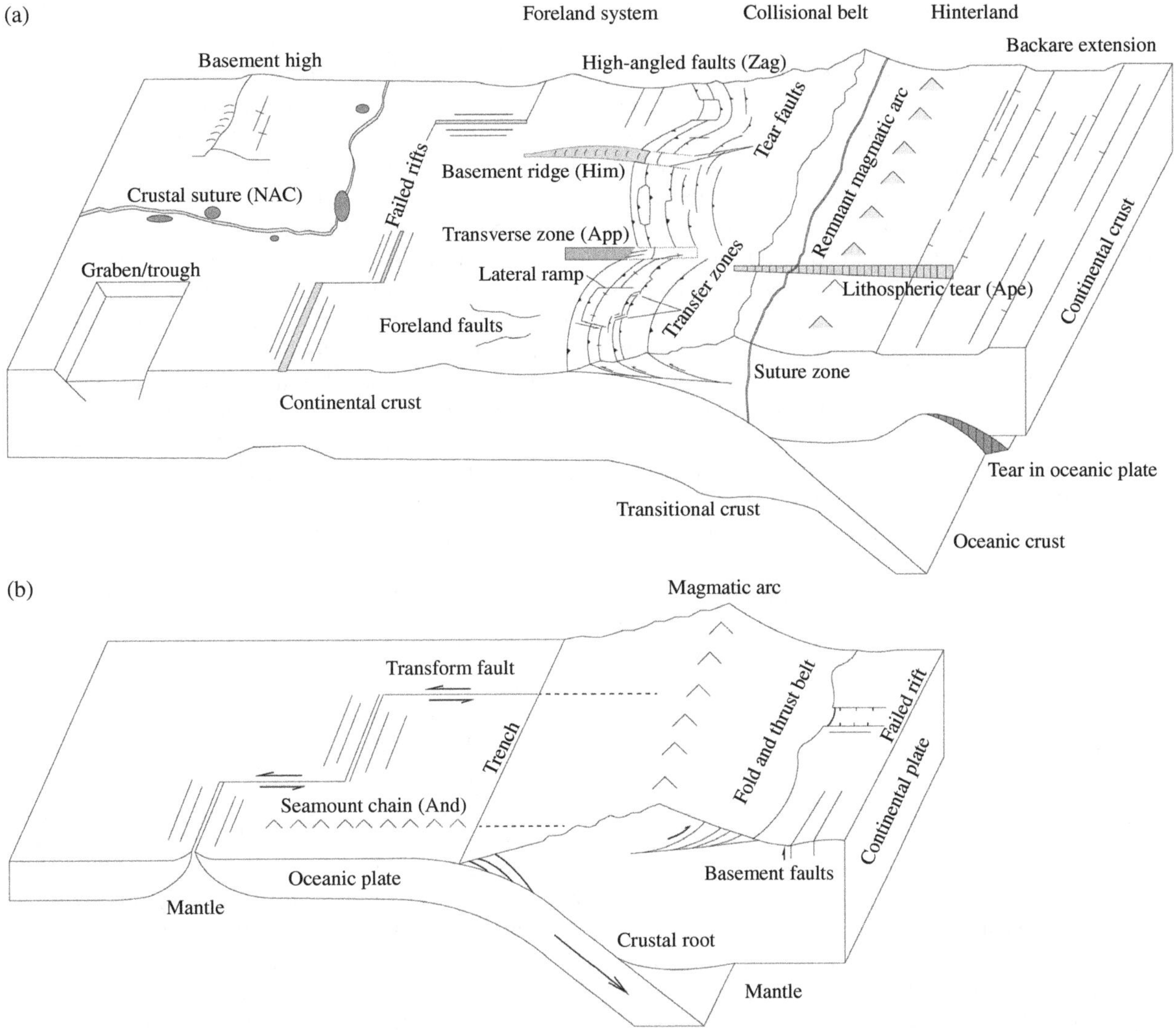

Figure 3.10 Various inherited basement structures and oceanic plate physiography in contractional settings. The gray ellipses in (a) represent igneous intrusions. The mountain belt that, in general, represents a certain structure are denoted in parentheses. Note: And=Andes; Ape=Apennines; App=Appalachians; Him = Himalaya; NAC = North American Cordillera; Zag = Zagros.

form weaker zones within the crust (Hope & Eaton, 2002; Whitmeyer & Karlstrom, 2007). These zones are known to be intruded by igneous bodies during the subsequent crustal stabilization (Whitmeyer & Karlstrom, 2007), and these intrusion boundaries can also behave as zones of weakness during various tectonic events (Simony & Carr, 1997; Bader, 2009).

Impacts of Basement Structures on Subsequent Sedimentation and Deformation

The presence of basement structures can greatly affect the foreland sedimentation and thereby have a strong control on the evolution of mountain belts as is seen in the Zagros, the Cordillera, and the Himalaya. Basement structures can isolate deposition centers in the foreland, which induces sharp lateral variations in both the thickness and facies of the sedimentary strata. A thicker foreland sedimentary sequence, when incorporated into the evolving taper wedge, will propagate much farther (salient) than a wedge consisting of a thinner sequence (recess) (e.g., Paulsen & Marshak, 1999). Differential tectonic transport between the adjacent segments (salients and recesses) is often accommodated by cross structures such as tear faults, lateral ramps, and displacement transfer zones (Paulsen & Marshak, 1999). Orogen-parallel faults and folds are generally truncated at, or dragged into, cross structures as seen in the Appalachians and the Cordillera (e.g., Thomas, 2007; Whisner et al., 2014).

Similarly, lateral variations in sedimentary facies will also have a significant impact on the wedge evolution. An abrupt change in the structural elevation of decoupling layers results in lateral ramps (Thomas, 1990). Since thrust-ramp geometry is governed by the nature of the sedimentary column, any lateral variation in sedimentary facies is likely to be reflected in the thrust geometry (e.g., Mitra, 1988). In a broader stratigraphic framework, if the presence of regional decollement horizons (units of salts, evaporites, and shales) varies laterally, then the deformation style in adjacent segments of the mountain belts will be very different. If a decollement horizon is present, then the taper angle will be low and propagate farther than adjacent areas. Likewise, in the absence of such a horizon, the taper will build up to a larger angle in a narrower belt as in the Zagros (e.g., Jadoon et al., 1994; Sobornov, 1996; Bahroudi & Koyi, 2003). The depth of these decollement horizons will also have a first-order control on the fold geometry, such that a deeper horizon will result in folds with larger amplitude (Sepehr et al., 2006). Furthermore, the stratigraphic variation boundaries will act as a lateral buttress to locally deflect the transport trajectories (Bahroudi & Koyi, 2003). If basement structures are reactivated during deposition, then synsedimentary faults and/or drape folds can form in the overlying sedimentary column (Thomas, 1990). These structures will also impact the fold-thrust belt evolution in a similar fashion to other cross-structures (e.g., Zerlauth et al., 2014).

Reactivation of Basement Structures

Regardless of their origin, inherited basement structures can be broadly categorized into three groups: orogen-parallel, oblique, and transverse/cross basement structures. Various styles of preferential reactivation and inversion of basement structures have been discussed in the Apennines (Tavarnelli et al., 2004; Butler et al., 2006), and are also seen in other mountain belts. When orogen-parallel basement structures/faults are present, it is possible for them to be reactivated as reverse faults during orogenic compression (Fig. 3.10a). However, it has been noted that basement faults can be deformed under compression without inversion or reactivation (Pantet et al., 2020). In a reactivation scenario, the deformation between the basement and the overlying sedimentary cover is partially decoupled, whereby the basement deforms in thick-skinned style and the cover deforms as thin-skinned, fold and thrust belts (e.g., Berberian, 1995; Lacombe et al., 2006). Discontinuities along the orogen-parallel basement structures will eventually result in laterally heterogeneous deformation style along the range. If the basement faults are steeply dipping, then they often cut into the thin-skinned decollement faults of the sedimentary cover and act as frontal ramps during the upper, brittle deformation (e.g., Bahroudi & Koyi, 2003). These orogen-parallel basement structures also accommodate convergence via strike-slip faulting under a favorable plate kinematic setting (e.g., Authemayou et al., 2006). Margin perpendicular and oblique structures, however, either serve as lateral/oblique ramps or get reactivated as strike-slip faults with a dip-slip component based on their orientation in the stress field. Based on their aerial extent, these inherited structures can affect a few thrust sheets, or they can partition the entire mountain belt (Scrocca, 2006; Pizzi & Galadini, 2009; Satolli et al., 2014). Cross faults in the Himalaya may have had an origin in reactivated, margin-perpendicular basement faults and basement highs such as the Delhi-Hardwar ridge (Sahoo et al., 2000; Godin & Harris, 2014; Hubbard et al., 2018; Hubbard et al., 2021).

3.9.3. Rheology of the Crust

Variation in crustal rheology is another significant element in inducing lateral heterogeneities in orogenic belts, which itself is a function of its composition and thermal structure. In general, a weaker crust is more likely to buckle under contraction resulting in a high crustal relief, whereas a stronger crust will distribute deformation in more brittle manner (Allmendinger et al., 1997; Coutand et al., 2001). It has also been proposed that rheological heterogeneities along a continental margin front could result in orogenic curvatures, such that a stronger segment better resists deformation and evolves as a salient (e.g., Malekzade et al., 2016). As seen in the Zagros, frictional strength in hinterland faults, which is governed by crustal rheology, can influence the mode of deformation partitioning in the orogenic belt and thereby have a control on the style and distribution of deformation (e.g., Vernant & Chéry, 2006). Thermal structure and rheology of the crust can vary laterally due to variations in arc magmatism. Therefore, previous tectonic events will also have an indirect impact on lateral heterogeneity of an orogen.

Giambiagi et al. (2012) proposed that a stronger coupling between the upper and lower crust deformation corresponds to a higher, surficial topography and greater crustal shortening. Crustal coupling is thought to be governed by crustal composition such that a thick and more felsic crust has a stronger coupling between the upper and lower units than a thin and mafic crust. Thus, if there is strong lateral variation in crustal properties due to tectonic events such as terrain accretion or rifting, then such variation will be manifested as laterally heterogenous deformation segments.

In the Alps, intensive folding and uplift of the Tauern window has been linked to the northward encroachment of the Dolomite acting as an indenter

(Rosenberg & Garcia, 2011), which indicates that rheology of both converging crustal blocks have an impact on the morpho-tectonic evolution of an orogen. As Butler et al. (2006) pointed out, the correlation between precise lithospheric strength profiles and orogenic deformation would further elucidate our understanding about the role of upper crustal rheology during orogenesis.

3.9.4. Plate Dynamics and Physiography of the Lower Plate

Plate dynamics between the upper and lower plate as well as the lower plate structures can disrupt the lateral homogeneity of orogenic belt (Fig. 3.10b). Effects of lateral change in the style of subduction in the deforming orogen have contributed to lateral heterogeneity in the Andes. Any transverse structure in the lower plate is likely to be inherited in the upper plate, such as the Colorado Mineral Belt and the lithospheric tear in the Apennines (e.g., Fig. 3.10a; Scrocca, 2006; Chapin, 2012). These structures in the lower plate impartially dissect the orogen rather than controlling the deformation style. The dynamics of the lower plate, however, generally differs across these transverse structures, which impacts the foreland sedimentation and fold-thrust belt evolution (Chapin, 2012). A similar phenomenon has been observed in cases where oceanic fracture zones (Fig. 3.10b), orthogonal to the subduction zone, get subducted and result in the occurrence of cross faults in the overriding plate. This scenario has been proposed on the Juan de Fuca plate subduction (Goldfinger et al., 1997), in Sumatra (Graindorge et al., 2008), and in the Andes (Robinson et al., 2006). Although not a cross structure, the Norumbega Fault System in the Appalachians has been regarded as a surface manifestation of a subducted oceanic ridge-transform system (Kuiper, 2016; Kuiper & Wakabayashi, 2018).

In orogens associated with oceanic subduction, variations in the relative rates of subduction versus convergence can create lateral heterogeneity in the overriding plates as is seen in the Andes. When subduction outpaces convergence such that the convergent boundary retreats, backarc extension occurs. These areas may have contrasting evolution when compared with adjacent areas where there is a better balance of the subduction versus convergence rates.

3.9.5. Obliquity of Plate Convergence

Lateral heterogeneity induced by oblique convergence is scale dependent. Obliquity of the plate convergence is generally accommodated by arrays of transtensional, transpressional, and strike-slip faults. In the Andes, transpressional cross faults are associated with mafic volcanism, while transtensional faults are linked with felsic magmatism as driven by magmatic viscosity difference (Petrinovic et al., 2006). At the scale of individual faults, these cross faults disrupt the continuity of the fold-thrust belt as noted in the southern Andes and the nNorthern Calcareous Alps (Zerlauth et al., 2014; Stanton-Yonge et al., 2016). Lateral variation in the degree of oblique convergence itself, however, will also induce lateral heterogeneity in the orogenic belt due to an overall change in the stress field (McQuarrie et al., 2003; Vernant et al., 2004; Vernant & Chéry, 2006).

3.9.6. Further Implications of Cross Structures

Besides their role in foreland sedimentation and tectonic evolution of orogens, cross faults also have implications in the seismicity, mineralization, igneous activity, numerical modeling, and topography of mountain belts. In the Zagros, a small seismic hazard has been linked to cross faults (Berberian, 1995; Authemayou et al., 2006). Cross faults in the Apennines are known to serve either as seismic segment boundaries or seismic loci (Valensise & Pantosti, 2001; Pizzi & Galadini, 2009). Similarly, in Taiwan, cross faults are known to act as earthquake nucleation sites and as rupture segment boundaries (Deffontaines et al., 1997; Ching et al., 2011). Microseismicity in the Himalaya has occurred along cross-strike trends, has terminated at cross-strike zones, and has made other spatial changes along cross-strike boundaries (Rajaure et al., 2013; Mugnier et al., 2017; Bilham, 2019; Mendoza, 2019). A transverse fault in the downgoing Nazca plate had an enormous impact during the 2001 Peru earthquake (Robinson et al., 2006). From the economic standpoint, cross faults form excellent conduits as well as reservoirs for magma and hydrothermal fluids (McMechan, 2012), which makes them target sites for economic mineral deposits. In the Papua New Guinea fold-thrust belt, transverse zones form sites of economically significant copper-gold deposits (Davies, 1991; Corbett, 1994; Hill et al., 2002). The Colorado Mineral Belt is another example of valuable mineral accumulation along a cross structure (Chapin, 2012). Moreover, cross structures can form potential traps for oil and gas accumulation, which in some areas makes them an important site for hydrocarbon exploration (Wheeler, 1980; Séjourné & Malo, 2007; Bader, 2009). Fault systems can form excellent conduits and reservoirs for magmatic fluids and therefore can control the locus of volcanic complexes in convergent settings (e.g., Petrinovic et al., 2006; Pérez-Flores et al., 2016; Roquer et al., 2017; Sielfeld et al., 2019; Lupi et al., 2020; Piquer et al., 2020). Thermo-mechanical tectonic models could potentially be rectified by changing the lithospheric parametric values along the strike such that they agree with the lateral variation in crustal deformation, as can be observed on the surface. Finally, cross structures also have a major control on the geomorphologic evolution of mountain belts. In the Himalayan front, cross faults spatially coincide with river channels (Sahoo et al., 2000; Srivastava et al., 2018). The anti-Apennine faults have distinct topographic signatures (Coltorti et al., 1996). Cross faults can be zones of weaknesses and, therefore, can also amplify erosional hazards, especially in active mountain belts.

3.10. CONCLUSIONS

While convergent mountain belts are dominated by spatially and temporally continuous orogen-parallel structures, geological and geophysical data show that various forms of lateral heterogeneity, often marked by cross structures, are ubiquitous in most orogenic settings. In general, lateral heterogeneities along orogens have been manifested as (1) along-strike changes in deformation style, (2) variation in igneous activity or metamorphic grade, (3) variation in seismic activity, (4) changes in topography and geomorphology, and (5) abrupt lateral stratigraphic changes. Common drivers behind these lateral heterogeneities include the geometry of the continental margin, inherited basement structures, lateral variation in stratigraphy of deforming sedimentary sequences, variation in crustal rheology, along-strike changes in plate tectonic setting, physiography of the lower plate, and obliquity of plate convergence. In most settings, these factors are interrelated and simultaneously influence the morphotectonic evolution of an orogen. Apart from their influence on foreland sedimentation and orogenic evolution, lateral heterogeneity and cross structures can have an impact on patterns of seismicity, natural resource occurrence, and natural hazards. We therefore stress the importance of documenting heterogeneity, mapping cross structures, and understanding the role these lateral changes play in mountain belt development along convergent margins.

ACKNOWLEDGMENTS

We would like to thank the anonymous reviewers for their suggestions, which have improved articulation and presentation of information in this contribution. Thanks to the editors of this book volume for their constant support. We acknowledge the National Geographic Society, the Explorer's Club, the Geological Society of America, and Montana State University for providing grant support toward Bibek Giri's dissertation research.

REFERENCES

Aleinikoff, J. N., Zartman, R. E., Walter, M., Rankin, D. W., Lyttle, P. T., & Burton, W. C. (1995). U-Pb ages of metarhyolites of the Catoctin and Mount Rogers Formations, central and southern Appalachians: Evidence for two pulses of Iapetan rifting. *American Journal of Science*, *295*(4), 428–454. https://doi.org/10.2475/ajs.295.4.428

Allen, M. B., & Armstrong, H. A. (2008). Arabia-Eurasia collision and the forcing of mid-Cenozoic global cooling. *Palaeogeography, Palaeoclimatology, Palaeoecology*, *265*(1–2), 52–58. https://doi.org/10.1016/j.palaeo.2008.04.021

Allmendinger, R. W., Jordan, T. E., Kay, S. M., & Isacks, B. L. (1997). The evolution of the Altiplano-Puna plateau of the Central Andes. *Annual Review of Earth and Planetary Sciences*, *25*(1), 139–174. https://doi.org/10.1146/annurev.earth.25.1.139

Anderson, H. E., & Davis, D. W. (1995). U-Pb geochronology of the Moyie sills, Purcell Supergroup, southeastern British Columbia: Implications for the Mesoproterozoic geological history of the Purcell (Belt) basin. *Canadian Journal of Earth Sciences*, *32*(8), 1180–1193. https://doi.org/10.1139/e95-097

Angermann, D., Klotz, J., & Reigber, C. (1999). Space-geodetic estimation of the Nazca-South America Euler vector. *Earth and Planetary Science Letters*, *171*(3), 329–334. https://doi.org/10.1016/S0012-821X(99)00173-9

Arriagada, C., Roperch, P., Mpodozis, C., & Cobbold, P. (2008). Paleogene building of the Bolivian Orocline: Tectonic restoration of the central Andes in 2D map view. *Tectonics*, *27*(6). https://doi.org/10.1029/2008TC002269

Audemard, F. A. (2009). Key issues on the post-Mesozoic southern Caribbean plate boundary. In K. James, M. Lorente, & J. Pindell (Eds.), *Origin and evolution of the Caribbean plate* (pp. 569–586). London: Geological Society, Special Publications, 328.

Audemard, F. A., Romero, G., Rendon, H., & Cano, V. (2005). Quaternary fault kinematics and stress tensors along the southern Caribbean from fault-slip data and focal mechanism solutions. *Earth-Science Reviews*, *69*(3–4), 181–233. https://doi.org/10.1016/j.earscirev.2004.08.001

Authemayou, C., Chardon, D., Bellier, O., Malekzadeh, Z., Shabanian, E., & Abbassi, M. R. (2006). Late Cenozoic partitioning of oblique plate convergence in the Zagros fold-and-thrust belt (Iran). *Tectonics*, *25*(3). https://doi.org/10.1029/2005TC001860

Avouac, J.-P. (2003). Mountain building, erosion, and the seismic cycle in the Nepal Himalaya. *Advances in Geophysics*, *46*, 1–80. https://doi.org/10.1016/S0065-2687(03)46001-9

Bader, J. W. (2008). Structural and tectonic evolution of the Cherokee Ridge arch, south-central Wyoming: Implications for recurring strike-slip along the Cheyenne Belt suture zone. *Rocky Mountain Geology*, *43*(1), 23–40. https://doi.org/10.2113/gsrocky.43.1.23

Bader, J. W. (2009). Structural and tectonic evolution of the Douglas Creek arch, the Douglas Creek fault zone, and environs, northwestern Colorado and northeastern Utah: Implications for petroleum accumulation in the Piceance and Uinta basins. *Rocky Mountain Geology*, *44*(2), 121–145. https://doi.org/10.2113/gsrocky.44.2.121

Bahroudi, A., & Koyi, H. A. (2003). Effect of spatial distribution of Hormuz salt on deformation style in the Zagros fold and thrust belt: an analogue modelling approach. *Journal of the Geological Society*, *160*(5), 719–733. https://doi.org/10.1144/0016-764902-135

Bartosch, T., Stüwe, K., & Robl, J. (2017). Topographic evolution of the Eastern Alps: The influence of strike-slip faulting activity. *Lithosphere*, *9*(3), 384–398. https://doi.org/10.1130/L594.1

Bayona, G., Thomas, W. A., & Van der Voo, R. (2003). Kinematics of thrust sheets within transverse zones: A structural and paleomagnetic investigation in the Appalachian thrust belt of Georgia and Alabama. *Journal of Structural Geology*, *25*(8), 1193–1212. https://doi.org/10.1016/S0191-8141(02)00162-1

Beck, R. A., Vondra, C. F., Filkins, J. E., & Olander, J. D. (1988). Syntectonic sedimentation and Laramide basement thrusting, Cordilleran foreland; timing of deformation. In C. J. Schmidt Jr. & W. J. Perry (Eds.), *Interaction of the Rocky Mountain foreland and the Cordilleran thrust belt* (pp. 465–488). Geological Society of America Memoir, 171.

Behrmann, J. H. (1988). Crustal-scale extension in a convergent orogen: The Sterzing-Steinach mylomte zone in the eastern Alps. *Geodinamica Acta*, *2*(2), 63–73. https://doi.org/10.1080/09853111.1988.11105157

Benvenuto, G. L., & Price, R. A. (1979). Structural evolution of the Hosmer thrust sheet, southeastern British Columbia. *Bulletin of Canadian Petroleum Geology*, *27*(3), 360–394.

Berberian, M. (1995). Master "blind" thrust faults hidden under the Zagros folds: Active basement tectonics and surface morphotectonics. *Tectonophysics*, *241*(3–4), 193–224. https://doi.org/10.1016/0040-1951(94)00185-C

Berberian, M., & King, G. C. P. (1981). Toward a paleogeography and tectonic evolution of Iran. *Canadian Journal of Earth Sciences*, *18*(2), 210–265. https://doi.org/10.1139/e81-019

Berger, Z., Boast, M., & Mushayandebvu, M. (2008). The contribution of integrated HRAM studies to exploration and exploitation of unconventional plays in North America. *Reservoir, Canadian Society of Petroleum Geologist*, *35*(10), 42–48.

Bernoulli, D. (2007). The pre-Alpine geodynamic evolution of the southern Alps: A short summary. *Bulletin für angewandte geologie*, *12*(2), 3–10.

Berra, F., & Carminati, E. (2010). Subsidence history from a backstripping analysis of the Permo-Mesozoic succession of the central southern Alps (northern Italy). *Basin Research*, *22*(6), 952–975. https://doi.org/10.1111/j.1365-2117.2009.00453.x

Bertotti, G., Picotti, V., Bernoulli, D., & Castellarin, A. (1993). From rifting to drifting: tectonic evolution of the South-Alpine upper crust from the Triassic to the Early Cretaceous. *Sedimentary Geology*, *86*(1–2), 53–76. https://doi.org/10.1016/0037-0738(93)90133-P

Bianchi, M., Heit, B., Jakovlev, A., Yuan, X., Kay, S., Sandvol, E., et al. (2013). Teleseismic tomography of the southern Puna plateau in Argentina and adjacent regions. *Tectonophysics*, *586*, 65–83. https://doi.org/10.1016/j.tecto.2012.11.016

Bielenstein, H. U. (1969). *The Rundle thrust sheet, Banff, Alberta: A structural analysis*. Ph.D. dissertation. Queen's University, Kingston, Ontario.

Bigi, S., Carminati, E., Aldega, L., Trippetta, F., & Kavoosi, M. A. (2018). Zagros fold and thrust belt in the Fars province (Iran) I: Control of thickness/rheology of sediments and prethrusting tectonics on structural style and shortening. *Marine and Petroleum Geology*, *91*, 211–224. https://doi.org/10.1016/j.marpetgeo.2018.01.005

Bilham, R. (2019). Himalayan earthquakes: A review of historical seismicity and early 21st century slip potential. In P. J. Treloar & M. P. Searle (Eds.), *Himalayan tectonics: A modern synthesis* (pp. 423–482). London: Geological Society of London Special Publication, 483.

Bilham, R., Gaur, V. K., & Molnar, P. (2001). Himalayan seismic hazard. *Science*, *293*. https://doi.org/10.1126/science.1062584

Boyer, S. E. (1995). Sedimentary basin taper as a factor controlling the geometry and advance of thrust belts. *American Journal of Science*, *295*(10), 1220–1254. https://doi.org/0.2475/ajs.295.10.1220

Boyer, S. E., & Elliott, D. (1982). Thrust systems. *AAPG Bulletin*, *66*(9), 1196–1230. https://doi.org/10.1306/03B5A77D-16D1-11D7-8645000102C1865D

Brack, P. (1981). Structures in the southwestern border of the Adamello intrusion (Alpi Bresciane, Italy). *Schweizerische mineralogische und petrographische mitteilungen*, *61(1)*, 37–50. https://doi.org/pascal-francis.inist.fr/vibad/index.php?action=getRecordDetail&idt=PASCALGEODEBRGM8220070062

Brewer, M. C. (2004). *Geometric and kinematic evolution of the Bessemer transverse zone, Alabama Alleghanian thrust belt*. Ph.D. dissertation. University of Kentucky.

Brown, D., Alvarez-Marrón, J., Perez-Estaun, A., Gorozhanina, Y., Baryshev, V., & Puchkov, V. (1997). Geometric and kinematic evolution of the foreland thrust and fold belt in the southern Urals. *Tectonics*, *16*(3), 551–562. https://doi.org/10.1029/97TC00815

Burchfiel, B. C., Cowan, D. S., & Davis, G. A. (1992). Tectonic overview of the Cordilleran orogen in the western United States. In B. C. Burchfiel, W. Lipman, & M. L. Zoback (Eds.), *The Cordilleran orogen: Conterminous US* (pp. 107–168). Geoogical Society of America, G3.

Burrard, S. G. (1915). Origin of the Indo-Gangetic trough, commonly called Himalayan foredeep. *Proceedings of the Royal Society of London, 91A*, 220–238. https://doi.org/10.1098/rspa.1915.0014

Butler, R. W. H., Tavarnelli, E., & Grasso, M. (2006). Structural inheritance in mountain belts: an Alpine-Apennine perspective. *Journal of Structural Geology*, *28*(11), 1893–1908. https://doi.org/10.1016/j.jsg.2006.09.006

Calais, E., Nocquet, J.-M., Jouanne, F., & Tardy, M. (2002). Current strain regime in the Western Alps from continuous Global Positioning System measurements, 1996–2001. *Geology*, *30*(7), 651–654. https://doi.org/10.1130/0091-7613(2002)030<0651:CSRITW>2.0.CO;2

Campani, M., Mancktelow, N., Seward, D., Rolland, Y., Müller, W., & Guerra, I. (2010). Geochronological evidence for continuous exhumation through the ductile-brittle transition along a crustal-scale low-angle normal fault: Simplon Fault Zone, central Alps. *Tectonics*, *29*(3). https://doi.org/10.1029/2009TC002582

Canavan, R. R., Carrapa, B., Clementz, M. T., Quade, J., DeCelles, P. G., & Schoenbohm, L. M. (2014). Early Cenozoic uplift of the Puna Plateau, Central Andes, based on stable isotope paleoaltimetry of hydrated volcanic glass. *Geology*, *42*(5), 447–450. https://doi.org/101130/G35239.1

Carosi, R., Montomoli, C., Rubatto, D., & Visoná, D. (2010). Late Oligocene high-temperature shear zones in the core of the Higher Himalayan Crystallines (lower Dolpo, Western Nepal). *Tectonics*, *29*. https://doi.org/10.1029/2008TC002400

Castellarin, A., & Cantelli, L. (2000). Neo-Alpine evolution of the southern Eastern Alps. *Journal of Geodynamics*, *30*(1–2), 251–274. https://doi.org/10.1016/j.tecto.2005.10.019

Castellarin, A., Vai, G. B., & Cantelli, L. (2006). The Alpine evolution of the Southern Alps around the Giudicarie faults: A Late Cretaceous to Early Eocene transfer zone. *Tectonophysics*, *414*(1), 203–223. https://doi.org/10.1016/j.tecto.2005.10.019

Cawood, P. A., & Botsford, J. W. (1991). Facies and structural contrasts across Bonne Bay cross-strike discontinuity, western Newfoundland. *American Journal of Science*, *291*(8), 737–759. https://doi.org/10.2475/ajs.291.8.737

Cembrano, J., Hervé, F., & Lavenu, A. (1996). The Liquiñe Ofqui fault zone: A long-lived intra-arc fault system in southern Chile. *Tectonophysics*, *259*(1–3), 55–66. https://doi.org/10.1016/0040-1951(95)00066-6

Chapin, C. E. (2012). Origin of the Colorado Mineral Belt. *Geosphere*, *8*(1), 28–43. https://doi.org/10.1130/GES00694.1

Chapple, W. M. (1978). Mechanics of thin-skinned fold-and-thrust belts. *Geological Society of America Bulletin*, *89*(8), 1189–1198. https://doi.org/10.1130/0016-7606(1978)89<1189:MOTFB>2.0.CO;2

Chatzaras, V., Xypolias, P., Kokkalas, S., & Koukouvelas, I. (2013). Tectonic evolution of a crustal-scale oblique ramp, Hellenides thrust belt, Greece. *Journal of Structural Geology*, *57*, 16–37. https://doi.org/10.1016/j.jsg.2013.10.003

Chauvet, A., Onézime, J., Charvet, J., & Barbanson, L. (2004). Syn-to late-tectonic stockwork emplacement within the Spanish section of the Iberian pyrite belt: Structural, textural, and mineralogical constraints in the Tharsis and la Zarza areas. *Economic Geology*, *99*(8), 1781–1792. https://doi.org/10.2113/gsecongeo.99.8.1781

Chen, Y.-W., Wu, J., & Suppe, J. (2019). Southward propagation of Nazca subduction along the Andes. *Nature*, *565*(7740), 441–447. https://doi.org/10.1038/s41586-018-0860-1

Ching, K.-E., Johnson, K. M., Rau, R.-J., Chuang, R. Y., Kuo, L.-C., & Leu, P.-L. (2011). Inferred fault geometry and slip distribution of the 2010 Jiashian, Taiwan, earthquake is consistent with a thick-skinned deformation model. *Earth and Planetary Science Letters*, *301*(1–2), 78–86. https://doi.org/10.1016/j.epsl.2010.10.021

Ching, K.-E., Rau, R.-J., Lee, J.-C., & Hu, J.-C. (2007). Contemporary deformation of tectonic escape in SW Taiwan from GPS observations, 1995–2005. *Earth and Planetary Science Letters*, *262*(3–4), 601–619. https://doi.org/10.1016/j.epsl.2007.08.017

Colman-Sadd, S. P. (1978). Fold development in Zagros simply folded belt, Southwest Iran. *AAPG Bulletin*, *62*(6), 984–1003. https://doi.org/10.1306/C1EA4F81-16C9-11D7-8645000102C1865D

Colman-Sadd, S. P., Cawood, P. A., Dunning, G. R., Hall, J. M., Kean, B. F., O'Brien, B. H., et al. (1992). *Lithoprobe East in Newfoundland (Burgeo Transect): A cross-section through the southwest Newfoundland Appalachians* (vol. 92-A7).

Coltorti, M., Farabollini, P., Gentili, B., & Pambianchi, G. (1996). Geomorphological evidence for anti-Apennine faults in the Umbro-Marchean Apennines and in the peri-Adriatic basin, Italy. *Geomorphology*, *15*(1), 33–45. https://doi.org/10.1016/0169-555X(95)00117-N

Coney, P. J., & Reynolds, S. J. (1977). Cordilleran benioff zones. *Nature*, *270*(5636), 403–406. https://doi.org/10.1038/270403a0

Constenius, K. N. (1996). Late Paleogene extensional collapse of the Cordilleran foreland fold and thrust belt. *Geological Society of America Bulletin*, *108*(1), 20–39. https://doi.org/10.1130/0016-7606(1996)108<0020:LPECOT>2.3.CO;2

Cook, B. S., & Thomas, W. A. (2009). Superposed lateral ramps in the Pell City thrust sheet, Appalachian thrust belt, Alabama. *Journal of Structural Geology*, *31*(9), 941–949. https://doi.org/10.1016/j.jsg.2009.06.001

Cooley, M. A., Price, R. A., Dixon, J. M., & Kyser, T. K. (2011). Along-strike variations and internal details of chevron-style, flexural-slip thrust-propagation folds within the southern Livingstone Range anticlinorium, a paleohydrocarbon reservoir in southern Alberta Foothills, Canada. *AAPG Bulletin*, *95*(11), 1821–1849. https://doi.org/10.1306/01271107097

Corbett, G. J. (1994). *Regional structural control of selected Cu/Au occurrences in Papua New Guinea*. Paper presented at the Proceedings of the Papua New Guinea Geology, Exploration and Mining Conference, Melbourne.

Cotton, J. T., & Koyi, H. A. (2000). Modeling of thrust fronts above ductile and frictional detachments: Application to structures in the Salt Range and Potwar Plateau, Pakistan. *Geological Society of America Bulletin*, *112*(3), 351–363. https://doi.org/10.1130/0016-7606(2000)112<351:MOTFAD>2.0.CO;2

Coutand, I., Cobbold, P. R., de Urreiztieta, M., Gautier, P., Chauvin, A., Gapais, D., et al. (2001). Style and history of Andean deformation, Puna plateau, northwestern Argentina. *Tectonics*, *20*(2), 210–234. https://doi.org/10.1029/2000TC900031

Dahl, P. S., Holm, D. K., Gardner, E. T., Hubacher, F. A., & Foland, K. A. (1999). New constraints on the timing of Early Proterozoic tectonism in the Black Hills (South Dakota), with implications for docking of the Wyoming province with Laurentia. *Geological Society of America Bulletin*, *111*(9), 1335–1349. https://doi.org/10.1130/0016-7606(1999)111<1335:NCOTTO>2.3.CO;2

Dahlen, F. A. (1990). Critical taper model of fold-and-thrust belts and accretionary wedges. *Annual Review of Earth and Planetary Sciences*, *18*(1), 55–99. https://doi.org/10.1146/annurev.ea.18.050190.000415

Davies, H. L. (1991). *Regional geologic setting of some mineral deposits of the New Guinea Orogen*. Paper presented at the Papua New Guinea Geology, Exploration and Mining Conference; Rabaul, Papua New Guinea, 1991.

DeCelles, P. G. (2004). Late Jurassic to Eocene evolution of the Cordilleran thrust belt and foreland basin system, western USA. *American Journal of Science*, *304*(2), 105–168. https://doi.org/10.2475/ajs.304.2.105

DeCelles, P. G., Carrapa, B., Ojha, T. P., Gehrels, G. E., & Collins, D. (2020). Structural and thermal evolution of the Himalayan thrust belt in midwestern Nepal. *Geological Society of America Special Paper*, *547*, 1–77. https://doi.org/10.1130/2020.2547(01)

DeCelles, P. G., Gray, M., Ridgway, K., Cole, R., Pivnik, D., Pequera, N., & Srivastava, P. (1991). Controls on synorogenic alluvial-fan architecture, Beartooth Conglomerate (Palaeocene), Wyoming and Montana. *Sedimentology*, *38*(4), 567–590. https://doi.org/10.1111/j.1365-3091.1991.tb01009.x

DeCelles, P. G., Robinson, D. M., Quade, J., Ojha, T. P., Garzione, C. N., Copeland, P., et al. (2001). Stratigraphy, structure, and tectonic evolution of the Himalayan fold-thrust belt in western Nepal. *Tectonics*, *20*, 487–509. https://doi.org/10.1029/2000TC001226

Deffontaines, B., Lacombe, O., Angelier, J., Chu, H., Mouthereau, F., Lee, C., Deramond, J., Lee, J., Yu, M., & Liew, P. (1997). Quaternary transfer faulting in the Taiwan Foothills: evidence from a multisource approach. *Tectonophysics*, *274*(1–3), 61–82. https://doi.org/10.1016/S0040-1951(96)00298-3

Dewey, J. F., & Bird, J. M. (1970). Mountain belts and the new global tectonics. *Journal of Geophysical Research*, *75*(14), 2625–2647. https://doi.org/10.1029/JB075i014p02625

Dewey, J. F., Holdsworth, R. E., & Strachan, R. A. (1998). Transpression and transtension zones. In R. E. Holdworth, R. A. Strachan, & J. F. Dewey (Eds.), *Continental transpressional and transtensional tectonics* (pp. 1–14). Geological Society, London, Special Publications, 135.

Dickinson, W. R. (2004). Evolution of the North American Cordillera. *Annual Reviwe of Earth and Planetary Sciences*, *32*, 13–45. https://doi.org/10.1146/annurev.earth.32.101802.120257

Dickinson, W. R., & Snyder, W. S. (1978). Plate tectonics of the Laramide orogeny. In V. W. Matthews (Ed.), *Laramide folding associated with block faulting in the Western United States* (pp. 355–366). Geoloical Society of America Memoir, 151.

Diehl, T., Singer, J., Hetényi, G., Grujic, D., Clinton, J., Giardini, D., et al. (2017). Seismotectonics of Bhutan: Evidence for segmentation of the Eastern Himalayas and link to foreland deformation. *Earth and Planetary Science Letters*, *471*, 54–65. https://doi.org/10.1016/j.epsl.2017.04.038

Dini, A., Westerman, D., Innocenti, F., & Rocchi, S. (2008). Magma emplacement in a transfer zone: The Miocene mafic Orano dyke swarm of Elba Island, Tuscany, Italy. *Geological Society, London, Special Publications*, *302*(1), 131–148. https://doi.org/10.1144/SP302.10

Divyadarshini, A., & Singh, V. (2019). Investigating topographic metrics to decipher structural model and morphotectonic evolution of the Frontal Siwalik Ranges, Central Himalaya, Nepal. *Geomorphology*, *337*, 31–52. doi:10.1016/j.geomorph.2019.03.028

Domeier, M. (2016). A plate tectonic scenario for the Iapetus and Rheic oceans. *Gondwana Research*, *36*, 275–295. https://doi.org/10.1016/j.gr.2015.08.003

Doutsos, T., Koukouvelas, I., & Xypolias, P. (2006). A new orogenic model for the External Hellenides. In A. H. F. Robertson & D. Mountrakis (Eds.), *Tectonic development of the Eastern Mediterrainan Region* (pp. 507–520). Geological Society, London, Special Publications, 260.

Drahovzal, J. A., & Thomas, W. A. (1976). *Cross-strike structural discontinuities in the Appalachians of Alabama: Geological Society of America Abstracts with Programs* (vol. *8*).

Drake, A. A. J., Sinha, A. K., Larid, J., & Guy, R. E. (1989). The Taconian Orogen. In R. D. Hatcher Jr, W. A. Thomas, & G. W. Viele (Eds.), *The Appalachian-Ouachita Orogen in the United States, the geology of North America* (pp. 101–177). Geological Society of America, F-2.

Drukpa, D., Velasco, A.A., & Doser, D. (2006). Seismicity in the Kingdom of Bhutan (1937–2003): Evidence for crustal transcurrent deformation. *Journal of Geophysical Research*, *111*, B06301. https://doi/org/10.1029/2004JB003087

Dubey, A. K., Mishra, R., & Bhakuni, S.S. (2001). Erratic shortening from balanced cross sections of the western Himalayan foreland basin causes and implications for basin evolution. *Journal of Asian Earth Sciences*, *19*, 765–777. https://doi.org/10.1016/S1367-9120(01)00010-4

Duncan, C., Masek, J., & Fielding, E. (2003). How steep are the Himalaya? Characteristics and implications of along-strike topographic variations. *Geology*, *31*(1), 75–78. https://doi.org/10.1130/0091-7613(2003)031<0075:HSATHC>2.0.CO;2

Dunkl, I., Kuhlemann, J., Reinecker, J., & Frisch, W. (2005). Cenozoic relief evolution of the Eastern Alps: Constraints from apatite fission track age-provenance of Neogene intramontane sediments. *Austrian Journal of Earth Sciences*, *98*, 92–105.

Duvall, M. J., Waldron, J. W. F., Godin, L., & Najman, Y. (2020). Active strike-slip faults and an outer frontal thrust in the Himalayan foreland basin. *Proceedings of the National Academy of Sciences*, *117*(30), 17615–17621. https://doi.org/10.1073/pnas.2001979117

Edey, A., Allen, M., & Nilfouroushan, F. (2020). Kinematic variation within the Fars Arc, eastern Zagros, and the development of fold-and-thrust belt curvature. *Tectonics*, *39*(8), e2019TC005941. https://doi.org/10.1029/2019TC005941

Egbue, O., & Kellogg, J. (2010). Pleistocene to present North Andean "escape." *Tectonophysics*, *489*(1–4), 248–257. https://doi.org/10.1016/j.tecto.2010.04.021

Elliott, D. (1976). The motion of thrust sheets. *Journal of Geophysical Research*, *81*(5), 949–963. https://doi.org/10.1029/JB081i005p00949

Elter, F. M., Piero, E., Claudio, E., Elena, E., Katharina, K. R., Matteo, P., et al. (2011). Strike-slip geometry inferred from the seismicity of the Northern-Central Apennines (Italy). *Journal of Geodynamics*, *52*(5), 379–388. https://doi.org/10.1016/j.jog.2011.03.003

Eugster, P., Thiede, R. C., Scherler, D., Stubner, K., Sobel, E. R., & Strecker, M. R. (2018). Segmentation of the main Himalayan thrust revealed by low-temperature thermochronometry in the western Indian Himalaya. *Tectonics*, *37*, 2710–2726. https://doi.org/10.1029/2017TC004752

Fakhari, M. D., Axen, G. J., Horton, B. K., Hassanzadeh, J., & Amini, A. (2008). Revised age of proximal deposits in the Zagros foreland basin and implications for Cenozoic evolution of the High Zagros. *Tectonophysics*, *451*(1–4), 170–185. https://doi.org/10.1016/j.tecto.2007.11.064

Falcon, N. L. (1974). Southern Iran: Zagros Mountains. *Geological Society, London, Special Publications*, *4*(1), 199–211. https://doi.org/10.1144/GSL.SP.2005.004.01.11

Farzipour-Saein, A., Nilfouroushan, F., & Koyi, H. (2013). The effect of basement step/topography on the geometry of the Zagros fold and thrust belt (SW Iran): An analog modeling approach. *International Journal of Earth Sciences*, *102*(8), 2117–2135. https://doi.org/10.1007/s00531-013-0921-5

Fermor, P. (1999). Aspects of the three-dimensional structure of the Alberta Foothills and Front Ranges. *GSA Bulletin*, *111*(3), 317–346. https://doi.org/10.1130/0016-7606(1999)111<0317:AOTTDS>2.3.CO;2

Festa, A., Balestro, G., Borghi, A., De Caroli, S., & Succo, A. (2020). The role of structural inheritance in continental break-up and exhumation of Alpine Tethyan mantle (Canavese Zone, Western Alps). *Geoscience Frontiers*, *11*(1), 167–188. https://doi.org/10.1016/j.gsf.2018.11.007

Fiigenschuh, B., Seward, D., & Mancktelow, N. (1997). Exhumation in a convergent orogen: The western Tauern Window. *Terra Nova*, *9*(5–6), 213–217. https://doi.org/10.1111/j.1365-3121.1997.tb00015.x

Folguera, A., Ramos, V. A., Hermanns, R. L., & Naranjo, J. (2004). Neotectonics in the foothills of the southernmost central Andes (37–38 S): Evidence of strike-slip displacement along the Antiñir-Copahue fault zone. *Tectonics*, *23*(5). https://doi.org/10.1029/2003TC001533

Fondriest, M., Aretusini, S., Di Toro, G., & Smith, S. A. (2015). Fracturing and rock pulverization along an exhumed seismogenic fault zone in dolostones: The Foiana Fault Zone (Southern Alps, Italy). *Tectonophysics*, *654*, 56–74. doi:http://dx.doi.org/10.1016/j.tecto.2015.04.015

Foo, W. K. (1979). *Evolution of transverse structures linking the Purcell Anticlinorium to the western Rocky Mountains near Canal Flats, British Columbia*. (MSc). Queen's University, Kingston.

Foster, D. A., Doughty, P. T., Kalakay, T. J., Fanning, C. M., Coyner, S., Grice, W. C., et al. (2007). Kinematics and timing of exhumation of metamorphic core complexes along the Lewis and Clark fault zone, northern Rocky Mountains, USA. *Special Papers-Geological Society of America*, *434*, 207. https://doi.org/10.1130/2007.2434(10)

Foster, D. A., Mueller, P. A., Mogk, D. W., Wooden, J. L., & Vogl, J. J. (2006). Proterozoic evolution of the western margin of the Wyoming craton: Implications for the tectonic and magmatic evolution of the northern Rocky Mountains. *Canadian Journal of Earth Sciences*, *43*(10), 1601–1619. https://doi.org/10.1139/E06-052

Frisch, W., Kuhlemann, J., Dunkl, I., & Brügel, A. (1998). Palinspastic reconstruction and topographic evolution of the Eastern Alps during late Tertiary tectonic extrusion. *Tectonophysics*, *297*(1–4), 1–15. https://doi.org/10.1016/S0040-1951(98)00160-7

Gaetani, M., Gianotti, R., Jadoul, F., Ciarapica, G., & Cirilli, S. (1986). Carbonifero superiore, Permiano et Triassico nell'area Lariana. *Memorie della Società Geologica Italiana*, *32*, 5–48. http://doi.org/pascal-francis.inist.fr/vibad/index.php?action=getRecordDetail&idt=7204188

Galliski, M., & Viramonte, J. (1988). The Cretaceous paleorift in northwestern Argentina: A petrologic approach. *Journal of South American Earth Sciences*, *1*(4), 329–342. https://doi.org/10.1016/0895-9811(88)90021-1

Gansser, A. (1964). *Geology of the Himalayas*. New York: Wiley Interscience.

Gaupp, R., & Batten, D. (1983). Depositional setting of Middle to Upper Cretaceous sediments in the Northern Calcareous Alps from palynological evidence. *Neues Jahrbuch für Geologie und Paläontologie-Monatshefte*, 585–600. https://doi.org/10.1127/njgpm/1983/1983/585

Ghiglione, M. C., Suarez, F., Ambrosio, A., Da Poian, G., Cristallini, E. O., Pizzio, M. F., et al. (2009). Structure and evolution of the Austral Basin fold-thrust belt, southern Patagonian Andes. *Revista de la Asociación Geológica Argentina*, *65*(1), 215–226.

Giambiagi, L., Mescua, J., Bechis, F., Tassara, A., & Hoke, G. (2012). Thrust belts of the southern Central Andes: Along-strike variations in shortening, topography, crustal geometry, and denudation. *Bulletin*, *124*(7–8), 1339–1351. https://doi.org/10.1130/B30609.1

Gifford, J. N., Mueller, P. A., Foster, D. A., & Mogk, D. W. (2014). Precambrian crustal evolution in the Great Falls tectonic zone: Insights from xenoliths from the Montana Alkali Province. *The Journal of Geology*, *122*(5), 531–548. https://doi.org/10.1086/677262

Gillis, R. J., Horton, B. K., & Grove, M. (2006). Thermochronology, geochronology, and upper crustal structure of the Cordillera Real: Implications for Cenozoic exhumation of the central Andean plateau. *Tectonics*, *25*(6). https://doi.org/10.1029/2005TC001887

Godin, L., & Harris, L. B. (2014). Tracking basement cross-strike discontinuities in the Indian crust beneath the Himalayan orogen using gravity data: Relationship to upper crustal faults. *Geophysical Journal International*, *198*, 198–215. https://doi.org/10.1093/gji/ggu131

Goldfinger, C., Kulm, L. D., Yeats, R. S., McNeill, L., & Hummon, C. (1997). Oblique strike-slip faulting of the central Cascadia submarine forearc. *Journal of Geophysical Research*, *102*(B4), 8217–8243. https://doi.org/10.1029/96JB02655

Graindorge, D., Klingelhoefer, F., Sibuiet, J. C., McNeill, L., Henstock, T., Dean, S., et al. (2008). Impact of lower plate structure on upper plate deformation at the NW Sumatran convergent margin from seafloor morphology. *Earth and Planetary Science Letters*, *275*, 201–210. https://doi.org/10.1016/j.epsl.2008.04.053

Grosjean, G., Sue, C., & Burkhard, M. (2004). Late Neogene extension in the vicinity of the Simplon fault zone (central Alps, Switzerland). *Eclogae Geologicae Helvetiae*, *97*(1), 33–46. https://doi.org/10.1007/s00015-004-1114-9

Grujic, D., Hollister, L.S., & Parrish, R. P. (2002). Himalayan metamorphic sequence as an orogenic channel: Insight from Bhutan. *Earth and Planetary Science Letters*, *198*, 177–191. https://doi.org/10.1016S0012-821X(02)00482-X

Guillot, S., Mahéo, G., De Sigoyer, J., Hattori, K., & Pecher, A. (2008). Tethyan and Indian subduction viewed from the Himalayan high-to-ultrahigh-pressure metmorphic rocks. *Tectonophysics*, *451*(1–4), 225–241. https://doi.org/10.1016/j.tecto.2007.11.059

Gutscher, M. A., Spakman, W., Bijwaard, H., & Engdahl, E. R. (2000). Geodynamics of flat subduction: Seismicity and tomographic constraints from the Andean margin. *Tectonics*, *19*(5), 814–833. https://doi.org/10.1029/1999TC001152

Handy, M. R., Schmid, S. M., Bousquet, R., Kissling, E., & Bernoulli, D. (2010). Reconciling plate-tectonic reconstructions of Alpine Tethys with the geological-geophysical record of spreading and subduction in the Alps. *Earth-Science Reviews*, *102*(3–4), 121–158. https://doi.org/10.1007/s00531-014-1060-3

Harrison, J., & Bally, A. (1988). Cross-sections of the Parry Islands fold belt on Melville Island, Canadian Arctic Islands: Implications for the timing and kinematic history of some thin-skinned decollement systems. *Bulletin of Canadian Petroleum Geology*, *36*(3), 311–332.

Harrison, J. E., Griggs, A. B., & Wells, J. D. (1974). *Tectonic features of the Precambrian Belt basin and their influcence on post-belt structures* (vol. 886). United States Geological Survey Professional Paper.

Harvey, J. E., Burbank, D. W., & Bookhagen, B. (2015). Along-strike changes in Himalayan thrust geometry: Topographic and tectonic discontinuities in western Nepal. *Lithosphere*, *7*(5), 511–518. https://doi.org/10.1130/L444.1

Hatcher, J. R. D. (1972). Developmental model for the southern Appalachians. *Geological Society of America Bulletin*, *83*(9), 2735–2760.

Hatcher, J. R. D. (2010). The Appalachian orogen: A brief summary. In R. P. Tollo, M. J. Bartholomew, J. P. Hibbard, & P. M. Karabinos (Eds.), *From Rodinia to Pangea: The lithotectonic record of the Appalachian region* (pp. 1–19). Geological Society of America Memoir 206.

Hatcher, J. R. D., Thomas, W. A., Geiser, P. A., Snoke, A. W., Mosher, S., & Wiltschko, D. V. (1989). Alleghanian orogen. In R. D. Hatcher Jr, W. A. Thomas, & G. W. Viele (Eds.), *The Appalachian-Ouachita Orogen in the United States, the geology of North America* (pp. 233–319). Geological Society of America, F-2.

Hauck, M. L., Nelson, K. D., Brown, L. D., Wenjin, Z., & Ross, A. R. (1998). Crustal structure of the Himalayan orogen at ~90° east longitude from Project INDEPTH deep reflection profiles. *Tectonics*, *17*, 481–500. https://doi.org/10.1029/98TC01314

Haynes, S. J., & McQuillan, H. (1974). Evolution of the Zagros suture zone, southern Iran. *Geological Society of America Bulletin*, *85*(5), 739–744. https://doi.org/10.1130/0016-7606(1974)85<739:EOTZSZ>2.0.CO;2

Haynes, S. J., & Reynolds, P. (1980). Early development of Tethys and Jurassic ophiolite displacement. *Nature*, *283*(5747), 561–563. https://doi.org/10.1038/283561a0

Hessami, K., Koyi, H. A., & Talbot, C. J. (2001). The significance of strike-slip faulting in the basement of the Zagros fold and thrust belt. *Journal of Petroleum Geology*, *24*(1), 5–28. https://doi.org/10.1111/j.1747-5457.2001.tb00659.x

Hibbard, J. P., van Staal, C. R., Rankin, D. W., & Williams, H. (2006). *Lithotectonic map of the Appalachian orogen, Canada-United States of America*. Geological Survey of Canada, Map 2096A, scale 1:1,500,000.

Hill, K. C., Kendrick, R. D., Crowhurst, P. V., & Gow, P. A. (2002). Copper-gold mineralisation in New Guinea: Tectonics, lineaments, thermochronology and structure. *Australian Journal of Earth Sciences*, *49*(4), 737–752. https://doi.org/10.1046/j.1440-0952.2002.00944.x

Hodges, K. V. (2000). Tectonics of the Himalaya and southern Tibet from two perspectives. *Geological Society of America Bulletin*, *112*(3), 324–350. https://doi.org/10.1130/0016-7606(2000)112<324:TOTHAS>2.0.CO;2

Hoffman, P. F. (1988). Early Proterozoic assembly. *Annual Review of Earth and Planetary Sciences*, *6*, 543–603. https://doi.org/10.1146/annurev.ea.16.050188.002551

Hofmann, F., Schlatter, R., & Weh, M. (2000). Erläuterungen zu Blatt 97: Beggingen (LK 1011) des Geologischen Atlas der Schweiz 1: 25,000. *Bundesamt für Wasser und Geologie, Bern*, *113*.

Hogan, J. P., & Gilbert, M. C. (1998). The Southern Oklahoma aulacogen: A Cambrian analog for Mid-Proterozoic AMCG (anorthosite-mangerite-charnockite-granite) complexes? In J. P. Hogan & M. C. Gilbert (Eds.), *Basement tectonics 12* (pp. 39–78). Proceedings of the International Conferences on Basement Tectonics (vol. 6). Springer.

Hope, J., & Eaton, D. (2002). Crustal structure beneath the western Canada sedimentary basin: Constraints from gravity and magnetic modelling. *Canadian Journal of Earth Sciences*, *39*(3), 291–312. https://doi.org/10.1139/e01-060

Horváth, F., & Cloetingh, S. A. P. L. (1996). Stress-induced late stage subsidence anomalies in the Pannonian Basin. *Tectonophysics*, *266*, 287–300.

Höy, T., & Heyden, P. (1988). Geochemistry, geochronology, and tectonic implications of two quartz monzonite intrusions, Purcell Mountains, southeastern British Columbia. *Canadian Journal of Earth Sciences*, *25*(1), 106–115. https://doi.org/10.1139/e88-011

Hubbard, J., Almeida, R., Foster, A., Sapkota, S. N., Burgi, P., & Tapponnier, P. (2016). Structural segmentation controlled the 2015 MW 7.8 Gorkha earthquake rupture in Nepal. *Geology*, *44*(8), 639–642. https://doi.org/10.1130/G38077.1

Hubbard, M., Gajurel, A. P., Mukul, M., & Seifert, N. (2018). *Cross faults and their role in Himalayan structural evolution*. Paper presented at the Geological Society of America Abstracts with Programs.

Hubbard, M., Mukul, M., Gajurel, A. P., Ghosh, A., Srivastava, V., Giri, B., et al. (2021). Orogenic segmentation and its role in Himalayan mountain building. *Frontiers in Earth Science*, *9*(259). https://doi.org/10.3389/feart.2021.641666

Hubbard, M. S. (1999). Norumbega fault zone: Part of an orogen-parallel strike-slip system, northern Appalachians. In A. Ludman & D. P. J. West (Eds.), *Normumberga fault system of the Northern Appalachians* (pp. 155–166). Geological Society of America Special Paper, 331.

Hubbard, M. S., & Mancktelow, N. S. (1992). Lateral displacement during Neogene convergence in the western and central Alps. *Geology*, *20*(10), 943–946. https://doi.org/10.1130/0091-7613(1992)020<0943:LDDNCI>2.3.CO;2

Hunziker, J. C., Frey, M., Clauer, N., Dallmeyer, R. D., Friedrichsen, H., Flehmig, W., et al. (1986). The evolution of illite to muscovite: mineralogical and isotopic data from the Glarus Alps, Switzerland. *Contributions to Mineralogy and Petrology*, *92*(2), 157–180. https://doi.org/10.1007/BF00375291

Husseini, M. I. (1988). The Arabian infracambrian extensional system. *Tectonophysics*, *148*(1–2), 93–103. https://doi.org/10.1016/0040-1951(88)90163-1

Hyndman, D. W., Alt, D., & Sears, J. W. (1988). Post-Archean metamorphic and tectonic evolution of western Montana and northern Idaho. In W. G. Ernst (Ed.), *Metamorphism and crustal evolution of the western United States* (pp. 332–361). Englewood Cliffs, New Jersey: Prentice Hall.

Hynes, A., & Rivers, T. (2010). Protracted continental collision: Evidence from the Grenville orogen. *Canadian Journal of Earth Sciences*, *47*(5), 591–620. https://doi.org/10.1139/E10-003

Isacks, B. L. (1988). Uplift of the central Andean plateau and bending of the Bolivian orocline. *Journal of Geophysical Research: Solid Earth*, *93*(B4), 3211–3231. https://doi.org/10.1029/JB093iB04p03211

Jackson, J. A. (1980). Reactivation of basement faults and crustal shortening in orogenic belts. *Nature*, *283*(5745), 343–346. https://doi.org/10.1038/283343a0

Jadoon, I. A. K., Lawrence, R. D., & Lillie, R. J. (1994). Seismic data, geometry, evolution, and shortening in the active Sulaiman fold-and-thrust belt of Pakistan, southwest of the Himalayas. *AAPG Bulletin*, *78*(5), 758–774. https://doi.org/10.1306/A25FE3AB-171B-11D7-8645000102C1865D

Japas, M. S., Ré, G. H., Oriolo, S., & Vilas, J. F. (2016). Basement-involved deformation overprinting thin-skinned deformation in the Pampean flat-slab segment of the southern Central Andes, Argentina. *Geological Magazine*, *153*(5–6), 1042–1065. https://doi.org/10.1017/S001675681600056X

Jessup, M. J., Newell, D. L., Cottle, J. M., Berger, A. L., & Spotila, J. A. (2008). Orogen-parallel extension and exhumation enhanced by denudation in the trans-Himalayan Arun River gorge, Ama Drime Massif, Tibet-Nepal. *Geology*, *36*(7), 587–590. https://doi.org/10.1130/g24722a.1

Jordan, T. E., Isacks, B., Ramos, V. A., & Allmendinger, R. W. (1983). Mountain building in the Central Andes. *Episodes*, *3*(3), 20–26.

Joudaki, M., Farzipour-Saein, A., & Nilfouroushan, F. (2016). Kinematics and surface fracture pattern of the Anaran basement fault zone in NW of the Zagros fold–thrust belt. *International Journal of Earth Sciences*, *105*(3), 869–883. https://doi.org/10.1007/s00531-015-1209-8

Karlstrom, K. E., & Houston, R. S. (1984). The Cheyenne belt: Analysis of a Proterozoic suture in southern Wyoming. *Precambrian Research*, *25*(4), 415–446. https://doi.org/10.1016/0301-9268(84)90012-3

Kendrick, E., Bevis, M., Smalley, R., Jr., Brooks, B., Vargas, R. B., Launa, E., & Fortes, L. P. S. (2003). The Nazca-South America Euler vector and its rate of change. *Journal of South American Earth Sciences*, *16*(2), 125–131. https://doi.org/10.1016/S0895-9811(03)00028-2

Kent, P. E. (1979). The emergent Hormuz salt plugs of southern Iran. *Journal of Petroleum Geology*, *2*(2), 117–144. https://doi.org/10.1111/j.1747-5457.1979.tb00698.x

Keppie, J. D., & Krogh, T. E. (1999). U-Pb Geochronology of Devonian granites in the Meguma terrain of Nova Scotia, Canada: Evidence for hotspot melting of a Neoproterozoic source. *The Journal of Geology*, *107*(5), 555–568. doi:10.1086/314369

Khadivi, S., Mouthereau, F., Larrasoaña, J. C., Vergés, J., Lacombe, O., Khademi, E., et al. (2010). Magnetochronology of synorogenic Miocene foreland sediments in the Fars arc of the Zagros Folded Belt (SW Iran). *Basin Research*, *22*(6), 918–932. https://doi.org/10.1111/j.1365-2117.2009.00446.x

Kley, J., Rossello, E. A., Monaldi, C. R., & Habighorst, B. (2005). Seismic and field evidence for selective inversion of Cretaceous normal faults, Salta rift, northwest Argentina. *Tectonophysics*, *399*(1–4), 155–172. https://doi.org/10.1016/j.tecto.2004.12.020

Kokkalas, S., & Doutsos, T. (2004). Kinematics and strain partitioning in the southeast Hellenides (Greece). *Geological Journal*, *39*(2), 121–140. https://doi.org/10.1002/gj.947

Koop, W. J., & Stoneley, R. (1982). Subsidence history of the Middle East Zagros basin, Permian to recent. *Philosophical Transactions of the Royal Society A Mathematical and Physical Sciences*, *305*(1489), 149–168. https://doi.org/10.1098/rsta.1982.0031

Kralik, M., Klima, K., & Riedmüller, G. (1987). Dating fault gouges. *Nature*, *327*(6120), 315–317. https://doi.org/10.1038/327315a0

Kuiper, Y. D. (2016). Development of the Norumbega fault system in mid-Paleozoic New England, USA: An integrated subducted oceanic ridge model. *Geology*, *44*(6), 455–458. https://doi.org/10.1130/G37599.1

Kuiper, Y. D., & Wakabayashi, J. (2018). A comparison between mid-Paleozoic New England, USA, and the modern western USA: Subduction of an oceanic ridge-transform fault system. *Tectonophysics*, *745*, 278–292. https://doi.org/10.1016/j.tecto.2018.08.020

Kwon, S., & Mitra, G. (2004). Strain distribution, strain history, and kinematic evolution associated with the formation of arcuate salients in fold-thrust belts: The example of the Provo salient, Sevier orogen, Utah. In A. J. Sussman & A. B. Weil (Eds.), *Orogenic curvature: Integrating Paleomagnetic and structural analyses* (pp. 205–223). Geological Society of America Special Paper, 383.

Kwon, S., & Mitra, G. (2006). Three-dimensional kinematic history at an oblique ramp, Leamington zone, Sevier belt, Utah. *Journal of Structural Geology*, *28*(3), 474–493. https://doi.org/10.1016/j.jsg.2005.12.011

Lacombe, O., Mouthereau, F., Angelier, J., & Deffontaines, B. (2001). Structural, geodetic and seismological evidence for tectonic escape in SW Taiwan. *Tectonophysics*, *333*(1–2), 323–345. https://doi.org/10.1016/S0040-1951(00)00281-X

Lacombe, O., Mouthereau, F., Kargar, S., & Meyer, B. (2006). Late Cenozoic and modern stress fields in the western Fars (Iran): Implications for the tectonic and kinematic evolution of central Zagros. *Tectonics*, *25*(1), 1–27. https://doi.org/10.1029/2005TC001831

Larson, K. P., & Cottle, J. M. (2014). Midcrustal discontinuities and the assembly of the Himalayan midcrust. *Tectonics*, *33*, 718–740. https://doi.org/10.1002/2013TC003452

Larson, K. P., Price, R. A., & Archibald, D. A. (2006). Tectonic implications of 40Ar/39Ar muscovite dates from the Mt. Haley stock and Lussier River stock, near Fort Steele, *British Columbia. Canadian Journal of Earth Sciences*, *43*(11), 1673–1684. https://doi.org/10.1139/E06-048

Laubscher, H. P. (1971). Das Alpen-Dinariden-Problem und die Palinspastik der südlichen tethys. *Geologische Rundschau*, *60*(3), 813–833. https://doi.org/10.1007/BF02046522

Laubscher, H. P. (1985). Large-scale, thin-skinned thrusting in the southern Alps: Kinematic models. *Geological Society of America Bulletin*, *96*(6), 710–718. https://doi.org/10.1130/0016-7606(1985)96<710:LTTITS>2.0.CO;2

Lawton, T. F., Sprinkel, D. A., Decelles, P. G., Mitra, G., Sussman, A. J., & Weiss, M. P. (1997). Stratigraphy and structure of the Sevier thrust belt and proximal foreland-basin system in central Utah: A transect from the Sevier Desert to the Wasatch Plateau. In K. P. Link & B. J. Kowallis (Eds.), *Brigham Young University geology studies field trip guide book, part 2* (pp.33–68). Geological Society of America Annual Meeting.

Le Garzic, E., Vergés, J., Sapin, F., Saura, E., Meresse, F., & Ringenbach, J. (2019). Evolution of the NW Zagros fold-and-thrust belt in Kurdistan region of Iraq from balanced and restored crustal-scale sections and forward modeling. *Journal of Structural Geology*, *124*, 51–69. https://doi.org/10.1016/j.jsg.2019.04.006

Leier, A., McQuarrie, N., Garzione, C., & Eiler, J. (2013). Stable isotope evidence for multiple pulses of rapid surface uplift in the Central Andes, Bolivia. *Earth and Planetary Science Letters*, *371*, 49–58. https://doi.org/10.1016/j.epsl.2013.04.025

Lemieux, S., Ross, G. M., & Cook, F. A. (2000). Crustal geometry and tectonic evolution of the Archean crystalline basement beneath the southern Alberta Plains, from new seismic reflection and potential-field studies. *Canadian Journal of Earth Sciences*, *37*(11), 1473–1491. https://doi.org/10.1139/e00-065

Li, Z. X., Bogdanova, S. V., Collins, A. S., Davidson, A., De Waele, B., Ernst, R. E., et al. (2008). Assembly, configuration, and break-up history of Rodinia: A synthesis. *Precambrian Research*, *160*(1), 179–210. https://doi.org/10.1016/j.precamres.2007.04.021

Likerman, J., Burlando, J. F., Cristallini, E. O., & Ghiglione, M. C. (2013). Along-strike structural variations in the Southern Patagonian Andes: Insights from physical modeling. *Tectonophysics*, *590*, 106–120. https://doi.org/10.1016/j.tecto.2013.01.018

Lin, S., Staal, C., & Dubé, B. (1994). Promontory-promontory collision in the Canadian Appalachians. *Geology*, *22*(10), 897–900. https://doi.org/10.1130/0091-7613(1994)022<0897:PPCITC>2.3.CO;2

Linzer, H.-G., Ratschbacher, L., & Frisch, W. (1995). Transpressional collision structures in the upper crust: The fold-thrust belt of the northern Calcareous Alps. *Tectonophysics*, *242*(1–2), 41–61. https://doi.org/10.1016/0040-1951(94)00152-Y

Llambías, E. J., Kleinman, L. E., & Salvarradi, J. A. (1993). El magmatismo Gondwánico. In V. Ramos (Ed.), *Relatorio geología y recursos naturales de Mendoza* (pp. 53–64). Buenos Aires: Servicio Geológico Minero Argentino (SEGEMAR).

Long, S., McQuarrie, N., Tobgay, T., & Grujic, D. (2011). Geometry and crustal shortening of the Himalayan fold-thrust belt, eastern and central Bhutan. *Geological Society of America Bulletin*, *123*(7/8), 1427–1447. https://doi.org/10.1130/B30203.1

Lupi, M., Trippanera, D., Gonzalez, D., D'amico, S., Acocella, V., Cabello, C., et al. (2020). Transient tectonic regimes imposed by megathrust earthquakes and the growth of NW-trending volcanic systems in the Southern Andes. *Tectonophysics*, *774*, 228204. https://doi.org/10.1016/j.tecto.2019.228204

Mahoney, L., Hill, K., McLaren, S., & Hanani, A. (2017). Complex fold and thrust belt structural styles: Examples from the Greater Juha area of the Papuan fold and thrust belt, Papua New Guinea. *Journal of Structural Geology*, *100*, 98–119. https://doi.org/10.1016/j.jsg.2017.05.010

Malekzade, Z., Bellier, O., Abbassi, M. R., Shabanian, E., & Authemayou, C. (2016). The effects of plate margin inhomogeneity on the deformation pattern within west-Central Zagros fold-and-thrust belt. *Tectonophysics*, *693*, 304–326. doi:http://dx.doi.org/10.1016/j.tecto.2016.01.030

Malusà, M. G. (2004). *Post-metamorphic evolution of the western Alps: Kinematic constraints from a multidisciplinary approach.* Ph.D. dissertation. University of Turin, Turin, Italy.

Mancktelow, N., Zwingmann, H., Campani, M., Fügenschuh, B., & Mulch, A. (2015). Timing and conditions of brittle faulting on the Silltal-Brenner fault zone, Eastern Alps (Austria). *Swiss Journal of Geosciences*, *108*(2–3), 305–326. https://doi.org/10.1007/s00015-015-0179-y

Mancktelow, N. S. (1992). Neogene lateral extension during convergence in the Central Alps: Evidence from interrelated faulting and backfolding around the Simplonpass (Switzerland). *Tectonophysics*, *215*(3–4), 295–317. https://doi.org/10.1016/0040-1951(92)90358-D

Mancktelow, N. S., Stöckli, D. F., Grollimund, B., Müller, W., Fügenschuh, B., Viola, G., et al. (2001). The DAV and Periadriatic fault systems in the eastern Alps south of the Tauern Window. *International Journal of Earth Sciences*, *90*(3), 593–622. https://doi.org/10.1007/s005310000190

Manea, V. C., Pérez-Gussinyé, M., & Manea, M. (2012). Chilean flat slab subduction controlled by overriding plate thickness and trench rollback. *Geology*, *40*(1), 35–38. https://doi.org/10.1130/g32543.1

Marshak, S., & Wilkerson, M. S. (1992). Effect of overburden thickness on thrust belt geometry and development. *Tectonics*, *11*(3), 560–566. https://doi.org/10.1029/92TC00175

McGugan, A. (1987). "Horses" and transverse faults in the Lewis thrust sheet, Elk range, Kananaskis valley, Rocky mountain front ranges, Southwestern Alberta. *Bulletin of Canadian Petroleum Geology*, *35*(3), 358–361. https://doi.org/10.35767/gscpgbull.35.3.358

McMechan, M. (2001). Large-scale duplex structures in the McConnell thrust sheet, Rocky Mountains, Southwest Alberta. *Bulletin of Canadian Petroleum Geology*, *49*(3), 408–425. https://doi.org/10.2113/49.3.408

McMechan, M. (2007). Nature, origin and tectonic significance of anomalous transverse structures, southeastern Skeena Fold Belt, British Columbia. *Bulletin of Canadian Petroleum Geology*, *55*(4), 262–274. https://doi.org/10.2113/gscpgbull.55.4.262

McMechan, M. E. (2000). Walker Creek fault zone, central Rocky Mountains, British Columbia: Southern continuation of the Northern Rocky Mountain Trench fault zone. *Canadian Journal of Earth Sciences*, *37*(9), 1259–1273. https://doi.org/10.1139/e00-038

McMechan, M. E. (2012). Deep transverse basement structural control of mineral systems in the southeastern Canadian Cordillera. *Canadian Journal of Earth Sciences*, *49*(5), 693–708. https://doi.org/10.1139/E2012-013

McMechan, M. E., & Price, R. A. (1982). Transverse folding and superposed deformation, Mount Fisher area, southern Canadian Rocky Mountain thrust and fold belt. *Canadian*

Journal of Earth Sciences, *19*(5), 1011–1024. https://doi.org/10.1139/e82-084

McQuarrie, N. (2004). Crustal scale geometry of the Zagros fold-thrust belt, Iran. *Journal of Structural Geology*, *26*(3), 519–535. https://doi.org/10.1016/j.jsg.2003.08.009

McQuarrie, N., & van Hinsbergen, D. J. J. (2013). Retrodeforming the Arabia-Eurasia collision zone: Age of collision versus magnitude of continental subduction. *Geology*, *41*(3), 315–318. https://doi.org/10.1130/G33591.1

McQuarrie, N., Stock, J. M., Verdel, C., & Wernicke, B. P. (2003). Cenozoic evolution of Neotethys and implications for the causes of plate motions. *Geophysical Research Letters*, *30*(20). https://doi.org/10.1029/2003GL017992

McQuillan, H. (1991). The role of basement tectonics in the control of sedimentary facies, structural patterns and salt plug emplacements in the Zagros fold belt of southwest Iran. *Journal of Southeast Asian Earth Sciences*, *5*(1–4), 453–463. https://doi.org/10.1016/0743-9547(91)90061-2

Mendoza, M. M., Ghosh, A., Karplus, M. S., Klemperer, S. L., Sapkota, S. N., Adhikari, L. B., et al. (2019). Duplex in the Main Himalayan Thrust illuminated by aftershocks of the 2015 Mw 7.8 Gorkha earthquake. *Nature Geoscience*, *12*, 1018–1022. https://doi.org/10.1038/s41561-019-0474-8

Mescua, J. F., & Giambiagi, L. B. (2012). Fault inversion vs. new thrust generation: a case study in the Malargüe fold-and-thrust belt, Andes of Argentina. *Journal of Structural Geology*, *35*, 51–63. https://doi.org/10.1016/j.jsg.2011.11.011

Mitra, S. (1988). Three-dimensional geometry and kinematic evolution of the Pine Mountain thrust system, southern Appalachians. *Geological Society of America Bulletin*, *100*(1), 72–95. https://doi.org/10.1130/0016-7606(1988)100<0072:TDGAKE>2.3.CO;2

Moffat, I. W., & Spang, J. H. (1984). Origin of transverse faulting, Rocky Mountain Front Ranges, Canmore, Alberta. *Bulletin of Canadian Petroleum Geology*, *32*(2), 147–161. https://doi.org/10.35767/gscpgbull.32.2.147

Mogk, D. W., Mueller, P. A., & Wooden, J. L. (1992). The nature of Archean terrain boundaries: an example from the northern Wyoming Province. *Precambrian Research*, *55*(1–4), 155–168. https://doi.org/10.1016/0301-9268(92)90020-O

Monod, B., Dhont, D., & Hervouët, Y. (2010). Orogenic float of the Venezuelan Andes. *Tectonophysics*, *490*(1–2), 123–135. https://doi.org/10.1016/j.tecto.2010.04.036

Mouthereau, F., & Lacombe, O. (2006). Inversion of the Paleogene Chinese continental margin and thick-skinned deformation in the Western Foreland of Taiwan. *Journal of Structural Geology*, *28*(11), 1977–1993. https://doi.org/10.1016/j.jsg.2006.08.007

Mouthereau, F., Deffontaines, B., Lacombe, O., & Angelier, J. (2002). Variations along the strike of the Taiwan thrust belt: Basement control on structural style, wedge geometry, and kinematics. In T. Byrne & C.-S. Liu (Eds.), *Geology and geophysics of an arc-continent collision, Taiwan, Republic of China* (pp. 35–58). Boulder, Colorado: Geological Society of America Special Paper, 358.

Mouthereau, F., Lacombe, O., & Vergés, J. (2012). Building the Zagros collisional orogen: Timing, strain distribution and the dynamics of Arabia/Eurasia plate convergence. *Tectonophysics*, *532*, 27–60. https://doi.org/10.1016/j.tecto.2012.01.022

Mpodozis, C., & Ramos, V. (1989). The Andes of Chile and Argentina. In G. Ericksen, M. Cañas, & J. Reinemund (Eds.), *Geology of the Andes and its relation to hydrocarbon and mineral resources* (pp. 59–90). Circum-Pacific Council for Energy and Mineral Resources.

Mueller, P. A., Heatherington, A. L., Kelly, D. M., Wooden, J. L., & Mogk, D. W. (2002). Paleoproterozoic crust within the Great Falls tectonic zone: Implications for the assembly of southern Laurentia. *Geology*, *30*(2), 127–130. https://doi.org/10.1130/0091-7613(2002)030<0127:PCWTGF>2.0.CO;2

Mugnier, J. L., Jouanne, F., Bhattarai, R., Cortes-Aranda, J., Gajurel, A., Leturmy, P., Robert, X., Upreti, B., & Vassallo, R. (2017). Segmentation of the Himalayan megathrust around the Gorkha earthquake (25 April 2015) in Nepal. *Journal of Asian Earth Sciences*, *141*, 236–252. https://doi.org/10.1016/j.jseaes.2017.01.015

Mugnier, J. L., Leturmy, P., Mascle, G., Huyghe, P., Chalaron, E., Vidal, G., Husson, L., & Delcaillau, B. (1999). The Siwaliks of western Nepal 1: Geometry and kinematics. *Journal of Asian Earth Sciences*, *17*, 629–642. https://doi.org/10.1016/S1367-9120(99)00038-3

Mukherjee, S., Talbot, C. J., & Koyi, H. A. (2010). Viscosity estimates of salt in the Hormuz and Namakdan salt diapirs, Persian Gulf. *Geological Magazine*, *147*(4), 497–507. https://doi/org/10.1017/S001675680999077X

Mukul, M. (2010). First-order kinematics of wedge-scale active Himalayan deformation: insights from Darjiling-Sikkim-Tibet (DaSiT) wedge. *Journal of Asian Earth Sciences*, *39*, 645–657. https://doi.org/10.1016/j.jseaes.2010.04.029

Mukul, M., Jade, S., Ansari, K., Matin, A, & Joshi, V. (2018). Structural insights from geodetic Global Positioning System measurements in the Darjiling-Sikkim Himalaya. *Journal of Structural Geology*, *114*, 346–356. https://doi.org/10.1016/j.jsg.2018.03.007

Müller, W., Mancktelow, N. S., & Meier, M. (2000). Rb-Sr microchrons of synkinematic mica in mylonites: An example from the DAV fault of the Eastern Alps. *Earth and Planetary Science Letters*, *180*(3–4), 385–397. https://doi.org/10.1016/S0012-821X(00)00167-9

Müller, W., Prosser, G., Mancktelow, N. S., Villa, I. M., Kelley, S. P., Viola, G., et al. (2001). Geochronological constraints on the evolution of the Periadriatic Fault System (Alps). *International Journal of Earth Sciences*, *90*(3), 623–653. https://doi.org/10.1007/s005310000187

Murphy, J. B., & Keppie, J. D. (2005). The Acadian orogeny in the northern Appalachians. *International Geology Review*, *47*(7), 663–687. https://doi.org/10.2747/0020-6814.47.7.663

Murphy, J. B., Waldron, J. W., Kontak, D. J., Pe-Piper, G., & Piper, D. J. (2011). Minas fault zone: Late Paleozoic history of an intracontinental orogenic transform fault in the Canadian Appalachians. *Journal of Structural Geology*, *33*(3), 312–328. https://doi.org/10.1016/j.jsg.2010.11.012

Nabelek, J., Hetenyi, G., Vergne, J., Sapkota, S., Kafle, B., Jiang, M., et al. (2009). Underplating in the Himalaya-Tibet collision zone revealed by the Hi-CL IMB experiment. *Science*, *325*, 1371–1374. https://doi.org/10.1126/science.1167719

Norris, D. (2001). Slickenlines and the kinematics of the Crowsnest Deflection in the southern Rocky Mountains of

Canada. *Journal of Structural Geology*, *23*(6–7), 1089–1102. https://doi.org/10.1016/S0191-8141(00)00180-2

O'Brien, T. M., & van der Pluijm, B. A. (2012). Timing of Iapetus Ocean rifting from Ar geochronology of pseudotachylytes in the St. Lawrence rift system of southern Quebec. *Geology*, *40*(5), 443–446. doi:10.1130/g32691.1

Oldham, R. D. (1917). Structure of the Himalayas and Indo-Gangetic plains. *Memoir of the Geological Survey of India*, *156*.

Oldow, J. S., Bally, A. W., Lallemant, H. G. A., & Leeman, W. P. (1989). Phanerozoic evolution of the North American Cordillera, United States and Canada. In A. W. Bally & A. R. Palmer (Eds.), *The geology of North America: An overview* (pp. 139–232). Geological Society of America, The Geology of North America.

O'Neill, J. M., & Lopez, D. A. (1985). Character and regional significance of Great Falls tectonic zone, east-central Idaho and west-central Montana. *AAPG Bulletin*, *69*(3), 437–447. https://doi.org/10.1306/AD462506-16F7-11D7-8645000102C1865D

Oriolo, S., Japas, M. S., Cristallini, E. O., & Giménez, M. (2014). Cross-strike structures controlling magmatism emplacement in a flat-slab setting (Precordillera, Central Andes of Argentina). *Geological Society, London, Special Publications*, *394*(1), 113–127. https://doi.org/10.1144/sp394.6

Oswald, P., Ortner, H., & Gruber, A. (2019). Deformation around a detached half-graben shoulder during nappe stacking (Northern Calcareous Alps, Austria). *Swiss Journal of Geosciences*, *112*(1), 23–37. https://doi.org/10.1007/s00015-018-0333-4

Pantet, A., Epard, J.-L., & Masson, H. (2020). Mimicking Alpine thrusts by passive deformation of synsedimentary normal faults: A record of the Jurassic extension of the European margin (Mont Fort nappe, Pennine Alps). *Swiss Journal of Geosciences*, *113*(1), 13. doi:10.1186/s00015-020-00366-2

Pash, R. R., Sarkarinejad, K., Ghoochaninejad, H. Z., Motamedi, H., & Yazdani, M. (2020). Accommodation of the different structural styles in the foreland fold-and-thrust belts: Northern Dezful Embayment in the Zagros belt, Iran. *International Journal of Earth Sciences*, 1–12. https://doi.org/10.1007/s00531-020-01844-6

Paul, H., Mitra, S., Bhattacharya, S. N., & Suresh, G. (2015). Active transverse faulting within underthrust Indian crust beneath the Sikkim Himalaya. *Geophysics Journal International*, *201*(2), 1070–1081. https://doi.org/10.1093/gji/ggv058

Paulsen, T., & Marshak, S. (1997). Structure of the Mount Raymond transverse zone at the southern end of the Wyoming salient, Sevier fold-thrust belt, Utah. *Tectonophysics*, *280*(3–4), 199–211. https://doi.org/10.1016/S0040-1951(97)00205-9

Paulsen, T., & Marshak, S. (1998). Charleston transverse zone, Wasatch Mountains, Utah: Structure of the Provo salient's northern margin, Sevier fold-thrust belt. *Geological Society of America Bulletin*, *110*(4), 512–522. https://doi.org/10.1130/0016-7606(1998)110<0512:CTZWMU>2.3.CO;2

Paulsen, T., & Marshak, S. (1999). Origin of the Uinta recess, Sevier fold-thrust belt, Utah: Influence of basin architecture on fold-thrust belt geometry. *Tectonophysics*, *312*(2–4), 203–216. https://doi.org/10.1016/S0040-1951(99)00182-1

Perello, P., Delle Piane, L., Piana, F., Morelli, M., Damiano, A., & Venturini, G. (2004). New constraints on late to post-Oligocene deformation history of the Western Alps: Data from Middle Susa valley and High Maurienne valley. *Abstract, 32nd IGC-Firenze*, *247*.

Perez-Estaun, A., Alvarez-Marron, J., Brown, D., Puchkov, V., Gorozhanina, Y., & Baryshev, V. (1997). Along-strike structural variations in the foreland thrust and fold belt of the southern Urals. *Tectonophysics*, *276*(1–4), 265–280. https://doi.org/10.1016/S0040-1951(97)00060-7

Pérez-Flores, P., Cembrano, J., Sánchez-Alfaro, P., Veloso, E., Arancibia, G., & Roquer, T. (2016). Tectonics, magmatism and paleo-fluid distribution in a strike-slip setting: Insights from the northern termination of the Liquiñe-Ofqui fault system, Chile. *Tectonophysics*, *680*, 192–210. https://doi.org/10.1016/j.tecto.2016.05.016

Perrone, G., Cadoppi, P., Tallone, S., & Balestro, G. (2011). Post-collisional tectonics in the Northern Cottian Alps (Italian Western Alps). *International Journal of Earth Sciences*, *100*(6), 1349–1373. https://doi.org/10.1007/s00531-010-0534-1

Petrinovic, I., Riller, U., Brod, J., Alvarado, G., & Arnosio, M. (2006). Bimodal volcanism in a tectonic transfer zone: Evidence for tectonically controlled magmatism in the southern Central Andes, NW Argentina. *Journal of Volcanology and Geothermal Research*, *152*(3–4), 240–252. https://doi.org/10.1016/j.jvolgeores.2005.10.008

Pfiffner, O. A. (2014). *Geology of the Alps*. John Wiley & Sons.

Picotti, V., Casolari, E., Castellarin, A., Mosconi, A., Cairo, E., Pessina, C., et al. (1997). Structural evolution of the eastern Lombardian Prealps: Alpine inversion of a Mesozoic rifted margin. *Agip SpA, San Donato Milanese*, *102*.

Pilger, R. H. J. (1981). Plate reconstructions, aseismic ridges, and low-angle subduction beneath the Andes. *GSA Bbulletin*, *92*(7), 448–456. doi:10.1130/0016-7606(1981)92<448:Praral>2.0.Co;2

Pilkington, M., Miles, W., Ross, G., & Roest, W. (2000). Potential-field signatures of buried Precambrian basement in the Western Canada Sedimentary Basin. *Canadian Journal of Earth Sciences*, *37*(11), 1453–1471. https://doi.org/10.1139/e00-020

Piquer, J., Rivera, O., Yañez, G., & Oyarzun, N. (2020). The Piuquencillo Fault System: A long-lived, *Andean-transverse fault system and its relationship with magmatic and hydrothermal activity. Solid Earth Discussions*, 1–34. https://doi.org/10.5194/se-2020-142

Pizzi, A., & Galadini, F. (2009). Pre-existing cross-structures and active fault segmentation in the northern-central Apennines (Italy). *Tectonophysics*, *476*(1–2), 304–319. https://doi.org/10.1016/j.tecto.2009.03.018

Plint, H. E., & Jamieson, R. A. (1989). Microstructure, metamorphism and tectonics of the western Cape Breton Highlands, Nova Scotia. *Journal of Metamorphic Geology*, *7*(4), 407–424. https://doi.org/10.1111/j.1525–1314.1989.tb00606.x

Polinski, R. K., & Eisbacher, G. H. (1992). Deformation partitioning during polyphase oblique convergence in the Karawanken Mountains, southeastern Alps. *Journal of Structural Geology*, *14*(10), 1203–1213. https://doi.org/10.1016/0191-8141(92)90070-D

Pomella, H., Stipp, M., & Fügenschuh, B. (2012). Thermochronological record of thrusting and strike-slip faulting along the Giudicarie fault system (Alps, Northern Italy). *Tectonophysics*, *579*, 118–130. https://doi.org/10.1016/j.tecto.2012.04.015

Prey, S. (1989). Ein steilstehendes Störungssystem als Westbegrenzung des Tauernfensters. *Journal of Geologische Bundesanstalt*, *132*(4), 745–749.

Price, R. A. (1981). The Cordilleran foreland thrust and fold belt in the southern Canadian Rocky Mountains. *Geological Society, London, Special Publications*, *9*(1), 427–448. https://doi.org/10.1144/GSL.SP.1981.009.01.39

Pride, K. R., Lecouteur, P. C., & Mawer, A. B. (1986). Geology and mineralogy of the Aley carbonatite, Ospika River area, British Columbia. *Canadian Institute of Mining Bulletin*, *79*(891), 32–32.

Prosser, G. (1998). Strike-slip movements and thrusting along a transpressive fault zone: The North Giudicarie line (Insubric line, northern Italy). *Tectonics*, *17*(6), 921–937. https://doi.org/10.1029/1998TC900010

Raiverman, V., Kunte, S.V., & Mukherjea, A. (1983). Basin geometry, Cenozoic sedimentation and hydrocarbon in northwestern Himalaya and Indo-Gangetic plains. *Petroleum Geology of Asia Journal*, *6*(4), 67–92.

Rajaure, S., Sapkota, S. N., Adhikari, L. B., Koirala, B., Bhattarai, M., Tiwari, D. R., et al. (2013). Double difference relocation of local earthquakes in the Nepal Himalaya. *Journal of the Nepal Geological Society*, *46*, 133–142.

Ramos, V. A. (1988). The tectonics of the Central Andes (30°–33°S latitude). In S. Clark, D. Burchfiel, & D. Suppe (Eds.), *Processes in continental lithospheric deformation* (pp. 31–54). Geological Society of America Special Paper, 218.

Ramos, V. A., & Folguera, A. (2009). Andean flat-slab subduction through time. *Geological Society, London, Special Publications*, *327*(1), 31–54. doi:10.1144/sp327.3

Ramos, V. A., Cristallini, E. O., & Pérez, D. J. (2002). The Pampean flat-slab of the Central Andes. *Journal of South American Earth Sciences*, *15*(1), 59–78. https://doi.org/10.1016/S0895-9811(02)00006-8

Rankin, D. W. (1976). Appalachian salients and recesses: Late Precambrian continental breakup and the opening of the Iapetus Ocean. *Journal of Geophysical Research*, *81*(32), 5605–5619. https://doi.org/10.1029/JB081i032p05605

Rao, M. B. R. (1973). The subsurface geology of the Indo-Gangetic plains. *Journal of the Geological Society of India*, *14*(3), 217–242.

Ratschbacher, L., Merle, O., Davy, P., & Cobbold, P. (1991). Lateral extrusion in the Eastern Alps; Part 1: Boundary conditions and experiments scaled for gravity. *Tectonics*, *10*(2), 245–256. https://doi.org/10.1029/90TC02622

Regard, V., Bellier, O., Thomas, J.-C., Bourles, D., Bonnet, S., Abbassi, M., et al. (2005). Cumulative right-lateral fault slip rate across the Zagros-Makran transfer zone: Role of the Minab-Zendan fault system in accommodating Arabia-Eurasia convergence in southeast Iran. *Geophysical Journal International*, *162*(1), 177–203. https://doi.org/10.1111/j.1365-246X.2005.02558.x

Reiter, F., Freudenthaler, C., Hausmann, H., Ortner, H., Lenhardt, W., & Brandner, R. (2018). Active seismotectonic deformation in front of the Dolomites indenter, Eastern Alps. *Tectonics*, *37*(12), 4625–4654. https://doi.org/10.1029/2017TC004867

Reynolds, M. W. (1979). Character and extent of Basin-Range faulting, Western Montana and east-central Idaho. In G. W. Newman & H. D. Goode (Eds.), *Basin and range symposium and great basin field conference* (pp. 185–193). Denver: Rocky Mountain Association of Geologists.

Ribes, C., Manatschal, G., Ghienne, J.-F., Karner, G. D., Johnson, C. A., Figueredo, P. H., et al. (2019). The syn-rift stratigraphic record across a fossil hyper-extended rifted margin: The example of the northwestern Adriatic margin exposed in the Central Alps. *International Journal of Earth Sciences*, *108*(6), 2071–2095. https://doi.org/10.1007/s00531-019-01750-6

Ring, U. (1994). The kinematics of the late Alpine Muretto fault and its relation to dextral transpression across the Periadriatic Line. *Eclogae Geologicae Helvetiae*, *87*(3), 811–831.

Ring, U., & Gerdes, A. (2016). Kinematics of the Alpenrhein-Bodensee graben system in the Central Alps: Oligocene/Miocene transtension due to formation of the Western Alps arc. *Tectonics*, *35*(6), 1367–1391. https://doi.org/10.1002/2015TC004085

Rivers, T. (2008). Assembly and preservation of lower, mid, and upper orogenic crust in the Grenville Province: Implications for the evolution of large hot long-duration orogens. *Precambrian Research*, *167*(3), 237–259. https://doi.org/10.1016/j.precamres.2008.08.005

Rivers, T. (2012). Upper-crustal orogenic lid and mid-crustal core complexes: Signature of a collapsed orogenic plateau in the hinterland of the Grenville Province. *Canadian Journal of Earth Sciences*, *49*(1), 1–42. https://doi.org/10.1139/e11-014

Rivers, T. (2015). Tectonic setting and evolution of the Grenville Orogen: An assessment of progress over the last 40 years. *Geoscience Canada*, *42*(1), 77–124. doi:10.12789/geocanj.2014.41.057

Robertson, A. H. F., Clift, P. D., Degnan, P. J., & Jones, G. (1991). Palaeogeographic and palaeotectonic evolution of the Eastern Mediterranean Neotethys. *Palaeogeography, Palaeoclimatology, Palaeoecology*, *87*(1–4), 289–343. https://doi.org/10.1016/0031-0182(91)90140-M

Robinson, D. P., Das, S., & Watts, A. B. (2006). Earthquake rupture stalled by a subducting fracture zone. *Science*, *312*(5777), 1203–1205. https://doi.org/10.1126/science.1125771

Robl, J., & Stüwe, K. (2005). Continental collision with finite indenter strength: 2. *European Eastern Alps. Tectonics*, *24*(4). https://doi.org/10.1029/2004TC001741

Rodgers, J. (1990). Fold-and-thrust belts in sedimentary rocks. Part 1: Typical examples. *American Journal of Science*, *290*(4), 321–359. https://doi.org/10.2475/ajs.290.4.321

Ronemus, C. B., Orme, D. A., Campbell, S., Black, S. R., & Cook, J. (2020). Mesoproterozoic-Early Cretaceous provenance and paleogeographic evolution of the Northern Rocky Mountains: Insights from the detrital zircon record of the Bridger Range, Montana, USA. *GSA Bulletin*. https://doi.org/10.1130/b35628.1

Root, K. G. (1987). *Geology of the Delphine Creek area, southeastern British Columbia: Implications for the Proterozoic and Paleozoic development of the Cordilleran Divergent Margin*. Ph.D. dissertation. University of Calgary, Calgary, Alberta.

Roquer, T., Arancibia, G., Rowland, J., Iturrieta, P., Morata, D., & Cembrano, J. (2017). Fault-controlled development of shallow hydrothermal systems: Structural and mineralogical insights from the Southern Andes. *Geothermics*, *66*, 156–173. doi:http://dx.doi.org/10.1016/j.geothermics.2016.12.003

Rosenberg, C. L., & Garcia, S. (2011). Estimating displacement along the Brenner Fault and orogen-parallel extension in the Eastern Alps. *International Journal of Earth Sciences*, *100*(5), 1129–1145. https://doi.org/10.1007/s00531-011-0645-3

Rosenberg, C. L., Brun, J.-P., & Gapais, D. (2004). Indentation model of the Eastern Alps and the origin of the Tauern Window. *Geology*, *32*(11), 997–1000. https://doi.org/10.1130/G20793.1

Ross, G., Parrish, R., Villeneuve, M., & Bowring, S. (1991). Geophysics and geochronology of the crystalline basement of the Alberta Basin, western Canada. *Canadian Journal of Earth Sciences*, *28*(4), 512–522. https://doi.org/10.1139/e91-045

Royden, L., Horváth, F., Nagymarosy, A., & Stegena, L. (1983). Evolution of the Pannonian basin system: 2. Subsidence and thermal history. *Tectonics*, *2*(1), 91–137. https://doi.org/10.1029/TC002i001p00091

Royden, L. H. (1993). Evolution of retreating subduction boundaries formed during continental collision. *Tectonics*, *12*(3), 629–638. https://doi.org/10.1029/92TC02641

Sadeghi, S., & Yassaghi, A. (2016). Spatial evolution of Zagros collision zone in Kurdistan, NW Iran: Constraints on Arabia-Eurasia oblique convergence. *Solid Earth*, *7*(2), 659–659. https://doi.org/10.5194/se-7-659-2016

Sahoo, P. K., Kumar, S., & Singh, R. P. (2000). Neotectonic study of Ganga and Yamuna tear faults, NW Himalaya, using remote sensing and GIS. *International Journal of Remote Sensing*, *21*(3), 499–518. https://doi.org/10.1080/014311600210713

Saleeby, J. B. (2003). Segmentation of the Laramide slab: Evidence from the southern Sierra Nevada region. *Geological Society of America Bulletin*, *115*(6), 655–668. https://doi.org/10.1130/0016-7606(2003)115<0655:SOTLSF>2.0.CO;2

Saleeby, J. B., Busby-Spera, C., Oldow, J. S., Dunne, G. C., Wright, J. E., Cowan, D. S., et al. (1992). Early Mesozoic tectonic evolution of the western US Cordillera. In B. C. Burchfiel, W. Lipman, & M. L. Zoback (Eds.), *The Cordilleran Orogen, conterminous U.S., the geology of North America G3* (pp. 107–168). Geological Society of America.

Salfity, J. A., & Marquillas, R. A. (1994). Tectonic and sedimentary evolution of the Cretaceous-Eocene Salta Group basin, Argentina. In J. A. Salfity (Ed.), *Cretaceous tectonics of the Andes* (pp. 266–315). Vieweg+Teubner Verlag, Wiesbaden.

Sarkarinejad, K., Pash, R. R., Motamedi, H., & Yazdani, M. (2018). Deformation and kinematic evolution of the subsurface structures: Zagros foreland fold-and-thrust belt, northern Dezful Embayment, Iran. *International Journal of Earth Sciences*, *107*(4), 1287–1304. https://doi.org/10.1007/s00531-017-1532-3

Sastri, V. V., Bhandari, L. L., Raju, A. T. R., & Datta, A. K. (1971). Tectonic framework and subsurface stratigraphy of the Ganga Basin. *Journal of The Geological Society of India*, *12*(3), 222–233.

Satolli, S., Pace, P., Viandante, M. G., & Calamita, F. (2014). Lateral variations in tectonic style across cross-strike discontinuities: An example from the Central Apennines belt (Italy). *International Journal of Earth Sciences*, *103*(8), 2301–2313. https://doi.org/10.1007/s00531-014-1052-3

Schepers, G., Van Hinsbergen, D. J., Spakman, W., Kosters, M. E., Boschman, L. M., & McQuarrie, N. (2017). South-American plate advance and forced Andean trench retreat as drivers for transient flat subduction episodes. *Nature Communications*, *8*, 15249. https://doi.org/10.1038/ncomms15249

Schmid, S. M., Aebli, H. R., Heller, F., & Zingg, A. (1989). The role of the Periadriatic Line in the tectonic evolution of the Alps. *Geological Society, London, Special Publications*, *45*(1), 153–171. https://doi.org/10.1144/gsl.Sp.1989.045.01.08

Schmid, S. M., Fügenschuh, B., Kissling, E., & Schuster, R. (2004). Tectonic map and overall architecture of the Alpine orogen. *Eclogae Geologicae Helvetiae*, *97*(1), 93–117. https://doi,org/10.1007/s00015-004-1113-x

Schmid, S. M., Pfiffner, O.-A., Froitzheim, N., Schönborn, G., & Kissling, E. (1996). Geophysical-geological transect and tectonic evolution of the Swiss-Italian Alps. *Tectonics*, *15*(5), 1036–1064. doi: https://doi.org/10.1029/96TC00433

Schönborn, G. (1992). Kinematics of a transverse zone in the Southern Alps, Italy. In K. R. McClay (Ed.), *Thrust tectonics* (pp. 299–310). Dordrecht: Springer Netherlands.

Schumacher, M. E. (1990). Alpine basement thrusts in the Eastern Seengebirge, Southern Alps (Italy Switzerland). *Eclogae Geologicae Helvetiae*, *83*(3), 645–663.

Scrocca, D. (2006). Thrust front segmentation induced by differential slab retreat in the Apennines (Italy). *Terra Nova*, *18*(2), 154–161. https://doi.org/10.1111/j.1365-3121.2006.00675.x

Searle, M. P., Simpson, R. L., Law, R. D., Parrish, R. R., & Waters, D. J. (2003). The structural geometry, metamorphic and magmatic evolution of the Everest massif, High Himalaya of Nepal–South Tibet. *Journal of the Geological Society*, *160*(3), 345–366. https://doi.org/10.1144/0016-764902-126

Sears, J. W., & Hendrix, M. S. (2004). Lewis and Clark line and the rotational origin of the Alberta and Helena salients, North American Cordillera. In A. Sussman & A. Weil (Eds.), *Orogenic curvature: Integrating Paleomagnetic and structural analyses* (pp. 173–186). Geological Society of America Special Paper, 383.

Sears, J. W., Hendrix, M., Waddell, A., Webb, B., Nixon, B., King, T., et al. (2000). Structural and stratigraphic evolution of the Rocky Mountain foreland basin in central-western Montana. In S. Roberts & D. Winston (Eds.), *Geologic field trips, western Montana and adjacent areas* (pp. 131–156). Rocky Mountain Section Geological Society of America Annual Meeting. Missoula: University of Montana.

Seifert, N. (2019). *Structural analysis of the Benkar Fault Zone, a cross structure in the higher Himalaya of the Khumbu Region, eastern Nepal*. MS. Montana State University, Bozeman, Montana.

Séjourné, S., & Malo, M. (2007). Pre-, syn-, and post-imbrication deformation of carbonate slices along the southern Quebec Appalachian front: Implications for hydrocarbon exploration. *Canadian Journal of Earth Sciences*, *44*(4), 543–564. https://doi.org/10.1139/E06-106

Selverstone, J. (1988). Evidence for east-west crustal extension in the Eastern Alps: Implications for the unroofing history of the Tauern Window. *Tectonics*, *7*(1), 87–105. https://doi.org/10.1029/TC007i001p00087

Sepehr, M., & Cosgrove, J. W. (2004). Structural framework of the Zagros fold-thrust belt, Iran. *Marine and Petroleum Geology*, *21*(7), 829–843. https://doi.org/10.1016/j.marpetgeo.2003.07.006

Sepehr, M., Cosgrove, J., & Moieni, M. (2006). The impact of cover rock rheology on the style of folding in the Zagros fold-thrust belt. *Tectonophysics*, *427*(1–4), 265–281. https://doi.org/10.1016/j.tecto.2006.05.021

Sielfeld, G., Ruz, J., Brogi, A., Cembrano, J., Stanton-Yonge, A., Pérez-Flores, P., et al. (2019). Oblique-slip tectonics in an active volcanic chain: A case study from the Southern Andes. *Tectonophysics*, *770*, 228221. https://doi.org/10.1016/j.tecto.2019.228221

Silver, C. R. P., Murphy, M. A., Taylor, M. H., Gosse, J., & Baltz, T. (2015). Neotectonics of the Western Nepal Fault System: Implications for Himalayan strain partitioning. *Tectonics*, *34*(12), 2494–2513. https://doi.org/10.1002/2014TC003730

Simony, P. S., & Carr, S. D. (1997). Large lateral ramps in the Eocene Valkyr shear zone: Extensional ductile faulting controlled by plutonism in southern British Columbia. *Journal of Structural Geology*, *19*(6), 769–784. https://doi.org/10.1016/S0191-8141(97)00011-4

Sims, P. K., Bankey, V., & Finn, C. (2001). *Preliminary Precambrian basement map of Colorado: A geologic interpretation of the aeromagnetic map*. U.S. Geological Survey.

Skourlis, K., & Doutsos, T. (2003). The Pindos fold-and-thrust belt (Greece): Inversion kinematics of a passive continental margin. *International Journal of Earth Sciences*, *92*(6), 891–903. https://doi.org/10.1007/s00531-003-0365-4

Sobornov, K. O. (1996). Lateral variations in structural styles of tectonic wedging in the northeastern Caucasus, Russia. *Bulletin of Canadian Petroleum Geology*, *44*(2), 385–399.

Sorgi, C., Deffontaines, B., Hippolyte, J. C., & Cadet, J. P. (1998). An integrated analysis of transverse structures in the northern Apennines, Italy. *Geomorphology*, *25*(3–4), 193–206. https://doi.org/10.1016/S0169-555X(98)00041-5

Soucy La Roche, R., & Godin, L. (2019). Inherited cross-strike faults and Oligocene-early Miocene segmentation of the main Himalayan thrust, West Nepal. *Journal of Geophysical Research Solid Earth*, *124*, 7429–7444. https://doi.org/10.1029/2019JB017467

Southworth, C. S. (1986). Side-looking airborne radar image map showing cross-strike structural discontinuities and lineaments of the central Appalachians. *In Miscellaneous Field Studies Map.* https://doi.org/10.3133/mf1891

Sperner, B., Ratschbacher, L., & Nemčok, M. (2002). Interplay between subduction retreat and lateral extrusion: Tectonics of the Western Carpathians. *Tectonics*, *21*(6), 1-1–1-24. https://doi.org/10.1029/2001TC901028

Srivastava, V., Mukul, M., Barnes, J. B., & Mukul, M. (2018). Geometry and kinematics of main frontal thrust-related fault propagation folding in the Mohand Range, northwest Himalaya. *Journal of Structural Geology*, *115*, 1–18. https://doi.org/10.1016/j.jsg.2018.06.022

Stampfli, G. M., & Borel, G. D. (2002). A plate tectonic model for the Paleozoic and Mesozoic constrained by dynamic plate boundaries and restored synthetic oceanic isochrons. *Earth and Planetary Science Letters*, *196*(1), 17–33. https://doi.org/10.1016/S0012-821X(01)00588-X

Stanton-Yonge, A., Griffith, W., Cembrano, J., St. Julien, R., & Iturrieta, P. (2016). Tectonic role of margin-parallel and margin-transverse faults during oblique subduction in the Southern Volcanic Zone of the Andes: Insights from boundary element modeling. *Tectonics*, *35*(9), 1990–2013. https://doi.org/10.1002/2016TC004226

Steck, A. (2008). Tectonics of the Simplon massif and Lepontine gneiss dome: Deformation structures due to collision between the underthrusting European plate and the Adriatic indenter. *Swiss Journal of Geosciences*, *101*(2), 515–546. https://doi.org//10.1007/s00015-008-1283-z

Stocklin, J. (1968). Structural history and tectonics of Iran: A review. *AAPG Bulletin*, *52*(7), 1229–1258. https://doi.org/10.1306/5D25C4A5-16C1-11D7-8645000102C1865D

Sue, C., & Tricart, P. (2003). Neogene to ongoing normal faulting in the inner western Alps: A major evolution of the late alpine tectonics. *Tectonics*, *22*(5). https://doi.org/10.1029/2002TC001426

Talbot, C. J., & Alavi, M. (1996). The past of a future syntaxis across the Zagros. *Geological Society, London, Special Publications*, *100*(1), 89–109.

Tavarnelli, E., Butler, R. W. H., Decandia, F. A., Calamita, F., Grasso, M., Alvarez, W., et al.(2004). *Implications of fault reactivation and structural inheritance in the Cenozoic tectonic evolution of Italy*. Paper presented at the International Geological Congress 32.

Tchalenko, J., & Braud, J. (1974). Seismicity and structure of the Zagros (Iran): The main recent fault between 33 and 35 N. *Philosophical Transactions of the Royal Society of London. Series A, Mathematical and Physical Sciences*, *277*(1262), 1–25. https://doi.org/10.1098/rsta.1974.0044

Thomas, W. A. (1977). Evolution of Appalachian-Ouachita salients and recesses from reentrants and promontories in the contineral margin. *American Journal of Science*, *277*, 1233–1278. https://doi.org/10.2475/ajs.277.10.1233

Thomas, W. A. (1990). Controls on locations of transverse zones in thrust belts. *Eclogae Geologicae Helvetiae*, *83*(3), 727–744.

Thomas, W. A. (1991). The Appalachian-Ouachita rifted margin of southeastern North America. *Geological Society of America Bulletin*, *103*(3), 415–431. https://doi.org/10.1130/0016-7606(1991)103<0415:TAORMO>2.3.CO;2

Thomas, W. A. (2007). Role of the Birmingham basement fault in thin-skinned thrusting of the Birmingham anticlinorium, Appalachian thrust belt in Alabama. *American Journal of Science*, *307*(1), 46–62. https://doi.org/10.2475/01.2007.03

Thomas, W. A. (2014). A mechanism for tectonic inheritance at transform faults of the Iapetan margin of Laurentia. *Geoscience Canada*, 321–344. doi:http://dx.doi.org/10.12789/geocanj.2014.41.048

Thomas, W. A., & Bayona, G. (2002). Palinspastic restoration of the Anniston transverse zone in the Appalachian thrust belt, Alabama. *Journal of Structural Geology*, *24*(4), 797–826. https://doi.org/10.1016/S0191-8141(01)00117-1

Thompson, R. I. (1989). Stratigraphy, tectonic evolution and structural analysis of the Halfway River map area (94B), northern Rocky Mountains, British Columbia. *Geological Survey of Canada*, *425*, 1–119.

Thomson, S. N. (2002). Late Cenozoic geomorphic and tectonic evolution of the Patagonian Andes between latitudes 42°S and 46° S: An appraisal based on fission-track results from the transpressional intra-arc Liquiñe-Ofqui fault zone. *Geological Society of America Bulletin*, *114*(9), 1159–1173. https://doi.org/10.1130/0016-7606(2002)114<1159:LCGATE>2.0.CO;2

Tricart, P. (1984). From passive margin to continental collision: A tectonic scenario for the Western Alps. *American Journal of Science*, *284*(2), 97–120. https://doi.org/10.2475/ajs.284.2.97

Tull, J. F., & Holm, C. S. (2005). Structural evolution of a major Appalachian salient-recess junction: Consequences of oblique collisional convergence across a continental margin transform fault. *Geological Society of America Bulletin*, *117*(3–4), 482–499. https://doi.org/10.1130/B25578.1

Urbanek, C., Frank, W., Grasemann, B., & Decker, K. (2002). *Eoalpine versus Tertiary deformation: Dating of heterogeneously partitioned strain (Tauern Window, Austria). Pangeo Austria: Erdwissenschaften in Osterreich 28.–30.6. 2002*. Institute for geology and paleontology University of Salzburg.

Valdiya, K. S. (1976). Himalayan transverse faults and folds and their parallelism with subsurface structures of North Indian plains. *Tectonophysics*, *32*, 353–386. https://doi.org/10.1016/0040-1951(76)90069-X

Valensise, G., & Pantosti, D. (2001). The investigation of potential earthquake sources in peninsular Italy: A review. *Journal of Seismology*, *5*(3), 287–306. https://doi.org/10.1023/A:1011463223440

Van der Beek, P., Litty, C., Baudin, M., Mercier, J., Robert, X., & Hardwick, E. (2016). Contrasting tectonically driven exhumation and incision patterns, western versus central Nepal Himalaya. *Geology*, *44*(4), 327–330. https://doi.org10.1130/G37579.1

Van Staal, C. R., Whalen, J. B., McNicoll, V. J., Pehrsson, S., Lissenberg, C. J., Zagorevski, et al. (2007). The Notre Dame arc and the Taconic orogeny in Newfoundland. In R. D. Hatcher Jr., M. P. Carlson, J. H. McBride, & J. R. Martinez Catalán (Eds.), *4-D framework of continental crust* (pp. 511–552). Geological Society of America Memoir, 200.

Vargas, G., Rebolledo, S., Sepúlveda, S. A., Lahsen, A., Thiele, R., Townley, B., et al. (2013). Submarine earthquake rupture, active faulting and volcanism along the major Liquiñe-Ofqui Fault Zone and implications for seismic hazard assessment in the Patagonian Andes. *Andean Geology*, *40*(1), 141–171. https://doi.org/10.5027/andgeoV40n1-a07

Vergani, G., Tankard, J., Belotti, J., & Welsink, J. (1995). Tectonic evolution and paleogeography of the Neuquén Basin, Argentina. In A. Tankard, R. Suárez, & H. Welsink (Eds.), *Petroleum basins of South America* (vol. *62*). American Association of Petroleum Geologists.

Vernant, P., & Chéry, J. (2006). Mechanical modelling of oblique convergence in the Zagros, Iran. *Geophysical Journal International*, *165*(3), 991–1002. https://doi.org/10.1111/j.1365-246X.2006.02900.x

Vernant, P., Nilforoushan, F., Hatzfeld, D., Abbassi, M., Vigny, C., Masson, F., et al. (2004). Present-day crustal deformation and plate kinematics in the Middle East constrained by GPS measurements in Iran and northern Oman. *Geophysical Journal International*, *157*(1), 381–398. https://doi.org/10.1111/j.1365-246X.2004.02222.x

Viola, G., Mancktelow, N. S., & Seward, D. (2001). Late Oligocene-Neogene evolution of Europe-Adria collision: New structural and geochronological evidence from the Giudicarie fault system (Italian Eastern Alps). *Tectonics*, *20*(6), 999–1020. https://doi.org/10.1029/2001tc900021

Waldron, J. W., Barr, S. M., Park, A. F., White, C. E., & Hibbard, J. (2015). Late Paleozoic strike-slip faults in maritime Canada and their role in the reconfiguration of the northern Appalachian orogen. *Tectonics*, *34*(8), 1661–1684. https://doi.org/10.1002/2015TC003882

Wallace, C. A., Lidke, D. J., & Schmidt, R. G. (1990). Faults of the central part of the Lewis and Clark line and fragmentation of the Late Cretaceous foreland basin in west-central Montana. *Geological Society of America Bulletin*, *102*(8), 1021–1037. https://doi.org/10.1130/0016-7606(1990)102<1021:FOTCPO>2.3.CO;2

Weil, A. B., Yonkee, A., & Schultz, M. (2016). Tectonic evolution of a Laramide transverse structural zone: Sweetwater Arch, south central Wyoming. *Tectonics*, *35*(5), 1090–1120. https://doi.org/10.1002/2016TC004122

Weinberg, R. F. (2016). Himalayan leucogranites and migmatites: nature, timing and duration of anatexis. *Journal of Metamorphic Geology*, *34*(8), 821–843. https://doi.org/10.1111/jmg.12204

West, D. P., Jr., & Hubbard, M. S. (1997). Progressive localization of deformation during exhumation of a major strike-slip shear zone: Norumbega fault zone, south-central Maine, USA. *Tectonophysics*, *273*(3–4), 185–201. https://doi.org/10.1016/S0040-1951(96)00306-X

Wheeler, R. L. (1980). Cross-strike structural discontinuities: Possible exploration tool for natural gas in Appalachian overthrust belt. *AAPG Bulletin*, *64*(12), 2166–2178. https://doi.org/10.1306/2F91975B-16CE-11D7-8645000102C1865D

Whisner, S. C., Schmidt, C. J., & Whisner, J. B. (2014). Structural analysis of the Lombard thrust sheet and adjacent areas in the Helena salient, southwest Montana, USA. *Journal of Structural Geology*, *69*, 351–376. http://doi.org/10.1016/j.jsg.2014.08.006

Whitmeyer, S. J., & Karlstrom, K. E. (2007). Tectonic model for the Proterozoic growth of North America. *Geosphere*, *3*(4), 220–259. https://doi.org/10.1130/GES00055.1

Williams, H., & Cawood, P. (1986). Relationships along the eastern margin of the Humber Arm allochthon between Georges Lake and Corner Brook, Newfoundland. *Current Research, Part A. Geological Survey of Canada, Paper*, *86*, 759–765.

Wooden, J., & Mueller, P. (1988). Pb, Sr, and Nd isotopic compositions of a suite of Late Archean, igneous rocks, eastern Beartooth Mountains: Implications for crust-mantle evolution. *Earth and Planetary Science Letters*, *87*(1–2), 59–72. https://doi.org/10.1016/0012-821X(88)90064-7

Yeats, R. S., & Lillie, R. J. (1991). Contemporary tectonics of the Himalayan frontal fault system: Folds, blind thrusts and the 1905 Kangra earthquake. *Journal of Structural Geology*, *13*(2), 215–225.

Yonkee, W. A., & Weil, A. B. (2015). Tectonic evolution of the Sevier and Laramide belts within the North American Cordillera orogenic system. *Earth-Science Reviews*, *150*, 531–593. https://doi.org/10.1016/j.earscirev.2015.08.001

Zanchetta, S., D'Adda, P., Zanchi, A., Barberini, V., & Villa, I. M. (2011). Cretaceous-Eocene compression in the central Southern Alps (N Italy) inferred from 40Ar/39Ar dating of pseudotachylytes along regional thrust faults. *Journal of Geodynamics*, *51*(4), 245–263. https://doi.org/10.1016/j.jog.2010.09.004

Zanchi, A., D'Adda, P., Zanchetta, S., & Berra, F. (2012). Synthrust deformation across a transverse zone: The Grem-Vedra fault system (central Southern Alps, N Italy). *Swiss Journal of Geosciences*, *105*(1), 19–38. https://doi.rog/10.1007/s00015-011-0089-6

Zerlauth, M., Ortner, H., Pomella, H., Pfiffner, O. A., & Fügenschuh, B. (2014). Inherited tectonic structures controlling the deformation style: An example from the Helvetic nappes of the Eastern Alps. *Swiss Journal of Geosciences*, *107*(2–3), 157–175. https://doi.org/10.1007/s00015-014-0167-7

Zhao, W., Nelson, K. D., & Project INDEPTH Team (1993). Deep seismic reflection evidence for continental underthrusting beneath southern Tibet. *Nature Geoscience*, *366*, 557–559. https://doi.org/10.1038/366557a0

4

A Review of the Dynamics of Subduction Zone Initiation in the Aegean Region

Elizabeth J. Catlos[1] and İbrahim Çemen[2]

ABSTRACT

The Hellenic arc, where the African (Nubian) slab subducts beneath the Aegean and Anatolian microplates, is a type-locality for understanding subduction dynamics. The subducting African slab is the driver for extension in the Aegean and Anatolian microplates and plays a significant role in accommodating the Anatolian microplate's westward extrusion. The Hellenic arc subduction zone initiation (SZI) age is central for deciphering ancient mantle flow, how plate tectonics is maintained, and mechanisms that triggered the onset of subduction. The SZI for the Hellenic arc is debated. A Late Cretaceous-Jurassic SZI age is proposed using tomography and timing of obducted ophiolite fragments thought to be related to the system. Alternatively, a Late Cenozoic (Eocene-Pliocene) SZI is proposed using the analysis of topography combined with estimates of slab age and depth, paleomagnetism, the timing of metamorphism, volcanic activity, and timing of sedimentation within its accretionary wedge. The younger SZI age is consistent with an induced-transference model, where a new subduction zone initiates following the jamming of an older one. The older SZI suggests induced-transference fails, and a single subduction zone persists. The presence of a long-lived subduction zone has implications for characterizing Earth's mantle dynamics and how plate tectonics operates.

4.1. INTRODUCTION

Subduction zones form when two lithospheric plates converge, and one plate abruptly descends beneath the other (Figs. 4.1 and 4.2) (e.g., White et al., 1970; Hayes, 2018; Stern & Gerya, 2018; Crameri et al., 2020). Large magnitude earthquakes, tsunamis, volcanic eruptions, and landslides occur near and are caused by this specific plate boundary. They are considered exceptional geological environments for recording significant ground-level changes that can trigger tsunamis and impact ground motion and climate change. Earthquakes that occur in such zones and those triggered by the subduction process far afield have global consequences. Understanding the dynamics of subduction zones involves diverse and multidisciplinary studies, which are critical for understanding their associated hazards and how they have influenced the dynamics of plate tectonics over Earth's history (e.g., Stern, 2004; Gerya, 2011; Le Pichon et al., 2019; Crameri et al., 2020).

The Aegean and Anatolian microplates (Fig. 4.1) are significantly impacted by the dynamics of the subducting northern portion of the African (Nubian) plate, which has emerged as the primary driver for extension and the development of metamorphic core complexes in the Aegean region (e.g., Jolivet & Faccenna, 2000; Çemen et al., 2006; Dilek & Sandvol, 2009; van Hinsbergen et al., 2010). The Hellenic and Cyprus arcs are the surface expression of the subducting Nubian plate and eastern

[1] *Jackson School of Geosciences, Department of Geological Sciences, The University of Texas at Austin, Austin, Texas, USA*

[2] *Department of Geological Sciences, The University of Alabama, Tuscaloosa, Alabama, USA*

Compressional Tectonics: Plate Convergence to Mountain Building, Geophysical Monograph 277, First Edition.
Edited by Elizabeth J. Catlos and İbrahim Çemen.

DOI:10.1002/9781119773856.ch04

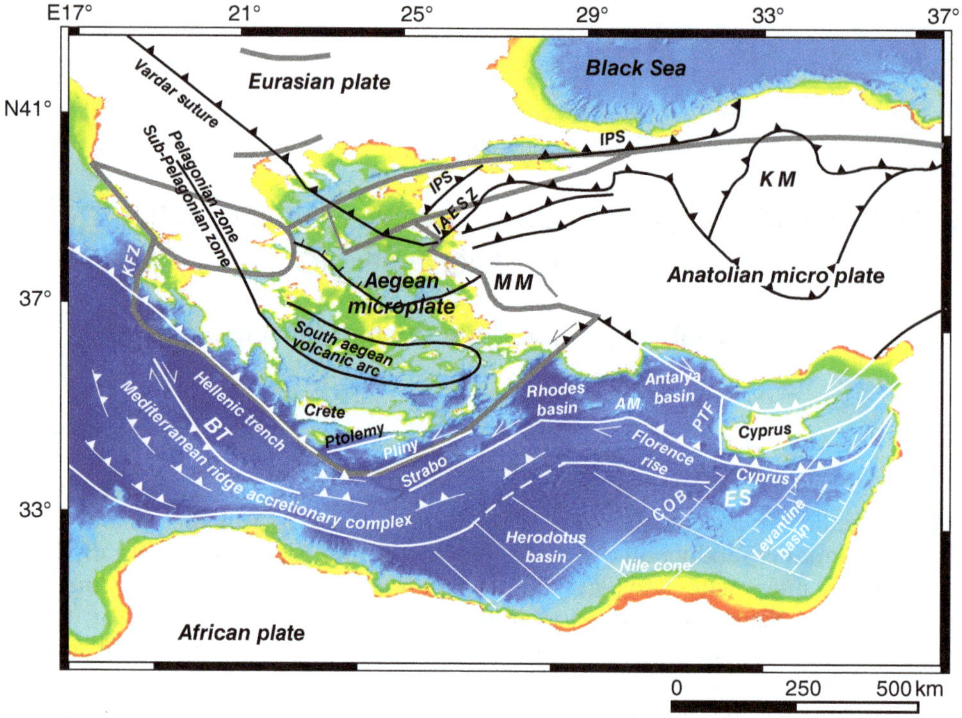

Figure 4.1 EMODnet Digital Bathymetry maps with some structures overlain. The Aegean and Anatolian microplate boundaries are shown in grey after Nyst and Thatcher (2004). Other structures after Hall et al. (1984) and (2009), Woodside et al. (2002), Peterek and Schwarze (2004), Meier et al. (2007), Kinnaird and Robertson (2012), and Symeou et al. (2018). Note: AM= Anaximander Mountains; BT= Backthrust; ES = Eratosthenes Seamount; IAESZ = Izmir-Ankara-Erzincan Suture Zone; IPS= Intra-Pontide Suture; KFZ = Kephalonia Fault Zone; KM= KirŞehir Massif; MP = Methana Peninsula; MM= Menderes Massif; NHSZ = North Hellenic Shear Zone; PTF = Paphos transform fault; SHSZ = South Hellenic Shear Zone.

Mediterranean lithosphere beneath the Aegean and Anatolian microplates, respectively (e.g., Le Pichon & Angelier, 1979; Angelier et al., 1982; Anastasakis & Keling, 1991; Papazachos et al., 2000; Ergün et al., 2005; Ganas & Parsons, 2009; Hall et al., 2009; Biryol et al., 2011; Royden & Papanikolaou, 2011; Hall et al., 2014; Symeou et al., 2018; Ventouzi et al., 2018).

Constraints regarding the subduction zone initiation (SZI) age of the present-day expression of the Hellenic arc developed from several independent approaches, including the timing of sedimentation within the intensely folded and faulted rocks of the Mediterranean Ridge accretionary prism (Figs. 4.1 and 4.2), paleomagnetism, the analysis of topography combined with estimates of slab age and depth, reconstructions of subducted slabs using tomography, and the timing of metamorphism and volcanic activity. SZI is the onset of downward plate motion forming a new slab, which later evolves into a self-sustaining subduction zone (Crameri et al., 2020). Some studies suggest a Cenozoic SZI age for the Hellenic arc, although estimates vary significantly, from the Eocene-Pliocene (e.g., Meulenkamp et al., 1988; Spakman et al., 1988; Papadopoulos, 1997; Brun & Sokoutis, 2010; Le Pichon et al., 2019) to Mesozoic (Late Cretaceous-Jurassic; Faccenna et al., 2003; van Hinsbergen et al., 2005; Royden & Papanikolaou, 2011; Jolivet et al., 2013; Malandri et al., 2017; Crameri et al., 2020; van Hinsbergen et al., 2021).

The disparity in the SZI age of onset of Nubian slab subduction along the Hellenic arc is significant as it impacts the tectonic history of the entire Aegean-Anatolian region, one of the most rapidly deforming regions across the Alpine-Himalayan chain. The region has emerged as the type-locality for understanding subduction zone dynamics, including slab tear, slab fragments, drips, and the role of transfer zones triggered by subduction. Understanding its SZI is also critical in deciphering ancient mantle flow, how plate tectonics is maintained, and the mechanisms involved in triggering the onset of subduction, among other factors (e.g., Crameri et al., 2020; van Hinsbergen et al., 2021). This chapter aims to summarize the approaches and results of studies that strive to constrain the SZI age of the African (Nubian) slab beneath the Aegean microplate that led to the formation of the Hellenic arc.

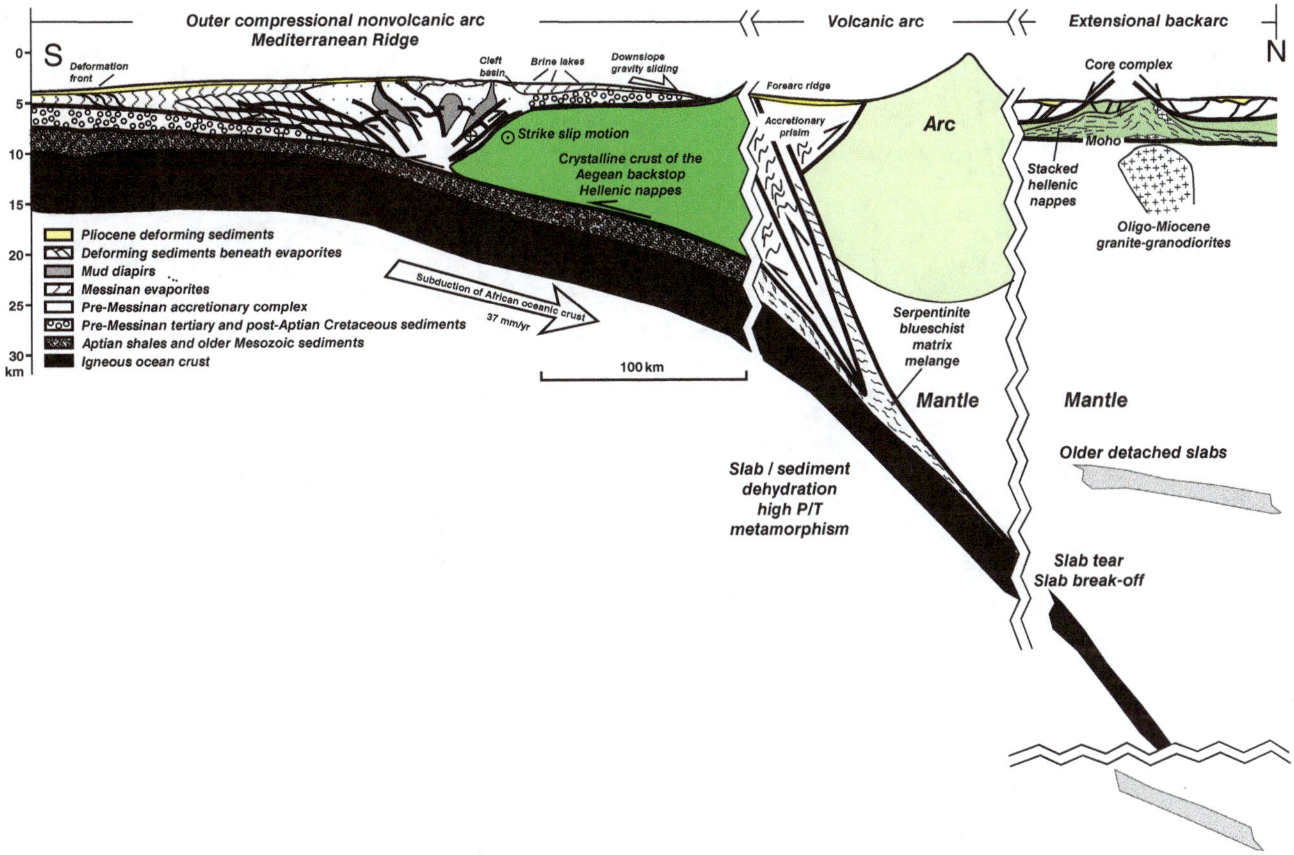

Figure 4.2 North-south generalized cross section through the Hellenic arc system showing the key structural elements. Map of the Mediterranean Ridge after Westbrook and Reston (2002).

4.2. GEOMETRY OF THE HELLENIC ARC (GREECE TO WESTERN TURKEY)

4.2.1. Definitions

The Hellenic subduction system extends ~1,200 km from approximately 37.5°N, 20.0°E offshore the island of Zakynthos to 36.0°N, 29.0°E offshore the island of Rhodes (Ganas & Parsons, 2009; Le Pichon et al., 2019; Papanikolaou, 2021) (Figs. 4.2 and 4.3). The system defines the boundary between the northern portion of the Nubian plate and the southern extent of the Aegean microplate within the central Mediterranean region (Pearce et al., 2012) and is sometimes referred to as the Aegean subduction zone (Wortel et al., 1990; Biryol et al., 2011; Bleier et al., 2007; Polat & Ozel, 2007; Taymaz et al., 2007; Crameri et al., 2020). This boundary between the Aegean microplate portion of Eurasia (Nyst & Thatcher, 2004) and the subducting Nubian slab is presently characterized by a strong curvature and fast slab rollback (e.g., Faccena et al., 2013; Evangelidis, 2017). Presently, the African plate advances toward Eurasia north-northwest at a rate of 5 mm/yr (Fernandes et al., 2006; Ganas & Parsons, 2009), but it subducts northward beneath Crete at a significantly faster rate of 35 mm/yr (McKenzie, 1972; Reilinger et al., 2006). The Aegean area also records the highest deformation rate along the entire Africa/Eurasia convergence zone (McClusky et al., 2000; Kassaras et al., 2005).

The Aegean and Anatolia microplates are sometimes classified as the single Aegean-Anatolian microplate (e.g., Jackson, 1994; Oral, 1995; Doutsos & Kokkalas, 2001; Le Pichon et al., 1995) with a Euler pole located north of the Sinai Peninsula (Cianetti et al., 2001). The Anatolian microplate itself is a distinct entity that includes over two-thirds of the country of Turkey (Fig. 4.1; Le Pichon et al., 1995; Oral et al., 1995; Reilinger et al., 2010; Papazachos, 1999). Şengör & Zabcı (2019) consider the whole of Turkey and the Balkan Peninsula within a plate boundary zone.

The Nubian plate includes the African continent. When Somalia is part of the definition, it is referred to as the African plate or the Nubia-Somalia plate (e.g., McCluksy et al., 2003). The African plate itself is defined by Nubia

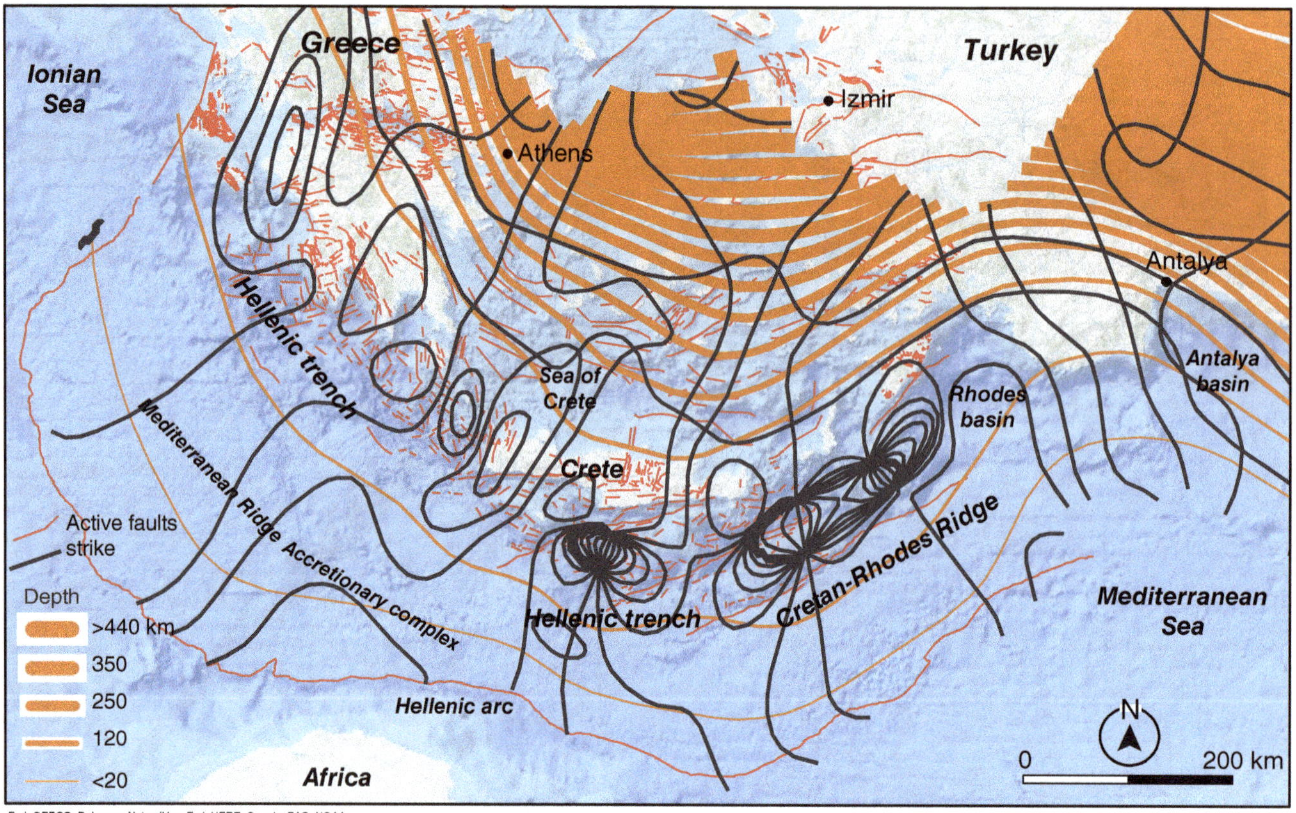

Figure 4.3 Map of the geometry of the Hellenic and part of the Cyprus subduction zones after Hayes (2018). The figure shows the depth (up to 440 km) and strike of the subducted African plate (African, Nubian, Aegean, or Hellenic slab) based on the Slab2 model. Overlain in red are fault systems from NOAFaults, a digital database of active faults in Greece (Ganas et al., 2013). Countries and some features are indicated.

to the west and Somalia to the east, both of which are separated by a diffuse plate boundary (Chase, 1978; Gordon & Stein, 1992; McCluksy et al., 2003; DeMets & Merkouriev, 2016). The African plate is geographically significant and is, at present, the third largest major tectonic plate (Gaina et al., 2013). A wide range of tectonic interactions characterizes its boundaries and internal dynamics, which were influenced by the collision between the plate and Eurasia (e.g., Meijer & Wortel, 1999; McCluksy et al., 2003; Fernandes et al., 2006; Catalano et al., 2008; DeMets & Merkouriev, 2016). The plate boundary between the Indian and African plates is over 12,000 km long and may have initiated around 105 Ma in the Early Cretaceous (e.g., van Hinsbergen et al., 2021). Rifting in the Gulf of Aden and the Red Sea signaled the break off of the Arabian peninsula from Africa at 29–24 Ma (e.g., Bosworth et al., 2006; Wolfenden et al., 2005), fragmentation of the Africa plate, and the onset of motion between Somalia and Nubia. Alternatively, the separation may have been more recent (11–10 Ma; Wolfenden et al., 2004; Keranen & Klemperer, 2008; Corti, 2009; see discussion in DeMets & Merkouriev, 2016).

The overriding Aegean microplate (Nyst & Thatcher, 2004) consists mainly of continental lithosphere with a similar thickness as the Anatolian plate (e.g., Zhu et al., 2006; Sodoudi et al., 2006). Significant crustal thickness variations exist from western Greece to eastern Anatolia (20–47 km), but thickness appears relatively uniform within specific regions of the Anatolian plate (Özacar et al., 2010; Karabulut et al., 2019). Moho depth across the Anatolian plate varies between 24 and 48 km, with the thinnest crustal thickness located on the coast of western Turkey and the thickest in eastern Turkey (Tezel et al., 2013). The Aegean Sea crustal thickness averages ~25 km (Zhu et al., 2006; Tirel et al., 2004; Kind et al., 2015), similar to the thickness beneath the central Menderes Massif at 25–30 km (Zhu et al., 2006; Karabulut et al., 2019). Estimates in the eastern Mediterranean show the Moho at an average 30 km depth (Marone et al., 2003). The Aegean microplate crustal thickness has thinner southern and central parts (20–22 km) compared with its northern portions (25–28 km) (Karagianni et al., 2005; Sodoudi et al., 2006). An exception to this observation is the crustal thickness beneath western Crete, 32.5–35 km or up to 45–50 km

(Bohnhoff et al., 2001; Meier et al., 2004; Snopek et al., 2007). The eastern part of Crete appears thinner (24–26 km) compared with its western portion (32.5 km; Bohnhoff et al., 2001). Crustal thickness also decreases to 15 km below the central Cretan Sea and 17 km, approximately 100 km off the southern coast of Crete (Bohnhoff et al., 2001). A crustal thickness of ~40 km is reported for western Greece (Karagianni et al., 2005), north of Antalya Bay (Karabulut et al., 2019), eastern Anatolia (Kind et al., 2015), and the Anatolian plateau (e.g., Saunders et al., 1998). In Eastern Anatolia, the crustal thickness increases toward the north, from 35 km at the Arabian foreland to 45 km farther north (Özacar et al., 2010).

The subducting portion of the African plate has multiple names based on the time frame that was active and differentiates it from other nearby subduction zone systems. In the Aegean and Western Anatolia region, the subducting plate is sometimes referred to as the Aegean slab (Wortel & Spakman, 2000; Widiyantoro et al., 2004; Vanderhaeghe et al., 2007; Biryol, 2009; van der Meer et al., 2018; El-Sharkawy et al., 2020), the African-Mediterranean slab (Sachpazi et al., 2016), the Hellenic plate or Hellenic slab (Chang et al., 2010; Vernant et al., 2014; Sachpazi et al., 2016; Confal et al., 2018; Hayes, 2018; Wei et al., 2019; Blom et al., 2019). The Aegean slab usually describes the system during the Late Cretaceous. In contrast, the other specific names distinguish the Aegean Sea and western Anatolia subduction system from that beneath Antalya and Cyprus. The subducting slab near Cyprus is referred to as the Cyprus slab (Confal et al., 2018), Cyprus/Tethys slab (Taylor et al., 2019), and Cyprus-Bitlis slab (Govers & Fichtner, 2016). The Antalya slab is recently defined between the Hellenic and Cyprus arcs (Güvercin et al., 2021).

The western edge of the Hellenic arc is bounded by the Kephalonia (also Kefalonia, Cephalonia) transform zone (KTZ) (Fig. 4.1; Bocchini et al., 2018; Hansen et al., 2019). This feature is a dextral strike-slip system that delineates the boundary between the Apulian platform (Adria microplate) and the Eurasian plate or Aegean microplate (Kokinou et al., 2006; Pearce et al., 2012). The KTZ has structural trends and seismicity that extend onshore, and it is the most active zone of shallow seismicity in the broader Aegean region (Kokinou et al., 2006). Over the last 1–4 Myr, the KTZ has linked to the North Anatolian Fault (NAF) system via the Central Hellenic Shear Zone (Papanikolaou & Royden, 2007; Reilinger et al., 2010; Vassilakis et al., 2011; Royden & Papanikolaou, 2011; Halpaap et al., 2018). The KTZ is considered one of the most seismically active zones in Europe (Pearce et al., 2012; Halpaap et al., 2018) and has been represented as a vertical tear between oceanic and continental lithosphere (Suckale et al., 2009), forming a subduction-transform-edge-propagator (STEP) fault (Govers & Wortel, 2005). The STEP fault may be in its initial stages of forming (Evangelidis, 2017; Özbakır et al., 2020), or the slab may have entirely detached (Wortel & Spakman, 2000). A smooth transition is also proposed between two segments without a tear, at least at depths shallower than 100 km (Pearce et al., 2012; Halpaap et al., 2018). Nine STEP structures may exist beneath southern Greece, segmenting the subducting African slab and contributing to seismicity and deformation (Sachpazi et al., 2016). Overall, the southwestward expansion and stretching of the Aegean microplate during Plio-Quaternary is accommodated by a northern right-lateral tectonic boundary defined by the KTZ and NASZ and a southern left-lateral tectonic boundary, marked by the Pliny and Strabo trenches (Sakellariou & Tsampouraki-Kraounaki, 2019).

The active right-lateral NAF and North Anatolian Shear Zone (NASZ) extends for ~1,200 from the Karlıova triple junction through the Sea of Marmara and Biga Peninsula (Fig. 4.1) (Ketin, 1948; Barka, 1992; Armijo et al., 1999; Şengör & Zabcı, 2019). The NASZ contains the NAF, which accommodated from 25 to 110 km of displacement, depending on location, since the Late Miocene (Westaway 1994; Yoshioka 1996; Armijo et al., 1999; Hubert-Ferrari et al., 2002; Şengör & Zabcı, 2019). The fault system disrupts the Aegean extensional domain in its northern section and may have been activated due to subduction rollback processes (e.g., Flerit et al., 2004; Faccenna et al., 2006). Current timing estimates suggest the NAF initiated in east Anatolia at 10–12 Ma, propagated westward through central Anatolia from 7 to 6 Ma, and was activated in the Aegean region by the Pliocene (Şengör et al., 2005; Faccenna et al., 2006). Its western and eastern terminations of the NASZ are poorly defined (Barbot & Weiss, 2021), but the system presently accommodates ~24 mm/yr of slip along northern Turkey (McClusky et al., 2000; Bulut et al., 2018).

Westward extrusion of the Anatolian plate accommodated by the NASZ and NAF dominates the present-day geodynamics of the Aegean microplate such that active extensional strain has been stated to be absent, except near the Corinth rift, south Viotia, south of Evia, and across the Sperchios-Kammena Vourla rift (Brooks & Ferentinos, 1984; Papanikolaou & Royden, 2007; Chousianitis et al., 2013, 2015). However, seismogenic faults in the internal Aegean domain associated with the Hellenic subduction arc are characterized by pure normal and strike-slip kinematics or by a combination, and active thrusting is limited to the central and western sectors of the Hellenic subduction zone and the offshore regions external to it (Maggini & Caputo, 2020). The nature of earthquakes varies with Hellenic arc depth, with normal senses of motion at depths up to 40 km and strike-slip faulting present parallel to strike for events deeper than 40 km (Benetatos et al., 2004). One of the strongest earthquakes of the 20th century, with an Mw 7.7 (or 7.8), occurred in the area of the south Aegean off the coast of the island of Amorgos on 9 July 1956 at 03:11 UTC (Fig. 4.1; Okal et al., 2009; Alatza et al., 2020). This event has debated focal mechanisms as either strike-slip or normal faulting geometries (Okal et al., 2009). An Mw 4.4 earthquake with normal motion recently occurred near its epicenter on 27 November 2018 (23:16 UTC, 36.7565°N, 25.877°E). A normal sense of motion also is found with some recent earthquakes near the NASZ, including 12 June 2017 Mw 6.2 (12:28 UTC, 38.8486°N, 26.313°E) and 2017 Mw 5.3 (10:58 UTC, 39.5275°N, 26.1373°E), likely associated with transtensional motion.

The Hellenic arc is separated into a western portion based on well-defined bathymetry. This portion is sometimes termed the *Western Hellenic Subduction Zone* or *Western Hellenic Arc and Trench* (WHA-T) system and extends ~400 km northwest from the central Adriatic Sea to the west coast of Crete (Figs. 4.1 and 4.3; Papadopoulos et al., 2010; Pearce et al., 2012; Hansen et al., 2019). This part of the Hellenic arc is divided into northern and southern sections termed the *Northern Hellenic arc and Southern Hellenic arc* (Royden & Papanikolaou, 2011). The Southern Hellenic arc exhibits classical features, but in the north, thick continental crust is subducted beneath northern Greece (Pearce et al., 2012; Halpaap et al., 2018). As a result, this area has significant variations in subduction rate, trench retreat, the occurrence of deep seismicity, and overriding plate extension, likely related to a change in convergence regime from the subduction of oceanic crust in the south to continental crust in the north (Papanikolaou & Royden, 2007; Halpaap et al., 2018). The southern portion of the Hellenic arc near the island of Crete is also affected by the subduction of continental crust (e.g., Meier et al., 2007).

Overall, the Hellenic arc has three distinct regions: an outer compressional nonvolcanic arc, a volcanic arc, and an extensional backarc region covering the broader Aegean Sea region (Figs. 4.1 and 4.2; McKenzie 1972; Papazachos, 2019). The Pliocene-Quaternary volcanic arc is located in the Methana peninsula, along the islands of Milos and Santorini in the Cycladic archipelago and the island of Nisryos in the Dodecanese (Fig. 4.1; Le Pichon & Angelier, 1979; McKenzie, 1972; Papazachos & Nolet, 1997, see review in Scoon, 2021). Most of the subducting Nubian plate is oceanic, except along the central sector of its southern margin, where the Mediterranean Ridge accretionary complex collides with the African continental margin (Chaumillon & Mascle, 1997; Westbrook & Reston, 2002). High-resolution scattered wave images of the northern portion of the Western Hellenic arc also show the subduction of thick continental crust beneath northern Greece (Pearce et al., 2012; Halpaap et al., 2018) and Crete (e.g., Meier et al., 2007).

The connection of the western portion of the Hellenic arc to its central and eastern sections is diffuse (Beißer et al., 1990; Shaw & Jackson, 2010; Biryol et al., 2011; Özbakır et al., 2013). The eastern portions of the Hellenic arc are sometimes referred to as the east Hellenic arc and trench system (HA-T) (Papadopoulos et al., 2007). The boundary between the Hellenic and Cyprus arcs in this area is obstructed by up to 300 km wide, 6–10 km thick section of sediments that compose the Mediterranean Ridge (Figs. 4.1–4.3; Heezen & Ewing, 1963; Emery et al., 1966; Le Pichon et al., 1982; Kenyon et al., 1982; Kastens et al., 1992; Foucher et al., 1993; Westbrook & Reston, 2002; Kopf et al., 2003). This accretionary complex is the most prominent structural unit of the eastern Mediterranean Sea, extending ~2,000 km from the Calabrian Rise east of Greece to the Florence Rise (Limonov et al., 1996; Cita et al., 1996). The front of subduction of the Hellenic arc is located south of the Mediterranean Ridge (Figs. 4.1–4.3; e.g., Le Pichon et al., 1995; Mascle & Chaumillon, 1997; Mascle et al., 1999; Jost et al., 2002; Westbrook & Reston, 2002; Meier et al., 2007; Jolivet et al., 2013).

The significant size of the Mediterranean Ridge indicates that the Hellenic arc is an accretionary as opposed to an erosive margin (e.g., Von Huene & Scholl, 1991, 1993; Cloos & Shreve, 1996; Clift & Vannucchi, 2004). In this case, material is transferred from the subducting plate to the forearc wedge and plate boundary zone (e.g., DeFranco et al., 2008; Scholl & Von Huene, 2009). This type of margin is favored when subduction rates are low and seismic moment release rates are high (Le Pichon et al., 1993; Clift & Vannuchi, 2004; DeFranco et al., 2008). However, the Mediterranean Ridge accretionary complex is unusual compared with others worldwide. It formed in a continent-continent collisional setting and has shallow, Messinian-age evaporites (e.g., Cita et al., 1996; Chaumillon & Mascle, 1997). These evaporites influence its deformation and growth rates due to their mechanical properties and effect on fluid flow and pressure (Kastens, 1991; Westbrook & Reston, 2002; Kopf et al., 2003). Extensive mud volcanism associated with the Mediterranean Ridge began in the early Pleistocene (1.75–1.25 Ma, Nikitas et al., 2021).

Understanding the development of the Mediterranean Ridge is critical to determining the Hellenic arc SZI because it is thought to grow via off scraping against a backstop formed by the Alpine nappes of the Hellenic arc (Kastens, 1991) at a rapid rate (10 km/Myr; Kastens, 1991; Kopf et al., 2003; Papanikolaou, 2021). The ridge varies in geometry along strike (Cita et al., 1996; Chaumillon & Mascle, 1997; Westbrook & Reston, 2002; Kopf et al., 2003; Andronikidis et al., 2017; Papanikolaou, 2021). The wedge accumulates sediments in its western and eastern portions, but the ridge behaves unlike a typical accretionary complex in the central portion between Libya and Crete, where it experienced maximum deformation and incipient collision had already caused uplift of the area (Chaumillon & Mascle, 1997; Kopf, 2002). Crete is situated on the central forearc of the Hellenic subduction zone and is underlain by material from two microcontinents, leading to larger observed crustal thicknesses (~50 km; e.g., Thomson et al., 1998; Stöckhert 1999; Meier et al., 2004, 2007).

South of Crete, the Hellenic trenches, Ptolemy, Pliny, and Strabo (Fig. 4.1), developed between the Mediterranean Ridge and volcanic arc. These trenches are not classical ocean trenches, as earthquakes beneath them originate along low-angle thrusts at 20–40 km (Taymaz et al., 1990; Shaw & Jackson, 2010). Instead, they develop due to back-thrusting beneath the northern edge of the accretionary complex (Galindo-Zaldivar et al., 1996; Westbrook & Reston, 2002) or the tearing of the Nubian slab (Özbakır et al., 2013). The Hellenic Trench has been described as the surface expression of a steep (~30°) reverse fault splaying off the deeper underlying thrust-fault interface of the subduction zone (Shaw et al., 2008; Shaw & Jackson, 2010).

Low-angle thrust faults along the Aegean coast associated with subduction zone tectonics pose significant tsunami hazards (e.g., Tinti et al., 2005; Howell et al., 2015; Bocchini et al., 2020). Offshore Crete Island is considered one of the most tsunamigenic areas in the entire Mediterranean Sea region (Papadopoulos et al., 2010; Triantafyllou et al., 2019). However, the complexity of the overall Hellenic arc plate boundary, combined with its aseismic nature, makes earthquake data alone a misleading guide for identifying the likely sources of tsunamigenic earthquakes (Yolsal et al., 2007; England et al., 2015; Howell et al., 2015). Tsunamigenic earthquakes infrequently occur in the eastern Mediterranean (Papazachos & Dimitriu, 1991; Papadopoulos et al., 2007). An evaluation of historical data, including the 1956 Amorgos event that generated the largest of the most recent tsunamis, indicates that a likely trigger of some past tsunamis in the region was submarine landslides generated by earthquakes (e.g., Dominey-Howes, 2002; Okal et al., 2009; Ebeling et al., 2012). Factors contributing to slope instability across portions of the Hellenic arc include its sloping bottom, thick accumulations and high rates of recent sedimentation, closely spaced active faults, active earthquakes, and diapirism (e.g., Ferentinos, 1990; Hooft et al., 2017). The eruption of Santorini (Fig. 4.1) in 1610 BCE generated a tsunami that affected civilizations throughout the eastern Mediterranean (Dominey-Howes, 2002; Hooft et al., 2017). Detailed bathymetry across the Mediterranean is critical in understanding tsunami propagation and mitigating its impacts (Fig. 4.1; e.g., CIESM, 2011).

The Rhodes Basin bounds the eastern portion of the Hellenic arc (Fig. 4.1; Ganas & Parsons, 2009) and is a deep depression lying in continuity with the Pliny and Strabo trenches. The depth of the basin is > 4,000 m and is one of the deepest portions of the Mediterranean Sea (Woodside et al., 2000). The Rhodes basin may represent an unsubducted portion of the deep Mesozoic Levantine basin (Rotstein & Ben-Avraham, 1985) or a former upper-Miocene subduction trench remnant that remained after a shift in the primary convergence zone (Mascle et al., 1986). Deep faults buried beneath the zone may mark the onset of extension (Woodside et al., 2000). Hall et al. (2009) suggest a two-part history of the basin. Following Miocene convergence, the basin experienced middle Pliocene-Quaternary sinistral transpression due to the actively curving Hellenic arc and change in the convergence vector of the African plate. Slab tear has been proposed to interpret the presence and structures within the deep Rhodes Basin (Woodside et al., 2000; Faccenna et al., 2014).

A STEP is also suggested to be located at the transition between the Cyprus and Hellenic arcs (Elitez et al., 2016). Trench-parallel tear affects the subducting African lithosphere between northern Greece and the Gulf of Corinth along the Western Hellenic arc (Hansen et al., 2019). Trench-perpendicular tear may accommodate the region between the Hellenic and Cyprian arcs, which differ in subduction steepness and material subducted (Dilek & Sandvol, 2009). The Cyprian arc involves shallower subduction dynamics with the Eratosthenes seamount and Anixamander Mountains (mud volcanoes; Lykousis et al., 2009) impinging on the trench (Kempler & Ben-Avraham 1987; Zitter et al., 2003; Biryol et al., 2011). This arc became effectively inactive during the onset of the westward extrusion of the Anatolian plate (Papanikolaou, 2021).

4.2.2. Geometry of the Hellenic Arc Subduction Zone

We must consider its present-day structure to understand when and why the Hellenic subduction zone was established. The Hayes (2018) Slab2 model uses active-source seismic data interpretations, receiver functions, local and regional seismicity catalogs, and seismic tomography, and models the subducting Nubian slab as uniformly northward dipping to > 440 km depths in its northern portion (Fig. 4.3). The Hellenic arc has a well-developed Wadati-Benioff zone at shallower depths but a debated slab geometry at intermediate depths (150–250 km; Suckale et al., 2009; Agostini et al., 2010; see

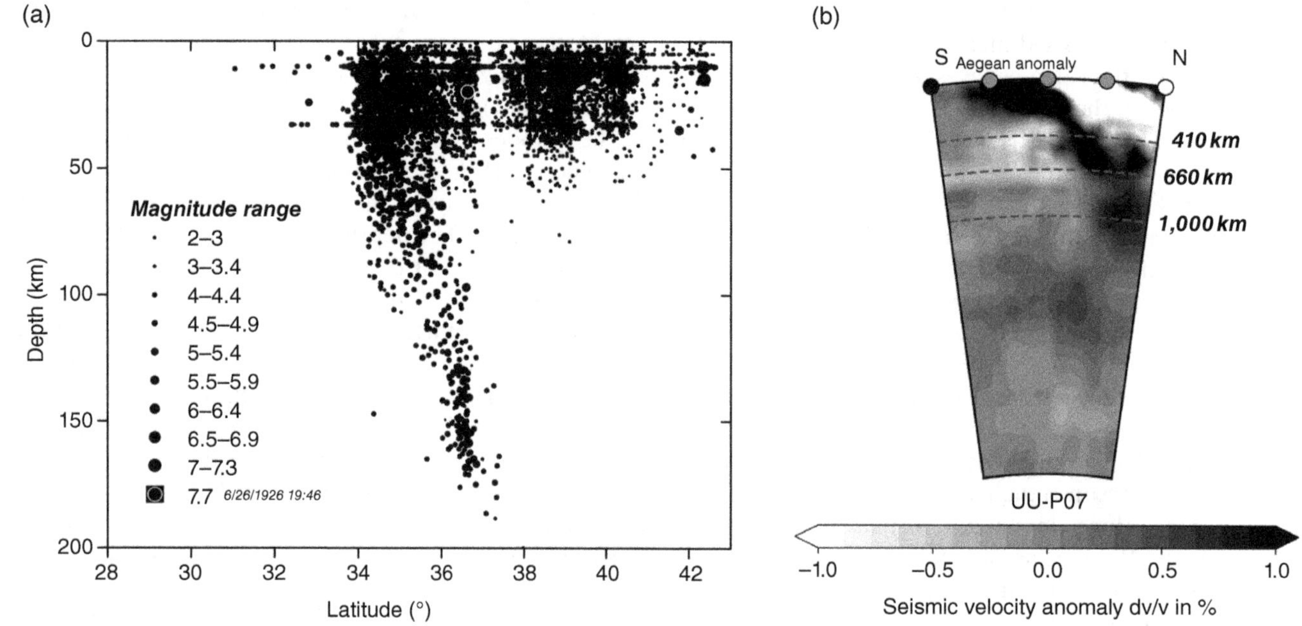

Figure 4.4 (a) Depth versus latitude of earthquakes taken from a line of the section of 28°–43° and longitude of 24°–28°. Events were extracted from the Turkish Ministry of the Interior, Disaster and Emergency Management Presidency, Earthquake Department Earthquake Catalog (M > = 4.0), 1900-20XX (https://deprem.afad.gov.tr/depremkatalogu) from 24 January 1900 to 17 June 2021. We indicate the largest event (26 June 1926, 19:46). The legend shows how the size of the earthquake correlates to the symbol. (b) Cross section of the Aegean anomaly interpreted as the African slab using the UUP07 P-wave model (Amaru, 2007). The depths of the dashed lines are 410, 660, and 1,000 km from the surface. Interpretations of the geology below 1,000 are debated and discussed in the text. Image created using Hosseini et al. (2018). See text for references to additional tomographic images.

review in Hansen et al., 2019). Figure 4.4a shows the slab clearly defined by earthquake depth versus latitude across the Hellenic subduction zone. Detailed analysis of the distribution of earthquakes indicates that the western part of the subduction zone dips under 20°–30° to the northeast and reaches the maximum depth of 180 km, and its eastern section dips under 40° to the northwest and reaches a maximum depth of 170 km (Papazachos & Comninakis, 1971; Vaněk et al., 1987; Papazachos et al., 2000; Suckale et al., 2009; Papazachos, 2019). At deeper levels (100–180 km), the Wadati-Benioff zone dips freely (without coupling) at a high angle (~45°) beneath the south Aegean trough and the volcanic arc (Mahatsente et al., 2017; Papazachos, 2019).

Seismic coupling is the ratio between the observed seismic moment release and the rate calculated from plate tectonic velocities (e.g., Ruff & Kanamori, 1983; Scholz & Campos, 2012). The plate interface coupling between the Hellenic trench fault and the Nubia-Aegean is low (<20%; Papadimitriou & Karakostas, 2008; Vernant et al., 2014) and only 10% in the southwestern portion of the arc (Papadimitriou & Karakostas, 2005). A low seismic coupling scenario tends to produce earthquakes with small rupture zones and high moment release that occur as small isolated patches on the subduction interface (Scholz, 1990; Messini et al., 2007). However, in the region near the Ionian islands (Fig. 4.1), complete seismic coupling of the subduction is most consistent with the observed moderate seismic moment release (Laigle et al., 2002, 2004). Higher degrees of seismic coupling are also found in specific regions in the western Hellenic arc (Jackson & McKenzie, 1988; Messini et al., 2007; Ganas et al., 2020).

Classical ray tomography using body and surface waves is often applied to the Mediterranean region to image the Hellenic arc subduction system (e.g., Piromallo & Morelli, 1997; Spakman et al., 1988; Snieder, 1988; Zielhuis & Nolet, 1994; Piromallo & Morelli, 2003; Marone et al., 2004a,b; Amaru, 2007; Schivardi & Morelli, 2009; Chang et al., 2010; Biryol et al., 2011; Salaün et al., 2012; Legendre et al., 2012; Portner et al., 2018; Blom et al., 2019; Wei et al., 2019; El-Sharkawy et al., 2020). We present a P-wave tomographic image of the Aegean anomaly as an example in Figure 4.4b. Although surface wave data improves ray coverage and allows the visualization of regions with low seismicity and few stations, body wave tomography affords better resolution yet suffers from vertical smearing and an unknown background velocity model compared with surface waves (Piromallo & Morelli 1997; Kassaras et al., 2005; Schivardi

& Morelli, 2009; Salaün et al., 2012). Body wave tomographic images across the Hellenic arc shows the geometry of the subduction zone clearly (e.g., Fig. 4.4b), with the slab sometimes exhibiting a strong velocity contrast (up to 6%) compared with the background model (e.g., Papazachos & Nolet, 1997; Kassaras et al., 2005). Some regional studies show the slab forms an asymmetrical shape, shallowly dipping and larger along the western side, and steeper and smaller in the eastern section (Papazachos & Nolet, 1997; Blom et al., 2019).

Segments of the African slab have long been imaged as subducting into the lower mantle (Fig. 4.4b; 1,400 ± 100 km depth; Spakman et al., 1988; Spakman, 1990, 1991; Spakman et al., 1993; Papadopoulos, 1997; Bijwaard et al., 1998; Kárason & van der Hilst, 2000; Wortel & Spakman, 2000; Romanowicz, 2003; Kassaras et al., 2005; Chang et al., 2010; Biryol et al., 2011; Legendre et al., 2012; Govers & Fichtner, 2016; van der Meer et al., 2018; Bocchini et al., 2018; Wei et al., 2019; El-Sharkawy et al., 2020). The width of the subducting slab differs depending on the approach applied to image it. For example, Chang et al. (2010) show the slab widening beneath 660 km in the lower mantle, whereas Piromallo & Morelli (2003) has widening within the transition zone. Widening can be associated with a slab avalanche where high viscosity occurs in the lower mantle (Capitanio et al., 2009; Chang et al., 2010). Slab avalanches form when large-volume subducted slabs that temporarily stagnate within the transition zone periodically penetrate the lower mantle (e.g., Solheim & Peltier, 1994; Deschamps & Tackley, 2009; Yang et al., 2018). Slab avalanches are unstable, controlled by mantle thermal instabilities, and will accelerate as slab sinking rates increase with time (e.g., Solheim & Peltier, 1994; Yang et al., 2018). More recent tomographic images show the Hellenic slab extends from the surface to the transition zone in a bent, arcuate shape with a high-velocity structure that flattens from the 410 km discontinuity and is not seen at deeper levels (Blom et al., 2019).

Mantle tomography has also shown that not all slabs in the Mediterranean region are connected to the lithosphere at the surface due to delamination (e.g., Spakman et al., 1988; Dilek & Sandvol, 2009; Wortel & Spakman, 2000). For example, gaps are visible in the slab at depths of 60–100 km just west of the south Hellenides, and slab tear may be visible at the 660 km discontinuity (Wei et al., 2019). Multiple remnant slabs are imaged beneath the North Hellenides (e.g., Wei et al., 2019) and most of Anatolia (e.g., Salaün et al., 2012). Vote maps, generated from stacking tomography models and identifying where the models agree based on an increasing vote count at a specified depth (Shephard et al., 2017), show a slab interpreted to be the African plate between 250 and 1,550 km depths (Crameri et al., 2020).

Although tomographic images are central in some studies of Hellenic arc SZI, their results are challenging to interpret, especially in locating regions of slab tear (e.g., Piromallo & Morelli 2003). The subduction zone's small size, its spatially highly variable nature, and the uneven distribution of its seismic stations make imaging the area more challenging (El-Sharkawy et al., 2020). The presence of slow velocity anomalies throughout the uppermost mantle beneath Anatolia and very slow velocities beneath eastern Anatolia has been explained by innovative ideas regarding how the mantle and lithosphere behave in terms of mantle upwelling, drip-style lithospheric delamination (GöğüŞ et al., 2017), or rollback (see discussion in Portner et al., 2018). Tomographic images of both the Hellenic and Cyrus arcs have transformed ideas regarding how subducting slabs operate as they enter the deep mantle and their impacts on seismicity and the creation of transfer zones (e.g., Meighan et al., 2013), extensive volume magmatism (e.g., Cocchi et al., 2017), and controlling ore-forming process and mineral deposits (e.g., de Boorder et al., 1998; Rabayrol et al., 2019; Rabayrol & Hart, 2021). Changes in the subducting slab dynamics significantly affect plate dynamics, including plate motion and mantle dynamics (e.g., Gianni et al., 2019).

4.3. GEOLOGICAL BACKGROUND OF AEGEAN-ANATOLIAN SUTURE ZONES

Determining the geological relationship between the present-day Hellenic arc and ancient subduction zones associated with the Aegean and Anatolian microplates is complicated due to the amalgamation of terrains that occurred over its tectonic history. The western portion of the Anatolian microplate exposes ophiolite remnants and other evidence of ancient subduction zones that may have influenced the onset of the present-day Hellenic arc. The Anatolian microplate's development is characterized by continental fragments separated by branches of the Paleo- and Neo-Tethyan oceans that collided and ultimately combined by the Late Cretaceous-Eocene. These exposures of ophiolitic and high-pressure/low-temperature (HP/LT) rock assemblages identify suture zones (Fig. 4.5; e.g., Şengör & Yılmaz, 1981; Okay, 2008; Moix et al., 2008; Okay & Tuysuz, 1999; Pourteau et al., 2016; Okay et al., 2020). This section describes the evidence of ancient subduction zones that may be linked to or may have influenced the development of the Hellenic arc.

4.3.1. Intra-Pontide Suture Zone

Western Anatolia contains the Pontide Mountains to the north and the Taurides Mountains to the south (Fig. 4.5; e.g., Şengör & Yılmaz, 1981; Yılmaz et al., 1997; Okay & Tuysuz, 1999; Pourteau et al., 2016). The Pontide

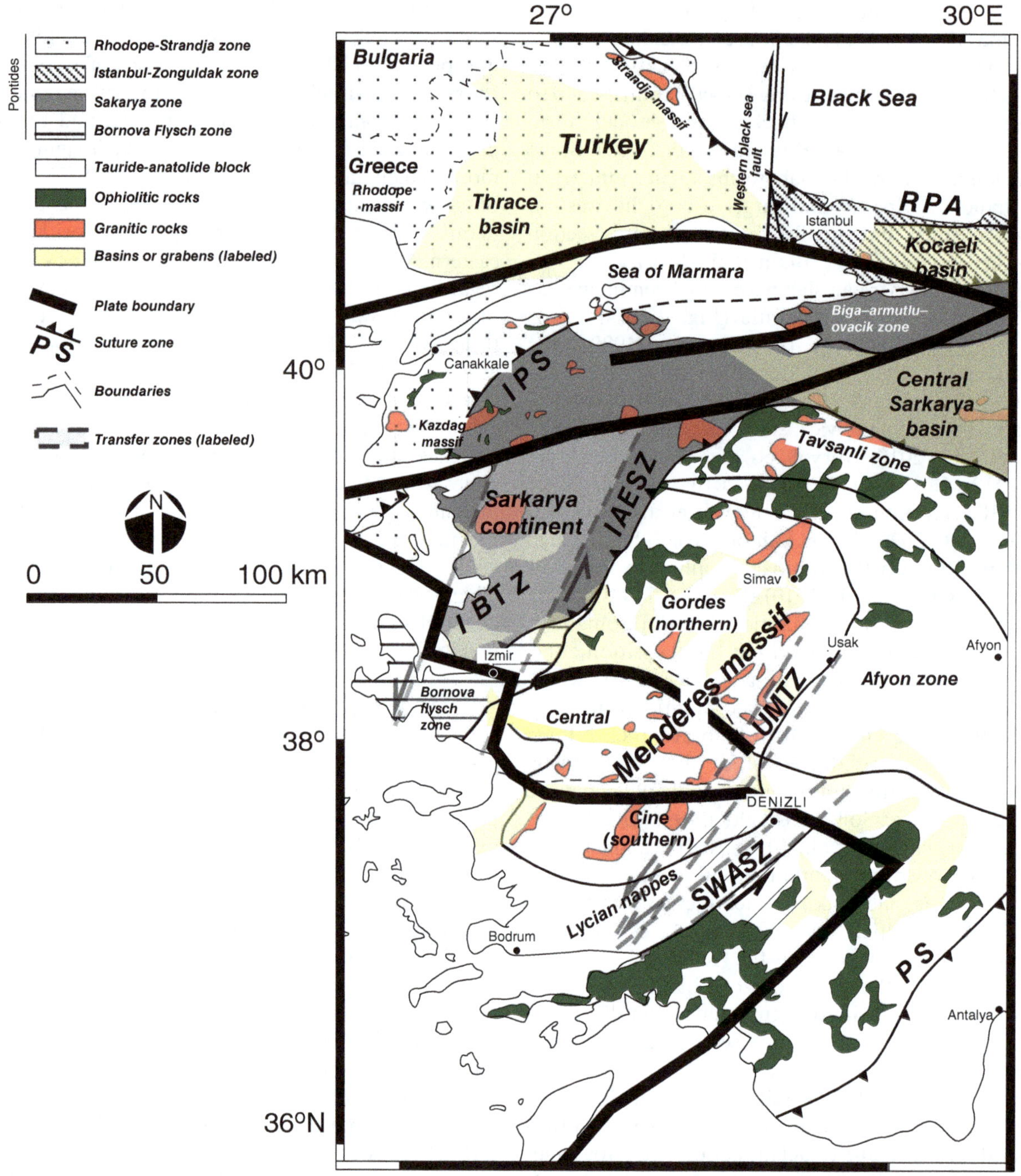

Figure 4.5 Geological map of western Anatolia focusing on the ophiolite and granite assemblages along the boundary between the Aegean and Anatolia microplates. Plate boundary after Nyst and Thatcher (2004). Terrain boundaries, major fault systems, and transfer zones after Okay (2008), Akbayram et al. (2016), Oner et al. (2010), and Karaoğlu and Helvacı (2014). Note: RPA = Rhodope -Pontide Arc; İBTZ = İzmir-Balıkesir Transfer Zone (also sometimes referred to as the Western Anatolia Transfer Zone; Gessner et al., 2013, 2017); SWASZ = Southwest Anatolian Shear Zone; IPS = Intra-Pontide Suture Zone; IAESZ = Izmir -Ankara -Erzincan Suture Zone; PS = Pamphylian suture zone; UMTZ = UŞak-Mugla Transfer Zone.

Mountains include a southern Eurasia margin affinity and are characterized by a Pan-African basement with Phanerozoic sedimentary cover units (Yılmaz et al., 1997; Moix et al., 2008; Pourteau et al., 2010; Okay et al., 2013). The Intra-Pontide Suture Zone (IPS) is mapped within the Pontide zone between the Sakarya continental zones and Istanbul-Zonguldak Unit (Istanbul-Zonguldak Zone, Istanbul Nappe, or Istanbul Zone; Yiğitbaş et al., 2004). The IPS has been interpreted as an accretionary complex, a suprasubduction zone, and a remnant of a former ocean basin that may have extended into eastern Europe (e.g., Okay et al., 1996; Robertson & Ustaömer, 2004; Göncüoğlu et al., 2012, 2014; Marroni et al., 2014; Akbayram et al., 2016; Sayit et al., 2016; Frassi et al., 2018). Its complex geology, composed of components from the Istanbul-Zonguldak and/or Sakarya zones, has led to a debate about its presence and utility in paleogeographic reconstructions (Moix et al., 2008). Geochemical data from mafic rocks collected along the IPS show signatures related to a suprasubduction zone that opened in the Middle Jurassic-Early Cretaceous, obducted during the Early-Late Cretaceous boundary, and had closed by the Late Paleocene (Göncüoğlu et al., 2014; Sayit et al., 2016).

In Greece, the Vardar suture connects to the IPS (Şengör & Yılmaz, 1981; Okay & Satir, 2000; Okay et al., 2001; Beccaletto & Jenny, 2004; Okay & Whitney, 2010; d'Atri et al., 2012; Di Rosa et al., 2019) or Meliata-Balkan suture (Stampfli, 2000). In Turkey's Biga Penisula, the IPS and Vardar connection may be recorded by an isolated ophiolite-bearing accretionary complex active until the Late Cretaceous (Fig. 4.5; e.g., Okay et al., 1991). However, several researchers have debated this connection (Altunkaynak & Genc, 2008; Burchfiel et al., 2008; Şengün et al., 2011). Furthermore, due to the uncertain link between IPS and Izmir-Ankara-Erzincan Suture Zone (IAESZ), the relationship of the Biga Peninsula basement assemblages to those in the Rhodope-Thrace Massif is also debated (e.g., Bonev & Beccaletto, 2007; Elmas, 2012). Both the IPS and IAESZ are thought to mark Late Cretaceous–earliest Tertiary closure of Neo-Tethyan ocean basins (e.g., Pourteau et al., 2010; Akbayram et al., 2016).

4.3.2. Izmir-Ankara-Erzincan Suture Zone (IAESZ)

The IAESZ separates the Pontide's Sakarya Composite Terrain in the north from the Anatolide-Tauride block to the south (Fig. 4.5; Şengör & Yılmaz, 1981; Okay & Tüysüz, 1999; Tekin et al., 2002; Göncüoğlu, 2010). In the Aegean microplate, the IAESZ is thought to record the closure of the Vardar ocean and link with the Vardar ophiolite (or Axios-Vardar Suture Zone) (Fig. 4.1; Channell & Kozur, 1997; Okay & Tuysuz, 1999; Tekin et al., 2002; Moix et al., 2008). However, the feature is not exposed due to the presence of the Aegean Sea (e.g., Burtman, 1994; Stampfli, 2000; Yılmaz et al., 2001; Burchfiel et al., 2008).

In Western Anatolia, blueschist assemblages exposed along the IAESZ are intruded by suture zone granitoids (SZGs) (Topuk, Orhaneli, Tepeldag (Gürgenyayla and Gürgenyayla)) with Paleocene (63.5 ± 2.8 Ma) to Oligocene (31.4 ± 0.6 Ma) ages but primarily characterized by early Eocene emplacement (~45–47 Ma; Okay & Satir, 2006; Altunkaynak, 2007). The SZGs intrude the western portion of the Tavşanlı zone, a blueschist sequence overlain by a Cretaceous accretionary complex and ophiolitic sheet. The TavŞanlı Zone is narrow (~50 km) and trends east to west for approximately 250–350 km (Okay & Whitney, 2010; Plunder et al., 2013). Its western and central portions contain blueschist facies metavolcanic and metasedimentary rocks with rare metabasalts (Okay, 1980a, 1980b, 1982; Okay & Kelley, 1994; see Seaton et al., 2009). The zone is thought to have formed as a result of northward-dipping subduction and represents the Mesozoic-Eocene closing of the northern branch of the Neo-Tethyan Ocean (Okay, 1986, 2008; Okay & Kelley, 1994; Sherlock et al., 1999; Moix et al., 2008; Shin et al., 2013; Plunder et al., 2013; Fornash & Whitney, 2020). The Sivrihisar Massif in the eastern portion of the TavŞanlı has eclogite, blueschist, and Barrovian sequences (Gautier, 1984; Seaton et al., 2009). Paleocene-Eocene ages from the TavŞanlı zone granites mark the timing of the closure of the IAESZ (e.g., Okay et al., 2020).

The Afyon zone (Afyon-Bolkardag Zone of Okay, 1986; Özdamar et al., 2013, or Ören-Afyon Zone of Pourteau et al., 2013) is considered the southward palaeogeographic extension of the TavŞanlı zone (Candan et al., 2005; Pourteau et al., 2010; Akal, 2013; Özdamar et al., 2013). Although its southern extent is unclear, the Afyon zone is mapped parallel to the TavŞanlı zone. A portion may also be exposed between the southern Menderes Massif and Lycian Nappes (Okay, 1986; Candan et al., 2005; Pourteau et al., 2013; Ustaömer et al., 2020). Afyon zone stratigraphy resembles the TavŞanlı Zone and consists of Palaeozoic-Mesozoic Tauride shelf sequences of metasedimentary and metavolcanic rocks that experienced regional greenschist to blueschist facies (Fe-Mg carpholite and glaucophane) metamorphism overlying Pan-African-related basement (Okay, 1984; Candan et al., 2005, Pourteau et al., 2010; Özdamar et al., 2013). HP/LT metamorphism in the Afyon zone is thought to have occurred coincident with the closure of the Neo-Tethyan Ocean at 70–65 Ma (Pourteau et al., 2010, 2013; Özdamar et al., 2013; Plunder et al., 2013). Portions of the Afyon zone may have subducted beneath the TavŞanlı zone during the Late Cretaceous based on zircon ages of granites that intrude TavŞanlı zone blueschist and altered ophiolitic assemblages (Speciale et al., 2012; Shin et al., 2013).

4.4. AGE CONSTRAINTS ON THE INITIATION OF SUBDUCTION

4.4.1. Cenozoic Estimates

Early estimates for the initiation age of Hellenic arc subduction were Late Miocene to Pliocene (13 ± 3–5 Ma, Le Pichon & Angelier, 1979; 5–10 Ma, McKenzie, 1978;

Mercier, 1981) based on interpretations of seismic activity coupled with assumptions regarding the age of subducted lithosphere and subduction depths. These ages are similar to the earliest volcanic activity in the South Aegean arc (Pliocene; Pe-Piper & Piper, 2005) and the onset of the KTZ based on geodynamic modeling and GPS data (6–8 Ma; Royden & Papanikolaou, 2011). The Late Miocene time frame is also cited as when the Western Hellenic arc divided into its northern and southern portions along the KTZ (Papanikolaou, 2010). Reconstructions of fault systems in the northern margin of the eastern Mediterranean Sea are also consistent with 15 Ma (Le Pichon et al., 2019).

Interpretations of the Aegean seismic velocity structure, tomography, and seismicity data in the Aegean area suggest an Eocene to Late Oligocene onset of the Hellenic arc subduction zone (40–26 Ma; Meulenkamp et al., 1988; Spakman et al., 1988; Papadopoulos, 1997; Brun & Sokoutis, 2010). These ages are consistent with the timing of granite crystallization throughout Western Anatolia and Mediterranean Ridge sedimentation (33–23.6 Ma, Fytikas et al., 1984; Kastens, 1991). Younger estimates from the ridge are also reported (~19 Ma, Kopf et al., 2003). Plate reconstructions suggest that the Northern Hellenic trench from the western portion of the Hellenic arc experienced the onset of subduction from 34 to 27 Ma. In contrast, the Southern Hellenic arc was active at 34 Ma (Royden & Papanikolaou, 2011).

The progressive deceleration in motion of Africa with respect to Europe in the Mediterranean region has been observed to occur since 35 Ma, and, in the eastern Mediterranean, from 35 Ma to 10 Ma to a convergence rate of a few millimeters per year (Savostin et al., 1986; Marsellos et al., 2010). The rate of trench retreat may have accelerated from ~0.6 cm/yr during the first 30 m.y. of subduction to 3.2 cm/yr during the past 15 m.y., perhaps due to slab tear during the Middle Miocene-Pliocene (Brun et al., 2017). Timing constraints on Aegean forearc curvature, due to opposite rotations, clockwise in the west and counterclockwise in the east, are Eocene and Middle Miocene (Morris & Robertson 1993; Cornée et al., 2018). Trench bending and rollback increased the obliquity of subduction over time, which was accommodated by strain partitioning within the upper Eurasian plate (Faccenna et al., 2013; Philippon et al., 2014; Brun et al., 2016; Cornée et al., 2018). By the Early Oligocene-Eocene, backarc basin formation was ongoing in the Aegean region as the African plate retreated as it subducted beneath Eurasia (Le Pichon & Angelier, 1981; Royden, 1993; Jolivet & Faccenna, 2000; Agostini et al., 2010; Carminati et al., 2012; Jolivet et al., 2018).

Figures 4.6 and 4.7 are north-south temporal cross sections across Western Anatolia that describe the region's tectonic evolution from the Late Cretaceous to the present. Figure 4.6a (from Dilek & Altunkaynak, 2007) illustrates the closure of a branch of the Neo-Tethyan Ocean (Izmir-Ankara Ocean), which occurred during the Late Cretaceous when the IPS already marked the division between the Sakarya and Rhodope continents. Break off of the subducting Neo-Tethyan slab occurred ca. 54 Ma along the IAESZ (Fig. 4.6b), concurrent with the development and emplacement of SZG and ophiolite obduction. The Hellenic arc forms due to subduction stepback processes completed by the Miocene (~25 Ma; Fig. 4.6c). Subduction rollback was most likely initiated soon after the complete formation of the Hellenic arc to start the Aegean backarc extension (Fig. 4.6d). Some variations of this model exist. For example, Shin et al. (2013) proposed an additional subduction system concurrent with the IAESZ but located further south between the IAESZ and the Hellenic arc (Fig. 4.7).

These models are the outcome of large data sets composed of field observations, geophysical, geochemical, petrological, geochronological, and structural data from multiple rock types but mainly affiliated with the lithosphere. They provide testable hypotheses for the timing and generation mechanisms of particular magmatic and structural features in the Aegean and western Anatolia to predict the presence of multiple suture zones and subducted slabs. They also delineate the affinity of specific crustal fragments north of the Hellenic arc to be distinct from the African plate in terms of their origin and geological history.

4.4.2. Mesozoic Estimates

Significantly older estimates for the onset of the Hellenic subduction exist and suggest the arc has been long-lived since the Jurassic–Late Cretaceous, with stipulations that Cenozoic age estimates are considering only the latest episode of its evolutionary history and disregard its earlier history (Fig. 4.8; e.g., Jolivet et al., 2018). In this model, the African slab has been part of a single, evolving subduction zone system that was initiated sometime in the Mesozoic during the initiation of the closure of the northern branch of the Neo-Tethys ocean (Faccenna et al., 2003; van Hinsbergen et al., 2005; Royden & Papanikolaou, 2011; Jolivet et al., 2013; Malandri et al., 2017; Jolivet et al., 2018; Crameri et al., 2020; van Hinsbergen et al., 2021). The model has been primarily applied to explain the evolution of the western Hellenic arc. The Vardar suture in Greece may be equivalent to the IAESZ (Channell & Kozur, 1997; Okay & Tuysuz, 1999; Moix et al., 2008). The Pindos Suture Zone could be equivalent to units within the Antalya domain and Dilek peninsula (Stampfli & Kozur, 2006) had buoyant microcontinents that entered and locked subduction. This process triggered southward slab

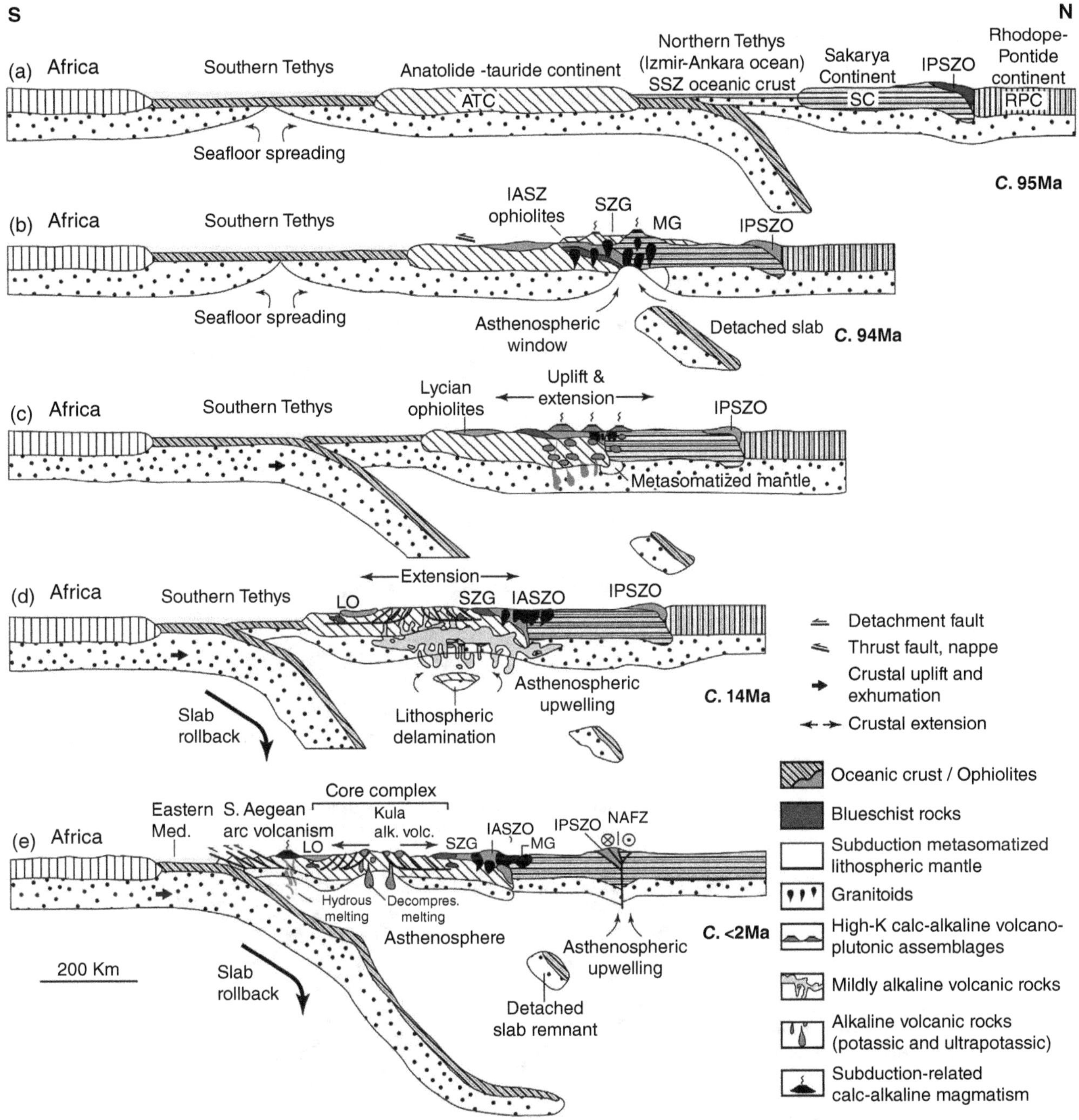

Figure 4.6 North-south cross section across the Aegean region from ca. 95 Ma to < 2 Ma. In this scenario, the Hellenic arc forms due to the transition from a divergent plate boundary to convergence sometime between 54 Ma and 25 Ma (Dilek & Altunkaynak, 2007). Scale is only a rough approximation. See Dilek and Altunkaynak (2007) for a color image of this figure. Note: ATC = Anatolide-Tauride Continent; IASZ = Izmir-Ankara Suture Zone; IPSZO = Intra-Pontide Suture Zone ophiolites; MG = Magmatic Granites; NAFZ= North Anatolian Fault Zone; SC = Sakarya Continent; SSZ = Supra-subduction zone; SZG = Suture Zone Granitoids; RPC = Rhodope-Pontide Continent.

rollback and migration of the volcanic arc (Fig. 4.8; van Hinsbergen et al., 2005; Brun & Faccenna, 2008; Jolivet & Brun, 2010; Jolivet et al., 2013; Cornée et al., 2018). In this model, the subduction zone accumulates tectonic fragments, terrains, and oceanic lithosphere, instead of following the induced transference subduction model (Figs. 4.6 and 4.7; e.g., Stern, 2004). Here, the Hellenic subduction zone front has behaved like an accretionary

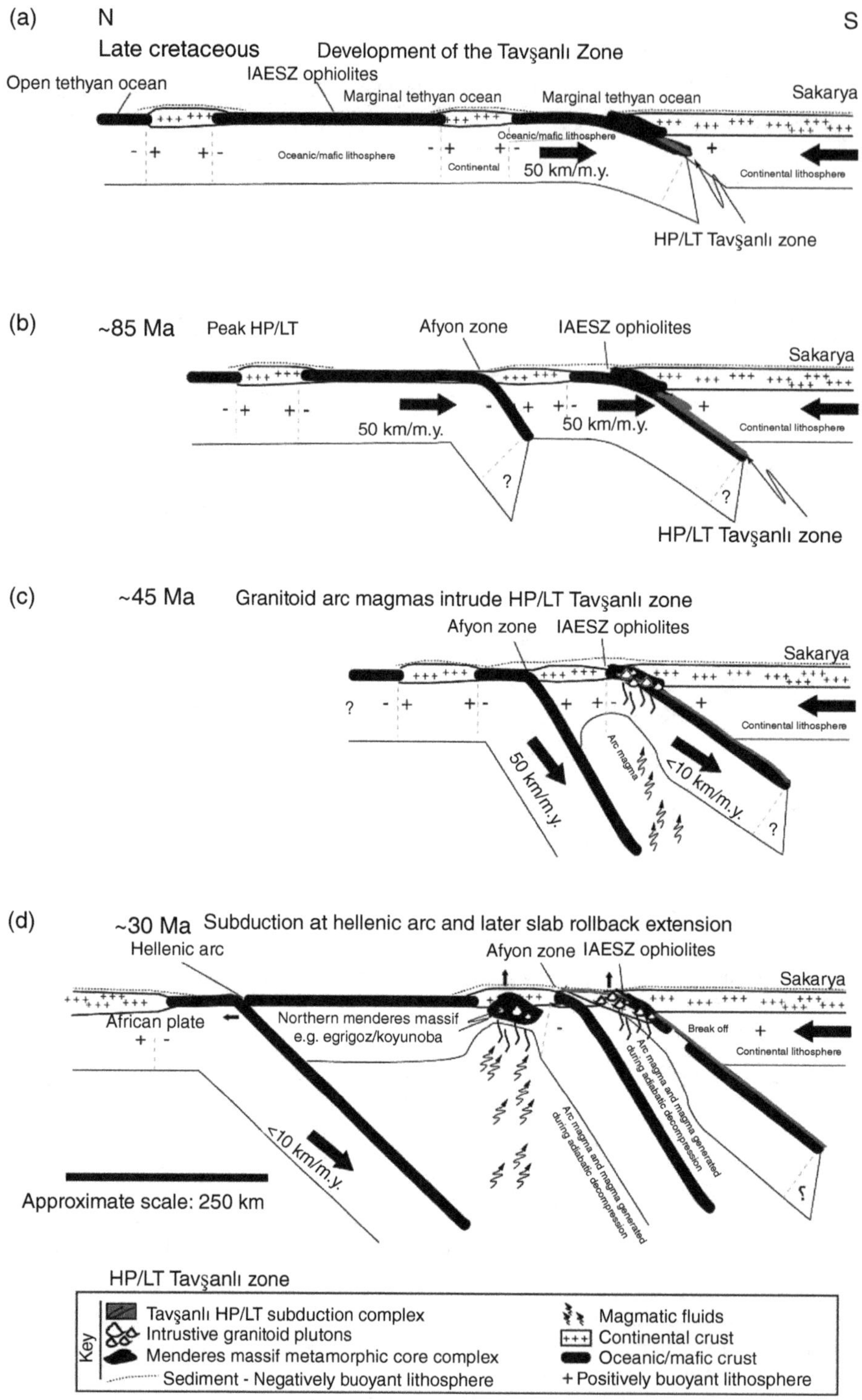

Figure 4.7 Lithospheric-scale cross sections of a model of the Aegean region through western Anatolia in which the region records and experiences progressively younger subduction zones after Shin et al. (2013). The subduction zone cartoon is adapted from Cloos et al. (2005). Scale is a rough approximation. Panel (a) shows the development of the IAESZ during the Early to Late Cretaceous. Panel (b) records the peak metamorphism of the zone and the development of the Afyon Zone subduction zone at ~85 Ma. Panel (c) shows granitoid arc magmas from the Afyon Zone intrude the IAESZ at ~45 Ma, emplacing granite bodies that intrude the ophiolite sequences. Panel (d) records the break off of the IAESZ slab and additional magma emplacement at 30 Ma. In this scenario, the Hellenic arc experiences ongoing subduction.

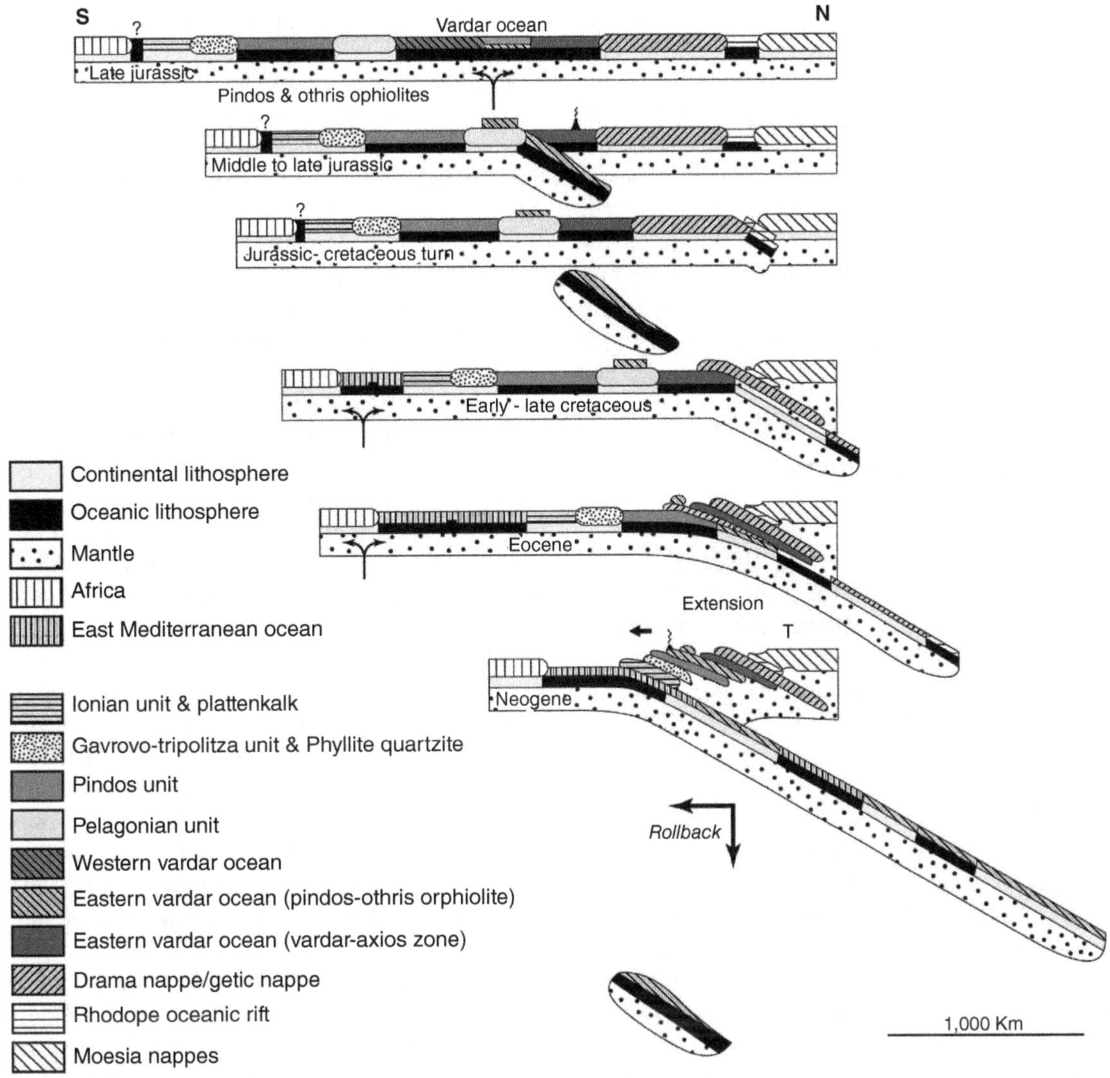

Figure 4.8 Schematic overview of the development of nappe stack and subduction during Alpine orogeny in Greece from van Hinsbergen et al. (2005). The cross section displays the single subduction zone model for the development of the Hellenic arc.

wedge during its entire development since the Jurassic–Late Cretaceous.

The model also requires a stationary trench from 100 to 45 Ma to allow for the penetration of the African slab into mantle depths (Bonneau, 1982; Stampfli & Borel, 2002; Jolivet et al., 2003; Capitanio et al., 2010). Penetration to 1,200–1,400 km depths is estimated to have occurred by ~50 Ma, which triggers the onset of extension in the region by 45 Ma (Jolivet & Faccenna, 2000; Brun & Faccenna, 2008; Capitanio et al., 2009, 2010). The present-day curvature of the Hellenic forearc represents oblique subduction that grew systematically (Huchon et al., 1982; Le Pichon et al., 1995; ten Veen & Kleinspehn, 2003; Gautier et al., 1999; Le Pichon et al., 2002; Wallace et al., 2005, 2008; van Hinsbergen & Schmid, 2012; Philippon et al., 2014; Cornée et al., 2018).

The single subduction model was developed primarily using interpretations of mantle tomography coupled with estimates of subduction rates over long periods of geological time. Thus, ages found in suture zones that would be entirely unrelated to the Hellenic arc in early SZI models now time the onset of subduction of the African slab. For example, the Late Cretaceous (104 Ma) estimate of Hellenic arc SZI is based on the ages of ophiolites exposed in the Tavsanli zone of Western Anatolia, part of the IAESZ (Peters et al., 2017; Pourteau et al., 2019; Crameri et al., 2020). The single subduction system also has implications for the affinity of Aegean crustal fragments and terrains in paleogeographical reconstructions. The model requires all the lower plate continental crust to be accreted into the upper plate while subducting continental lithosphere. It also requires the entire Aegean crust from the Vardar (Izmir-Ankara) suture to the Mediterranean ridge to be derived from the lower plate (Fig. 4.6–4.8). The model also eliminates the need for multiple sutures and subducted slabs to be present beneath western Turkey and the Aegean and simplifies the evolution of the Aegean microplate to a single evolving, long-lived subduction system (van Hinsbergen et al., 2005).

4.5. DISCUSSION

Plate interactions are thought to be most clear at convergent plate boundaries (Silver et al., 1998). However, several outstanding issues remain regarding the timing of the onset of convergence between the Nubian Plate and Aegean microplate. Several observations indicate that the Aegean microplate does not behave like a typical backarc basin (Agostini et al., 2010; Doglioni et al., 2002). It lacks an ocean floor and associated ocean crust and is underlain by a thick layer of continental crust (e.g., Makris, 1978). The Aegean crust comprises stacked carbonate and oceanic nappes, metamorphic massifs, and core complexes (Jolivet et al., 2013; Papazachos, 2019). Models of deformation along the Hellenic arc intended to capture the large-scale 3D structure of the Nubian plate suggest the region more closely resembles a continental thrust than a typical oceanic subduction zone (Ganas & Parsons, 2009).

The Aegean microplate is also bound to the north by the active strike-slip systems (NAF, NASZ, KTZ, and Central Hellenic Shear Zone; Brooks & Ferentinos, 1984; Lyberis, 1984; Nyst & Thatcher, 2004; Gürer et al., 2006; Kreemer et al., 2004; Kokkalas et al., 2006; Papanikolaou & Royden, 2007; Reilinger et al., 2010; Royden & Papanikolaou, 2011; Vassilakis et al., 2011; Halpaap et al., 2018). This results in a complex tensional regime where crustal stretching is inconsistent with the geometry and direction of the subducting Hellenic slab (e.g., Mantovani et al., 1997; Agostini et al., 2010). Differences in the timing of initiation and rate of subduction exist between segments along the Western Hellenic arc exist, and should also be expected to occur along other portions of the Hellenic and Cyprus arcs (Royden & Papanikolaou, 2011). Subduction zones behave chaotically and may retreat, advance, or remain stationary at different stages, especially if the incoming lithosphere is heterogeneous (e.g., Royden & Husson, 2009; Husson et al., 2009). The extent and role of several ancient fault systems, including older Neo-Tethyan suture zones that likely played a significant role in its development, are thus debated.

The Mesozoic SZI age for the Hellenic arc requires all of the lower plate continental crust to be accreted into the upper plate while subducting continental lithosphere. The entire Aegean crust from the Vardar suture to the Mediterranean ridge would be derived from the lower plate (Fig. 4.8). This process requires a redefinition of components associated with ancient supercontinents and would impact paleogeographic reconstructions of the region. Oceans between the accreted domains were of significant size (500 km in some cases). The process would lead not only to changes in elevation and crustal thicknesses but also critical changes in the zone of subduction transitions occurring from oceanic to continental shear zones (see discussion in Le Pichon et al., 2019). Not all units record blueschist facies conditions, and some experienced Barrovian prograde (burial) P-T paths, such as on the island of Naxos (e.g., Lamont et al., 2019).

Subduction zones with limited trench-parallel lengths on the order of the Hellenic arc (600–800 km) and narrow slabs (<1,500 km) typically have rapid retreat rates (Schellart et al., 2007; Bolhar et al., 2010). The requirement of a stationary trench from 100 to 45 Ma to allow for the African slab to enter mantle depths (Bonneau, 1982; Stampfli & Borel, 2002; Jolivet et al., 2003; Capitanio et al., 2010) seems unusual in that most of the slab today is characterized by significant tear. Sites of slab tear vary in scale from regional to local. These locations include the boundary between the Hellenic and Cyprus arcs (Wortel & Spakman, 1992; Biryol et al., 2011), south of Crete at the Pliny-Strabo shear zone (Özbakır et al., 2013), the İzmir-Balıkesir transfer zone (e.g., Kaya, 1981; Gessner et al., 2013), and beneath the Menderes Massif (Biryol et al., 2011; Rabayrol & Hart, 2021), and the eastern Aegean (Jolivet et al., 2015).

At present, it is unclear how the subducting African (Nubian) slab could survive through the Earth's transition zone, suggesting that its composition and convergent rate, and the material properties of the mantle were critical constraints to allow this process to occur (e.g., Kárason & van der Hilst, 2000). Interpretations of the deeper images are debated: the subducted slab may be a single folded body that overturned in the lower mantle (Faccenna et al., 2003) or two different slabs between 2,000 and 1,500 km and from 1,500 km to the surface (van Hinsbergen et al., 2005; van der Meer et al., 2018). The unhindered slab penetration into lower mantle depths is unexpected due to the relationship between slab morphology and trench migration (e.g., Widiyantoro et al., 2004). However, the process of slabs reaching depths of 1,700 km is not uncommon (e.g., Kárason & van der Hilst, 2000).

Numerous examples also suggest significant slab tear across the Aegean and Western Anatolia (e.g., Wortel & Spakman, 1992; Woodside et al., 2000; Biryol et al., 2011; Gessner et al., 2013; Faccenna et al., 2014; Roche et al., 2019; Rabayrol & Hart, 2021). Legendre et al. (2012) note that body wave studies show slab segments are trapped within the mantle transition zone in the western Mediterranean and Alpine region and do not dive deeper as is observed in the Aegean region. The Mediterranean Sea exhibits lateral variations in crustal and upper mantle structures (Marone et al., 2004a).

Long-lived Aegean subduction would have affected the mantle's thermal and chemical convection and impacted surface conditions, including climate, over time (e.g., Sigl et al., 2015). The region contains active and quiescent volcanoes located in the Methana Peninsula (Peloponnese) and islands of the Cycladic archipelago (Fig. 4.1; see review in Scoon, 2021). Holocene volcanism along the Hellenic arc allows for identifying and characterizing high-frequency climatic changes during the Holocene in particular (e.g., Zanchetta et al., 2011). Fluids released from past and present subducting lithosphere have affected not only volcanism but also grain mobility and seismic anisotropy throughout the Mediterranean and elsewhere (e.g., Karato, 1995; Van der Meijde et al., 2003; Marone et al., 2004b; Faccenda et al., 2008). This impact over the length of planetary history may have been significant in terms of triggering other subduction zones, controlling sites of seismicity and magmatism, and generating ore deposits. A Jurassic-age subduction zone that is presently active and capable of generating significant hazards has implications for our understanding of lithospheric strength, the driving force of plate tectonics over geological time, and the tectonic history of the Aegean region.

The single subduction system has its roots in tomography and the interpretation of large low shear-wave velocity anomalies in the lowermost mantle. These compositional anomalies thus appear as crucial as continents are for the surface, yet they are inaccessible and have limited data regarding their thermal histories and chemical compositions. Their role as reservoirs and their impact on the evolution of the Hellenic arc over time is unclear. The interpretation that the seismic tomography images reveal a single, long-lived subduction system may be incorrect. The Aegean-Anatolian microplates have long been known to have an extensive and complex orogenic history based on surface geology. With the presence and data from mapped suture zones, it is expected that the region would record multiple subducted slabs from earlier collisional events.

Determining the Hellenic arc SZI is also important because the timing constraints shed light on the mode of initiation (i.e., induced or spontaneous; from compression or extension; Gurnis et al., 2004; Stern, 2004; Gerya, 2011; Stern & Gerya, 2018) based on our understanding of the geology at that time. Subduction initiation requires breaking a zone of weakness in the lithosphere (e.g., Stern & Gerya, 2018; Baes et al., 2020a; Cloetingh et al., 2021). The process may be accommodated along passive margins (e.g., Cloetingh et al., 1984), although the conversion of a passive margin into an active one has not been observed to have occurred during the Cenozoic (Gurnis et al., 2004; Cloetingh et al., 2021). Subduction initiation could also occur due to motion along transform or fracture zones associated with midocean ridges due to large-scale changes in plate motion (e.g., Hall et al., 2003). This process could occur across extinct midocean ridges or within backarc regions within mature subduction zones, as well as mantle plume interactions with old oceanic lithosphere (Baes et al., 2020a,b; Cloetingh et al., 2021). No evidence or hypothesis exists for a mantle plume triggering the development of the Hellenic arc. A young SZI initiation age suggests the Hellenic arc developed due to the Late Cenozoic collision of the African plate with Eurasia (e.g., Meulenkamp et al., 1988). An older SZI links its origination to near and parallel to a spreading ridge associated with the western portion of the Neo-Tethyan ocean (e.g., Maffione & van Hinsbergen, 2018).

Younger estimates for the Hellenic arc are consistent with the observation that nearly half of Earth's active subduction zones were initiated during the Cenozoic (Gurnis et al., 2004). Crameri et al. (2020) indicate that SZI has happened at least 13 times during the last hundred million years, with global occurrences of SZI occurring within 55 to 40 Ma and 16 to 6 Ma. These times mark higher SZI activity on Earth and are remarkably consistent with Cenozoic estimates of the onset of the Hellenic arc (McKenzie, 1978; Le Pichon & Angelier, 1979; Mercier, 1981; Meulenkamp et al., 1988; Spakman et al., 1988; Papadopoulos, 1997; Brun & Sokoutis, 2010; Royden & Papanikolaou, 2011; Le Pichon et al., 2019). A significant slowdown in Africa's northward motion relative to surrounding plates has also been attributed to the Eocene-Oligocene collision of Africa and Eurasia. The deceleration altered the kinematics and tectonic evolution of the Aegean region and the restructuring, reorganization, and fracturing of the African plate elsewhere (e.g., DeMets & Merkouriev, 2016).

4.6. CONCLUSIONS

The Hellenic arc is the result of the convergence of the northern margin of the African (Nubian) plate with the southern portion of the Aegean microplate. The region has emerged as the type locality for subduction zone tectonics, and many novel ideas regarding the development of this particular plate boundary have emerged from the area. Two end-member models for the SZI age of the Hellenic arc evolved using disparate datasets. Seismic tomography images help determine the present-day subduction-related structure beneath the Aegean microplate, and some interpretations suggest a long-lived subduction zone from the Jurassic to Late Cretaceous. However, tomographic images can be limited in their interpretations of deeper depths and our ability to observe slab tears or structures that disrupt the subducting slab. Surface geological data sets suggest a complex, long-lived history of the Aegean region with multiple suture zones. These observations are more consistent with multiple detached slabs, even some unrelated to the African plate. Geological, geochemical, and geochronological constraints from the continental lithosphere affected by subduction help constrain the pathways of rocks affected by subduction dynamics. However, these are limited by the currently available data sets and interpretation within an existing framework with limited exposure. The Hellenic arc is a significant plate boundary that would likely be influenced by differential retreat, inherited lower plate lithospheric heterogeneities, and mantle upwelling. Multidisciplinary studies that couple detailed field mapping of available outcrops in the Western Anatolian and Aegean microplates with geophysical observations, geochemistry, and geochronology are needed to determine its SZI age and which model, a long-lived single subduction zone or multiple subduction zones with the latest SZI in Cenozoic, applies to the region.

ACKNOWLEDGMENTS

No real or perceived financial conflicts of interests exist for the authors. We appreciate the comments from two anonymous reviewers. We thank Jeffrey S. Horowitz for assistance with drafting figures.

REFERENCES

Agostini, S., Doglioni, C., Innocenti, F., Manetti, P., & Tonarini, S. (2010). On the geodynamics of the Aegean Rift. In İ. Çemen (Ed.), Extensional tectonics in the Basin and Range, the Aegean, and western Anatolia. *Tectonophysics*, *488*, 7–21. https://doi.org/10.1016/j.tecto.2009.07.025

Akal, C. (2013). Coeval shoshonitic-ultrapotassic dyke emplacements within the Kestanbol Pluton, Ezine-Biga Peninsula (NW Anatolia). *Turkish Journal of Earth Sciences*, *21*, 1–20.

Akbayram, K., Şengör, A. M. C., & Ozcan, E. (2016). The evolution of the intra-Pontide suture; implications of the discovery of late Cretaceous-early Tertiary melanges. In R. Sorkhabi (Ed.), *Tectonic evolution, collision, and seismicity of southwest Asia: In honor of Manuel Berberian's forty-five years of research contributions. Special Paper, Geological Society of America*, 525. https://doi.org/10.1130/2016.2525(18)

Alatza, S., Papoutsis, I., Paradissis, D., Kontoes, C., & Papadopoulos, G.A. (2020). Multi-Temporal InSAR analysis for monitoring ground deformation in Amorgos Island, *Greece. Sensors*, *20*(2), 338. https://doi.org/10.3390/s20020338

Altunkaynak, Ş. (2007). Collision-driven slab breakoff magmatism in northwestern Anatolia, Turkey. *Journal of Geology*, *115*, 63–82

Altunkaynak, Ş., & Genc, S. (2008). Petrogenesis and time-progressive evolution of the Cenozoic continental volcanism in the Biga Peninsula, NW Anatolia (Turkey). In Y. Xu et al. (Eds.), Continental volcanism and chemistry of the Earth's interior. *Lithos*, *102*, 316–340.

Amaru, M.L. (2007). *Global travel time tomography with 3-D reference models*. Ph.D. Dissertation. Utrecht University. http://dspace.library.uu.nl/bitstream/handle/1874/19338/index.htm;jsessionid=88B6AA4941C5E76FD65034D89D4065E0?sequence=17

Anastasakis, G., & Keling, G. (1991). Tectonic connection of the Hellenic and Cyprus arcs and related geotectonic elements. *Marine Geology*, *97*(3–4), 261–277. https://doi.org/10.1016/0025-3227(91).90120-S

Andronikidis, N., Kokinou, E., Vafidis, A., Kamberis, E., & Manoutsoglou, E. (2017). Deformation patterns in the southwestern part of the Mediterranean Ridge (South Matapan Trench, Western Greece). *Marine Geophysical Research*. https://doi.org/10.1007/s11001-017-9337-0

Angelier, J., Lybéris, N., Le Pichon, X., Barrier, E., & Huchon, P. (1982). The tectonic development of the Hellenic arc and the sea of Crete: A synthesis. *Tectonophysics*, *86*(1–3), 159–196. https://doi.org/10.1016/0040-1951(82).90066-X

Armijo, R., Meyer, B., Hubert, A., & Barka, A. (1999). Westward propagation of the North Anatolian Fault into the northern Aegean: Timing and kinematics. *Geology*, *27*, 267–270. https//doi.org/10.1130/0091-7613(1999)027<0267:WPOTNA>2.3.CO;2

Baes, M., Sobolev, S., Gerya, T., & Brune, S. (2020a). Subduction initiation by plume-plateau interaction: Insights from numerical models. *Geochemistry, Geophysics, Geosystems*, *21*, e2020GC009119. https://doi.org/10.1029/2020GC009119

Baes, M., Sobolev, S., Gerya, T., & Brune, S. (2020b). Plume-induced subduction initiation: Single-slab or multi-slab subduction?. *Geochemistry, Geophysics, Geosystems*, *21*, e2019GC008663. https://doi.org/10.1029/2019GC008663

Barbot, S., & Weiss, J. R. (2021). Connecting subduction, extension and shear localization across the Aegean Sea and Anatolia. *Geophysical Journal International*, *226*(1), 422–445. https://doi.org/10.1093/gji/ggab078

Barka, A.A. (1992). The north Anatolian fault zone. *Annales Tectonicae*, *6*, 64–195

Beccaletto, L., & Jenny, C. (2004). Geology and correlation of the Ezine zone: A Rhodope fragment in NW Turkey? *Turkish Journal of Earth Sciences*, *13*, 45–176.

Beißer, M., Wyss, M., & Kind, R. (1990). Inversion of source parameters for subcrustal earthquakes in the Hellenic Arc. *Geophysics Journal International*, *103*, 439–450.

Benetatos, C., Kiratzi, A., Papazachos, C., & Karakaisis, G. (2004). Focal mechanisms of shallow and intermediate depth earthquakes along the Hellenic Arc. *Journal of Geodynamics*, *37*(2), 253–296. https://doi.org/10.1016/j.jog.2004.02.002

Bijwaard, H., Spakman, W., & Engdahl, E. R. (1998). Closing the gap between regional and global travel time tomography. *Journal of Geophysical Research*, *103*, 30,055 – 30,078.

Biryol, C. (2009). *Complex rupture processes of the Solomon Islands subduction zone earthquake and subduction controlled upper mantle structure beneath Anatolia*. Ph.D. Dissertation. University of Arizona.

Biryol, C. B., Beck, S. L., Zandt, G., & Ozacar, A. A. (2011). Segmented African lithosphere beneath the Anatolian region inferred from teleseismic P-wave tomography. *Geophysics Journal International*, *184*, 1037–1057. https://doi.org/0.1111/j.1365-246X.2010.04910.x

Bleier, N. Bruhns, O., Friederich, W., Stoeckhert, B. (2007). Time variation of the stress field in the Aegean subduction zone. In W. Friederich (Ed.), *Subduction dynamics, bridging the scales. Bochumer Geowissenschaftliche Arbeiten*, *10*, 98.

Blom, N., Gokhberg, A., & Fichtner, A. (2019). Seismic waveform tomography of the central and eastern Mediterranean upper mantle. *Solid Earth*, *11*, 669–690. https://doi.org/10.5194/se-11-669-2020

Bocchini, G. M., Brüstle, A., Becker, D., Meier, T., van Keken, P. E., Ruscic, M., et al. (2018). Tearing, segmentation, and backstepping of subduction in the Aegean: New insights from seismicity. *Tectonophysics*, *734–735*, 96–118. https://doi.org/10.1016/j.tecto.2018.04.002

Bocchini, G. M., Novikova, T., Papadopoulos, G. A., et al. (2020). Tsunami potential of moderate earthquakes: The July 1, 2009 earthquake (Mw 6.45) and its associated local tsunami in the Hellenic Arc. *Pure Applied Geophysics*, *177*, 1315–1333. https://doi.org/10.1007/s00024-019-02246-9

Bohnhoff, M., Makris, J., Papanikolaou, D., & Stavrakakis, G. (2001). Crustal investigation of the Hellenic subduction zone using wide aperture seismic data. *Tectonophysics*, *343*(3–4), 239–262. https://doi.org/10.1016/S0040-1951(01)00264-5

Bolhar, R., Ring, U., & Allen, C. M. (2010). An integrated zircon geochronological and geochemical investigation into the Miocene plutonic evolution of the Cyclades, Aegean Sea, Greece: Part 1: Geochronology. *Contributions to Mineralogy and Petrology*, *160*, 719–742. https://doi.org/10.1007/s00410-010-0504-4

Bonev, N., & Beccaletto, L. (2007). From syn- to post-orogenic Tertiary extension in the north Aegean region: Constraints on the kinematics in the eastern Rhodope-Thrace, Bulgaria-Greece and the Biga Peninsula, NW Turkey. *Geological Society, London, Special Publications*, *291*, 113–142. https://doi.org/10.1144/SP291.6

Bonneau, M. (1982). Tectonic evolution of central Aegean Hellenides from Upper Jurassic to Miocene times. *Bulletin Society Geology France*, *24*, 229–242.

Bosworth, W., Huchon, P., & McClay, K. (2006). The Red Sea and Gulf of Aden basins. *Journal of African Earth Sciences*, *43*(1–3), 334–378. https://doi.org/10.1016/j.jafrearsci.2005.07.020

Brooks, M., & Ferentinos, G. (1984). Tectonics and sedimentation in the Gulf of Corinth and the Zakynthos and Kefallinia channels, *Western Greece. Tectonophysics*, *101*(1–2). 25–54. https://doi.org/10.1016/0040-1951(84)90040-4

Brun, J.-P., & Faccenna, C. (2008). Exhumation of high-pressure rocks driven by slab rollback. *Earth and Planetary Science Letters*, *272*, 1–7. https://doi.org/10.1016/j.epsl.2008.02.038

Brun, J-P., & Sokoutis, D. (2010).45 m.y. of Aegean crust and mantle flow driven by trench retreat. *Geology*, *38*(9), 815–818. https://doi.org/10.1130/G30950.1

Brun, J-P., Faccenna, C., Gueydan, F., Sokoutis, D., Philippon, M., Kydonakis, K., et al. (2016). The two-stage aegean extension, from localized to distributed, a result of slab rollback acceleration. *Canadian Journal of Earth Sciences, National Research Council Canada*, *53*(11), 1142–1157. https://doi.org/10.1139/cjes-2015-0203ff. ffinsu-01271296f

Brun, J-P., Faccenna, C., Gueydan, F., Sokoutis, D., Philippon, M., Kydonakis, K., et al. (2017). Effects of slab rollback acceleration on Aegean extension. *Bulletin of the Geological Society of Greece*, *50*(1), 5–23. https://doi.org/10.12681/bgsg.11697

Bulut, F., Özener, H., Doğru, A., Aktuğ, B., & Yaltırak, C. (2018). Structural setting along the western North Anatolian Fault and its influence on the 2014 North Aegean earthquake (Mw 6.9). *Tectonophysics*, *745*, 382–394. https://doi.org/10.1016/j.tec.2018.07.006

Burchfiel, B. C., Nakov, R., Dumurdzanov, N., Papanikolaou, D., Tzankov, T., Serafimovski, T., et al. (2008). Evolution and dynamics of the Cenozoic tectonics of the South Balkan extensional system. *Geosphere*, *4*, 918–938.

Burtman, V. S. (1994). Meso-Tethyan oceanic sutures and their deformation. *Tectonophysics*, *234*, 305–327.

Candan, O., Çetinkaplan, M., Oberhänsli, R., Rimmelé, G., & Akal, C. (2005). Alpine high-P/low-T metamorphism of the Afyon zone and implications for the metamorphic evolution of Western Anatolia, Turkey. *Lithos*, *84*, 102–124.

Capitanio, F. A., Faccenna, C., & Funiciello, R. (2009). Opening of Sirte Basin: Result of slab avalanche? *Earth and Planetary Science Letters*, *285*, 210–216. https://doi.org/10.1016/j.epsl.2009.06.019

Capitanio, F. A., Stegman, D.R., Moresi, L., & Sharples W. (2010). Upper plate controls on deep subduction, trench migrations and deformations at convergent margins. *Tectonophysics*, *483*, 80–92. https://doi.org/10.1016/j.tecto.2009.08.020

Carminati, E., Lustrino, M., & Doglioni, C. (2012). Geodynamic evolution of the central and western Mediterranean: Tectonics vs. igneous petrology constraints. *Tectonophysics*, *579*, 173–192. https://doi.org/10.1016/j.tecto.2012.01.026

Catalano, S., De Guidi, G., Romagnoli, G., Torrisi, S., Tortorici, G., & Tortorici, L. (2008). The migration of plate boundaries in SE Sicily: Influence on the large-scale kinematic model of the African promontory in southern Italy. *Tectonophysics*, *449*(1–4), 41–62. https://doi.org/10.1016/j.tecto.2007.12.003

Çemen, İ, Catlos, E. J., Gogus, O., & Ozerdem, C. (2006). Postcollisional extensional tectonics and exhumation of the Menderes Massif in the western Anatolia extended terrain, Turkey. *Special Paper, Geological Society of America*, *409*, 353–379.

Chang, S.-J., van der Lee, S., Flanagan, M. P., Bedle, H., Marone, F., Matzel, E. M., et al. (2010). Joint inversion for three-dimensional *S* velocity mantle structure along the Tethyan margin. *Journal of Geophysical Research*, *115*, B08309. https://doi.org/10.1029/2009JB007204

Channell, J. E. T., & Kozur H. W. (1997). How many oceans? Meliata, Vardar and Pindos oceans in Mesozoic Alpine paleogeography. *Geology*, *25*(2), 183–186. https://doi.org/10.1130/0091-7613(1997)025<0183:HMOMVA>2.3.CO;2

Chase, C. G. (1978). Plate kinematics: The Americas, East Africa, and the rest of the world. *Earth and Planetary Science Letters*, *37*(3), 355–368. https://doi.org/10.1016/0012-821X(78)90051-1

Chaumillon, E., & Mascle, J. (1997). From foreland to forearc domains: New multichannel seismic reflection survey of the Mediterranean ridge accretionary complex (Eastern Mediterranean). *Marine Geology*, *138*, 237–259. https://doi.org/10.1016/S0025-3227(97)00002-9

Chousianitis, K., Ganas, A., & Evangelidis, C. P. (2015). Strain and rotation rate patterns of mainland Greece from continuous GPS data and comparison between seismic and geodetic moment release. *Journal of Geophysical Research, Solid Earth*, *120*, 3909– 3931. https://doi.org/10.1002/2014JB011762

Chousianitis, K., Ganas, A., & Gianniou, M. (2013). Kinematic interpretation of present-day crustal deformation in central Greece from continuous GPS measurements. *Journal of Geodynamics*, *71*, 1–13. https://doi.org/10.1016/j.jog.2013.06.004

Cianetti, S., Gasperini, P., Giunchi, C., & Boschi, E. (2001). Numerical modeling of the Aegean-Anatolian region: Geodynamical constraints from observed rheological heterogeneities. *Geophysics Journal International*, *146*, 760–780.

CIESM (2011). *Marine geohazards in the Mediterranean*. N°42. In F. Briand (Ed), CIESM Workshop Monographs, Marine Geohazards in the Mediterranean, The Mediterranean Science Commission Workshop in Nicosia, 2–5 February 2011.

Cita, M. B., Erba, E., Lucchi, R., Pott, M., van der Meer, R., & Nieto, L. (1996). Stratigraphy and sedimentation in the Mediterranean Ridge diapiric belt. *Marine Geology*, *132*(1–4), 131–150. https://doi.org/10.1016/0025-3227(96)00157-0

Clift, P., & Vannucchi, P. (2004). Controls on tectonic accretion versus erosion in subduction zones: Implications for the origin and recycling of the continental crust. *Reviews in Geophysics*, *42*, RG2001. https://doi.org/10.1029/2003RG000127

Cloetingh, S., Koptev, A., Kovács, I., Gerya, T., Beniest, A., Willingshofer, E., et al. (2021). Plume-induced sinking of intracontinental lithospheric mantle: An overlooked mechanism of subduction initiation? *Geochemistry, Geophysics, Geosystems*, *22*, e2020GC009482. https://doi.org/10.1029/2020GC009482

Cloetingh, S. A. P. L., Wortel, M. J. R., & Vlaar, N. J. (1984). Passive margin evolution, initiation of subduction and the Wilson cycle. *Tectonophysics*, *109*(1–2), 147– 163.

Cloos, M., & Shreve, R. L. (1996). Shear-zone thickness and the seismicity of Chilean- and Marianas-type subduction zones. *Geology*, *24* (2), 107–110. https://doi.org/10.1130/0091-7613(1996)024<0107:SZTATS>2.3.CO;2

Cloos, M., Sapiie, B., van Ufford, A., Weiland, R. J., Warren, P. Q., & McMahon, T. P. (2005). Collisional delamination in New Guinea: The geotectonics of subducting slab breakoff. *Geological Society of America Special Papers, 400*. https://doi.org/10.1130/SPE400

Cocchi, L., Passaro, S., Tontini, F. C., & Ventura, G. (2017). Volcanism in slab tear faults is larger than in island-arcs and back-arcs. *Nature Communications*, *8*, 1451. https://doi.org/10.1038/s41467-017-01626-w

Confal, J. M., Faccenda, M., Eken, T., & Taymaz, T. (2018). Numerical simulation of 3-D mantle flow evolution in subduction zone environments in relation to seismic anisotropy beneath the eastern Mediterranean region. *Earth and Planetary Science Letters*, *497*, 50–61. https://doi.org/10.1016/j.epsl.2018.06.005

Cornée, J-J., Quillévéré, F., Moissette, P., Fietzke, J., López-Otálvaro, G. E., Melinte-Dobrinescu, M., et al. (2018). Tectonic motion in oblique subduction forearcs: insights from the revisited Middle and Upper Pleistocene deposits of Rhodes, Greece. *Journal of the Geological Society*, *176*, 78–96. https://doi.org/10.1144/jgs2018-090

Corti, G. (2009). Continental rift evolution: From rift initiation to incipient break-up in the Main Ethiopian Rift, *East Africa. Earth-Science Reviews*, *96*(1–2), 1–53. https://doi.org/10.1016/j.earscirev.2009.06.005.

Crameri, F., Magni, V., Domeier, M., et al. (2020). A transdisciplinary and community-driven database to unravel subduction zone initiation. *Nature Communications*, *11*, 3750. https://doi.org/10.1038/s41467-020-17522-9

d'Atri, A., Zuffa, G. G., Cavazza, W., Okay, A. I., & Di Vincenzo, G. (2012). Detrital supply from subduction/accretion complexes to the Eocene-Oligocene post-collisional southern Thrace Basin (NW Turkey and NE Greece). *Sedimentary Geology*, *243–244*, 117–129. https://doi.org/10.1016/j.sedgeo.2011.10.008

de Boorder, H., Spakman, W., White, S. H., & Wortel, M. J. R. (1998). Late Cenozoic mineralization, orogenic collapse and slab detachment in the European Alpine Belt. *Earth and Planetary Science Letters*, *164*(3–4), 569–575. https://doi.org/10.1016/S0012-821X(98)00247-7

DeFranco, R., Govers, R., & Wortel, R. (2008). Nature of the plate contact and subduction zones diversity. *Earth and Planetary Science Letters*, *271*(1–4), 245–253. https://doi.org/10.1016/j.epsl.2008.04.019

DeMets, C., & Merkouriev, S. (2016). High-resolution reconstructions of Pacific–North America plate motion: 20 Ma to present. *Geophysical Journal International*, *207*(2), 741–773. https://doi.org/10.1093/gji/ggw305

Deschamps, F., & Tackley, P. J. (2009). Searching for models of thermo-chemical convection that explain probabilistic tomography. II: Influence of physical and compositional parameters. *Physics of Earth and Planetary Interiors*, *176*, 1–18.

Dilek, Y., & Altunkaynak, Ş. (2007). Cenozoic crustal evolution and mantle dynamics of post-collisional magmatism in Western Anatolia. *International Geology Review*, *49*(5), 431–453. https://doi.org/10.2747/0020-6814.49.5.431

Dilek, Y., & Sandvol, E. (2009). Seismic structure, crustal architecture and tectonic evolution of the Anatolian-African plate boundary and the Cenozoic orogenic belts in the eastern Mediterranean region. *Geological Society, London, Special Publications*, *327*, 127–160. https://doi.org/10.1144/SP327.8

Di Rosa, M., Farina, F., Marroni, M., Pandolfi, L., Göncüoglu, M. C., Ellero, A., et al. (2019). U-Pb zircon geochronology of intrusive rocks from an exotic block in the Late Cretaceous–Paleocene Tarakli Flysch (northern Turkey): Constraints on the tectonics of the Intrapontide suture zone. *Journal of Asian Earth Sciences*, *171*, 277–288.

Doglioni, C., Agostini, S., Crespi, M., Innocenti, F., Manetti, P., Riguzzi, F., et al. (2002). On the extension in western Anatolia and the Aegean sea. In G. Rosenbaum & G. S. Lister (Eds.), Reconstruction of the evolution of the Alpine-Himalayan orogen. *Journal of the Virtual Explorer*, *8*, 161–176.

Dominey-Howes, D. (2002). Documentary and geological records of tsunamis in the Aegean Sea region of Greece and their potential value to risk assessment and disaster management. *Natural Hazards*, *25*, 195–224. https://doi.org/10.1023/A:1014808804611

Doutsos, T., & Kokkalas, S. (2001). Stress and deformation patterns in the Aegean region. *Journal of Structural Geology*, *23*, 455–472.

Ebeling, C. W., Okal, E. A., Kalligeris, N., & Synolakis, C. E. (2012). Modern seismological reassessment and tsunami simulation of historical Hellenic Arc earthquakes. *Tectonophysics*, *530–531*, 225–239. https://doi.org/10.1016/j.tecto.2011.12.036

Elitez, I., Yaltırak, C., & Aktuğ, B. (2016). Extensional and compressional regime driven left-lateral shear in southwestern Anatolia (eastern Mediterranean): The Burdur-Fethiye Shear Zone. *Tectonophysics*, *688*, 26–35. https://doi.org/10.1016/j.tecto.2016.09.024

Elmas, A. (2012). Basement types of the Thrace Basin and a new approach to the pre-Eocene tectonic evolution of the northeastern Aegean and northwestern Anatolia: A review of data and concepts. *International Journal of Earth Sciences*, *101*, 1895–1911. https://doi.org/10.1007/s00531-012-0756-5

El-Sharkawy, A., Meier, T., Lebedev, S., Behrmann, J. H., Hamada, M., Cristiano, L., et al. (2020). The slab puzzle of the Alpine-Mediterranean region: Insights from a new, high-resolution, shear wave velocity model of the upper mantle. *Geochemistry, Geophysics, Geosystems*, *21*, e2020GC008993. https://doi.org/10.1029/2020GC008993

Emery, K. O., Heezen, B., & Allan, T. D. (1966). Bathymetry of the Eastern Mediterranean sea. *Deep Sea Research*, *13*, 173–192.

England, P., Howell, A., Jackson, J., & Synolakis, C. (2015). Palaeotsunamis and tsunami hazards in the Eastern Mediterranean. *Philosophical Transactions of the Royal Society A: Mathematical, Physical and Engineering Sciences*, *373*, 2053. https://doi.org/10.1098/rsta.2014.0374

Ergün, M., Okay, S., Sari, C., Oral, E. Z., Ash, M., Hall, J., & Miller, H. (2005). Gravity anomalies of the Cyprus Arc and their tectonic implications. *Marine Geology*, *221*(1–4), 349–358. https://doi.org/10.1016/j.margeo.2005.03.004

Evangelidis, C. P. (2017). Seismic anisotropy in the Hellenic subduction zone: Effects of slab segmentation and subslab mantle flow. *Earth and Planetary Science Letters*, *480*, 97–106. https://doi.org/10.1016/j.epsl.2017.10.003

Faccenda, M., Burlini, L., Gerya, T. V., & Mainprice, D. (2008). Fault-induced seismic anisotropy by hydration in subducting oceanic plates. *Nature*, *455*(7216), 1097–1100. https://doi.org/10.1038/nature07376

Faccenna, C., Becker, T. W., Auer, L., Billi, A., Boschi, L., Brun, J.-P., Capitanio, F. A., et al. (2014). Mantle dynamics in the Mediterranean. *Review of Geophysics*, *52*, 283– 332. https://doi.org/10.1002/2013RG000444

Faccenna, C., Becker, T. W., Jolivet, L., & Keskin, M. (2013). Mantle convection in the Middle East: Reconciling Afar upwelling, Arabia indentation and Aegean trench rollback. *Earth and Planetary Science Letters*, *375*, 254–269. https://doi.org/10.1016/j.epsl.2013.05.043

Faccenna, C., Bellier, O., Martinod, J., Piromallo, C., & Regard, V. (2006). Slab detachment beneath eastern Anatolia: A possible cause for the formation of the North Anatolian fault. *Earth and Planetary Science Letters*, *242*(1–2), 85–97. https://doi.org/10.1016/j.epsl.2005.11.046.

Faccenna, C., Jolivet, L., Piromallo, C., & Morelli A. (2003). Subduction and the depth of convection in the Mediterranean mantle. *Journal of Geophysical Research*, *108*(B2), https://doi.org/10.1029/2001JB001690

Ferentinos, G. (1990). Offshore geological hazards in the Hellenic arc. *Marine Geotechnology*, *9*(4), 261–277. https://doi.org/10.1080/10641199009388244

Fernandes, R. M. S., Bastos, L., Miranda, J. M., Lourenço, N., Ambrosius, B. A. C., Noomen, R., et al. (2006). Defining the plate boundaries in the Azores region. *Journal of Volcanology and Geothermal Research*, *156*(1–2), 1–9. https://doi.org/10.1016/j.jvolgeores.2006.03.019.

Flerit, F., Armijo, R., King, G., & Meyer, B. (2004). The mechanical interaction between the propagating North Anatolian Fault and the back-arc extension in the Aegean. *Earth and Planetary Science Letters*, *224*(3–4), 347–362. https://doi.org/10.1016/j.epsl.2004.05.028

Fornash, K. F., & Whitney, D. L. (2020). Lawsonite-rich layers as records of fluid and element mobility in subducted crust (Sivrihisar Massif, Turkey). *Chemical Geology*, *533*, 119356, https://doi.org/10.1016/j.chemgeo.2019.119356

Foucher, J. P., Chamot-Roocke, N., Alexandry, S., Augustin, J. M., Monti, S., Pavlakis, P., & Voisset, M. (1993). Multibeam bathymetry and seabed reflectivity maps of the MEDRIFT corridor across the eastern Mediterranean Ridge. *Terra Cognita*, 278–279.

Frassi, C., Marroni, M., Pandolfi, L., Göncüoğlu, M. C., Ellero, A., Ottria, G., et al. (2018). Burial and exhumation history of the Daday Unit (central Pontides, Turkey): Implications for the closure of the Intra-Pontide oceanic basin. In G. Capponi, A. Festa, & G. Rebay (Eds.), *Birth and death of oceanic basins: Geodynamic processes from rifting to continental collision in Mediterranean and circum-Mediterranean orogeny. Geological Magazine*, *155*, 356–376.

Fytikas, M., Innocenti, F., Manetti, P., Mazzuoli, R., Peccerillo, A., & Villari, L. (1984). Tertiary to quaternary evolution of volcanism in the Aegean region. In J. E. Dixon & A. H. F. Robertson (Eds.), *The geological evolution of the eastern Mediterranean* (pp. 687–699). Geological Society of London, Special Publications, 17.

Gaina, C., Torsvik, T. H., van Hinsbergen, D. J. J., Medvedev, S., Werner, S. C., & Labails, C. (2013). The African plate: A history of oceanic crust accretion and subduction since the Jurassic. *Tectonophysics*, *604*, 4–25. https://doi.org/10.1016/j.tecto.2013.05.037

Galindo-Zaldivar, J., Nieto, L., & Woodside, J. (1996). Structural features of mud volcanoes and the fold system of the Mediterranean Ridge, south of Crete. *Marine Geology*, *132*(1–4), 95–112. https://doi.org/10.1016/0025-3227(96)00155-7

Ganas, A., & Parsons, T. (2009). Three-dimensional model of Hellenic Arc deformation and origin of the Cretan uplift. *Journal of Geophysical Research*, *114*(B06404), 1–14. https://doi.org/10.1029/2008JB005599

Ganas, A., Briole, P., Bozionelos, G., Barberopoulou, A., Elias, P., Tsironi, V., et al. (2020). The 25 October 2018 Mw = 6.7

Zakynthos earthquake (Ionian Sea, Greece): A low-angle fault model based on GNSS data, relocated seismicity, small tsunami and implications for the seismic hazard in the west Hellenic Arc. *Journal of Geodynamics*, *137*, https://doi.org/10.1016/j.jog.2020.101731

Ganas A., Marinou, A., Anastasiou, D., Paradissis, D., Papazissi, K., Tzavaras, P., et al. (2013). GPS-derived estimates of crustal deformation in the central and north Ionian Sea, Greece: 3-yr results from NOANET continuous network data. *Journal of Geodynamics*, *67*, 62–71.

Gautier, P., Brun, J-P., Moriceau, R., Sokoutis, D., Martinod, J., & Jolivet, L. (1999). Timing, kinematics and cause of Aegean extension: a scenario based on a comparison with simple analogue experiments. *Tectonophysics*, *315*(1–4), 31–72. https://doi.org/10.1016/S0040-1951(99)00281-4

Gautier, Y. (1984). Déformations et métamorphismes associés à la fermeture téthysienne en Anatolie centrale (région de Sivrihisar, Turquie). Ph.D. dissertation. University Paris-Sud, France, 236.

Gerya, T. (2011). Future directions in subduction modeling. *Journal of Geodynamics*, *52*, 344–378.

Gessner, K., Gallardo, L. A., Markwitz, V., Ring, U., & Thomson, S. N. (2013). What caused the denudation of the Menderes Massif: Review of crustal evolution, lithosphere structure, and dynamic topography in southwest Turkey. *Gondwana Research*, *24*(1), 243–274.

Gessner, K., Markwitz, V., & Güngör, T. (2017). Crustal fluid flow in hot continental extension: tectonic framework of geothermal areas and mineral deposits in western Anatolia. *Geological Society, London, Special Publications*, *453*, 289–311. https://doi.org/10.1144/SP453.7

Gianni, G. M., Navarrete, C., & Spagnotto, S. (2019). Surface and mantle records reveal an ancient slab tear beneath Gondwana. *Scientific Reports*, *9*, 19774. https://doi.org/10.1038/s41598-019-56335-9

Göğüş, O. H., Pysklywec, R. N., Şengör, A. M. C., et al. (2017). Drip tectonics and the enigmatic uplift of the Central Anatolian plateau. *Nature Communication*, *8*, 1538. https://doi.org/10.1038/s41467-017-01611-3

Göncüoğlu, M. C. (2010). Introduction to the geology of Turkey: Geodynamic evolution of the Pre-Alpine and Alpine terrains. *General Directorate of Mineral Research and Exploration Monography Series*, *5*, 1–66.

Göncüoğlu, M. C., Marroni, M., Pandolfi, L., Ellero, A., Ottria, G., Catanzariti, R., et al. (2014). The Arkot Dag Melange in Arac area, central Turkey; evidence of its origin within the geodynamic evolution of the Intra-Pontide suture zone. *Journal of Asian Earth Sciences*, *85*, 117–139. https://doi.org/10.1016/j.jseaes.2014.01.013

Göncüoğlu, M. C., Marroni, M., Sayit, K., Tekin, U. K., Ottria, G., Pandolfi, L., et al. (2012). The Ayli Dag ophiolite sequence (central-northern Turkey): A fragment of Middle Jurassic oceanic lithosphere within the Intra-Pontide suture zone. *Ofioliti*, *37*, 77–92.

Gordon, R. G., & Stein, S. (1992). Global tectonics and space geodesy. *Science*, *17*, 333–342.

Govers, R., & Fichtner, A. (2016). Signature of slab fragmentation beneath Anatolia from full-waveform tomography. *Earth and Planetary Science Letters*, *450*, 10–19. https://doi.org/10.1016/j.epsl.2016.06.014

Govers, R., & Wortel, M. J. R. (2005). Lithosphere tearing at STEP faults: Response to edges of subduction zones. *Earth and Planetary Science Letters*, *236*(1–2), 505–523. https://doi.org/10.1016/j.epsl.2005.03.022

Gürer, Ö., Sangu, E., & Özburan, M. (2006). Neotectonics of the SW Marmara region, NW Anatolia, *Turkey. Geological Magazine*, *143* (2), 229–241. https://doi.org/10.1017/S0016756805001469

Gurnis, M., Hall, C., & Lavier, L. (2004). Evolving force balance during incipient subduction. *Geochemistry, Geophysics, Geosystems*, *5*, Q07001. https://doi.org/10.1029/2003GC000681

Güvercin, S.E., Konca, A. Ö., Özbakır, A. D., Ergintav, S., & Karabulut, H. (2021). New focal mechanisms reveal fragmentation and active subduction of the Antalya slab in the Eastern Mediterranean. *Tectonophysics*, *805*, 228792. https://doi.org/10.1016/j.tecto.2021.228792

Hall, C. E., Gurnis, M., Sdrolias, M., Lavier, L. L., & Müller, R. D. (2003). Catastrophic initiation of subduction following forced convergence across fracture zones. *Earth and Planetary Science Letters*, *212*(1–2), 15–30.

Hall, J., Aksu, A. E., Elitez, I., Yaltırak, C., & Çifçi, G. (2014). The Fethiye-Burdur Fault Zone: A component of upper plate extension of the subduction transform edge propagator fault linking Hellenic and Cyprus Arcs, Eastern Mediterranean. *Tectonophysics*, *635*, 80–99. https://doi.org/10.1016/j.tecto.2014.05.002

Hall, J., Aksu, A. E., Yaltırak, C., & Winsor, J. D. (2009). Structural architecture of the Rhodes Basin: A deep depocentre that evolved since the Pliocene at the junction of Hellenic and Cyprus Arcs, eastern Mediterranean. *Marine Geology*, *258*(1–4), 1–23. https://doi.org/10.1016/j.margeo.2008.02.007

Hall, R., Audley-Charles, M. G., & Carter, D. J. (1984). The significance of Crete for the evolution of the Eastern Mediterranean. *Geological Society, London, Special Publications*, *17*, 499–516. https://doi.org/10.1144/GSL.SP.1984.017.01.37

Halpaap, F., Rondenay, S., & Ottemöller, L. (2018). Seismicity, deformation, and metamorphism in the Western Hellenic Subduction Zone: New constraints from tomography. *Journal of Geophysical Research: Solid Earth*, *123*, 3000–3026. https://doi.org/10.1002/2017JB015154

Hansen, S. E., Evangelidis, C. P., & Papadopoulos, G. A. (2019). Imaging slab detachment within the Western Hellenic Subduction Zone. *Geochemistry, Geophysics, Geosystems*, *20*, 895–912. https://doi.org/10.1029/2018GC007810

Hayes, G. (2018). Slab2: A comprehensive subduction zone geometry model: U.S. *Geological Survey data release.* https://doi.org/10.5066/F7PV6JNV

Heezen, B. C., & Ewing, M. (1963). The midoceanic ridge. In M. N. Hill (Ed.), *The seas* (vol. *3*, pp. 388–410). New York: Interscience.

Hooft, E. E. E., Nomikou, P., Toomey, D. R., Lampridou, D., Getz, C., Christopoulou, M-E., et al. (2017). Backarc tectonism, volcanism, and mass wasting shape seafloor morphology in the Santorini-Christiana-Amorgos region of the Hellenic Volcanic Arc. *Tectonophysics*, *712–713*, 396–414. https://doi.org/10.1016/j.tecto.2017.06.005

Hosseini, K., Matthews, K.J., Sigloch, K., Shephard, G.E., Domeier, M., & Tsekhmistrenko, M. (2018). SubMachine: Web-Based tools for exploring seismic tomography and other models of Earth's deep interior. *Geochemistry, Geophysics, Geosystems*, *19*. https://doi.org/10.1029/2018GC007431

Howell, A., Jackson, J., England, P., Higham, T., & Synolakis, C. (2015). Late Holocene uplift of Rhodes, Greece: Evidence for a large tsunamigenic earthquake and the implications for the tectonics of the eastern Hellenic Trench System. *Geophysical Journal International*, *203*(1), 459–474.

Hubert-Ferrari, A., Armijo, R., King, G., Meyer, B., & Barka, A. (2002). Morphology, displacement, and slip rates along the North Anatolian Fault, *Turkey. Journal of Geophysical Research*, *107*(B10), 2235. https://doi.org/10.1029/2001JB000393

Huchon, P., Lybéris, N., Angelier, J., Le Pichon, X., & Renard, V. (1982). Tectonics of the hellenic trench: A synthesis of sea-beam and submersible observations. *Tectonophysics*, *86*(1–3), 69–112. https://doi.org/10.1016/0040-1951(82)90062-2

Husson, L., Brun, J-P., Yamato, P., & Faccenna, C. (2009). Episodic slab rollback fosters exhumation of HP-UHP rocks. *Geophysics Journal International*, *179*, 1292–1300

Jackson, J. (1994). Active tectonics of the Aegean region. *Annual Review of Earth and Planetary Sciences*, *22*, 239–271.

Jackson, J., & McKenzie, D. (1988). The relationship between plate motions and seismic moment tensors and the rates of active deformation in the Mediterranean and Middle East. *Geophysical Journal*, *93*, 45–73.

Jolivet, L., & Brun, J.-P. (2010). Cenozoic geodynamic evolution of the Aegean. *International Journal of Earth Sciences*, *99*(1), 109–138. https://doi.org/10.1007/s00531-008-0366-4

Jolivet, L., & Faccenna, C. (2000). Mediterranean extension and the Africa-Eurasia collision. *Tectonics*, *19*(6), 1095– 106. https://doi.org/10.1029/2000TC900018

Jolivet, L., Faccenna, C., Goffé, B., Burov, E., & Agard, P. (2003). Subduction tectonics and exhumation of high-pressure metamorphic rocks in the Mediterranean orogens. *American Journal of Science*, *303*, 353–409. https//doi.org/10.2475/ajs.303.5.353

Jolivet, L., Faccenna, C., Huet, B., Labrousse, L., Le Pourhiet, L., et al. (2013). Aegean tectonics: Strain localisation, slab tearing and trench retreat. *Tectonophysics*, *597–598*, 1–33.

Jolivet, L., Menant, A., Clerc, C., Sternai, P., Bellahsen, N., Leroy, S., Pik, R., et al. (2018). Extensional crustal tectonics and crust-mantle coupling, a view from the geological record. *Earth-Science Reviews*, *185*, 1187–1209. https://doi.org/10.1016/j.earscirev.2018.09.010

Jolivet, L., Menant, A., Sternai, P., Rabillard, A., Arbaret, L., Augier, R., et al. (2015). The geological signature of a slab tear below the Aegean, *Tectonophysics*, *659*, 166–182. https://doi.org/10.1016/j.tecto.2015.08.004

Jost, M. L., Knabenbauer, O., Cheng, J., & Harjes, H-P. (2002). Fault plane solutions of microearthquakes and small events in the Hellenic arc. *Tectonophysics*, *356* (1–3), 87–114. https://doi.org/10.1016/S0040-1951(02)00378-5

Karabulut, H., Paul, A., Özbakır, A. D., Ergün, T., & Şentürk, S. (2019). A new crustal model of the Anatolia-Aegean domain: Evidence for the dominant role of isostasy in the support of the Anatolian plateau. *Geophysical Journal International*, *218*(1), 57–73. https://doi.org/10.1093/gji/ggz147

Karagianni, E. E., Papazachos, C. B., Panagiotopoulos, D. G., Suhadolc, P., Vuan, A., & Panza, G. F. (2005). Shear velocity structure in the Aegean area obtained by inversion of Rayleigh waves. *Geophysical Journal International*, *160*(1), 127–143. https://doi.org/10.1111/j.1365-246X.2005.02354.x

Karaoğlu, Ö, & Helvacı, C. (2014). Isotopic evidence for a transition from subduction to slab-tear related volcanism in western Anatolia, Turkey. *Lithos*, *192–195*, 226–239. https://doi.org/10.1016/j.lithos.2014.02.006

Kárason, H., & van der Hilst R. (2000). Constraints on mantle convection from seismic tomography. *Geophysical Monograph*, *121*, 277–288.

Karato, S.-I. (1995). Effects of water on seismic wave velocities in the upper mantle. *Proceedings of the Japan Academy, Ser. B*, *71*, 61–66.

Kassaras, I., Makropoulos, K., Bourova, E., Pedersen, H., & Hatzfeld, D. (2005). 3: Upper mantle structure of the Aegean derived from two-station phase velocities of fundamental mode Rayleigh waves. In M. Fytikas & G. E. Vougioukalakis (Eds). *Developments in Volcanology*, *7*, 19–45. https://doi.org/10.1016/S1871-644X(05)80031-2

Kastens, K. A. (1991). Rate of outward growth of the Mediterranean Ridge accretionary complex. *Tectonophysics*, *199*, 25–50.

Kastens, K. A., Breen, N. A., & Cita, M. B. (1992). Progressive deformation of an evaporite-bearing accretionary complex: SeaMARC I, SeaBeam and piston-core observations from the Mediterranean Ridge. *Marine Geophysics Research*, *14*, 249–298. https://doi.org/10.1007/BF01203620

Kaya, O. (1981). Miocene reference section for the coastal parts of West Anatolia. *Newsletters on Stratigraphy*, *10*, 164–191.

Kempler, D., & Ben-Avraham, Z. (1987). The tectonic evolution of the Cyprean Arc. *Annales Tectonicae*, *1*, 58–71.

Kenyon, N. H., Belderson, R. H., & Stride, A. H. (1982). Detailed tectonic trends on the central part of the Hellenic outer ridge and in the Hellenic trench system. *Geological Society of London*, *10*, 335–343.

Keranen, K., & Klemperer, S. L. (2008). Discontinuous and diachronous evolution of the Main Ethiopian Rift: Implications for development of continental rifts. *Earth and Planetary Science Letters*, *265*(1–2), 96–111. https://doi.org/10.1016/j.epsl.2007.09.038

Ketin, I. (1948). Über die tektonisch-mechanischen Folgerungen aus den großen anatolischen Erdbeben des letzten Dezenniums. *International Journal of Earth Sciences*, *36*, 77–83. https://doi.org/10.1007/BF01791916

Kind, R., Eken, T., Tilmann, F., Sodoudi, F., Taymaz, T., Bulut, F., et al. (2015). Thickness of the lithosphere beneath Turkey and surroundings from S-receiver functions. *Solid Earth*, *6*, 971–984. https://doi.org/10.5194/se-6-971-2015

Kinnaird, T., & Robertson, A. (2012). Tectonic and sedimentary response to subduction and incipient continental collision in southern Cyprus, easternmost Mediterranean region. *Geological Society, London, Special Publications*, *372*, 585–614. https://doi.org/10.1144/SP372.10

Kokinou, E., Papadimitriou, E., Karakostas, V., Kamberis, E., & Vallianatos, F. (2006). The Kefalonia Transform Zone (offshore Western Greece) with special emphasis to its prolongation toward the Ionian Abyssal Plain. *Marine Geophysical Researches, Dordrecht*, *27*(4), 241–252. https://doi.org/10.1007/s11001-006-9005-2

Kokkalas, S., Paraskevas, X., Koukouvelas, I., & Doutsos, T. (2006). Postcollisional contractional and extensional deformation in the Aegean region. *Special Paper, Geological Society of America*, *409*, 97–123. https://doi.org/10.1130/0-8137-2409-0.97

Kopf, A., Mascle, J., & Klaeschen, D. (2003). The Mediterranean ridge: A mass balance across the fastest growing accretionary complex on Earth. *Journal of Geophysical Research*, *108*(B8), 2372. https://doi.org/10.1029/2001JB000473

Kopf, A. J. (2002). Significance of mud volcanism. *Reviews in Geophysics*, *40*(2), 1005. https://doi.org/10.1029/2000RG000093

Kreemer, C., Chamot-Rooke, N., & Le Pichon, X. (2004). Constraints on the evolution and vertical coherency of deformation in the Northern Aegean from a comparison of geodetic, geologic and seismologic data. *Earth and Planetary Science Letters*, *225*(3–4), 329–346. https://doi.org/10.1016/j.epsl.2004.06.018

Laigle, M., Hirn, A., Sachpazi, M., & Clément, C. (2002). Seismic coupling and structure of the Hellenic subduction zone in the Ionian Islands region. *Earth and Planetary Science Letters*, *200*(3–4), 243–253. https://doi.org/10.1016/S0012-821X(02)00654-4

Laigle, M., Sachpazi, M., & Him, A. (2004). Variation of seismic coupling with slab detachment and upper plate structure along the western Hellenic subduction zone. *Tectonophysics*, *391*, 85–95.

Lamont, T. N., Searle, M. P., Roberts, N. M. W., Waters, D. J., Palin, R. M., Smye, A. J., et al. (2019). Compressional origin to the Naxos metamorphic core complex, Greece: structure, petrography and thermobarometry. *Geological Society of America Bulletin*, *132*, 149–197. https://doi.org/10.1130/B31978.1

Legendre, C. P., Meier, T., Lebedev, S., Friederich, W., & Viereck-Götte, L. (2012). A shear wave velocity model of the European upper mantle from automated inversion of seismic shear and surface waveforms. *Geophysical Journal International*, *191*(1), 282–304. https://doi.org/10.1111/j.1365-246X.2012.05613.x

Le Pichon, X., & Angelier, J. (1979). The Hellenic arc and trench system: A key to the neotectonic evolution of the eastern Mediterranean area. *Tectonophysics*, *60*, 1–42.

Le Pichon, X., & Angelier, J. (1981). The Aegean Sea. *Philosophical Transactions of the Royal Society of London. Series A., Mathematical and Physical Sciences*, *300*, 357–372, http://doi.org/10.1098/rsta.1981.0069

Le Pichon, X., Chamot-Rooke, N., & Lallemant, S. (1995). Geodetic determination of the kinematics of central Greece with respect to Europe: Implications for eastern Mediterranean tectonics. *Journal of Geophysical Research*, *100*, 12,675-12,690. https://doi.org/10.1029/95JB00317

Le Pichon, X., Henry, P., & Lallemand, S. (1993). Accretion and erosion in subduction zones: The role of fluids. *Annual Reviews in Earth and Planetary Science*, *21*, 307–331.

Le Pichon, X., Lallemant, S. J., Chamot-Rooke, N., Lemeur, D., & Pascal, G. (2002). The Mediterranean Ridge backstop and the Hellenic nappes, *Marine Geology*, *186*(1–2), 111–125. https://doi.org/10.1016/S0025-3227(02)00175-5

Le Pichon, X., Lybéris, N., Angelier, J., & Renard, V. (1982). Strain distribution over the East Mediterrranean Ridge: A synthesis incorporating new sea-beam data. *Tectonophysics*, *86*, 243–274. https://doi.org/10.1016/0040-1951(82)90069-5

Le Pichon, X., Şengör, A. M. C., & İmren. C. (2019). A new approach to the opening of the eastern Mediterranean Sea and the origin of the Hellenic subduction zone. Part 2: The Hellenic subduction zone. *Canadian Journal of Earth Sciences*, *56*(11), 1144–1162. https://doi.org/10.1139/cjes-2018-0315

Limonov, A. F., Woodside, J. M., Cita, M. B., & Ivanov, M. K. (1996). The Mediterranean Ridge and related mud diapirism: A background. *Marine Geology*, *132*(1–4), 7–19. https://doi.org/10.1016/0025-3227(96)00150-8

Lyberis, N. (1984). Tectonic evolution of the North Aegean trough. *Geological Society, London, Special Publications*, *17*, 709–725.

Lykousis, V., Alexandri, S., Woodside, J., de Lange, G., Dählmann, A., Perissoratis, C., et al. (2009). Mud volcanoes and gas hydrates in the Anaximander mountains (Eastern Mediterranean Sea). *Marine and Petroleum Geology*, *26*(6), 854–872. https://doi.org/10.1016/j.marpetgeo.2008.05.002

Maffione, M., & van Hinsbergen, D. J. J. (2018). Reconstructing plate boundaries in the Jurassic Neo-Tethys from the East and West Vardar Ophiolites (Greece and Serbia). *Tectonics*, *37*, 858– 887. https://doi.org/10.1002/2017TC004790

Maggini, M., & Caputo, R. (2020). Sensitivity analysis for crustal rheological profiles: Examples from the Aegean Region. *Annals of Geophysics*, *63*(3), GT334. http://dx.doi.org/10.4401/ag-8244

Mahatsente, R., Alemdar, S., & Çemen, İ. (2017). Effect of Slab-Tear on Crustal Structure in Southwestern Anatolia. In İ. Çemen & Y. Yılmaz (Eds.), *Active global seismology* (103–119). Geophysical Monograph Series. https://doi.org/10.1002/9781118944998.ch4

Makris, J. (1978). The crust and upper mantle of the Aegean region from deep seismic soundings. *Tectonophysics*, *46*(3–4), 269–284. https://doi.org/10.1016/0040-1951(78)90207-X

Malandri, C., Soukis, C., Maffione, M., Özkaptan, M., Vassilakis, E., Lozios, S., et al. (2017). Vertical-axis rotations accommodated along the Mid-Cycladic lineament on Paros Island in the extensional heart of the Aegean orocline (Greece). *Lithosphere*, *9*(1), 78–99. https://doi.org/10.1130/L575

Mantovani, E., Albarello, D., Tamburelli, C., Babbucci, D., & Viti, M. (1997). Plate convergence, crustal delamination, extrusion tectonics and minimization of shortening work as main controlling factors of the recent Mediterranean deformation pattern. *Annals of Geophysics*, *40*, 611–643.

Marone, F., van der Lee, S., & Giardini, D. (2004a). Three-dimensional upper-mantle S-velocity model for the Eurasia-Africa plate boundary region. *Geophysical Journal International*, *158*(1), 109–130. https://doi.org/10.1111/j.1365-246X.2004.02305.x

Marone, F., van der Lee, S., & Giardini, D. (2004b). Shallow anisotropy in the Mediterranean mantle from surface waves. *Geophysical Research Letters*, *31*(6). https://doi.org/10.1029/2003GL018948

Marone, F., van der Meijde, M., van der Lee, S., & Giardini, D. (2003). Joint inversion of local, regional and teleseismic data for crustal thickness in the Eurasia-Africa plate boundary region. *Geophysical Journal International*, *154*(2), 499–514. https://doi.org/10.1046/j.1365-246X.2003.01973.x

Marroni, M., Frassi, C., Göncüoğlu, M. C., Di Vincenzo, G., Pandolfi, L., Rebay, G., et al. (2014). Late Jurassic amphibolite-facies metamorphism in the Intra-Pontide Suture Zone (Turkey): An eastward extension of the Vardar Ocean from the Balkans into Anatolia? *Journal of the Geological Society*, *171*(5), 605–608. https://doi.org/10.1144/jgs2013-104

Marsellos, A. E., Kidd, W. S. F., & Garver, J. I. (2010). Extension and exhumation of the HP/LT rocks in the Hellenic forearc ridge. *American Journal of Science*, *310*(1), 1–36. https://doi.org/10.2475/01.2010.01

Mascle, J., & Chaumillon, E. (1997). Pre-collisional geodynamics of the Mediterranean Sea: The Mediterranean Ridge and the Tyrrhenian Sea. *Annales Geophysicae*, *40*(3), https://www.annalsofgeophysics.eu/index.php/annals/article/view/3891

Mascle, J., Huguen, C., Benkhelil, J., ChamotRooke, N., Chaumillon, E., Foucher, J. P., et al. (1999). Images may show start of European-African plate collision. *Eos Transations AGU*, *80*(37), 421–436.

Mascle, J., Le Cleac'h, A., & Jongsma, D. (1986). The eastern Hellenic margin from Crete to Rhodes: Example of progressive collision. *Marine Geology*, *73*(1–2), 145–168. https://doi.org/10.1016/0025-3227(86)90116-7

McClusky, S., et al. (2000). Global positioning system constraints on plate kinematics and dynamics in the eastern Mediterranean and Caucasus. *Journal of Geophysical Research*, *105*(B3), 5695–5719. https://doi.org/10.1029/1999JB900351

McClusky, S., Reilinger, R., Mahmoud, S., Ben Sari, D., & Tealeb, A. (2003). GPS constraints on Africa (Nubia) and Arabia plate motions. *Geophysical Journal International*, *155*(1), 126–138.

McKenzie, D. (1972). Active tectonics of the Mediterranean region. *Geophysical Journal of the Royal Astronomical Society*, *30*, 109–185. https://doi.org/10.1111/j.1365-246X.1972.tb02351.x

McKenzie, D. (1978). Active tectonics of the Alpine-Himalayan belt: The Aegean Sea and surrounding regions. *Geophysics Journal of the Royal Astronomical Society*, *55*, 217–254.

Meier, T., Becker, D., Endrun, B., Rische, M., Bohnhoff, M., Stöckhert, B.,et al. (2007). A model for the Hellenic subduction zone in the area of Crete based on seismological investigations. *Geological Society, London, Special Publications*, *291*, 183–199. https://doi.org/10.1144/SP291.9

Meier, T., Dietrich, K., Stöckhert, B., & Harjes, H-P. (2004). One-dimensional models of shear wave velocity for the eastern Mediterranean obtained from the inversion of Rayleigh wave phase velocities and tectonic implications. *Geophysical Journal International*, *156*(1), 45–58. https://doi.org/10.1111/j.1365-246X.2004.02121.x

Meighan, H. E., ten Brink, U., & Pulliam, J. (2013). Slab tears and intermediate-depth seismicity. *Geophysical Research Letters*, *40*, 4244– 4248. https://doi.org/10.1002/grl.50830

Meijer, P. T., & Wortel, M. J. R. (1999). Cenozoic dynamics of the African plate with emphasis on the Africa-Eurasia collision. *Journal of Geophysical Research*, *104*(B4), 7405–7418. https://doi.org/10.1029/1999JB900009

Mercier, J. L. (1981). Extensional-compressional tectonics associated with the Aegean Arc: comparison with the Andean Cordillera of south Peru-north Bolivia. *Philosophical Transactions of the Royal Society of London A, Mathematical and Physical Sciences*, *300*(1454), 337–355.

Messini, A., Papadimitriou, E., Karakostas, V., & Baskoutas, I. (2007). Stress interaction between thrust faults along the SW Hellenic arc (Greece). *Bulletin of the Geological Society of Greece*, *40*(1), 386–398. https://doi.org/10.12681/bgsg.16625

Meulenkamp, J. E., Wortel, M. J. R., van Wamel, W. A., Spakman, W., & Hoogerduyn, S. E. (1988). On the Hellenic subduction zone and the geodynamic evolution of Crete since the late middle Miocene. *Tectonophysics*, *146*, 203–215.

Moix, P., Beccaletto, L., Kozur, H. W., Hochard, C., Rosselet, F., & Stampfli, G. M. (2008). A new classification of the Turkish terrains and sutures and its implication for the paleotectonic history of the region. *Tectonophysics*, *451*(1–4), 7–39. https://doi.org/10.1016/j.tecto.2007.11.044

Morris, A., & Robertson, A. H. F. (1993). Miocene remagnetisation of carbonate platform and Antalya Complex units within the Isparta angle, *SW Turkey. Tectonophysics*, *220*(1–4), 243–266. https://doi.org/10.1016/0040-1951(93)90234-B

Nikitas, A., Triantaphyllou, M. V., Rousakis, G., Panagiotopoulo, S. I., Pasadakis, N., Hatzianestis, I., & Gogou, A. (2021). Pre-Messinian deposits of the Mediterranean Ridge: biostratigraphic and geochemical evidence from the Olimpi mud volcano field. *Water*, *13*(10), 1367. https://doi.org/10.3390/w13101367

Nyst, M., & Thatcher, W. (2004). New constraints on the active tectonic deformation of the Aegean. *Journal of Geophysical Research*, *109*(11), 23. https://doi.org/10.1029/2003JB002830

Okal, E. A., Synolakis, C. E., Uslu, B., Kalligeris, N., & Voukouvalas, E. (2009). The 1956 earthquake and tsunami in Amorgos, *Greece. Geophysical Journal International*, *178*(3), 1533–1554.https://doi.org/10.1111/j.1365-246X.2009.04237.x

Okay, A. I. (1980a). Mineralogy, petrology and phase relations of glaucophane-lawsonite zone blueschists from the TavŞanlı region, northwest Turkey. *Contributions to Mineralogy and Petrology*, *72*, 243–255.

Okay, A. I. (1980b). Lawsonite zone blueschists and a sodic amphibole producing reaction in the TavŞanlı region, northwest Turkey. *Contributions to Mineralogy and Petrology*, *75*, 179–186.

Okay, A. I. (1982). Incipient blueschist metamorphism and metasomatism in the Tavşanlı region, northwest Turkey. *Contributions to Mineralogy and Petrology*, *79*, 361–367.

Okay, A. I. (1984). Distribution and characteristics of the north-west Turkish blueschists. In J. E. Dixon & A. H. F. Robertson (Eds.), *The geological evolution of the eastern Mediterranean* (pp. 455–466). Geological Society of London Special Publications, 17.

Okay, A. I. (1986). High-pressure/low-temperature metamorphic rocks of Turkey. In B. W. Evans & E. H. Brown (Eds.), *Blueschists and eclogites* (pp. 333–347). Memoir-Geological Society of America, 164.

Okay, A. I. (2008). *Geology of Turkey: A synopsis. Anschnitt*, *21*, 19–42.

Okay, A. I., & Kelley, S. P. (1994). Tectonic setting, petrology and geochronology of jadeite + glaucophane and chloritoid

+ glaucophane schists from north-west Turkey. *Journal of Metamorphic Geology*, *12*, 455–466.

Okay A. I., & Satir, M. (2000). Coeval plutonism and metamorphism in a latest Oligocene metamorphic core complex in northwest Turkey. *Geological Magazine*, *137*, 495–516.

Okay, A. I., & Satır, M. (2006). Geochronology of Eocene plutonism and metamorphism in northwest Turkey: Evidence for a possible magmatic arc. *Geodinamica Acta*, *19*, 251–265.

Okay A. I., & Tüysüz, O. (1999). Tethyan sutures of northern Turkey. In B. Durand, et al. (Eds.), *The Mediterranean basins: Tertiary extension with the Alpine orogen* (pp. 475–515). Geological Society, London, Special Publications, 156. https://doi.org/10.1144/GSL.SP.1999.156.01.22

Okay, A. I., & Whitney, D. L. (2010). Blueschists, eclogites, ophiolites and suture zones in northwest Turkey: A review and a field excursion guide. *Ofioliti*, *35*(2), 131–172.

Okay, A. I., Satır, M., Maluski, H., Siyako, M., Monie, P., Metzger, R., & Akyüz, S. (1996). Paleo- and Neo-Tethyan events in northwestern Turkey: Geologic and geochronologic constraints. In T. M. Harrison (Ed.), *The tectonic evolution of Asia* (pp. 420–441). Cambridge: Cambridge University Press.

Okay, A. I., Satır, M., Tuysuz, O., Akyuzm S., & Chen, F. (2001). The tectonics of Strandja Massif: late-Variscan and mid-Mesozoic deformation and metamorphism in the northern Aegean. *International Journal of Earth Sciences*, *90*, 217–233.

Okay, A. I., Siyako, M., & Burkan, K. A. (1991). Geology and tectonic evolution of the Biga Peninsula, northwest Turkey. *Bulletin of Technical University Istanbul*, *44*, 191–256.

Okay, A. I., Sunal, G., Sherlock, S., Altner, D., Tuysuz, O., Kylander-Clark, A. R. C., et al. (2013). Early Cretaceous sedimentation and orogeny on the active margin of Eurasia; southern-central Pontides, Turkey. *Tectonics*, *32*, 1247–1271.

Okay, A. I., Sunal, G., Sherlock, S., Kylander-Clark, A. R. C., & Özcan, E. (2020). İzmir-Ankara suture as a Triassic to Cretaceous plate boundary: Data from central Anatolia. *Tectonics*, *39*, e2019TC005849. https://doi.org/10.1029/2019TC005849

Oner, Z., Dilek, Y., & Kadioglu, Y. K. (2010). Geology and geochemistry of the synextensional Salihli granitoid in the Menderes core complex, western Anatolia, *Turkey. International Geology Review*, *52*(2–3), 336–368. https://doi.org/10.1080/00206810902815871

Oral, M. B., Reilinger, R., Toksöz, M., King, R.W., Barka, A., Kinik, I., et al. (1995). Global positioning system offers evidence of plate motions in eastern Mediterranean. *Eos*, *76*, 9–11.

Özacar, A. A., Zandt, G., Gilbert, H., & Beck, S. L. (2010). Seismic images of crustal variations beneath the East Anatolian plateau (Turkey) from teleseismic receiver functions. *Geological Society, London, Special Publications*, *340*, 485–496. https://doi.org/10.1144/SP340.21

Özbakır A. D., Govers, R., & Fichtner, A. (2020). The Kefalonia Transform Fault: A STEP fault in the making. *Tectonophysics*, *787*, 228471. https://doi.org/10.1016/j.tecto.2020.228471.

Özbakır, A. D., Şengör, A. M. C., Wortel, M. J. R., & Govers, R. (2013). The Pliny-Strabo trench region: A large shear zone resulting from slab tearing. *Earth and Planetary Science Letters*, *375*, 188–195. https://doi.org/10.1016/j.epsl.2013.05.025

Özdamar, S., Billor, M. Z., Sunal, G., Esenli, F., & Roden, N. F. (2013). First U-Pb SHRIMP zircon and 40Ar/39Ar ages of metarhyolites from the Afyon-Bolkardag zone, SW Turkey: Implications for the rifting and closure of the Neo-Tethys. *Gondwana Research*, *24*(1), 377–391. https://doi.org/10.1016/j.gr.2012.10.006

Papadimitriou, E. E., & Karakostas, V. G. (2005). *Faulting geometry and seismic coupling of the southwest part of the Hellenic subduction zone*. Abstract in 33d IASPEI General Assembly, Santiago, Chile, 2–8 October 2005.

Papadimitriou, E. E., & Karakostas, V. G. (2008). Rupture model of the great AD 365 Crete earthquake in the southwestern part of the Hellenic Arc. *Acta Geophysica*, *56*(2), 293–312. https://doi.org/10.2478/s11600-008-0001-6

Papadopoulos, G. A. (1997). On the interpretation of large-scale seismic tomography images in the Aegean sea area. *Annals of Geophysics*, *40*(1), 37–42. https://doi.org/10.4401/ag-3933

Papadopoulos, G. A., Daskalaki, E., Fokaefs, A., & Giraleas, N. (2007). Tsunami hazards in the Eastern Mediterranean: Strong earthquakes and tsunamis in the East Hellenic Arc and Trench system. *Natural Hazards Earth System Science*, *7*, 57–64. https://doi.org/10.5194/nhess-7-57-2007

Papadopoulos, G. A., Daskalaki, E., Fokaefs, A., & Giraleas, N. (2010). Tsunami hazard in the Eastern Mediterranean Sea: Strong earthquakes and tsunamis in the West Hellenic arc and Trench System. *Journal of Earthquake and Tsunami*, *4*(3), 145–179.

Papanikolaou, D. (2010). Major paleogeographic, tectonic and geodynamic changes from the last stage of the Hellenides to the actual Hellenic arc and trench system. *Bulletin of the Geological Society of Greece*, *43*(1), 72–85. https://doi.org/10.12681/bgsg.11161

Papanikolaou, D. I. (2021). The Mediterranean. In *The geology of Greece. regional geology reviews*. Springer, Cham. https://doi.org/10.1007/978-3-030-60731-9_3

Papanikolaou, D. J., & Royden, L. H. (2007). Disruption of the Hellenic arc: Late Miocene extensional detachment faults and steep Pliocene-Quaternary normal faults: Or what happened at Corinth? *Tectonics*, *26*, TC5003. https://doi.org/10.1029/2006TC002007

Papazachos, B. C., & Comninakis, P. E. (1971). Geophysical and tectonic features of the Aegean Arc. *Journal of Geophysical Research*, *76*(8517).

Papazachos, B. C., & Dimitriu, P. P. (1991). Tsunamis in and near Greece and their relation to the earthquake focal mechanisms. *Natural Hazards*, *4*, 161–170. https://doi.org/10.1007/BF00162785

Papazachos, B. C., Karakostas, V. G., Papazachos, C. B., & Scordilis, E. M. (2000). The geometry of the Wadati-Benioff zone and lithospheric kinematics in the Hellenic arc. *Tectonophysics*, *319*(4), 275–300. https://doi.org/10.1016/S0040-1951(99)00299-1

Papazachos, C. B. (1999). Deep structure and active tectonics of the South Aegean volcanic arc. *Elements*, *15* (3), 153–158. https://doi.org/10.2138/gselements.15.3.153

Papazachos, C. B. (2019). Deep structure and active tectonics of the south Aegean volcanic arc. *Elements*, *15*(3), 153–158. https://doi.org/10.2138/gselements.15.3.153

Papazachos, C. B., & Nolet, G. (1997). P and S deep velocity structure of the Hellenic area obtained by robust nonlinear inversion of travel times. *Journal of Geophysical Research*, *102*(B4), 8349–8367.

Pearce, F. D., Rondenay, S., Sachpazi, M., Charalampakis, M., & Royden, L. H. (2012). Seismic investigation of the transition from continental to oceanic subduction along the western Hellenic subduction zone. *Journal of Geophysical Research*, *117*, B07306. https://doi.org/10.1029/2011JB009023

Pe-Piper, G., & Piper, D. J. W. (2005). The south Aegean active volcanic arc: Relationships between magmatism and tectonics. *Developments in Volcanology*, *7*, 113–133. https://doi.org/10.1016/S1871-644X(05)80034-8.

Peterek, A., & Schwarze, J. (2004). Architecture and Late Pliocene to recent evolution of outer-arc basins of the Hellenic subduction zone (south-central Crete, Greece). *Journal of Geodynamics*, *38*(1), 19–55. https://doi.org/10.1016/j.jog.2004.03.002

Peters, K., Smit, M., van Hinsbergen, D. J. J., van Roermund, H., & Brouwer, F. (2017). *The longevity of Neotethyan metamorphic soles from Lu-Hf garnet chronology*. 19th EGU General Assembly, EGU2017, Vienna, Austria. https://ui.adsabs.harvard.edu/abs/2017EGUGA.1915673P/abstract

Philippon, M., Brun, J.-P., Gueydan, F., & Sokoutis, D. (2014). The interaction between Aegean back-arc extension and Anatolia escape since Middle Miocene, *Tectonophysics*, *631*, 176– 188. https://doi.org/10.1016/j.tecto.2014.04.039

Piromallo, C., & Morelli, A. (1997). Imaging the Mediterranean upper mantle by P-wave travel time tomography. *Annali Di Geofisica*, *XL(4)*, 963–979. https://doi.org/10.4401/ag-3890

Piromallo, C., & Morelli, A. (2003). P wave tomography of the mantle under the Alpine-Mediterranean area. *Journal of Geophysical Research*, *108*, 2065. https://doi.org/10.1029/2002JB001757

Plunder, A., Agard, P., Chopin, C., & Okay, A. I. (2013). Geodynamics of the TavŞanlı zone, western Turkey: Insights into subduction/obduction processes. *Tectonophysics*, *608*, 884–903. https://doi.org/10.1016/j.tecto.2013.07.028

Polat, G., & Ozel, M. N. (2007). *Deep azimuthal seismic anisotropy in the western Anatolia and Aegean subduction zone*. International Union of Geodesy and Geophysics General *Assembly*, *24*. http://kocaeli2007.kocaeli.edu.tr/kocaeli2007/TAM_METIN_NUMARALI-SIRALI-PDF/141-145.pdf

Portner, D. E., Delph, J. R., Biryol, C. B., Beck, S. L., Zandt, G., Özacar, A. A., et al. (2018). Subduction termination through progressive slab deformation across Eastern Mediterranean subduction zones from updated P-wave tomography beneath Anatolia. *Geosphere*, *14*(3), 907–925. https://doi.org/10.1130/GES01617.1

Pourteau, A., Candan, O., & Oberhänsli, R. (2010). High-pressure metasediments in central Turkey: Constraints on the Neotethyan closure history. *Tectonics*, *29*, 1–18.

Pourteau, A., Oberhänsli, R., Candan, O., Barrier, E., & Vrielynck, B. (2016). Neotethyan closure history of western Anatolia: A geodynamic discussion. *International Journal of Earth Sciences*, *105*, 203–224. https://doi.org/10.1007/s00531-015-1226-7

Pourteau, A., Scherer, E. E., Schorn, S., Bast, R., Schmidt, A., & Ebert, L. (2019). Thermal evolution of an ancient subduction interface revealed by Lu-Hf garnet geochronology, Halilbağı Complex (Anatolia). *Geoscience Frontiers*, *10*(1), 127–148.

Pourteau, A., Sudo, M., Candan, O., Lanari, P., Vidal, O., & Oberhaensli, R. (2013). Neotethys closure history of Anatolia; insights from 40Ar-39Ar geochronology and P-T estimation in highpressure metasedimentary rocks. *Journal of Metamorphic Geology*, *31*(6), 585–606.

Rabayrol, F., & Hart, C. J. R. (2021). Petrogenetic and tectonic controls on magma fertility and the formation of post-subduction porphyry and epithermal mineralization along the late Cenozoic Anatolian Metallogenic Trend, Turkey. *Mineralium Deposita*, *56*, 279–306. https://doi.org/10.1007/s00126-020-00967-9

Rabayrol, F., Hart, C. J. R., & Creaser, R. A. (2019). Tectonic Triggers for postsubduction magmatic-hydrothermal gold metallogeny in the late Cenozoic Anatolian metallogenic trend, *Turkey. Economic Geology*, *114* (7), 1339–1363. https://doi.org/10.5382/econgeo.4682

Reilinger, R., et al. (2006). GPS constraints on continental deformation in the Africa-Arabia-Eurasia continental collision zone and implications for the dynamics of plate interactions, *Journal of Geophysical Research*, *111*, B05411. https://doi.org/10.1029/2005JB004051

Reilinger, R., McClusky, S., Paradissis, D., Ergintav, S., & Vernant, P. (2010). Geodetic constraints on the tectonic evolution of the Aegean region and strain accumulation along the Hellenic subduction zone. *Tectonophysics*, *488*(1–4), 22–30. https://doi.org/10.1016/j.tecto.2009.05.027

Robertson, A. H. F., & Ustaömer, T. (2004). Tectonic evolution of the Intra-Pontide suture zone in the Armutlu Peninsula, *NW Turkey. Tectonophysics*, *381*(1–4), 175–209. https://doi.org/10.1016/j.tecto.2002.06.002

Romanowicz, B. (2003). 3D structure of the Earth's lower mantle. *Comptes Rendus Geoscience*, *335*(1), 23–35. https://doi.org/10.1016/S1631-0713(03)00012-9

Rotstein, Y., &., Ben-Avraham, Z. (1985). Accretionary processes at subduction zones in the eastern Mediterranean. *Tectonophysics*, *112*, 551–561.

Royden, L. H. (1993). The tectonic expression slab pull at continental convergent boundaries. *Tectonics*, *12*(2), 303–325. https://doi.org/10.1029/92TC02248

Royden L. H., & Husson L. (2009). Subduction with variations in slab buoyancy: Models and application to the Banda and Apennine systems. In S. Lallemand & F. Funiciello (Eds.), *Subduction zone geodynamics. frontiers in earth sciences*. Berlin, Heidelberg: Springer. https://doi.org/10.1007/978-3-540-87974-9_2

Royden, L. H., & Papanikolaou, D. J. (2011). Slab segmentation and late Cenozoic disruption of the Hellenic arc. *Geochemistry, Geophysics, Geosystems*, *12*(3). https://doi.org/10.1029/2010GC003280

Ruff, L., & Kanamori, H. (1983). Seismic coupling and uncoupling at subduction zones. *Tectonophysics*, *99*(2–4), 99–117. https://doi.org/10.1016/0040-1951(83)90097-5

Sachpazi, M., Laigle, M., Charalampakis, M., Diaz, J., Kissling, E., Gesret, A., et al. (2016). Segmented Hellenic slab rollback driving Aegean deformation and seismicity. *Geophysical Research Letters*, *43*, 65–658. https://doi.org/10.1002/2015GL066818

Sakellariou, D., & Tsampouraki-Kraounaki, K. (2019). Plio-Quaternary extension and strike-slip tectonics in the Aegean. In J. C. Duarte (Ed.), *Transform plate boundaries and fracture zones* (pp. 339–374). Elsevier. https://doi.org/10.1016/B978-0-12-812064-4.00014-1

Salaün, G., Pedersen, H. A., Paul, A., Farra, V., Karabulut, H., Hatzfeld, D., et al. (2012). High-resolution surface wave tomography beneath the Aegean-Anatolia region: Constraints on upper-mantle structure. *Geophysical Journal International*, *190*(1), 406–420. https://doi.org/10.1111/j.1365-246X.2012.05483.x

Saunders, P., Priestley, K., & Taymaz, T. (1998). Variations in the crustal structure beneath western Turkey. *Geophysical Journal International*, *134*(2), 373–389. https://doi.org/10.1046/j.1365-246x.1998.00571.x

Savostin, L. A., Sibuet, J-C., Zonenshain, L. P., Le Pichon, X., & Roulet, M-J. (1986). Kinematic evolution of the Tethys belt from the Atlantic ocean to the pamirs since the Triassic. *Tectonophysics*, *123*(1–4), 1–35. https://doi.org/10.1016/0040-1951(86)90192-7

Sayit, K., Marroni, M., Göncüoğlu, M. C., Pandolfi, L., Ellero, A., Ottria, G., et al. (2016). Geological setting and geochemical signatures of the mafic rocks from the Intra-Pontide Suture Zone: Implications for the geodynamic reconstruction of the Mesozoic Neotethys. *International Journal of Earth Sciences*, *105*(1), 39–64.

Schellart, W., Freeman, J., Stegman, D. et al. (2007). Evolution and diversity of subduction zones controlled by slab width. *Nature*, *446*, 308–311. https://doi.org/10.1038/nature05615

Schivardi R., & Morelli A. (2009). Surface wave tomography in the European and Mediterranean region. *Geophysical Journal International*, *177*(3), 1050–1066.

Scholl, D.W., & von Huene, R. (2009). Implications of estimated magmatic additions and recycling losses at the subduction zones of accretionary (non–collisional) and collisional (suturing) orogens. *Geological Society, London, Special Publications*, *318*, 105–125. https://doi.org/10.1144/SP318.4

Scholz, C. H. (1990). *The mechanics of earthquakes and faulting*. Cambridge: Cambridge University Press.

Scholz, C. H., & Campos, J. (2012). The seismic coupling of subduction zones revisited. *Journal of Geophysical Research*, *117*, B05310. https://doi.org/10.1029/2011JB009003

Scoon R. N. (2021). Volcanoes of the Hellenic volcanic arc, Greece. In *The geotraveller*. Springer, Cham. https://doi.org/10.1007/978-3-030-54693-9_12

Seaton, N.C.A., Whitney, D. L., Teyssier, C., Toraman, E., & Heizler, M. T. (2009). Recrystallization of high-pressure marble (Sivrihisar, Turkey). *Tectonophysics*, *479*, 241–253. https://doi.org/10.1016/j.tecto.2009.08.015

Şengör, A. M. C., & Yılmaz, Y. (1981). Tethyan evolution of Turkey: A plate tectonic approach. *Tectonophysics*, *75*, 181–241.

Şengör, A. M. C., & Zabcı, C. (2019). The North Anatolian fault and the North Anatolian shear zone. In C. Kuzucuoğlu et al. (Eds.), *Landscapes and landforms of Turkey*. Springer International Publishing, World Geomorphological Landscapes. https://doi.org/10.1007/978-3-030-03515-0_27

Şengün, F., YiğitbaŞ, E., & Tunç, I. O. (2011). Geology and tectonic emplacement of eclogite and blueschists, Biga Peninsula, northwest Turkey. *Turkish Journal of Earth Sciences*, *20*, 273–285.

Shaw, B., & Jackson, J. (2010). Earthquake mechanisms and active tectonics of the Hellenic subduction zone. *Geophysical Journal International*, *181*, 966–984. https://doi.org/10.1111/j.1365-246X.2010.04551.x

Shaw, B., Ambraseys, N., England, P., et al. (2008). Eastern Mediterranean tectonics and tsunami hazard inferred from the AD 365 earthquake. *Nature Geoscience*, *1*, 268–276. https://doi.org/10.1038/ngeo151

Shephard, G. E., Matthews, K. J., Hosseini, K., & Domeier, M. (2017). On the consistency of seismically imaged lower mantle slabs. *Scientific Reports*, *7*, 10976. https://doi.org/10.1038/s41598-017-11039-w

Sherlock, S., Kelley, S., Inger, S., Harris, N., & Okay, A. (1999). 40Ar-39Ar and Rb-Sr geochronology of high-pressure metamorphism and exhumation history of the TavŞanlı Zone, NW Turkey. *Contributions to Mineralogy and Petrology*, *137*, 46–58.

Shin, T. A., Catlos, E. J., Jacob, L., & Black, K. (2013). Relationships between very high pressure subduction complex assemblages and intrusive granitoids in the TavŞanlı Zone, Sivrihisar Massif, central Anatolia. *Tectonophysics*, *595–596*, 183–197.

Sigl, M., Winstrup, M., McConnell, J., et al. (2015). Timing and climate forcing of volcanic eruptions for the past 2,500 years. *Nature*, *523*, 543–549. https://doi.org/10.1038/nature14565

Silver, P. G., Russo, R. M., & Lithgow-Bertelloni, C. (1998). Coupling of South American and African plate motion and plate deformation. *Science*, *279*, 60–63.

Snieder, R. (1988). Large-scale waveform inversions of surface waves for lateral heterogeneity: 2. Application to surface waves in Europe and the Mediterranean. *Journal of Geophysical Research, Solid Earth*, *93*, 12067–12080.

Snopek, K., Meier, T., Endrun, B., Bohnhoff, M., & Casten, U. (2007). Comparison of gravimetric and seismic constraints on the structure of the Aegean lithosphere in the forearc of the Hellenic subduction zone in the area of Crete. *Journal of Geodynamics*, *44*(3–5), 173–185. https://doi.org/10.1016/j.jog.2007.03.005

Sodoudi, F., Kind, R., Hatzfeld, D., Priestley, K., Hanka, W., Wylegalla, K., et al. (2006). Lithospheric structure of the Aegean obtained from P and S receiver functions. *Journal of Geophysical Research*, *111*, B12307. https//doi.org/10.1029/2005JB003932

Solheim, L. P., & Peltier, W. R. (1994). Avalanche effects in phase transition modulated thermal convection: a model of Earth's mantle. *Journal of Geophysical Research*, *99*, 6997–7018.

Spakman, W. (1990). Tomographic images of the upper mantle below central Europe and the Mediterranean. *Terra Nova*, *2*, 542–553. https://doi.org/10.1111/j.1365-3121.1990.tb00119.x

Spakman, W. (1991). Delay-time tomography of the upper mantle below Europe, the Mediterranean, and Asia Minor. *Geophysical Journal International*, *107*(2), 309–332. https://doi.org/10.1111/j.1365-246X.1991.tb00828.x

Spakman, W., van der Lee, S., & van der Hilst, R. (1993). Travel-time tomography of the European-Mediterranean mantle down to 1,400 km. *Physics of the Earth and Planetary Interiors*, *79*(1–2), 3–74. https://doi.org/10.1016/0031-9201(93)90142-V

Spakman, W., Wortel, M. J. R., & Vlaar, N. J. (1988). The Hellenic subduction zone: A tomographic image and its geodynamic implications. *Geophysical Research Letters*, *15*, 60–63.

Speciale, P. A., Catlos, E. J., Yıldız, G. O., Shin, T. A., & Black, K. N. (2012). Zircon ages from the Beypazarı granitoid pluton (north central Turkey): tectonic implications. *Geodinamica Acta*, *25*(3–4), 162–182. https://doi.org/10.1080/09853111.2013.858955

Stampfli, G. M. (2000). Tethyan oceans. In E. Bozkurt et al. (Eds.), *Tectonics and magmatism in Turkey and surrounding area* (pp. 1–23). Geological Society of London, Special Publication, London (vol. *173*). https://doi.org/10.1144/GSL.MEM.2006.032.01.04

Stampfli, G. M., & Borel, G. D. (2002). A plate tectonic model for the Paleozoic and Mesozoic constrained by dynamic plate boundaries and restored synthetic oceanic isochrons. *Earth and Planetary Science Letters*, *196*, 17–33. https://doi.oirg/10.1016/S0012-821X(01)00588-X

Stampfli, G. M., & Kozur, H. W. (2006). Europe from the Variscan to the Alpine cycles. *Geological Society, London, Memoirs*, *32*, 57–82.

Stern, R. J. (2004). Subduction initiation: Spontaneous and induced. *Earth and Planetary Science Letters*, *226*, 275–292.

Stern, R. J., & Gerya, T. (2018). Subduction initiation in nature and models: A review. *Tectonophysics*, *746*, 173–198. https://doi.org/10.1016/j.tecto.2017.10.014

Stöckhert, B. (1999). *The Hellenic subduction zone, a world site to study the mechanics of roll back*. CIESM Workshop Series 6, Workshop on Mediterranean Scientific Drilling Prospectives, Granada 10–12 June 1999, 103–108.

Suckale, J., Rondenay, S., Sachpazi, M., Charalampakis, M., Hosa, A., & Royden, L. H. (2009). High-resolution seismic imaging of the western Hellenic subduction zone using teleseismic scattered waves. *Geophysical Journal International*, *178*(2), 775–791. https://doi.org/10.1111/j.1365-246X.2009.04170.x

Symeou, V., Homberg, C., Nader, F. H., Darnault, R., Lecomte, J. C., & Papadimitriou, N. (2018). Longitudinal and temporal evolution of the tectonic style along the Cyprus Arc system, assessed through 2-D reflection seismic interpretation. *Tectonics*, *37*, 30–47. https://doi.org/10.1002/2017TC004667

Taylor, G., Thompson, D. A., Cornwell, D., & Rost, S. (2019). Interaction of the Cyprus/Tethys slab with the mantle transition zone beneath Anatolia. *Geophysical Journal International*, *216*(3), 1665–1674. https://doi.org/10.1093/gji/ggy514

Taymaz, T., Jackson, J., & Westaway, R. (1990). Earthquake mechanisms in the Hellenic Trench near Crete. *Geophysics Journal International*, *102*, 695–731.

Taymaz, T., Yılmaz, Y., & Dilek, Y. (2007). The geodynamics of the Aegean and Anatolia: introduction. *Geological Society, London, Special Publications*, *291*(1–16). https://doi.org/10.1144/SP291.1

Tekin, U. K., Göncüoğlu, M. C., & Turhan, N. (2002). First evidence of Late Carnian radiolarians from the Izmir-Ankara suture complex, central Sakarya, Turkey: Implications for the opening age of the Izmir-Ankara branch of Neo-Tethys. *Geobios*, *35*(1), 127–135. https://doi.org/10.1016/S0016-6995(02)00015-3

Ten Veen, J. H., & Kleinspehn, K. L. (2002). Geodynamics along an increasingly curved convergent plate margin: Late Miocene-Pleistocene Rhodes, Greece. *Tectonics*, *21*(3), https://doi.org/10.1029/2001TC001287

Tezel, T., Shibutani, T., & Kaypak, B. (2013). Crustal thickness of Turkey determined by receiver function. *Journal of Asian Earth Sciences*, *75*, 36–45. https://doi.org/10.1016/j.jseaes.2013.06.016

Thomson, S. N., Stöckhert N., & Brix M. R. (1998). Thermochronology of the high-pressure metamorphic rocks of Crete, Greece: Implications for the speed of tectonic processes. *Geology*, *26*, 259–262.

Tinti, S., Armigliato, A., Pagnoni, G., & Zaniboni, F. (2005). Scenarios of giant tsunamis of tectonic origin in the Mediterranean. *ISET Journal of Earthquake Technology*, *42*(4), 171–188.

Tirel, C., Gueydan, F., Tiberi, C., & Brun, J-P. (2004). Aegean crustal thickness inferred from gravity inversion. Geodynamical implications. *Earth and Planetary Science Letters*, *228*(3–4), 267–280. https://doi.org/10.1016/j.epsl.2004.10.023

Triantafyllou, I., Novikova, T., Charalampakis, M., et al. (2019). Quantitative tsunami risk assessment in terms of building replacement cost based on tsunami modelling and GIS methods: The case of Crete Island, Hellenic Arc. *Pure Applied Geophysics*, *176*, 3207–3225. https://doi.org/10.1007/s00024-018-1984-9

Ustaömer, T., Ustaömer, P. A., Robertson, A. H. F., & Gerdes, A. (2020). U-Pb-Hf isotopic data from detrital zircons in late Carboniferous and Mid-Late Triassic sandstones, and also Carboniferous granites from the Tauride and Anatolide continental units in S Turkey: Implications for Tethyan palaeogeography. *International Geology Review*, *62*(9), 1159–1186. https://doi.org/10.1080/00206814.2019.1636415

Vanderhaeghe, O., Hibsch, C., Siebenaller, L., Duchêne, S., de St. Blanquat, M., Kruckenberg, S., et al. (2007). Penrose Conference, Extending a Continent: Naxos Field Guide. In G. Lister et al. (Eds.), *Inside the Aegean metamorphic core complexes. Journal of the Virtual Explorer*, *27*(4). https://doi.org/10.3809/jvirtex.2007.00175

van der Meer, D. G., van Hinsbergen, D. J. J., & Spakman, W. (2018). Atlas of the underworld: Slab remnants in the mantle, their sinking history, and a new outlook on lower mantle viscosity. *Tectonophysics*, *723*, 309–448. https://doi.org/10.1016/j.tecto.2017.10.004.

van der Meijde, M., Marone, F., Giardini, D., & van der Lee, S. (2003). Seismic evidence for water deep in Earth's upper mantle. *Science*, *300*, 1556–1558.

Vaněk, J., Hanuš, V., Christova, C., & Simeonova, S. (1987). Morphology of the wadati-benioff zone in the hellenic arc. *Journal of Geodynamics*, *8*(1), 79–93. https://doi.org/10.1016/0264-3707(87)90059-7

van Hinsbergen, D. J. J., & Schmid, S. M. (2012). Map view restoration of Aegean-West Anatolian accretion and extension since the Eocene. *Tectonics*, *31*, TC5005. https://doi.org/10.1029/2012TC003132

van Hinsbergen, D. J. J., Hafkenscheid, E., Spakman, W., Meulenkamp, J. E., & Wortel, R. (2005). Nappe stacking resulting from subduction of oceanic and continental lithosphere below Greece. *Geology*, *33* (4), 325–328. https://doi.org/10.1130/G20878.1

van Hinsbergen, D. J. J., Kaymakçi, N., Spakman, W., & Torsvik, T. H. (2010). Reconciling the geological history of western Turkey with plate circuits and mantle tomography. *Earth and Planetary Science Letters*, *297*(3–4), 674–686.

van Hinsbergen, D. J. J., Steinberger, B., Guilmette, C., et al. (2021). A record of plume-induced plate rotation triggering subduction initiation. *Nature Geoscience*, *14*, 626–630. https://doi.org/10.1038/s41561-021-00780-7

Vassilakis, E., Royden, L., & Papanikolaou, D. (2011). Kinematic links between subduction along the Hellenic trench and extension in the Gulf of Corinth, Greece: A multidisciplinary analysis. *Earth and Planetary Science Letters*, *303*(1–2), 108–120.

Ventouzi, C., Papazachos, C., Hatzidimitriou, P., Papaioannou, C., & EGELADOS Working Group (2018). Anelastic P- and S- upper mantle attenuation tomography of the southern Aegean Sea subduction area (Hellenic Arc) using intermediate-depth earthquake data. *Geophysical Journal International*, *215*(1), 635–658. https://doi.org/10.1093/gji/ggy292

Vernant, P., Reilinger, R., & McClusky, S. (2014). Geodetic evidence for low coupling on the Hellenic subduction plate interface. *Earth and Planetary Science Letters*, *385*, 122–129. https://doi.org/10.1016/j.epsl.2013.10.018

Von Huene, R., & Scholl, D. W. (1991). Observations at convergent margins concerning sediment subduction, subduction erosion, and the growth of continental-crust. *Reviews in Geophysics*, *29*(3), 279–316.

Von Huene, R., & Scholl, D. W. (1993). The return of sialic material to the mantle indicated by terrigeneous material subducted at convergent margins. *Tectonophysics*, *219*(1–3), 163–175.

Wallace, L. M., Ellis, S., & Mann, P. (2008). Tectonic block rotation, arc curvature, and back-arc rifting: insights into these processes in the Mediterranean and the western Pacific. *IOP Conference Series: Earth and Environmental Sciences*, *2*, 012010. https://doi.org/10.1088/1755-1307/2/1/012010

Wallace, L. M., McCaffrey, R., Beavan, J., & Ellis, S. (2005). Rapid microplate rotations and backarc rifting at the transition between collision and subduction. *Geology*, *33* (11), 857–860. https://doi.org/10.1130/G21834.1

Wei, W., Zhao, D., Wei, F., Bai, X., & Xu, J. (2019). Mantle dynamics of the Eastern Mediterranean and Middle East: Constraints from P-wave anisotropic tomography. *Geochemistry, Geophysics, Geosystems*, *20*, 4505–4530. https://doi.org/10.1029/2019GC008512

Westaway, R. (1994). Present-day kinematics of the Middle East and eastern Mediterranean. *Journal of Geophysical Research*, *99*(B6), 12071–12090. https://doi.org/10.1029/94JB00335

Westbrook, G. K., & Reston, T. J. (2002). The accretionary complex of the Mediterranean Ridge: tectonics, fluid flow and the formation of brine lakes: An introduction to the special issue of *Marine Geology*. *Marine Geology*, *186*, 1–8.

White, D. A., Roeder, D. H., Nelson, T. H., & Crowell, J. C. (1970). Subduction. *Geological Society of America Bulletin*, *81*, 3431–3432. https://doi.org/10.1130/0016-7606(1970)81[3431:S]2.0.CO;2

Widiyantoro, S., van der Hilst, R. D., & Wenzel, F. (2004). Deformation of the Aegean slab in the mantle transition zone. *International Journal of Tomography & Statistics*, *D04*, 1–14.

Wolfenden, E., Ebinger, C., Yirgu, G., Deino, A., & Ayalew, D. (2004). Evolution of the northern Main Ethiopian rift: Birth of a triple junction. *Earth and Planetary Science Letters*, *224*(1–2), 213–228. https://doi.org/10.1016/j.epsl.2004.04.022

Wolfenden, E., Ebinger, C., Yirgu, G., Renne, P. R., & Kelley, S. P. (2005). Evolution of a volcanic rifted margin: Southern Red Sea, *Ethiopia. Geological Society of America Bulletin*, *117* (7–8), 846–864. https://doi.org/10.1130/B25516.1

Woodside, J. M., Mascle, J., Huguen, C., & Volkonskaia, A. (2000). The Rhodes basin, a post-Miocene tectonic trough. *Marine Geology*, *165*, 1–12.

Woodside, J. M., Mascle, J., Zitter, T. A. C., Limonov, A. F., Ergün, M., & Volkonskaia, A. (2002). The Florence Rise, the western bend of the Cyprus Arc. *Marine Geology*, *185*(3–4), 177–194. https://doi.org/10.1016/S0025-3227(02)00194-9

Wortel, M. J. R., & Spakman, W. (1992). Structure and dynamics of subducted lithosphere in the Mediterranean region. *Proceedings of the Koninklijke Nederlandse Akademie van Wetenschappen*, *95*(3), 325–347.

Wortel, M. J. R., & Spakman, W. (2000). Subduction and slab detachment in the Mediterranean-Carpathian region. *Science*, *290*, 1910–1917. https://doi.org/10.1126/science.290.5498.1910

Wortel, M. J. R., Goes, S. D. B., & Spakman, W. (1990). Structure and seismicity of the Aegean subduction zone. *Terra Nova*, *2*, 554–562.

Yang, T., Gurnis, M., & Zahirovic, S. (2018). Slab avalanche-induced tectonics in self-consistent dynamic models. *Tectonophysics*, *746*, 251–265. https://doi.org/10.1016/j.tecto.2016.12.007

YiğitbaŞ, E., Kerrich, R., Yılmaz, Y., Elmas, A., & Xie, Q. (2004). Characteristics and geochemistry of Precambrian ophiolites and related volcanics from the Istanbul-Zonguldak Unit, Northwestern Anatolia, Turkey: Following the missing chain of the Precambrian south European suture zone to the east. *Precambrian Research*, *132*(1–2), 179–206. https://doi.org/10.1016/j.precamres.2004.03.003

Yılmaz, Y., Tüysüz, O., Yigitbas, E., Genç, C., & Şengör, A. M. C. (1997). Geology and tectonic evolution of the Pontides. In A. G. Robinson (Ed.), *Regional and petroleum geology of the Black Sea and surrounding region* (pp. 183–226). American Association of Petroleum Geology Memiors, 68. Tulsa, Oklahoma.

Yolsal, S., Taymaz, T., & Yalçiner, A. C. (2007). Understanding tsunamis, potential source regions and tsunami-prone mechanisms in the Eastern Mediterranean. *Geological Society, London, Special Publications*, *291*, 201–230. https://doi.org/10.1144/SP291.10

Yoshioka, T. (1996). Evolution of fault geometry and development of strike-slip basins: Comparative studies on the transform zones in Turkey and Japan. *Island Arc*, *5*, 407–419. https://doi.org/10.1111/j.1440-1738.1996.tb00162.x

Zanchetta, G, Sulpizio, R, Roberts, N, et al. (2011). Tephrostratigraphy, chronology and climatic events of the Mediterranean basin during the Holocene: An overview. *The Holocene*, *21*(1), 33–52. https://doi.org/10.1177/0959683610377531

Zhu, L., Mitchell, B. J., Akyol, N., Çemen, İ., & Kekovali, K. (2006). Crustal thickness variations in the Aegean region and implications for the extension of continental crust, *Journal of Geophysical Research*, *111*, B01301. https//doi.org/10.1029/2005JB003770

Zielhuis, A., & Nolet, G. (1994). Shear-wave velocity variations in the upper-mantle beneath Central Europe. *Geophysics Journal International*, *117*(3), 695–715.

Zitter, T. A. C., Woodside, J. M., & Mascle, J. (2003). The Anaximander Mountains: A clue to the tectonics of southwest Anatolia. *Geological Journal*, *38*, 375–394. https://doi.org/10.1002/gj.961

Part II
Alpine-Himalayan Collision

5

Genesis of Himalayan Stratigraphy and the Tectonic Development of the Thrust Belt

Delores M. Robinson[1] and Aaron J. Martin[2]

ABSTRACT

This chapter is a reference for researchers either to begin their studies in the Himalayan thrust belt or to deepen their knowledge of the region. Integration of the plethora of data, all with their own assumptions and conventions, is difficult. The Himalaya is composed of rocks that are at least as old as Paleoproterozoic and have been recycled and deformed multiple times. We begin in Paleoproterozoic time and continue to review the data and models up to the present day. Our caveat is that this review is destined to change with the acquisition of new data and the development of new ideas that will tilt the balance to one side or the other of the many controversies and poorly-known issues presented here.

5.1. INTRODUCTION

The Himalayan-Tibetan orogenic system is an active continent-continent collision with an ~5 km plateau bordered to the south by an ~2,400 km long thrust belt. As such, researchers flock to the system to unravel pressing questions about how continent-continent collisions evolve through time. Since ~60 Ma, India and Asia have been involved in a slow-motion continent crash at ~5.5 cm per year. Hidden within the mountain range today are answers to questions fundamental to all orogenic collisional systems, including the following: (1)How do preexisting heterogeneities in the continents influence the development of the mountain range? (2)What were the sizes and configurations of the continents before collision and what effects did those factors have on the system? (3) How are metamorphism and orogeny related? (4)Why did the thrust belt form a duplex and what causes its varying formation in different places? (5)How did surficial processes drive landscape evolution, and what were the effects of tectonics and climate in different parts of the mountain belt?

We present a review of the development of the Himalayan thrust belt, which extends from the Indus-Yarlung suture southward to the Indo-Gangetic Plain and along the ~2,400 km thrust belt from Namcha Barwa in southeastern Tibet to Nanga Parbat in Pakistan. Where appropriate, we discuss the geologic history of Pakistan, although correlation with the rest of the Himalayan arc is still uncertain. We review from Paleoproterozoic time to the present because the tectonic and stratigraphic history determine how the modern thrust belt developed. We integrate tectonic elements within and surrounding the orogen to teach the fundamentals of the Himalayan thrust belt to geoscientists beginning investigations there.

[1] *Department of Geological Sciences, and Center for Sedimentary Basin Studies, The University of Alabama, Tuscaloosa, Alabama, USA*

[2] *Division de Geociencias Aplicadas, Instituto Potosino de Investigacion Cientifica y Tecnologica, San Luis Potosi, S.L.P., Mexico*

Compressional Tectonics: Plate Convergence to Mountain Building, Geophysical Monograph 277, First Edition.
Edited by Elizabeth J. Catlos and İbrahim Çemen.

DOI:10.1002/9781119773856.ch05

5.2. GEOLOGIC FRAMEWORK OF THE HIMALAYA

The Cenozoic Himalaya is the result of a collection of processes that began in Paleoproterozoic time, evolving to be the deformed southern edge of the modern 5 km high Tibetan Plateau (Fig. 5.1). Because these two components are intimately tied together, we refer to it as the Himalayan-Tibetan orogenic system. Prior to India colliding with Asia, multiple continental fragments and magmatic arcs collided with the Asian margin, creating a collage of terranes, some of which are within the Tibetan Plateau. Kapp and DeCelles (2019) review the evolution of the Tibetan Plateau including possible scenarios for ocean closures. Final closure of the Neo-Tethys Ocean started with the initial contact and mixing between Indian and Asian sediment at ~61–58 Ma in the Indus-Yarlung suture zone (Wang et al., 2011; DeCelles et al., 2014; Wu et al., 2014; Hu et al., 2015a; Wang et al., 2017; An et al., 2021). However, some argue for two-stage collision, with the first collision between India and an intra-oceanic arc (e.g., Klootwijk et al., 1979; Aitchison et al., 2000, 2007; Bouilhol et al., 2013; Baxter et al., 2016; Martin et al., 2020) or India and the rifted southern margin of Asia (Kapp & DeCelles, 2019), followed by a second collision with the unrifted southern margin of Asia. Each of these two-stage collision models requires subduction of another ocean basin after initial collision at ca. 60 Ma. Final closure began the southward deformation of rocks that extended northward from cratonic India, called Greater India, which includes the remnants of prior tectonic activity since Paleoproterozoic time. The tectonic history of the Himalaya includes several collision and magmatic events, which have resulted in different metamorphic events. Kohn (2014) and Waters (2019) contain a review of these metamorphic events.

Because tectonostratigraphic and high strain zone (fault) names are used in the following subsections, here we give a brief introduction to the Himalayan orogen and go into depth in the appropriate sections with accompanying citations. Himalayan stratigraphy is contentious and confusing, with different researchers using their preferred scheme. The classification of Himalayan tectonostratigraphy that we use is based on protolith characteristics and follows the work of Heim and Gansser (1939) and Gansser (1964). We do not subscribe to schemes that name high strain zones based on the age of motion (Webb et al., 2013) or the locations of exposed fossil brittle-ductile transitions combined with metamorphic grade (e.g., Goscombe et al., 2006, 2018; Searle et al., 2008; cf. Martin, 2017a). Figure 5.1a shows the Himalayan stratigraphy divided by age (Webb, 2013). Martin (2017b) uses a scheme that divides the rocks into two parts, Himalayan Assemblage A (equivalent to Lesser and Sub-Himalayan rocks in our scheme) and Himalayan Assemblage B (equivalent to our Greater and Tethyan Himalayan rocks), based on the nature of contacts within and between the assemblages (Fig. 5.1b). Rocks within each assemblage share observed depositional or intrusive contacts, whereas rocks of the different assemblages presently are in contact across a high strain zone. Although this division is simpler, we retain a four tectonostratigraphic zone breakdown for learning purposes because that is how most of the literature is divided (Fig. 5.1c). We use Tethyan Himalayan, Greater Himalayan, Lesser Himalayan, and Sub-Himalayan rocks as our breakdown, which refers to rocks that are defined and mapped according to their depositional or intrusion age, stratigraphic relationships, and geochemical attributes. As a cautionary note, when reading literature from the Himalaya, be sure to understand how the authors are dividing the stratigraphy and how they define the location of each high strain zone. We do not use "Zone" after Tethyan, Greater, and Lesser Himalaya because zone refers to elevations and geography within the Himalaya. Some researchers use the word "sequence" instead of "rocks." Those researchers are also referring to the stratigraphic characteristics of a unit.

Beginning at the Indus-Yarlung suture within southern Tibet, the Neoproterozoic to Cenozoic Tethyan Himalayan rocks were deformed into the Tethyan fold-thrust belt (TFTB) southward to the South Tibetan detachment system. This part of the Himalayan thrust belt north of the South Tibetan detachment system is not as well documented as the southern part. In general, Neoproterozoic-Cambrian Greater Himalayan rocks crop out from the South Tibetan detachment system south to the Main Central Thrust. This is not always true but a good place to start to understand the Himalaya. Greater Himalayan rocks are the highest-grade metamorphic rocks exposed in the mountain belt and contain intra-Greater Himalayan shear zones. The southern part of the Himalayan thrust belt is located south of the Main Central Thrust within Lesser and Sub-Himalayan rocks. Lesser Himalayan rocks were thrusted and folded into long-traveled thrust sheets and duplexes (Ramgarh-Munsiari Thrust and Lesser Himalayan duplex) as far southward as the Main Boundary Thrust. The age of what strata to include as Lesser Himalayan rocks is debated. Most of the rock is Paleoproterozoic-Mesoproterozoic strata. However, a Carboniferous-Paleocene Gondwanan unit and an Eocene-Miocene foreland basin unit are sometimes mapped as Lesser Himalayan rocks and sometimes separated into the Gondwana unit (or sequence) and foreland basin unit (or sequence). Sub-Himalayan rocks are as old as upper Paleocene-middle Eocene in northwestern India (Subathu Formation; e.g., Colleps et al., 2020) and as old as ~16 Ma in Nepal (Siwalik Group, Ojha et al., 2000, 2009).

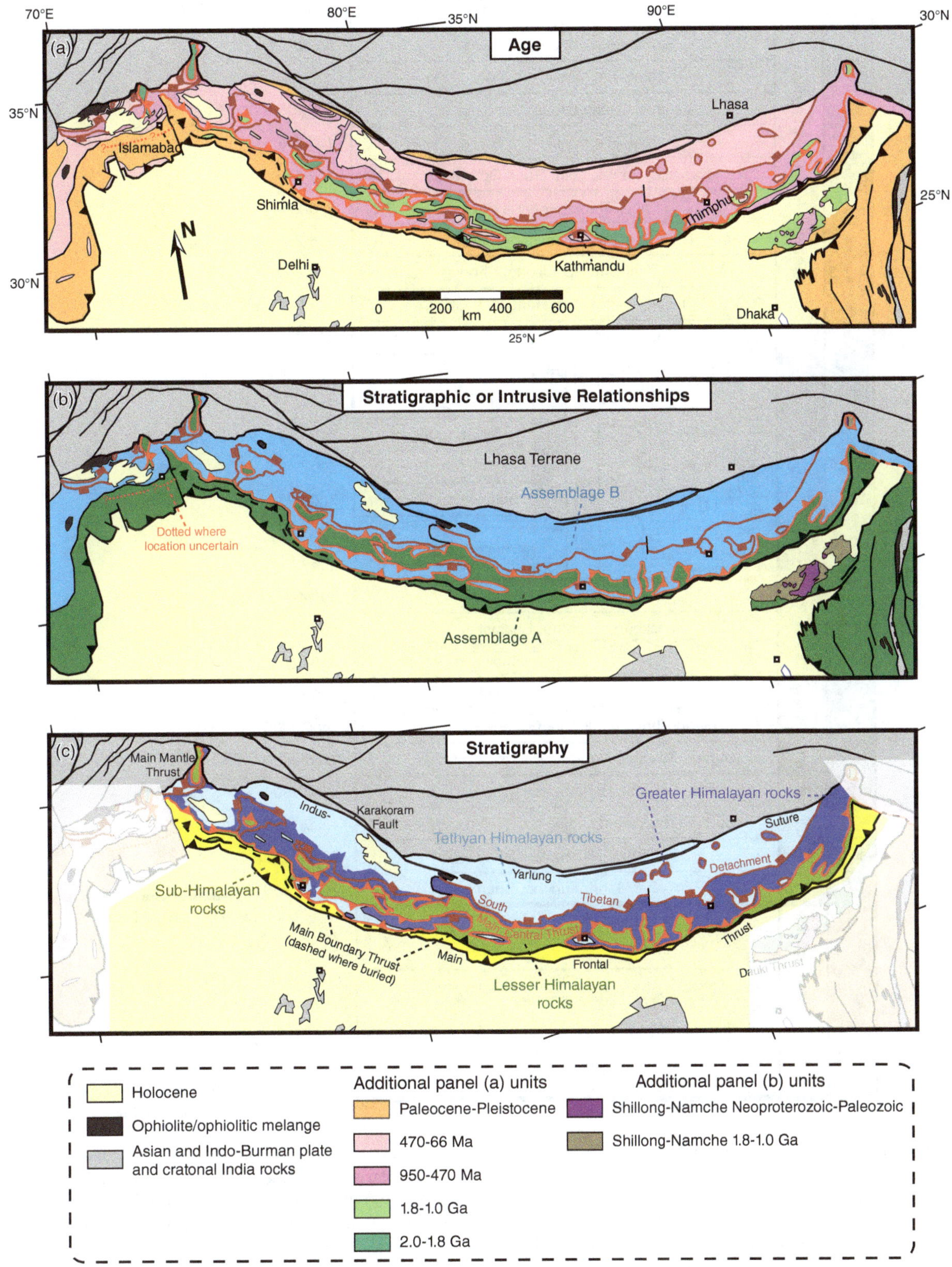

Figure 5.1 Generalized geologic maps of the Himalaya divided according to three different criteria: (a) Depositional or intrusion age. Locations are the same for (b) and (c). (b) Relationships set at the time of deposition or intrusion. Members of an assemblage share depositional or intrusive relationships, but the contact between assemblages is currently a high strain zone. (c) Stratigraphic characteristics. Used in this way, the four-unit breakdown is a division of the two-unit breakdown shown in panel (b). This four-part division applies only between the syntaxes, so the western and eastern parts of the orogen are blocked out. Note that Martin (2017b) defines the four units based on structural position, not stratigraphic characteristics (modified from Webb, 2013; Martin, 2017b; and Fan & Murphy, 2021).

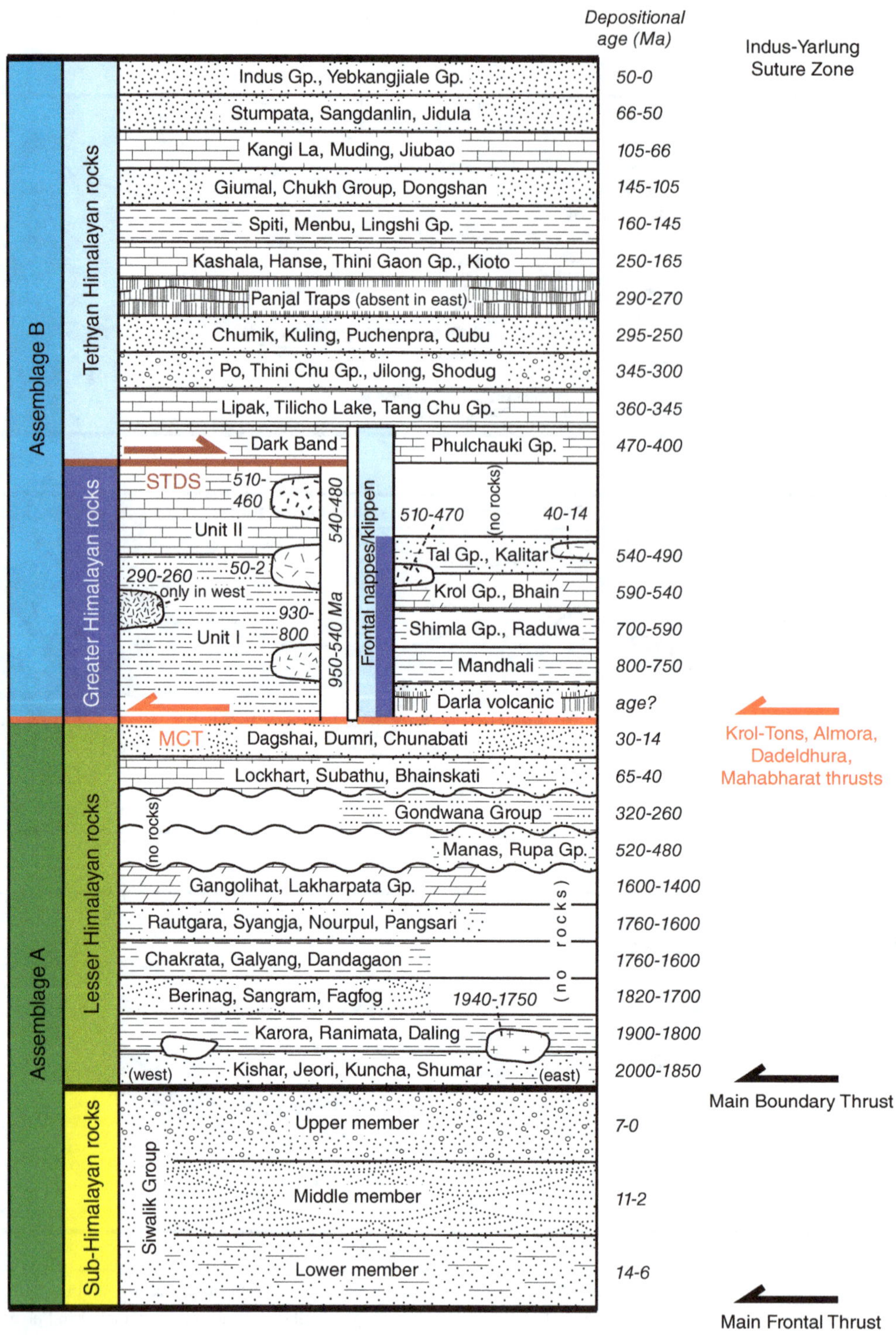

Figure 5.2 Highly generalized tectonostratigraphic column for the Himalayan fold-thrust belt between the syntaxes showing relationships between lithologic units and package-bounding high strain zones. Protolith lithologies are shown; metamorphic effects are not depicted. Labels on lithologic units are examples from along the orogen; they do not include all the names that have been assigned to the rocks. The east and west labels apply only to Lesser Himalayan (meta)sedimentary rocks. Different frontal nappes and klippen contain Greater or Tethyan Himalayan rocks, or both, so both shades of blue are used for the frontal nappes/klippen. Wavy lines indicate unconformities that span more than 100 m.y. Uncertainties in depositional and intrusion ages are not included. This compilation does not include the following units because they are exposed in small areas relative to the size of the entire orogen: (1) ca. 990 Ma metamafic rocks in Lesser Himalayan rocks in the Ama Drime Massif (Cottle et al., 2009); (2) ca. 600 Ma granite in Greater Himalayan rocks in Sikkim (Mottram et al., 2014); and (3) the Cretaceous Taltung Formation and < 120 Ma Amile Formation in the Gondwana Group in central and western Nepal (Sakai, 1983; Stickroth et al., 2019). Note: MCT=Main Central thrust, STDS=South Tibetan detachment system, Bhain=Bhainsedobhan Formation. Darla volcanic unit sensu Webb et al. (2011a).

Sub-Himalayan rocks are deformed into an imbricate thrust system between the Main Boundary and Main Frontal thrusts. All faults sole down into the basal décollement of the thrust system, called the Main Himalayan Thrust.

Because the Himalayan Mountains are ~500 km across strike and >2,400 km along strike, researchers generalize the structure and stratigraphy, as we just did, and often assume that tectonic and stratigraphic elements are continuous and uniform along strike. However, the structure and stratigraphy are variable (Fig. 5.2), and fault timing and stratigraphic position both change along strike. In fact, because of its length and protracted tectonic history, it is remarkable that major structures and stratigraphic patterns can be correlated along much of the Himalaya.

5.3. PALEOPROTEROZOIC TIME (2.5–1.6 GA)

Archean (3.5–2.5 Ga) rocks compose the Indian craton onto which Paleoproterozoic Lesser Himalayan strata were deposited. Lesser Himalayan rocks are younger than 2.0 Ga. In the Himalayan thrust belt, there is only one location where possible Archean rocks are exposed. DiPietro and Isachsen (2001) date 15 individual zircon crystals from a rock collected in the tectonic window of the Indus River anticlinorium in northern Pakistan by dissolution of each grain followed by thermal ionization mass spectrometry. Two grains, interpreted to be detrital, yield $^{207}Pb/^{206}Pb$ dates near 2.93 Ga, providing an Archean maximum depositional age. The remaining grains are interpreted to be metamorphic, including seven crystals with $^{207}Pb/^{206}Pb$ dates between 2.831 and 2.136 Ga. If instead these grains are detrital, the actual maximum depositional age would be Paleoproterozoic. The need for more dating of rocks in the Indus River anticlinorium keeps open the possibility for Archean depositional ages. In India, deep wells drilled to a maximum of 5 km in the Indo-Gangetic Plain indicate Archean basement rock covered by the Vindhyan Group (Proterozoic), which is covered by the sedimentary rock deposited as the Cenozoic Himalaya grew (Sastri et al., 1971). Thus, underneath Paleoproterozoic rocks deposited on the Indian craton as the eroded detritus from the craton, Archean rocks are the depositional basement. These Archean rocks are deformed into extensional and transtensional basins and were filled with Paleoproterozoic sediment (compiled in Martin, 2017b). In fact, the location of one such Paleoproterozoic basin in front of the thrust belt in central Nepal corresponds spatially with surface outcrops of the 2.5–3.3 km thick Kuncha Formation (1.9–1.85 Ga) within the thrust belt (Robinson et al., 2021), which is some of the oldest rock from Namcha Barwa in southeastern Tibet to Nanga Parbat in Pakistan. Northwestern India also has some of the oldest rock exposed in the Sutlej River valley (Himachal Pradesh, India), where Webb et al. (2011a) found that the Jeori Formation has a maximum depositional age of ~1.9 Ga.

The motif of Paleoproterozoic sandstone and mudstone intruded by ca. 1.9–1.83 Ga granitoid, all subsequently metamorphosed, occurs along the entire Himalayan thrust belt (Fig. 5.2; see Robinson & Pearson, 2013; Martin, 2017b, for different formation names along the orogen). In several locations, granitic magmatism occurred as late as 1.77 Ga, and in northeastern India it continued until ca. 1.75 Ga. The tectonic setting during deposition and intrusion of these rocks is disputed (Fig. 5.3a). Some researchers argue for a magmatic arc setting, either the main part of the arc or the backarc, largely based on the chemical compositions of the granitic rocks and the zircon they contain (Kohn et al., 2010; Rao & Sharma, 2011; Spencer et al., 2012; Sen et al., 2013; Mandal et al., 2016; Phukon et al., 2018). Other geologists infer a continental rift setting (Sakai et al., 2013; Martin, 2017b; Larson et al., 2019). Larson et al. (2019) show that Paleoproterozoic granitic gneiss in eastern Nepal has an A-type affinity, indicating that the rocks are from an upper crustal source consistent with a continental rift setting. A caveat in this debate is that it is a mistake to project modern-style plate tectonics into the deep past. Perhaps Paleoproterozoic A-type granite represents locations of upwelling and melting of the mantle during an intermediate stagnant lid episode, as suggested by Stern (2020). Extension as early as 1.8 Ga is recorded by abundant mafic sills and dikes intruding into Paleoproterozoic strata that was deposited in a backarc basin (Mandal et al., 2016) or perhaps a continental rift environment (Larson et al., 2019). Siliciclastic sediment was deposited before 1.6 Ga into the arc- or rift-related basins, which perhaps transition up section into a passive margin. These units now form the Paleoproterozoic (1.9–1.6 Ga) lower Lesser Himalayan rock present all along the Himalayan orogen (Figs. 5.1 and 5.2; e.g., Ahmad et al., 2000; Kohn et al., 2010, and references therein; McQuarrie et al., 2008, 2013; McKenzie et al., 2011; Robinson & Pearson, 2013; Webb et al., 2013). These rocks subsequently were metamorphosed to greenschist to lower amphibolite facies conditions (e.g., Beyssac et al., 2004; Chakraborty et al., 2016; Catlos et al., 2020; Shrestha et al., 2020).

The supercontinent during Paleoproterozoic-Mesoproterozoic time was Columbia (or Nuna; Rogers & Santosh, 2002; Brown & Johnson, 2018). The reconstructions of Rogers and Santosh (2002), Zhao et al. (2004), and Cawood et al. (2018) do not show a subduction zone outboard of northern India as do the models of Phukon et al. (2018). If the Paleoproterozoic rocks were deposited and intruded in a backarc basin, there must have been an

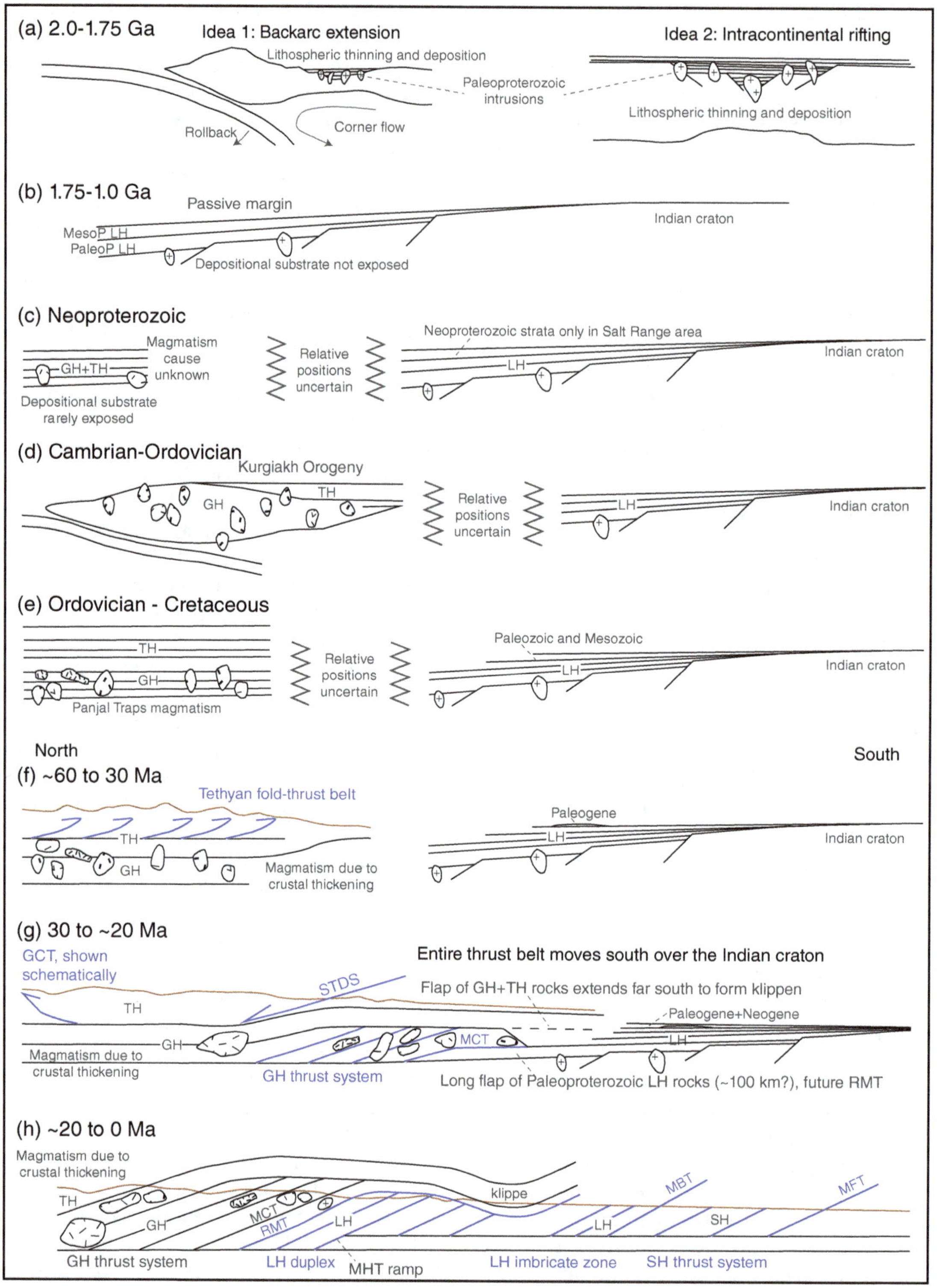

Figure 5.3 Summary of the geologic history of Himalayan rocks: (a) Two ideas for the development of Paleoproterozoic strata and intrusions; (b) Mesoproterozoic margin; (c) Neoproterozoic margin with no assumptions on the relationship between TH/GH and LH rocks; (d) Cambro-Ordovician Kurgiakh orogeny; (e) Ordovician-Cretaceous margin; (f) Collision to 30 Ma margin; (g) 30–20 Ma margin generalized for the central Himalayan fold-thrust belt. In Pakistan, the Main Mantle Thrust may serve the same role as the STDS and the Kohistan arc is directly north instead of the TFTB. (h) 20–0 Ma margin. Active faults are in blue. Hypothesized erosion levels are represented by solid brown lines. Intrusion patterns are the same as in Figure 5.2. Note: TH=Tethyan Himalaya; GH=Greater Himalaya; LH=Lesser Himalaya; SH=Sub-Himalaya; STDS=South Tibetan detachment system; GCT=Great Counter Thrust; MCT=Main Central Thrust; RMT=Ramgarh-Munsiari Thrust; MBT=Main Boundary Thrust; MFT =Main Frontal Thrust.

associated arc and subduction zone. However, if instead they were the result of a continental rift, no subduction is needed; the rift setting thus is compatible with reconstructions that show northern India facing an open margin at that time. Much is unknown about the Paleoproterozoic evolution of strata now composing the Himalayan thrust belt. With better dating of these deposits in the future, we will be able to better piece together the puzzle. We do know that these Paleoproterozoic deposits are present between Afghanistan and the Namcha Barwa syntaxis, so the environment in which these strata were deposited must have been at least 3,000 km long in one dimension.

5.4. MESOPROTEROZOIC TIME (1.6–1.0 GA)

Rocks deposited in Mesoproterozoic time are Lesser Himalayan (Fig. 5.2). In and west of northwestern India as well as in and east of Sikkim, Mesoproterozoic units exposed at the surface are thin. More Mesoproterozoic rocks could have been deposited and were eroded, or perhaps they were never deposited, or perhaps they remain buried under thrust sheets. In northwestern India, the Mesoproterozoic rocks are limited to the ~ 1.6 Ga Gangolihat Formation (see Martin 2017b, Appendix A for discussion of Meso- and Neoproterozoic carbonate-dominated formations along the thrust belt). In central and western Nepal, the Mesoproterozoic Lakharpata Group makes up 3-4 km of the Lesser Himalayan stratigraphy. In central Nepal, the top of the Mesoproterozoic succession is the Malekhu Formation with $\delta^{13}C$ values that suggest deposition prior to 1.3 Ga (Martin et al., 2011). Thus, we know that these rocks are older than 1.3 Ga but we are uncertain how much older. Martin (2017b) correlates these units with the ~1.6 Ga Gangolihat Formation, whereas DeCelles et al. (2020) infer a depositional age of 1.5–1.3 Ga. Clearly, there is much work still to be done to understand the stratigraphic correlations.

Cawood et al. (2020) reconstruct the Mesoproterozoic northern Indian margin as a tectonically inactive margin. The supercontinent Columbia (Nuna) initially began to break up at 1.45 Ga and ended breakup at ~1.3 Ga. It is not clear whether breakup of the supercontinent moved the continents around significantly. O'Neill et al. (2015) found slow plate-motion velocities through Early and Middle Mesoproterozoic time while Earth perhaps remained in a single-lid tectonic regime (Stern, 2020). The supercontinent Rodinia assembled between 1.1 and 0.9 Ga (Zhao et al., 2018). The northern Indian margin was located at the edge of Columbia and remained along the continental periphery into Neoproterozoic time through breakup of Columbia and assembly of Rodinia (Cawood et al., 2020). The time frame from 1.85 to 0.85 Ga has been informally termed the "boring billion" (Holland, 2006) because it was a period of environmental stability even though one supercontinent was torn apart and reassembled into another (Fig. 5.3b).

5.5. NEOPROTEROZOIC TO EARLY ORDOVICIAN TIME (1.0–0.46 GA)

Neoproterozoic-Cambrian rocks are found in the Tethyan, Greater, and Lesser Himalayan sequences. This is a case in which the four-unit breakdown of the tectonostratigraphy is not sufficient (Fig. 5.2). These tectonostratigraphic units have rocks that were deposited at the same time but perhaps were not connected spatially at the time of deposition (Fig. 5.3c). Martin (2017b) contains a review of the different ideas about when and how these rock packages may have been connected.

An unmodified depositional contact between Neoproterozoic and Paleoproterozoic/lower Mesoproterozoic strata may crop out in two locations in the orogen; in all other locations, this boundary lies within a high strain zone. South and west of Tarbela Lake in northern Pakistan, Pogue et al. (1999) and DiPietro et al. (1999) map the Neoproterozoic-Cambrian Tanawal Formation in depositional contact with the Gandaf Formation. These authors infer a Paleoproterozoic depositional age for the Gandaf Formation there based on correlation with dated deposits farther north. However, the depositional age was not determined directly on the strata depositionally below the Tanawal Formation, leading to uncertainty about their actual depositional age. If the true depositional age of the Gandaf Formation near Tarbela Lake were Neoproterozoic, then actually no depositional contact between Paleoproterozoic and Neoproterozoic strata is exposed there. In the Tons River valley in northwestern India, 20 km north of the Main Boundary Thrust, McKenzie et al. (2011) describe shale and sandstone of the Neoproterozoic Mandhali Formation resting depositionally on Mesoproterozoic or Neoproterozoic dolostone (photograph in Hughes et al., 2020, their Fig. 1). However, the depositional age of the dolostone is poorly known (discussion and references in McKenzie et al., 2011). If the dolostone beneath the contact were deposited during the Neoproterozoic Era, then this locality does not expose a depositional contact between Neoproterozoic and lower Mesoproterozoic strata.

Overlying the Mandhali Formation are the Shimla, Krol, and Tal Groups, which have depositional ages from Early Neoproterozoic to Cambrian (Singh et al., 1999; Hughes et al., 2005; McKenzie et al., 2011; Hughes, 2016). Eastern Bhutan does not have Mesoproterozoic rocks; the Neoproterozoic-Cambrian Baxa Group directly overlies the Paleoproterozoic section (McQuarrie et al., 2019). It is possible that as more stratigraphy in the Himalaya is analyzed, Neoproterozoic rock will be found structurally below the Main Central Thrust, perhaps by using detrital zircon to determine a maximum age of deposition.

Myrow et al. (2016) state that Tethyan Himalayan rocks are Neoproterozoic-Cenozoic strata; however, including the Neoproterozoic-Cambrian strata within the Tethyan Himalayan rocks can be contentious. Some researchers consider the Tethyan Himalayan rocks to begin only in Ordovician time (e.g., Liu & Einsele, 1994; Jadoul et al., 1998; Garzanti, 1999; Hu et al., 2008; Sciunnach & Garzanti, 2012), which is the perspective of those who have worked mainly in southern Tibet in the central Himalaya and consider all rock under the South Tibetan detachment system to be something other than Tethyan Himalayan rocks. Myrow et al. (2016) review these Neoproterozoic to Cambrian rocks in the northern part of the thrust belt in the western and central Himalaya and show that in some parts of the Himalaya, these rocks are separated from the overlying Ordovician rocks by an unconformity and in other places by a younger detachment fault (South Tibetan detachment system), and that the South Tibetan detachment system is at different stratigraphic levels along the thrust belt. Throughout the Himalayan orogen, researchers state that there is/was an unconformity that marks the contact of Ordovician strata with the strata underneath (Hayden, 1904; Garzanti et al., 1986; Wiesmayr & Grasemann, 2002; Gehrels et al., 2003, 2006a, 2006b; Cawood et al., 2007). In some cases, the strata underneath are unmetamorphosed whereas in others they are metamorphosed. In Zanskar, Garzanti et al. (1986) report an angular unconformity between Cambrian carbonate rock and Middle Ordovician conglomerate, and hypothesize about an orogenic event at the Cambrian-Ordovician boundary called the Kurgiakh orogeny. Martin (2017b) and Kapp and DeCelles (2019) term the metamorphosed rocks stratigraphically under the unequivocal Tethyan Himalayan rocks, as well as Neoproterozoic to Cambrian rocks found in klippen within the thrust belt, Greater Himalayan Protolith. For identification, basal Tethyan Himalayan rocks have youngest detrital zircon ages of ca. 510–470 Ma (DeCelles et al., 2000, 2004; Gehrels et al., 2003, 2006a, 2006b, 2011; Aikman et al., 2008). Greater Himalayan and Greater Himalayan Protolith rocks have youngest detrital zircon grains that range in age from ca. 830 to 540 Ma with the spectra dominated by 1,100–750 Ma zircon grains (DeCelles et al., 2000, 2020; Myrow et al., 2010; Gehrels et al., 2011).

In general, Greater Himalayan rocks crop out between the South Tibetan detachment system and the Main Central Thrust (see 30–15 Ma section 5.8) as well as in isolated klippen south of the Main Central Thrust surrounded by Lesser Himalayan rocks (e.g., Dadeldhura/Almora klippe, Jarjarkot klippe, Kathmandu klippe). Greater Himalayan rocks are mostly sedimentary rocks metamorphosed from garnet to sillimanite grades (Kohn, 2014), including quartzite, calcsilicate, mica-quartz-garnet ± aluminosilicate schist and gneiss, and migmatite, which range in age from ~830 to 460 Ma (e.g., Gehrels et al., 2011), with the youngest age determined by the youngest intrusions into Greater Himalayan rock (460 Ma; Horton et al., 2015). Neoproterozoic to Early Ordovician intrusion ages within Greater Himalayan rocks are 930–800 Ma, which occur only in the western and eastern Himalaya, and 510–460 Ma, which are widespread around the Himalayan orogen. Some researchers divide Greater Himalayan rocks into different units; however, the stratigraphic order of these units varies along strike. In general, unit I is pelitic and psammitic, unit II is calcsilicate and marble, and unit III is the metagranite that intruded at ca. 500 Ma (e.g., Hodges et al., 1996). Mukherjee and Koyi (2010a) divide the Greater Himalayan rocks into a lower or southern nonmigmatitic unit and an upper or northern migmatitic unit in the Sutlej section (Himachal Pradesh, India). Greater Himalayan rocks are not Indian Archean basement because they are too young and do not have the same Nd isotopic signatures (e.g., Parrish & Hodges, 1996; DeCelles et al., 2000; Robinson et al., 2001; Mukherjee et al., 2019). Martin (2017b) contains a discussion on possibilities of where Greater and Tethyan Himalayan rocks were positioned with respect to the Indian craton during deposition and intrusion. Modern ideas fall into two categories.

1. Deposition and intrusion mostly east of the Indian craton (Brookfield, 1993; Martin, 2017b). In this scenario, the eastern and central parts of the Greater and Tethyan Himalayan rocks intruded and were deposited north of western Australia while the western portion intruded and was deposited north of eastern India.

2. Contiguous deposition outboard of India forming a passive margin, with Lesser, Greater, and Tethyan Himalayan rocks deposited contiguously outboard of one another from south to north (Burrard & Hayden, 1908; Frank et al., 1973; Colchen et al., 1982; Searle, 1986; Myrow et al., 2003). In this scenario, the passive margin changed to a convergent margin during Cambrian time, during the Cambrian-Ordovician Kurgiakh orogeny (e.g., Garzanti et al., 1986; Fig. 5.3d). After this orogeny, the margin switched back to a passive margin in Late Ordovician time (Fig. 5.3e). Hybrid scenarios that combine elements of the endmember models are possible.

Neoproterozoic granitic plutons are present in the western and eastern Himalaya. Most of these are located north of the Main Central Thrust in Greater Himalayan rocks (DiPietro & Pogue, 2004; Richards et al., 2006; Yin et al., 2010; Webb et al., 2011a; Mottram et al., 2014; Clarke et al., 2016). If found north of the Main Central Thrust, these rocks give a maximum age for Greater Himalayan rocks, which could be as old as 930 Ma. In northwestern India, there are ~930 and ~830 Ma crystallization ages from plutons in the hanging wall of the Jutogh Thrust (Webb et al., 2011a; Singh et al., 2020). This thrust lies south of the position of the Main Central Thrust interpreted by most workers. The affinity of the rocks carried by the Jutogh Thrust could be Greater Himalayan (e.g., Richards et al., 2005), Lesser Himalayan (e.g., Dubey & Bhakuni, 2007), or Tethyan (e.g., Webb et al., 2011a). Singh et al. (2020) address this problem by placing the ~930–830 Ma plutons in a separate thrust sheet because the plutons are surrounded by shear zones. It is possible that most of the evidence of a Neoproterozoic magmatic arc or rift was wiped out by subsequent Cambrian-Ordovician tectonic activity (Spencer et al., 2019).

In Pakistan, Cambrian-Ordovician granite (~520–475 Ma; Naeem et al., 2016; Sajid et al., 2018) intruded both the ca. 823 Ma Black Mountain Complex (DiPietro & Isachsen, 2001) and the Neoproterozoic-Cambrian Tanawal Formation. Peak metamorphism south of the Main Mantle Thrust in kyanite- and sillimanite-bearing gneiss in the Tanawal Group indicate Ordovician ages (ca. 482–464 Ma), and Permian ages (ca. 272–267) in the staurolite-grade schist (Palin et al., 2018).

The Cambrian-Ordovician orogenic event, which deformed equatorial Gondwana, is called the Kurgiakh (or Bhimphedi) orogeny (Fig. 5.3d; Garzanti et al., 1986; Argles et al., 1999; Gehrels et al., 2003, 2006a, 2006b; Kohn et al., 2005; Cawood et al., 2007; Yin et al., 2010; Xu et al., 2014a, 2014b). Deformation associated with the Kurgiakh orogeny affected Greater and Tethyan Himalayan rocks, including rocks now found in the frontal klippen, between ~500 Ma and ~460 Ma based on preorogenic and postorogenic granite emplacement and an angular unconformity between the sedimentary successions (Garzanti et al., 1986; Le Fort et al., 1986; Gehrels et al., 2003, 2006a; Myrow et al., 2016). Myrow et al. (2016) infer that a 22–36 m.y. stratigraphic gap near the Cambrian-Ordovician boundary, first described by Hayden (1904), was caused by deformation associated with this orogeny. The absence of Lesser Himalayan deformation, metamorphism, and magmatism at 930–800 Ma, 500–460 Ma, and 290–260 Ma requires an explanation. One possible explanation is that Lesser Himalayan rocks that were affected by this event have been eroded or covered by other rocks. Alternatively, it could be that Greater and Tethyan Himalayan rocks were far away from Lesser Himalayan rocks at this time (Fig. 5.3d), perhaps laterally offset to the east (Martin, 2017b). One result of this uncertainty is that when reconstructing the Greater Indian margin by palinspastically restoring Cenozoic offsets on major high strain zones, the preorogenic northward extent of Lesser Himalayan rocks is unknown. These Proterozoic rocks may have extended much farther north than reconstructions indicate (see discussion in Robinson & Martin, 2014) resulting in a wider Greater India. Balanced cross-section reconstructions commonly place Greater Himalayan rock north of Lesser Himalayan rock at the time of initiation of Cenozoic orogeny; however, the reconstructions offer no judgment on the location of Tethyan and Greater Himalayan rocks during the pre-Cenozoic construction of Greater India. We simply do not have enough data to know.

Some researchers argue that the Earth's modern plate tectonic regime began in Neoproterozoic time (e.g., Stern, 2020). At this time, some reconstructions of Gondwana do not leave much space between the northern edge of India and the South China terranes for rocks north of cratonal India (Merdith et al., 2017). In that space, the Lesser, Greater, and Tethyan Himalayan rocks, as well as the Lhasa/South Qiangtang terrane, may have been present (Spencer al., 2019). Alternatively, the Greater and Tethyan Himalayan rocks could have been to the east of that space (Martin, 2017b). The pre-Cenozoic location of the Greater and Tethyan Himalaya is debated in Martin (2017b) and remains an open question (Hughes et al., 2019, 2020; Martin, 2020).

Martin (2017b) places uppermost Ediacaran to Cretaceous strata found in and near the Salt Range in Pakistan (Potwar Plateau) in Himalayan Assemblage B (Tethyan strata). Hughes et al. (2019) point out that the Ediacaran and Cambrian Salt Range strata are more like rocks deposited on the Indian craton to the south than Assemblage B (Tethyan and Greater Himalayan) (meta)sedimentary rocks farther north in Pakistan. Similarities between the Salt Range and cratonal India deposits include deposition of uppermost Ediacaran evaporite- and carbonate-rich strata and lower Cambrian clastic strata (Gee, 1989; Hughes et al., 2019). In contrast, there are at least three differences between the Salt Range and Assemblage B (Tethyan and Greater Himalayan) deposits.

1. The Salt Range contains thick uppermost Ediacaran salt and gypsum deposits as well as lower Cambrian deposits with salt pseudomorphs (Wynne, 1878; Gee, 1989), but evaporite deposits in northern Pakistan strata of the same age are limited to uncommon, thin gypsum layers (Qasim et al., 2015; cf. Latif, 1973).

2. Uppermost Ediacaran or lowermost Cambrian volcanic rocks crop out in the Salt Range (Fleming, 1853; Jan & Faruqi, 1995) but are absent in northern Pakistan deposits of the same age (DiPietro et al., 1999; Pogue et al., 1999).

3. Assemblage B (Tethyan and Greater Himalayan) rocks in northern Pakistan include Permian mafic metavolcanic rocks and dikes (DiPietro et al., 1999), but Permian Salt

Range strata contain neither volcanic nor volcaniclastic deposits, and mafic dikes did not intrude the Salt Range rocks during Permian time (Gee, 1989; Jan et al., 2016).

Accordingly, we place the Salt Range strata in Himalayan Assemblage A (Lesser Himalayan rocks). The location of the boundary between assemblages A and B (Lesser and Greater Himalayan rocks, respectively) in northern Pakistan is uncertain, so we show it with a dotted line in Figure 5.1b.

5.6. MIDDLE ORDOVICIAN TO CRETACEOUS TIME (470–66 MA)

Post-Middle Ordovician rocks are mostly unmetamorphosed and are found in both the Tethyan and Lesser Himalayan rocks (e.g., Fuchs, 1977; Stocklin, 1980; Colchen et al., 1986; Fuchs et al., 1988; Garzanti, 1999; see also Strachey, 1851). Tethyan Himalayan rocks are Paleozoic to Cenozoic carbonate and clastic sedimentary rocks deposited in a mostly passive margin setting (Searle et al., 1987; Liu & Einsele, 1994; Garzanti, 1999). In Lesser Himalayan rocks, a large unconformity exists between the Proterozoic rocks below and Paleozoic rocks above. In Bhutan, the unconformity spans only about 200 m.y., separating the Manas or Rupa Formation below (520–480 Ma) from the Gondwana unit above (320–260 Ma; Fig. 5.2). In central Nepal, the unconformity in Lesser Himalayan rocks is between the >1.3 Ga Malekhu Formation and the Upper Carboniferous-Permian Sisne Formation (Gondwana unit) and spans ~1.0–1.3 Ga (Martin et al., 2011; summary in Robinson & Martin, 2014). The Gondwana unit contains glacial dropstones from the Gondwanan icecaps and are Late Carboniferous in age in Bhutan and central Nepal. In most places, the Gondwana unit is entirely Late Carboniferous to Permian in age; however, in central Nepal, it includes about 200 m of the Cretaceous Taltung Formation (Sakai, 1983).

In the Tethyan Himalaya, Carboniferous rifting between Gondwana and Peri-Gondwanan blocks (Cimmeria) formed the Neo-Tethys Ocean and terrigenous rocks were deposited in the resulting rift basins (Fig. 5.2; Garzanti et al., 1999). Liu and Einsele (1994) state that this rifting was likely perpendicular to the modern Tethyan fold-thrust belt based on the stratigraphy, creating normal faults in a favorable orientation for reactivation during Cenozoic orogeny. In Early Permian time, rifting ended and basalt and granite were intruded into western Tethyan Himalayan rocks (Spring et al., 1993; Vannay & Spring, 1993). The Tethyan Himalayan Panjal Traps in Pakistan and Kashmir are from the rifting (Shellnutt et al., 2015) and are similar in composition to the Tethyan Himalayan Nar Tsum Spilite and Bhote Kosi Basalt in the central Himalaya (Garzanti et al., 1999).

In northwestern India, there are isolated klippen of the Jurassic-Cretaceous Spongtang oceanic arc within the Tethyan Himalaya (Reuber et al., 1987; Garzanti et al., 1987; Pedersen et al., 2001; Buckman et al., 2018; Catlos et al., 2019). Late Jurassic rifting of Argoland from Gondwana is often shown extending from Java across northern India to Arabia (e.g., Gibbons et al., 2015) with no Jurassic detrital zircon signal in Tethyan Himalayan strata. Early Cretaceous Indo-Australia rifting occurred along the eastern margin of India. Some authors argue for a rifted northern margin of Greater India at this time (e.g., Chen et al., 2018b), while others propose a transform margin (Ali & Aitchison, 2005, 2014; Gibbons et al., 2012).

In central Nepal, it is uncertain whether deposition of the 50–200 m thick Amile Formation (Sakai, 1983) was unrelated to the Himalayan orogeny or if it was deposited in a distal part of the foreland basin created by the initial stages of the orogeny (cf. DeCelles et al., 2014; Martin, 2017b). If depositional basin formation was not related to the Himalayan orogeny, the Amile Formation should be included in the Gondwana unit within the Lesser Himalaya strata, whereas if the Amile Formation were deposited in a Himalayan foreland basin, these deposits instead should be grouped with the other foreland basin strata described in the next section. The Amile Formation has a maximum depositional age of 120 Ma, taken from a prominent peak of youngest detrital zircon grains (DeCelles et al., 2014). During the same time frame that Lesser Himalayan rocks were being deposited in the southern part of the thrust belt, Tethyan Himalayan rocks were being deposited in the northern part, with some unconformities, particularly during Devonian time (Fig. 5.2). Correlation of strata between the Lesser and Tethyan Himalaya is difficult, limited to a possible link in central Nepal between Lower Cretaceous volcaniclastic rocks of the Taltung Formation in the Gondwana unit and the Chukh Group in the Tethyan strata. However, we do not know if the Lesser and Tethyan Himalayan depositional basins were contiguous or separated by 2,000 km in the east-west direction (Fig. 5.3e). The Lower Cretaceous strata of the Tethyan Himalaya are locally interbedded with intraplate volcanic rocks (Hu et al., 2015b), and were deposited due to thermal subsidence associated with the Rajmahal Traps, which consist of ca. 118 Ma basalt currently located west of the Shillong Plateau (Zhu et al., 2008). The strata near the Cretaceous-Paleogene contact contain detritus associated with the Deccan Traps (Garzanti & Hu, 2015). During the development of two large igneous provinces (145–115 Ma) in northeastern India (Hu et al., 2015b), India and Australia rifted away from each other (~130 Ma).

In Pakistan, rocks that fall into this time range are the Carboniferous-Jurassic Alpurai Group and related rocks found in the footwall of the Main Mantle Thrust. The

Alpurai Group contains mostly metasedimentary rocks but also includes Permian metabasalt correlated with the Panjal Traps in northwestern India (DiPietro et al., 1999). This volcanism is associated with the intrusion of mafic dikes and granite (DiPietro et al., 1999; DiPietro & Isachsen, 2001). The Peshawar Plain alkaline igneous province extends ~200 km, contains alkaline granite, syenite, and carbonatite (e.g., Ahmad et al., 2013). It was emplaced during three different episodes (Early Silurian, Late Carboniferous, and Late Cretaceous) related to different tectonic events (Khan, 2021). The Late Cretaceous carbonatite is associated with the passage over the Réunion hotspot mantle plume (Khan, 2021). In western Pakistan, Garzanti and Hu (2015) state that the oldest basalt and mafic alkalic rocks chemically and isotopically similar to the modern Réunion lavas have an age of ~70 Ma.

5.7. PALEOCENE TO EARLY OLIGOCENE TIME (66–30 MA)

Greater India is a generic term for the lithosphere that extended north of the northern Indian craton immediately prior to the India-Asia collision (Veevers et al., 1975; Klootwijk et al., 1979; Ali & Aitchison, 2005, and references therein; Robinson et al., 2006; Guillot et al., 2008; Long et al., 2011a; van Hinsbergen et al., 2011, 2012, 2019; Ma et al., 2016; Meng et al., 2019; Rowley, 2019; Parsons et al., 2020) and includes all of the rocks discussed in the prior sections. Greater India is currently exposed from the Indus-Yarlung suture to the Main Boundary Thrust. Using a collision time of 58 Ma, plate reconstructions suggest that Greater India was ~2,600–3,400 km wide prior to initial collision (van Hinsbergen et al., 2019), much wider than the largest amount of minimum shortening found across the entire Himalayan thrust belt of ~1,000 km (Robinson et al., 2006). This discrepancy has led to debate about whether Greater India was all continental crust (Fuchs & Willems, 1990; Dupont-Nivet et al., 2010) or contained ocean basins between continental fragments (van Hinsbergen et al., 2011). One location for the "missing" shortening could be north of the South Tibetan detachment system within the northern part of the Himalayan thrust belt, the Tethyan fold-thrust belt (TFTB). The northernmost Indian paleomagnetic sites, located in the southern TFTB (Huang et al., 2015), require at least 1,800 km of convergence across the TFTB (van Hinsbergen et al., 2019). If Greater India were all continental crust, then the total convergence should be reflected in the TFTB shortening. However, if the shortening in the TFTB is drastically less than the ~1,800 km predicted, then it is likely parts of Greater India were oceanic crust, perhaps with intervening subduction zones and associated magmatic arcs that were eventually destroyed during final closure. Unfortunately, not enough work has been done on the TFTB to make a determination. Shortening estimates exist but need updating (cf., Ratschbacher et al., 1994). Prior to the final continent-continent collision, perhaps southern Asia was home to a Mesozoic protoplateau (Ding et al., 2014; Ingalls et al., 2018). Kapp and DeCelles (2019) have a thorough discussion on the development of the Tibetan Plateau prior to 45 Ma and the different closure scenarios as India docked with the southern margin of Asia. Li and Robinson (2023) detail two possible pre-collisional configurations of northern Greater India prior to collision.

Initial continental contact leading to mixing between sediment derived from India and Asia occurred at ~61–58 Ma in the Indus-Yarlung suture zone in the central Tibetan Plateau (Wang et al., 2011; DeCelles et al., 2014; Wu et al., 2014; Hu et al., 2015a; Wang et al., 2017; An et al., 2021). In the Indus basin in northwestern India, mixing between Indian and Asian sediment first occurred at ca. 50 Ma (e.g., Bhattacharya et al., 2021). As the India-Asia collision began, the rocks on the northernmost Greater Indian margin began forming the TFTB, which we assume developed between collision and ~30 Ma (e.g., Ratschbacher et al., 1994; Murphy & Yin, 2003; Zhang & Guo, 2007; Montomoli et al., 2017; Figs. 5.3, 5.4f). North of the South Tibetan detachment system, rocks that are currently structurally underneath the TFTB crop out in structural windows as gneiss domes that experienced metamorphic zircon growth at 48–45 Ma (Ding et al., 2016a) and yield 54–49 Ma Lu-Hf garnet dates (Smit et al., 2014). South of the South Tibetan detachment system, rocks structurally underneath the TFTB are Greater Himalayan rocks, in which metamorphism initiated at ~45 Ma and continued until at least Early Miocene time (see Fig. 5.4 for timing variations; Vannay & Hodges, 1996; Guillot, 1999; Godin et al., 2001; Catlos et al., 2002; Shrestha et al., 2020; Benetti et al., 2021, and many other articles). An undeformed intrusion cut the northern part of the TFTB at 44 Ma (Aikman et al., 2008). In the southern part of the TFTB, metamorphic crystallization ages of 46–42 Ma are interpreted using synkinematic illite analyses (Wiesmayr & Grasemann, 2002). A phase of rapid cooling is suggested between 41 and 37 Ma from detrital thermochronology data (Shen et al., 2016). In western and central Nepal, peak pressures of 10–13 kbar and temperatures of 700°C indicate Greater Himalayan rocks were buried 35–30 km between 41 and 30 Ma (Kohn et al., 2004; Kohn, 2008; Yakymchuk & Godin, 2012; Montomoli et al., 2013; Carosi et al., 2015; Soucy La Roche et al., 2016; Shrestha et al., 2020), presumably by the overburden of the TFTB. Ultrahigh pressure metamorphism of Greater Indian rocks in northwestern India and Pakistan occurred at the same time as shortening in the TFTB and metamorphism of Greater Himalayan rocks between 55 and 45 Ma (de

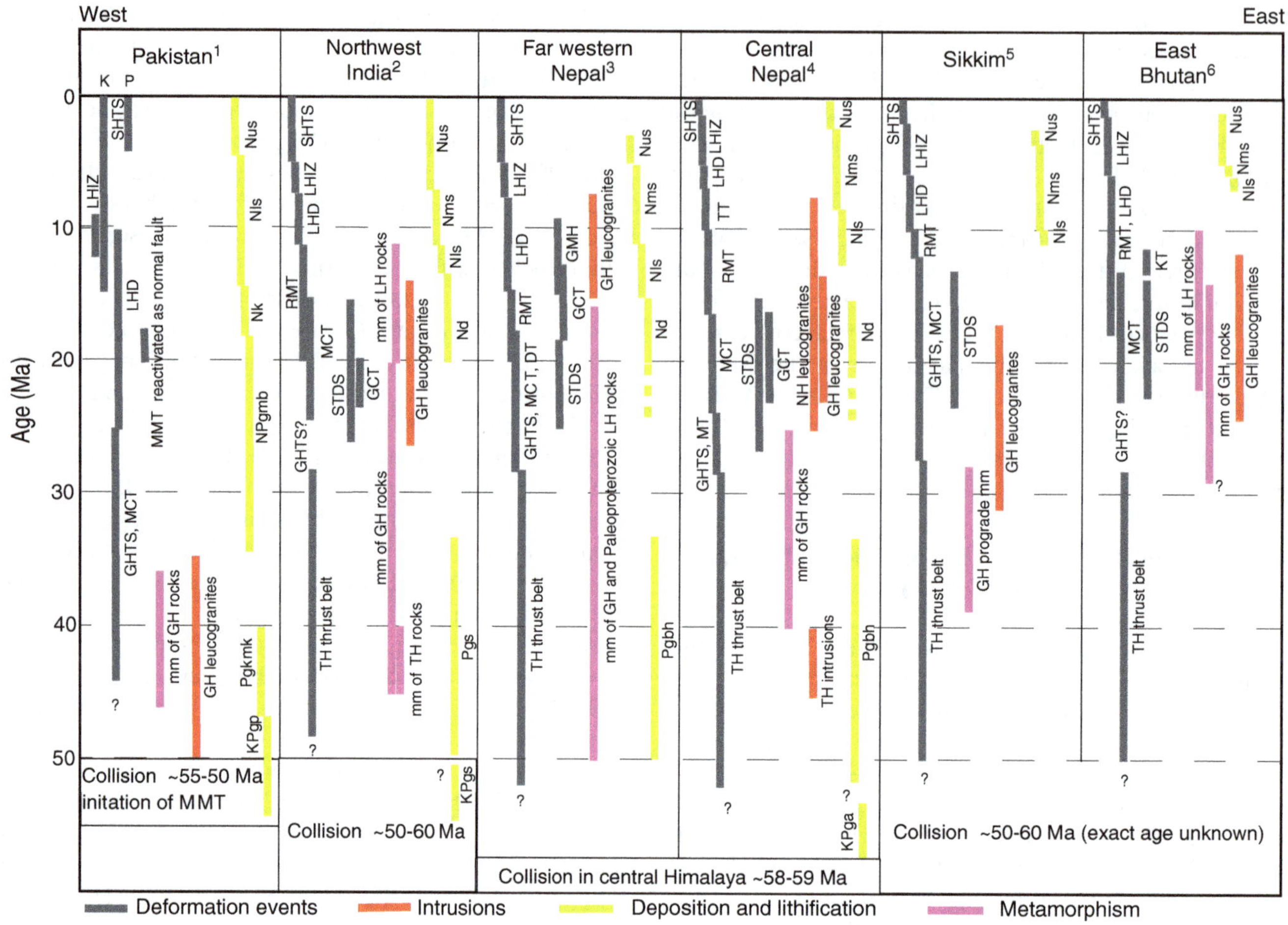

Figure 5.4 Along-strike comparison of tectonic events in the Himalaya. Note: Nus=Neogene upper Siwalik unit; Nms=Neogene middle Siwalik unit; Nls=Neogene lower Siwalik unit; Nk=Neogene Kamlial Fm.; Ns=Neogene Subathu Fm.; Nd=Neogene Dumri and Dharmsala Fms.; NPgmb=Neogene-Paleogene Murree and Balakot Fms.; Pgbh=Paleogene Bhainskati Fm.; Pgkmk=Paleogene Kahat, Mamikhel, and Kuldana Fms.; KPgs=Cretaceous -Paleogene Singtali Fm.; KPga=Cretaceous-Paleogene Amile Fm.; KPgp=Cretaceous-Paleogene Panoba Fm.; SHTS=Sub-Himalayan thrust system; LHIZ=Lesser Himalayan imbricate zone; LHD=Lesser Himalayan duplex; MCT=Main Central Thrust; GHTS=Greater Himalayan thrust system; MMT=Main Mantle Thrust; RMT=Ramgarh-Munsiari Thrust; STDS=South Tibetan detachment system; GCT=Great Counter Thrust; DT =Dadeldhura Thrust; MT=Mahabarat Thrust; KT=Kaktang Thrust; GMH=Gurla Mandata-Humla fault system; TH=Tethyan Himalaya; GH=Greater Himalaya; LH=Lesser Himalaya; NH=North Himalaya; mm=metamorphism; K=Kohat plateau; P=Potwar plateau (based on [1]Treloar & Rex,1990; Zeitler & Chamberlain, 1991; Pivnik & Wells, 1996; Foster et al., 2002; Ding et al., 2016b, and references therein; Treloar et al., 2019; Ghani et al., 2021a, 2021b; [2]Clift et al., 2002; Najman et al., 2004; Vannay et al., 2004; Finch et al., 2014; Weinberg, 2016, and references therein; Iaccarino et al., 2017; Colleps et al., 2019, and references therein; [3]Ojha et al., 2000, 2009; DeCelles et al., 2004, 2011; Murphy & Copeland, 2005; Robinson & McQuarrie, 2012; Carosi et al., 2013; Montomoli et al., 2013; McCallister et al., 2014; Hu et al., 2015a; [4]DeCelles et al., 2004; Ojha et al., 2009; Wang et al., 2016; Laskowski et al., 2017, and references therein; Ghoshal et al., 2020; [5]Ojha et al., 2009; Kellett et al., 2013; McQuarrie et al., 2014; Mottram et al., 2015; Weinberg, 2016, and references therein; Chakraborty et al., 2019; [6]Kellett et al., 2010; Coutand et al., 2016; Weinberg, 2016, and references therein; Gilmore et al., 2018; Grujic et al., 2020).

Sigoyer et al., 2000; Kaneko et al., 2003; Leech et al., 2005; Parrish et al., 2006; Wilke et al., 2010; Donaldson et al., 2013; Rehman et al., 2013; St-Onge et al., 2013). Thus, we know that the TFTB buried Greater Himalayan rocks and attained significant surficial height between 60 and 30 Ma. Currently, we cannot identify all the thrusts necessary to accomplish this burial and thickening, only that it must have been during Paleocene-Eocene time.

Leucogranite intruded into the upper part of the Greater Himalayan rocks consistently along the Himalayan arc. The leucogranite is two-mica ± tourmaline ± garnet and cordierite-bearing (e.g., Godin et al., 2006; Visona et al., 2012). In Pakistan in the Kaghan valley, the leucogranite intruded from 50–35 Ma, and from northwestern India to Bhutan, the leucogranite is younger, from ~32 to 7 Ma (see references in Fig. 5.4 caption; reviews in Leech, 2008, and Weinberg, 2016). As Greater Himalayan rocks were buried, the leucogranite formed by the dehydration melting of muscovite and biotite (Cottle et al., 2015), which could produce 15–20 volume % melt (Groppo et al., 2012).

The Indian plate moved northward under Tibet as collision continued. Webb et al. (2017) and Kapp and DeCelles (2019), and more generally Garzanti et al. (2018), discuss breakoff of the subducting Indian lithospheric slab from ~45 to 38 Ma. The TFTB is intruded by 45–40 Ma igneous bodies that can be interpreted as being derived from the asthenosphere (e.g., Aikman et al., 2008, 2012; Pullen et al., 2011); thus, the subducting slab must have retreated far enough south to allow this magmatism to occur within the TFTB. North of the TFTB in the Lhasa terrane, the youngest granite is ~38 Ma (Sanchez et al., 2013), which can be interpreted as indicating the beginning of the Indian continental lithospheric slab flattening, shutting off magmatism.

In the northwestern Himalaya, collision may have occurred at 60–50 Ma (Green et al., 2008; Wu et al., 2014; Ding et al., 2016b; Hu et al., 2016). Garzanti et al. (1987) and Li et al. (2017) interpret the forebulge unconformity in the Zanskar Range and in South Tibet at ca. 56 Ma. In the upper Kaghan valley of Pakistan, ultra-high-pressure coesite-eclogite metamorphism occurred at ~47 Ma (Pognante & Spencer, 1991; Tonarini et al., 1993; Kaneko et al., 2003; Treloar et al., 2003; Parrish et al., 2006; Wilke et al., 2010; Rehman et al., 2013; Rehman, 2019). Treloar et al. (2019) suggest that collision was no older than 47 Ma and no younger than 40 Ma based on hornblende cooling ages. The metasedimentary rocks contain evidence for Cenozoic metamorphism prior to cooling through 500°C by 40 Ma (Treloar, 1997) in the Kaghan valley. In the Swat valley, this schist was metamorphosed to ca. 10–12 kbar with temperatures of 650–700°C (e.g., Treloar et al., 2019). These authors state that isothermal decompression from peak metamorphic conditions of 11 kbar and 700°C occurred prior to 40 Ma. Thus, collision timing is poorly known there but must be before these metamorphic ages.

5.7.1. Sedimentary Record

In the TFTB near the Indus-Yarlung suture in the central Himalaya, the foreland basin record gets younger southward with an ~60 Ma trench, ~56 Ma forebulge disconformity, and ~51 Ma distal foredeep strata (Garzanti et al., 1987; Ding et al., 2005; DeCelles et al., 2014; Hu et al., 2016; Li et al., 2017). In the southern TFTB, strata record Middle Maastrichtian to Paleocene flexural bulge migration signaling the downflexure of the northern Indian passive margin and Eocene foreland basin deposition (e.g., Willems et al., 1996; Zhang et al., 2012; Li et al., 2015a). Sediment accumulation southward distal from the frontal mountain belt includes deposition of the Bhainskati Formation of central Nepal in a back-bulge to distal forebulge depozone at ~45 Ma (DeCelles et al., 1998a, 2004; Najman et al., 2005). Erosion of Tethyan Himalayan rocks in the growing TFTB is recorded in the distal foreland basin rocks of the Bhainskati Formation (DeCelles et al., 1998a, 2004; Najman et al., 2005; Jain et al., 2009; Ravikant et al., 2011; Stickroth et al., 2019). At ~38 Ma, the Bengal basin experienced an increased flux of sand with short detrital thermochronological lag times (Najman et al., 2005) indicating increased erosion from the growing Himalayan mountain belt. In Bhutan, no strata of Paleogene age are preserved.

In the Indus-Yarlung suture in Ladakh, northwestern India, rocks of this age are found in the Indus basin in the Upper Cretaceous-Lower Eocene marine Tar Group and the Lower Eocene-Lower Miocene Lower Indus Group. Deposition began in an arc-bounded or a forearc marine basin and evolved into a continental intermontane basin after collision began (Garzanti & Van Haver, 1988; Henderson et al., 2010). Bhattacharya et al. (2020, 2021) contain more data on the age and evolution of the Indus basin. Toward the foreland in northwestern India, the shallow marine Subathu Formation is Late Paleocene-Middle Eocene in age (Mathur, 1978; Batra, 1989), and detrital zircon data indicate the foreland basin system was in place by at least 50 Ma (e.g., Colleps et al., 2020). Burbank et al. (1996), Najman (2006), and Garzanti (2019) contain reviews of synorogenic rocks along the Himalayan orogen that record the evolution of the orogenic system.

In Pakistan, Ding et al. (2016b) show a correlation of synorogenic sedimentary rocks of these ages with those in Himachal and central Nepal. In some locations, sedimentation is continuous throughout Cenozoic time, while in other places, there are disconformities that last anywhere between ~20 and 5 m.y. in Eocene to Oligocene time. Correlation with the units in Pakistan is not direct because the western depositional setting in Pakistan was dominated by Paleocene-Middle Eocene carbonate strata (Pivnik & Wells, 1996), rather than clastic synorogenic deposits. However, in the eastern part of Pakistan, Ding et al. (2016b) interpret a disconformity in the ca. 57–55 Ma (Shah, 2009) Patala Formation formed by the migration of a flexural forebulge; thus, sediment in the uppermost part of the Patala Formation was deposited in a distal foredeep. Bossart and Ottiger (1989) and Critelli and Garzanti (1994) identify the first Asian-derived detritus in the upper Patala Formation. Pivnik and Wells (1996) have an extensive review of the Paleocene – Eocene stratigraphy in the Kohat Plateau of Pakistan, south of the Main Boundary Thrust.

5.8. EARLY OLIGOCENE TO MIDDLE MIOCENE TIME (30–15 MA)

The southern part of the Tibetan Plateau may have reached its modern height by 25 Ma (Currie et al., 2005, 2016). In the TFTB from 26–23 Ma, the north Himalayan domes experienced crustal thinning and north-south stretching (e.g., Lee et al., 2004, 2006). The Leo Pargil dome in northwestern India was active at 23 Ma (Langille et al., 2012; Jessup et al., 2016). Kapp and DeCelles (2019) state that beginning about 21 Ma, the Indian lithosphere began to return to a gentler underthrusting angle. This may or may not correlate with motion on the Great Counter Thrust in the northern TFTB (Heim & Gansser, 1939), which occurred at ~21–15 Ma in the central Himalaya (Ratschbacher et al., 1994; Yin et al., 1994; Quidelleur et al., 1997; DeCelles et al., 2011; Zhang et al., 2011; Carrapa et al., 2014; Leary et al., 2016; Laskowski et al., 2018; Orme, 2019). In Ladakh, northwestern India, motion on the Great Counter Thrust system was indirectly determined to be 23–20 Ma (Searle et al., 1992; Sinclair & Jaffey, 2001; Clift et al., 2002). At about the same time, first motion on the South Tibetan detachment system (Powell & Conaghan, 1973; Burg & Chen, 1984; Burg et al., 1984; Herren, 1987; Burchfiel et al., 1992) was at ca. 26–25 Ma in most locales (Carosi et al., 2018; Kellett et al., 2019; Zhang et al., 2020). However, caution is needed because the system did not initiate coevally along the orogen (Fig. 5.4). The South Tibetan detachment system consists of one or more top-to-the-north sheared low-angle high strain zones that in some places separate weakly metamorphosed Tethyan strata from underlying Greater Himalayan rocks as in central Nepal and northwestern India and in others separate unmetamorphosed Tethyan strata from underlying Greater Himalayan rocks as near Mount Everest (e.g., Carosi et al., 1998; Kellett et al., 2019). The thermal blanket created by the Tethyan thrust belt pushed Greater Himalayan rocks to peak temperatures between 700–800°C at 28–23 Ma (Kohn, 2014).

South of the TFTB, the first recorded thrusts within the southern part of the Himalayan thrust belt are the intra-Greater Himalayan shear zones, sometimes referred to as the High Himalayan Discontinuities or Greater Himalayan thrust belt (Shrestha et al., 2020) (Fig. 5.3g). These high strain zones are inferred to have been active beginning at 28–25 Ma and continuing until 18–16 Ma, depending on the location (Fig. 5.4; Kohn et al., 2004; Goscombe et al., 2006; Carosi et al., 2007, 2010, 2018; Corrie & Kohn, 2011; Imayama et al., 2012; Montomoli et al., 2013; Chakraborty et al., 2019; Benetti et al., 2021), with one report of motion at 41–36 Ma in central Nepal (Carosi et al., 2016). Shrestha et al. (2020) determine pressure-temperature-time paths for rocks above and below multiple shear zones in the Modi Khola valley of central Nepal, finding that the emplacement of three Greater Himalayan thrust sheets began at ~24 Ma, and ended with emplacement of the southernmost thrust that carried high-grade metamorphic rocks, the Main Central Thrust, at ~17 Ma. The end of activity on the Greater Himalayan shear zones was concurrent with motion on the south-dipping Great Counter Thrust system in the northern TFTB and Indus-Yarlung suture, which overprinted the older south-verging structures with north-verging folds and faults (Heim & Gansser, 1939; Gansser, 1964; Burg et al., 1987; Ratschbacher et al., 1994; Yin et al., 1994, 1999; Quidelleur et al., 1997; Yin & Harrison, 2000; Murphy & Yin, 2003; Montomoli et al., 2017).

The identification of the Main Central Thrust is a major point of disagreement (see Martin et al., 2017a, for a review). The definition is contentious for several reasons. First, offset on thrusts brought rocks that were metamorphosed at depth and stacked them atop other rocks, creating a package of ductilely deformed rocks 10–20 km thick, depending on location. There are thrusts above, within, and below this package, and it can be difficult to decide which of these high strain zones to correlate with other thrusts along strike to compose the orogen-spanning Main Central Thrust. Second, geologists have created new definitions or adapted old ones to accord with their favored models for the tectonic development of the orogen, resulting in multiple active definitions. Identifying the Main Central Thrust based on a gradient in ductile strain in quartz is problematic for multiple reasons, as discussed in Martin et al. (2017a). Two of these reasons follow. (1) This definition does not work in quartz-poor rocks such as limestone. (2) The location of a gradient in ductile strain is not necessarily the position of a gradient in total strain, that is, the edge of a high strain zone. Instead, a decrease in ductile strain could be matched by a corresponding increase in brittle strain, producing no change in total strain. Without measurements of contemporaneous ductile and brittle strain, it is impossible to know whether a gradient in total strain exists at any particular location. In this chapter, we identify the Main Central Thrust as the Cenozoic foreland-vergent thrust that separates Greater Himalayan from Lesser Himalayan rocks, with Greater and Lesser Himalayan rocks defined by their stratigraphy (Fig. 5.2). The Main Central Thrust was active after 25 Ma but not after 10 Ma in most places, with most ages clustering about 23–20 Ma (e.g., Kohn et al., 2004; Shrestha et al., 2020; reviews in Li et al., 2015b; Mukherjee, 2015). However, Montemagni et al. (2019, 2020) interpret 9–6 Ma motion on the Main Central Thrust using mica $^{40}Ar/^{39}Ar$ ages in Garhwal, northwestern India. The timing of motion varies along the Himalayan arc (Fig. 5.4), and it is unclear in some of the earlier literature whether these dates are for the initiation of intra-Greater Himalayan shear zones or the Main Central Thrust at the base of the Greater Himalayan rocks.

As the Main Himalayan thrust propagated southward, the development of the Lesser Himalayan duplex folded these overlying Tethyan and Greater Himalayan rocks and preserved them in synclines now exposed as frontal klippen (Fig. 5.3h). Some of these synclines contain both Greater and Tethyan Himalayan rocks and others contain only Greater Himalayan rocks. In the cores of others, the strata have the same detrital zircon signatures as Tethyan Himalayan deposits and are only weakly metamorphosed (Gehrels et al., 2006a, 2006b, 2011); thus, most researchers classify them as Tethyan Himalayan rocks. The contact between the Greater Himalayan and overlying Tethyan Himalayan rocks in these synclines is debated. One argument is that the contact was originally depositional, and it was subsequently sheared during folding, creating a high strain zone (Gehrels et al., 2006a, 2006b). Another argument is that the contact is the top-to-the-north sheared South Tibetan detachment system (Antolin et al. 2013; Soucy La Roche et al., 2016). Yet another argument is that it is a top-to-the-north fault that functioned as a passive roof thrust (e.g., He et al., 2016). It should be a simple prospect to look at these contacts and decide between the three models; however, there is a mix of top-north and top-south shear sense indicators and it is difficult to determine if a particular shear zone was the result of local folding or orogen-scale faulting because of the rheologic contrasts in different lithologies. As a result, the debate continues.

In Pakistan, the Main Mantle Thrust, which is tectonically equivalent to the Indus-Yarlung suture, was reactivated as a top-to-the-north normal fault in Oligocene-Miocene time (Treloar et al., 1991; Burg et al, 1996; Vince & Treloar, 1996; Anczkiewicz et al., 2001). This may be the western expression of the collapse of the orogenic system that the South Tibetan detachment system indicates toward the east.

The deformation then propagated southward into the Lesser Himalayan rocks (Fig. 5.3h). From northwestern India to Bhutan, the first intraformational thrust within the Lesser Himalayan rocks is the Ramgarh-Munsiari Thrust (review in Robinson & Pearson, 2013). The timing of this thrust also varies from 5–4 Ma in northwestern India (Montemagni et al., 2020) to 20–16 Ma in western Nepal (Robinson & McQuarrie, 2012) to 10 Ma in central Nepal (Kohn et al., 2004) to 20–15 Ma in eastern Bhutan (Long et al., 2012). Robinson and Pearson (2013) outline that it is possible that transfer of slip from the Main Central Thrust to the Ramgarh-Munsiari Thrust may have served as the trigger for motion on the South Tibetan detachment system. The age of initiation of motion on the South Tibetan detachment system is 26–25 Ma in most places (Carosi et al., 2018) with most ages clustering between 22–18 Ma, although there are younger ages (reviews in Robinson & Pearson, 2013; Kellett et al., 2019). Hence initial timing is similar to first motion on the Main Central Thrust and other timing constraints make the South Tibetan detachment system synchronous with the Ramgarh-Munsiari Thrust. Thus, depending on the location in the Himalayan thrust belt, the South Tibetan detachment system was synchronous with the Great Counter Thrust, Greater Himalayan thrust system, and Main Central Thrust, and/or the Ramgarh-Munsiari Thrust. It remains to be determined which faults were synchronous and which were diachronous.

5.8.1. Sedimentary Record

Synorogenic sedimentary strata document the unroofing of the Himalayan thrust belt. The strata are the Dumri Formation of Nepal (DeCelles et al., 2004) and the Dharamsala (Dagshai/Kasauli) Formation in northwestern India (Najman, 2006). Deposition began ca. 21 Ma (Figs. 5.2, 5.4). In Pakistan, Ding et al. (2016b) show that the Murree (Balakot) Formation is as old as 34 Ma. Oligocene age strata from the Kohat Plateau in Pakistan are reviewed in Pivnik and Wells (1996). Strata that are Oligocene–Middle Miocene in age do not exist in Bhutan.

The input of sediment derived from metamorphic rocks suggests that Greater and Tethyan Himalayan rocks were at the surface contributing sediment into the fluvial systems draining the Himalaya. The lack of high-grade metamorphic grains indicates that perhaps the lower Tethyan Himalayan rocks provided most of the sediment in Nepal (e.g., DeCelles et al., 2004). However, Najman (2006) indicates that there are temporal and spatial variations along the orogen regarding exposure of Greater Himalayan rocks. From 20–15 Ma, the data show that Tethyan and possibly Greater Himalayan rocks were at the surface providing detritus into the fluvial system draining the Himalaya (e.g., DeCelles et al., 1998a, b, 2004; Robinson et al., 2001; Najman et al., 2005). Burbank et al. (1996), Najman (2006), and Garzanti (2019) contain reviews of synorogenic rocks along the Himalayan orogen.

5.9. MIDDLE MIOCENE TO PRESENT TIME (15–0 MA)

From Namcha Barwa in the east to Nanga Parbat in the west, the ~20–16 Ma Ramgarh-Munsiari Thrust is the roof thrust for the Lesser Himalayan duplex, which consists of multiple thrusts in an antiformal-hinterland dipping duplex composed entirely of Lesser Himalayan strata (Srivastava & Mitra, 1994; DeCelles et al., 1998a, 2001, 2016; Robinson et al., 2003, 2006; McQuarrie et al., 2008; Mitra et al., 2010; Long et al., 2011b; Khanal & Robinson, 2013; Robinson & Pearson, 2013; Webb, 2013; Robinson & Martin, 2014; Mandal et al., 2019; McQuarrie et al., 2019). All along the Himalayan arc, the duplex began building at 15–13 Ma (Fig. 5.4) and had a profound effect, driving thickening, exhumation, and the attainment of high elevations (Long & Robinson, 2021). During building of the Lesser Himalayan duplex, multiple thin, smaller-displacement thrust sheets were clipped off the footwall of the Main Himalayan Thrust and incorporated into the hanging wall (Fig. 5.3h). The Lesser Himalayan duplex was exposed at the surface in erosional windows through Greater Himalayan rocks at 11–9 Ma (e.g., Robinson et al., 2001) indicating that the duplex was stacking and attaining vertical height (i.e., building taper in the critical wedge). From eastern Nepal to Bhutan, Greater Himalayan rocks cover most of the surface, with small windows into the footwall exposing some Lesser Himalayan rocks. Either the duplex was not vertically thick or there was less erosion due to another factor such as less precipitation. McQuarrie et al. (2014) found that the Lesser Himalayan duplex in Bhutan was built from 12 to 9 Ma. South of Bhutan, the Shillong Plateau in the foreland of the Himalayan thrust belt was emplaced from 14–8 Ma (Clark & Bilham, 2008). Synchronous with the building of the duplex, north-south trending grabens and low-angle normal faults formed in the highest part of the mountain belt on the Tibetan Plateau, and cut through the South Tibetan detachment system, Main Central Thrust, and possibly the Ramgarh-Munsiari Thrust sheet.

Exhumation of the north Himalayan domes occurred mostly at 12 ± 4 Ma (e.g., Lee et al., 2000, 2006; King et al., 2011; Aikman et al., 2012) with a few a little older at 18–15 Ma (Chen et al., 2018a). Coevally, grabens formed along the crest of the Himalaya that have east-west directed extension and north-south striking normal faults. In eastern Nepal, Ama Drime formed prior to 11 Ma and was accompanied by lower crustal material being brought to the surface from 40–50 km depth (Jessup et al., 2008; Cottle et al., 2009; Kali et al., 2010). These authors show that the switch from east-striking (South Tibetan detachment system) to north-striking (Ama Drime) normal faulting occurred between 13 and 12 Ma. North of far western Nepal in Tibet, motion on the Gurla Mandata metamorphic core complex occurred after 15 Ma (Murphy, 2007). The Thakkhola graben has north-south striking high-angle faults with less exhumation and extension. Garzione et al. (2000, 2003) found an initiation age of 11 Ma. In northwestern India, the Leo Pargil dome may also be a metamorphic core complex with a protracted history from 16–6 Ma (Thiede et al., 2006). This east-west directed extension is along the high peaks of the Himalaya and also across the Tibetan Plateau. Ideas for the causes of extension are reviewed in Kapp and DeCelles (2019).

Along the southern edge of the Tibetan Plateau, the data in the previous paragraphs document the southward building of the thrust belt during the 15–10 Ma time frame. This Himalaya-wide southward building occurred when Tibet grew outward on its other margins. It is unclear what caused the outward growth of Tibet. One idea is that it is a natural outward growth of the highest part of the Tibetan Plateau. Another idea is that the lower crust of the Tibetan Plateau is moving around to cause the outward growth (Clark & Royden, 2000; Royden et al., 2008). One more idea is that the lithospheric root of the Tibetan Plateau delaminated (e.g., Molnar et al., 1993), either all at once or in a series of smaller delamination events.

The first stage of deformation in the Lesser Himalayan rock created antiformal duplex culminations that increase in size westward along the arc (Long & Robinson, 2021). These culminations folded the overlying Greater and Tethyan Himalayan rock, which are now preserved in synclinal klippen toward the front of the thrust belt. The second stage of deformation in the Lesser Himalayan rock is the emplacement of the Lesser Himalayan imbricate zone (e.g., Robinson et al., 2006) near the front of the thrust belt (Fig. 5.3h). This is a complexly faulted zone that extends 10–20 km north of the Main Boundary Thrust. Motion on the Main Boundary Thrust occurred at ~5 Ma (DeCelles et al., 1998a). In the footwall of the Main Boundary Thrust in Nepal, the Siwalik Group is deformed into imbricate thrust sheets (Sub-Himalayan thrust system) that end with the recent emplacement of the Main Frontal Thrust (2–0 Ma). In northwestern India, both the Siwalik Group and the older Neogene synorogenic sedimentary rock are involved in the Sub-Himalayan thrust system.

In Pakistan, the 15 Ma to present history is different because the deformation moved south of the Main Boundary Thrust earlier than in the rest of the Himalaya. Two regions are present south of the Main Boundary Thrust, the Kohat Plateau to the west and Potwar Plateau to the east. In the Kohat Plateau, the Main Frontal Thrust (Surghar Range Thrust) was active at ~15 Ma and then out-of-sequence thrusting occurred on the Main Boundary Thrust between ~12–9 Ma (Ghani et al., 2021a). In the Potwar Plateau to the east, the Main Frontal Thrust (Salt Range Thrust) was active since at least 4 Ma (Ghani et al., 2021b). The two regions deform differently because the Potwar Plateau has an Ediacaran salt layer, with less internal deformation and a far traveled decollement, whereas the Kohat Plateau lacks thick evaporite deposits and has more internal deformation with a double decollement system (Ghani et al., 2018).

5.9.1. Sedimentary Record

The sedimentary record for 15–0 Ma is stored along the arc in the Siwalik Group. In general, the Siwalik Group coarsened upward as the frontal thrust of the mountain belt approached. One reason geologists study the Siwalik Group is that it yields the unroofing pattern, which is at what time rock units were exposed in the thrust belt (see Najman et al., 2009). For example, at the base of the Siwalik Group, Tethyan and Greater Himalayan detritus was shed into the foreland basin. Upsection in the Siwalik Group deposits, the growing Lesser Himalayan duplex forced Lesser Himalayan rocks to breach the surface and shed sediment into the basin. This exposure is found using whole-rock ε_{Nd} values because the Paleoproterozoic Lesser Himalayan rocks have more negative ε_{Nd} values than the other Himalayan units (e.g., Robinson et al., 2001). Robinson and McQuarrie (2012) contains an example of combining foreland basin data with deformation in the thrust belt in far western Nepal and details how other indicators from the foreland basin can be used. From paleomagnetic data, we know the depositional age of the Siwalik strata. Examining when Paleoproterozoic sediment was added to the Siwalik fluvial system has determined that parts of the Ramgarh-Munsiari Thrust sheet and Lesser Himalayan duplex were exposed and shedding sediment into the foreland basin at ~11–9 Ma in Nepal. Sediment cores from the Bengal fan indicate that erosion of the Himalaya peaked around 15 Ma and remained high until 10.5 Ma and subsequently slowed gradually until 3.5 Ma (Clift et al., 2008). Sedimentation increased once again in the late Pliocene and Pleistocene epochs (Clift et al., 2008). Reviews of the Siwalik Group are in Burbank et al. (1996), Najman (2006), and Garzanti (2019).

5.10. DISCUSSION

5.10.1. Naming Confusion

Figure 5.1 shows a geologic map of the Himalayan orogen divided three ways: (1) by age (Fig. 5.1a); (2) by type of contact between rock units, making a two-unit breakdown (Fig. 5.1b); and (3) by stratigraphic relationships (plus metamorphic grade for the Greater versus Tethyan Himalayan distinction), which produces the historic four-unit breakdown (Fig. 5.1c). Defined this way, the two- and four-unit breakdowns are based on essentially the same criteria, the four-unit division simply shows more detail. Figure 5.2 is an integrated tectonostratigraphic diagram that shows how these three ways of dividing stratigraphy compare with each other. We provide the information this way to not exclude viewpoints and to inform those who are learning about the Himalaya that there are controversies surrounding the names and definitions of these units.

The labeling of tectonostratigraphic units has been debated for decades, which has led to confusion. Considering how Himalayan geologic studies evolved gives some perspective. Over the past 200 yr of investigation of Himalayan geology, most studies have covered small spatial scales (tens of km) relative to the size of the entire orogen, with uncommon attempts to integrate information over larger areas. One advantage of working at small scales is that it facilitates detailed understanding; however, a drawback is that uncertain connections with other Himalayan regions led geologists to create a myriad of local stratigraphic and fault names. Gansser (1964) makes one of the first attempts to integrate information throughout much of the orogen. Subsequently, the opening of geographic regions that previously were difficult to access as well as the development of new and improved techniques in disciplines such as geochronology, metamorphic petrology, and structural geology permitted the publication of abundant new geologic data from the orogen. The availability of innovative ideas such as plate tectonics and critical wedge theory prompted researchers to interpret Himalayan geologic data in the new contexts provided by the novel ideas. Such applications typically involved integrating local- and regional-scale results into interpretations of the tectonic development of large sectors of the orogen, or its entirety. In the process, the four-unit breakdown of Gansser (1964) was modified and expanded to cover new information. As pointed out in previous sections, there are places where the four-unit breakdown does not work well. Except for the Sub-Himalayan deposits, all the tectonostratigraphic units partially overlap in depositional or intrusion age, which means that decisions must be made about whether a unit is Tethyan, Greater, or Lesser Himalayan based on other criteria in addition to age. Inevitably, that means that some researchers make different decisions than others, hence the confusion. One person's Lesser Himalayan rock is not another person's Lesser Himalayan rock. Webb (2013) provides a perspective in which rocks are coded by age instead. Martin (2017b) simplifies the four-unit breakdown to just two units to remove the confusion.

The problem with simplifying is that details are lost in the process and again, decisions are made. For this paper, the results are in Figures 5.1 and 5.2, which integrate the different types of information.

5.10.2. Development of the Himalayan Thrust Belt

Hand in hand with the naming confusion is the resulting confusion in interpretations. In recent years, many papers have been written that add support for a particular model for development of the Himalayan thrust belt. Many papers exist that explain the different models so we will not review these models in depth. Simply, there is the critical taper model (e.g., Robinson et al., 2006; Kohn, 2008), channel flow model (e.g., Beaumont et al., 2001, 2004; Jamieson et al., 2004, 2006; Mukherjee & Koyi, 2010b; Mukherjee et al., 2012; Mukherjee, 2013), and tectonic wedging model (e.g., Webb et al., 2011a,b). Torvela (2017) contains a recent review of channel flow models. What we know after 20 yr of comparing the different models is that they are endmembers, but reality incorporates aspects of all the models (e.g., Cottle et al., 2015). Using a large-scale frame of reference, the models are not that different from each other.

Figure 5.4 is one way to view the temporal development of the Himalayan thrust belt. In this diagram (inspired by a seminal researcher in continental tectonics, Peter Coney), the thrust belt is organized from west to east with time from 60 to 0 Ma on the vertical axis. Deformational events, metamorphism, intrusion, and deposition are indicated for each region of the Himalayan thrust belt. Timing of events is diachronous along strike, which leads to confusion when a researcher is focusing on the geological evolution in one area and expanding the significance to the entire orogen.

5.10.3. Integration of New Techniques

The long-term average shortening rate in the Himalaya is ~20 mm/yr, which agrees with the active shortening rate across the Himalaya of 15–20 mm/yr derived from satellite positioning (e.g., Bettinelli et al., 2006) as well as the Holocene slip rate of 21.5 ± 2 mm/yr on the Main Frontal Thrust (Lave & Avouac, 2000). These rates average shortening over time, and imply a constant shortening rate. However, Robinson and McQuarrie (2012) show that temporally variable shortening rates match the foreland basin and hinterland data in far western Nepal. Shortening rates were as high as 33 mm/yr from 12.99–11 Ma, and 27 mm/yr from 10.99–10 Ma, and as low as 14 mm/yr from 15.99–13 Ma and 13 mm/yr from 9.99 to 5 Ma. Bhutan is another place where variable shortening rates have been interpreted. Long et al. (2012) find rapid rates (25–60 mm/yr) from 15 to 9 Ma. McQuarrie and Ehlers (2015) use different time frames and find rates of 10–11 mm/yr from ~17 to 11 Ma and fast rates of shortening (55–75 mm/yr) from ~11 to 8.5 Ma. In the past 5 years, more Himalayan studies have emerged that tie low-temperature thermochronology to movement over faults at depth, particularly the Main Himalayan Thrust (McQuarrie & Ehlers, 2017; Gilmore et al., 2018; Olsen et al., 2019; Ghoshal et al., 2020; Robinson et al., 2021; Ghoshal et al., 2023). These efforts are herculean in terms of how much time goes into producing one viable model. Samples and data are collected and processed, two or three low-temperature thermochronologic systems are analyzed, cross sections are balanced and forward modeled, and, finally, all data are put into a thermokinematic model to predict the low-temperature cooling ages. If the modeled ages do not match the measured cooling ages, the balanced cross section must be redone until a close fit is reached. The results have shown diachronous development of the thrust system along the arc and provided detailed information about the rates and timing of shortening. Increased rates of shortening and subsequent erosion occurred at different times and locations along the Himalayan thrust belt, yielding a pulsed shortening pattern.

As in other parts of the world, we struggle with keeping up with the amount of data being produced and with integrating the information into an understanding of the entire Himalayan system. Thermokinematic modeling (e.g., Ghoshal et al., 2020; Grujic et al., 2020) and finite element modeling (e.g., Kelly et al., 2020) are techniques that can help with this task. Caution should be used in comparing different models because the same parameters are not used in each study. In the past 5 years, much progress has been made in understanding the shear zones within Greater Himalayan rocks using petrochronological investigations (e.g., Carosi et al., 2018; Shrestha et al., 2020; Benetti et al., 2021), and the importance of the Greater Himalayan Thrust system has come into focus. Other studies have focused on along-strike variations in the major thrust systems and possible causes (e.g., Webb et al., 2017; Mottram et al., 2019; Fan & Murphy, 2021; Long & Robinson, 2021). The ability to determine accurate deformation temperatures has increased using Raman spectroscopy of carbonaceous material, Ti-in-quartz thermobarometry, and quartz c-axis textures, yielding more detailed information about the thermal evolution of the thrust belt, particularly the Lesser Himalayan rocks (e.g., Beyssac et al., 2004; Bollinger et al., 2004; Celerier et al., 2009; Long et al., 2011c, 2016; Bose & Mukherjee, 2020; Grujic et al., 2020; Starnes et al., 2020; Larson et al., 2021). The publication of more analyses (e.g., Colleps et al., 2018; DeCelles et al., 2020; Ghoshal et al., 2020) has influenced how we think about the Himalayan thrust belt. Conversely, more data are more difficult to manage, integrate, and understand. The challenge for researchers who want to start working in the Himalaya is that it is a huge orogenic system with factions who interpret the data differently, and data production never stops.

This review contains only a sample of the available references for each subject. Treat this work as a general overview and use the citations within the references to dive deeply into a subject. Throughout most sections of this paper, we have stated problems that still need to be resolved in the Himalaya and where advances are needed. It is our hope that future researchers will use this review as a starting point to begin their work advancing our understanding of the evolution of the Himalayan thrust belt and the entire Himalayan-Tibetan orogenic system.

ACKNOWLEDGMENTS

We thank İbrahim Çemen and Elizabeth Catlos for inviting this contribution. We thank Eduardo Garzanti, Soumyajit Mukherjee, and three anonymous reviewers for comments that helped us improve the manuscript. We thank Alex Webb for sharing Figure 5.1a in vector file format. This work was funded by the following grants: NSF-EAR0809405, NSF-EAR1524320, and National Academy of Sciences Prime Cooperative Agreement No. AID-391-A-17-00001.

REFERENCES

Ahmad, I., Khan, S., Lapen, T., Burke, K., & Jehan, N. (2013). Isotopic ages for alkaline igneous rocks, including a 26Ma ignimbrite, from the Peshawar plain of northern Pakistan and their tectonic implications. *Journal of Asian Earth Sciences*, *62*, 414–424. https://doi.org/10.1016/j.jseaes.2012.10.025

Ahmad, T., Harris, N., Bickle, M., Chapman, H., Bunbury, J., & Prince, C. (2000). Isotopic constraints on the structural relationships between the Lesser Himalayan Series and the High Himalayan Crystalline Series, Garhwal Himalaya. *Geological Society of America Bulletin*, *112*, 467–477. https://doi.org/10.1130/0016-7606(2000)112<467:ICOTSR>2.0.CO;2

Aikman, A. B., Harrison, T. M., & Hermann, J. (2012). Age and thermal history of Eo- and Neohimalayan granitoids, eastern Himalaya. *Journal of Asian Earth Sciences*, *51*, 85–97. https://doi.org/10.1016/j.jseaes.2012.01.011

Aikman, A. B., Harrison, T. M., & Lin, D. (2008). Evidence for early (> 44 Ma) Himalayan crustal thickening, Tethyan Himalaya, southeastern Tibet. *Earth and Planetary Science Letters*, *274*, 14–23. https://doi.org/10.1016/j.epsl.2008.06.038

Aitchison, J. C., Ali, J. R., & Davis, A. M. (2007). When and where did India and Asia collide? *Journal of Geophysical Research*, *112*, B05423. https://doi.org/10.1029/2006JB004706

Aitchison, J. C., Badengzhu, A., Liu, J., Luo, H., Malpas, J. G., McDermid, I. R. C., et al. (2000). Remnants of a Cretaceous intra-oceanic subduction system within the Yarlung-Zangbo suture (southern Tibet). *Earth and Planetary Science Letters*, *183*, 231–244. https://doi.org/10.1016/S0012-821X(00)00287-9

Ali, J. R., & Aitchison, J. C. (2005). Greater India. *Earth-Science Reviews*, *72*, 169–188. https://doi.org/10.1016/j.earscirev.2005.07.005

Ali, J. R., & Aitchison, J. C. (2014). Greater India's northern margin prior to its collision with Asia. *Basin Research*, *26*, 73–84. https://doi.org/10.1111/bre.12040

An, W., Hu, X., Garzanti, E., Wang, J., & Liu, Q. (2021). New precise dating of the India-Asia collision in the Tibetan Himalaya at 61 Ma. *Geophysical Research Letters*, *48*, e2020GL090641. https://doi.org/10.1029/2020GL090641.

Anczkiewicz, R., Oberli, F., Burg, J. P., Villa, I. M., Gunther, D., & Meier, M. (2001). Timing of normal faulting along the Indus Suture in Pakistan Himalaya and a case of major 231 Pa/ 235 U initial disequilibrium in zircon. *Earth and Planetary Science Letters*, *191*, 101–114. https://doi.org/10.1016/S0012-821X(01)00406-X

Antolin, B., Godin, L., Wemmer, K., & Nagy, C. (2013). Kinematics of the Dadeldhura klippe shear zones (W Nepal): Implications for the foreland evolution of the Himalayan metamorphic core. *Terra Nova*, *25*, 282–291. https://doi.org/10.1111/ter.12034

Argles, T. W., Prince, C. I., Foster, G. L., & Vance, D. (1999). New garnets for old? Cautionary tales from young mountain belts. *Earth and Planetary Science Letters*, *172*, 301–309. https://doi.org/10.1016/S0012-821X(99)00209-5

Batra, R. S. (1989). A reinterpretation of the geology and biostratigraphy of the Lower Tertiary formations exposed along the Bilaspur-Shimla Highway, Himachal Pradesh, India. *Journal of the Geological Society of India*, *33*, 503–523.

Baxter, A. T., Aitchison, J. C., Ali, J. R., Chan, J. S-L., & Chan, G. H. N. (2016). Detrital chrome spinel evidence for a Neotethyan intra-oceanic island arc collision with India in the Paleocene. *Journal of Asian Earth Sciences*, *128*, 90–104. https://doi.org/10.1016/j.jseaes.2016.06.023

Beaumont, C., Jamieson, R. A., Nguyen, M. H., & Lee, B. (2001). Himalayan tectonics explained by extrusion of a low-viscosity crustal channel coupled to focused surface denudation. *Nature*, *414*, 738–742. https://doi.org/10.1038/414738a

Beaumont, C., Jamieson, R. A., Nguyen, M. H., & Medvedev, S. (2004). Crustal channel flows: 1. Numerical models with applications to the tectonics of the Himalayan-Tibetan orogen. *Journal of Geophysical Research, Solid Earth*, *109*, B06406. https://doi.org/10.1029/2003JB002809

Benetti, B., Montomoli, C., Iaccarino, S., Langone, A., & Carosi, R. (2021). Mapping tectono-metamorphic discontinuities in orogenic belts: Implications for mid-crust exhumation in NW Himalaya. *Lithos*, *392–393*, 106129. https://doi.org/10.1016/j.lithos.2021.106129

Bettinelli, P., Avouac, J.-P., Flouzat, M., Jouanne, F., Bollinger, L., Willis, P., et al. (2006). Plate motion of India and interseismic strain in the Nepal Himalaya from GPS and DORIS measurements. *Journal of Geodesy*, *80*, 567–589. https://doi.org/10.1007/s00190-006-0030-3

Beyssac, O., Bollinger, L., Avouac, J-P., & Goffe, B. (2004). Thermal metamorphism in the lesser Himalaya of Nepal determined from Raman spectroscopy of carbonaceous material. *Earth and Planetary Science Letters*, *225*, 233–241. https://doi.org/10.1016/j.epsl.2004.05.023

Bhattacharya, G., Robinson, D. M., & Wielicki, M. M. (2021). Detrital zircon provenance of the Indus Group, Ladakh, NW India: Implications for the timing of the India-Asia collision and other syn-orogenic processes. *GSA Bulletin*, *133*, 1007–1020. https://doi.org/10.1130/B35624.1

Bhattacharya, G., Robinson, D. M., Orme, D. A., Najman, Y., & Carter, A. (2020). Low-temperature thermochronology of the Indus Basin in central Ladakh, northwest India: Implications of Miocene-Pliocene cooling in the India-Asia collision zone. *Tectonics*, *39*, e2020TC006333. https://doi.org/10.1029/2020TC006333

Bollinger, L., Avouac, J. P., Beyssac, O., Catlos, E. J., Harrison, T. M., Grove, M., et al. (2004). Thermal structure and exhumation history of the Lesser Himalaya in central Nepal. *Tectonics*, *23*, TC5015. https://doi.org/10.1029/2003TC001564

Bose, N., & Mukherjee, S. (2020). Estimation of deformation temperatures, flow stresses & strain rates from an intra-continental shear zone: The Main Boundary Thrust, NW Himalaya (Uttarakhand, India). *Marine and Petroleum Geology*, *112*, 104094. https://doi.org/10.1016/j.marpetgeo.2019.104094

Bossart, P., & Ottiger, R. (1989). Rocks of the Murree Formation in northern Pakistan: Indicators of a descending foreland basin of late Paleocene to middle Eocene age. *Eclogae Geologicae Helvetiae*, *82*, 133–165. https://doi.org/10.5169/SEALS-166369

Bouilhol, P., Jagoutz, O., Hanchar, J. M., & Dudas, F. O. (2013). Dating the India-Eurasia collision through arc magmatic records. *Earth and Planetary Science Letters*, *366*, 163–175. https://doi.org/10.1016/j.epsl.2013.01.023

Brookfield, M. E. (1993). The Himalayan passive margin from Precambrian to Cretaceous times. *Sedimentary Geology*, *84*, 1–35. https://doi.org/10.1016/0037-0738(93)90042-4

Brown, M., & Johnson, T. (2018). Secular change in metamorphism and the onset of global plate tectonics. *American Mineralogist*, *103*, 181–196. https://doi.org/10.2138/am-2018-6166

Buckman, S., Aitchison, J. C., Nutman, A. P., Bennett, V. C., Saktura, W. M., Walsh, J. M. J., et al. (2018). The Spongtang Massif in Ladakh, NW Himalaya: An Early Cretaceous record of spontaneous, intra-oceanic subduction initiation in the Neotethys. *Gondwana Research*, *63*, 226–249. https://doi.org/10.1016/j.gr.2018.07.003

Burbank, D. W., Beck, R. A., & Mulder, T. (1996). The Himalayan foreland basin. In A. Yin & T. M. Harrison (Eds.), *The tectonic evolution of Asia* 149–188. Cambridge: Cambridge University Press.

Burchfiel, B. C., Chen, Z., Hodges, K. V., Liu, Y., Royden, L. H., Deng, C., et al. (1992). The South Tibetan detachment system, Himalayan orogen: Extension contemporaneous with and parallel to shortening in a collisional mountain belt. *Boulder, Geological Society of America, Special Paper 269*. https://doi.org/10.1130/SPE269

Burg, J. P., & Chen, G. M. (1984). Tectonics and structural zonation of southern Tibet, China. *Nature*, *311*, 219–223. https://doi.org/10.1038/311219a0

Burg, J. P., Chaudhry, M. N., Ghazanfar, M., Anczkiewicz, R., & Spencer, D. (1996). Structural evidence for back sliding of the Kohistan arc in the collisional system of northwest Pakistan. *Geology*, *24*, 739–742. https://doi.org/10.1130/0091-7613(1996)024<0739:SEFBSO>2.3.CO;2

Burg, J. P., Guiraud, M., Chen, G. M., & Li, G. C. (1984). Himalayan metamorphism and deformations in the North Himalayan Belt (southern Tibet, China). *Earth and Planetary Science Letters*, *69*, 391–400. https://doi.org/10.1016/0012-821X(84)90197-3

Burg, J. P., Leyreloup, A., Girardeau, J., & Chen, G. M. (1987). Structure and metamorphism of a tectonically thickened continental crust: The Yalu Tsangpo suture zone (Tibet). *Philosophical Transactions of the Royal Society of London. Series A, Mathematical and Physical Sciences*, *321*, 67–86. https://doi.org/10.1098/rsta.1987.0005

Burrard, S. G., & Hayden, H. H. (1908). *A sketch of the geography and geology of the Himalaya Mountains and Tibet*. Calcutta: Superintendent Government Printing, India.

Carosi, R., Lombardo, B., Molli, G., Musumeci, G., & Pertusati, P. C. (1998). The South Tibetan detachment system in the Rongbuk valley, Everest region. Deformation features and geological implications. *Journal of Asian Earth Sciences*, *16*, 299–311. https://doi.org/10.1016/S0743-9547(98)00014-2

Carosi, R., Montomoli, C., & Iaccarino, S. (2018). 20 years of geological mapping of the metamorphic core across central and eastern Himalayas. *Earth-Science Reviews*, *177*, 124–138. https://doi.org/10.1016/j.earscirev.2017.11.006

Carosi, R., Montomoli, C., & Visona, D. (2007). A structural transect in the Lower Dolpo: Insights on the tectonic evolution of Western Nepal. *Journal of Asian Earth Sciences*, *29*, 407–423. https://doi.org/10.1016/j.jseaes.2006.05.001

Carosi, R., Montomoli, C., Iaccarino, S., Massonne, H.-J., Rubatto, D., Langone, A., et al. (2016). Middle to late Eocene exhumation of the Greater Himalayan Sequence in the central Himalayas: Progressive accretion from the Indian plate. *Geological Society of America Bulletin*, *128*, 1571–1592. https://doi.org/10.1130/B31471.1

Carosi, R., Montomoli, C., Langone, A., Turina, A., Cesare, B., Iaccarino, S., et al. (2015). Eocene partial melting recorded in peritectic garnets from kyanite-gneiss, Greater Himalayan Sequence, central Nepal. In S. Mukherjee et al. (Eds.), *Tectonics of the Himalaya* (pp. 111–129). London, Geological Society, Special Publication, 412. https://doi.org/10.1144/SP412.1

Carosi, R., Montomoli, C., Rubatto, D., & Visona, D. (2010). Late Oligocene high-temperature shear zones in the core of the Higher Himalayan Crystallines (Lower Dolpo, western Nepal). *Tectonics*, *29*, TC4029. https://doi.org/10.1029/2008TC002400

Carosi, R., Montomoli, C., Rubatto, D., & Visona, D. (2013). Leucogranite intruding the South Tibetan Detachment in western Nepal: Implications for exhumation models in the Himalayas. *Terra Nova*, *25*, 478–489. https://doi.org/10.1111/ter.12062

Carrapa, B., Orme, D. A., DeCelles, P. G., Kapp, P., Cosca, M. A., & Waldrip, R. (2014). Miocene burial and exhumation of the India-Asia collision zone in southern Tibet: Response to slab dynamics and erosion. *Geology*, *42*, 443–446. https://doi.org/10.1130/G35350.1

Catlos, E. J., Harrison, T. M., Manning, C. E., Grove, M., Rai, S. M., Hubbard, M. S., et al. (2002). Records of the evolution of the Himalayan orogen from in situ Th-Pb ion microprobe dating of monazite: Eastern Nepal and western Garhwal. *Journal of Asian Earth Sciences*, *20*, 459–479. https://doi.org/10.1016/S1367-9120(01)00039-6

Catlos, E. J., Pease, E. C., Dygert, N., Brookfield, M., Schwarz, W. H., Bhutani, R., et al. (2019). Nature, age and emplacement of the Spongtang ophiolite, Ladakh, NW India. *Journal of the Geological Society*, *176*, 284–305. https://doi.org/10.1144/jgs2018-085

Catlos, E. J., Perez, T. J., Lovera, O. M., Dubey, C. S., Schmitt, A. K., & Etzel, T. M. (2020). High-resolution P-T-time paths across Himalayan faults exposed along the Bhagirathi Transect NW India: Implications for the construction of the Himalayan Orogen and ongoing deformation. *Geochemistry, Geophysics, Geosystems*, *21*, e2020GC009353. https://doi.org/10.1029/2020GC009353

Cawood, P. A., et al. (2020). Deconstructing South China and consequences for reconstructing Nuna and Rodinia. *Earth-Science Reviews*, *204*, 103169. https://doi.org/10.1016/j.earscirev.2020.103169

Cawood, P. A., Johnson, M. R. W., & Nemchin, A. A. (2007). Early Palaeozoic orogenesis along the Indian margin of Gondwana: Tectonic response to Gondwana assembly. *Earth and Planetary Science Letters*, *255*, 70–84. https://doi.org/10.1016/j.epsl.2006.12.006

Cawood, P. A., Zhao, G., Yao, J., Wang, W., Xu, Y., & Wang, Y. (2018). Reconstructing South China in Phanerozoic and Precambrian supercontinents. *Earth-Science Reviews*, *186*, 173–194. https://doi.org/10.1016/j.earscirev.2017.06.001

Celerier, J., Harrison, T. M., Beyssac, O., Herman, F., Dunlap, W. J., & Webb, A. A. G. (2009). The Kumaun and Garwhal Lesser Himalaya, India: Part 2. Thermal and deformation histories. *Geological Society of America Bulletin*, *121*, 1281–1297. https://doi.org/10.1130/B26343.1

Chakraborty, S., Anczkiewicz, R., Gaidies, F., Rubatto, D., Sorcar, N., Faak, K., et al. (2016). A review of thermal history and timescales of tectonometamorphic processes in Sikkim Himalaya (NE India) and implications for rates of metamorphic processes. *Journal of Metamorphic Geology*, *34*, 785–803. https://doi.org/10.1111/jmg.12200

Chakraborty, S., Mukul, M., Mathew, G., & Pande, K. (2019). Major shear zone within the Greater Himalayan Sequence and sequential evolution of the metamorphic core in Sikkim, India. *Tectonophysics*, *770*, 228183. https://doi.org/10.1016/j.tecto.2019.228183

Chen, J., Carosi, R., Cao, H., Montomoli, C., Iaccarino, S., Langone, A., & Li, G. (2018a). Structural setting of the Yalaxiangbo dome, SE Tibet (China). *Italian Journal of Geosciences*, *137*, 330–347. https://doi.org/10.3301/IJG.2018.18

Chen, S.-S., Fan, W.-M., Shi, R.-D., Liu, X.-H., & Zhou, X.-J. (2018b). 118–115 Ma magmatism in the Tethyan Himalaya igneous province: Constraints on Early Cretaceous rifting of the northern margin of Greater India. *Earth and Planetary Science Letters*, *491*, 21–33. https://doi.org/10.1016/j.epsl.2018.03.034

Clark, M. K., & Bilham, R. (2008). Miocene rise of the Shillong Plateau and the beginning of the end for the Eastern Himalaya. *Earth and Planetary Science Letters*, *269*, 336–350. https://doi.org/10.1016/j.epsl.2008.01.045

Clark, M. K., & Royden, L. H. (2000). Topographic ooze: Building the eastern margin of Tibet by lower crustal flow. *Geology*, *28*, 703–706. https://doi.org/10.1130/0091-7613(2000)28<703:TOBTEM>2.0.CO;2

Clarke, G. L., Bhowmik, S. K., Ireland, T. R., Aitchison, J. C., Chapman, S. L., & Kent, L. (2016). Inverted Oligo-Miocene metamorphism in the Lesser Himalaya Sequence, Arunachal Pradesh, India: Age and grade relationships. *Journal of Metamorphic Geology*, *34*, 805–820. https://doi.org/10.1111/jmg.12202

Clift, P. D., Carter, A., Krol, M., & Kirby, E. (2002). Constraints on India-Eurasia collision in the Arabian Sea region taken from the Indus Group, Ladakh Himalaya, India. In P. Clift, et al. (Eds.), *The tectonic and climatic evolution of the Arabian Sea* (pp. 97–116). London, Geological Society, 195. https://doi.org/10.1144/GSL.SP.2002.195.01.07

Clift, P. D., Hodges, K. V., Heslop, D., Hannigan, R., Van Long, H., & Calves, G. (2008). Correlation of Himalayan exhumation rates and Asian monsoon intensity. *Nature Geoscience*, *1*, 875–880. https://doi.org/10.1038/ngeo351

Colchen, M., Bassoullet, J. P., & Mascle, G. (1982). La paleogeographie des orogenes, l'example de l'Himalaya. *Memoires Geologiques de l'Universite de Dijon*, *7*, 453–471.

Colchen, M., Le Fort, P., & Pecher, A. (1986). *Recherches geologiques dans l'Himalaya du Nepal: Annapurna-Manaslu-Ganesh Himal*. Paris, Centre National de la Recherche Scientifique.

Colleps, C. L., McKenzie, N. R., Horton, B. K., Webb, A. A. G., Ng, Y. W., & Singh, B. P. (2020). Sediment provenance of pre- and post-collisional Cretaceous-Paleogene strata from the frontal Himalaya of northwest India. *Earth and Planetary Science Letters*, *534*, 116079. https://doi.org/10.1016/j.epsl.2020.116079

Colleps, C. L., McKenzie, N. R., Stockli, D. F., Hughes, N. C., Singh, B. P., Webb, A. A. G., et al. (2018). Zircon (U-Th)/He thermochronometric constraints on Himalayan Thrust Belt exhumation, bedrock weathering, and Cenozoic seawater chemistry. *Geochemistry, Geophysics, Geosystems*, *19*, 257–271. https://doi.org/10.1002/2017GC007191

Colleps, C. L., Stockli, D. F., McKenzie, N. R., Webb, A. A. G., & Horton, B. K. (2019). Neogene kinematic evolution and exhumation of the NW India Himalaya: Zircon geo and thermochronometric insights from the fold-thrust belt and foreland basin. *Tectonics*, *38*, 2059–2086. https://doi.org/10.1029/2018TC005304

Corrie, S. L., & Kohn, M. J. (2011). Metamorphic history of the central Himalaya, Annapurna region, Nepal, and implications for tectonic models. *Geological Society of America Bulletin*, *123*, 1863–1879. https://doi.org/10.1130/B303761

Cottle, J. M., Jessup, M. J., Newell, D. L., Horstwood, M. S. A., Noble, S. R., Parrish, R. R., Waters, D. J., et al. (2009). Geochronology of granulitized eclogite from the Ama Drime

Massif: Implications for the tectonic evolution of the South Tibetan Himalaya. *Tectonics*, *28*, TC1002. https://doi.org/10.1029/2008TC002256

Cottle, J. M., Larson, K. P., & Kellett, D. A. (2015). How does the mid-crust accommodate deformation in large, hot collisional orogens? A review of recent research in the Himalayan orogen. *Journal of Structural Geology*, *78*, 119–133. https://doi.org/10.1016/j.jsg.2015.06.008

Coutand, I., Barrier, L., Govin, G., Grujic, D., Hoorn, C., Dupont-Nivet, G., et al. (2016). Late Miocene-Pleistocene evolution of India-Eurasia convergence partitioning between the Bhutan Himalaya and the Shillong Plateau: New evidences from foreland basin deposits along the Dungsam Chu section, eastern Bhutan. *Tectonics*, *35*, 2963–2994. https://doi.org/10.1002/2016TC004258

Critelli, S., & Garzanti, E. (1994). Provenance of the Lower Tertiary Murree redbeds (Hazara-Kashmir Syntaxis, Pakistan) and initial rising of the Himalayas. *Sedimentary Geology*, *89*, 265–284. https://doi.org/10.1016/0037-0738(94)90097-3

Currie, B. S., Polissar, P. J., Rowley, D. B., Ingalls, M., Li, S., Olack, G., & Freeman, K. H. (2016). Multiproxy paleoaltimetry of the Late Oligocene-Pliocene Oiyug Basin, southern Tibet. *American Journal of Science*, *316*, 401–436. https://doi.org/10.2475/05.2016.01

Currie, B. S., Rowley, D. B., & Tabor, N. J. (2005). Middle Miocene paleoaltimetry of southern Tibet: Implications for the role of mantle thickening and delamination in the Himalayan orogen. *Geology*, *33*, 181–184. https://doi.org/10.1130/G21170.1

DeCelles, P. G., Carrapa, B., Gehrels, G. E., Chakraborty, T., & Ghosh, P. (2016). Along-strike continuity of structure, stratigraphy, and kinematic history in the Himalayan thrust belt: The view from Northeastern India. *Tectonics*, *35*, 2995–3027. https://doi.org/10.1002/2016TC004298

DeCelles, P. G., Carrapa, B., Ojha, T. P., Gehrels, G. E., & Collins, D. (2020). Structural and thermal evolution of the Himalayan Thrust Belt in midwestern Nepal. *Boulder, Geological Society of America, Special Paper*, *547*. https://doi.org/10.1130/2020.2547(01)

DeCelles, P. G., Gehrels, G. E., Najman, Y., Martin, A. J., Carter, A., & Garzanti, E. (2004). Detrital geochronology and geochemistry of Cretaceous–Early Miocene strata of Nepal: Implications for timing and diachroneity of initial Himalayan orogenesis. *Earth and Planetary Science Letters*, *227*, 313–330. https://doi.org/10.1016/j.epsl.2004.08.019

DeCelles, P. G., Gehrels, G. E., Quade, J., & Ojha, T. P. (1998a). Eocene-early Miocene foreland basin development and the history of Himalayan thrusting, western and central Nepal. *Tectonics*, *17*, 741–765. https://doi.org/10.1029/98TC02598

DeCelles, P. G., Gehrels, G. E., Quade, J., LaReau, B., & Spurlin, M. (2000). Tectonic implications of U-Pb zircon ages of the Himalayan Orogenic Belt in Nepal. *Science*, *288*, 497–499. https://doi.org/10.1126/science.288.5465.497

DeCelles, P. G., Gehrels, G. E., Quade, J., Ojha, T. P., Kapp, P. A., & Upreti, B. N. (1998b). Neogene foreland basin deposits, erosional unroofing, and the kinematic history of the Himalayan fold-thrust belt, western Nepal. *Geological Society of America Bulletin*, *110*, 2–21. https://doi.org/10.1130/0016-7606(1998)110<0002:NFBDEU>2.3.CO;2

DeCelles, P. G., Kapp, P., Gehrels, G. E., & Ding, L. (2014). Paleocene-Eocene foreland basin evolution in the Himalaya of southern Tibet and Nepal: Implications for the age of initial India-Asia collision. *Tectonics*, *33*, 824–849. https://doi.org/10.1002/2014TC003522

DeCelles, P. G., Kapp, P., Quade, J., & Gehrels, G. E. (2011). Oligocene-Miocene Kailas basin, southwestern Tibet: Record of postcollisional upper-plate extension in the Indus-Yarlung suture zone. *Geological Society of America Bulletin*, *123*, 1337–1362. https://doi.org/10.1130/B30258.1

DeCelles, P. G., Robinson, D. M., Quade, J., Ojha, T. P., Garzione, C. N., Copeland, P., & Upreti, B. N. (2001). Stratigraphy, structure, and tectonic evolution of the Himalayan fold-thrust belt in western Nepal. *Tectonics*, *20*, 487–509. https://doi.org/10.1029/2000TC001226

de Sigoyer, J., Chavagnac, V., Blichert-Toft, J., Villa, I. M., Luais, B., Guillot, S., et al. (2000). Dating the Indian continental subduction and collisional thickening in the northwest Himalaya: Multichronology of the Tso Morari eclogites. *Geology*, *28*, 487–490. https://doi.org/10.1130/0091-7613(2000)28<487:DTICSA>2.0.CO;2

Ding, H., Zhang, Z., Dong, X., Tian, Z., Xiang, H., Mu, H., Gou, Z., Shui, X., Li, W., & Mao, L. (2016a). Early Eocene (c. 50 Ma) collision of the Indian and Asian continents: Constraints from the North Himalayan metamorphic rocks, southeastern Tibet. *Earth and Planetary Science Letters*, *435*, 64–73. https://doi.org/10.1016/j.epsl.2015.12.006

Ding, L., Kapp, P., & Wan, X. (2005). Paleocene-Eocene record of ophiolite obduction and initial India-Asia collision, south central Tibet. *Tectonics*, *24*, TC3001. https://doi.org/10.1029/2004TC001729

Ding, L., Qasim, M., Jadoon, I. A. K., Khan, M. A., Xu, Q., Cai, F., Wang, H., Baral, U., & Yue, Y. (2016b). The India–Asia collision in north Pakistan: Insight from the U-Pb detrital zircon provenance of Cenozoic foreland basin. *Earth and Planetary Science Letters*, *455*, 49–61. https://doi.org/10.1016/j.epsl.2016.09.003

Ding, L., Xu, Q., Yue, Y., Wang, H., Cai, F., & Li, S. (2014). The Andean-type Gangdese Mountains: Paleoelevation record from the Paleocene–Eocene Linzhou Basin. *Earth and Planetary Science Letters*, *392*, 250–264. https://doi.org/10.1016/j.epsl.2014.01.045

DiPietro, J. A., & Isachsen, C. E. (2001). U-Pb zircon ages from the Indian plate in northwest Pakistan and their significance to Himalayan and pre-Himalayan geologic history. *Tectonics*, *20*, 510–525. https://doi.org/10.1029/2000TC001193

DiPietro, J. A., & Pogue, K. R. (2004). Tectonostratigraphic subdivisions of the Himalaya: A view from the west. *Tectonics*, *23*, TC5001. https://doi.org/10.1029/2003TC001554

DiPietro, J. A., Pogue, K. R., Hussain, A., & Ahmad, I. (1999). Geologic map of the Indus syntaxis and surrounding area, northwest Himalaya, Pakistan. In Himalaya and Tibet: Mountain roots to mountain tops. *Geological Society of America, Special Paper*, *328*. https://doi.org/10.1130/0-8137-2328-0.159

Donaldson, D. G., Webb, A. A. G., Menold, C. A., Kylander-Clark, A. R. C., & Hacker, B. R. (2013). Petrochronology of Himalayan ultrahigh-pressure eclogite. *Geology*, *41*, 835–838. https://doi.org/10.1130/G33699.1

Dubey, A. K., & Bhakuni, S. S. (2007). Younger hanging wall rocks along the Vaikrita thrust of the high Himalaya: A model based on inversion tectonics. *Journal of Asian Earth Sciences*, *29*, 424–429. https://doi.org/10.1016/j.jseaes.2005.10.005

Dupont-Nivet, G., Lippert, P. C., Van Hinsbergen, D. J. J., Meijers, M. J. M., & Kapp, P. (2010). Palaeolatitude and age of the Indo-Asia collision: Palaeomagnetic constraints. *Geophysical Journal International*, *182*, 1189–1198. https://doi.org/10.1111/j.1365-246X.2010.04697.x

Fan, S., & Murphy, M. A. (2021). Three-dimensional strain accumulation and partitioning in an arcuate orogenic wedge: An example from the Himalaya. *Geological Society of America Bulletin*, *133*, 3–18. https://doi.org/10.1130/B35528.1

Finch, M., Hasalova, P., Weinberg, R. F., & Fanning, C. M. (2014). Switch from thrusting to normal shearing in the Zanskar shear zone, NW Himalaya: Implications for channel flow. *Geological Society of America Bulletin*, *126*, 892–924. https://doi.org/10.1130/B30817.1

Fleming, A. (1853). On the Salt Range of the Punjaub. *Quarterly Journal of the Geological Society*, *9*, 189–200. https://doi.org/10.1144/GSL.JGS.1853.009.01-02.34

Foster, G., Gibson, H. D., Parrish, R., Horstwood, M., Fraser, J., & Tindle, A. (2002). Textural, chemical and isotopic insights into the nature and behaviour of metamorphic monazite. *Chemical Geology*, *191*, 183–207. https://doi.org/10.1016/S0009-2541(02)00156-0

Frank, W., Hoinkes, G., Miller, C., Purtscheller, F., Richter, W., & Thoni, M. (1973). Relations between metamorphism and orogeny in a typical section of the Indian Himalayas. *Tschermaks Mineralogische und Petrographische Mitteilungen*, *20*, 303–332. https://doi.org/10.1007/BF01081339

Fuchs, G. (1977). The geology of the Karnali and Dolpo Regions, western Nepal. *Jahrbuch der Geologischen Bundesanstalt Wien*, *120*, 165–217.

Fuchs, G., & Willems, H. (1990). The final stages of sedimentation in the Tethyan zone of Zanskar and their geodynamic significance (Ladakh-Himalaya). *Jahrbuch der Geologischen Bundesanstalt Wien*, *133*, 259–273.

Fuchs, G., Widder, R. W., & Tuladhar, R. (1988). Contributions to the geology of the Annapurna Range (Manang area, Nepal). *Jahrbuch der Geologischen Bundesanstalt Wien*, *131*, 593–607.

Gansser, A. (1964). *Geology of the Himalayas*. London, Wiley.

Garzanti, E. (1999). Stratigraphy and sedimentary history of the Nepal Tethys Himalaya passive margin. *Journal of Asian Earth Sciences*, *17*, 805–827. https://doi.org/10.1016/S1367-9120(99)00017-6

Garzanti, E. (2019). The Himalayan Foreland Basin from collision onset to the present: A sedimentary-petrology perspective. In P. J. Treloar & M. Searle (Eds.), *Himalayan tectonics: A modern synthesis* (65–122). London, Geological Society, Special Publication, 483. https://doi.org/10.1144/SP483.17

Garzanti, E., & Hu, X. (2015). Latest Cretaceous Himalayan tectonics: Obduction, collision or Deccan-related uplift? *Gondwana Research*, *28*, 165–178. https://doi.org/10.1016/j.gr.2014.03.010

Garzanti, E., & Van Haver, T. (1988). The Indus clastics: Forearc basin sedimentation in the Ladakh Himalaya (India). *Sedimentary Geology*, *59*, 237–249. https://doi.org/10.1016/0037-0738(88)90078-4

Garzanti, E., Baud, A., & Mascle, G. (1987). Sedimentary record of the northward flight of India and its collision with Eurasia (Ladakh Himalaya, India). *Geodinamica Acta*, *1*, 297–312. https://doi.org/10.1080/09853111.1987.11105147

Garzanti, E., Casnedi, R., & Jadoul, F. (1986). Sedimentary evidence of a Cambro-Ordovician orogenic event in the northwestern Himalaya. *Sedimentary Geology*, *48*, 237–265. https://doi.org/10.1016/0037-0738(86)90032-1

Garzanti, E., Le Fort, P., & Sciunnach, D. (1999). First report of Lower Permian basalts in South Tibet: Tholeiitic magmatism during break-up and incipient opening of Neotethys. *Journal of Asian Earth Sciences*, *17*, 533–546. https://doi.org/10.1016/S1367-9120(99)00008-5

Garzanti, E., Radeff, G., & Malusa, M. G. (2018). Slab breakoff: a critical appraisal of a geological theory as applied in space and time. *Earth-Science Reviews*, *177*, 303–319. https://doi.org/10.1016/j.earscirev.2017.11.012

Garzione, C. N., DeCelles, P. G., Hodkinson, D. G., Ojha, T. P., & Upreti, B. N. (2003). East-west extension and Miocene environmental change in the southern Tibetan plateau: Thakkhola graben, central Nepal. *Geological Society of America Bulletin*, *115*, 3–20. https://doi.org/10.1130/0016-7606(2003)115<0003:EWEAME>2.0.CO;2

Garzione, C. N., Dettman, D. L., Quade, J., DeCelles, P. G., & Butler, R. F. (2000). High times on the Tibetan Plateau: Paleoelevation of the Thakkhola graben, *Nepal. Geology*, *28*, 339–342. https://doi.org/10.1130/0091-7613(2000)28<339:HTOTTP>2.0.CO;2

Gee, E. R. (1989). Overview of the geology and structure of the Salt Range, with observations on related areas of northern Pakistan. In *Tectonics of the western Himalayas* (pp. 95–112). *Geological Society of America, Special Paper, 232*. https://doi.org/10.1130/SPE232-p95

Gehrels, G., et al. (2011). Detrital zircon geochronology of pre-Tertiary strata in the Tibetan-Himalayan orogen. *Tectonics*, *30*, TC5016. https://doi.org/10.1029/2011TC002868

Gehrels, G. E., DeCelles, P. G., Martin, A., Ojha, T. P., Pinhassi, G., & Upreti, B. N. (2003). Initiation of the Himalayan Orogen as an early Paleozoic thin-skinned thrust belt. *GSA Today*, *13*, 4–9. https://doi.org/10.1130/1052-5173(2003)13<4:IOTHOA>2.0.CO;2

Gehrels, G. E., DeCelles, P. G., Ojha, T. P., & Upreti, B. N. (2006a). Geologic and U-Pb geochronologic evidence for early Paleozoic tectonism in the Dadeldhura thrust sheet, far-west Nepal Himalaya. *Journal of Asian Earth Sciences*, *28*, 385–408. https://doi.org/10.1016/j.jseaes.2005.09.012

Gehrels, G. E., DeCelles, P. G., Ojha, T. P., & Upreti, B. N. (2006b). Geologic and U-Th-Pb geochronologic evidence for early Paleozoic tectonism in the Kathmandu thrust sheet, central Nepal Himalaya. *Geological Society of America Bulletin*, *118*, 185–198. https://doi.org/10.1130/B25753.1

Ghani, H., Sobel, E. R., Zeilinger, G., Glodny, J., Irum, I., & Sajid, M. (2021a). Spatio-temporal structural evolution of the Kohat fold and thrust belt of Pakistan. *Journal of Structural Geology*, *145*, 104310. https://doi.org/10.1016/j.jsg.2021.104310

Ghani, H., Sobel, E. R., Zeilinger, G., Glodny, J., Zapata, S., & Irum, I. (2021b). Palaeozoic and Pliocene tectonic evolution of the Salt Range constrained by low-temperature thermochronology. *Terra Nova*, *33*, 293–305. https://doi.org/10.1111/ter.12515

Ghani, H., Zeilinger, G., Sobel, E. R., & Heidarzadeh, G. (2018). Structural variation within the Himalayan fold and thrust belt: a case study from the Kohat-Potwar Fold Thrust Belt of Pakistan. *Journal of Structural Geology*, *116*, 34–46. https://doi.org/10.1016/j.jsg.2018.07.022

Ghoshal, S., McQuarrie, N., Robinson, D. M., Adhikari, D. P., Morgan, L. E., & Ehlers, T. A. (2020). Constraining central Himalayan (Nepal) fault geometry through integrated thermochronology and thermokinematic modeling. *Tectonics*, *39*, e2020TC006399. https://doi.org/10.1029/2020TC006399

Ghoshal, S., McQuarrie, N., Huntington, K. W., Robinson, D. M., & Ehlers, T. A. (2023). Testing erosional and kinematic drivers in the central Himalaya, *Earth and Planetary Science Letters, 609*, https://doi.org/10.1016/j.epsl.2023.118087

Gibbons, A. D., Barckhausen, U., van den Bogaard, P., Hoernle, K., Werner, R., Whittaker, J. M., et al. (2012). Constraining the Jurassic extent of Greater India: Tectonic evolution of the West Australian margin. *Geochemistry, Geophysics, Geosystems*, *13*, Q05W13. https://doi.org/10.1029/2011GC003919

Gibbons, A. D., Zahirovic, S., Muller, R. D., Whittaker, J. M., & Yatheesh, V. (2015). A tectonic model reconciling evidence for the collisions between India, Eurasia and intra-oceanic arcs of the central-eastern Tethys. *Gondwana Research*, *28*, 451–492. https://doi.org/10.1016/j.gr.2015.01.001

Gilmore, M. E., McQuarrie, N., Eizenhofer, P. R., & Ehlers, T. A. (2018). Testing the effects of topography, geometry, and kinematics on modeled thermochronometer cooling ages in the eastern Bhutan Himalaya. *Solid Earth*, *9*, 599–627. https://doi.org/10.5194/se-9-599-2018

Godin, L., Grujic, D., Law, R. D., & Searle, M. P. (2006). Channel flow, ductile extrusion and exhumation in continental collision zones: An introduction. In R. D. Law et al. (Eds.), *Channel flow, ductile extrusion and exhumation in continental collision zones* (pp. 1–23). London, Geological Society, Special Publication, 268. https://doi.org/10.1144/GSL.SP.2006.268.01.01

Godin, L., Parrish, R. R., Brown, R. L., & Hodges, K. V. (2001). Crustal thickening leading to exhumation of the Himalayan metamorphic core of central Nepal: insight from U-Pb geochronology and ^{40}Ar/^{39}Ar thermochronology. *Tectonics*, *20*, 729–747. https://doi.org/10.1029/2000TC001204

Goscombe, B., Gray, D., & Foster, D. A. (2018). Metamorphic response to collision in the central Himalayan Orogen. *Gondwana Research*, *57*, 191–265. https://doi.org/10.1016/j.gr.2018.02.002

Goscombe, B., Gray, D., & Hand, M. (2006). Crustal architecture of the Himalayan metamorphic front in eastern Nepal. *Gondwana Research*, *10*, 232–255. https://doi.org/10.1016/j.gr.2006.05.003

Green, O. R., Searle, M. P., Corfield, R. I., & Corfield, R. M. (2008). Cretaceous-Tertiary carbonate platform evolution and the age of the India-Asia collision along the Ladakh Himalaya (northwest India). *The Journal of Geology*, *116*, 331–353. https://doi.org/10.1086/588831

Groppo, C., Rolfo, F., & Indares, A. (2012). Partial melting in the Higher Himalayan Crystallines of eastern Nepal: the effect of decompression and implications for the "channel flow" model. *Journal of Petrology*, *53*, 1057–1088. https://doi.org/10.1093/petrology/egs009

Grujic, D., Ashley, K. T., Coble, M. A., Coutand, I., Kellett, D. A., Larson, K. P., et al. (2020). Deformational temperatures across the Lesser Himalayan Sequence in eastern Bhutan and their implications for the deformation history of the Main Central Thrust. *Tectonics*, *39*, e2019TC005914. https://doi.org/10.1029/2019TC005914

Guillot, S. (1999). An overview of the metamorphic evolution in Central Nepal. *Journal of Asian Earth Sciences*, *17*, 713–725. https://doi.org/10.1016/S1367-9120(99)00045-0

Guillot, S., Maheo, G., de Sigoyer, J., Hattori, K. H., & Pecher, A. (2008). Tethyan and Indian subduction viewed from the Himalayan high- to ultrahigh-pressure metamorphic rocks. *Tectonophysics*, *451*, 225–241. https://doi.org/10.1016/j.tecto.2007.11.059

Hayden, H. H. (1904). The geology of Spiti, with parts of Bashahr and Rupshu. *Memoirs of the Geological Survey of India*, *36*, 1–129.

He, D., Webb, A. A. G., Larson, K. P., & Schmitt, A. K. (2016). Extrusion vs. duplexing models of Himalayan mountain building 2: The South Tibet detachment at the Dadeldhura klippe. *Tectonophysics*, *667*, 87–107. https://doi.org/10.1016/j.tecto.2015.11.014

Heim, A., & Gansser, A. (1939). Central Himalaya: Geological observations of the Swiss expedition 1936. *Zurich, Memoirs of the Swiss Society of Natural Sciences*, *73*.

Henderson, A. L., Najman, Y., Parrish, R., BouDagher-Fadel, M., Barford, D., Garzanti, E., et al. (2010). Geology of the Cenozoic Indus Basin sedimentary rocks: Paleoenvironmental interpretation of sedimentation from the western Himalaya during the early phases of India-Eurasia collision. *Tectonics*, *29*, TC6015. https://doi.org/10.1029/2009TC002651

Herren, E. (1987). Zanskar shear zone: Northeast-southwest extension within the Higher Himalayas (Ladakh, India). *Geology*, *15*, 409–413. https://doi.org/10.1130/0091-7613(1987)15<409:ZSZNEW>2.0.CO;2

Hodges, K. V., Parrish, R. R., & Searle, M. P. (1996). Tectonic evolution of the central Annapurna Range, Nepalese Himalayas. *Tectonics*, *15*, 1264–1291. https://doi.org/10.1029/96TC01791

Holland, H. D. (2006). The oxygenation of the atmosphere and oceans. *Philosophical Transactions of the Royal Society B: Biological Sciences*, *361*, 903–915. https://doi.org/10.1098/rstb.2006.1838

Horton, F., Lee, J., Hacker, B., Bowman-Kamaha'o, M., & Cosca, M. (2015). Himalayan gneiss dome formation in the middle crust and exhumation by normal faulting: New geochronology of Gianbul dome, northwestern India. *Geological Society of America Bulletin*, *127*, 162–180; https://doi.org/10.1130/B31005.1

Hu, X., Garzanti, E., Moore, T., & Raffi, I. (2015a). Direct stratigraphic dating of India-Asia collision onset at the Selandian (middle Paleocene, 59 ± 1 Ma). *Geology*, *43*, 859–862. https://doi.org/10.1130/G36872.1

Hu, X., Garzanti, E., Wang, J., Huang, W., An, W., & Webb, A. (2016). The timing of India-Asia collision onset: Facts, theories, controversies. *Earth-Science Reviews*, *160*, 264–299. https://doi.org/10.1016/j.earscirev.2016.07.014

Hu, X., Jansa, L., & Wang, C. (2008). Upper Jurassic-Lower Cretaceous stratigraphy in south-eastern Tibet: A comparison with the western Himalayas. *Cretaceous Research*, *29*, 301–315. https://doi.org/10.1016/j.cretres.2007.05.005

Hu, X.-M., Garzanti, E., An, W., & Hu, X.-F. (2015b). Provenance and drainage system of the Early Cretaceous volcanic detritus in the Himalaya as constrained by detrital zircon geochronology. *Journal of Palaeogeography*, *4*, 85–98. https://doi.org/10.3724/SP.J.1261.2015.00069

Huang, W., van Hinsbergen, D. J. J., Lippert, P. C., Guo, Z., & Dupont-Nivet, G. (2015). Paleomagnetic tests of tectonic reconstructions of the India-Asia collision zone: reconstructing India-Asia collision. *Geophysical Research Letters*, *42*, 2642–2649. https://doi.org/10.1002/2015GL063749

Hughes, N. C. (2016). The Cambrian palaeontological record of the Indian subcontinent. *Earth-Science Reviews*, *159*, 428–461. https://doi.org/10.1016/j.earscirev.2016.06.004

Hughes, N. C., Myrow, P. M., Ghazi, S., McKenzie, N. R., Stockli, D. F., & DiPietro, J. A. (2019). Cambrian geology of the Salt Range of Pakistan: linking the Himalayan margin to the Indian craton. *GSA Bulletin*, *131*, 1095–1114. https://doi.org/10.1130/B35092.1

Hughes, N. C., Myrow, P. M., Ghazi, S., McKenzie, N. R., Stockli, D. F., & DiPietro, J. A. (2020). Cambrian geology of the Salt Range of Pakistan: Linking the Himalayan margin to the Indian craton: reply. *Geological Society of America Bulletin*, *132*, 446–448. https://doi.org/10.1130/B35365.1

Hughes, N. C., Peng, S., Bhargava, O. N., Ahluwalia, A. D., Walia, S., Myrow, P. M., et al. (2005). Cambrian biostratigraphy of the Tal Group, Lesser Himalaya, India, and early Tsanglangpuan (late early Cambrian) trilobites from the Nigali Dhar syncline. *Geological Magazine*, *142*, 57–80. https://doi.org/10.1017/S0016756804000366

Iaccarino, S., Montomoli, C., Carosi, R., Montemagni, C., Massonne, H.-J., Langone, A., et al. (2017). Pressure-temperature-deformation-time constraints on the South Tibetan Detachment System in the Garhwal Himalaya (NW India). *Tectonics*, *36*, 2281–2304. https://doi.org/10.1002/2017TC004566

Imayama, T., et al. (2012). Two-stage partial melting and contrasting cooling history within the Higher Himalayan Crystalline Sequence in the far-eastern Nepal Himalaya. *Lithos*, *134*, 1–22. https://doi.org/10.1016/j.lithos.2011.12.004

Ingalls, M., et al. (2018). Paleocene to Pliocene low-latitude, high-elevation basins of southern Tibet: Implications for tectonic models of India-Asia collision, Cenozoic climate, and geochemical weathering. *GSA Bulletin*, *130*, 307–330. https://doi.org/10.1130/B31723.1

Jadoul, F., Berra, F., & Garzanti, E. (1998). The Tethys Himalayan passive margin from Late Triassic to Early Cretaceous (south Tibet). *Journal of Asian Earth Sciences*, *16*, 173–194. https://doi.org/10.1016/S0743-9547(98)00013-0

Jain, A. K., Lal, N., Sulemani, B., Awasthi, A. K., Singh, S., Kumar, R., & Kumar, D. (2009). Detrital-zircon fission-track ages from the Lower Cenozoic sediments, NW Himalayan foreland basin: clues for exhumation and denudation of the Himalaya during the India-Asia collision. *Geological Society of America Bulletin*, *121*, 519–535. https://doi.org/10.1130/B26304.1

Jamieson, R. A., Beaumont, C., Medvedev, S., & Nguyen, M. H. (2004). Crustal channel flows: 2. Numerical models with implications for metamorphism in the Himalayan-Tibetan orogen. *Journal of Geophysical Research*: Solid *Earth*, *109*, B06407. https://doi.org/10.1029/2003JB002811

Jamieson, R. A., Beaumont, C., Nguyen, M. H., & Grujic, D. (2006). Provenance of the Greater Himalayan Sequence and associated rocks: predictions of channel flow models. In R. D. Law et al. (Eds.), *Channel flow, ductile extrusion and exhumation in continental collision zones* (pp. 165–182). London, Geological Society, Special Publication, 268. https://doi.org/10.1144/GSL.SP.2006.268.01.07

Jan, I. U., Shah, A., Stephenson, M. H., Iqbal, S., Hanif, M., Wagreich, M. et al. (2016). The sedimentology of the Lower Permian Dandot Formation: a component of the Gondwana deglaciation sequence of the Salt Range, Pakistan. *Rivista Italiana di Paleontologia e Stratigrafia*, *122*, 75–90. https://doi.org/10.13130/2039-4942/6948

Jan, M. Q., & Faruqi, S. H. (1995). Khewra Trap: An unusual ultrapotassic rock in the Salt Range of Pakistan. *Journal of Nepal Geological Society*, *11*, 237–252.

Jessup, M. J., Langille, J. M., Cottle, J. M., & Ahmad, T. (2016). Crustal thickening, Barrovian metamorphism, and exhumation of midcrustal rocks during doming and extrusion: Insights from the Himalaya, NW India. *Tectonics*, *35*, 160–186. https://doi.org/10.1002/2015TC003962

Jessup, M. J., Newell, D. L., Cottle, J. M., Berger, A. L., & Spotila, J. A. (2008). Orogen-parallel extension and exhumation enhanced by denudation in the trans-Himalayan Arun River gorge, Ama Drime Massif, Tibet-Nepal. *Geology*, *36*, 587–590. https://doi.org/10.1130/G24722A.1

Kali, E., Leloup, P. H., Arnaud, N., Maheo, G., Liu, D., Boutonnet, E., Van der Woerd, J., Liu, X., Liu-Zeng, J., & Li, H. (2010). Exhumation history of the deepest central Himalayan rocks, Ama Drime range: key pressure-temperature-deformation-time constraints on orogenic models. *Tectonics*, *29*, TC2014. https://doi.org/10.1029/2009TC002551

Kaneko, Y., Katayama, I., Yamamoto, H., Misawa, K., Ishikawa, M., Rehman, H. U., Kausar, A. B., & Shiraishi, K. (2003). Timing of Himalayan ultrahigh-pressure metamorphism: sinking rate and subduction angle of the Indian continental crust beneath Asia. *Journal of Metamorphic Geology*, *21*, 589–599. https://doi.org/10.1046/j.1525-1314.2003.00466.x

Kapp, P., & DeCelles, P. G. (2019). Mesozoic–Cenozoic geological evolution of the Himalayan-Tibetan orogen and working tectonic hypotheses. *American Journal of Science*, *319*, 159–254. https://doi.org/10.2475/03.2019.01

Kellett, D. A., Cottle, J. M., & Larson, K. P. (2019). The South Tibetan Detachment System: History, advances, definition and future directions. In P. J. Treloar & M. Searle (Eds.),

Himalayan tectonics: A modern synthesis (pp. 377–400). London, Geological Society, Special Publication, 483. https://doi.org/10.1144/SP483.2

Kellett, D. A., Grujic, D., Coutand, I., Cottle, J., & Mukul, M. (2013). The South Tibetan detachment system facilitates ultra rapid cooling of granulite-facies rocks in Sikkim Himalaya. *Tectonics*, *32*, 252–270. https://doi.org/10.1002/tect.20014

Kellett, D. A., Grujic, D., Warren, C., Cottle, J., Jamieson, R., & Tenzin, T. (2010). Metamorphic history of a syn-convergent orogen-parallel detachment: the South Tibetan detachment system, Bhutan Himalaya. *Journal of Metamorphic Geology*, *28*,785–808.https://doi.org/10.1111/j.1525-1314.2010.00893.x

Kelly, S., Beaumont, C., & Butler, J. P. (2020). Inherited terrane properties explain enigmatic post-collisional Himalayan-Tibetan evolution. *Geology*, *48*, 8–14. https://doi.org/10.1130/G46701.1

Khan, A. (2021). *Petrochronology, geochemistry, and rare earth element potential of Himalayan carbonatite complexes, northwest Pakistan*. Ph. D. Dissertation. University of Peshawar.

Khanal, S., & Robinson, D. M. (2013). Upper crustal shortening and forward modeling of the Himalayan thrust belt along the Budhi-Gandaki River, central Nepal. *International Journal of Earth Sciences*, *102*, 1871–1891. https://doi.org/10.1007/s00531-013-0889-1

King, J., Harris, N., Argles, T., Parrish, R., & Zhang, H. (2011). Contribution of crustal anatexis to the tectonic evolution of Indian crust beneath southern Tibet. *Geological Society of America Bulletin*, *123*, 218–239. https://doi.org/10.1130/B30085.1

Klootwijk, C., Sharma, M. L., Gergan, J., Tirkey, B., Shah, S. K., & Agarwal, V. (1979). The extent of Greater India, II. Palaeomagnetic data from the Ladakh Intrusives at Kargil, northwestern Himalayas. *Earth and Planetary Science Letters*, *44*, 47–64. https://doi.org/10.1016/0012-821X(79)90007-4

Kohn, M. J. (2008). P-T-t data from central Nepal support critical taper and repudiate large-scale channel flow of the Greater Himalayan Sequence. *Geological Society of America Bulletin*, *120*, 259–273. https://doi.org/10.1130/B26252.1

Kohn, M. J. (2014). Himalayan metamorphism and its tectonic implications. *Annual Review of Earth and Planetary Sciences*, *42*, 381–419. https://doi.org/10.1146/annurev-earth-060313-055005

Kohn, M. J., Paul, S. K., & Corrie, S. L. (2010). The lower Lesser Himalayan sequence: A Paleoproterozoic arc on the northern margin of the Indian plate. *Geological Society of America Bulletin*, *122*, 323–335. https://doi.org/10.1130/B26587.1

Kohn, M. J., Wieland, M. S., Parkinson, C. D., & Upreti, B. N. (2004). Miocene faulting at plate tectonic velocity in the Himalaya of central Nepal. *Earth and Planetary Science Letters*, *228*, 299–310. https://doi.org/10.1016/j.epsl.2004.10.007

Kohn, M. J., Wieland, M. S., Parkinson, C. D., & Upreti, B. N. (2005). Five generations of monazite in Langtang gneisses: Implications for chronology of the Himalayan metamorphic core. *Journal of Metamorphic Geology*, *23*, 399–406. https://doi.org/10.1111/j.1525-1314.2005.00584.x

Langille, J. M., Jessup, M. J., Cottle, J. M., Lederer, G., & Ahmad, T. (2012). Timing of metamorphism, melting and exhumation of the Leo Pargil dome, northwest India. *Journal of Metamorphic Geology*, *30*, 769–791. https://doi.org/10.1111/j.1525-1314.2012.00998.x

Larson, K., Piercey, S., & Cottle, J. (2019). Preservation of a Paleoproterozoic rifted margin in the Himalaya. Insight from the Ulleri-Phaplu-Melung orthogneiss. *Geoscience Frontiers*, *10*, 873–883. https://doi.org/10.1016/j.gsf.2017.05.010

Larson, K. P., Shrestha, S., Soret, M., & Smit, M. (2021). The P-T-t-D evolution of the Mahabharat, east-central Nepal. the out-of-sequence development of the Himalaya. *Geoscience Frontiers*. https://doi.org/10.1016/j.gsf.2020.08.001

Laskowski, A. K., Kapp, P., & Cai, F. (2018). Gangdese culmination model: Oligocene-Miocene duplexing along the India-Asia suture zone, Lazi region, southern Tibet. *GSA Bulletin*, *130*, 1355–1376. https://doi.org/10.1130/B31834.1

Laskowski, A. K., Kapp, P., Ding, L., Campbell, C., & Liu, X. (2017). Tectonic evolution of the Yarlung suture zone, Lopu Range region, southern Tibet. *Tectonics*, *36*, 108–136. https://doi.org/10.1002/2016TC004334

Latif, M. A. (1973). Partial extension of the evaporite facies of the Salt Range to Hazara, Pakistan. *Nature Physical Science*, *244*, 124–125. https://doi.org/10.1038/physci244124a0

Lave, J., & Avouac, J. P. (2000). Active folding of fluvial terraces across the Siwaliks Hills, Himalayas of central Nepal. *Journal of Geophysical Research: Solid Earth*, *105*, 5735–5770. https://doi.org/10.1029/1999JB900292

Leary, R. J., DeCelles, P. G., Quade, J., Gehrels, G. E., & Waanders, G. (2016). The Liuqu Conglomerate, southern Tibet: Early Miocene basin development related to deformation within the Great Counter Thrust system. *Lithosphere*, *8*, 427–450. https://doi.org/10.1130/L542.1

Lee, J., Hacker, B., & Wang, Y. (2004). Evolution of North Himalayan gneiss domes: structural and metamorphic studies in Mabja Dome, southern Tibet. *Journal of Structural Geology*, *26*, 2297–2316. https://doi.org/10.1016/j.jsg.2004.02.013

Lee, J., Hacker, B. R., Dinklage, W. S., Wang, Y., Gans, P., Calvert, A., et al. (2000). Evolution of the Kangmar Dome, southern Tibet: Structural, petrologic, and thermochronologic constraints. *Tectonics*, *19*, 872–895. https://doi.org/10.1029/1999TC001147

Lee, J., McClelland, W., Wang, Y., Blythe, A., & McWilliams, M. (2006). Oligocene-Miocene middle crustal flow in southern Tibet: Geochronology of Mabja Dome. In R. D. Law et al. (Eds.), *Channel flow, ductile extrusion and exhumation in continental collision zones* (pp. 445–469). London, Geological Society, Special Publication, 268. https://doi.org/10.1144/GSL.SP.2006.268.01.21

Leech, M. (2008). Does the Karakoram fault interrupt mid-crustal channel flow in the western Himalaya? *Earth and Planetary Science Letters*, *276*, 314–322. doi:10.1016/j.epsl.2008.10.006

Leech, M. L., Singh, S., Jain, A. K., Klemperer, S. L., & Manickavasagam, R. M. (2005). The onset of India-Asia continental collision: Early, steep subduction required by the timing of UHP metamorphism in the western Himalaya. *Earth and Planetary Science Letters*, *234*, 83–97. https://doi.org/10.1016/j.epsl.2005.02.038

Le Fort, P., Debon, F., Pecher, A., Sonet, J., et al. (1986). The 500 Ma magmatic event in Alpine southern Asia, a thermal episode at Gondwana scale. *Sciences de la Terre Memoire*, *47*, 191–209.

Li, G., Kohn, B., Sandiford, M., Xu, Z., & Wei, L. (2015a). Constraining the age of Liuqu Conglomerate, southern Tibet: Implications for evolution of the India-Asia collision zone. *Earth & Planetary Science Letters*, *426*, 259–266. https://doi.org/10.1016/j.epsl.2015.06.010.

Li, J., Hu, X., Garzanti, E., & BouDagher-Fadel, M. (2017). Shallow-water carbonate responses to the Paleocene-Eocene thermal maximum in the Tethyan Himalaya (southern Tibet): Tectonic and climatic implications. *Palaeogeography, Palaeoclimatology, Palaeoecology*, *466*, 153–165. https://doi.org/10.1016/j.palaeo.2016.11.026

Li, Y., and Robinson, D. M. (2023). Two Possible Pre-collisional Crustal Configurations of Northern Greater India: Implications for the India-Asia Collision. *Earth and Planetary Science Letters*, *610*, https://doi.org/10.1016/j.epsl.2023.118098.

Li, Y., Wang, C., Dai, J., Xu, G., Hou, Y., & Li, X. (2015b). Propagation of the deformation and growth of the Tibetan-Himalayan orogen: A review. *Earth-Science Reviews*, *143*, 36–61. https://doi.org/10.1016/j.earscirev.2015.01.001

Liu, G., & Einsele, G. (1994). Sedimentary history of the Tethyan basin in the Tibetan Himalayas. *Geologische Rundschau*, *83*, 32–61. https://doi.org/10.1007/BF00211893

Long, S., McQuarrie, N., Tobgay, T., & Grujic, D. (2011b). Geometry and crustal shortening of the Himalayan fold-thrust belt, eastern and central Bhutan. *Geological Society of America Bulletin*, *123*, 1427–1447. https://doi.org/10.1130/B30203.1

Long, S., McQuarrie, N., Tobgay, T., & Hawthorne, J. (2011c). Quantifying internal strain and deformation temperature in the eastern Himalaya, Bhutan: Implications for the evolution of strain in thrust sheets. *Journal of Structural Geology*, *33*, 579–608. https://doi.org/10.1016/j.jsg.2010.12.011

Long, S., McQuarrie, N., Tobgay, T., Rose, C., Gehrels, G., & Grujic, D. (2011a). Tectonostratigraphy of the Lesser Himalaya of Bhutan: Implications for the along-strike stratigraphic continuity of the northern Indian margin. *Geological Society of America Bulletin*, *123*, 1406–1426. https://doi.org/10.1130/B30202.1

Long, S. P., & Robinson, D. M. (2021). Construction of the Lesser Himalayan–Subhimalayan thrust belt: The primary driver of thickening, exhumation, and high elevations in the Himalayan orogen since the middle Miocene. *Geology*, *49*, 1283–1288. https://doi.org/10.1130/G48967.1

Long, S. P., Gordon, S. M., Young, J. P., & Soignard, E. (2016). Temperature and strain gradients through Lesser Himalayan rocks and across the Main Central thrust, south central Bhutan: implications for transport-parallel stretching and inverted metamorphism. *Tectonics*, *35*, 1863–1891. https://doi.org/10.1002/2016TC004242

Long, S. P., McQuarrie, N., Tobgay, T., Coutand, I., Cooper, F. J., Reiners, P. W., et al. (2012). Variable shortening rates in the eastern Himalayan thrust belt, Bhutan: Insights from multiple thermochronologic and geochronologic data sets tied to kinematic reconstructions. *Tectonics*, *31*, TC5004. https://doi.org/10.1029/2012TC003155

Ma, Y., Yang, T., Bian, W., Jin, J., Zhang, S., Wu, H., et al. (2016). Early Cretaceous paleomagnetic and geochronologic results from the Tethyan Himalaya: insights into the Neotethyan paleogeography and the India-Asia collision. *Scientific Reports*, *6*, 21605. https://doi.org/10.1038/srep21605

Mandal, S., Robinson, D. M., Kohn, M. J., Khanal, S., & Das, O. (2019). Examining the tectono-stratigraphic architecture, structural geometry, and kinematic evolution of the Himalayan fold-thrust belt, Kumaun, northwest India. *Lithosphere*, *11*, 414–435. https://doi.org/10.1130/L1050.1

Mandal, S., Robinson, D. M., Kohn, M. J., Khanal, S., Das, O., & Bose, S. (2016). Zircon U-Pb ages and Hf isotopes of the Askot klippe, Kumaun, northwest India: Implications for Paleoproterozoic tectonics, basin evolution and associated metallogeny of the northern Indian cratonic margin. *Tectonics*, *35*, 965–982. https://doi.org/10.1002/2015TC004064

Martin, A. J. (2017a). A review of definitions of the Himalayan Main Central Thrust. *International Journal of Earth Sciences*, *106*, 2131–2145. https://doi.org/10.1007/s00531-016-1419-8

Martin, A. J. (2017b). A review of Himalayan stratigraphy, magmatism, and structure. *Gondwana Research*, *49*, 42–80. https://doi.org/10.1016/j.gr.2017.04.031

Martin, A. J. (2020). Cambrian geology of the Salt Range of Pakistan: Linking the Himalayan margin to the Indian craton: Comment. *Geological Society of America Bulletin*, *132*, 444–445. https://doi.org/10.1130/B35329.1

Martin, A. J., Burgy, K. D., Kaufman, A. J., & Gehrels, G. E. (2011). Stratigraphic and tectonic implications of field and isotopic constraints on depositional ages of Proterozoic Lesser Himalayan rocks in central Nepal. *Precambrian Research*, *185*, 1–17. https://doi.org/10.1016/j.precamres.2010.11.003

Martin, C. R., Jagoutz, O., Upadhyay, R., Royden, L. H., Eddy, M. P., Bailey, E., Nichols, C. I. O., and Weiss, B. P. (2020). Paleocene latitude of the Kohistan -Ladakh arc indicates multistage India -Eurasia collision. *Proceedings of the National Academy of Sciences*, *117*, 29487–29494. doi:10.1073/pnas.2009039117.

Mathur, N. S. (1978). Biostratigraphical aspects of the Subathu Formation, Kumaun Himalaya. *Recent Researches in Geology*, *5*, 96–112.

McCallister, A. T., Taylor, M. H., Murphy, M. A., Styron, R. H., & Stockli, D. F. (2014). Thermochronologic constraints on the late Cenozoic exhumation history of the Gurla Mandhata metamorphic core complex, southwestern Tibet. *Tectonics*, *33*, 27–52. https://doi.org/10.1002/2013TC003302

McKenzie, N. R., Hughes, N. C., Myrow, P. M., Xiao, S., & Sharma, M. (2011). Correlation of Precambrian-Cambrian sedimentary successions across northern India and the utility of isotopic signatures of Himalayan lithotectonic zones. *Earth and Planetary Science Letters*, *312*, 471–483. https://doi.org/10.1016/j.epsl.2011.10.027

McQuarrie, N., & Ehlers, T. A. (2015). Influence of thrust belt geometry and shortening rate on thermochronometer cooling ages: Insights from thermokinematic and erosion modeling of the Bhutan Himalaya. *Tectonics*, *34*, 1055–1079. https://doi.org/10.1002/2014TC003783

McQuarrie, N., & Ehlers, T. A. (2017). Techniques for understanding fold-and-thrust belt kinematics and thermal evolution. In R. D. Law et al. (Eds.), Linkages and feedbacks in orogenic systems. *Geological Society of America, Memoir, 213*, 25–54. https://doi.org/10.1130/2017.1213(02)

McQuarrie, N., Eizenhofer, P. R., Long, S. P., Tobgay, T., Ehlers, T. A., Blythe, A. E., et al. (2019). The influence of foreland structures on hinterland cooling: Evaluating the drivers of exhumation in the eastern Bhutan Himalaya. *Tectonics, 38*, 3282–3310. https://doi.org/10.1029/2018TC005340

McQuarrie, N., Long, S. P., Tobgay, T., Nesbit, J. N., Gehrels, G., & Ducea, M. N. (2013). Documenting basin scale, geometry and provenance through detrital geochemical data: Lessons from the Neoproterozoic to Ordovician Lesser, Greater, and Tethyan Himalayan strata of Bhutan. *Gondwana Research, 23*, 1491–1510. https://doi.org/10.1016/j.gr.2012.09.002

McQuarrie, N., Robinson, D., Long, S., Tobgay, T., Grujic, D., Gehrels, G., et al. (2008). Preliminary stratigraphic and structural architecture of Bhutan: implications for the along strike architecture of the Himalayan system. *Earth and Planetary Science Letters, 272*, 105–117. https://doi.org/10.1016/j.epsl.2008.04.030

McQuarrie, N., Tobgay, T., Long, S. P., Reiners, P. W., & Cosca, M. A. (2014). Variable exhumation rates and variable displacement rates: Documenting recent slowing of Himalayan shortening in western Bhutan. *Earth and Planetary Science Letters, 386*, 161–174. https://doi.org/10.1016/j.epsl.2013.10.045

Meng, J., Gilder, S. A., Wang, C., Coe, R. S., Tan, X., Zhao, X., & He, K. (2019). Defining the limits of Greater India. *Geophysical Research Letters, 46*, 4182–4191. https://doi.org/10.1029/2019GL082119

Merdith, A. S., et al. (2017). A full-plate global reconstruction of the Neoproterozoic. *Gondwana Research, 50*, 84–134. https://doi.org/10.1016/j.gr.2017.04.001

Mitra, G., Bhattacharyya, K., & Mukul, M. (2010). The Lesser Himalayan Duplex in Sikkim: Implications for variations in Himalayan shortening. *Journal of the Geological Society of India, 75*, 289–301.

Molnar, P., England, P., & Martinod, J. (1993). Mantle dynamics, uplift of the Tibetan Plateau, and the Indian Monsoon. *Reviews of Geophysics, 31*, 357–396. https://doi.org/10.1029/93RG02030

Montemagni, C., Carosi, R., Fusi, N., Iaccarino, S., Montomoli, C., Villa, I. M., et al. (2020). Three-dimensional vorticity and time-constrained evolution of the Main Central Thrust zone, Garhwal Himalaya (NW India). *Terra Nova, 32*, 215–224. https://doi.org/10.1111/ter.12450

Montemagni, C., Montomoli, C., Iaccarino, S., Carosi, R., Jain, A. K., Massonne, H.-J., et al. (2019). Dating protracted fault activities: microstructures, microchemistry and geochronology of the Vaikrita Thrust, Main Central Thrust zone, Garhwal Himalaya, NW India. In R. Sharma et al. (Eds.), *Crustal architecture and evolution of the Himalaya-Karakoram-Tibet orogen* (pp. 127–146). London, Geological Society, Special Publication, 481. https://doi.org/10.1144/SP481.3

Montomoli, C., Iaccarino, S., Antolin, B., Appel, E., Carosi, R., Dunkl, I., et al. (2017). Tectono-metamorphic evolution of the Tethyan Sedimentary Sequence (Himalayas, SE Tibet). *Italian Journal of Geosciences, 136*, 73–88.

Montomoli, C., Iaccarino, S., Carosi, R., Langone, A., & Visona, D. (2013). Tectonometamorphic discontinuities within the Greater Himalayan Sequence in western Nepal (central Himalaya): Insights on the exhumation of crystalline rocks. *Tectonophysics, 608*, 1349–1370. https://doi.org/10.1016/j.tecto.2013.06.006

Mottram, C. M., Argles, T. W., Harris, N. B. W., Parrish, R. R., Horstwood, M. S. A., Warren, C. J., et al. (2014). Tectonic interleaving along the Main Central Thrust, Sikkim Himalaya. *Journal of the Geological Society, 171*, 255–268. https://doi.org/10.1144/jgs2013-064

Mottram, C. M., Cottle, J. M., & Kylander-Clark, A. R. C. (2019). Campaign-style U-Pb titanite petrochronology: along-strike variations in timing of metamorphism in the Himalayan metamorphic core. *Geoscience Frontiers, 10*, 827–847. https://doi.org/10.1016/j.gsf.2018.09.007

Mottram, C. M., Parrish, R. R., Regis, D., Warren, C. J., Argles, T. W., Harris, N. B. W., et al. (2015). Using U-Th-Pb petrochronology to determine rates of ductile thrusting: time windows into the Main Central Thrust, Sikkim Himalaya. *Tectonics, 34*, 1355–1374. https://doi.org/10.1002/2014TC003743

Mukherjee, P. K., Jain, A. K., Singhal, S., Singha, N. B., Singh, S., Kumud, K., et al. (2019). U-Pb zircon ages and Sm-Nd isotopic characteristics of the Lesser and Great Himalayan sequences, Uttarakhand Himalaya, and their regional tectonic implications. *Gondwana Research, 75*, 282–297. https://doi.org/10.1016/j.gr.2019.06.001

Mukherjee, S. (2013). Channel flow extrusion model to constrain dynamic viscosity and Prandtl number of the Higher Himalayan Shear Zone. *International Journal of Earth Sciences, 102*, 1811–1835. https://doi.org/10.1007/s00531-012-0806-z

Mukherjee, S. (2015). A review on out-of-sequence deformation in the Himalaya. In S. Mukherjee, et al. (Eds.), *Tectonics of the Himalaya* (pp. 67–109). London, Geological Society, Special Publication, 412. https://doi.org/10.1144/SP412.13

Mukherjee, S., & Koyi, H. A. (2010a). Higher Himalayan Shear Zone, Sutlej section: Structural geology and extrusion mechanism by various combinations of simple shear, pure shear and channel flow in shifting modes. *International Journal of Earth Sciences, 99*, 1267–1303. https://doi.org/10.1007/s00531-009-0459-8

Mukherjee, S., & Koyi, H. A. (2010b). Higher Himalayan Shear Zone, Zanskar Indian Himalaya: Microstructural studies and extrusion mechanism by a combination of simple shear and channel flow. *International Journal of Earth Sciences, 99*, 1083–1110. https://doi.org/10.1007/s00531-009-0447-z

Mukherjee, S., Koyi, H. A., & Talbot, C. J. (2012). Implications of channel flow analogue models for extrusion of the Higher Himalayan Shear Zone with special reference to the out-of-sequence thrusting. *International Journal of Earth Sciences, 101*, 253–272. https://doi.org/10.1007/s00531-011-0650-6

Murphy, M. A. (2007). Isotopic characteristics of the Gurla Mandhata metamorphic core complex: Implications for the architecture of the Himalayan orogen. *Geology, 35*, 983–986. https://doi.org/10.1130/G23774A.1

Murphy, M. A., & Copeland, P. (2005). Transtensional deformation in the central Himalaya and its role in accommodating growth of the Himalayan orogen. *Tectonics*, *24*, TC4012. https://doi.org/10.1029/2004TC001659

Murphy, M. A., & Yin, A. (2003). Structural evolution and sequence of thrusting in the Tethyan fold-thrust belt and Indus-Yalu suture zone, southwest Tibet. *Geological Society of America Bulletin*, *115*, 21–34. https://doi.org/10.1130/0016-7606(2003)115<0021:SEASOT>2.0.CO;2

Myrow, P. M., Hughes, N. C., Goodge, J. W., Fanning, C. M., Williams, I. S., Peng, S., et al. (2010). Extraordinary transport and mixing of sediment across Himalayan central Gondwana during the Cambrian-Ordovician. *Geological Society of America Bulletin*, *122*, 1660–1670. https://doi.org/10.1130/B30123.1

Myrow, P. M., Hughes, N. C., McKenzie, N. R., Pelgay, P., Thomson, T. J., Haddad, E. E., et al. (2016). Cambrian-Ordovician orogenesis in Himalayan equatorial Gondwana. *Geological Society of America Bulletin*, *128*, 1679–1695. https://doi.org/10.1130/B31507.1

Myrow, P. M., Hughes, N. C., Paulsen, T. S., Williams, I. S., Parcha, S. K., Thompson, K. R., et al. (2003). Integrated tectonostratigraphic analysis of the Himalaya and implications for its tectonic reconstruction. *Earth and Planetary Science Letters*, *212*, 433–441. https://doi.org/10.1016/S0012-821X(03)00280-2

Naeem, M., Burg, J.-P., Ahmad, N., Chaudhry, M. N., & Khalid, P. (2016). U-Pb zircon systematics of the Mansehra Granitic Complex: implications on the early Paleozoic orogenesis in NW Himalaya of Pakistan. *Geosciences Journal*, *20*, 427–447. https://doi.org/10.1007/s12303-015-0062-x

Najman, Y. (2006). The detrital record of orogenesis: A review of approaches and techniques used in the Himalayan sedimentary basins. *Earth-Science Reviews*, *74*, 1–72. https://doi.org/10.1016/j.earscirev.2005.04.004

Najman, Y., Bickle, M., Garzanti, E., Pringle, M., Barfod, D., Brozovic, N., et al. (2009). Reconstructing the exhumation history of the Lesser Himalaya, NW India, from a multitechnique provenance study of the foreland basin Siwalik Group. *Tectonics*, *28*, TC5018. https://doi.org/10.1029/2009TC002506

Najman, Y., Carter, A., Oliver, G., & Garzanti, E. (2005). Provenance of Eocene foreland basin sediments, Nepal: Constraints to the timing and diachroneity of early Himalayan orogenesis. *Geology*, *33*, 309–312. https://doi.org/10.1130/G21161.1

Najman, Y., Johnson, K., White, N., & Oliver, G. (2004). Evolution of the Himalayan foreland basin, NW India. *Basin Research*, *16*, 1–24. https://doi.org/10.1111/j.1365-2117.2004.00223.x

Ojha, T. P., Butler, R. F., DeCelles, P. G., & Quade, J. (2009). Magnetic polarity stratigraphy of the Neogene foreland basin deposits of Nepal. *Basin Research*, *21*, 61–90. https://doi.org/10.1111/j.1365-2117.2008.00374.x

Ojha, T. P., Butler, R. F., Quade, J., DeCelles, P. G., Richards, D., & Upreti, B. N. (2000). Magnetic polarity stratigraphy of the Neogene Siwalik Group at Khutia Khola, far western Nepal. *Geological Society of America Bulletin*, *112*, 424–434. https://doi.org/10.1130/0016-7606(2000)112<424:MPSOTN>2.0.CO;2

Olsen, J. E. S., McQuarrie, N., & Robinson, D. M. (2019). Determining kinematic order and relative age of faulting via flexural-kinematic restoration: A case study in far western Nepal. *Basin Research*, *31*, 1153–1177. https://doi.org/10.1111/bre.12362

O'Neill, C., Lenardic, A., & Condie, K. C. (2015). Earth's punctuated tectonic evolution: Cause and effect. *Geological Society, London, Special Publications*, *389*, 17–40. https://doi.org/10.1144/SP389.4

Orme, D. A. (2019). Burial and exhumation history of the Xigaze forearc basin, Yarlung suture zone, Tibet. *Geoscience Frontiers*, *10*, 895–908. https://doi.org/10.1016/j.gsf.2017.11.011

Palin, R. M., Treloar, P. J., Searle, M. P., Wald, T., White, R. W., & Mertz-Kraus, R. (2018). U-Pb monazite ages from the Pakistan Himalaya record pre-Himalayan Ordovician orogeny and Permian continental breakup. *GSA Bulletin*. https://doi.org/10.1130/B31943.1

Parrish, R. R., & Hodges, K. V. (1996). Isotopic constraints on the age and provenance of the Lesser and Greater Himalayan sequences, Nepalese Himalaya. *Geological Society of America Bulletin*, *108*, 904–911. https://doi.org/10.1130/0016-7606(1996)108<0904:ICOTAA>2.3.CO;2

Parrish, R. R., Gough, S. J., Searle, M. P., & Waters, D. J. (2006). Plate velocity exhumation of ultrahigh-pressure eclogites in the Pakistan Himalaya. *Geology*, *34*, 989–992. https://doi.org/10.1130/G22796A.1

Parsons, A. J., Hosseini, K., Palin, R. M., & Sigloch, K. (2020). Geological, geophysical and plate kinematic constraints for models of the India-Asia collision and the post-Triassic central Tethys oceans. *Earth-Science Reviews*, *208*, 103084. https://doi.org/10.1016/j.earscirev.2020.103084

Pedersen, R. B., Searle, M. P., & Corfield, R. I. (2001). U-Pb zircon ages from the Spontang Ophiolite, Ladakh Himalaya. *Journal of the Geological Society*, *158*, 513–520. https://doi.org/10.1144/jgs.158.3.513

Phukon, P., Sen, K., Srivastava, H. B., Singhal, S., & Sen, A. (2018). U-Pb geochronology and geochemistry from the Kumaun Himalaya, NW India, reveal Paleoproterozoic arc magmatism related to formation of the Columbia supercontinent. *GSA Bulletin*, *130*, 1164–1176. https://doi.org/10.1130/B31866.1

Pivnik, D. A., & Wells, N. A. (1996). The transition from Tethys to the Himalaya as recorded in northwest Pakistan. *Geological Society of America Bulletin*, *108*, 1295–1313. https://doi.org/10.1130/0016-7606(1996)108<1295:TTFTTT>2.3.CO;2

Pognante, U., & Spencer, D. A. (1991). First report of eclogites from the Himalayan belt, Kaghan valley (northern Pakistan). *European Journal of Mineralogy*, *3*, 613–618. https://doi.org/10.1127/ejm/3/3/0613

Pogue, K. R., Hylland, M. D., Yeats, R. S., Khattak, W. U., & Hussain, A. (1999). Stratigraphic and structural framework of Himalayan foothills, northern Pakistan. In *Himalaya and Tibet: Mountain roots to mountain tops* (pp. 257–274). Boulder, Geological Society of America, Special Paper, 328. https://doi.org/10.1130/0-8137-2328-0.257

Powell, C., & Conaghan, P. J. (1973). Plate tectonics and the Himalayas. *Earth and Planetary Science Letters*, *20*, 1–12. https://doi.org/10.1016/0012-821X(73)90134-9

Pullen, A., Kapp, P., DeCelles, P. G., Gehrels, G. E., & Ding, L. (2011). Cenozoic anatexis and exhumation of Tethyan Sequence rocks in the Xiao Gurla Range, southwest Tibet. *Tectonophysics*, *501*, 28–40. https://doi.org/10.1016/j.tecto.2011.01.008

Qasim, M., Khan, M. A., & Haneef, M. (2015). Stratigraphic characterization and structural analysis of the northwestern Hazara Ranges across Panjal Thrust, northern Pakistan. *Arabian Journal of Geosciences*, *8*, 7811–7829. https://doi.org/10.1007/s12517-015-1787-6

Quidelleur, X., Grove, M., Lovera, O. M., Harrison, T. M., Yin, A., & Ryerson, F. J. (1997). Thermal evolution and slip history of the Renbu Zedong Thrust, southeastern Tibet. *Journal of Geophysical Research: Solid Earth*, *102*, 2659–2679. https://doi.org/10.1029/96JB02483

Rao, D., & Sharma, R. (2011). Arc magmatism in eastern Kumaun Himalaya, India. A study based on geochemistry of granitoid rocks. *Island Arc*, *20*, 500–519. https://doi.org/10.1111/j.1440-1738.2011.00781.x

Ratschbacher, L., Frisch, W., Liu, G., & Chen, C. (1994). Distributed deformation in southern and western Tibet during and after the India-Asia collision. *Journal of Geophysical Research, Solid Earth*, *99*, 19917–19945. https://doi.org/10.1029/94JB00932

Ravikant, V., Wu, F.-Y., & Ji, W.-Q. (2011). U-Pb age and Hf isotopic constraints of detrital zircons from the Himalayan foreland Subathu sub-basin on the Tertiary palaeogeography of the Himalaya. *Earth and Planetary Science Letters*, *304*, 356–368. https://doi.org/10.1016/j.epsl.2011.02.009

Rehman, H. U. (2019). Geochronological enigma of the HP-UHP rocks in the Himalayan orogen. In L. Zhang et al. (Eds.), *HP-UHP metamorphism and tectonic evolution of orogenic belts* (pp. 183–207). London, Geological Society, Special Publication, 474. https://doi.org/10.1144/SP474.14

Rehman, H. U., et al. (2013). Ion microprobe U-Th-Pb geochronology and study of micro-inclusions in zircon from the Himalayan high- and ultrahigh-pressure eclogites, Kaghan Valley of Pakistan. *Journal of Asian Earth Sciences*, *63*, 179–196. https://doi.org/10.1016/j.jseaes.2012.04.025

Reuber, I., Colchen, M., & Mevel, C. (1987). The geodynamic evolution of the South-Tethyan margin in Zanskar, NW-Himalaya, as revealed by the Spongtang ophiolitic melanges. *Geodinamica Acta*, *1*, 283–296. https://doi.org/10.1080/09853111.1987.11105146

Richards, A., Argles, T., Harris, N., Parrish, R., Ahmad, T., Darbyshire, F., et al. (2005). Himalayan architecture constrained by isotopic tracers from clastic sediments. *Earth and Planetary Science Letters*, *236*, 773–796. https://doi.org/10.1016/j.epsl.2005.05.034

Richards, A., Parrish, R., Harris, N., Argles, T., & Zhang, L. (2006). Correlation of lithotectonic units across the eastern Himalaya, Bhutan. *Geology*, *34*, 341–344. https://doi.org/10.1130/G22169.1

Robinson, D. M., & Martin, A. J. (2014). Reconstructing the Greater Indian margin: A balanced cross section in central Nepal focusing on the Lesser Himalayan duplex. *Tectonics*, *33*, 2143–2168. https://doi.org/10.1002/2014TC003564

Robinson, D. M., & McQuarrie, N. (2012). Pulsed deformation and variable slip rates within the central Himalayan thrust belt. *Lithosphere*, *4*, 449–464. https://doi.org/10.1130/L204.1

Robinson, D. M., & Pearson, O. N. (2013). Was Himalayan normal faulting triggered by initiation of the Ramgarh-Munsiari thrust and development of the Lesser Himalayan duplex? *International Journal of Earth Sciences*, *102*, 1773–1790. https://doi.org/10.1007/s00531-013-0895-3

Robinson, D. M., DeCelles, P. G., & Copeland, P. (2006). Tectonic evolution of the Himalayan thrust belt in western Nepal: implications for channel flow models. *Geological Society of America Bulletin*, *118*, 865–885. https://doi.org/10.1130/B25911.1

Robinson, D. M., DeCelles, P. G., Garzione, C. N., Pearson, O. N., Harrison, T. M., & Catlos, E. J. (2003). Kinematic model for the Main Central thrust in Nepal. *Geology*, *31*, 359–362. https://doi.org/10.1130/0091-7613(2003)031<0359:KMFTMC>2.0.CO;2

Robinson, D. M., DeCelles, P. G., Patchett, P. J., & Garzione, C. N. (2001). The kinematic evolution of the Nepalese Himalaya interpreted from Nd isotopes. *Earth and Planetary Science Letters*, *192*, 507–521. https://doi.org/10.1016/S0012-821X(01)00451-4

Robinson, D. M., Khanal, S., Olree, E., Bhattacharya, G., & Mandal, S. (2021). Controls of stratigraphic architecture on along strike cooling age patterns. *Terra Nova*, *33*, 129–136. https://doi.org/10.1111/ter.12498

Rogers, J. J. W., & Santosh, M. (2002). Configuration of Columbia, a Mesoproterozoic supercontinent. *Gondwana Research*, *5*, 5–22. https://doi.org/10.1016/S1342-937X(05)70883-2

Rowley, D. B. (2019). Comparing paleomagnetic study means with apparent wander paths: a case study and paleomagnetic test of the Greater India versus Greater Indian Basin hypotheses. *Tectonics*, *38*, 722–740. https://doi.org/10.1029/2017TC004802

Royden, L. H., Burchfiel, B. C., & van der Hilst, R. D. (2008). The geological evolution of the Tibetan Plateau. *Science*, *321*, 1054–1058. https://doi.org/10.1126/science.1155371

Sajid, M., Andersen, J., Rocholl, A., & Wiedenbeck, M. (2018). U-Pb geochronology and petrogenesis of peraluminous granitoids from northern Indian plate in NW Pakistan: Andean type orogenic signatures from the early Paleozoic along the northern Gondwana. *Lithos*, *318–319*, 340–356. https://doi.org/10.1016/j.lithos.2018.08.024

Sakai, H. (1983). Geology of the Tansen Group of the Lesser Himalaya in Nepal: Memoirs of the faculty of science, Kyushu University, Series D. *Geology*, *25*, 27–74.

Sakai, H., Iwano, H., Danhara, T., Takigami, Y., Rai, S. M., Upreti, B. N., et al. (2013). Rift-related origin of the Paleoproterozoic Kuncha Formation, and cooling history of the Kuncha nappe and Taplejung granites, eastern Nepal Lesser Himalaya: A multichronological approach. *Island Arc*, *22*, 338–360. https://doi.org/10.1111/iar.12021

Sanchez, V. I., Murphy, M. A., Robinson, A. C., Lapen, T. J., & Heizler, M. T. (2013). Tectonic evolution of the India-Asia suture zone since Middle Eocene time, Lopukangri area, south-central Tibet. *Journal of Asian Earth Sciences*, *62*, 205–220. https://doi.org/10.1016/j.jseaes.2012.09.004

Sastri, V. V., Bhandari, L. L., Raju, A. T. R., & Datta, A. K. (1971). Tectonic framework and subsurface stratigraphy of the Ganga Basin. *Journal of the Geological Society of India*, *12*, 222–233.

Sciunnach, D., & Garzanti, E. (2012). Subsidence history of the Tethys Himalaya. *Earth-Science Reviews*, *111*, 179–198. https://doi.org/10.1016/j.earscirev.2011.11.007

Searle, M. P. (1986). Structural evolution and sequence of thrusting in the High Himalayan, Tibetan-Tethys and Indus suture zones of Zanskar and Ladakh, Western Himalaya. *Journal of Structural Geology*, *8*, 923–936. https://doi.org/10.1016/0191-8141(86)90037-4

Searle, M. P., et al. (1987). The closing of Tethys and the tectonics of the Himalaya. *Geological Society of America Bulletin*, *98*, 678–701. https://doi.org/10.1130/0016-7606(1987)98<678:TCOTAT>2.0.CO;2

Searle, M. P., Law, R. D., Godin, L., Larson, K. P., Streule, M. J., Cottle, J. M., et al. (2008). Defining the Himalayan Main Central Thrust in Nepal. *Journal of the Geological Society*, *165*, 523–534. https://doi.org/10.1144/0016-76492007-081

Searle, M. P., Waters, D. J., Rex, D. C., & Wilson, R. N. (1992). Pressure, temperature and time constraints on Himalayan metamorphism from eastern Kashmir and western Zanskar. *Journal of the Geological Society*, *149*, 753–773. https://doi.org/10.1144/gsjgs.149.5.0753

Sen, K., Tripathi, K., & Dubey, A. K. (2013). Is the North Indian continental margin a Palaeo-proterozoic magmatic arc? Insights from magnetomineralogy and geochemistry of the Wangtu Gneissic Complex, Himachal Lesser Himalaya. *Current Science*, *104*, 1527–1533.

Shah, S. I. (2009). Stratigraphy of Pakistan. Government of Pakistan Ministry of Petroleum and Natural Resources. *Geological Survey of Pakistan*.

Shellnutt, J. G., Bhat, G. M., Wang, K.-L., Yeh, M.-W., Brookfield, M. E., & Jahn, B.-M. (2015). Multiple mantle sources of the early Permian Panjal Traps, Kashmir, India. *American Journal of Science*, *315*, 589–619. https://doi.org/10.2475/07.2015.01

Shen, T., Wang, G., Leloup, P. H., van der Beek, P., Bernet, M., Cao, K., et al. (2016). Controls on Cenozoic exhumation of the Tethyan Himalaya from fission-track thermochronology and detrital zircon U-Pb geochronology in the Gyirong Basin area, southern Tibet. *Tectonics*, *35*, 1713–1734. https://doi.org/10.1002/2016TC004149

Shrestha, S., Larson, K. P., Martin, A. J., Guilmette, C., Smit, M. A., & Cottle, J. M. (2020). The Greater Himalayan Thrust Belt: Insight into the assembly of the exhumed Himalayan metamorphic core, Modi Khola Valley, central Nepal. *Tectonics*, *39*, e2020TC006252. https://doi.org/10.1029/2020TC006252

Sinclair, H. D., & Jaffey, N. (2001). Sedimentology of the Indus Group, Ladakh, northern India: Implications for the timing of initiation of the palaeo-Indus River. *Journal of the Geological Society*, *158*, 151–162. https://doi.org/10.1144/jgs.158.1.151

Singh, P., Singhal, S., & Das, A. N. (2020). U-Pb (zircon) geochronologic constraint on tectono-magmatic evolution of Chaur granitoid complex (CGC) of Himachal Himalaya, NW India: Implications for the Neoproterozoic magmatism related to Grenvillian orogeny and assembly of the Rodinia supercontinent. *International Journal of Earth Sciences*, *109*, 373–390. https://doi.org/10.1007/s00531-019-01808-5

Singh, S. K., Trivedi, J. R., & Krishnaswami, S. (1999). Re-Os isotope systematics in black shales from the Lesser Himalaya: Their chronology and role in the 1870s/1880s evolution of seawater. *Geochimica et Cosmochimica Acta*, *63*, 2381–2392. https://doi.org/10.1016/S0016-7037(99)00201-X

Smit, M. A., Hacker, B. R., & Lee, J. (2014). Tibetan garnet records early Eocene initiation of thickening in the Himalaya. *Geology*, *42*, 591–594. https://doi.org/10.1130/G35524.1

Soucy La Roche, R., Godin, L., Cottle, J. M., & Kellett, D. A. (2016). Direct shear fabric dating constrains early Oligocene onset of the South Tibetan detachment in the western Nepal Himalaya. *Geology*, *44*, 403–406. https://doi.org/10.1130/G37754.1

Spencer, C. J., Dyck, B., Mottram, C. M., Roberts, N. M. W., Yao, W. -H., & Martin, E. L. (2019). Deconvolving the pre-Himalayan Indian margin: Tales of crustal growth and destruction. *Geoscience Frontiers*, *10*, 863–872. https://doi.org/10.1016/j.gsf.2018.02.007

Spencer, C. J., Harris, R. A., & Dorais, M. J. (2012). Depositional provenance of the Himalayan metamorphic core of Garhwal region, India: Constrained by U-Pb and Hf isotopes in zircons. *Gondwana Research*, *22*, 26–35. https://doi.org/10.1016/j.gr.2011.10.004

Spring, L., Bussy, F., Vannay, J.-C., Huon, S., & Cosca, M. A. (1993). Early Permian granitic dykes of alkaline affinity in the Indian High Himalaya of Upper Lahul and SE Zanskar: geochemical characterization and geotectonic implications. In P. J. Treloar & M. Searle (Eds.), *Himalayan tectonics* (pp. 251–264). London, Geological Society, Special Publication, 74. https://doi.org/10.1144/GSL.SP.1993.074.01.18

Srivastava, P., & Mitra, G. (1994). Thrust geometries and deep structure of the outer and lesser Himalaya, Kumaon and Garhwal (India): Implications for evolution of the Himalayan fold-and-thrust belt. *Tectonics*, *13*, 89–109. https://doi.org/10.1029/93TC01130

Starnes, J. K., Long, S. P., Gordon, S. M., Zhang, J., & Soignard, E. (2020). Using quartz fabric intensity parameters to delineate strain patterns across the Himalayan Main Central thrust. *Journal of Structural Geology*, *131*, 103941. https://doi.org/10.1016/j.jsg.2019.103941

Stern, R. (2020). The Mesoproterozoic single-lid tectonic episode: Prelude to modern plate tectonics. *GSA Today*, *30*, 4–10. https://doi.org/10.1130/GSATG480A.1

Stickroth, S. F., Carrapa, B., DeCelles, P. G., Gehrels, G. E., & Thomson, S. N. (2019). Tracking the growth of the Himalayan fold-and-thrust belt from lower Miocene foreland basin strata: Dumri Formation, western Nepal. *Tectonics*, *38*, 3765–3793. https://doi.org/10.1029/2018TC005390

Stocklin, J. (1980). Geology of Nepal and its regional frame: Thirty-third William Smith lecture. *Journal of the Geological Society*, *137*, 1–34. https://doi.org/10.1144/gsjgs.137.1.0001

St-Onge, M. R., Rayner, N., Palin, R. M., Searle, M. P., & Waters, D. J. (2013). Integrated pressure-temperature-time constraints for the Tso Morari dome (northwest India): Implications for the burial and exhumation path of UHP units in the western Himalaya. *Journal of Metamorphic Geology*, *31*, 469–504. https://doi.org/10.1111/jmg.12030

Strachey, R. (1851). On the geology of part of the Himalaya Mountains and Tibet. *Quarterly Journal of the Geological Society*, *7*, 292–310. https://doi.org/10.1144/GSL.JGS.1851.007.01-02.54

Thiede, R. C., Arrowsmith, J. R., Bookhagen, B., McWilliams, M., Sobel, E. R., & Strecker, M. R. (2006). Dome formation and extension in the Tethyan Himalaya, Leo Pargil, northwest India. *Geological Society of America Bulletin*, *118*, 635–650. https://doi.org/10.1130/B25872.1

Tonarini, S., Villa, I. M., Oberli, F., Meier, M., Spencer, D. A., Pognante, U., et al. (1993). Eocene age of eclogite metamorphism in Pakistan Himalaya: Implications for India-Eurasia collision. *Terra Nova*, *5*, 13–20. https://doi.org/10.1111/j.1365-3121.1993.tb00221.x

Torvela, T. (2017). How (not) to recognize a midcrustal channel from outcrop patterns. In R. D. Law et al. (Eds.), *Linkages and feedbacks in orogenic systems* (pp. 129–148). Boulder, Geological Society of America, Memoir, 213. https://doi.org/10.1130/2017.1213(07)

Treloar, P. J. (1997). Thermal controls on early-Tertiary, short-lived, rapid regional metamorphism in the NW Himalaya, Pakistan. *Tectonophysics*, *273*, 77–104. https://doi.org/10.1016/S0040-1951(96)00289-2

Treloar, P. J., & Rex, D. C. (1990). Cooling and uplift histories of the crystalline thrust stack of the Indian Plate internal zones west of Nanga Parbat, Pakistan Himalaya. *Tectonophysics*, *180*, 323–349. https://doi.org/10.1016/0040-1951(90)90317-2

Treloar, P. J., O'Brien, P. J., Parrish, R. R., & Khan, M. A. (2003). Exhumation of early Tertiary, coesite-bearing eclogites from the Pakistan Himalaya. *Journal of the Geological Society*, *160*, 367–376. https://doi.org/10.1144/0016-764902-07

Treloar, P. J., Palin, R. M., & Searle, M. P. (2019). Towards resolving the metamorphic enigma of the Indian Plate in the NW Himalaya of Pakistan. In P. J. Treloar, & M. Searle (Eds.), *Himalayan tectonics: A modern synthesis* (pp. 255–279). London, Geological Society, Special Publication, 483. https://doi.org/10.1144/SP483-2019-22

Treloar, P. J., Rex, D. C., & Williams, M. P. (1991). The role of erosion and extension in unroofing the Indian Plate thrust stack, Pakistan Himalaya. *Geological Magazine*, *128*, 465–478. https://doi.org/10.1017/S0016756800018628

van Hinsbergen, D. J. J., Kapp, P., Dupont-Nivet, G., Lippert, P. C., DeCelles, P. G., & Torsvik, T. H. (2011). Restoration of Cenozoic deformation in Asia and the size of Greater India. *Tectonics*, *30*, TC5003. https://doi.org/10.1029/2011TC002908

van Hinsbergen, D. J. J., Lippert, P. C., Dupont-Nivet, G., McQuarrie, N., Doubrovine, P. V., Spakman, W., et al. (2012). Greater India Basin hypothesis and a two-stage Cenozoic collision between India and Asia. *Proceedings of the National Academy of Sciences*, *109*, 7659–7664. https://doi.org/10.1073/pnas.1117262109

van Hinsbergen, D. J. J., Lippert, P. C., Li, S., Huang, W., Advokaat, E. L., & Spakman, W. (2019). Reconstructing Greater India: Paleogeographic, kinematic, and geodynamic perspectives. *Tectonophysics*, *760*, 69–94. https://doi.org/10.1016/j.tecto.2018.04.006

Vannay, J. C., & Hodges, K. V. (1996). Tectonometamorphic evolution of the Himalayan metamorphic core between the Annapurna and Dhaulagiri, central Nepal. *Journal of Metamorphic Geology*, *14*, 635–656. https://doi.org/10.1046/j.1525-1314.1996.00426.x

Vannay, J. C., & Spring, L. (1993). Geochemistry of the continental basalts within the Tethyan Himalaya of Lahul-Spiti & SE Zanskar, northwest India. In P. J. Treloar & M. Searle (Eds.), *Himalayan tectonics* (pp. 237–249). London, Geological Society, Special Publication, 74. https://doi.org/10.1144/GSL.SP.1993.074.01.17

Vannay, J. C., Grasemann, B., Rahn, M., Frank, W., Carter, A., Baudraz, V., et al. (2004). Miocene to Holocene exhumation of metamorphic crustal wedges in the NW Himalaya: evidence for tectonic extrusion coupled to fluvial erosion. *Tectonics*, *23*, TC1014. https://doi.org/10.1029/2002TC001429

Veevers, J. J., Powell, C., & Johnson, B. D. (1975). Greater India's place in Gondwanaland and in Asia. *Earth and Planetary Science Letters*, *27*, 383–387. https://doi.org/10.1016/0012-821X(75)90056-4

Vince, K. J., & Treloar, P. J. (1996). Miocene, north-vergent extensional displacements along the Main Mantle Thrust, NW Himalaya, Pakistan. *Journal of the Geological Society*, *153*, 677–680. https://doi.org/10.1144/gsjgs.153.5.0677

Visona, D., Carosi, R., Montomoli, C., Tiepolo, M., & Peruzzo, L. (2012). Miocene andalusite leucogranite in central-east Himalaya (Everest–Masang Kang area): Low-pressure melting during heating. *Lithos*, *144–145*, 194–208. https://doi.org/10.1016/j.lithos.2012.04.012

Wang, H.-Q., Ding, L., Cai, F.-L., Xu, Q., Li, S., Fu, J.-J., et al. (2017). Early Tertiary deformation of the Zhongba-Gyangze Thrust in central southern Tibet. *Gondwana Research*, *41*, 235–248. https://doi.org/10.1016/j.gr.2015.02.017

Wang, J., Hu, X., Jansa, L., & Huang, Z. (2011). Provenance of the Upper Cretaceous-Eocene deep-water sandstones in Sangdanlin, Southern Tibet: Constraints on the timing of initial India-Asia collision. *Journal of Geology*, *119*, 293–309. https://doi.org/10.1086/659145

Wang, J.-M., Zhang, J.-J., Liu, K., Zhang, B., Wang, X.-X., Rai, S., et al. (2016). Spatial and temporal evolution of tectonometamorphic discontinuities in the central Himalaya: Constraints from P-T paths and geochronology. *Tectonophysics*, *679*, 41–60. https://doi.org/10.1016/j.tecto.2016.04.035

Waters, D. J. (2019). Metamorphic constraints on the tectonic evolution of the High Himalaya in Nepal: The art of the possible. In P. J. Treloar & M. Searle (Eds.), *Himalayan tectonics: A modern synthesis* (pp. 325–375). London, Geological Society, Special Publication, 483. https://doi.org/10.1144/SP483-2018-187

Webb, A. A. G. (2013). Preliminary balanced palinspastic reconstruction of Cenozoic deformation across the Himachal Himalaya (northwestern India). *Geosphere*, *9*, 572–587. https://doi.org/10.1130/GES00787.1

Webb, A. A. G., Guo, H., Clift, P. D., Husson, L., Muller, T., Costantino, D., et al. (2017). The Himalaya in 3D: Slab dynamics controlled mountain building and monsoon intensification. *Lithosphere*, *9*, 637–651. https://doi.org/10.1130/L636.1

Webb, A. A. G., Schmitt, A. K., He, D., & Weigand, E. L. (2011b). Structural and geochronological evidence for the leading edge of the Greater Himalayan Crystalline complex in the central Nepal Himalaya. *Earth and Planetary Science Letters*, *304*, 483–495. https://doi.org/10.1016/j.epsl.2011.02.024

Webb, A. A. G., Yin, A., & Dubey, C. S. (2013). U-Pb zircon geochronology of major lithologic units in the eastern Himalaya: Implications for the origin and assembly of Himalayan rocks. *Geological Society of America Bulletin*, *125*, 499–522. https://doi.org/10.1130/B30626.1

Webb, A. A. G., Yin, A., Harrison, T. M., Celerier, J., Gehrels, G. E., Manning, C. E., et al. (2011a). Cenozoic tectonic history of the Himachal Himalaya (northwestern India) and its constraints on the formation mechanism of the Himalayan orogen. *Geosphere*, *7*, 1013–1061. https://doi.org/10.1130/GES00627.1

Weinberg, R. F. (2016). Himalayan leucogranites and migmatites: nature, timing and duration of anatexis. *Journal of Metamorphic Geology*, *34*, 821–843. https://doi.org/10.1111/jmg.12204

Wiesmayr, G., & Grasemann, B. (2002). Eohimalayan fold and thrust belt: Implications for the geodynamic evolution of the NW-Himalaya (India). *Tectonics*, *21*, 1058. https://doi.org/10.1029/2002TC001363

Wilke, F. D. H., O'Brien, P. J., Gerdes, A., Timmerman, M. J., Sudo, M., & Khan, M. A. (2010). The multistage exhumation history of the Kaghan Valley UHP series, NW Himalaya, Pakistan from U-Pb and 40Ar/39Ar ages. *European Journal of Mineralogy*, *22*, 703–719. https://doi.org/10.1127/0935-1221/2010/0022-2051

Willems, H., Zhou, Z., Zhang, B., & Grafe, K.-U. (1996). Stratigraphy of the Upper Cretaceous and Lower Tertiary strata in the Tethyan Himalayas of Tibet (Tingri area, China). *Geologische Rundschau*, *85*, 723–754. https://doi.org/10.1007/BF02440107

Wu, F.-Y., Ji, W.-Q., Wang, J.-G., Liu, C.-Z., Chung, S.-L., & Clift, P. D. (2014). Zircon U-Pb and Hf isotopic constraints on the onset time of India-Asia collision. *American Journal of Science*, *314*, 548–579. https://doi.org/10.2475/02.2014.04

Wynne, A. B. (1878). On the geology of the Salt Range in the Punjab. *Memoirs of the Geological Survey of India*, *14*, 1–313.

Xu, Y., Cawood, P. A., Du, Y., Huang, H., & Wang, X. (2014a). Early Paleozoic orogenesis along Gondwana's northern margin constrained by provenance data from South China. *Tectonophysics*, *636*, 40–51. https://doi.org/10.1016/j.tecto.2014.08.022

Xu, Y., Cawood, P. A., Du, Y., Zhong, Z., & Hughes, N. C. (2014b). Terminal suturing of Gondwana along the southern margin of South China Craton: Evidence from detrital zircon U-Pb ages and Hf isotopes in Cambrian and Ordovician strata, Hainan Island. *Tectonics*, *33*, 2490–2504. https://doi.org/10.1002/2014TC003748

Yakymchuk, C., & Godin, L. (2012). Coupled role of deformation and metamorphism in the construction of inverted metamorphic sequences: An example from far-northwest Nepal. *Journal of Metamorphic Geology*, *30*, 513–535. https://doi.org/10.1111/j.1525-1314.2012.00979.x

Yin, A., & Harrison, T. M. (2000). Geologic evolution of the Himalayan-Tibetan orogen. *Annual Review of Earth and Planetary Sciences*, *28*, 211–280. https://doi.org/10.1146/annurev.earth.28.1.211

Yin, A., Dubey, C. S., Kelty, T. K., Webb, A. A. G., Harrison, T. M., Chou, C. Y., et al. (2010). Geologic correlation of the Himalayan orogen and Indian craton: Part 2. Structural geology, geochronology, and tectonic evolution of the Eastern Himalaya. *Geological Society of America Bulletin*, *122*, 360–395. https://doi.org/10.1130/B26461.1

Yin, A., Harrison, T. M., Murphy, M. A., Grove, M., Nie, S., Ryerson, F. J., et al. (1999). Tertiary deformation history of southeastern and southwestern Tibet during the Indo-Asian collision. *Geological Society of America Bulletin*, *111*, 1644–1664. https://doi.org/10.1130/0016-7606(1999)111<1644:TDHOSA>2.3.CO;2

Yin, A., Harrison, T. M., Ryerson, F. J., Chen, W., Kidd, W. S. F., & Copeland, P. (1994). Tertiary structural evolution of the Gangdese Thrust System, southeastern Tibet. *Journal of Geophysical Research: Solid Earth*, *99*, 18175–18201. https://doi.org/10.1029/94JB00504

Zeitler, P. K., & Chamberlain, C. P. (1991). Petrogenetic and tectonic significance of young leucogranites from the northwestern Himalaya, Pakistan. *Tectonics*, *10*, 729–741. https://doi.org/10.1029/91TC00168

Zhang, J., & Guo, L. (2007). Structure and geochronology of the southern Xainza-Dinggye rift and its relationship to the south Tibetan detachment system. *Journal of Asian Earth Sciences*, *29*, 722–736. https://doi.org/10.1016/j.jseaes.2006.05.003

Zhang, L., et al. (2020). Activity of the South Tibetan detachment system: Constraints from leucogranite ages in the eastern Himalayas. *Geological Journal*, *55*, 5540–5573. https://doi.org/10.1002/gj.3756

Zhang, Q., Willems, H., Ding, L., Grafe, K.-U., & Appel, E. (2012). Initial India-Asia continental collision and foreland basin evolution in the Tethyan Himalaya of Tibet: Evidence from stratigraphy and paleontology. *The Journal of Geology*, *120*, 175–189. https://doi.org/10.1086/663876

Zhang, R., Murphy, M. A., Lapen, T. J., & Heizler, M. (2011). Late Eocene crustal thickening followed by Early-Late Oligocene extension along the India-Asia suture zone: Evidence for cyclicity in the Himalayan orogen. *Geosphere*, *7*, 1249–1268. https://doi.org/10.1130/GES00643.1

Zhao, G., Sun, M., Wilde, S. A., & Li, S. (2004). A Paleo-Mesoproterozoic supercontinent: Assembly, growth and breakup. *Earth-Science Reviews*, *67*, 91–123. https://doi.org/10.1016/j.earscirev.2004.02.003

Zhao, G., Wang, Y., Huang, B., Dong, Y., Li, S., Zhang, G., et al. (2018). Geological reconstructions of the East Asian blocks: From the breakup of Rodinia to the assembly of Pangea. *Earth-Science Reviews*, *186*, 262–286. https://doi.org/10.1016/j.earscirev.2018.10.003

Zhu, D., Mo, X., Pan, G., Zhao, Z., Dong, G., Shi, Y., et al. (2008). Petrogenesis of the earliest Early Cretaceous mafic rocks from the Cona area of the eastern Tethyan Himalaya in south Tibet: Interaction between the incubating Kerguelen plume and the eastern Greater India lithosphere? *Lithos*, *100*, 147–173. https://doi.org/10.1016/j.lithos.2007.06.024

6

Records of Himalayan Metamorphism and Contractional Tectonics in the Central Himalayas (Darondi Khola, Nepal)

Elizabeth J. Catlos

ABSTRACT

The Himalayan orogen exposes assemblages from low-grade Indian shelf sediments of the Tethyan Formation to eclogite and ultra-high-pressure rocks from the suture zone between the Indian craton and Asian subcontinent. Barrovian-grade pelites in the Himalayan core comprise the Greater Himalayan Crystallines and Lesser Himalayan formations. These units are separated by the Main Central Thrust (MCT), which accommodated a significant amount of convergence. We describe and apply isopleth thermobarometry and high-resolution pressure-temperature (P-T) path modeling to decipher the metamorphic history of garnet-bearing rocks collected across the MCT in central Nepal. Results are compared to with previous reports of conventional rim P-T conditions and P-T paths that used Gibb's method on the same data and assemblages. Predictions of the paths on garnet zoning are also presented for the high-resolution P-T path modeling and Gibb's method using the program TheriaG. Although the approaches yield different absolute conditions and P-T path shapes, all are consistent with the development of the MCT shear zone due to the imbrication of distinct rock packages. Greater Himalayan Crystalline garnets experienced higher grade conditions, making extracting its P-T conditions and paths challenging. Lesser Himalayan garnets appear to behave as closed systems and are ideally suited for thermodynamic approaches.

6.1. INTRODUCTION

Nineteenth-century geologists studying the Himalayas perceived an anomalous geologic relationship that contradicted two commonly accepted principles: the oldest rocks in a sedimentary succession are found at the base of the pile, and metamorphosed strata are older than unmetamorphosed. The Main Central Thrust (MCT), located at the base of the Himalayan break in slope, places lower-grade Lesser Himalayan Formation (LHF) metasediments beneath high-grade gneisses of the Greater Himalayan Crystallines (GHC) (Figs. 6.1 and 6.2). Exploration of the Himal Pradesh region in northern India surprised the pioneering geologists (Medicott, 1864; Oldham, 1883; Middlemiss, 1887) who saw the highest grade, and thus supposedly oldest, rocks form the tallest peaks:

> Now the belief which is at present so rapidly gaining ground that metamorphic strata are presumably older than unmetamorphosed strata makes one at first glance assume the strong probability in favor of the inner schistose series (Greater Himalayan Crystallines) being of much greater age than the outer zone of formations (Lesser Himalaya). But no sooner has this a priori probability obtained a firm hold of the mind than a rude shock is given to it by the discovery that at every point round the schistose area the Outer formations appear to dip towards and under the schistose series at steep angles (50°–60°), whilst the schistose series itself is disposed apparently. . .upon the top of the Outer Formations, and culminating in a capping of gneissose rock on the summit of Kalogarhi mountain, the highest point in the neighborhood. (Middlemiss, 1887, p. 34)

Jackson School of Geosciences, Department of Geological Sciences, The University of Texas at Austin, Austin, Texas, USA

Compressional Tectonics: Plate Convergence to Mountain Building, Geophysical Monograph 277, First Edition.
Edited by Elizabeth J. Catlos and İbrahim Çemen.

DOI:10.1002/9781119773856.ch06

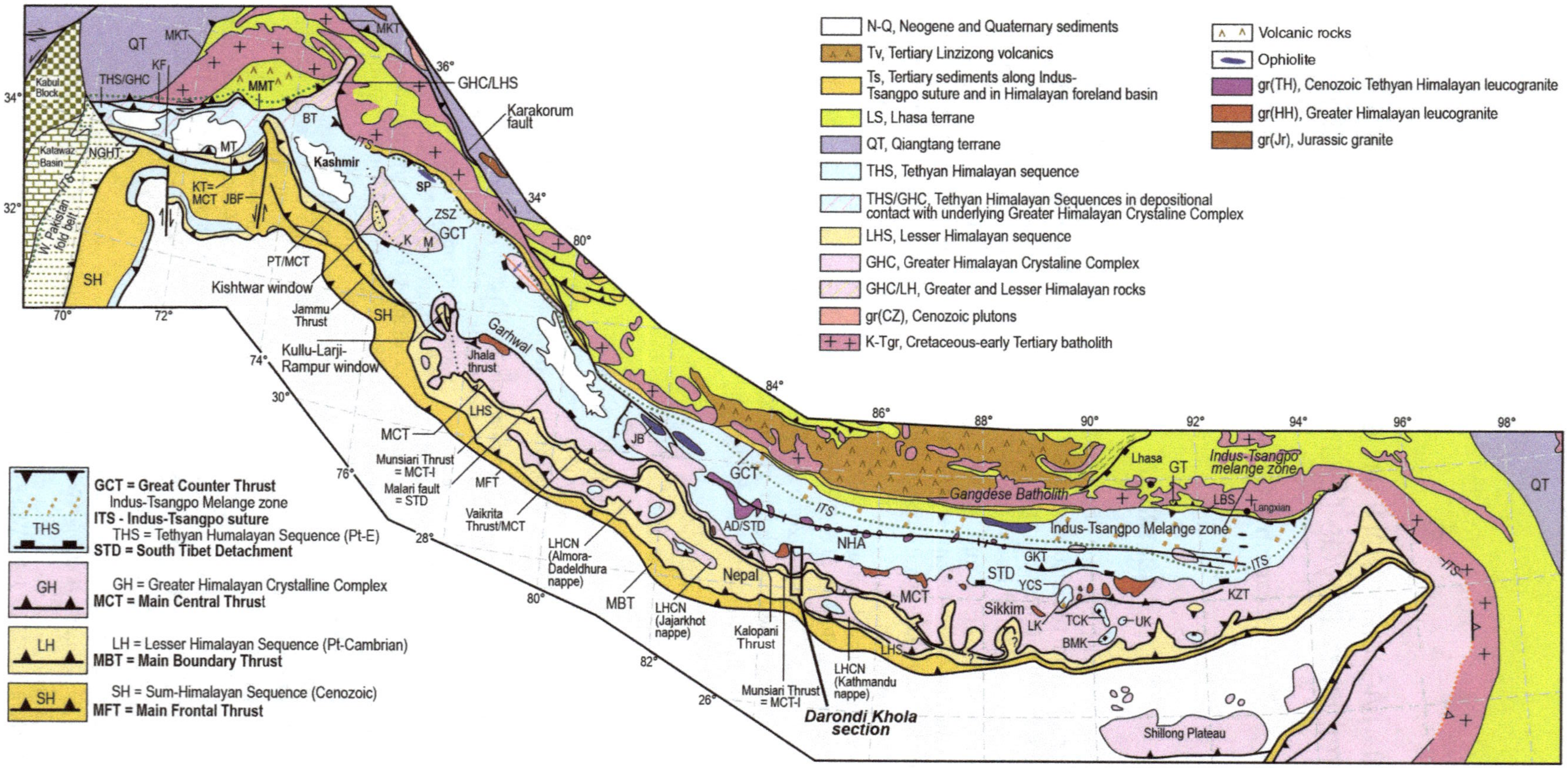

Figure 6.1 Geological map of the Himalayas after Yin (2006). See Figure 6.2 for a cross section through central Nepal and Figures 6.3, 6.4, and 6.5 for the sample transect taken across the MCT shear zone along the Darondi Khola. Abbreviations: Faults, AD - Annapurna detachment; BT - Batal thrust; GT -Gangdese Thrust; JBF - Jhelum-Balakot fault; K - Kilar shear zone; KZT - Kakhtang-Zimithang thrust; KF - Kohistan fault; KT - Khairabad thrust LHCN - Lesser Himalayan Crystalline Nappes; M - Miyar thrust; MKT - Main Karakorum Thrust; MMT - Main Mantle Thrust; MT - Mansehra thrust; NGHT - Nathia Gali-Hissartang thrust; NHA - North Himalayan antiform; PT - Panjal Thrust = MCT; YCS - Yadong cross-structure; ZSZ - Zanskar shear zone. Ophiolites: LBS - Luobusha, JB - Jungbwa, SP - Spongtang, XG - Xigaze.

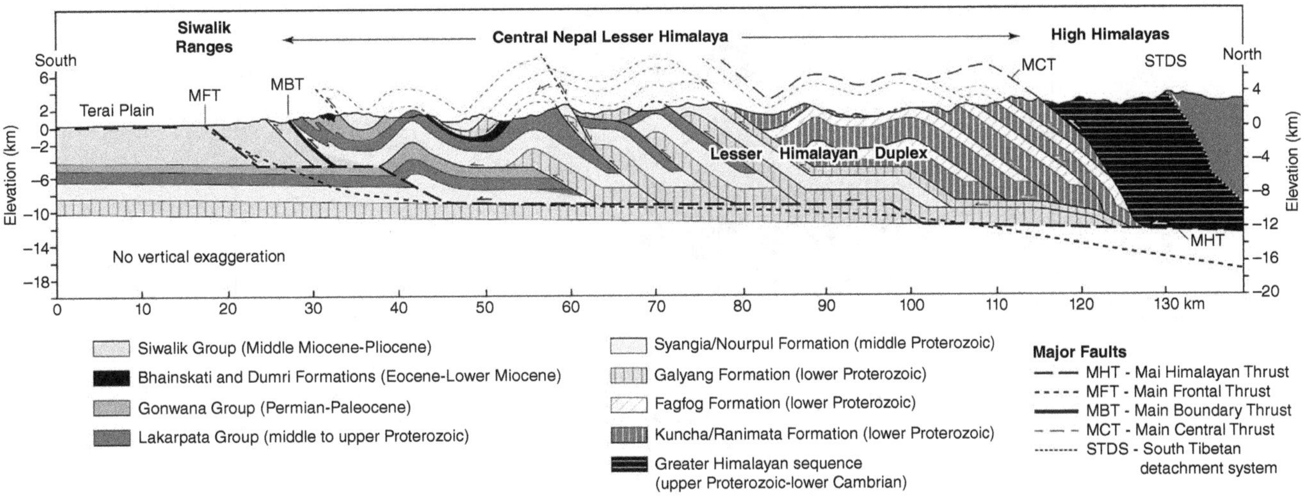

Figure 6.2 Generalized cross section through the Himalayas in central Nepal after DeCelles (2015) and Robinson et al. (2006). See additional cross sections in the range in DeCelles et al. (2020).

The recognition of widespread thrusting within the Himalayas was primarily based on the observations (von Loczy, 1907; Pilgrim & West, 1928; Auden, 1937; Heim & Gansser, 1939) and the orogen's "inverted metamorphism," an increase in metamorphic intensity toward higher structural levels, appeared resolved by invoking compressional/contractional tectonics and the thrusting of the GHC unit on top of the LHF. However, subsequent studies indicated that the kyanite isograd remained unbroken across the MCT, and its footwall was characterized by an inverted geotherm (Fig. 6.3; Ray, 1947; Gansser, 1964; Le Fort, 1975; Pêcher, 1989; England et al., 1992). The idea of Himalayan inverted metamorphism reemerged but was relocated to the MCT footwall. In other words, the observer after a hasty examination is almost driven to the conclusions that there is an upper metamorphic series lying normally upon the comparatively unmetamorphosed zone of Outer formations (a counterpart of the opinion long held with regard to the strata of the Scotch Highlands).

Inverted metamorphic gradients suggest the presence of wholly overturned strata or heat sources that counteract the influence of the asthenosphere (e.g., England & Molnar, 1993; Jamieson et al., 1996; Grasemann & Vannay, 1999; Kidder et al., 2013). The cause of the phenomenon has implications for establishing mechanisms of heat transfer within collisional belts and the role of heat sources, such as shear heating in fault zones, heat advection by magmas, radiogenic heating, and asthenospheric input. Inverted metamorphism has long been associated with areas of extensive thrust faulting where heat is thought to flow from a hot upper plate to a colder lower plate (Ernst, 1973; Graham & England, 1976; Spear et al., 1995). Some models of Himalayan orogenesis link the anomalous geothermal gradient spatially and temporally with motion along the MCT, whereas others suggest a juxtaposition of previously metamorphosed sequences (Searle & Rex, 1989; Harrison et al., 1999; Hodges, 2000; Dasgupta et al., 2004; Larson et al., 2015).

The MCT has accommodated a significant amount of Indo-Asian convergence (e.g., Schelling & Arita, 1991; Yin & Harrison, 2000; Yin, 2006; Tobgay et al., 2012; Roberts et al., 2020), but other large-scale structures, including crustal-scale strike-slip faults to the north, active thrusts to the south, and the Main Himalayan Thrust (MHT) decollement compete for strain accommodation (Fig. 6.2). The MCT has long been considered quiescent (Ye et al., 1981; Ni & Barazangi, 1984; Schelling & Arita, 1991; England et al., 1992). However, some earthquakes, like the 2015 Nepal (Gorkha) and 1991 Uttarkashi (Garhwal) events, challenge the notion, as their epicenters and epicenters of their aftershocks are modeled to zones within the LHF (Thakur & Kumar, 1994; Jain & Chander, 1995; Kayal, 1996; Gupta et al., 2015; Bai et al., 2016; Catlos et al., 2020). The MCT has also been linked to the generation of numerous geologic elements that characterize the Himalayan range, including hanging wall anatectic granitoids and crustal-scale extension found in association with the GHC (e.g., Burg et al., 1984; Valdiya, 1988; Burchfiel et al., 1992; Harrison et al., 1997; Kawakami et al., 2019).

Studies of the MCT hanging wall indicate the unit has discontinuities possibly related to internal structures (e.g., western Nepal: Carosi et al., 2010; Montomoli et al., 2013; Braden et al., 2017; central Nepal: Cottle et al., 2015; Wang et al., 2013, 2016; Rapa et al., 2018; Larson et al., 2015; Sikkim: Chakraborty et al., 2019; NW India: Iaccarino et al., 2020; Benetti et al., 2021; see discussions in Mukherjee et al., 2012; Larson et al., 2013; Montomoli et al., 2015). The nature of these cryptic discontinuities is

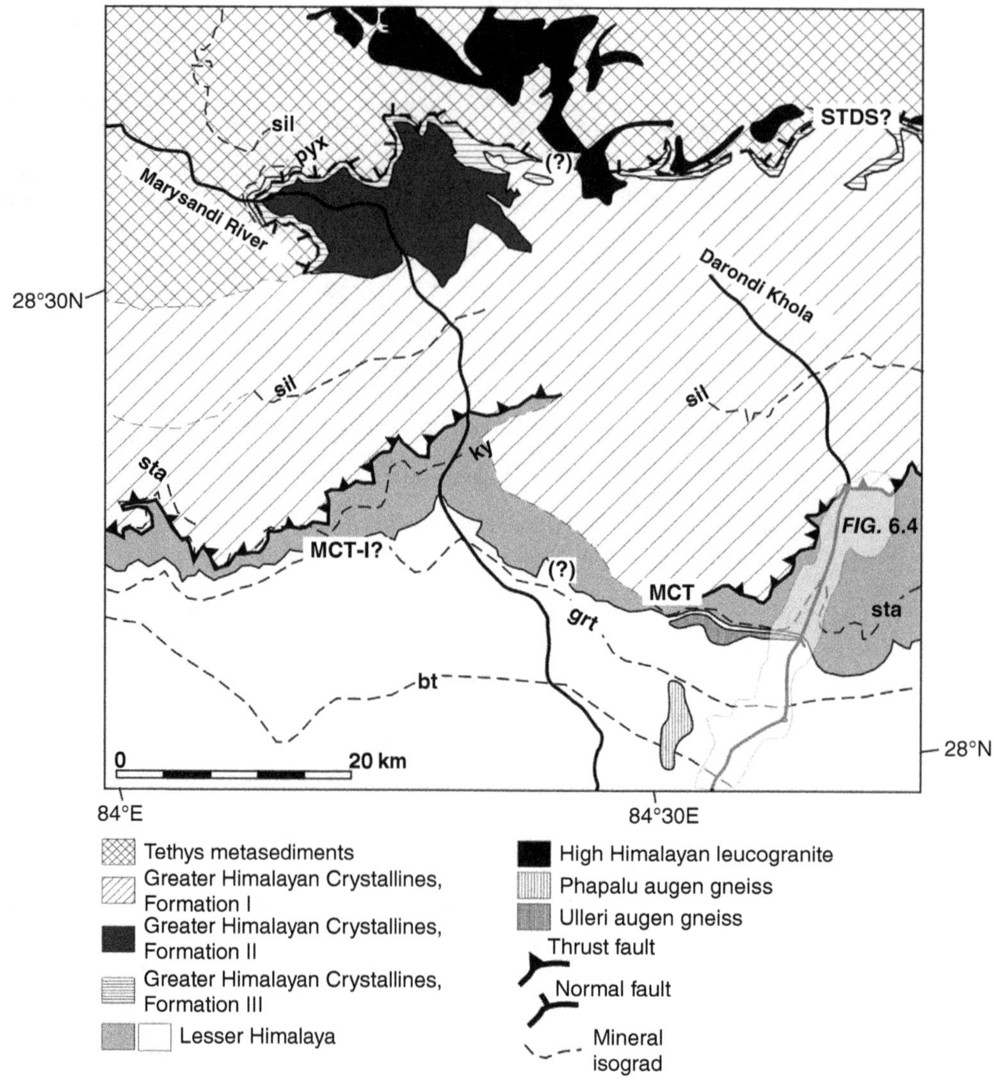

Figure 6.3 Generalized geological map of the Annapurna-Manaslu-Ganesh region of central Nepal after Colchen et al. (1980). Isograds are dashed and labeled: bt = biotite, grt = garnet; ky = kyanite; sta = staurolite; sil = sillimanite; pyx = pyroxene, carbonate lithologies; STDS? = the presence of the South Tibetan Detachment is debated; MCT= Main Central Thrust; MCT-I?= the presence of Arita's (1983) thrust at the base of the MCT shear zone is debated.

unclear and could be due to unmapped faults, shear zones, or inheritance of preexisting basement structures (e.g., Cottle et al., 2015). Their presence enhances the complexity of the Himalayan orogenic system as their activity could have significantly disrupted geotherms within the GHC core during its exhumation. This scenario is not accounted for in the channel flow model for the extrusion of the Himalayan core (e.g., Benetti et al., 2021; Maiti & Mandal, 2021). Understanding when Himalayan fault systems were active is critical for deciphering the processes involved during convergence.

The metamorphic history of the Himalayas has been the focus of sustained attention for almost 75 yr (e.g., Ray, 1947). Garnet-bearing assemblages have long been used to test hypotheses proposed for the origin of MCT inverted metamorphism, understand the slip history of the MCT, and the dynamics of Himalayan convergence. The focus includes generating the peak pressure-temperature (P-T) conditions and paths that rocks followed as they were metamorphosed during the Indo-Asia collision (e.g., Brunel & Kienast, 1986; Hodges et al., 1988; Hodges & Silverberg, 1988; Hubbard, 1989; Inger & Harris, 1992; Hodges et al., 1993; Metcalfe, 1993; Pognante & Benna, 1993; Kaneko, 1995; Macfarlane, 1995; Coleman, 1996; Vannay & Hodges, 1996; Vannay & Grasemann, 1998; Manickavasagam et al., 1999; Catlos et al., 2001; Kohn et al., 2001; Kohn, 2008; Phukon et al., 2019; Waters, 2019; Iaccarino et al., 2020; Catlos et al., 2018, 2020).

As a universal outcome, models for the development of the Himalayas predict the P-T paths that rocks follow as they track the conditions they experience during displacement. Common approaches to generating Himalayan P-T paths have included connecting peak metamorphic conditions of individual rocks, inferring from mineral assemblages, pseudosections, or Gibb's method thermodynamic modeling. However, some rocks yield problematic P-T estimates based on (1) a lack of evidence of phases in equilibrium among phases, (2) the application of barometers to inappropriate (uncalibrated) mineral compositions, and (3) calculated conditions that appear at odds with observed mineral assemblages and structural data (e.g., Kohn & Spear, 2000; Kohn, 2008). As a result, P-T paths and absolute peak P-T conditions may not diagnose the tectonic processes involved (e.g., Gervais & Brown, 2011). In addition, low-resolution P-T paths can be limited in their ability to test ideas regarding lithospheric response to perturbations, including motion within fault zones. However, this information can supplement other data, such as the timing of deformation or strain recorded in microstructures (see Kohn, 2016; Rolfo et al., 2014).

This chapter has two goals. The first is to review the geological framework of the Himalayas with a focus on the protolith of its metamorphosed assemblages. The second is to describe and apply particular thermobarometric approaches to decipher the metamorphic history of garnet-bearing rocks collected from the central portion of the range across the MCT using data published in the literature (Darondi Khola, Figs. 6.3–6.5; Kohn et al., 2001). Various approaches are available to obtain P-T and time garnet growth (P-T-t data; see review by Waters, 2019). The Darondi Khola case study compares conventional and isopleth thermobarometry in terms of

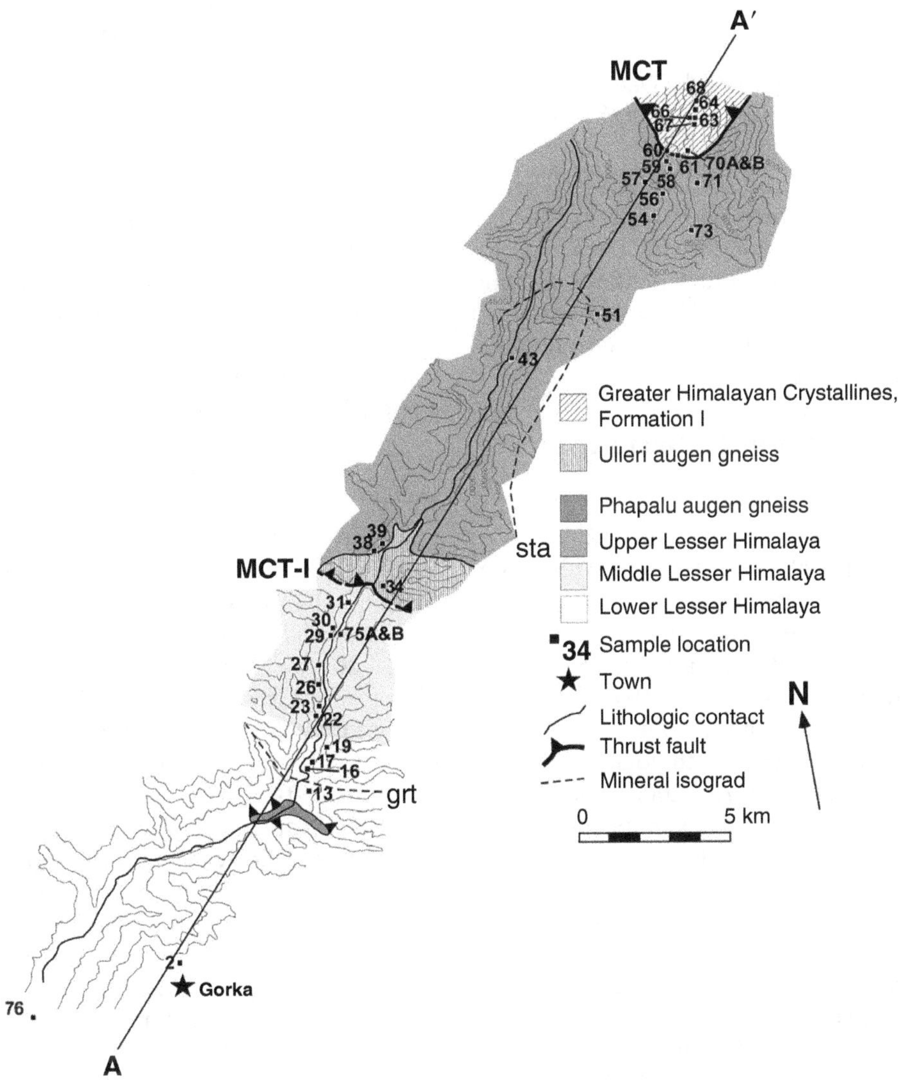

Figure 6.4 Sample location map from rocks collected along the Darondi Khola. The contour interval is 500 ft. See Figure 6.3 for the location of this transect on the geological map of the Himalayas. Samples are indicated by "DH#" in the text. Isograds are dashed and labeled (grt = garnet; stau = staurolite). See Figure 6.5 for a cross section along A-A′.

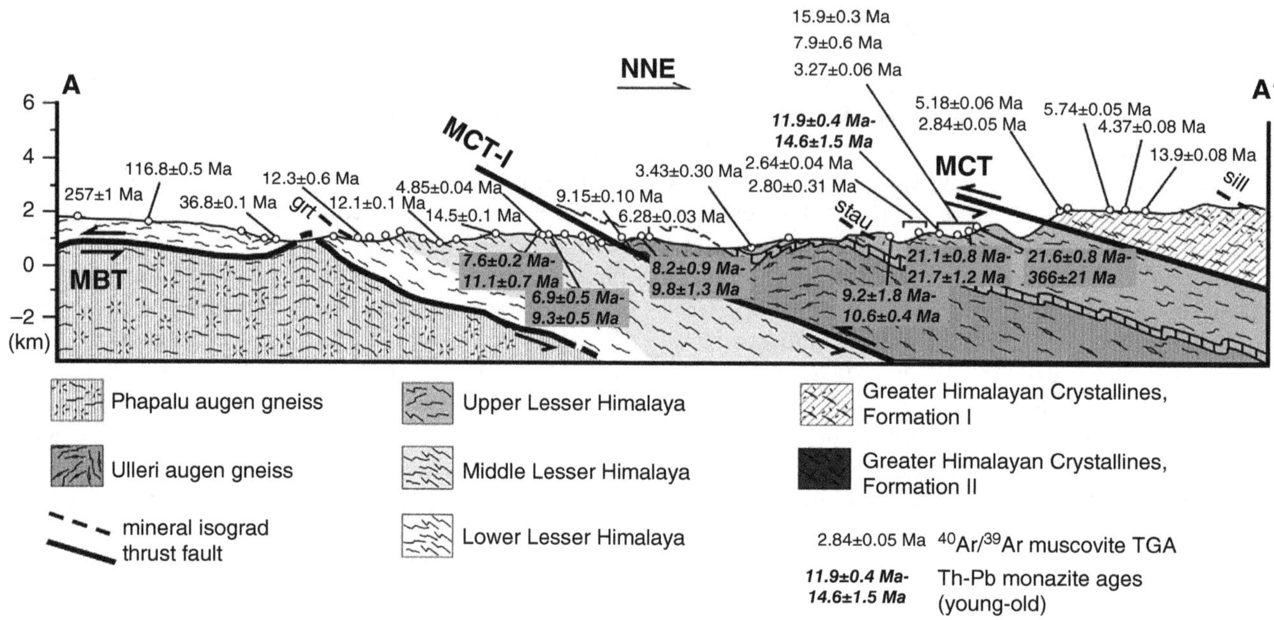

Figure 6.5 Cross section across the Darondi Khola section showing available muscovite $^{40}Ar/^{39}Ar$ and monazite Th-Pb ages after Catlos et al. (2001). See Figure 6.4 for line of section.

their outcomes and insights and presents new P-T paths from metamorphosed garnet-bearing rocks using previously collected data. Garnet-bearing pelitic assemblages exposed in units across the Himalayas contain information regarding their history that can be extracted and applied to test models developed to decipher the crustal response during orogeny.

6.2. GEOLOGICAL BACKGROUND

The Himalayan arc extends ~2,400 km from Nanga Parbat (8,138 m) in the west to Namche Barwa (7,756 m) in the east (e.g., Le Fort, 1996; Figs. 6.1, 6.2). This region includes the independent kingdoms of Nepal and Bhutan and parts of Pakistan, India, and China. The orogen forms a sharp transition between the average ~5 km high, arid Tibetan plateau and the warmer, wetter Indian lowlands and is composed of roughly parallel, crustal-scale fault systems that bound distinctive lithologic units along strike (DiPietro & Pogue, 2004; Yin, 2006). These units have Indian affinity and experienced variable degrees of metamorphism before their assembly with Asia.

6.2.1. Geological Framework Before the Collision

The Indian subcontinent initiated rifting from other continents and fragments associated with Gondwana during the Early Cretaceous (140–130 Ma; e.g., Scotese et al., 1988; Jadoul et al., 1998; Hu et al., 2010). Evidence for Early Cretaceous rift- and plume-related alkaline and basaltic volcanism exists within the LHF, which extends the entire length of the Himalayas, and is the oldest stratigraphically lowest unit (e.g., Sakai et al., 2013; Bhandari et al., 2019). The LHF is considered the MCT footwall and Main Boundary Thrust (MBT) sheet (Figs. 6.1 and 6.2; Dey et al., 2020). It is mainly composed of Paleoproterozoic Gondwana-associated sediments that experienced deposition and granite intrusion centered around 1,800 Ma (e.g., Trivedi et al., 1984; Tripathi & Singh, 1987; Parrish & Hodges, 1996; Miller et al., 2000; DeCelles et al., 2004; Kohn et al., 2010; Martin et al., 2011; McKenzie et al., 2011; Long et al., 2011; Sakai et al., 2013; Khanal et al., 2014; Mandal et al., 2016). The depositional environment is a passive-margin, shallow-water coastal, fluvial, volcanoclastic (e.g., Parrish & Hodges, 1996; Ahmad et al., 1999; Martin et al., 2011; Sakai et al., 2013; Bhandari et al., 2019). However, the lower section of the LHF may have formed in a continental arc on the northern margin of the Indian plate (Kohn et al., 2010; Mandal et al., 2016). In some locations, rift-related alkaline trachytic lava and pillow-bearing volcanic rocks are interbedded with pebble conglomerates and black shales, and signatures of significant asthenosphere upwelling based on geochemical proxies (e.g., Ahmad et al., 1999; Larson et al., 2019; Bhandari et al., 2019).

The stratigraphic classification of the Proterozoic succession of Lesser Himalaya is yet to be defined following the Code of Stratigraphic Nomenclature, and orogen-scale correlations along strike are hindered by uncertainties (Myrow et al., 2006; Long et al., 2011; Mandal et al., 2016; Dey et al., 2020). Sedimentation was long lasting, but a lack of fossiliferous assemblages makes correlating specific units problematic (Upreti, 1999; Martin et al., 2011; Long et al., 2011). Detrital zircon geochronology and stable isotopic analysis of suitable assemblages assist in this regard (e.g., Long et al., 2011; Martin et al., 2011; Sakai et al., 2013). A series of augen gneisses (the Melung Salleri or Phaplu Augen Gneiss of eastern Nepal and northwest India, the Ulleri of central Nepal, or Chainpurplay in western Nepal) distinguish different levels of the MCT shear zone and LHF stratigraphy (Figs. 6.3–6.5; Kohn et al., 2010; Dyck et al., 2019; Jharendra & Paudyal, 2019). These gneisses have ages from 2.2 Ga to 900 Ma (e.g., Le Fort & Rai, 1999; DeCelles et al., 2000; Catlos et al., 2002; Kohn et al., 2010) and define the base of the MCT shear zone in some interpretations. A Permian-Cambrian unconformity is recognized in northwest India, Nepal, and Bhutan (e.g., Bhargava et al., 2011; Martin et al., 2011; Long et al., 2011). Some have related its presence and Cambro-Ordovician granites found in GHC and LHF units to a pre-Tertiary orogeny (Gehrels et al., 2003, 2006; Cawood et al., 2007; Bhargava et al., 2011).

The GHC protolith is a clastic sedimentary sequence intruded by Cambro-Ordovician granitoids (e.g., Trivedi et al., 1984; Bhargava & Bassi, 1994; Parrish & Hodges, 1996; Upreti & Le Fort, 1999; DeCelles et al., 2004; Dyck et al., 2019). The depositional age is bracketed between the age of the youngest detrital zircons (~800–600 Ma) and granite intrusions (~500–460 Ma) (Ahmad et al., 2000; DeCelles et al., 2000; DeCelles et al., 2004; Martin et al., 2005; Dyck et al., 2019). Whole-rock Nd isotopes distinguish GHC [εNd(0) –19 to –12, average –16] from LHF affinities [εNd(0) –20 to –26, average –21.5] (Robinson et al., 2001; Martin et al., 2005). The GHC detrital zircons may have originated from the East African portion of the Pan-African orogeny (Arabian-Nubian Shield), uplifted during the Neoproterozoic (DeCelles et al., 2000, 2004), or from the late Mesoproterozoic terrains of Western Australia and East Antarctica (Circum-East Antarctic Orogen, Upreti & Yoshida, 2005; Yoshida & Upreti, 2006). The GHC has been modeled as unconformably deposited on the LHF (Parrish & Hodges, 1996) or tectonically juxtaposed (Upreti & Le Fort, 1999; DeCelles et al., 2000). The similarity in Cambrian ages between the LHF and GHC suggests they may be part of a shared depositional environment, with the LHF as proximal and the GHC as distal (Brookfield, 1993; Parish & Hodges, 1996; Corfield & Searle, 2000; Myrow et al., 2003; Myrow et al., 2006; Long et al., 2011). Alternatively, the GHC may have been a distinct basement unit separating the LHF from the Tethyan metasediments (Saxena, 1971; Aharon et al., 1987) or an exotic terrain involved in a pre-Himalayan collision (DeCelles et al., 2000; Gehrels et al., 2003).

Tethyan sedimentary and metasedimentary rocks are bounded to the north by the Indus Yarlung-Tsangpo suture zone (also Yarlung Zangbo Ophiolite Zone, Liu et al., 2010) (Gansser, 1964; Dewey & Bird, 1970; Burg et al., 1984; Yin et al., 1994; Quidelleur et al., 1997; Zhang et al., 2004; Yin, 2006; Zyabrev et al., 2008), and to the south by the GHC or South Tibetan Detachment System (STDS) (Figs. 6.1 and 6.2) (e.g., Makovsky et al., 1996; Wu et al., 1998; Searle, 2010; Long et al., 2017; Montomoli et al., 2017; Hughes et al., 2018; Kellett et al., 2018; Long et al., 2019). The Great Counter Thrust (or the Renbu-Zedong Thrust or Himalayan Backthrust) is mapped south of the zone and works to accommodate crustal shortening (Yin et al., 1999; Yin, 2006; Aikman et al., 2008). The Tethyan unit has long been considered a contiguous stratigraphic cover of the GHC (Bodenhausen et al., 1964; Bordet et al., 1971; Stöcklin, 1980; Garzanti & Pagni Frette, 1991; Brookfield, 1993; Liu & Einsele, 1994; Fuchs & Linner, 1995; Vannay & Steck, 1995; Garzanti, 1999; Dyck et al., 2019), and the suture zone marks the geological boundary separating rocks of Indian and Asian affinity (e.g., Gansser, 1981; Yin & Harrison, 2000). In northwest India, the Tethyan Formation may be in thrust contact with the LHF (Webb et al., 2007). The Tethyan Formation consists of Paleoproterozoic to Eocene Indian shelf sediments (marine, fossiliferous strata) interbedded with Paleozoic and Mesozoic volcanic assemblages (Bassoullet et al., 1980; Brookfield, 1993; Yin, 2006; Bhargava & Singh, 2020). The unit has been divided into four sequences: a Proterozoic to Devonian prerift, a Carboniferous–Lower Jurassic rift and postrift, a Jurassic–Cretaceous passive continental margin sequence, and an uppermost Cretaceous–Eocene syn-collision sequence (Liu & Einsele, 1994; Garzanti, 1999; Yin, 2006). Tethyan Formation lithostratigraphy changes along and perpendicular to the Himalayan orogeny (Brookfield, 1993; Yin, 2006). In some locations, the unit has undergone pre-Himalaya low-grade to greenschist facies metamorphism (e.g., Crouzet et al., 2007; Dunkl et al., 2011; Montomoli et al., 2017), and the latest Cretaceous to Paleocene sequence of the formation records the obduction of ophiolitic material (Allègre et al., 1984; Burg et al., 1987; Willems et al., 1996; Gnos et al., 1997; Makovsky et al., 1999; Aitchison et al., 2000; Ding et al., 2005).

6.2.2. Timing of Major Metamorphic Events and Fault Systems

Collision and Metamorphism in the Tethyan Formation

The Indian subcontinent moved over 60° latitude north toward Asia during the mid-Mesozoic to Eocene, closing the ancient Neo-Tethyan Ocean (e.g., Burg, 2011).

Remnants of Neo-Tethyan ophiolites are present across the Himalayas and provide information regarding the timing and processes involved during Indo-Asia collision and the nature of Neo-Tethyan Ocean crust and upper mantle (e.g., Hébert et al., 2012; Hu et al., 2016; Catlos et al., 2019). Based on data from these ophiolites, a Late Cretaceous intraoceanic arc has been suggested to be present within the Neo-Tethyan Ocean near the paleo-equator (e.g., Reuber, 1986; Abrajevitch et al., 2005; Metcalfe, 2009; Dai et al., 2011; Siddiqui et al., 2012; Siddiqui et al., 2017) or ~30°N (Zhu et al., 2013). The arc is recorded by the Samail (Oman), Zagros (Iran), Chagai-Raskoh (western Pakistan, southern Iran, eastern Afghanistan), Kandahar and Kohistan-Ladakh, Dazhuqu, and Zhongba arcs (Brookfield & Reynolds, 1981; Bhutani et al., 2004; Abrajevitch et al., 2005; Dai et al., 2011; Siddiqui et al., 2012; Baxter et al., 2016). Paleogeographic reconstructions of Lawver et al. (2018) restrict the location of the intraoceanic arc to the south of the Lhasa Terrain, although others suggest the Lhasa Terrain had already accreted onto other Tibetan-related continental fragments by the Late Jurassic–Early Cretaceous (Rolland et al., 2002; Kapp et al., 2003, 2007; Guynn et al., 2006; Zhu et al., 2013). Alternatively, the Lhasa Terrain accreted just before the final collisional event in the Paleocene (54.9 ± 2.3 Ma and 40.0 ± 3.3 Ma; Yang et al., 2015; 61 Ma and ~53–48 Ma; Yuan et al., 2020). Besides the Lhasa Terrain, multiple arcs may have been present (e.g., Zyabrev et al., 2008). The oldest portion of the Neo-Tethyan domain is the Late Triassic to Late Cretaceous (Sinha-Roy, 1982; Şengör & Atayman, 2009; Cao et al., 2018) and is sometimes termed the Ceno-Tethyan Ocean (Metcalfe, 1999; Matsuoka et al., 2002; Wakita & Metcalfe, 2005). The Yarlung-Tsangpo suture zone is mapped as the zone of the closure of the Ceno-Tethyan Ocean (e.g., Metcalfe, 1999, 2009, 2013).

The timing of initial collision varies along strike of the range but is often cited as during the Paleocene (Patriat & Achache, 1984; Klootwijk et al., 1992; Rowley, 1996; Yin & Harrison, 2000; Najman et al., 2001, 2002, 2003; Zhu et al., 2005; Ding et al., 2005; Yang et al., 2015; Hu et al., 2016). Much younger constraints are also suggested (e.g., Eocene/Oligocene boundary: Aitchison et al., 2007) and a division between a soft (Paleocene) and hard collision (25–20 Ma: van Hinsbergen et al. 2012; see review in Parsons et al., 2020). In the hard scenario of the Indo-Asia collision, the Tethyan unit represents the northern extension of the Indian subcontinent (Brookfield, 1993; Yin & Harrison, 2000; Myrow et al., 2003, Hughes et al., 2005; Myrow et al., 2009; Myrow et al., 2015; Hughes, 2016). Others suggest the Tethyan Formation may have been an independent terrain in the Mesozoic (DeCelles et al., 2000; see review in Yuan et al., 2020). Parsons et al. (2020) note that little progress has been made to resolve the differences between models of Indo-Asia collision. Gehrels et al. (2003) indicate that ascertaining the relative contributions of early Paleozoic versus Tertiary tectonism poses a significant challenge in understanding the Himalayan orogen. Determining the relevant model requires the metamorphic and timing history of active fault systems during the collision. The Great Counter Thrust, which bounds the upper portion of the Tethyan Formation along much of its strike (Fig. 6.1), is significantly younger than the Paleocene ages of the Indo-Asia collision. This structure shows activity primarily during the Miocene (20–13 Ma) across the western and central Himalayas (see review in Yin, 2006).

Compilations regarding the Tethyan Formation's metamorphic history collected along strike suggest the Tethyan sequences have experienced multiple (4–5) deformation events, although the timing of these episodes is poorly constrained (Aikman et al., 2008; Dunkl et al., 2011). Studies focusing on the low-grade history reveal Early Cretaceous precollisional metamorphism (e.g., Crouzet et al., 2007; Dunkl et al., 2011). Eocene greenschist to amphibolite facies metamorphism is recorded in portions of the unit and records conditions likely related to the onset of collision (580–600°C, 5–8 kbar; Dunkl et al., 2011; Catlos et al., 2020). Paleocene radiometric ages related to the collision are found in the Spongtang ophiolite (Fig. 6.1) (64.3 ± 0.8 Ma and 42.4 ± 0.5 Ma, zircon U-Pb ages; Catlos et al., 2019) and in a Tethyan Formation garnet (50.3 ± 0.6 Ma; Catlos et al., 2020). The collisional event is also recorded by eclogite, high-pressure (HP) and ultra-high-pressure (UHP) rocks documented within the Tethyan Formation near the suture zone (e.g., Guillot et al., 2008; Laskowski et al., 2016). In some locations, HP assemblages record multiple metamorphic stages (e.g., Chen et al., 2021). Oligo-Miocene anchi- to epizonal metamorphism and alteration are related to crustal shortening during this time, whereas rocks at the base of the unit record Miocene STDS motion and gneiss dome exhumation (e.g., Dunkl et al., 2011). Oligocene- to Miocene-age contact metamorphism marks the onset of the intrusion of some North Himalayan granitic bodies and gneiss domes (e.g., Guillot et al., 1995, 1999; Liu et al., 2016; Gao et al., 2016; Lihter et al., 2020). The onset of the north-dipping STDS and its associated imbrications (Fig. 6.1) is often constrained to the Miocene (Carosi et al., 1998, 1999a, 1999b; Searle, 1999; Sachan et al., 2010; Iaccarino et al., 2017; Long et al., 2017; Montomoli et al., 2017; Kellett et al., 2018; Iaccarino et al., 2020). However, at some locations, the GHC shows a transitional relationship with limestone of the Tethys sediments, and metamorphic grade remains unchanged across the STDS with stratigraphy and lithology excluding a distinct fault boundary (e.g., Bordet et al., 1975; Stöcklin, 1980; Fuchs et al., 1988; Schneider & Masch, 1993; Vannay & Steck, 1995; Cooper et al., 2012; Long et al., 2017). Long et al. (2019) suggest that extensional motion associated with the STDS in Bhutan occurs within both the GHC and Tethyan units.

The GHC Eohimalayan and Neohimalayan Events

The GHC is mainly composed of kyanite- to sillimanite-grade gneisses intruded by High Himalayan leucogranites in structurally higher levels (e.g., Upreti, 1999; Searle et al., 2006; Sachan et al., 2010; Wu et al., 2020). The GHC is generally divided into different units, although the assemblages that compose its sections differ along strike (Le Fort, 1975; Myrow et al., 2003; Yin, 2006). In central Nepal (Guillot, 1999), the upper Formation III consists of augen orthogneisses, whereas the middle Formation II are calc-silicate gneisses and marbles, and the basal Formation I is kyanite- and sillimanite-bearing metapelites, gneisses, and metagreywackes with abundant quartzite. The division of the package into three units mirrors the nomenclature ~250 km east (Lombardo et al., 1993; Pognante & Benna, 1993; Carosi et al., 1999a), although the assemblages differ from that of central Nepal. In eastern Nepal, the upper Black Gneiss is composed of biotite-sillimanite paragneisses with metaconglomerates and quartzite layers. The middle Namche Migmatite Orthogneiss contains granite-granodiorite sillimanite-bearing orthogneisses. The Barun Gneiss at the base is a migmatized paragneiss with minor metabasites, calc-silicate rocks, and marbles. In the Garhwal Himalaya, the Vaikrita Group is analogous to the GHC (e.g., Ahmad et al., 2000). Early observations, the unit's complex metamorphic history, and the possibility of out-of-sequence thrusting at higher levels led Stöcklin (1980) to doubt the assumption of a three-tiered, laterally continuous, tectonostratigraphic framework. More recent thermobarometric data and P-T-t paths from garnet-bearing assemblages from the GHC unit indicate the unit itself has discontinuities related to internal structures that overlap in age with MCT motion (Carosi et al., 2010; Montomoli et al., 2013; Wang et al., 2013; Larson et al., 2015; Montomoli et al., 2015; Carosi et al., 2016; Wang et al., 2016; Braden et al., 2017; Rapa et al., 2018; Chakraborty et al., 2019; Benetti et al., 2021).

A Tertiary history involving two metamorphic episodes has been proposed for the GHC (see Pêcher & Le Fort, 1986; Metcalfe, 1993; Pognante & Benna, 1993; Wiesmayr & Grasemann, 2002; Carosi et al., 1999b; Lombardo & Rolfo, 2000; Hodges, 2000; Wiesmayr & Grasemann, 2002; Cottle et al., 2009; Kohn, 2014; Robyr & Lanari, 2020). The first stage (Eocene-Oligocene) of Barrovian-type metamorphism, termed the *Eohimalayan event*, corresponds to the nappe's burial beneath Asia and is best preserved near the unit's base (e.g., Pêcher, 1989; Hodges et al., 1994; Wiesmayr & Grasemann, 2002). During this stage, the base of the unit reached garnet-grade conditions (550°C–750°C and at least 8–10 kbar; e.g., Pognante & Benna, 1993; Hodges et al., 1994). Thrusting may have been accommodated by a fault system considered a proto-STDS that reactivated as a normal fault during the second stage in the Miocene (e.g., Vannay & Hodges, 1996; Wiesmayr & Grasemann, 2002; DiPietro & Pogue, 2004). The locations where this scenario has been proposed show mylonitic contractional fabrics overprinted by extensional structures and include the Himachal Pradesh, northwest India (Vannay & Grasemann, 1998), Zanskar, northwest India (Patel et al., 1993), and the Kali Gandaki, central Nepal (Vannay & Hodges, 1996). Even earlier events may have preceded Himalayan orogenesis and can be challenging to distinguish from Cenozoic tectonics (Gehrels et al., 2003).

The Eohimalayan event occurred between 50 and 35 Ma and coincides with a dramatic decrease in the convergence rate between India and Asia from ~15 to ~4 cm/yr (Copley et al., 2010). During the Miocene Neohimalayan event, the base of the GHC experienced 550°C–600°C, and the top records lower pressures and/or temperatures (3–7 kbar, 575°C–850°C) (e.g., Hodges et al., 1994; Simpson et al., 2000; Daniel et al., 2003). The Neohimalayan event has been associated with MCT slip and with the development of Miocene-age High Himalayan leucogranites found in close association with STDS (Pêcher, 1989; Harris et al., 1993; Metcalfe, 1993; Hopkinson et al., 2017; Liu et al., 2018; Wu et al. 2020).

The duration and onset of MCT movement vary along strike and is often attributed to sometime during the Miocene. To time fault activity, monazite [(Ce,La,Th)PO4] is often targeted, a mineral that appears in these rocks at garnet grades and is ideal for timing metamorphism (e.g., Catlos, 2013). In western Bhutan, the duration of MCT displacement is bracketed by monazite ages between 20.8 ± 1.1 Ma and 15.0 ± 2.4 Ma (Togbay et al., 2012). However, in eastern Bhutan, prograde metamorphism and deformation are reported to have been underway along the structure by ca. 23 Ma (Daniel et al., 2003). Monazite ages from the lower portions of the GHC in western Nepal yield younger ages timed to fault motion at 17–13 Ma (Montomoli et al., 2013). Depth profiling of unpolished monazite indicates that the MCT hanging wall was deforming in central Nepal between 24–22 Ma (Harrison et al., 1995). The 22 Ma age is also attributed to MCT activity in central Nepal (Hodges et al., 1996; Coleman & Hodges, 1998)

In northwest India (Garhwal region), the MCT shear zone is thought to be active from 20 Ma to 5 Ma (Iaccarino et al., 2020) or even younger to 1 Ma (Catlos et al., 2007, 2020). Mukhopadhyay et al. (2017) suggest that activity on the MCT itself in the Sikkim region occurred in pulses over an extended period from 26 to 11 Ma. Monazite grains from the MCT shear zone in the Sikkim indicate the structure was active at c. 22 Ma, 14–15 Ma, and 12–10 Ma (Catlos et al., 2007) and from 21–13 Ma (Mottram et al., 2014). Younger ages are attributed to structures south of the GHC-LHF contact and are often found at lower structural levels within the shear zone elsewhere. For example, Figure 6.5 shows the distribution of in situ secondary ion mass spectrometry (SIMS) Th-Pb monazite and muscovite $^{40}Ar/^{39}Ar$ total gas ages along the Darondi Khola (Catlos et al., 2001). The oldest early Miocene monazite ages are found in upper LHF units (21.7 ± 1.2 Ma to 21.1 ± 0.8 Ma), but they decrease towards lower structural levels to 7.6 ± 0.2 Ma to 6.9 ± 0.5 Ma. In the Arun valley of eastern Nepal Himalayas, Oligocene activity within the upper portions of the MCT shear zone has been reported (~31 Ma: Groppo et al., 2010). Late Eocene to Oligocene monazite ages have been found in GHC samples from central Nepal (37.6 ± 3.8 Ma to 30.4 ± 0.5 Ma: Catlos et al., 2001; see also Gibson et al., 2016 for the same transect) and northwest India (37.9 ± 0.9 Ma to 34.3 ± 0.8 Ma: Catlos et al., 2007). The Oligocene ages are consistent with the events related to the Eohimalayan time frame of Himalayan orogenesis and may not be related to activity along the MCT.

Distinguishing between the Eohimalayan and Neohimalayan events is challenging, and the Eohimalayan event may only be recorded in GHC and Tethyan units in particular locations (Aikman et al., 2008; Stickroth et al., 2019). Northeast verging folds of the Tethys Formation may have developed during the Eohimalayan event (e.g., Godin et al. 1999a, 1999b), whereas these and others reported in the GHC (Bhargava & Bassi, 1994; Carosi et al., 1999b) may be the result of gravity sliding along the STDS (e.g., Burchfiel et al., 1992; Hodges et al., 1996; Vannay & Hodges, 1996; Searle, 2010). Large-scale folding that aligns preexisting isograds into an apparent inverted metamorphic position was attributed as a significant factor in the metamorphic history of the GHC (Searle & Rex, 1989). In a petrochronology study of GHC monazite-bearing assemblages, Gibson et al. (2016) suggest that Eocene-Oligocene monazite ages represent prograde burial, followed by Miocene retrograde metamorphism and Miocene-Pliocene exhumation. Wu et al. (2020) suggest the High Himalayan leucogranites can be classified as Eohimalayan (46–25 Ma), Neohimalayan (25–14 Ma), and post-Himalayan (< 14 Ma) based on their relationship with particular detachment systems, ages, and compositions. Some North Himalayan granites that intrude Indian shelf sediments are similar in age to their post-Himalayan stage (Harrison et al., 1997, 1998; Lee et al., 2004; Zhang et al., 2004).

Metamorphism in the LHF

At their base, the GHC is thrust over Middle Proterozoic phyllites, metaquartzites, and mylonitic augen gneisses of LHF along the broadscale 8–12 km thick MCT shear zone (Le Fort, 1975; Gansser, 1981; Arita, 1983; Brunel & Kienast, 1986; Pêcher, 1989; Searle & Rex, 1989; England et al., 1992; Schelling, 1992; Le Fort, 1996; Henry et al., 1997; Harrison et al., 1998; Kohn et al., 2004; Bollinger et al., 2006; Carosi et al., 2013; Parsons et al., 2016; Martin, 2017a,b; Mukhopadhyay et al., 2017; Yin & Harrison, 2000; Catlos et al., 2001; Catlos et al., 2018, 2020). The onset of MCT activity occurred during the early Miocene (Pêcher, 1991; Vannay et al., 2004; Yin, 2006), at a time when Indo-Asia convergence slowed (> 40% between 20 and 10 Ma: Molnar & Stock, 2009) and had a significant impact on the extrusion of the GHC orogenic wedge (Maiti et al., 2020). The MCT fault system plays a central role in many models for the evolution of the Himalayas (e.g., Le Fort, 1975; Searle & Rex, 1989; England et al., 1992; Henry et al., 1997; Harrison et al., 1998; Bollinger et al., 2006; Kohn, 2008; Carosi et al., 2013; Beaumont et al., 2001, 2004; Jamieson et al., 2004, 2006; Searle et al., 2006; Long & McQuarrie, 2010; Wang et al., 2013; Cottle et al., 2015).

The lack of an apparent break in metamorphic grade between the GHC and LHF makes the placement of the boundaries of the MCT shear zone challenging to discern. Definitions of the structure center around age, metamorphic history, rheology, and role as separating units of different depositional environments (see Martin, 2017a). As the GHC and LHF evolved in different depositional environments, geochemical and geochronological evidence aids in their placement (e.g., Parrish & Hodges, 1996; Ahmad et al., 2000; Martin, 2017a,b; Khanal et al., 2014). The original Pêcher (1989) MCT definition involved three criteria in identifying the MCT in the field: (1) The boundary between hanging wall gneisses and upper carbonate-rich formations of the Lesser Himalaya, (2) where Lesser Himalaya shear fabric (L-S) is replaced by the flattening fabric of the Greater Himalayan Crystallines, and (3) where the rotational deformation that increases progressively through the Lesser Himalaya reaches a maximum. In central Nepal, Arita (1983) places two thrusts (MCT-I and MCT-II) on each side of the MCT shear zone. The MCT-II corresponds to that described by Pêcher (1989), whereas the MCT-I separates a mylonitic augen gneiss from other Lesser Himalaya metasedimentary rocks. The MCT-I correlates to the Ramgarh Thrust, which accommodates the Ramgarh Thrust sheet within the LHF duplex (DeCelles et al., 2001; Robinson et al., 2003; Pearson & DeCelles, 2005; Robinson et al., 2006; Matin & Mukul, 2010; Khanal et al., 2014; Mandal et al., 2015). Some researchers do not recognize the MCT-I anywhere in the Nepal Himalayas (see Upreti, 1999). Along the Dudh Kosi-Everest transect, the MCT corresponds to the contact between the GHC gneisses and the upper LHF pelitic schists. In contrast, the MCT-I separates the mylonitic Phaplu augen gneiss from low-grade Lesser Himalaya metasedimentary rocks. Along the Bhagirathi River, Garhwal region, the Vaikrita Thrust (= MCT) and Munsiari Thrust (= MCT-I) bound the MCT shear zone, but no equivalent to the Phaplu augen gneiss is exposed (e.g., Valdiya, 1980; Pêcher, 1991; Metcalfe, 1993; Searle et al., 1993; Ahmad et al., 2000).

The MCT footwall is characterized and defined by inverted metamorphism, where metamorphic grade increases toward structurally shallower levels (e.g., Ray, 1947; Pêcher, 1989; Larson et al., 2015; Chakraborty et al., 2016; Searle et al., 2008; Iaccarino et al., 2020). Understanding the origin of this phenomenon has implications for establishing the role of various crustal heat sources and mechanisms of heat transfer within collisional belts (e.g., radiogenic, asthenospheric input, shear heating, the addition of melts) (e.g., England et al., 1992) and is facilitated by obtaining metamorphic P-T conditions and paths from shear zone garnet-bearing assemblages (Stäubli, 1989; Metcalfe, 1993; Kaneko, 1995; Vannay & Hodges, 1996; Vannay & Grasemann, 1998; Manickavasagam et al., 1999; Kohn et al., 2001; Catlos et al., 2001; Imayama et al., 2010; Corrie et al., 2010; Larson et al., 2013; Anczkiewicz et al., 2014; Kohn, 2014; Mottram et al., 2014; Mukhopadhyay et al., 2017; Catlos et al., 2018; Waters, 2019; Catlos et al., 2020).

The MCT has also been considered to be an expression of the Main Himalayan Thrust (MHT) (Bollinger et al., 2004; Mahajan et al., 2010), a pervasive decollement that separates the downgoing Indian plate from the Himalayan orogenic wedge (Fig. 6.2; Bilham et al., 1997; Nelson et al., 1996; Subedi et al., 2018; Zhao et al., 1993). Other surface expressions of the MHT include the MBT and Main Frontal Thrust (MFT; Figs. 6.1 and 6.2). The STDS may be a local phenomenon, occurring when the Tethys dissociated from the GHC and slid along north-dipping planes due to gravity following uplift (e.g., Pêcher & Le Fort, 1986; Fuchs, 1987), although some models for channel flow connect this fault system to the MHT (e.g., Beaumont et al., 2001, 2004; Jamieson et al., 2004, 2006). The MHT is one of the largest and fastest slipping continental megathrusts on Earth (e.g., Duputel et al., 2016; Rajendran et al., 2017; Searle et al., 2017). Understanding its geometry, history, and fault systems that splay into the structure has implications for assessing and predicting the initiation, propagation, and termination of major event Himalayan earthquakes (e.g., Wang et al., 2017). The topography of the MHT is uncertain (e.g., Caldwell et al., 2013; Denolle et al., 2015; Elliott et al., 2016; Hazarika et al., 2017; Hubbard et al., 2016; Nábělek et al., 2009; Whipple et al., 2016; Wang et al., 2017; Zhou et al., 2019), due in part to the lack of recognition that fault systems within the LHF duplex or MCT shear zone have the potential to accommodate present-day slip (Catlos et al., 2020). The LHF duplex has long been known to be a ~50 km wide seismogenic zone of predominately moderate earthquakes (Bai et al., 2016; Cattin & Avouac, 2000; Khattri & Tyagi, 1983; Mahajan et al., 2010). Bai et al. (2016) suggest that a thrust system within the LHF is the most seismically active region in the Himalayas and accommodates most of its elastic strain accumulation. Alternatively, the shallower events are explained by a segmented MHT that includes a ramp (He et al., 2018; Hubbard et al., 2016; Pandey et al., 1995).

The MBT and MFT

South of the MCT, the MBT separates the Lesser Himalaya from Neogene molasse, the Siwalik Formation (Fig. 6.1; Valdiya, 1992; Meigs et al., 1995; Mukul, 2000; Thakur et al., 2010; Goswami & Deopa, 2017; Dhamodharan et al., 2020). South of the MBT, the MFT is the boundary between the Siwalik Formation and the northern Indo-Gangetic Plains (e.g., Mugnier et al., 1999; Mukul et al., 2007; Burgess et al., 2012; Bollinger et al., 2014). The MBT has few timing constraints but is primarily thought to have been initiated during the Late Miocene (13–10 Ma: Meigs et al., 1995; Chirouze et al., 2012; Patra & Saha, 2019), although it may have been active since 5 Ma in Nepal (DeCelles et al. 2020). The Siwalik Formation overall is a 7–10 km thick succession of dominantly fluvial sedimentary rocks located along the entire length of the Himalaya from the Potwar plateau to the Brahmaputra in the east, likely due to the evolution of the Ganga River system (Bora & Shukla, 2005; Sanyal & Sinha, 2010; Khan et al., 2019; Dhamodharan et al., 2020). The Siwalik unit is divided into several sectors due to lineaments related to normal faulting in the Indian basement that reactivated as thrust faults during the Himalayan collision (Raiverman et al., 1983; Dubey, 1997; Sanyal & Sinha, 2010). The Siwalik formation records critical information regarding Himalaya erosional history, paleoclimate, transitions in paleobotany, and exhumation rates (e.g., Quade et al., 1989, 1995; Acharyya, 1994; Sanyal & Sinha, 2010; Najman et al., 2017; Ghosh et al., 2018; Khan et al., 2019). In some locations, the MBT has nearby active steep faults that show normal or strike-slip senses of motion as they accommodate a critical taper (Mugnier et al., 1994; Patra & Saha, 2019). The MFT cuts Siwalik strata in places and is often manifested as growing anticlines (Yeats et al., 1992; Powers et al., 1998; Srivastava et al., 2018). These crustal-scale faults sole into the MHT (Fig. 6.2; Zhao et al., 1993; Nelson et al., 1996).

6.3. MODELS FOR THE EXTRUSION OF THE HIMALAYAN CORE

The MCT fault system has long played a central role in models for the evolution of the Himalayas. Initial models presume that activity along the MCT occurred only during the Early Miocene (Fig. 6.6; Le Fort, 1975; Searle & Rex, 1989; Hubbard, 1996; England et al., 1992; Hodges et al., 1993; Harris & Massey, 1994). Le Fort (1975) conceived that the inverted metamorphism in the LHF footwall was caused by the transfer of thermal energy due to the large-scale underthrusting of the LHF beneath the GHC (the hot-iron model, Fig. 6.6a). In this scenario, fluids released from the MCT footwall migrate through the hanging wall and flux a leucogranite belt. The alternative was proposed by Hubbard (1996), in which the inverted mineral isograds result from ductile shearing of a preexisting zone of right-way-up metamorphism (Fig. 6.6b). Searle and Rex (1989) delegated a significant role to STDS and suggested that the present-day distribution of metamorphic facies and leucogranite bodies are caused by overprinting earlier isograds during MCT-related anatexis and folding (Fig. 6.6c). Fluids released from the MCT footwall assist with the formation of leucogranite. The STDS has appeared prominently in almost every model since. Hodges et al. (1993) proposed that the wedge GHC extruded via synchronous STDS and MCT movement (Fig. 6.6d), and Harris and Massey (1994) suggested extensional collapse led to rapid GHC exhumation and decompression melting of kyanite-bearing schists that led to melts emplaced near the STDS (Fig. 6.6e).

As more data became available, more numerical solutions and quantitative models appeared (Fig. 6.7). Molnar and England (1990) matched P-T conditions obtained by Hubbard (1996) by solving a one-dimensional time-dependent heat equation for an inclined fault (Fig. 6.7a). The model derived thermal energy from three sources: (1) radioactive nuclides, (2) the asthenosphere, and (3) frictional heating along the MCT, and suggested that shear stresses >100 MPa can account for peak T < 600°C at the MCT and contributes ~13°C/km to the inverted geotherm. Huerta et al. (1998) introduced that MCT inverted metamorphism was caused by accretion and erosion acting on a crust enriched with radiogenic elements (Fig. 6.7b). The presence of post-Miocene mineral crystallization ages in the LHF and MCT shear zone led Harrison et al. (1998) to model footwall inverted metamorphism as the accretion of tectonic slivers of the LHF to the hanging wall (Figs. 6.7 and 6.8a). This model was one of the first to suggest the concept of critical taper as an essential control in developing inverted metamorphism and the geometry of the fold-and-thrust belt. In this scenario, anatexis is linked to shear heating along the MHT. This model is also the first in a series that suggests the Himalaya is an outcome of stacked thrust systems within and below the GHC (Fig. 6.8; e.g., Kohn, 2008; Corrie & Kohn, 2011; Cottle et al., 2015; Catlos et al., 2020).

Large-scale imaging of the Himalayan and Tibet crust led Nelson et al. (1996) to advocate that thrusting within the Himalayas results from anatexis (Fig. 6.7d). The Tibetan middle crust is assumed to be partially molten

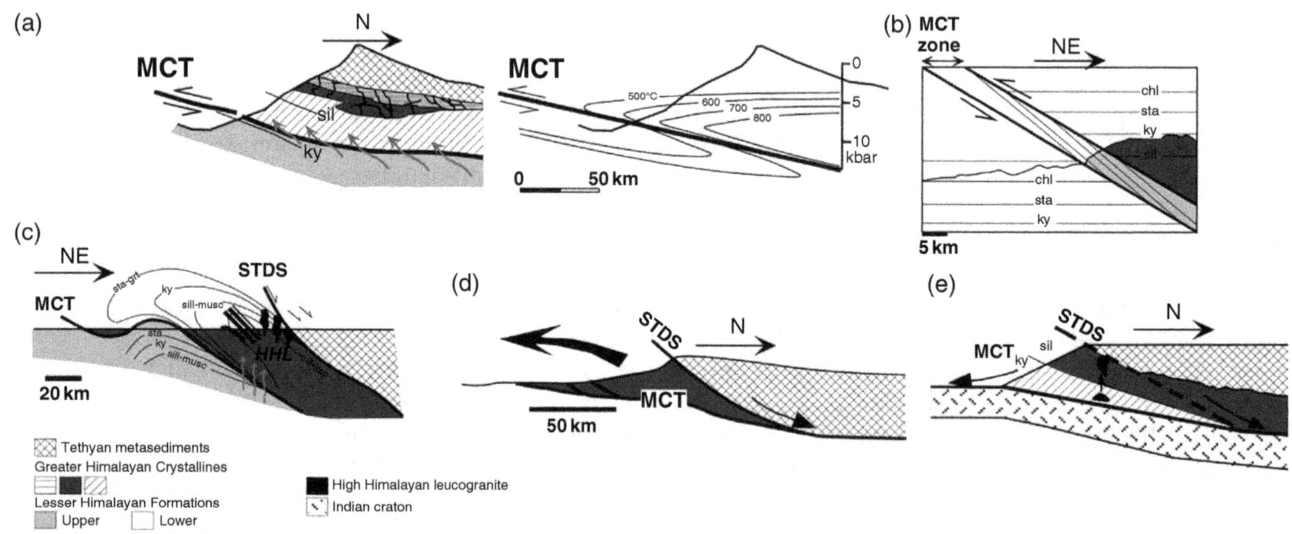

Figure 6.6 Cross sections of early models of Himalayan inverted metamorphism and leucogranite formation.: (a) The hot-iron model of Le Fort (1975); (b) juxtaposing right-way-up metamorphic isograds after Hubbard (1996); (c) a combination of models in panels (a) and (b) after Searle and Rex (1989); wedge extrusion models after (d) Hodges et al. (1993) and (e) Harris and Massey (1994).

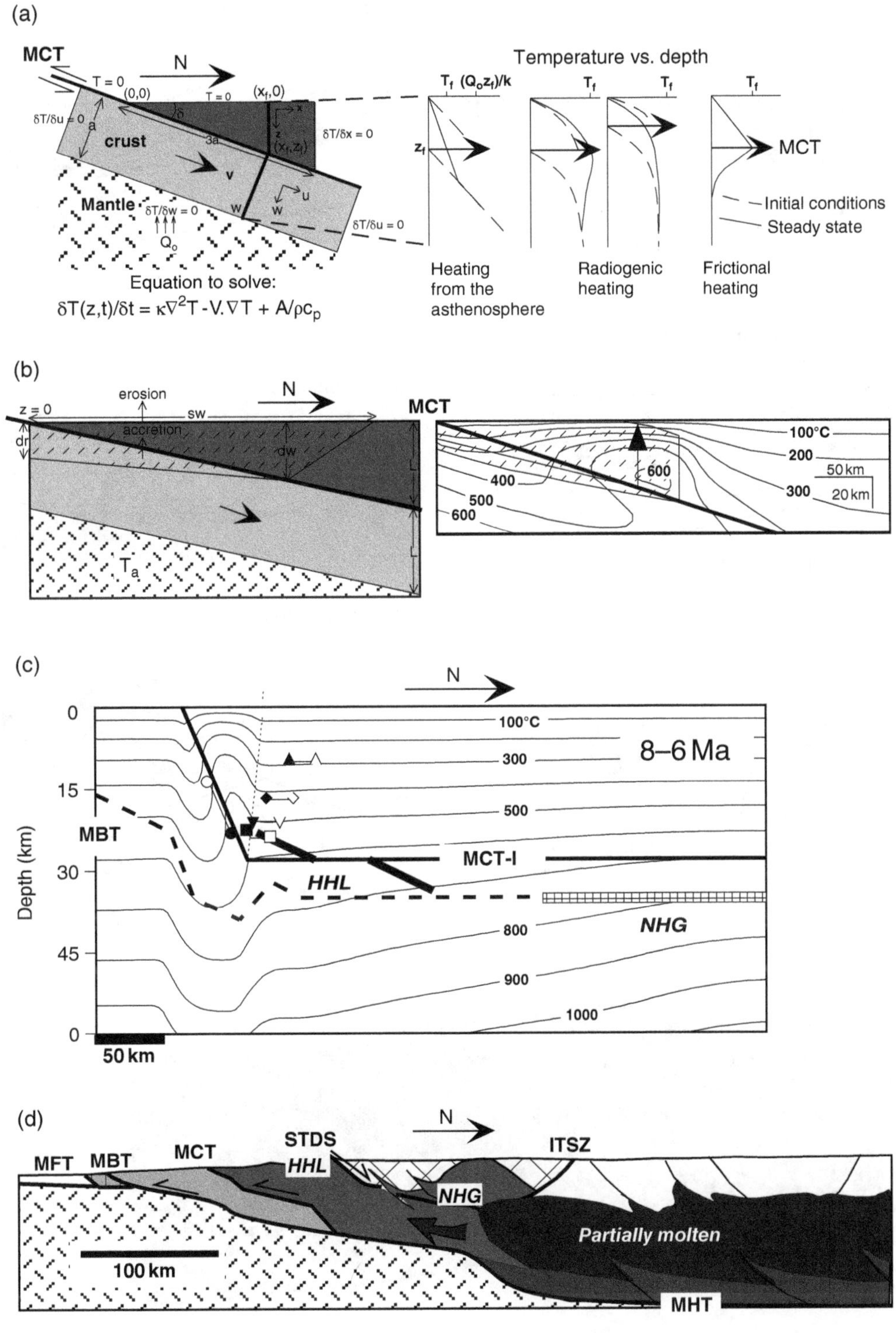

Figure 6.7 Cross sections of early kinematic models of Himalayan inverted metamorphism and leucogranite formation: (a) After Molnar and England (1990); (b) after Huerta et al. (1998); (c) after Harrison et al. (1998); (d) after Nelson et al. (1996).

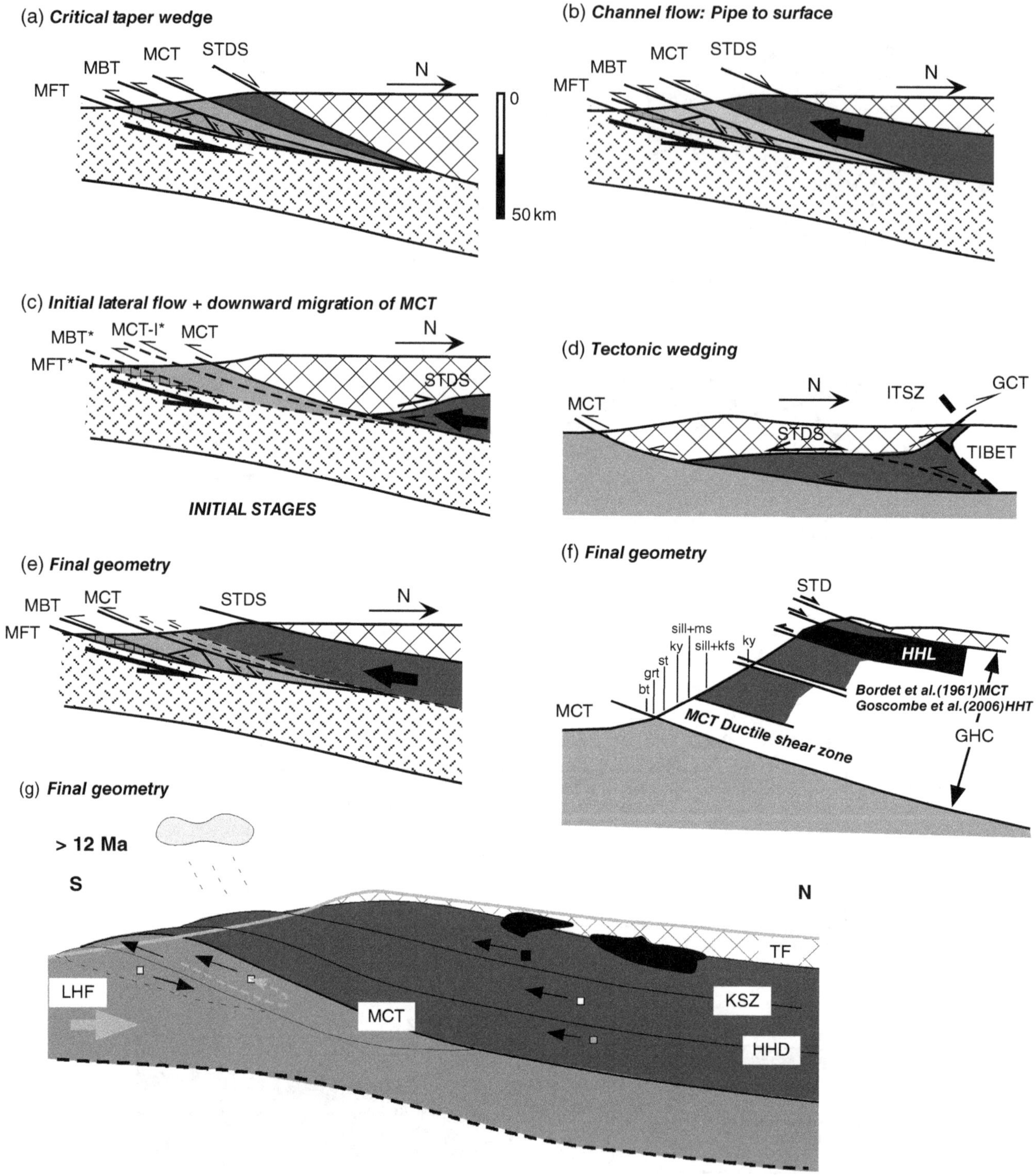

Figure 6.8 (a) An illustration of the critical taper model; (b) schematic of channel flow and wedge extrusion; (c) a combination of critical taper and tectonic wedging of the GHC. A "*" indicates a developing fault; (d) illustration of GHC tectonic wedging after Webb et al. (2011). Panels (e) and (f) show the final geometry of the range, which has affected the position and definition of the MCT. Panels (a), (b), (c), and (e) are after Cottle et al. (2015), panel (f) is after Searle et al. (2008), and panel (g) is after Carosi et al. (2016).

today, and the region between the MCT and STDS is earlier extruded equivalent. This model is the foundation of others that suggest the GHC experienced gravity-driven lateral midcrustal flow (Grujic et al., 1996; Beaumont et al., 2001; Jamieson et al., 2006; Searle, 2010; Webb et al., 2011; Cottle et al., 2015). Channel flow was initially conceived as a Neohimalayan event, with the extrusion of the GHC due to synchronous MCT and STDS activity with focused erosion along the topographic front (e.g., Beaumont et al., 2001, 2004; Jamieson et al., 2004, 2006; Robinson et al., 2006; Searle et al., 2006; Long & McQuarrie, 2010; Wang et al., 2013; Cottle et al., 2015). This model can be combined with the scenario where LHF footwall slivers are accreted to the hanging wall as the GHC channel flow progresses (Fig. 6.8b). LHF footwall inverted metamorphism may also be an outcome of channel flow experienced by the GHC (Daniel et al., 2003; Searle et al., 2008).

Channel flow has also been invoked to model the inverted metamorphic sequence within the MCT footwall, where discrete fault systems within the LHF and the MCT or other faults within the GHC are active during the Miocene (Imayama et al., 2010; Goswami-Banerjee et al., 2014). The alternative to channel flow is the critical taper model, where the GHC is exhumed through a series of stacked thrust systems within and below the GHC (Fig. 6.8a; Kohn, 2008; Corrie & Kohn, 2011). Critical taper has also been applied to explain thrusting with fault systems associated with the MFT (Mukul et al., 2007; Hirschmiller et al., 2014). In addition, the recognition that in northwest India, the LHF and Tethyan Formation appear in thrust contact suggests that the GHC may have acted as a tectonic wedge (Fig. 6.8c,d; Webb et al., 2007; Webb et al., 2011; Cottle et al., 2015). One outcome of these models has been a wholesale redistribution and redefining of the MCT from a fault system that separates the GHC and LHF to one in which portions of the GHC are involved in large-scale thrust movement (Searle et al., 2008; Carosi et al., 2018). The inverted metamorphic sequence is redelegated from the LHF and MCT footwall to lower levels of the GHC (Fig. 6.8e,f). Thermal heating along faults within the GHC may have played a role in accommodating wedge extrusion and developing inverted metamorphism (e.g., Goscombe et al., 2006; Searle et al., 2008). Fig. 6.8g shows a scenario where the MCT shear zone develops due to imbrication, and fault systems accommodate activity within the GHC (Carosi et al., 2016). GHC structures include the Kalopani shear zone (KSZ) at 41–30 Ma (Eohimalayan time frame) and High Himalayan Discontinuity (HHD) at 26–24 Ma (Neohimalayan).

Although lines of evidence exist that the LHF inverted metamorphism involved multiple episodes of post-Miocene imbrication and deformation with the MCT shear zone or LHF duplex (e.g., Fig. 6.5; Caddick et al., 2007; Catlos et al., 2001; Kohn et al., 2001; Groppo et al., 2009, 2010; Herman et al., 2010; Mosca et al., 2012; Montomoli et al., 2013; Mottram et al., 2014; Carosi et al., 2016; Braden et al., 2018; Catlos et al., 2018; Montemagni et al., 2018; Catlos et al., 2020; Montemagni et al., 2020), the idea of tectonic inversion of a coherent rock package that experienced a single Barrovian Neohimalayan-related metamorphic event remains (Hubbard, 1996; Martin et al., 2010; Gaidies et al., 2015). Multiple episodes of ductile overthrusts of the GHC over the LHF are proposed as an explanation (Goscombe & Hand, 2000), as well as post and tectonic overpressure that changed fundamental rock properties (Thakur et al., 2015; model of Schmalholz & Podladchikov, 2013). The originally proposed hot-iron model in which the primary heat source for inverted metamorphism is GHC thermal energy transferred as Miocene MCT emplacement occurred (England & Molnar, 1993) has an alternative end-member where no contribution of dissipative to downward conductive heating from the GHC is required (Stephenson et al., 2000).

The Himalaya is often framed as a large-hot orogen that may have grown by distributed extrusion (channel flow) or discrete thrusting (critical taper) (Beaumont et al., 2006; Jamieson & Beaumont, 2013; Mukherjee, 2013; Iaccarino et al., 2020; review in Wang et al., 2013). Both scenarios may be relevant to the Himalayan core, depending on the location within the range and the time frame of metamorphism (Larson et al., 2010; Mukherjee, 2013; Cottle et al., 2015). The Himalaya may represent a scenario in which the two end-member models apply and are not mutually exclusive (Beaumont & Jamieson, 2010; Larson et al., 2010, 2013; Corrie et al., 2012; Jamieson & Beaumont, 2013; Cottle et al., 2015). Evaluating models for GHC extrusion requires an understanding of the P-T-t paths of its rocks as they experienced the transition from convergence and subduction to their final exhumation (e.g., Catlos et al., 2001; Caddick et al., 2007; Kohn, 2008; Corrie et al., 2010; Goswami-Banerjee et al., 2014; Catlos et al., 2018, 2020).

Kohn (2008) presents particle paths predicted by end-member critical taper and channel flow models. Channel flow predicts the GHC experienced isothermal exhumation, and the LHF experiences isobaric heating. Critical taper predictions imply isobaric cooling for GHC rocks and hairpin LHF P-T paths (Kohn, 2008). The metamorphic field gradients predicted by the models also differ. Complicating this scenario is the observation that P-T conditions from garnet-bearing assemblages collected from the MCT hanging wall indicate the unit has cryptic discontinuities related to unmapped faults, shear zones, or inheritance of preexisting basement structures (see discussions in Mukherjee et al., 2012; Larson et al., 2013; Montomoli et al., 2015; Cottle et al., 2015; Carosi et al., 2016). Understanding when these possible fault systems, the High Himalayan Discontinuities (HHD), were active is critical for deciphering the processes involved during convergence. Their age may be similar to activity along the MCT (28–18 Ma; e.g., Carosi et al., 2010; Montomoli et al., 2013, Larson et al., 2015; Montomoli et al., 2015; Carosi et al., 2016; Carosi et al., 2018; Benetti et al., 2021). This scenario is not accounted for in the channel flow model (e.g., Benetti et al., 2021; Maiti & Mandal, 2021).

6.4. HIMALAYAN METAMORPHISM AND CONTRACTIONAL TECTONICS (DARONDI KHOLA, CENTRAL NEPAL)

As indicated in the previous sections, garnet-bearing assemblages are valuable recorders of compressional/contractional metamorphism and help constrain models for the Himalayas' uplift history. P-T data from garnet-bearing Himalayan assemblages can be generated using several approaches (see review in Waters, 2019). This section compares conditions generated using conventional thermobarometers and Gibb's method (Kohn et al., 2001) to those obtained using isochemical phase diagrams, isopleth thermobarometry, and the garnet zoning method (e.g., de Capitani & Petrakakis, 2010; Moynihan & Pattison, 2013; Catlos et al., 2018, 2020). The approaches were applied to the same samples and data from rocks collected from the GHC and LHF units along the Darondi Khola in central Nepal (Figs. 6.3–6.5). The TheriaG model of Gaides et al. (2008) is also used to predict garnet zoning based on Gibb's P-T paths for some samples. The reason for any discrepancies in the P-T paths and conditions within the context of equilibrium thermodynamics is evaluated, and the impact of the results on models for Himalayan orogenesis is explored. Regardless of calibrations used, the P-T conditions and paths coupled with previously reported timing constraints from Darondi Khola assemblages suggest the MCT shear zone developed during pulses of movement that resulted in a progressive transfer of rock packages as the MCT shear zone developed (Catlos et al., 2001; Kohn et al., 2001).

6.4.1. Methods, Samples, and Assumptions

Samples

Kohn et al. (2001) report rim P-T conditions from 18 garnet-bearing assemblages exposed along the Darondi Khola and divides LHF samples into lowermost (n = 3), lower to mid (n = 6), and upper MCT zones (n = 4) (Figs. 6.4 and 6.5, Table 6.1). Five GHC rocks were also

Table 6.1 Conventional P-T data from samples collected along the Darondi Khola

Sample[a]	Gibbs Core P (±1kbar)	Gibbs Core T (±25°C)	Gibbs Rim P (± 1kbar)	Gibbs Rim T (±25°C)	Conv. Rim P (kbar ±1σ)	Conv. Rim T (°C ±1σ)
			Lower LHF			
DH16	–[b]	–	–	–	5.0±1.0	520±25
DH17	6.0	510	7.0	525	6.8±1.0	525±15
DH19	5.5	475	6.0	525	6.2±0.5	527±12
			Middle LHF			
DH22	6.0	498	5.0	525	5.2±1.0	520±25
DH23	7.0	525	7.3	575	7.2±2.0	540±35
DH26	6.5	500	5.5	520	5.5±1.0	524±25
DH30	–	–	–	–	8.5±2.0	570±20
DH75A	8.5	530	9.0	560	8.5±2.0	570±20
DH75B	8.0	500	8.0	560	8.5±2.0	570±20
			Upper LHF			
DH38	–	–	–	–	8.2±1.0	578±20
DH57	–	–	–	–	11.5±2.0	640±25
DH58	–	–	–	–	11.5±2.0	640±25
DH71	–	–	–	–	11.5±1.0	640±20
			GHC			
DH60	–	–	–	–	10.5±1.0	652±10
DH61	–	–	–	–	9.1±2.0	625±12
DH63	–	–	–	–	11.5±1.5	705±25
DH66	–	–	–	–	7.6±1.0	660±12
DH67	–	–	–	–	10.2±2.0	715±5

[a] See Figure 6.3 for samples location. Conditions estimated from Kohn et al. (2001).
[b] "–" not reported.

analyzed. We use the definition of the MCT shear zone following this nomenclature but also note that based on the placement of the Phaplu and Ulleri augen gneisses, others would allocate all of these samples squarely in the GHC unit (e.g., Searle et al., 2008).

Mineral assemblages, X-ray element maps, and compositional transects were made across garnets and are reported in Kohn et al. (2001). Petrographic images of some samples are shown in Figure 6.9. All samples contain prograde garnet, quartz, muscovite, biotite, and ilmenite. All are metamorphosed pelites, except DH38, a metabasite containing hornblende and plagioclase. All lower LHF rocks contain prograde chlorite and plagioclase, but DH75A and DH75B also have retrograde chlorite (Fig. 6.9). The upper LHF and GHC rocks do not have chlorite (DH38, DH60, DH61, DH63, DH66, DH67) or contain only retrograde chlorite (DH57, DH58, DH71). All samples, except upper LHF rock DH57, contain plagioclase. Most plagioclase show core-rim zoning with higher-Ca cores and lower-Ca rims, expected if garnet grew in a closed chemical system (Spear, 1993). Samples DH26, DH75A, and DH75B have plagioclase with relict albite cores overgrown by an oligoclase mantle that is either unzoned or slightly zoned to lower Ca towards its rim. The mantle was assumed to be metamorphic and reflects garnet growth and uptake of Ca. Sample DH57 has both staurolite and kyanite, whereas DH63 has staurolite. Other rocks for this study do not show any of these index minerals. No chloritoid or carbonate minerals are reported in the samples. All samples appear syntectonic, and the presence of strain shadows is common. Figure 6.9 shows micas and quartz appear deflected around the garnet porphyroblasts. Some garnets are inclusion free; however, the middle LHF sample DH30 has quartz inclusions in the center of the garnet that appear thinner and elongated compared with the larger, anhedral grains in its outer-rim region. GHC samples DH61, DH63, and DH67 have inclusion-filled cores and inclusion-free outer rims. Inclusions of ilmenite and quartz in middle LHF samples DH51 and DH75B are aligned in relatively straight tracks into the rock matrix, consistent with syntectonic growth.

Figure 6.10 shows garnet compositional transects for LHF samples in which high-resolution P-T paths (DH17, DH19, DH22, DH23, DH26, DH75B) were obtained, and Figure 6.11 shows compositional transects across garnets in two GHC samples (DH61 and DH66). LHF samples show typical prograde bell-shaped profiles in Mn, with no evidence of retrogression at the rims. The garnets show a smooth decrease in grossular and increase in pyrope and almandine from core to rim. GHC garnets, however, show evidence of retrogression with sharp increases in Mn contents at the rim. These garnets also show significant grossular, pyrope, and almandine fluctuations from core to rim. GHC samples are not ideal candidates for high-resolution P-T path modeling because their zoning profiles suggest modification of prograde garnet compositions after growth, open-system behavior, and the potential of changes in rock bulk composition after growth. These situations are assumed not to occur when modeling and developing high-resolution P-T paths.

Conventional Thermobarometry and Gibbs Method P-T Paths

The garnet-biotite thermometer calibrated by Ferry and Spear (1978) with the Berman (1990) Ca-in garnet solution model and the garnet-muscovite-biotite-plagioclase barometer of Hoisch (1990) were applied to the assemblages with pelitic bulk compositions (Kohn et al., 2001). Sample DH38 is a metabasite, so the garnet-hornblende thermometer of Graham and Powell (1984) and the garnet amphibolite barometer of Kohn and Spear (1990) were used. Results are summarized in Table 6.1. The conditions are internally consistent, and Kohn et al. (2001) emphasize that different calibrations would change estimated P and T by ± 25°C and ± 0.5–1 kbar and would not alter overall trends. The conditions were estimated using minimum garnet Mn contents. Note that these garnet compositions are the same used for estimating the garnet rim conditions using isopleth thermobarometry as described in the next section (Table 6.2; polygons in Figs. 6.12–6.17).

Gibb's method to generate the P-T paths from zoned garnets and their coexisting matrix mineral assemblages is outlined in Spear and Selverstone (1983), Spear (1986, 1993), Spear et al. (1984), and Spear and Rumple (1986). P-T paths were only generated from LHF rocks with pelitic bulk compositions (DH16, DH17, DH22, DH23, DH26, DH75A, and DH75B) that preserved prograde compositional zoning best. The rocks were modeled with the observed solid assemblage garnet + biotite + chlorite + muscovite + plagioclase + quartz in the MnNCKFMASH system. Compositional changes were derived from garnet and plagioclase compositional zoning produced as the garnet grew. Garnet core and rim conditions were selected for samples DH17, DH19, DH22, and DH26, whereas an intermediate point was also included for samples DH23, DH75A, and DH75B. This intermediate garnet composition was used to better account for nonlinearity in the garnet zoning pattern. In the case of DH23 and DH75B, inclusions influence the development of the hairpin P-T paths (Figs. 6.13a, 6.14c, and 6.17). A pure H_2O fluid at lithostatic pressure was assumed to have been present. Kohn et al. (2001) emphasize that activity models would not significantly affect the trends of retrieved P-T paths.

The P-T results obtained using conventional and Gibb's method show an inverted metamorphic signature in the

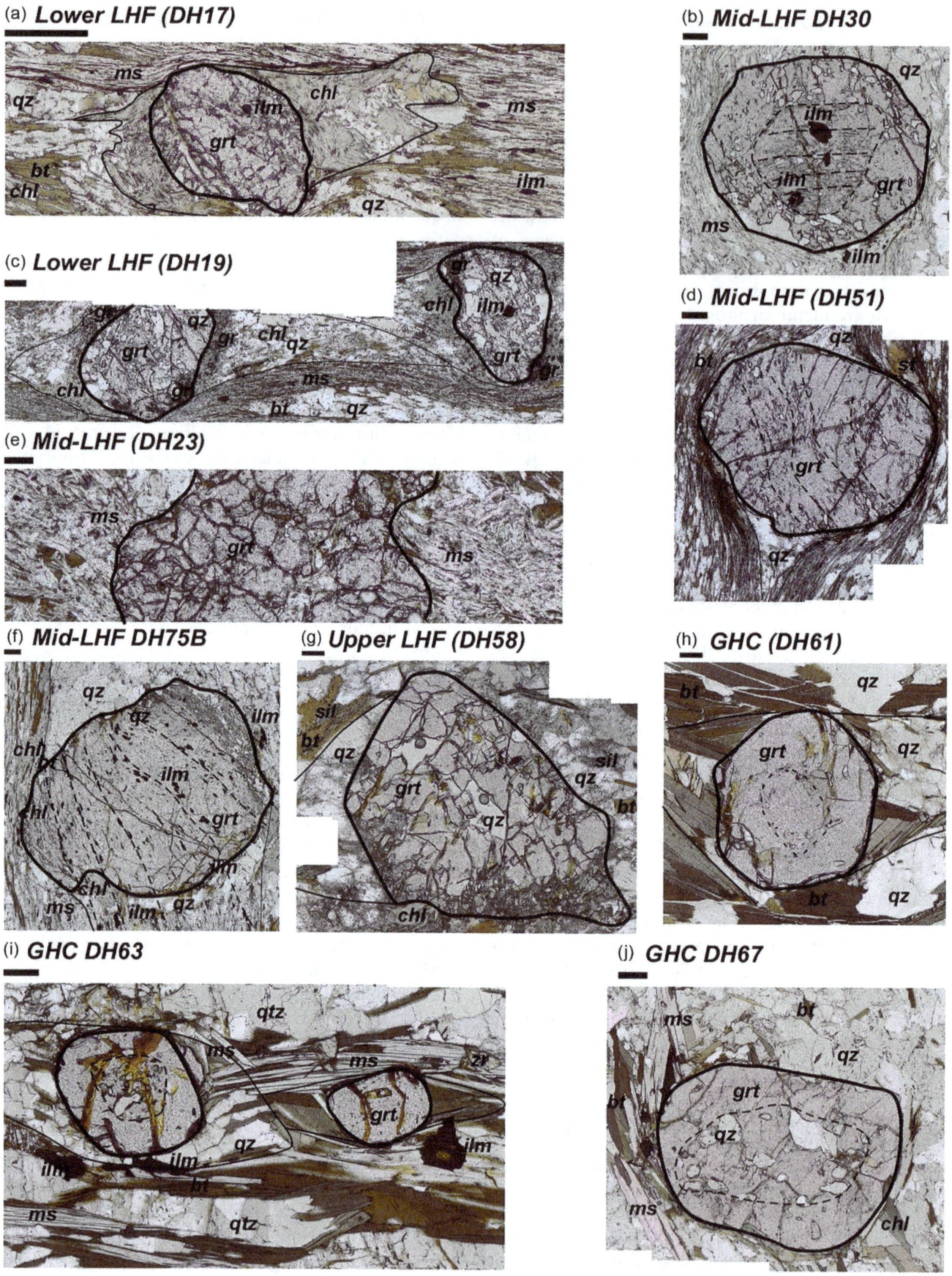

Figure 6.9 Selected petrographic (plane-polarized light) images of samples along the Darondi Khola show the relationship between the garnet porphyroblasts and rock textures. Garnets are outlined using bold lines. Lighter and dashed lines indicate pressure shadows and inclusion trails, respectively; mineral abbreviations after Whitney and Evans (2010). Panels are labeled with the sample number. The scale bar for each image is 200 μm. See Figures 6.4 and 6.5 for sample locations.

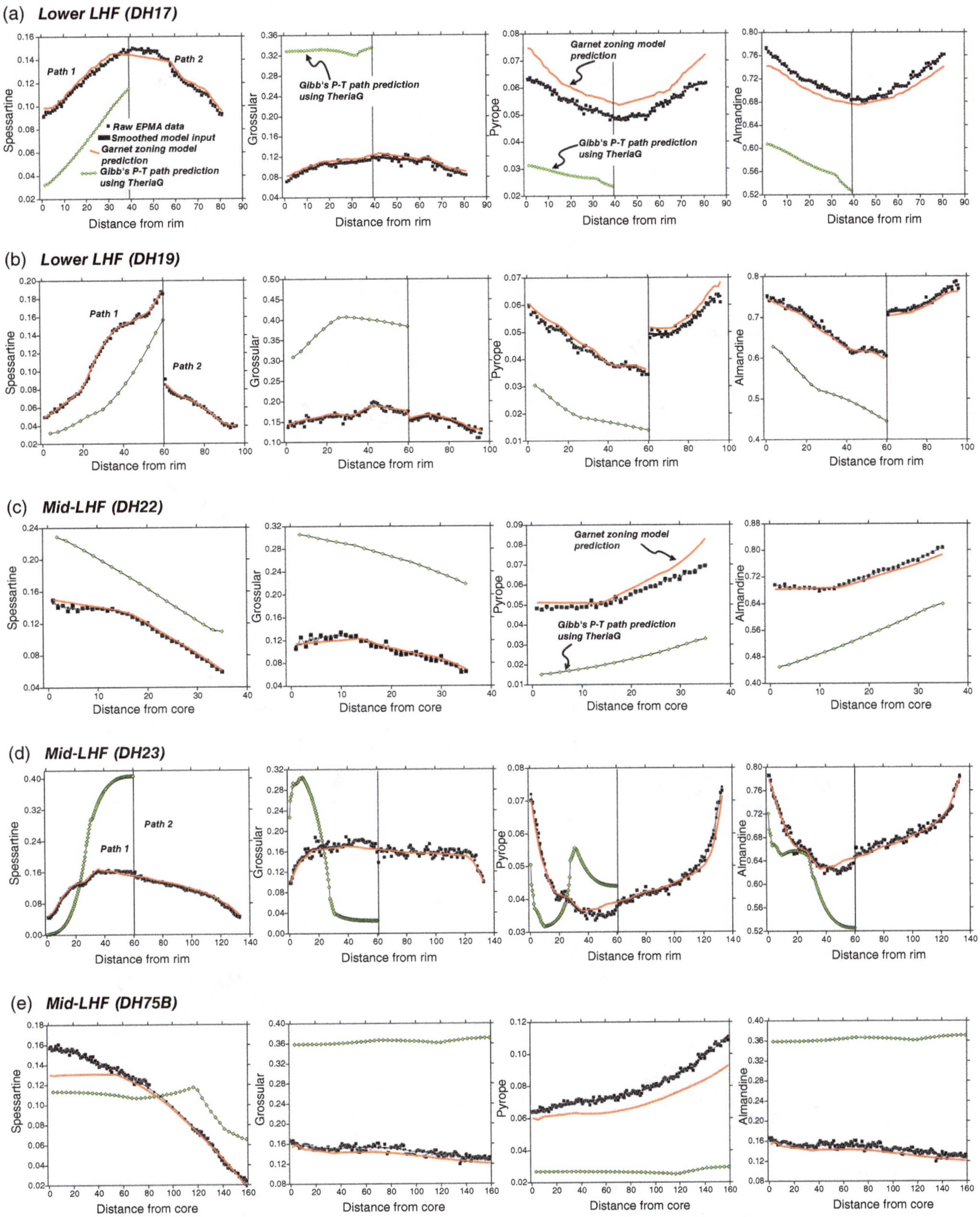

Figure 6.10 Compositional transects across garnets in lower LHF samples: (a) DH17 and (b) DH19, and middle LHF samples (c) DH22, (d) DH23, and (e) DH75B. Distance is in analytical points, and the spacing between the points is ~20 μm. The larger black squares are the raw electron microprobe data, whereas the smaller black squares near the EPMA data points are the smoothed data used for input into the model to generate the high-resolution P-T paths. The red lines show garnet zoning based on predictions made by the high-resolution P-T paths. TheriaG was used to predict the garnet zoning for Gibb's P-T paths, which are also indicated as green symbols.

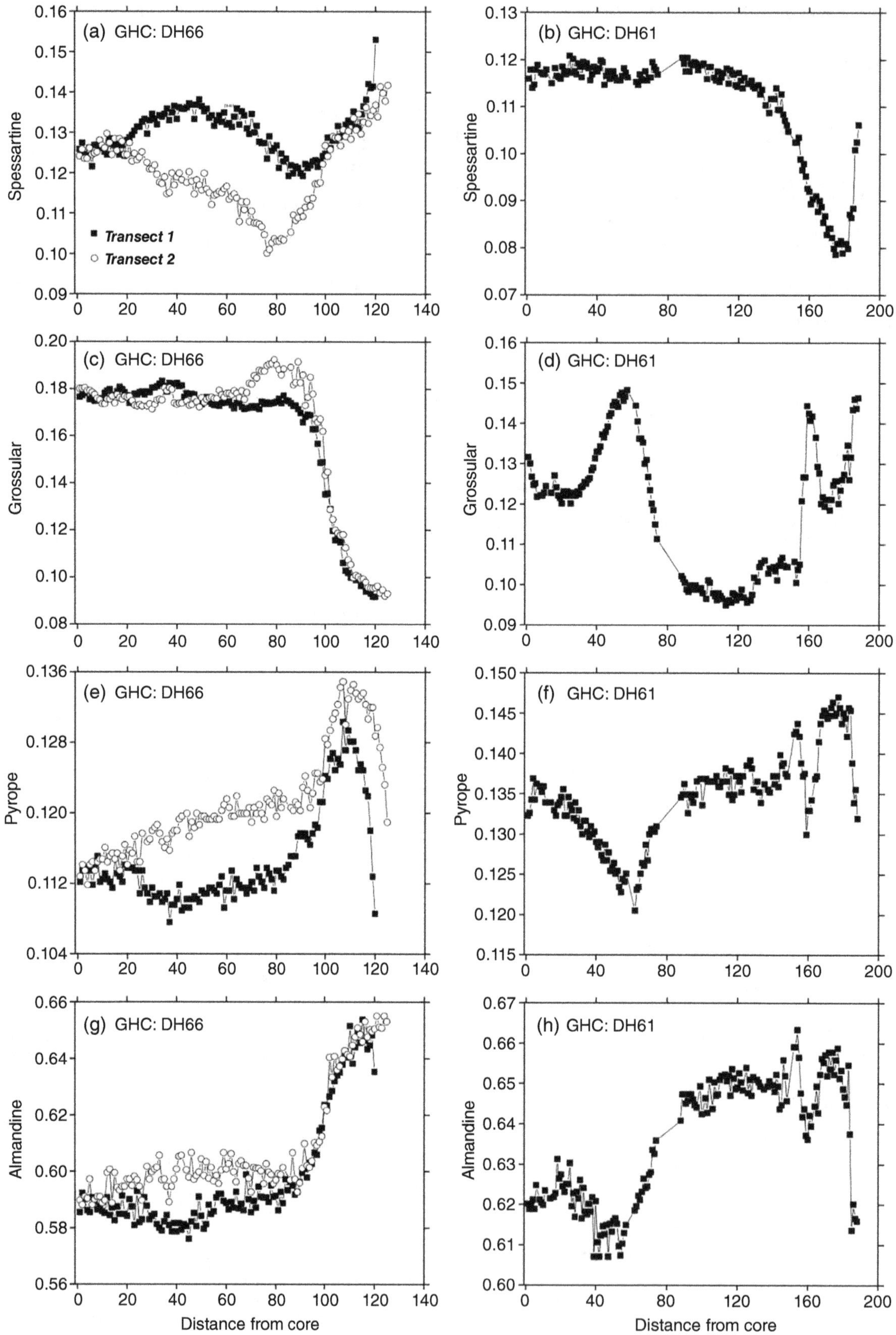

Figure 6.11 Compositional transects across garnets in GHC samples DH66 (two transects) in (a) spessartine, (c) grossular, (e) pyrope, and (g) almandine. Panels (b), (d), (f), and (h) are compositional transects in spessartine, grossular, pyrope, and almandine, respectively, across a garnet in sample DH61. Distance from the garnet core is in analytical points, and the spacing between the points is ~20 μm.

Table 6.2 P-T data generated from samples collected along the Darondi Khola using garnet isopleth thermobarometry

Sample[a]	Isopleth Core P (±1 kbar)	Isopleth Core T (±25°C)	Isopleth Rim P (±1 kbar)	Isopleth Rim T (±25°C)
		Lower LHF		
DH17	4.5	540	4.5	560
DH19	4.0	520	4.8	550
		Middle LHF		
DH22	4.8	540	6.0	580
DH23	5.1	525	6.0	580
DH26	6.5	550	5.5	560
DH75A	6.8	540	6.5	580
DH75B	7.5	550	8.8	590
		Upper LHF		
DH51	–[b]	–	7.0	650
DH58	6.0	550	–	-
		GHC		
DH60	–	–	7.5	570
DH61	7.0	580	7.5	600
DH63	–	–	10.5	650
DH66			7.5	550

[a] See Figure 6.3 for sample locations.
[b] "–" not reported.

LHF, with P-T conditions increasing up section from near the garnet isograd at 520°C ± 25°C and 5.0 ± 1.0 kbar (DH16) to 640 ± 25°C and 11.5 ± 2 kbar (samples DH57 and DH58) just below the mapped MCT. GHC rocks were collected near the mapped MCT (Figs. 6.4 and 6.5) and recorded the highest P-T conditions (660°C–715°C and 7.6–11.5 kbar). Samples DH17, DH19, and DH75A have P-T paths consistent with burial (increasing in both P and T), whereas those from DH22 and DH26 yield exhumation paths (decreasing P with increasing T). P-T paths from DH75B and DH23 are hairpin and interpreted as recording both burial and exhumation during the MCT shear zone imbrication. Monazite inclusions in DH75B garnets ranged from 11.1 ± 0.7 Ma to 7.6 ± 0.2 Ma (Th-Pb secondary ion mass spectrometry, ±1σ). These ages constrain Late Miocene reactivation of the MCT shear zone (Fig. 6.5). Late Miocene monazite inclusions in garnet are also found in other samples along the transect, and young (Pliocene-Late Miocene) $^{40}Ar/^{39}Ar$ muscovite total gas ages are also reported (Catlos et al., 2001; Kohn et al., 2001).

Isopleth Core Thermobarometry

The approach to obtain isopleth P-T conditions and paths is the same as that outlined in Catlos et al. (2018, 2020) and Etzel et al. (2019). Bulk rock compositions were obtained from rock chips of the DH samples using inductively coupled plasma spectrometry (ICP) at Activation Laboratories (Canada; Tables 6.3 and 6.4). No modifications in these compositions were made for the approach. Samples DH22 and DH23 were collected nearby, and the same bulk composition was used for both rocks. Compositions vary from low SiO_2 (~45 wt % samples DH17, DH22, and DH23) to higher SiO_2 (69–76 wt% DH26, DH75A). Upper LHF and GHC rocks generally have mid to high SiO_2 contents (56–80 wt%). Their molar values are used as direct input for the effective bulk composition needed to create the isochemical phase diagrams using the software package Theriak-Domino (de Capitani & Brown, 1987; de Capitani & Petrakakis, 2010) (Figs. 6.12–6.17). The Holland and Powell (1998), with updates to solution models through 2010, thermodynamic data set and appropriate mixing models in the system MnO-Na_2O-CaO-K_2O-FeO-MgO-Al_2O_3- SiO_2-H_2O-TiO_2 were used. The specific solid solution models used are the same as in Catlos et al. (2018). H_2O saturation is assumed for these samples (i.e., H_2O (100) in Theriak-Domino), as is appropriate for these assemblages. Fe^{3+} was not estimated but did not significantly affect results.

For samples where compositional data were available from the garnet's central section, isopleths of ± 0.01–0.02 mole fraction spessartine, almandine, pyrope, grossular, and ± 0.01–0.02 Mg-# (Mg/Fe+Mg), are plotted on the phase diagram as grey-shaded bars. Their intersection corresponds with our closest approximation of garnet core P-T conditions as indicated by a polygon. Although isopleth intersections occurred over small regions, the uncertainty in the conditions is likely approaching that of conventional methods (i.e., ± 25°C and ± 1 kbar). Overlapping garnet core isopleths are found only for LHF samples DH17, DH19, DH22, DH23, DH26, DH75A, DH75B, DH58, and GHC sample DH61 (Figs. 6.12–6.16). The core was defined as the portion of the garnet with the highest Mn content and was the best approximation of the chemical system when the garnet began growth. In all cases, the intersections are located in mineral stability fields consistent with their assemblages (feldspar + garnet + biotite + phengite + ilmenite ± rutile ± chlorite + quartz + H_2O) and anticipated conditions. Core metamorphic conditions increase up section over a north-south distance of ~5 km from a low of 4–4.5 kbar and 520°C–540°C in lower LHF samples DH17 and DH19 to 6.8–7.5 kbar and 540°C–550°C in middle LHF samples DH75A and DH75B (Table 6.2). Upper LHF sample DH58 collected directly beneath the MCT yields overlapping core isopleths at ~550°C and 6.0 kbar. Core isopleths for GHC sample DH61 overlap at ~7 kbar and 580°C, similar to those from some LHF rocks (Table 6.2).

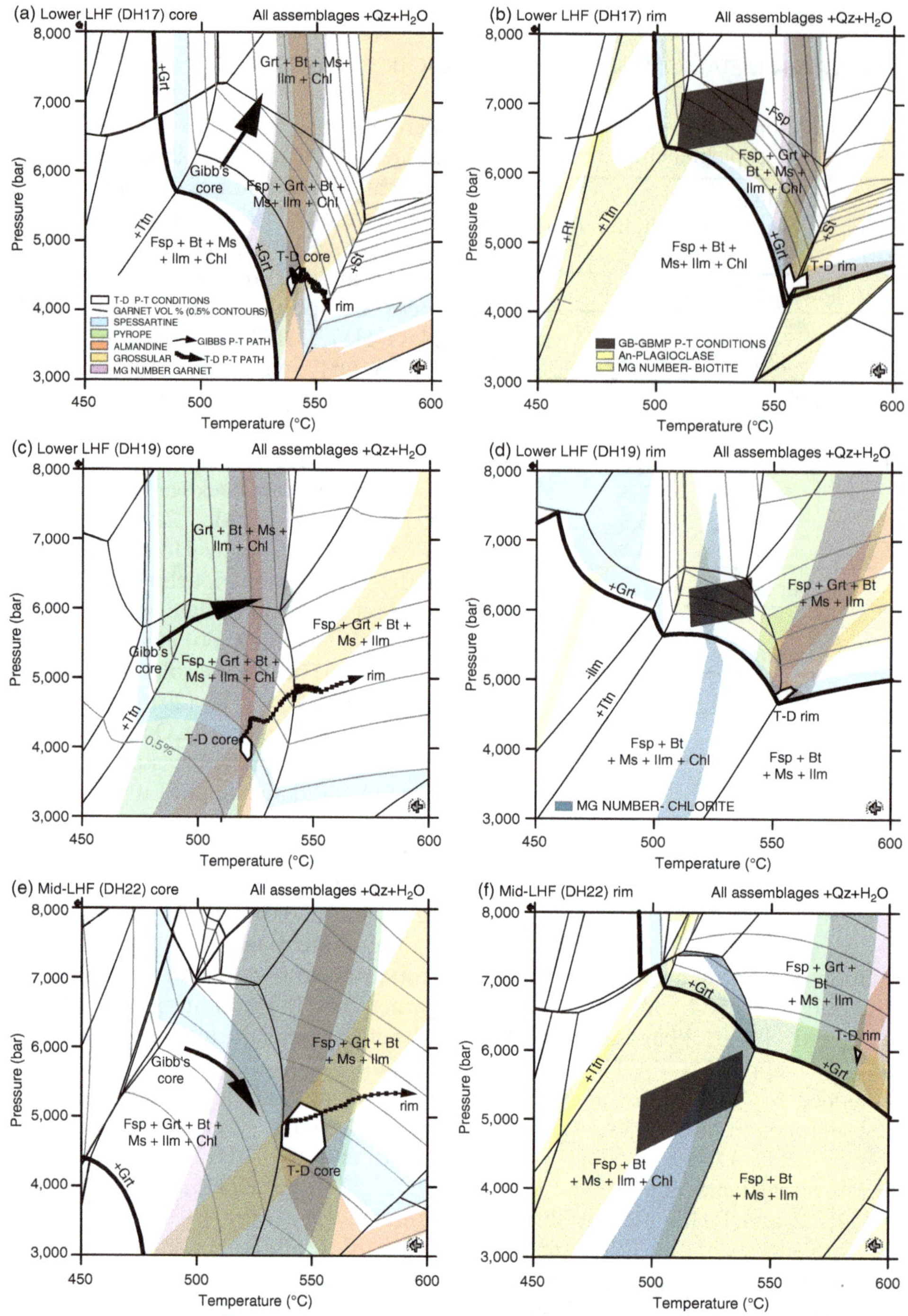

Figure 6.12 Isochemical phase diagrams from lower LHF samples DH17: (a) garnet core and (b) garnet rim, DH19 (c) core and (d) rim, and mid-LHF sample DH22 (e) core and (f) rim. See Figure 6.4 for sample locations. Garnet-in reaction line (+Grt) and garnet growth contours (volume 0.5% increments) are overlaid on each diagram. Each core diagram was created using the rock bulk compositions reported in Table 6.3 and the software program Theriak-Domino. Some mineral stability fields are labeled using abbreviations after de Capitani and Petrakakis (2010) and include quartz and H_2O. For the core diagrams, isopleths of ± 0.01–0.02 mole fraction spessartine (XMn), grossular (XCa), pyrope (XMg), almandine (XFe), and ± 0.01–0.02 Mg-number (Mg/Fe + Mg, Mg#) of the compositional data point selected from the garnet's highest Mn content are overlain (Kohn et al., 2001). They intersect as indicated by the polygon and are labeled "T-D core." High-resolution P-T paths (DH17, n = 2; DH19, n = 2; DH22, n = 1) were generated from garnet core-to-rim transects after the approach of Moynihan and Pattison (2013). These are compared with the P-T paths obtained using Gibb's method for the samples using the same data by Kohn et al. (2001). The starting point of Gibb's P-T paths is labeled with "Gibb's core." Isochemical phase diagrams for the garnet rim were created using the final effective bulk composition generated by Theriak-Domino. In this case, isopleths for the garnet compositions reported for garnet rims are overlain on the diagram, as well as matrix mineral composition isopleths for An-content for plagioclase, Mg-number for biotite and chlorite (Mg/Fe + Mg, Mg# - bt and Mg# - chl; Kohn et al. 2001), when available. They intersect as indicated by the polygon and are labeled as "T-D rim." These are compared with the rim P-T conditions [garnet-biotite (GB) thermometry and garnet-biotite-muscovite-plagioclase (GBMP) barometry] for the samples using the same data and approaches by Kohn et al. (2001).

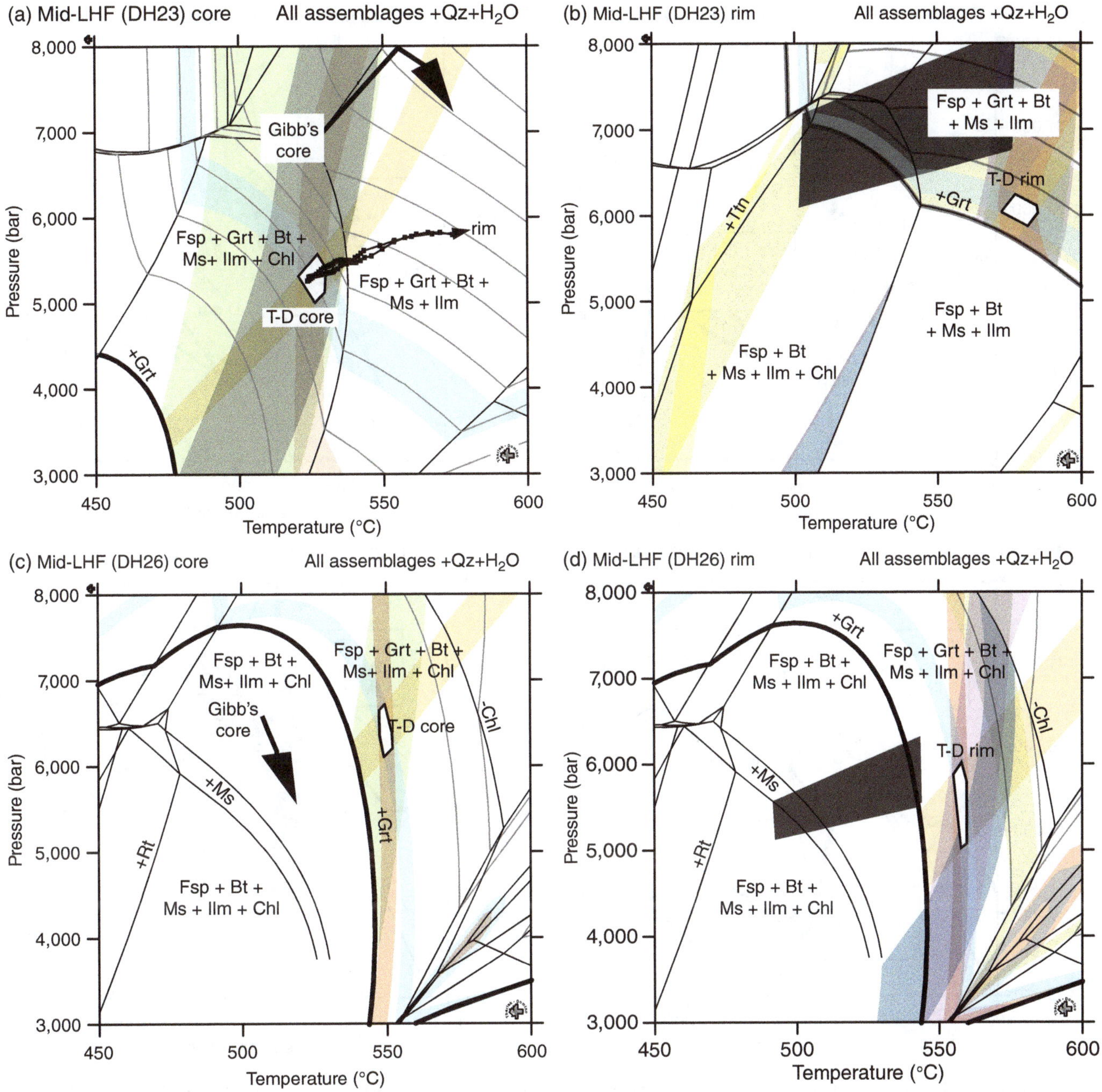

Figure 6.13 Isochemical phase diagrams from middle LHF samples DH23: (a) garnet core and (b) rim and DH26 (c) core, and (d) rim. See Figure 6.4 for sample locations. The detailed caption is the same as in Figure 6.12. High-resolution P-T paths for sample DH23 were generated from two garnet core-to-rim transects after the approach of Moynihan and Pattison (2013). These are compared with the P-T paths obtained using Gibb's method for the samples using the same data by Kohn et al. (2001). No data were available for the garnet transect for sample DH26, so a high-resolution P-T path was not created. In this case, the rock bulk composition (Table 6.3) was used for both core and rim panels.

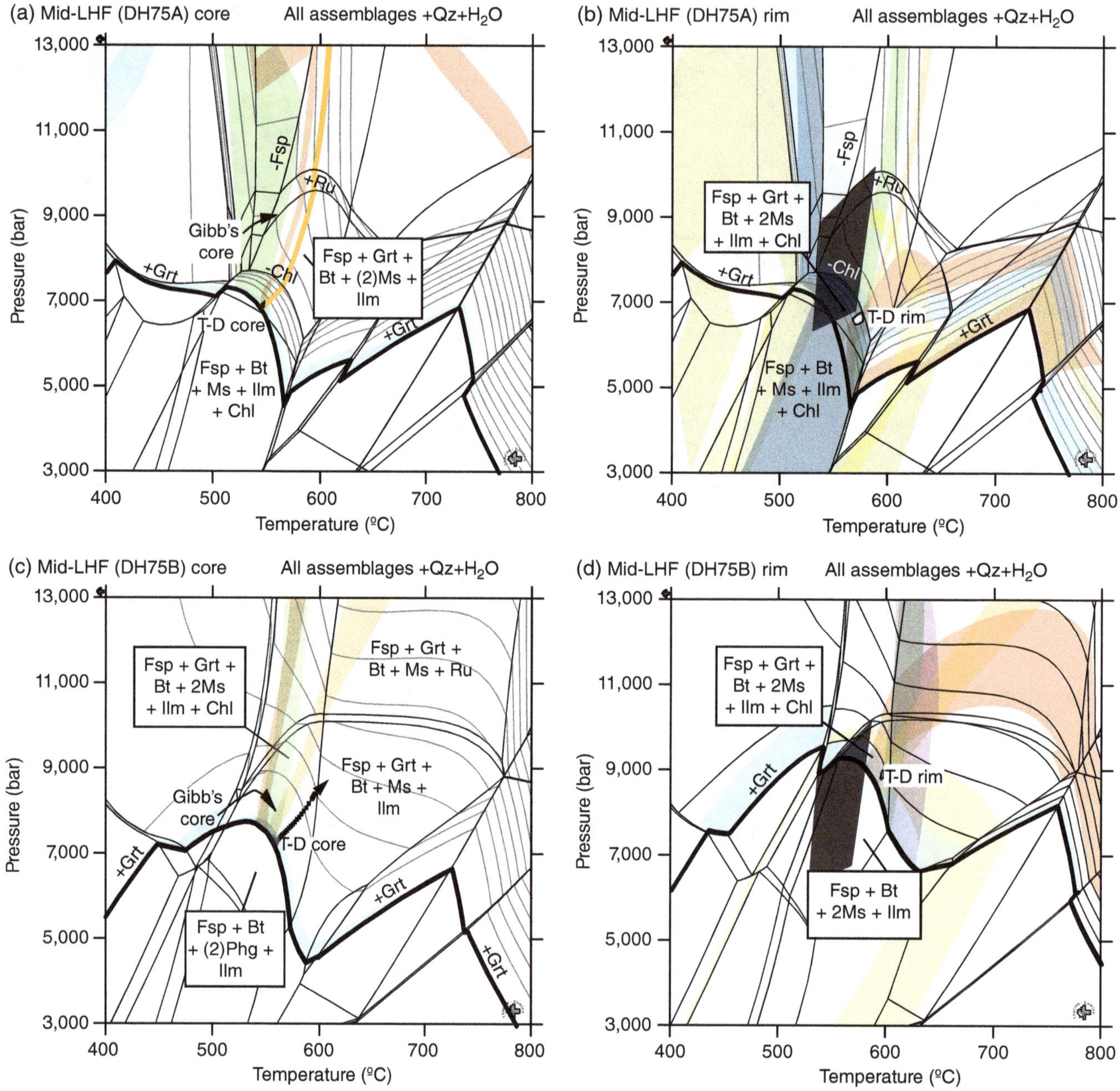

Figure 6.14 Isochemical phase diagrams from middle LHF samples DH75A (a) garnet core and (b) rim and DH75B (c) core and (d) rim. See Figure 6.4 for sample locations. The detailed caption is the same as in Figure 6.12. High-resolution P-T paths for sample DH75B were generated from garnet core-to-rim transects after the approach of Moynihan and Pattison (2013). These are compared with the P-T paths obtained using Gibb's method for the samples using the same data by Kohn et al. (2001). Gibb's path was also reported for sample DH75A. The starting point of Gibb's P-T paths is labeled "Gibb's core." The isochemical phase diagram for the garnet rim in sample DH75B was created using the final effective bulk composition generated by Theriak-Domino. In this case, isopleths for the garnet compositions reported for garnet rims are overlain on the diagram and matrix mineral compositions isopleths for ± 0.01 An-content for plagioclase (Kohn et al., 2001). The polygon indicates intersecting isopleths and is labeled "T-D rim." No data were available for the garnet transect for sample DH75A, so a high-resolution P-T path was not created, and the rock bulk composition (Table 6.3) was used for both core and rim panels. In the rim panels, data are compared with the conventional P-T conditions for the samples (grey polygon) using the same data reported by Kohn et al. (2001).

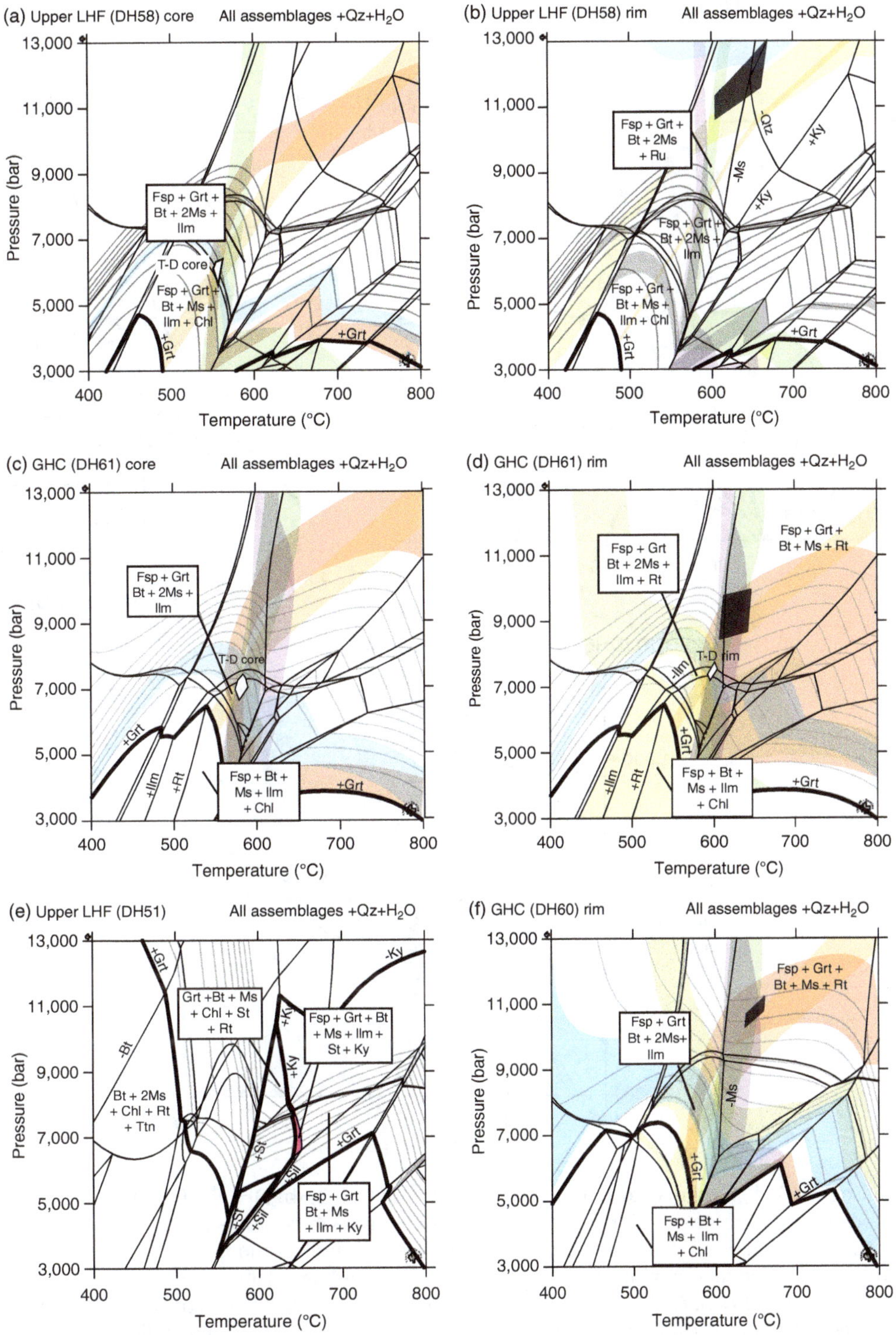

Figure 6.15 Isochemical phase diagrams from upper LHF samples DH58 (a) garnet core and (b) rim, and (e) DH51 show the mineral reactions only. The figure also includes isochemical phase diagrams from GHC sample DH61 (c) core and (d) rim, and DH60 (f) garnet rim data only. See Figure 6.4 for sample locations. Garnet-in reaction line (+ Grt) and garnet growth contours (volume 0.5% increments) are overlaid on each diagram. Each diagram was created using the rock bulk compositions reported in Table 6.4 and the software program Theriak-Domino. The detailed caption is the same as in Figure 6.12. All labeled stability fields include quartz and H_2O, except in panel (b), where the quartz-out reaction line is noted (-Qtz). The isopleths did not intersect in panels (b) and (f) but did in panel (d), as indicated by the polygon labeled "T-D rim." In the rim panels, data are compared with the rim P-T conditions for the samples (labeled polygon) using the same data and conventional thermobarometric approaches by Kohn et al. (2001). No mineral data are available for sample DH51, so the isochemical phase diagram in panel (e) shows only reaction lines with staurolite, sillimanite, garnet, and kyanite reaction lines in bold. This sample has coexisting kyanite and staurolite, so the colored shaded area indicates the field where both of these minerals are stable.

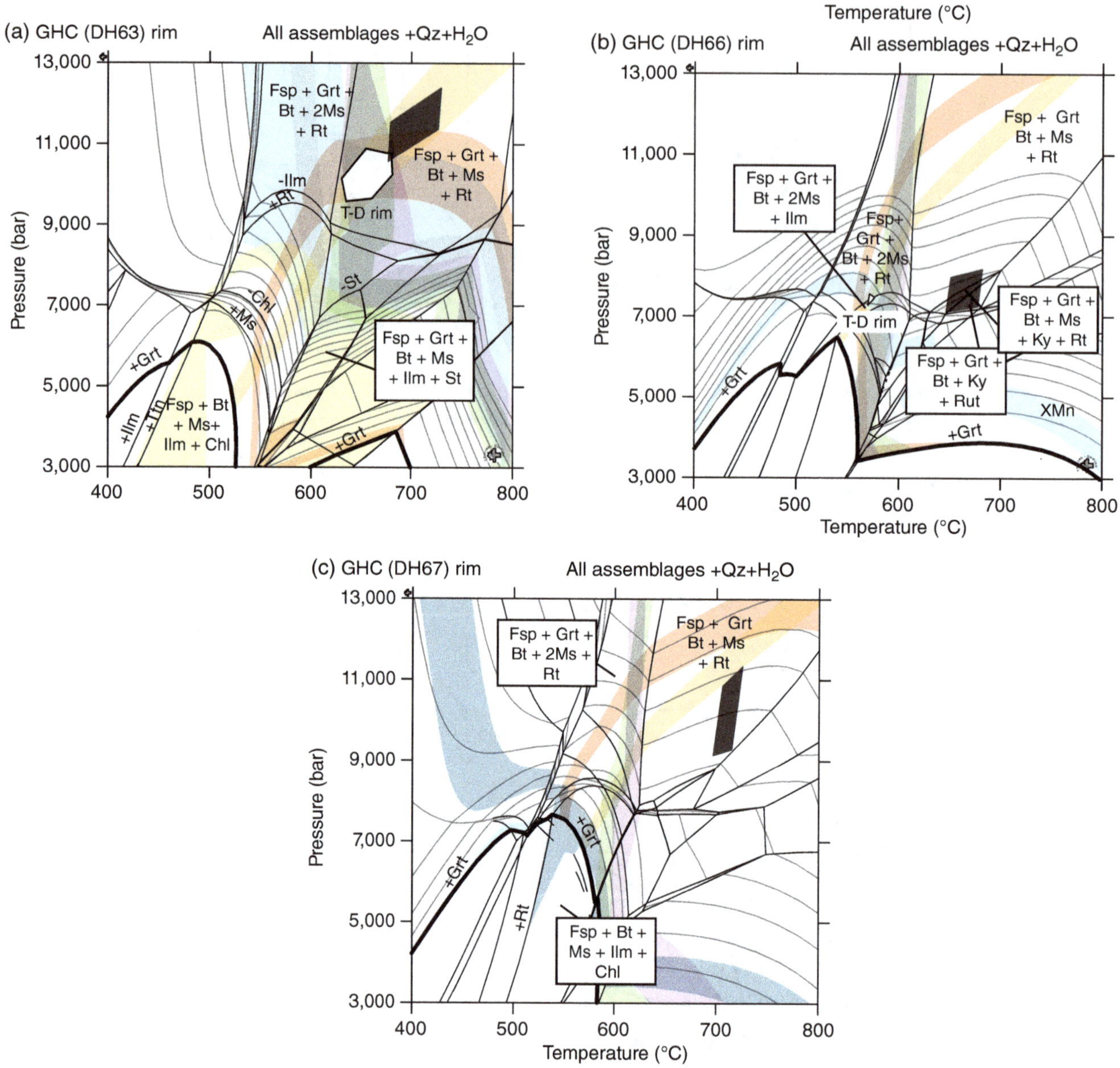

Figure 6.16 Isochemical phase diagrams from GHC samples (a) DH63, (b) DH66, and (c) DH67. See Figure 6.4 for sample locations. Garnet-in reaction line (+ Grt) and garnet growth contours (volume 0.5% increments) are overlaid on each diagram. Each diagram was created using the rock bulk compositions reported in Table 6.4 and the software program Theriak-Domino. Some mineral stability fields are labeled using abbreviations after de Capitani and Petrakakis (2010). All labeled stability fields include quartz and H_2O. For each diagram, isopleths of ± 0.01–0.02 mole fraction spessartine (XMn), grossular (XCa), pyrope (XMg), almandine (XFe), and ± 0.01–0.02 Mg number (Mg/Fe + Mg, Mg#) of the compositional data point selected from the garnet's rim are overlain (Kohn et al., 2001). Isopleths of An-content for matrix plagioclase (a) and Mg# for matrix chlorite are also overlaid on the diagrams in panels (a) and (c), respectively. Mineral isopleths intersect in panels (a) and (b) as indicated by the polygon, and the intersection is labeled "T-D rim." These isopleths did not intersect in panel (c). In each panel, data are compared with the rim P-T conditions for the samples (grey polygon) using the same data and conventional thermobarometric approaches by Kohn et al. (2001).

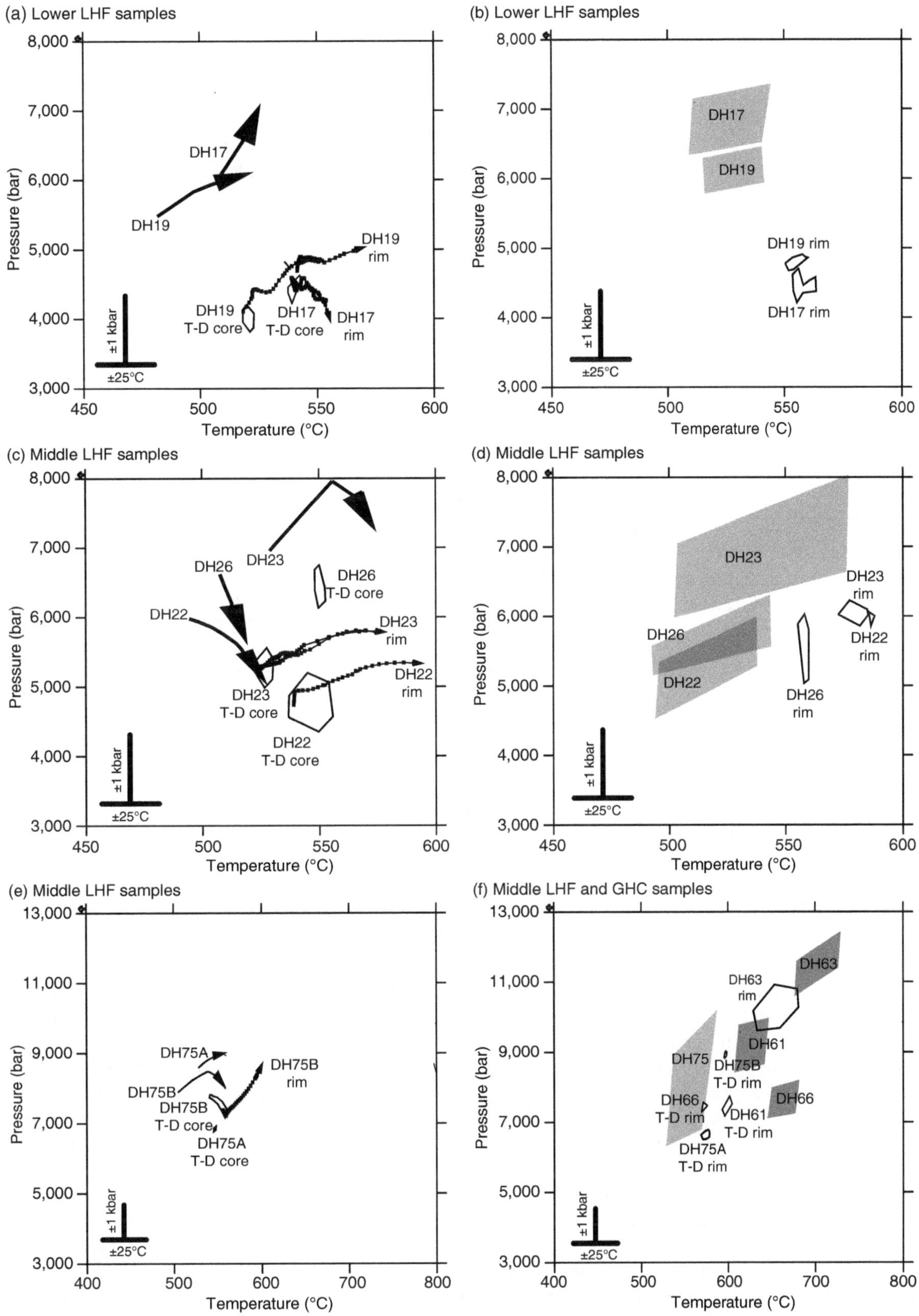

Figure 6.17 Summary of the P-T conditions and paths reported in Figures 6.12–6.16. In panels (a), (c), and (e), Gibb's P-T paths are shown as bold arrows, and high-resolution P-T paths are labeled with core and rim points. In panels (b), (d), and (f), rim data generated using isopleth (white polygons) and conventional thermobarometry (grey polygons) are compared. Each panel shows uncertainty scales in T (± 25°C) and P (± 1 kbar) as insets.

Table 6.3 Bulk rock compositions (wt%) from lower and middle LHF samples

Analyte/sample[a]	DH17	DH19	DH22/23	DH26	DH30	DH75A	DH75B
SiO_2	45.39	66.04	45.22	76.04	50.89	70.24	68.75
Al_2O_3	27.31	16.36	28.09	11.79	27.80	12.79	8.16
Fe_2O_3(T)	9.18	7.12	6.37	3.52	6.11	4.98	3.78
MnO	0.10	0.17	0.20	0.05	0.02	0.02	0.02
MgO	2.81	1.88	1.50	1.24	2.03	1.78	1.52
CaO	0.30	0.44	0.51	0.58	0.20	0.39	0.46
Na_2O	1.07	0.45	1.57	3.54	1.44	1.14	1.51
K_2O	7.72	5.14	7.26	1.22	6.12	2.97	1.99
TiO_2	0.86	0.52	0.86	0.40	0.79	0.49	0.35
P_2O_5	0.11	0.16	0.13	0.11	0.14	0.09	0.07
LOI	n.m.	2.15	n.m.	2.26	5.21	4.89	n.m.
Total	94.8	100.4	91.7	100.7	100.7	99.8	86.6

[a] See Figure 6.3 for sample locations.

Table 6.4 Bulk rock compositions (wt%) from upper LHF and GHC samples

	DH51	DH58B	DH60	DH61	DH63	DH66	DH67
Analyte/sample[a]	LHF samples				GHC samples		
SiO_2	66.13	55.66	71.51	55.47	69.01	68.35	79.92
Al_2O_3	15.14	23.82	12.62	19.54	13.88	10.47	9.01
Fe_2O_3(T)	5.29	4.48	5.80	7.76	6.69	5.44	3.57
MnO	0.03	0.19	0.04	0.14	0.10	0.07	0.02
MgO	2.31	1.66	2.41	4.71	1.99	2.00	1.65
CaO	0.21	3.38	1.02	2.59	1.15	1.07	1.37
Na_2O	0.58	6.46	2.55	4.23	2.46	1.52	2.18
K_2O	3.78	2.08	2.68	3.84	2.32	2.53	1.22
TiO_2	0.52	0.67	0.65	0.99	0.68	0.54	0.62
P_2O_5	0.12	0.01	0.15	0.03	0.15	0.09	0.05
LOI	5.89	n.m.	n.m.	n.m.	n.m.	n.m.	1.28
Total	100	98.4	99.4	99.3	98.4	92.1	100.9

[a] See Figure 6.3 for sample locations.

The garnet-in reaction line and garnet growth contours (volume 0.5% increments) are overlaid on each isochemical phase diagram (Figs. 6.12–6.16). The topology of the diagrams and the location of the garnet core suggest that garnet appears in all samples through chlorite dehydration. Only samples DH75A and DH75B yield core P-T conditions that overlap the garnet-in reaction line (Fig. 6.14). The other samples yield overlapping isopleths on or near the 0.5% (DH17, Fig. 6.12a), 1% (DH19, DH61, DH58, Figs. 6.12c, 6.15a, and 6.15c), and 1.5–2% volume contours (DH22 and DH23, Figs. 6.12e and 6.13a). The results suggest that the true core was missed while analyzing these garnets. The lack of retrograde zoning in the LHF samples suggests that the effect of diffusional homogenization is minimal (Fig. 6.10), although this process likely modified GHC DH61 garnet compositions (Fig. 6.11b).

High-Resolution and Gibb's P-T Path Modeling

For samples DH17, DH19, DH22, DH23, DH26, DH75A, and DH75B, the garnet's core isopleth P-T conditions can be directly compared with those obtained using Gibb's approach (Fig. 6.17). Sample DH26 yields the same pressure using both methods (~6.5 kbar), but the thermal conditions are lower using Gibb's method by 50°C. Sample DH23 yields the same thermal conditions using both approaches (~525°C), but the isopleth barometry suggests the sample experienced lower pressures by ~2 kbar (5.1 kbar compared with 7.0 kbar). In all other samples, isopleth thermobarometry results in the samples recording lower P (from 0.5 kbar to 1.7 kbar, samples DH75A and DH75B) and higher T (from 10°C to 50°C, samples DH75A and DH75B) than those generated by Gibb's method. The discrepancies of 10°C and 0.5 kbar are within the stated uncertainties of both methods (Kohn, 1993), but those approaching 50°C and 1.5 kbar are significant differences that would influence the understanding of their depth of exhumation and metamorphic reaction history. For example, in sample DH26, Gibb's core P-T condition shows similar P, but the low Gibb's T starting conditions lie in a region where chlorite, not garnet, is stable (Fig. 6.13c).

High-resolution P-T paths (Moynihan & Pattison, 2013) were generated for lower LHF samples where garnet transect data were available, suggested prograde conditions, and minimal modification since growth. In generating the high-resolution P-T paths, zoning profiles are smoothed using a Savitzky-Golay function to minimize the impact of missing analyses due to inclusions or cracks (Fig. 6.10). The Matlab script starts with the bulk composition and initial smoothed core garnet composition and calls Theriak-Domino to calculate an isochemical phase diagram. A Matlab optimization function searches the P-T grid for the smallest misfit between a modeled garnet composition and the smoothed composition. It then calculates the portion of the bulk composition sequestered in the first step of garnet growth. Sequestered components are subtracted from the bulk composition to estimate an effective bulk composition for the next growth step. A new diagram is calculated from the effective bulk composition, and the process is repeated for all steps along the garnet zoning profile from core to rim. Each step yields an estimate of the P-T conditions of incremental garnet growth, culminating in a P-T path. These data are available in the repository.

Using the garnet zoning method, two independently estimated high-resolution P-T paths can be obtained from a rim-to-rim compositional transect across the garnet. The expectation is that the P-T paths from the same garnet should record similar trajectories, and this was the case for samples DH17, DH19, and DH23 (Figs. 6.12a,c , 6.13a). However, the starting and endpoints may differ due to the proximity of the initial condition near the garnet core and the extent of garnet rim preservation. In addition, garnet zoning can be predicted by the high-resolution P-T path and is indicated as a bolder grey line in Figure 6.10. The high-resolution P-T paths reproduce the original garnet zoning to ± 0.01 mole fraction in most cases and for most compositions, which is expected if the garnet behaved in a closed system and had no significant changes in bulk rock composition as it grew.

As with the core conditions, the shapes of the high-resolution P-T paths for samples DH17, DH19, DH22, DH23, DH75A, and DH75B can be directly compared with those obtained using Gibb's approach (Fig. 6.17). The length of the high-resolution and Gibb's P-T paths are similar (ranges from 14°C to 52°C) but do not show the same trajectories. For example, instead of hairpin trajectories (increase in P followed a P decrease as T increases) predicted by DH23 and DH75B Gibb's P-T paths, their high-resolution P-T paths suggest that these rocks only experience an increase in P as T increases. Gibb's P-T path in sample DH17 shows an increase in P, but the high-resolution P-T path from the same rock using the same data decreases in P as T increases. Minor fluctuations in P in the high-resolution P-T paths (± 10–50 bars) in sample DH19 and near the core of sample DH22 are likely due to small compositional changes as the program seeks the best fit and should not be interpreted tectonically. DH19 shows an overall similar P-T path shape as the one reported using Gibb's approach, but the conditions are different, with the high-resolution P-T path showing much lower P (by ~2 kbar) and higher T (by ~50°C). The high-resolution P-T paths from samples DH22 and DH23 are similar in that they show minor increases in P as T increases, but Gibb's P-T path from sample DH22 suggests decompression, a result not observed with this sample. A P-T path from sample DH26 could not be generated due to a lack of garnet zoning data, but core and rim isopleth conditions suggest it could have followed the decompression path as T increased. A decrease in P as T increases is the same trajectory suggested by Gibb's P-T path but at higher T conditions.

Gibb's P-T paths were used as input in the program TheriaG (Gaides et al., 2008) using the bulk composition indicated in Table 6.3 to gauge how well the paths reproduced garnet zoning. The time of garnet growth over the P-T path is regularly spaced over durations of 5 m.y., 10 m.y., and 15 m.y. Modeling parameters are available in Catlos (2022). In each case, the predicted Gibb's P-T path was identical and would predict garnet zoning inconsistent with what is observed in the sample (Fig. 6.10). Results may change if different size classes than those used were selected (10 m successive shells in 2,000 m garnet radius), if garnet growth did not occur over regular space durations, or if the selected exhumation path were different. Using the chosen parameters, Gibb's P-T path predictions for samples DH17, DH19, and DH22 show similar trends with the garnet zoning but significantly different compositions (Fig. 6.10). The Gibb's P-T paths predictions for those samples that produced hairpin P-T paths (DH75B and DH23) resulted in significantly different garnet zoning and compositions and did not replicate observed garnet compositions.

Isopleth Rim Thermobarometry

Garnet rim isopleth conditions were estimated using the same garnet compositions used to generate conventional P-T conditions. Rim isopleths also include those from the sample's average composition of matrix plagioclase (± 0.01 mole fraction Ca), chlorite, and biotite [± 0.01 Mg-number, Mg# = Mg/(Mg + Fe)], when available. The last effective bulk composition generated by the high-resolution P-T paths was used for LHF samples

DH17, DH19, DH22, DH23, and DH75B. For other LHF samples where garnet transect data are unavailable and for the GHC rocks, the bulk composition was used (DH26, DH75A, DH58, DH60, DH61, DH63, DH66, DH67) (Tables 6.3 and 6.4). Only three GHC samples (DH61, DH63, and DH66) yield overlapping garnet compositional isopleths (Figs. 6.15d and 6.16a,b), and those from upper LHF sample DH58 did not overlap.

The extent of overlap of the garnet rim with matrix mineral compositions varies for the GHC and LHF rocks. Garnet rim isopleths for GHC sample DH61 rim overlapped with all matrix mineral compositions, but DH66 shows no overlap of the garnet rim with any of the matrix mineral compositions. The garnet rim isopleths for GHC sample DH63 overlapped with ± 0.01 Mg-number chlorite but not with the biotite or plagioclase compositions. As seen with some footwall samples, the overlaps for the GHC assemblages are located far from the garnet-in reaction line, as seen by the garnet growth contours (> 1.5 vol%) (Figs. 6.15 and 6.16). The garnet rim isopleths for LHF samples DH17, DH19, DH22, DH23, and DH75B overlap with Mg# biotite but not plagioclase. No matrix mineral isopleths overlap with the garnet rim isopleths for samples DH26 and DH75A within the compositional ranges applied here (± 0.01 mole fraction Ca and Mg#).

For samples where garnet and matrix mineral isopleths overlap, conditions are consistent with their mineral assemblages. They are similar to the garnet core assemblages (feldspar + garnet + biotite + phengite + ilmenite ± rutile ± chlorite + quartz + H_2O). As with the core conditions, the rim P-T conditions increase up section over a north-south distance of ~5 km from a low of 4.5–4.8 kbar and 550°C–560°C in lower LHF samples DH17 and DH19 to 5.5–8.8 kbar and 560°C–590°C in middle LHF samples (Table 6.2). Although garnet compositional data are not available for Upper LHF sample DH51, its mineral assemblage of coexisting staurolite and kyanite allows for an approximation of rock conditions using only its bulk rock composition and observations conditions where these minerals coexist (Fig. 6.15e), which appears at ~7.0 kbar and ~650°C. Rim isopleths for GHC samples DH60, DH61, and DH66 yield similar P of ~7 kbar, but T ranges from 550°C–600°C. GHC sample DH63 yields the highest P-T isopleth conditions of ~10.5 kbar and 650°C.

Comparisons are made between the conventional rim P-T conditions and isopleth rim conditions. As seen in Figure 6.17b, the lower LHF samples yield higher T (by 25–30°C) and lower P (by 1.4–2.3 kbar). All middle LHF samples (Fig. 6.17d, f) overlap in P conditions within uncertainty, but the isopleth T for samples DH22 and DH26 is higher than the conventional results by 5°C–85°C, depending on how uncertainty is applied. For GHC sample DH61, the approaches yield similar T conditions, but P differs by 1–2 kbar, depending on uncertainty (Fig. 6.17f). The opposite observation is seen with GHC sample DH66, where P is similar, but the isopleth conditions suggest significantly lower T (Fig. 6.17f). Finally, some overlap is seen with GHC sample DH63, but the conventional results suggest higher P-T than the isopleth results.

6.5. DISCUSSION

Using the same samples and data, Darondi Khola MCT footwall P-T paths using Gibb's method and high-resolution garnet modeling do not yield the same conditions or shapes (Fig. 6.17), even within the estimated uncertainties of Gibb's method (e.g., Kohn, 1993). In addition, the lowest-grade footwall samples record higher T and lower P isopleth rim P-T conditions than those generated using conventional thermometers and barometers. Conventional garnet rim P-T conditions and isopleth thermobarometry for GHC samples yield differing absolute conditions, although overlap exists within uncertainty (± 25°C and ± 1 kbar). An important check on the feasibility of the P-T conditions generated using any approach is if the results seem geologically reasonable and consistent with mineral assemblages (e.g., Moynihan & Pattison, 2013; Kelly et al., 2015; Catlos et al., 2018; Etzel et al., 2019; Craddock Affinati et al., 2020). However, this is the case with all conditions reported for the Darondi Khola samples, regardless of approach.

Several assumptions underlie many P-T estimates generated using thermodynamic modeling. For all thermobarometric methods applied here, a critical assumption is that the samples' minerals experienced equilibrium, which can never be proven for any rock system (e.g., Spear & Peacock, 1989; Lanari & Duesterhoeft, 2019). The samples are also assumed to have experienced closed system behavior, and the original compositions of the mineral phases and the bulk rock have not changed significantly since metamorphism (e.g., Lanari & Engi, 2017). LHF assemblages appear to have preserved their original garnet compositions, as shown by their prograde zoning profiles (Fig. 6.10). Garnets with preserved divalent cation zoning based on previously reported thermal conditions of generally < 600°C (e.g., Carlson, 1989; Spear, 1993; Carlson, 2002) are consistent with the results shown here. GHC samples show fluctuations in garnet compositions from core to rim and have evidence of diffusional modification by an increase of Mn at the rims (Fig. 6.11).

Multiple sources of error are inherent in conventional P-T conditions. They include uncertainty in the accuracy of end-member reactions, electron microprobe analyses, calibration errors, variations in activity models, and compositional heterogeneity (e.g., Kohn & Spear, 1991).

The precise uncertainty with approaches that involve isochemical phase diagrams is likewise challenging to determine due to the same factors incorporated into their creation as well as the uncertainty associated with the thermodynamic properties inherent in the choice of internally consistent database (e.g., molar enthalpy of formation, molar entropy, molar volume, heat capacity, bulk modulus, Landau parameters, and Margules parameters; e.g., White et al., 2014; Lanari & Duesterhoeft, 2019). The error suggested by the grid created due to overlapping mineral compositional isopleths likely underestimates the actual uncertainty in the identified conditions. Therefore, applying standard values of uncertainty (± 25°C and ± 1 kbar) to the overlapping isopleth conditions as those used for conventional results appears appropriate and is commonly reported (e.g., Spear & Peacock, 1989; Kohn, 1993; Kohn et al., 2001).

Ultimately, each approach to generating P-T conditions discussed here transforms the sample into a model representing the true rock and mineral assemblage. However, it restricts its behavior as if it were in a closed system that experienced particular boundary conditions. Confidence in conventional and Gibb's P-T paths increases when conditions agree with mineral assemblages and if the P-T paths reproduce broadscale trends in garnet zoning from core to rim. Samples collected from the same outcrop or nearby should yield similar P-T conditions and paths. Although Kohn et al. (2001) report only one Gibb's P-T path per sample, the expectation is that multiple paths collected from the same garnet or garnets in the same rock would agree in terms of shapes and conditions. The high-resolution P-T path approach and the garnet isopleth thermobarometry use these criteria to evaluate the estimated result's appropriateness. However, they have two additional values in critically evaluating results. First, a user can gauge the extent of overlapping mineral isopleths in P-T space. Second, a user can identify how well the high-resolution P-T paths predict the trends and values of garnet compositional zoning (Fig. 6.10). A significant value of the high-resolution P-T path and isopleth approaches is that a user can detect when systems stray from the equilibrium and closed system assumptions.

These samples illustrate that not all garnets are suitable candidates for high-resolution P-T path modeling and isopleth thermobarometry. Garnets with significant changes in composition over short distances from the core to the rim and those affected by diffusion cannot be modeled. Garnets in samples that experienced significant changes in bulk composition or multiple deformation episodes resulting in modification of composition are also unable to be modeled. Not all field areas are ideal candidates, and the GHC samples show that they often fail assumptions required for isopleth thermobarometry and high-resolution P-T path modeling. For example, overlapping garnet core isopleths were found for only one GHC sample, DH61, which was located far from the garnet-in reaction line (Fig. 6.15c). The intersections for all samples, except DH75A and DH75B, are far from the garnet-in reaction line (> 1 vol%), although all overlap mineral stability fields are consistent with rock assemblages. The compositional core may not coincide with the geometric garnet center (e.g., Spear & Daniel, 1998), as shown for most samples. Overlapping garnet compositional rim isopleths were found for three GHC samples (DH61, DH63, DH66), but only GHC sample DH61 appears ideal as garnet rim isopleths also intersect those of the matrix minerals (± 0.01 mole fraction Ca in anorthite and ± 0.01 Mg# chlorite and biotite). Confidence in isopleth conditions increases when matrix mineral compositions overlap the garnet rim conditions, as these mineral compositions are independent.

The high-resolution P-T paths should be considered approximations of how a garnet with a specific type of compositional zoning would behave in a closed system of a known bulk composition as it evolves during increasing T. Rocks are open systems, but LHF garnet-bearing assemblages appear as if they approach an ideal scenario of a closed system. This appearance of equilibrium is shown by overlapping isopleths of compositions from the garnet core and those of the garnet rim with matrix minerals. In addition, predictions of garnet zoning made by the high-resolution P-T paths closely match the original garnet for these samples (Fig. 6.10). Multiple paths from the same sample yield similar conditions and shapes. The inability to reproduce garnet zoning using Gibb's P-T path trajectories using TheriaG modeling suggests these paths may not be relevant to the samples using the applied parameters.

Regardless of calibrations used, the P-T conditions and paths, along with previously reported timing constraints, are consistent with an imbrication model that suggest the MCT shear zone developed as rock packages within the LHF were progressively transferred (Catlos et al., 2001; Kohn et al., 2001). For example, Figure 6.18 shows P-T path predictions for one such imbrication model described in Catlos et al. (2018 and 2020). This model calculates thermobarometric histories using a two-dimensional finite-difference solution to the diffusion-advection equation. Samples within the LHF travel along the MCT at a 5 km/Ma speed rate from 25 to 18 Ma (Fig. 6.18a). The hanging wall speed rate is 10 km/Ma, and topography progressively accumulates until a maximum height of 3.5 km. The increase in topography is required to accommodate the pressure changes recorded by the garnets while matching their thermal histories. Once the topography is achieved at 18 Ma, a period of cessation is applied to the MCT between 18 and 15 Ma, and

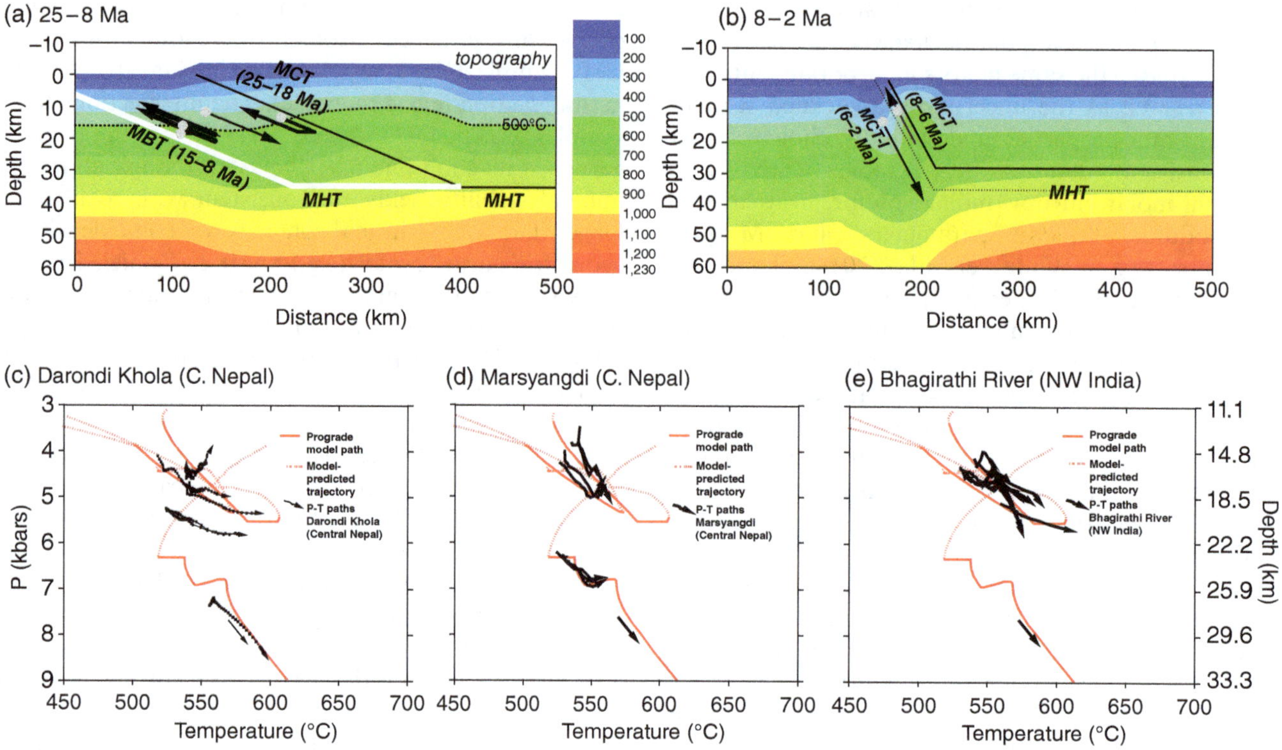

Figure 6.18 (a) Thermal-kinematic model cross section after Catlos et al. (2018) showing the MCT (dark line) and MBT (white line) from 25 to 8 Ma. The MCT and MBT soles into the MHT at depth. Isothermal sections in degree increments are indicated by the scale bar. The isotherms show the thermal situation at 18 Ma after MCT slip. Example sample trajectories on the diagram are represented by arrows with dots at the initial and heads at the final position. The MCT is active from 25 to 18 Ma, whereas slip transfers to the MBT from 15 to 8 Ma. (b) The model cross section of the reactivation of the MCT shear zone from 8 to 2 Ma. Both the MCT and MCT-I sole into the MHT at depth. This panel represents the thermal situation at 6 Ma right before the development of MCT shear zone inverted metamorphism. Example sample trajectories are shown. (c) P-T diagram showing the trajectories of the model predictions for samples panels (a) and (b) and high-resolution P-T paths for the Darondi Khola samples. Sample DH75B is identified. Panels (d) and (e) show the same model predictions, but high-resolution P-T paths from the Marsyangdi River (Catlos et al., 2018) and Bhagirathi River transects (Catlos et al., 2020).

topography is reduced at a rate of 1.5 km/Ma. The model returns to activity within the MCT shear zone with the activation of the MCT footwall slivers from 8 to 2 Ma (Fig. 6.18b). P-T changes recorded by the footwall garnets are the direct result of thermal advection combined with alterations in topography. Changes in the timing of fault motion would affect the model outcomes. However, the model's current constraints and boundary conditions match the high-resolution P-T paths. For example, the P-T diagram in Figure 6.18 c–e are model predictions for samples that experienced imbrication in the MCT footwall. High-resolution P-T paths are also plotted in these panels from samples collected from the LHF along the Darondi (Fig. 6.18c) and Marsyangdi (Fig. 6.18d) rivers in central Nepal and from along the Bhagirathi River in northwest India (Fig. 6.18e). For most samples, the P-T paths match the model predictions remarkably well. P-T paths for sample DH75B (Panel 6.18c) suggest the possibility of very high exhumation rates (> 12 mm/yr) within the MCT shear zone since the Pliocene, which is a scenario predicted by this imbrication model.

6.6. CONCLUSIONS

This paper reviews the geological framework of the Himalayas. It describes and applies particular thermobarometric approaches to decipher the metamorphic history of garnet-bearing rocks collected across the MCT along the Darondi Khola in central Nepal using previously reported data (Kohn et al., 2001). A comparison is made between conventional and isopleth thermobarometry for all samples and high-resolution and Gibb's P-T paths for MCT footwall rocks only. A significant value of the high-resolution P-T path and isopleth approaches is that users can detect when systems stray from the equilibrium assumption. Confidence in conditions exists when minerals assemblages predicted by thermodynamic modeling appear consistent with the actual rock and when the P-T paths reproduce broadscale trends in garnet zoning from core to rim. The expectation is that multiple paths collected from the same garnet or multiple garnets in the same rock would agree in terms of shapes and conditions and that samples collected from the same outcrop or nearby should record similar P-T conditions and paths. Using isopleth thermobarometry, a user can gauge the extent of overlapping mineral compositions and where the overlap occurs with respect to the garnet-in reaction line and garnet volume % growth contours. MCT footwall garnet compositions predicted by Gibb's P-T paths using the software package TheriaG fail to reproduce the original garnet zoning. However, high-resolution P-T paths reproduce the original garnet zoning to ± 0.01 mole fraction in most cases and for most compositions, expected if the garnet behaved in a closed system and had no significant changes in bulk rock composition as it grew. Although the assumption of equilibrium has long been known can never be proven for any rock system (e.g., Spear & Peacock, 1989), isopleth thermobarometry and high-resolution P-T path modeling applied to garnet-grade Himalayan MCT footwall assemblages show they appear to behave as if they evolved in a closed system that experienced particular P-T path trajectories. Ultimately, the P-T conditions and paths generated for rocks across the MCT along the Darondi Khola, regardless of calibrations used, are consistent with the imbrication model that suggests the MCT shear zone developed as rock packages within the LHF were progressively transferred (Catlos et al., 2001; Kohn et al., 2001).

ACKNOWLEDGMENTS

I appreciate Matt Kohn (Boise State University) for supplying the data from the Darondi Khola samples and Mark Harrison (UCLA) for access the rock samples. Discussions and assistance from Eric D. Kelly helped to refine the ideas in the manuscript. I thank Theresa Perez (UT Austin), who helped generate some of the P-T diagrams, and Jeffrey S. Horowitz (UT Austin) for drafting assistance. Finally, comments from three reviewers improved the original version. Data Availability Statement: Supplementary data used in the paper are available in the Texas Data Repository, a platform for publishing and sharing data sets. The repository is Catlos, Elizabeth, 2022, "Replication Data for: Records of Himalayan Metamorphism and Contractional Tectonics in the central Himalayas (Darondi Khola, Nepal)," https://doi.org/10.18738/T8/OLZIJM, Texas Data Repository, V1.

REFERENCES

Abrajevitch, A. V., Ali, J., Aitchison, J. C., Badengzhu, Davis, A. M., & Liu, J., et al. (2005). Neotethys and the India-Asia collision: Insights from a palaeomagnetic study of the Dazhuqu ophiolite, southern Tibet. *Earth and Planetary Science Letters*, *233*, 87–102.

Acharyya, S. K. (1994). The Cenozoic foreland basin and tectonics of the eastern sub-Himalaya: Problems and prospects. *Himalayan Geology*, *15*, 3–21.

Aharon, P., Schidlowski, M., & Singh, I. (1987). Chronostratigraphic markers in the end-Precambrian carbon isotope record of the Lesser Himalaya. *Nature*, *327*, 699–702. https://doi.org/10.1038/327699a0

Ahmad, T., Harris, N., Bickle, M., Chapman, H., Bunbury, J., & Prince, C. (2000). Isotopic constraints on the structural relationships between the Lesser Himalayan Series and High Himalayan Crystalline Series, Garhwal Himalaya. *Geological Society of America Bulletin*, *112*, 467–477.

Ahmad, T., Mukherjee P. K., & Trivedi, J. R. (1999). Geochemistry of Precambrian mafic magmatic rocks of the Western Himalaya, India: Petrogenetic and tectonic implications. *Chemical Geology*, *160*, 103–119.

Aikman, A. B., Harrison, T. M., & Lin, D. (2008). Evidence for Early (> 44 Ma) Himalayan crustal thickening, Tethyan Himalaya, southeastern Tibet. *Earth and Planetary Science Letters*, *274*, 14–23. https://doi.org/10.1016/j.epsl.2008.06.038

Aitchison, J. C, Ali, J. R., & Davis, A. M. (2007). When and where did India and Asia collide? *Journal of Geophysical Research*, *112*, B05423. https://doi.org/10.1029/2006JB004706

Aitchison, J. C., Badengzhu, Davis, A. M., Liu, J., Luo, H., Malpas J. G., et al. (2000). Remnants of a Cretaceous intraoceanic subduction system within the Yarlung-Zangbo suture (southern Tibet). *Earth and Planetary Science Letters*, *183*, 231–244.

Allègre, C. J., et al. (1984). Structure and evolution of the Himalayan-Tibet orogenic belt. *Nature*, *307*, 17–22.

Anczkiewicz, R., Chakraborty, S., Dasgupta, S., Mukhopadhyay, D., & Kołtonik, K. (2014). Timing, duration and inversion of prograde Barrovian metamorphism constrained by high resolution Lu-Hf garnet dating: A case study from the Sikkim Himalaya, NE India. *Earth and Planetary Science Letters*, *407*, 70–81. https://doi.org/10.1016/j.epsl.2014.09.035

Arita, K. (1983). Origin of the inverted metamorphism of the Lower Himalaya, central Nepal. *Tectonophysics*, *95*, 43–60.

Auden, M. A. (1937). The structure of the Himalaya in Garhwal. *Records of the Geological Survey of India*, *71*, 407–433.

Bai, L., Liu, H. B., Ritsema, J. Mori, J., Zhang, T. Z., Ishikawa, Y., et al. (2016). Faulting structure above the Main Himalayan Thrust as shown by relocated aftershocks of the 2015 Mw7.8 Gorkha, Nepal, earthquake. *Geophysical Research Letters*, *43*, 637–642. https://doi.org/101002/2015GL066473

Bassoullet, J. P., Colchen, M., Juteau, T., Marcoux, J., & Mascle, G. (1980). L'edifice de nappes du Zanskar (Ladakh, Himalaya). *Comptes rendus de l'Académie des Sciences, Series D*, *290*, 389–392.

Baxter, A. T., Aitchison, J. C., Ali, J. R., Chan, J. S.-L., & Chan, G. H. N. (2016). Detrital chrome spinel evidence for a

Neotethyan intra-oceanic island arc collision with India in the Paleocene. *Journal of Asian Earth Sciences*, *128*, 90–104.

Beaumont, C., & Jamieson, R. A. (2010). Himalayan-Tibetan orogeny: Channel flow versus (critical) wedge models, afalse dichotomy. In M. L. Leech et al. (Eds.), *Proceedings for the 25th Himalaya-Karakoram-Tibet workshop*. U.S. Geological Survey, Open-File Report 2010-1099. http://pubs.usgs.gov/of/2010/1099/beaumont

Beaumont, C., Jamieson, R. A., Nguyen, M. H., & Lee, B. (2001). Himalayan tectonics explained by extrusion of a low-viscosity crustal channel coupled to focused surface denudation. *Nature*, *414*, 738– 742.

Beaumont, C., Jamieson, R. A., Nguyen, M. H., & Medvedev, S. (2004). Crustal channel flows: 1. Numerical models with applications to the tectonics of the Himalayan-Tibetan orogen. *Journal of Geophysical Research*, *109*. https://doi.org/10.1029/2003JB002809

Beaumont, C., Nguyen, M., Jamieson, R., & Ellis, S. (2006). Crustal flow modes in large hot orogens. In R. D. Law et al. (Eds), *Channel flow, ductile extrusion and exhumation in continental collision zones* (pp. 91–146). Geological Society, London, Special Publications, 268.

Benetti, B., Montomoli, C., Iaccarino, S., Langone, A., & Carosi, R. (2021). Mapping tectono-metamorphic discontinuities in orogenic belts: Implications for mid-crust exhumation in NW Himalaya. *Lithos*, *392–393*. https://doi.org/10.1016/j.lithos.2021.106129.

Berman, R.G. (1990). Mixing properties of Ca-Mg-Fe-Mn garnets. *American Mineralogist*, *75*, 328–344.

Bhandari, S., Xiao, W., Ao, S., Windley, B. F., Zhu, R., Li, R., et al. (2019). Rifting of the northern margin of the Indian craton in the Early Cretaceous: Insight from the Aulis Trachyte of the Lesser Himalaya (Nepal). *Lithosphere*, *11*(5), 643–651. https://doi.org/10.1130/L1058.1

Bhargava, O. N., & Bassi, U. K. (1994). The crystalline thrust sheets of the Himachal Himalaya and the age of amphibolite facies metamorphism. *Journal of the Geological Society of India*, *43*, 343–352.

Bhargava, O. N., & Singh, B. P. (2020). Geological evolution of the Tethys Himalaya. *Episodes*, *43*(1), 404–416. https://doi.org/10.18814/epiiugs/2020/020025

Bhargava, O. N., Frank, W., & Bertle, R. (2011). Late Cambrian deformation in the Lesser Himalaya. *Journal of Asian Earth Sciences*, *40*, 201–212.

Bhutani, R., Pande, K., & Venkatesan, T. R. (2004). Tectonothermal evolution of India-Asia collision zone based on 40Ar-39Ar thermochronology in Ladakh, India. *Proceedings of the Indian Academy of Sciences (Earth and Planetary Sciences)*, *113*, 737–754.

Bilham, R., Larson, K., Freymueller, J., & Project Idylhim members (1997). GPS measurements of present-day convergence across the Nepal Himalaya. *Nature*, *386*, 61–64.

Bodenhausen, J. W. A., DeBooy, T., Egelar, C. G., & Nijhuis, H. J. (1964). On the geology of Central west Nepal: A preliminary note. 22nd International Geological Congress, New Delhi. *Special Publication*, *11*, 101–122.

Bollinger, L., Avouac, J., Cartin, R., & Pandey, M. R. (2004). Stress buildup in the Himalaya. *Journal of Geophysical Research*, *109*, 1–8. https://doi.org/10.1029/2003JB002911

Bollinger, L., Henry, P., & Avouac, J. P. (2006). Mountain building in the Nepal Himalaya: Thermal and kinematic model. *Earth and Planetary Science Letters*, *244*, 58–71.

Bollinger, L., Soma Nath Sapkota, S. N., Tapponnier, P., Yann Klinger, Y., Magali Rizza, M., et al. (2014). Estimating the return times of great Himalayan earthquakes in eastern Nepal: Evidence from the Patu and Bardibas strands of the Main Frontal Thrust. *Journal of Geophysical Research*, *119*(9), 7123–7163.

Bora, D. S., & Shukla, U. K. (2005). Petrofacies implication for the Lower Siwalik Foreland Basin evolution, Kumaun Himalaya, India. *Special Publication of the Palaeontological Society of India*, *2*, 163–179.

Bordet, P., Colchen, M., & Le Fort, P. (1971). Esquisse géologique = Geological sketch map Nyi-Shang (Nepal central) / Mission géologiques du centre national de la recherche scientifique (R.C.P. Nepal) avec le patronage de la Fédération française de la montagne et du Club Alpin; Les contours géologiques ont été levés par P. Bordet, M. Colchen, P. Le Fort, ca. 1:75 000 (E 83°50'–E 84°15'/N 28°49'–N 28°32'). https://www.sudoc.fr/131377973

Bordet, P., Colchen, M., & Le Fort. P. (1975). *Recherches geologiques dans l'Himalaya du Nepal, region du Nyi-Shang*. Paris Centre National de la Recherche Scientifique.

Braden, Z., Godin, L., & Cottle, J. M. (2017). Segmentation and rejuvenation of the Greater Himalayan sequence in western Nepal revealed by in situ U-Th/Pb monazite petrochronology. *Lithos*, *284–285*, 751–765. https://doi.org/10.1016/j.lithos.2017.04.023

Braden, Z., Godin, L., Cottle, J., & Yakymchuk, C. (2018). Renewed late Miocene (< 8 Ma), hinterland ductile thrusting, western Nepal Himalaya. *Geology*, *46*, 503–506.

Brookfield, M. E. (1993). The Himalayan passive margin from Precambrian to Cretaceous. *Sedimentary Geology*, *84*, 1–35.

Brookfield, M. E., & Reynolds, P. H. (1981). Late Cretaceous emplacement of the Indus suture zone ophiolitic melanges and an Eocene-Oligocene magmatic arc on the northern edge of the Indian plate. *Earth and Planetary Science Letters*, *55*, 157–162.

Brunel, M., & Kienast, J. R. (1986). Etude pétro-structurale des chevauchements ductiles himalayens sur la transversale de l'Everest-Makalu (Nepal oriental). *Canadian Journal of Earth Sciences*, *23*, 1117–1137.

Burchfiel, C. B., Zhiliang, C., Hodges, K. V., Yuping, L., Royden, L. H., Changrong, D., et al. (1992). The South Tibetan detachment system, Himalayan orogen: Extension contemporaneous with and parallel to shortening in a collisional mountain belt. *USGS Special Paper*, *269*, 1–40.

Burg, J.-P. (2011). The Asia-Kohistan-India collision: Review and discussion. In D. Brown & P. D. Ryan (Eds.), *Arc-continent collision* (279–309). Frontiers in Earth Sciences. Berlin: Springer.

Burg, J.-P., Brunel, M., Gapais, D., Chen G. M., & Liu, G. H. (1984). Deformation of leucogranites of the crystalline Main Central Sheet in southern Tibet (China). *Journal of Structural Geology*, *6*, 535–542.

Burg, J.-P., Leyreloup, A., Girardeau, J., & Chen, G. M. (1987). Structure and metamorphism of a tectonically thickened continental crust: The Yalu Tsangpo suture zone (Tibet).

Philosophical Transactions of the Royal Society of London Series A, *321*, 67– 86.

Burgess, W. P., Yin, A., Dubey, C. S., Shen, Z-K., & Kelty, T. K. (2012). Holocene shortening across the Main Frontal Thrust zone in the eastern Himalaya. *Earth and Planetary Science Letters*, *357–358*, 152–167. https://doi.org/10.1016/j.epsl.2012.09.040

Caddick, M. J., Bickle, M. J., Harris, N. B. W., Holland, T. J. B., Horstwood, M. S. A., Parrish, R. R., et al. (2007). Burial and exhumation history of a Lesser Himalayan schist: Recording the formation of an inverted metamorphic sequence in NW India. *Earth and Planetary Science Letters*, *264*, 375–90. https://doi.org/10.1016/j.epsl.2007.09.011

Caldwell, W. B., Klemperer, S. L., Lawrence, J. F., & Rai, S. S. (2013). Characterizing the Main Himalayan Thrust in the Garhwal Himalaya, India with receiver function CCP stacking. *Earth and Planetary Science Letters*, *367*, 15–27. https://doi.org/10.1016/j.epsl.2013.02.009

Cao, H., Huang, Y., Li, G., Zhang, L., Wu, J., Dong, L., et al. (2018). Late Triassic sedimentary records in the northern Tethyan Himalaya: Tectonic link with Greater India. *Geoscience Frontiers*, *9*(1), 273–291. https://doi.org/10.1016/j.gsf.2017.04.001

Carlson, W. D. (1989). The significance of intergranular diffusion to the mechanisms and kinetics of porphyroblast crystallization. *Contributions to Mineralogy and Petrology*, *103*, 1–24.

Carlson, W. D. (2002). Scales of disequilibrium and rates of equilibration during metamorphism. *American Mineralogist*, *87*, 185–204.

Carosi, R., Lombardo, B., Molli, G., Musumeci, G., & Pertusati, P. C. (1998).The South Tibetan Detachment System in the Rongbuk valley, Everest region: Deformation and geological implications. *Journal of Asian Earth Sciences*, *16*, 299–311.

Carosi, R., Lombardo, B., Musumeci, G., & Pertusati, P. C. (1999a). Geology of the Higher Himalayan Crystallines in Khumbu Himal (eastern Nepal). *Journal of Asian Earth Sciences*, *17*, 785–803.

Carosi, R., Montomoli, C., & Iaccarino, S. (2018). 20 years of geological mapping of the metamorphic core across Central and Eastern Himalayas. *Earth-Science Reviews*, *177*, 124–138. https://doi.org/10.1016/j.earscirev.2017.11.006

Carosi, R., Montomoli, C., Iaccarino, S., Massonne, H. J., Rubatto, D., Langone, A., et al. (2016). Middle to late Eocene exhumation of the Greater Himalayan sequence in the central Himalayas: Progressive accretion from the Indian plate. *Bulletin of the Geological Society of America*, *128*, 1571–1592. https://doi.org/10.1130/B31471.1

Carosi, R., Montomoli, C., Rubatto, D., & Visonà, D. (2010). Late Oligocene high-temperature shear zones in the core of the Higher Himalayan Crystallines (Lower Dolpo, western Nepal). *Tectonics*, *29*, TC4029. https://doi.org/10.1029/2008TC002400

Carosi, R., Montomoli, C., Rubatto, D., & Visonà, D. (2013). Leucogranite intruding the South Tibetan Detachment in western Nepal: implications for exhumation models in the Himalayas. *Terra Nova*, *25*, 478– 489, 2013

Carosi, R., Musumeci, G., & Pertusati, P. C. (1999b). Extensional tectonics in the higher Himalayan crystallines of Khumbu Himal, Eastern Nepal. In A. Macfarlane et al. (Eds.), *Himalaya and Tibet: Mountain roots to mountain tops* (pp. 211–223). Geological Society of America Special Paper, 328.

Catlos, E. J. (2013). Versatile Monazite, resolving geological records and solving challenges in materials science: Generalizations about monazite, implications for geochronologic studies. *American Mineralogist*, *98* (5–6), 819–832. https://doi.org/10.2138/am.2013.4336

Catlos, E. J. (2022). *Replication data for records of Himalayan metamorphism and contractional tectonics in the central Himalayas (Darondi Khola, Nepal)*. *Texas Data Repository*, *V1*. https://doi.org/10.18738/T8/OLZIJM

Catlos, E. J., Dubey, C. S., Marston, R. A., & Harrison, T. M. (2007). Geochronologic constraints across the Main Central Thrust shear zone, Bhagirathi River (NW India): Implications for Himalayan tectonics. In M. Cloos et al. (Eds.), Convergent margin terranes and associated regions: A tribute to W. G. Ernst. *Geological Society of America Bulletin*, *419*, 135–151. https://doi.org/10.1130/SPE419

Catlos, E. J., Harrison, T. M., Kohn, F. J., Grove, Z. M., Ryerson, F. J., Manning, C. E., et al. (2001). Geochronologic and thermobarometric constraints on the evolution of the Main Central Thrust, central Nepal Himalaya. *Journal of Geophysical Research*, *106*, 16177–16204. https://doi.org/10.1029/2000JB900375

Catlos, E. J., Harrison, T. M., Manning, C. E., Grove, M., Rai, S. M., Hubbard, M. S., et al. (2002). Records of the evolution of the Himalayan Orogen from in situ Th-Pb ion microprobe dating of monazite; eastern Nepal and western Garhwal. *Journal of Asian Earth Sciences*, *20*, 459–479. https://doi.org/10.1016/S1367-9120(01)00039-6

Catlos, E. J., Lovera, O. M., Kelly, E. D., Ashley, K. T., Harrison, T. M., & Etzel, T. (2018). Modeling high-resolution pressure-temperature paths across the Himalayan Main Central Thrust (central Nepal): Implications for the dynamics of collision. *Tectonics*, *37*, 2363–2388.

Catlos, E. J., Pease, E. C., Dygert, N., Brookfield, M., Schwarz, W. H., Bhutani, R., et al. (2019). Nature, age and emplacement of the Spongtang ophiolite, Ladakh, NW India. *Journal of the Geological Society*, *176*, 284–305. https://doi.org/10.1144/jgs2018-085

Catlos, E. J., Perez, T. J., Lovera, O. M., Dubey, C. S., Schmitt, A. K., & Etzel, T. M. (2020). High-resolution P-T-Time paths across Himalayan faults exposed along the Bhagirathi transect NW India: Implications for the construction of the Himalayan orogen and ongoing deformation. *Geochemistry, Geophysics, Geosystems*, *21*, e2020GC009353. https://doi.org/10.1029/2020GC009353

Cattin, R., & Avouac, J. (2000). Modeling mountain building and the seismic cycle in the Himalaya of Nepal. *Journal of Geophysical Research*, *105*, 13389–13407. https://doi.org/10.1029/2000JB900032

Cawood, P. A., Johnson, M. R. W., & Nemchin, A. A. (2007). Early Palaeozoic orogenesis along the Indian margin of Gondwana: Tectonic response to Gondwana assembly. *Earth and Planetary Science Letters*, *255*, 70–84. https://doi.org/10.1016/j.epsl.2006.12.006

Chakraborty, S., Anczkiewicz, R., Gaidies, F., Rubatto, D., Sorcar, N., Faak, K., et al. (2016). A review of thermal history and timescales of tectonometamorphic processes in Sikkim Himalaya (NE India).and implications for rates of metamorphic processes. *Journal of Metamorphic Geology*, *34*, 785–803. https://doi.org/10.1111/jmg.12200

Chakraborty, S., Mukul, M., Mathew, G., & Pande, K. (2019). Major shear zone within the Greater Himalayan sequence and sequential evolution of the metamorphic core in Sikkim, India. *Tectonophysics*, *770*, 228183. https://doi.org/10.1016/j.tecto.2019.228183

Chen, X., Schertl, H-P., Gu, P., Zheng, Y., Xu, R., Zhang, J., et al. (2021). Newly discovered MORB-Type HP garnet amphibolites from the Indus-Yarlung Tsangpo suture zone: Implications for the Cenozoic India–Asia collision. *Gondwana Research*, *90*, 102–117. https://doi.org/10.1016/j.gr.2020.11.006

Chirouze, F., Dupont-Nivet, G., Huyghe, P., van der Beek, P., Chakraborti, T., Bernet, M., et al. (2012). Magnetostratigraphy of the Neogene Siwalik Group in the far eastern Himalaya: Kameng section, Arunachal Pradesh, India. *Journal of Asian Earth Sciences*, *44*, 117–135. https://doi.org/10.1016/j.jseaes.2011.05.016

Colchen, M., Le Fort, P., & Pêcher, A. (1980). *Annapurna-Manaslu-Ganesh Himal*. Centre National de la Recherches Scientifiques, Paris.

Coleman, M. E. (1996). Orogen-parallel and orogen-perpendicular extension in the central Nepalese Himalayas. *Geological Society of America Bulletin*, *108*, 1594–1607.

Coleman, M. E., & Hodges, K. V. (1998). Contrasting Oligocene and Miocene thermal histories from the hanging wall and footwall of the South Tibetan detachment in the central Himalaya from $^{40}Ar/^{39}Ar$ thermochronology, Marsyandi Valley, central Nepal. *Tectonics*, *17*(5), 726– 740. https://doi.org/10.1029/98TC02777

Cooper, F. J., Adams, B. A., Edwards, C. S., & Hodges, K. V. (2012). Large normal-sense displacement on the South Tibetan fault system in the eastern Himalaya. *Geology*, *40* (11), 971–974. https://doi.org/10.1130/G33318.1

Copley, A., Avouac, J.-P., & Royer, J.-Y. (2010). India-Asia collision and the Cenozoic slowdown of the Indian plate: Implications for the forces driving plate motions. *Journal of Geophysical Research*, *115*, B03410. https://doi.org/10.1029/2009JB006634

Corfield, R. L., & Seale, M. P. (2000). Crustal shortening across the north Indian continental margin, Ladakh, India. In M. A. Khan et al. (Eds.), *Tectonics of the Nanga Parbat syntaxis and the Western Himalaya* (pp. 395– 410). The Geological Society Special Publication, vol. *170*.

Corrie, S. L., & Kohn, M. J. (2011). Metamorphic history of the central Him alaya, Annapurna region, Nepal, and implications for tectonic models. *Geological Society of America Bulletin*, *123*(9–10), 1863–1879.

Corrie, S. L., Kohn, M. J., & Vervoort, J. D. (2010). Young eclogite from the Greater Himalayan Sequence, Arun Valley, eastern Nepal; P-T-t path and tectonic implications. *Earth and Planetary Science Letters*, *289*, 406–416. https://doi.org/10.1016/j.epsl.2009.11.029

Corrie, S. L., Kohn, M. J., McQuarrie, N., & Long, S. (2012). Flattening the Bhutan Himalaya. *Earth and Planetary Science Letters*, *349–350*, 67–74.

Cottle, J. M., Larson, K. P., & Kellett, D. A. (2015). How does the mid-crust accommodate deformation in large, hot collisional orogens? A review of recent research in the Himalayan orogen. *Journal of Structural Geology*, *78*, 119–133. https://doi.org/10.1016/j.jsg.2015.06.008

Cottle, J. M., Searle, M. P., Horstwood, M. S. A., & Waters, D. (2009). Timing of midcrustal metamorphism, melting, and deformation in the Mount Everest region of south Tibet revealed by U(-Th)-Pb geochronology. *The Journal of Geology*, *117*, 643–664.

Craddock Affinati, A., Hoisch, T. D., Wells, M. L., & Vervoort, J. D. (2020). Pressure-temperature-time paths from the Funeral Mountains, California, reveal Jurassic retroarc underthrusting during early Sevier orogenesis. *Geological Society of America Bulletin*, *132* (5–6), 1047–1065. https://doi.org/10.1130/B35095.1

Crouzet, C., Dunkl, I., Paudel, L., Árkai, P., Rainer, T. M., Balogh, K., et al. (2007). Temperature and age constraints on the metamorphism of the Tethyan Himalaya in Central Nepal: A multidisciplinary approach. *Journal of Asian Earth Sciences*, *30*(1), 113–130. https://doi.org/10.1016/j.jseaes.2006.07.014

Dai, J.-G., Wang, C.-S., Hébert, R., Santosh, M., Li, Y.-L., & Xu, J.-Y. (2011). Petrology and geochemistry of peridotites in the Zhongba ophiolite, Yarlung Zangbo Suture Zone: Implications for the Early Cretaceous intra-oceanic subduction zone within the Neo-Tethys. *Chemical Geology*, *288*, 133–148.

Daniel, C. G., Hollister, L. S., Parrish, R. R., & Grujic, D. (2003). Exhumation of the Main Central Thrust from lower crustal depths, eastern Bhutan Himalaya. *Journal of Metamorphic Geology*, *21*(4), 317–334. https://doi.org/10.1046/j.1525-1314.2003.00445.x

Dasgupta, S., Ganguly, J., & Neogi, S. (2004). Inverted metamorphic sequence in the Sikkim Himalayas: crystallization history, P-T gradient and implications. *Journal of Metamorphic Geology*, *22*, 395–412. https://doi.org/10.1111/j.1525–1314.2004.00522.x

de Capitani, C., & Brown, T. H. (1987).The computation of chemical equilibrium in complex systems containing non-ideal solutions. *Geochimica et Cosmochimica Acta*, *51*, 2639–2652.

de Capitani, C., & Petrakakis, K. (2010). The computation of equilibrium assemblage diagrams with Theriak/Domino software. *American Mineralogist*, *95*, 1006–1016. https://doi.org/10.2138/am.2010.3354

DeCelles, P. G. (2015). *Structural-kinematic setting of the 2015 Gorkha, Nepal earthquakes: Lessons from a critically tapered orogenic wedge*. 2015 GSA Annual Meeting in Baltimore, Maryland, USA (1–4 November 2015), Paper No. 105-9. https://gsa.confex.com/gsa/2015AM/webprogram/Paper266639.html

DeCelles, P. G., Carrapa, B., Ojha, T. P., Gehrels, G. E., & Collins, D. (2020). Structural and thermal evolution of the Himalayan thrust belt in midwestern Nepal. *Geological Society of America Special Paper*, *547*, 1–77. https://doi.org/10.1130/2020.2547(01)

DeCelles, P. G., Gehrels, G. E., Najman, Y., Martin, A. J., Carter, A., & Garzanti, E. (2004). Detrital geochronology and geochemistry of Cretaceous–Early Miocene strata of Nepal: Implications for timing and diachroneity of initial Himalayan orogenesis. *Earth and Planetary Science Letters*, *227*, 313– 330.

DeCelles, P. G., Gehrels, G. E., Quade, J., LaReau, B., & Spurlin, M. (2000). Tectonic implications of U-PL zircon ages of the Himalayan orogenic belt in Nepal. *Science*, *288*, 497–499.

DeCelles, P. G., Robinson, D. M., Quade, J., Ojha, T. P., Garzione, C. N., Copeland, P., et al. (2001). Stratigraphy, structure, and tectonic evolution of the Himalayan fold-thrust belt in western Nepal. *Tectonics*, *20*, 487–509.

Denolle, M. A., Fan, W., & Shearer, M. (2015). Dynamics of the 2015 M7.8 Nepal earthquake. *Geophysical Research Letters*, *42*, 7467–7475. https://doi.org/10.1002/2015GL065336

Dewey, J. F., & Bird, J. M. (1970). Mountain belts and new global tectonics. *Journal of Geophysical Research*, *75*, 2625–2685.

Dey, S., Dasgupta, P., Das, K., & Matin, A. (2020). Neoproterozoic Blaini Formation of Lesser Himalaya, India: Fiction and fact. *Geological Society of America Bulletin*, *132* (11–12), 2267–2281. https://doi.org/10.1130/B35483.1

Dhamodharan, S., Rawat, G., Kumar, S., & Bagri, D. S. (2020). Sedimentary thickness of the northern Indo-Gangetic plain inferred from magnetotelluric studies. *Journal of Earth System Sciences*, *129*, 156–168. https://doi.org/10.1007/s12040-020-01422-z

Ding, L., Kapp, P., & Wan, X. Q. (2005). Paleocene-Eocene record of ophiolite obduction and initial India-Asia collision, south central Tibet. *Tectonics*, *24*, TC3001. https://doi.org/10.1029/2004TC001729

DiPietro, J. A., & Pogue, K. R. (2004). Tectonostratigraphic subdivisions of the Himalaya: A view from the west. *Tectonics*, *23*, TC5001. https://doi.org/10.1029/2003TC001554

Dubey, A. K. (1997). Simultaneous development of noncylindrical folds, frontal ramps and transfer faults in a compressional regime-experimental investigation of Himalayan examples. *Tectonics*, *16*, 336–346.

Dunkl, I., Antolín, B., Wemmer, K., Rantitsch, G., Kienast, M., Montomoli, C., et al. (2011). Metamorphic evolution of the Tethyan Himalayan flysch in SE Tibet. *Geological Society, London, Special Publications*, *353*, 45–69. https://doi.org/10.1144/SP353.4

Duputel, Z., Vergne, J., Rivera, L., Wittlinger, G., Farra, V., & Hetényi, G. (2016). The 2015 Gorkha earthquake: A large event illuminating the Main Himalayan Thrust fault. *Geophysical Research Letters*, *43*, 2517–2525. https://doi.org/10.1002/2016GL068083

Dyck, B., St-Onge, M., & Searle, M. P., Rayner, N., Waters, D., & Weller, O. M. (2018). Protolith lithostratigraphy of the Greater Himalayan Series in Langtang, Nepal: Implications for the architecture of the northern Indian margin. *Geological Society, London, Special Publications*, *483*, 281–304. https://doi.org/10.1144/SP483.9

Elliott, J., Jolivet, R., González, P. J., Avouac, J-P., Hollingsworth, J., Searle, M., & Stevens, L. (2016). Himalayan megathrust geometry and relation to topography revealed by the Gorkha earthquake. *Nature Geoscience*, *9*, 174–180. https://doi.org/10.1038/ngeo2623

England, P. C., & Molnar, P. (1993). The interpretation of inverted metamorphic isograds using simple physical calculations. *Tectonics*, *12*, 145–157.

England, P. C., Le Fort P., Molnar P., & Pêcher, A. (1992). Heat sources for Tertiary metamorphism and anatexis in the Annapurna-Manaslu region, central Nepal. *Journal of Geophysical Research*, *97*, 2107–2128.

Ernst, W. G. (1973). Blueschist metamorphism and P-T regimes in active subduction zones. *Tectonophysics*, *17*, 255.

Etzel, T. M., Catlos, E. J., Ataktürk, K., Lovera, O. M., Kelly, E. D., Çemen, I., & Diniz, E. (2019). Implications for thrust-related shortening punctuated by extension from P-T paths and geochronology of garnet-bearing schists, Southern (Çine) Menderes Massif, SW Turkey. *Tectonics*, *38*, 1974– 1998. https://doi.org/10.1029/2018TC005335

Ferry, J. M., & Spear, F. S. (1978). Experimental calibration of partitioning of Fe and Mg between biotite and garnet. *Contributions to Mineralogy and Petrology*, *66*, 113–117.

Fuchs, G. (1987). The Geology of southern Zanskar (Ladakh) - Evidence for the autochthony of the Tethys Zone of the Himalaya. *Jahrbuch der Geologischen Bundesanstalt-A*, *130*, 465–491.

Fuchs, G., & Linner, M. (1995). Geological traverse across the western Himalaya- a contribution to the geology of eastern Ladakh, Lahul, and Chamba. *Jahrbuch der Geologischen Bundesanstalt-A.*, *138*, 655–685.

Fuchs, G., Widder, R. W., & Tuladhar, R. (1988). Contributions to the Geology of the Annapurna Range (Manang Area, Nepal). *Jahrbuch der Geologischen Bundesanstalt-A*, *131*, 593–607.

Gaidies, F., de Capitani, C., & Abart, R. (2008). THERIA_G: a software program to numerically model prograde garnet growth. *Contributions to Mineralogy and Petrology*, *155*, 657–671. https://doi.org/10.1007/s00410-007-0263-z

Gaidies, F., Petley-Ragan, A., Chakraborty, S., Dasgupta, S., & Jones, P. (2015). Constraining the conditions of barrovian metamorphism in Sikkim, India: P-T-t paths of garnet crystallization in the Lesser Himalayan Belt. *Journal of Metamorphic Geology*, *33*(1), 23–44. https://doi.org/10.1111/jmg.12108

Gansser, A. (1964). *The geology of the Himalayas*. New York: Wiley Interscience.

Gansser, A. (1981). The geodynamic history of the Himalaya. In H. K. Gupta, & F. M. Delany, (Eds.), *Zagros, Hindu Kush, Himalayan Geodynamic Evolution* (pp. 111–121). American Geophysical Union, 3.

Gao, L-E., Zeng, L., Gao, J., Shang, Z., Hou, K.,, & Wang, Q. (2016). Oligocene crustal anatexis in the Tethyan Himalaya, southern Tibet. *Lithos*, *264*, 201–209. https://doi.org/10.1016/j.lithos.2016.08.038

Garzanti, E. (1999). Stratigraphy and sedimentary history of the Nepal Tethys Himalaya passive margin. *Journal of Asian Earth Science*, *17*, 805–827.

Garzanti, E.,, & Pagni Frette, M. (1991). Stratigraphic succession of the Thakkhola region (central Nepal)-Comparison with the Northwestern Tethys Himalaya. *Rivista Italiana di Paleontologia e Stratigrafia*, *97*(1). https://doi.org/10.13130/2039-4942/8980

Gehrels, G. E., DeCelles, P. G., Martin, A., Ojha, T. P., Pinhassi, G.,, & Upreti, B. N. (2003). Initiation of the Himalayan

Orogen as an early Paleozoic thin-skinned thrust belt. *Geological Society of America Today*, *13*, 4–9.

Gehrels, G. E., DeCelles, P. G., Ojha, T. P., & Upreti, B. N. (2006). Geologic and U-Th-Pb geochronologic evidence for early Paleozoic tectonism in the Kathmandu thrust sheet, central Nepal Himalaya. *Geological Society of America Bulletin*, *118*, 185–198.

Gervais, F.,, & Brown, R. L. (2011). Testing modes of exhumation in collisional orogens: Synconvergent channel flow in the southeastern Canadian Cordillera. *Lithosphere*, *3*, 55–75. https://doi.org/10.1130/L98.1

Ghosh, S., Sanyal, P., Sangode, S. J.,, & Nanda, A. C. (2018). Substrate control of C4 plant abundance in the Himalayan foreland: A study based on inter-basinal records from Plio-Pleistocene Siwalik Group sediments. *Palaeogeography, Palaeoclimatology, Palaeoecology*, *511*, 341–351.

Gibson, R., Godin, L., Kellett, D. A., Cottle, J. M.,, & Archibald, D. (2016). Diachronous deformation along the base of the Himalayan metamorphic core, west-central Nepal. *GSA Bulletin*, *128* (5–6), 860–878. https://doi-org/10.1130/B31328.1

Gnos, E., Immenhauser, A.,, & Peters, T. (1997). Late Cretaceous/early Tertiary convergence between the Indian and Arabian plates recorded in ophiolites and related sediments. *Tectonophysics*, *271*(1–2), 1–19. https://doi.org/10.1016/S0040-1951(96)00249-1

Godin, L., Brown, R. L., & Hanmer, S. (1999b). High strain zone in the hanging wall of the Annapurna detachment, central Nepal Himalaya. In A. Macfarlane et al. (Eds.), *Himalaya and Tibet: Mountain roots to mountain tops* (pp. 199–210). Geological Society of America Special Paper, 328.

Godin, L., Brown, R. L., Hanmer, S.,, & Parrish, R. R. (1999a). Back folds in the Himalayan orogen: An alternative interpretation. *Geology*, *27*, 151–154.

Goscombe, B.,, & Hand, M. (2000). Contrasting P-T paths in the eastern Himalaya, Nepal: Inverted isograds in a paired metamorphic mountain belt. *Journal of Petrology*, *41*(12), 1673–1719. https://doi.org/10.1093/petrology/41.12.1673

Goscombe, B., Gray, D.,, & Hand, M. (2006). Crustal architecture of the Himalayan metamorphic front in eastern Nepal. *Gondwana Research*, *10*(3–4), 232–255. https://doi.org/10.1016/j.gr.2006.05.003.

Goswami, P. K.,, & Deopa, T. (2017). Petrotectonic setting of the provenance of Lower Siwalik sandstones of the Himalayan foreland basin, southeastern Kumaun Himalaya, India. *Island Arc*, *27*, e12242. https://doi.org/10.1111/iar.12242

Goswami-Banerjee, S., Bhowmik, S. K., Dasgupta, S., & Pant, N. C. (2014). Burial of thermally perturbed Lesser Himalayan mid-crust: Evidence from petrochemistry and P-T estimation of the western Arunachal Himalaya, India. *Lithos*, *208–209*, 298–311. https://doi.org/10.1016/j.lithos.2014.09.015

Graham, C. M., & England, P. C. (1976). Thermal regimes and regional metamorphism in the vicinity of overthrust faults: an example of shear heating and inverted metamorphic zonation from southern California. *Earth and Planetary Science Letters*, *31*, 142–152.

Graham, C. M., & Powell, R. (1984). A garnet-hornblende geothermometer; calibration, testing, and application to the Pelona Schist, Southern California. *Journal of Metamorphic Geology*, *2*, 13–31.

Grasemann B., & Vannay J-C. (1999). Flow controlled inverted metamorphism in shear zones. *Journal of Structural Geology*, *21*, 743–750.

Groppo, C., Rolfo, F., & Lombardo, B. (2009). P-T evolution across the Main Central Thrust Zone (eastern Nepal): Hidden discontinuities revealed by petrology. *Journal of Petrology*, *50*(6), 1149–1180. https://doi.org/10.1093/petrology/egp036

Groppo, C., Rubatto, D., Rolfo, F., & Lombardo, B. (2010). Early Oligocene partial melting in the Main Central Thrust Zone (Arun valley, eastern Nepal Himalaya). *Lithos*, *118*(3–4),287–301.https://doi.org/10.1016/j.lithos.2010.05.003

Grujic, D., Casey, M., Davidson, C., Hollister, L. D., Kündig, R., Pavlis, T., & Schmid, S. (1996). Ductile extrusion of the Higher Himalayan Crystalline in Bhutan: evidence from quartz microfabrics. *Tectonophysics*, *260*(1–3), 21–43. https://doi.org/10.1016/0040-1951(96)00074-1

Guillot, S. (1999). An overview of the metamorphic evolution in Central Nepal. *Journal of Asian Earth Sciences*, *17*, 713–725.

Guillot, S., Cosca, M., Allemand, P., & Le Fort, P. (1999). Contrasting metamorphic and geochronologic evolution along the Himalayan belt. In A. Macfarlane et al. (Eds.), *Himalaya and Tibet: Mountain roots to mountain tops* (pp. 117–128). Geological Society of America Special Papers, 328.

Guillot, S., Le Fort, P., Peçher, A., Barman, M. R., & Aprahmaian, J. (1995). Contact-metamorphism and depth of emplacement of the Manaslu granite (Central Nepal)—implications for Himalayan orogenesis. *Tectonophysics*, *241*, 99– 119.

Guillot, S., Mahéo, G., de Sigoyer, J., Hattori, K. H., & Pêcher, A. (2008). Tethyan and Indian subduction viewed from the Himalayan high- to ultrahigh-pressure metamorphic rocks. *Tectonophysics*, *451*, 225–241.

Gupta, D. K., Bhowmick, D., & Roy, N.S. (2015). Himalayan hazard study on the basis of stress and strain state of 1991 Uttarkashi earthquake using Coulomb stress transfer model. *Geomatics, Natural Hazards and Risk*, *6*, 131–148. https://doi.org/10.1080/19475705.2013.820797

Guynn, J. H., Kapp, P., Pullen, A., Gehrels, G., Heizler, M., & Ding, L. (2006). Tibetan basement rocks near Amdo reveal "missing" Mesozoic tectonism along the Bangong suture, central Tibet. *Geology*, *34*, 505–508.

Harris, N., & Massey, J. (1994).Decompression and anatexis of Himalayan metapelites. *Tectonics*, *13*(6), 1537– 1546. https://doi.org/10.1029/94TC01611

Harris, N., Inger, S., & Massey, J. (1993). The role of fluids in the formation of High Himalayan leucogranites. In P.J. Trealor & M. P. Searle (Eds.), *Himalayan tectonics* (pp. 391–400). Geological Society Special Publication, 74.

Harrison, T. M., Grove M., & Lovera, O. M. (1997). New insights into the origin of two contrasting Himalayan granite belts. *Geology*, *25*, 899–902.

Harrison, T. M., Grove, M., Lovera, O. M., & Catlos, E. J. (1998). A model for the origin on Himalayan anatexis and inverted metamorphism. *Journal of Geophysical Research*, *103*, 27017–27032.

Harrison, T. M., Grove, M., Lovera, O. M., Catlos, E. J., & D'Andrea, J. (1999). The origin of Himalayan anatexis and inverted metamorphism: Models and constraints. *Journal of Asian Earth Sciences*, *17*, 755–772.

Harrison, T. M., McKeegan, K. D., & Le Fort, P. (1995). Detection of inherited monazite in the Manaslu leucogranite by 208Pb232Th ion microprobe dating: Crystallization age and tectonic implications. *Earth and Planetary Science Letters*, *133*(3–4), 271–282. https://doi.org/10.1016/0012-821X(95)00091-P.

Hazarika, D., Wadhawan M., Paul, A., Kumar, N., & Borah, K. (2017). Geometry of the Main Himalayan Thrust and Moho beneath Satluj valley, northwest Himalaya: Constraints from receiver function analysis. *Journal of Geophysical Research*, *122*, 2929–2945. https://doi.org/10.1002/2016JB013783

He, P., Lei, J., Yuan, X., Xu, X., Xu, Q., Liu, Z., Mi, Q., & Zhou, L. (2018). Lateral Moho variations and the geometry of the Main Himalayan Thrust beneath the Nepal Himalayan orogen revealed by teleseismic receiver functions. *Geophysical Journal International*, *214*, 1004–1017. https://doi.org/10.1093/gji/ggy192

Hébert, R., Bezard, R., Guilmette, C., Dostal, J., Wang, C. S., & Liu, Z. F. (2012). The Indus-Yarlung Zangbo ophiolites from Nanga Parbat to Namche Barwa syntaxes, southern Tibet: First synthesis of petrology, geochemistry, and geochronology with incidences on geodynamic reconstructions of Neo-Tethys. *Gondwana Research*, *22*, 337–397.

Heim, A. A., & Gansser, A. (1939). *The throne of the gods; An account of the first Swiss expedition to the Himalayas*. New York: The Macmillan Company.

Henry, P., Le Pichon, X., & Goffe, B. (1997). Kinematic, thermal and petrological model of the Himalayas: Constraints related to metamorphism within the underthrust Indian crust and topographic elevation. *Tectonophysics*, *273*, 31–56.

Herman, F., Copeland, P., Avouac, J. , Bollinger, L., Mahéo, G., Le Fort, P., et al. (2010). Exhumation, crustal deformation, and thermal structure of the Nepal Himalaya derived from the inversion of thermochronological and thermobarometric data and modeling of the topography. *Journal of Geophysical Research*, *115*, B06407. https://doi.org/10.1029/2008JB006126

Hirschmiller, J., Grujic, D., Bookhagen, B., Coutand, I., Huyghe, P., Mugnier, J-L., & Ojha, T. (2014). What controls the growth of the Himalayan foreland fold-and-thrust belt? *Geology*, *42* (3), 247–250. https://doi.org/10.1130/G35057.1

Hodges, K. V. (2000). Tectonics of the Himalaya and southern Tibet from two perspectives. *Geological Society of America Bulletin*, *112*, 324–350.

Hodges, K. V., & Silverberg, D.S. (1988). Thermal evolution of the Greater Himalaya, Garhwal, India. *Tectonics*, *7*, 583–600.

Hodges, K. V., Burchfiel, B. C., Royden, L. H., Chen, Z., & Liu, Y. (1993). The metamorphic signature of contemporaneous extension and shortening in the central Himalayan orogen: data from the Nyalam transect, southern Tibet. *Journal of Metamorphic Geology*, *11*, 721–737.

Hodges, K. V., Hames, W. E., Olszewski, W. J., Burchfiel, B. C., Royden, L. H., & Chen, Z. (1994). Thermobarometric and $^{40}Ar/^{39}Ar$ geochronologic constraints on Eohimalayan metamorphism in the Dinggy area, southern Tibet. *Contributions to Mineralogy Petrology*, *117*, 151– 163.

Hodges, K. V., Le Fort, P., & Pêcher, A. (1988). Possible thermal buffering by crustal anatexis in collisional orogens: thermobarometric evidence from the Nepalese Himalaya. *Geology*, *16*, 707–710.

Hodges, K. V., Parrish, R. R., & Searle, M. P. (1996). Tectonic evolution of the central Annapurna Range, Nepalese Himalayas. *Tectonics*, *15*, 1264–1291.

Hoisch, T. D. (1990). Empirical calibration of six geobarometers for the mineral assemblage quartz + muscovite + biotite + plagioclase + garnet. *Contributions to Mineralogy and Petrology*, *104*, 225–234.

Holland, T. J. B., & Powell, R. (1998). An internally consistent thermodynamic data set for phases of petrological interest. *Journal of Metamorphic Geology*, *16*, 309–343. https://doi.org/10.1111/j.1525-1314.1998.00140.x

Hopkinson, T. N., Harris, N. B. W., Warren, C. J., Spencer, C. J., Roberts, N. M. W., Horstwood, M. S. A., et al. (2017). The identification and significance of pure sediment-derived granites. *Earth and Planetary Science Letters*, *467*, 57–63. https://doi.org/10.1016/j.epsl.2017.03.018

Hu, X., Garzanti, E., Wang, J., Huang, W., An, Q., & Webb, A. (2016). The timing of India-Asia collision onset: Facts, theories, controversies. *Earth-Science Reviews*, *160*, 264–299.

Hu, X., Jansa, L., Chen, L., Griffin, W. L., O'Reilly, S. Y. & Wang, J. (2010). Provenance of Lower Cretaceous Wölong volcaniclastics in the Tibetan Tethyan Himalaya: Implications for the final breakup of Eastern Gondwana. *Sedimentary Geology*, *223*, 193–205. https://doi.org/10.1016/j.sedgeo.2009.11.008

Hubbard, J., Almeida, R., Foster, A., Sapkota, S. N., Bürgi, P., & Tapponnier, P. (2016) Structural segmentation controlled the 2015 Mw 7.8 Gorkha earthquake rupture in Nepal. *Geology*, *44* (8), 639–642. https://doi.org/10.1130/G38077.1

Hubbard, M. S. (1989). Thermobarometric constraints on the thermal history of the Main Central Thrust Zone and Tibet Slab, eastern Nepal Himalaya. *Journal of Metamorphic Geology*, *7*, 19–30.

Hubbard, M. S. (1996). Ductile shear as a cause of inverted metamorphism: example from the Nepal Himalaya. *Journal of Geology*, *104*, 493–499.

Huerta, A. D., Royden, L. H., and Hodges, K. V. (1998), The thermal structure of collisional orogens as a response to accretion, erosion, and radiogenic heating. *Journal of Geophysical Research*, *103*(B7), 15287–15302. https://doi.org/10.1029/98JB00593.

Hughes, N. C. (2016). The Cambrian palaeontological record of the Indian subcontinent. Earth-Science Reviews, *159*, 428–461. https://doi.org/10.1016/j.earscirev.2016.06.004,

Hughes, N. C., Myrow, P. M., Peng, S., Banerjee, D. M. (2018). The Parahio Formation of the Tethyan Himalaya: The type section, thickness, lithostratigraphy and biostratigraphy of the best characterised Cambrian succession in the Indian subcontinent. *Journal of the Palaeontological Society of India*, *63*(1), 1–18.

Hughes, N. C., Peng, S., Bhargava, O. N., Ahulwalia, A. D., Walia, S., Myrow, P. M., & Parcha, S. K. (2005). The Cambrian biostratigraphy of the Tal Group, Lesser Himalaya, India, and early Tsanglangpuan (late early Cambrian) trilobites from the Nigali Dhar Syncline. *Geological Magazine*, *142*, 57–80.

Iaccarino, S., Montomoli, C., Carosi, R., Massonne, H.-J., & Visonà, D. (2017). Geology and tectono-metamorphic evolution of the Himalayan metamorphic core: Insights from the Mugu Karnali transect, western Nepal (central Himalaya). *Journal of Metamorphic Geology*, *35*(3), 301–325. https://doi.org/10.1111/jmg.12233

Iaccarino, S., Montomoli, C., Montemagni, C., Massonne, H-J., Langone, A., Jain, A. K., Visonà, D., & Carosi, R. (2020). The Main Central Thrust zone along the Alaknanda and Dhauli Ganga valleys (Garhwal Himalaya, NW India): Insights into an inverted metamorphic sequence. *Lithos*, *372–373*. https://doi.org/10.1016/j.lithos.2020.105669

Imayama, T., Takeshita, T., & Arita, K. (2010). Metamorphic P-T profile and P-T path discontinuity across the far-eastern Nepal Himalaya: Investigation of channel flow models. *Journal of Metamorphic Geology*, *28*, 527–549. https://doi.org/10.1111/j.1525-1314.2010.00879.x

Inger, S., & Harris, N. B. W. (1992). Tectonothermal evolution of the High Himalaya Crystalline sequence, Langtang Valley, northern Nepal. *Journal of Metamorphic Geology*, *10*, 439–452

Jadoul, F., Berra, F., & Garzanti, E. (1998). The Tethys Himalayan passive margin from Late Triassic to Early Cretaceous (South Tibet): *Journal of Asian Earth Sciences*, *16*, 173–194. https://doi.org/10.1016/S0743-9547(98)00013-0

Jain, A. K., & Chander, R. (1995). *Geodynamic models for Uttarkashi earthquake of October 20, 1991. Memoirs of the Geological Society of India*, Bangalore, 225–233.

Jamieson, R. A., & Beaumont, C. (2013). On the origin of orogens. *Geological Society of America Bulletin*, *11–12*, 1671–1702.

Jamieson, R. A., Beaumont, C., Hamilton, J., & Fullsack, P. (1996). Tectonic assembly of inverted metamorphic sequences. *Geology*, *24*, 839–842.

Jamieson, R. A., Beaumont, C., Medvedev, S., Nguyen, M. H. (2004). Crustal channel flows: 2. Numerical models with implications for metamorphism in the Himalayan-Tibetan orogen. *Journal of Geophysical Research*, *109*. https://doi.org/10.1029/2003JB002811

Jamieson, R. A., Beaumont, C., Nguyen, M. H., & Grujic, D. (2006). Provenance of the Greater Himalayan Sequence and associated rocks: Predictions of channel flow models. *Geological Society, London, Special Publications*, *268*, 165–182. https://doi.org/10.1144/GSL.SP.2006.268.01.07

Jharendra, K. C., & Paudyal, K. R. (2019). Characteristics and field relation of Ulleri Augen Gneiss to country rocks in the Lesser Himalaya: A case study from Syaprubesi-Chhyamthali area, central Nepal. *Journal of Nepal Geological Society*, *58*, 89–96. https://doi.org/10.3126/jngs.v58i0.24577

Kaneko, Y. (1995). Thermal structure in the Annapurna region, central Nepal Himalaya: implication for the inverted metamorphism. *Journal of Mineralogical and Petrological Sciences*, *90*, 143–154.

Kapp, P., DeCelles, P. G., Gehrels, G. E., Heizler, M., & Ding, L. 2007. Geological records of the Lhasa-Qiangtang and Indo-Asian collisions in the Nima area of central Tibet. *Geological Society of America Bulletin*, *119*, 917–932.

Kapp, P., Murphy, M. A., Yin, A., Harrison, T. M., Ding, L. & Guo, J. R. 2003. Mesozoic and Cenozoic tectonic evolution of the Shiquanhe area of western Tibet. *Tectonics*, *22*. https://doi.org/10.1029/2001TC001332

Kawakami, T., Sakai, H., & Sato, K. (2019). Syn-metamorphic B-bearing fluid infiltrations deduced from tourmaline in the Main Central Thrust zone, Eastern Nepal Himalayas. *Lithos*, *348–349*, 105175. https://doi.org/10.1016/j.lithos.2019.105175

Kayal, J. R. (1996). Precursor seismicity, foreshocks and aftershocks of the Uttarkashi earthquake of October 20, 1991 at Garhwal Himalaya. *Tectonophysics*, *263*, 339–345.

Kellett, S. A., Cottle, J. M., & Larson, K. P. (2018). The South Tibetan Detachment System: history, advances, definition and future directions. *Geological Society, London, Special Publications*, *483*, 377–400. https://doi.org/10.1144/SP483.2

Kelly, E. D., Hoisch, T. D., Wells, M. L., Vervoort, J. D., & Beyene, M. A. (2015). An Early Cretaceous garnet pressure-temperature path recording synconvergent burial and exhumation from the hinterland of the Sevier orogenic belt, Albion Mountains, Idaho. *Contributions to Mineralogy and Petrology 170*, 1–22.

Khan, M. A., Bera, M., Spicer, R. A., Spicer, T. E. V., & Bera, S. (2019). Palaeoclimatic estimates for a latest Miocene-Pliocene flora from the Siwalik Group of Bhutan: Evidence for the development of the South Asian Monsoon in the eastern Himalaya, *Palaeogeography, Palaeoclimatology, Palaeoecology*, *514*, 326–335. https://doi.org/10.1016/j.palaeo.2018.10.019

Khanal, S., Robinson, D. M., Mandal, S., & Simkhada, P. (2014). Structural, geochronological and geochemical evidence for two distinct thrust sheets in the 'Main Central thrust zone', the Main Central thrust and Ramgarh-Munsiari thrust: implications for upper crustal shortening in central Nepal. *Geological Society, London, Special Publications*, *412*, 221–245. https://doi.org/10.1144/SP412.2

Khattri, K. N., & Tyagi, A. K. (1983). Seismicity patterns in the Himalayan plate boundary and identification of areas of high seismic potential. *Tectonophysics*, *96*, 281–297.

Kidder, S. B., Herman, F., Saleeby, J., Avouac, J-P., Ducea, M. N., & Chapman, A. (2013). Shear heating not a cause of inverted metamorphism. *Geology*, *41*, 899–902. https://doi.org/10.1130/G34289.1

Klootwijk, C.T., Gee, J. S., Peirce, J. W., Smith, G. M., & McFadden, P. L. (1992). An early India-Asia contact: Paleomagnetic constraints from Ninetyeast Ridge, ODP Leg 121. *Geology*, *20*(5), 395–398. https://doi.org/10.1130/0091-7613(1992)020<0395:AEIACP>2.3.CO;2

Kohn, M. J. (1993). Uncertainties in differential thermodynamic (Gibbs' method) P-T paths. *Contributions to Mineralogy and Petrology*, *113*, 24–39. https://doi.org/10.1007/BF00320829

Kohn, M. J. (2008). P-T-t data from central Nepal support critical taper and repudiate large-scale channel flow of the Greater Himalayan Sequence. *Geological Society of America Bulletin*, *120*, 259–273. https://doi.org/10.1130/B26252.1

Kohn, M. J. (2014). Himalayan metamorphism and its tectonic implications. *Annual Review of Earth and Planetary Sciences*, *42*(1), 381–419.

Kohn, M. J. (2016). Metamorphic chronology—a tool for all ages: Past achievements and future prospects. *American Mineralogist*, *101*, 25–42. https://doi.org/10.2138/am-2016-5146

Kohn, M. J., & Spear, F. (2000). Retrograde net transfer reaction insurance for pressure-temperature estimates. *Geology*, *28*, 1127–1130. https://doi.org/10.1130/0091-7613(2000)28<1127:RNTRIF>2.0.CO;2

Kohn, M. J., & Spear, F. S. (1990). Two new barometers for garnet amphibolites with applications to southeastern Vermont: *American Mineralogist*, *75*, 89–96.

Kohn, M. J., & Spear, F. S. (1991). Error propagation for barometers: 2. Application to rocks. *American Mineralogist*, *76* (1–2), 138–147.

Kohn, M. J., Catlos, E. J., Ryerson, F. J., & Harrison, T. M. (2001). Pressure-temperature-time path discontinuity in the Main Central thrust zone, central Nepal. *Geology*, *29* (7), 571–574. https://doi.org/10.1130/0091-7613(2001)029<0571:PTTPDI>2.0.CO;2

Kohn, M. J., Paul, S. K. , & Corrie S. L. (2010). The lower Lesser Himalayan sequence: A Paleoproterozoic arc on the northern margin of theIndian plate, *Geological Society of America Bulletin*, *122*, 323–335.

Kohn, M. J., Wieland, M. Parkinson, C. D. , & Upreti, B. N. (2004). Miocene faulting at plate tectonic velocity in the Himalaya of central Nepal, *Earth and Planetary Science Letters*, *228*, 299–310.

Lanari, P., & Duesterhoeft, E. (2019). Modeling metamorphic rocks using equilibrium thermodynamics and internally consistent databases: Past achievements, problems and perspectives. *Journal of Petrology*, *60*, 19–56. https://doi.org/10.1093/petrology/egy105

Lanari, P., & Engi, M. (2017). Local bulk composition effects on metamorphic mineral assemblages. *Reviews in Mineralogy and Geochemistry*, *83*, 55–102. https://doi.org/10.2138/rmg.2017.83.3

Larson, K., Gervais, F., & Kellett, D. A. (2013). A P-T-t-D discontinuity in east-central Nepal: Implications for the evolution of the Himalayan mid-crust. *Lithos*, *179*, 275–292. https://doi.org/10.1016/j.lithos.2013.08.012

Larson, K. P., Ambrose, T. K., Webb, A. A. G., Cottle, J. M., & Shrestha, S. (2015). Reconciling Himalayan midcrustal discontinuities: The Main Central thrust system. *Earth and Planetary Science Letters*, *429*, 139–146. https://doi.org/10.1016/j.epsl.2015.07.070

Larson, K. P., Godin, L., & Price, R. A. (2010). Relationships between displacement and distortion in orogens: Linking the Himalayan foreland and hinterland in central Nepal. *Geological Society of America Bulletin*, *122* (7–8), 1116–1134. https://doi.org/10.1130/B30073.1

Larson K. P., Kellet, D. A., Cottle, J. M., Camacho, A., & Brubacher, A. D. (2019). Mid-Miocene initiation of E-W extension and recoupling of the Himalayan Orogen. *Terra Nova*, *12443*, 1–8.

Laskowski, A. K., Kapp, P., Vervoort, J. D., & Ding, L. (2016). High-pressure Tethyan Himalaya rocks along the India-Asia suture zone in southern Tibet. *Lithosphere*, *8* (5), 574–582. https://doi.org/10.1130/L544.1

Lawver, L. A., Norton, I. O., Dalziel, I. W.D., Davis, J. K., & Gahagan, L. M. (2018). The PLATES 2017 Atlas of Plate Reconstructions (550 Ma to Present Day). *PLATES Progress Report*, 390–0318.

Lee, J., Hacker, B., & Wang, Y. (2004). Evolution of North Himalayan gneiss domes: structural and metamorphic studies in Mabja Dome, southern Tibet. *Journal of Structural Geology*, *26*, 2297–316.

Le Fort, P. (1975). Himalaya, the collided range, Present knowledge of the continental arc. *American Journal of Science*, *275A*, 1–44.

Le Fort, P. (1996). Evolution of the Himalaya. In A. Yin & T. M. Harrison (Eds.), *The tectonic evolution of Asia* (pp. 95–109). Cambridge University Press.

Le Fort, P., & Rai, S. M. (1999). Pre-Tertiary felsic magmatism of the Nepal Himalaya: recycling of continental crust. *Journal of Asian Earth Sciences*, *17*, 607– 628.

Lihter, I., Larson, K. P., Shrestha, S., Cottle, J. M., & Brubacher, A. D. (2020). Contact metamorphism of the Tethyan Sedimentary Sequence, Upper Mustang region, west-central Nepal. *Geological Magazine*, *157*(11), 1917–1932. https://doi.org/10.1017/S0016756820000229

Liu, G., & Einsele, G. (1994). Sedimentary history of the Tethyan basin in the Tibetan Himalayas. *Geologische Rundschau*, *82*, 32– 61.

Liu, J-H., Xie, C-M., Li, C., Wang, M., Wu, H., Li, X-K., et al. (2018). Early Carboniferous adakite-like and I-type granites in central Qiangtang, northern Tibet: Implications for intra-oceanic subduction and back-arc basin formation within the Paleo-Tethys Ocean. *Lithos*, *296–299*, 265–280. https://doi.org/10.1016/j.lithos.2017.11.005

Liu, X., Ju, Y., Wei, L, & Li, G. (2010) An alternative tectonic model for the Yarlung Zangbo suture zone. *Science in China Series D: Earth Sciences*, *53*, 27–41. https://doi.org/10.1007/s11430-009-0177-x

Liu, Z. C., Wu, F. Y., Ding, L., Liu, X. C., Wang, J. G., & Ji, W. Q. (2016). Highly fractionated Late Eocene (~ 35 Ma). leucogranites in the Xiaru Dome, Tethyan Himalaya, *South Tibet. Lithos*, *240–243*, 337–354.

Lombardo, B., & Rolfo, F. (2000). Two contrasting eclogite types in the Himalayas: Implications for the Himalayan orogeny. *Journal of Geodynamics*, *30*, 37–60.

Lombardo B., Pertusati, P., & Borghi, S. (1993). Geology and tectonomagmatic evolution of the eastern Himalaya along the Chomolungma-Makalu transect. In P. J. Trealor & M. P. Searle (Eds.), *Himalayan tectonics* (pp. 341–355). Geological Society Special Publication, 74.

Long, S., & McQuarrie, N. (2010). Placing limits on channel flow: Insights from the Bhutan Himalaya. *Earth and Planetary Science Letters*, *290*, 475–390.

Long, S., McQuarrie, N., Tobgay, T., & Grujic, D. (2011). Geometry and crustal shortening of the Himalayan fold-thrust belt, eastern and central Bhutan. *Geological Society of America Bulletin*, *123* (7–8), 1427–1447. https://doi.org/10.1130/B30203.1

Long, S. P., Gordon, S. M., & Soignard, E., 2017, Distributed north-vergent shear and flattening through Greater and Tethyan Himalayan rocks: Insights from metamorphic and strain data from the Dang Chu region, central Bhutan. *Lithosphere*, *9*, 774–795. https://doi.org/10.1130/L655.1

Long, S. P., Mullady, C. L., Starnes, J. K., Gordon, S. M., Larson, K. P., Pianowski, L. S., Miller, R. B., & Soignard, E. (2019). A structural model for the South Tibetan detachment system in northwestern Bhutan from integration of temperature, fabric, strain, and kinematic data. *Lithosphere*, *11* (4), 465–487. https://doi.org/10.1130/L1049.1

Macfarlane, A. M. (1995). An evaluation of the inverted metamorphic gradient at Langtang National Park, central Nepal Himalaya. *Journal of Metamorphic Geology*, *13*, 595–612.

Mahajan, A. K., Thakur, C., Sharma, M. L., & Chauhan, M. (2010). Probabilistic seismic hazard map of NW Himalaya and its adjoining area, India. *Natural Hazards*, *53*, 443–457. https://doi.org/10.1007/s11069-009-9439-3

Maiti, G., & Mandal, N. (2021) Early Miocene exhumation of high-pressure rocks in the Himalaya: A response to reduced India-Asia convergence velocity. *Frontiers in Earth Science.* https://doi.org/10.3389/feart.2021.632806

Maiti, G., Mandal, N., & Misra, S. (2020). Insights into the dynamics of an orogenic wedge from lubrication theory: Implications for the Himalayan tectonics. *Tectonophysics*, *776*, 228335. https://doi.org/10.1016/j.tecto.2020.228335

Makovsky, Y., Klemperer, S. L., Huang, L., Lu, D., & Project INDEPTH Team (1996). Structural elements of the southern Tethyan Himalaya crust from wide-angle seismic data, *Tectonics*, *15*(5), 997– 1005. https://doi.org/10.1029/96TC00310

Makovsky, Y., Klemperer, S. L., Ratschbacher, L., & Alsdorf, D. (1999). Midcrustal reflector on INDEPTH wide-angle profiles: an ophiolitic slab beneath the India-Asia suture in southern Tibet? *Tectonics*, *18*, 793–808.

Mandal, S., Robinson, D. M., Khanal, S., & Das, O. (2015). Redefining the tectonostratigraphic and structural architecture of the Almora klippe and the Ramgarh-Munsiari thrust sheet in NW India. In S. Mukherjee et al. (Eds.), *Tectonics of the Himalaya* (pp. 247–269). Geological Society of America Special Publications, 412.

Mandal, S.. Robinson, D. M., Kohn, M. J., Khanal, S., Das, O., & Bose, S. (2016). Zircon U-Pb ages and Hf isotopes of the Askot klippe, Kumaun, northwest India: Implications for Paleoproterozoic tectonics, basin evolution and associated metallogeny of the northern Indian cratonic margin. *Tectonics*, *35*. https://doi.org/10.1002/2015TC004064

Manickavasagam, R. M., Jain, A. K., Singh, S., & Asokan, A. (1999). Metamorphic evolution of the northwest Himalaya, India: Pressure-temperature data, inverted metamorphism, and exhumation in the Kashmir, Himachal, and Garhwal Himalayas. In A. Macfarlane et al. (Eds.), *Himalaya and Tibet: Mountain roots to mountain tops* (pp. 179-198). Geological Society of America Special Paper, 328.

Martin, A. J. (2017a). A review of Himalayan stratigraphy, magmatism, and structure. *Gondwana Research*, *49*, 42–80. https://doi.org/10.1016/j.gr.2017.04.031

Martin, A. J. (2017b). A review of definitions of the Himalayan Main Central Thrust. *International Journal of Earth Sciences*, *106*, 2131-45. https://doi.org/10.1007/s00531-016-1419-8

Martin, A. J., Burgy, K. D., Kaufman, A. J., & Gehrels, G. E. (2011). Stratigraphic and tectonic implications of field and isotopic constraints on depositional ages of Proterozoic Lesser Himalayan rocks in central Nepal. *Precambrian Research*, *185*, 1–17.

Martin, A. J., DeCelles, P. G., Gehrels, G. E., Patchett, P. J., & Isachsen, C. (2005). Isotopic and structural constraints on the location of the MainCentral Thrust in the Annapurna Range, central Nepal Himalaya. *Geological Society of America Bulletin*, *117*(7–8), 926–944.

Martin, A. J., Ganguly, J., & DeCelles, P. G. (2010). Metamorphism of Greater and Lesser Himalayan rocks exposed in the Modi Khola valley, central Nepal. *Contributions to Mineralogy and Petrology*, *159*, 203–223.

Matin, A., & Mukul, M. (2010). Phases of deformation from cross-cutting structural relationships in external thrust sheets: Insights from small-scale structures in the Ramgarh thrust sheet, Darjiling Himalaya, *West Bengal. Current Science*, *99*(10), 1369–1377.

Matsuoka, A., Yang, Q., Kobayashi, K., Takei, M., Nagahashi, T., Zeng, Q., et al. (2002). Jurassic-Cretaceous radiolarian biostratigraphy and sedimentary environments of the Ceno-Tethys: records from the Xialu Chert in the Yarlung-Zangbo Suture Zone, southern Tibet. *Journal of Asian Earth Sciences*, *20*(3), 277–287. https://doi.org/10.1016/S1367-9120(01)00044-X

McKenzie, N. R., Hughes, N. C., Myrow, P. M., Xiao, S., & Sharma, M. (2011). Correlation of Precambrian-Cambrian sedimentary successions across northern India and the utility of isotopic signatures of Himalayan lithotectonic zones. *Earth and Planetary Science Letters*, *312*(3–4), 471–483. https://doi.org/10.1016/j.epsl.2011.10.027

Medicott, H. B. (1864). On the geologic structure and relations of the southern portion of the Himalayan range between the rivers Ganges and Ravee. *Memoirs of the Geological Survey of India*, *3*, 1–206.

Meigs, A. J., Burbank, D. W., & Beck, R. A. (1995). Middle-late Miocene [>10 Ma] formation of the Main Boundary Thrust in the Western Himalaya. *Geology*, *23*, 423–426.

Metcalfe, I. (1999). The Tethys; How many? How old? How deep? How wide? *Proceedings of the International Symposium on Shallow Tethys*, *5*, 1–15.

Metcalfe, I. (2009). Late Palaeozoic and Mesozoic tectonic and palaeogeographic evolution of SE Asia. *Geological Society of London, Special Publication*, *315*, 7–23.

Metcalfe, I. (2013). Gondwana dispersion and Asian accretion: Tectonic and palaeogeographic evolution of eastern Tethys. *Journal of Asian Earth Sciences*, *66*, 1–33.

Metcalfe, R. P. (1993). Pressure, temperature and time constraints on metamorphism across the Main Central Thrust zone and High Himalaya Slab in the Garhwal Himalaya. In P. J. Trealor & M. P. Searle, (Eds.), *Himalayan tectonics* (pp. 485–509). Geological Society Special Publication, 74.

Middlemiss, C. S. (1887). Physical geology of West British Garhwal, with notes on a route traverse through Jaunsar Bawar and Tiri-Garhwal. *Records of the Geological Survey of India*, *20*, 26–40.

Miller, C., Klotzli, U., Frank, W., Thoni, M., & Grasemann, B. (2000). Proterozoic crustal evolution in the NW Himalaya (India) as recorded by circa 1.80 Ga mafic and 1.84 Ga granitic magmatism. *Precambrian Research*, *103* (3–4), 191– 206.

Molnar, P., & England, P. (1990). Late Cenozoic uplift of mountain ranges and global climate change: Chicken or egg? *Nature*, *346*, 29–34. https://doi.org/10.1038/346029a0

Molnar, P., & Stock, J. M. (2009). Slowing of India's convergence with Eurasia since 20 Ma and its implications for

Tibetan mantle dynamics. *Tectonics*, *28*. https://doi.org/10.1029/2008TC002271

Montemagni, C., Carosi, R., Fusi, N., Iaccarino, S., Montomoli, C., Villa, I. M., et al. (2020) Three-dimensional vorticity and time-constrained evolution of the Main Central Thrust zone, Garhwal Himalaya (NW India). *Terra Nova*, *32*, 215– 224. https://doi.org/10.1111/ter.12450

Montemagni, C., Montomoli, C., Iaccarino, S., Carosi, R., Jain, A. K., Massonne, et al. (2018). Dating protracted fault activities: Microstructures, microchemistry and geochronology of the Vaikrita Thrust, Main Central Thrust zone, Garhwal Himalaya, NW India. *Geological Society, London, Special Publications*, *481*, 127–146. https://doi.org/10.1144/SP481.3

Montomoli, C., Carosi, R., & Iaccarino, S. (2015). Tectonometamorphic discontinuities in the Greater Himalayan Sequence: A local or a regional feature? *Geological Society Special Publication*, *412*, 25–41.

Montomoli, C., Carosi, R., Rubatto, D., Visonà, D., & Iaccarino, S. (2017). Tectonic activity along the inner margin of the South Tibetan Detachment constrained by syntectonic leucogranite emplacement in Western Bhutan. *Italian Journal of Geosciences*, *136*(1), 5–14. https://doi.org/10.3301/IJG.2015.26

Montomoli, C., Iaccarino, S., Carosi, R., Langone, A., & Visonà, D. (2013). Tectonometamorphic discontinuities within the Greater Himalayan sequence in Western Nepal (Central Himalaya): Insights on the exhumation of crystalline rocks. *Tectonophysics*, *608*, 1349–1370.

Mosca, P., Groppo, C., & Rolfo, F. (2012). Structural and metamorphic features of the Main Central Thrust Zone and its contiguous domains in the eastern Nepalese Himalaya. In M. Zucali (Eds.), Multiscale structures and tectonic trajectories in active margins. *Journal of the Virtual Explorer*, *41*, 1–34. https://doi.org/10.3809/jvirtex.2011.00294

Mottram, C. M., Warren, C. J., Regis, D., Roberts, N. M. W., Harris, N. B. W. Argles, T. W., et al. (2014). Developing an inverted Barrovian sequence: Insights from monazite petrochronology. *Earth and Planetary Science Letters*, *403*, 418–431. https://doi.org/10.1016/j.epsl.2014.07.006

Moynihan, D., & Pattison, D. R. M. (2013). An automated method for the calculation of P-T paths from garnet zoning, with application to metapelitic schist from the Kootenay Arc, British Columbia, Canada. *Journal of Metamorphic Geology*, *31*, 525–548. https://doi.org/10.1111/jmg.12032

Mugnier, J. L., Huyghe, P., Chalaron, E., & Mascle, G. (1994). Recent movements along the Main Boundary Thrust of the Himalayas: Normal faulting in an over-critical thrust wedge? *Tectonophysics*, *238*(1–4), 199–215. https://doi.org/10.1016/0040-1951(94)90056-6

Mugnier, J. L., Leturmy, P., Mascle, G., Huyghe, P., Chalaron, E., Vidal, G., et al. (1999). The Siwaliks of western Nepal: I. Geometry and kinematics. *Journal of Asian Earth Sciences*, *17* (5–6), 629–642. https://doi.org/10.1016/S1367-9120(99)00038-3

Mukherjee, S. (2013). Higher Himalaya in the Bhagirathi section (NW Himalaya, India): its structures, backthrusts and extrusion mechanism by both channel flow and critical taper mechanisms. *International Journal of Earth Sciences (Geol Rundsch)*, *102*, 1851–1870. https://doi.org/10.1007/s00531-012-0861-5

Mukherjee, S., Koyi, H. A., & Talbot, C. J. (2012). Implications of channel flow analogue models for extrusion of the Higher Himalayan Shear Zone with special reference to the out-of-sequence thrusting. *International Journal of Earth Sciences (Geol Rundsch)*, *101*, 253–272. https://doi.org/10.1007/s00531-011-0650-6

Mukhopadhyay, D. K., Chakraborty, S., Trepmann, C., Rubatto, D., Anczkiewicz, R., Gaidies, F., et al. (2017). The nature and evolution of the Main Central Thrust: Structural and geochronological constraints from the Sikkim Himalaya, NE India. *Lithos*, *282–283*, 447–463. https://doi.org/10.1016/j.lithos.2017.01.015

Mukul, M. (2000). The geometry and kinematics of the Main Boundary Thrust and related neotectonics in the Darjiling Himalayan Fold-and-Thrust belt, West Bengal, India. *Journal of Structural Geology*, *22*, 1261–1283.

Mukul, M., Jaiswal, M., & Singhvi, A. K. (2007). Timing of recent out-of-sequence active deformation in the frontal Himalayan wedge: Insights from the Darjiling sub-Himalaya, India. *Geology*, *35* (11), 999–1002. https://doi.org/10.1130/G23869A.1

Myrow, P. M., Hughes, N. C., Derry, L. A., McKenzie, N. R., Jiang, G., Webb, A. A. G., et al. (2015). Neogene marine isotopic evolution and the erosion of Lesser Himalayan strata: Implications for Cenozoic tectonic history. *Earth Planetary Science Letters*, *417*, 142–150.

Myrow, P. M., Hughes, N. C., Paulsen, T. S., Williams, I. S., Parcha, S.K., Thompson, K. R., et al. (2003). Integrated tectonostratigraphic analysis of the Himalaya and implications for its tectonic reconstruction. *Earth and Planetary Science Letters*, *212*, 433–441.

Myrow, P. M., Hughes, N. C., Searle, M. P., Fanning, C. M., Peng, S., & Parcha, S. K. (2009). Stratigraphic correlation of Cambrian-Ordovician deposits along the Himalaya: Implications for the age and nature of rocks in the Mt. Everest region. *Geological Society of America Bulletin*, *120*, 323–332.

Myrow, P. M., Snell, K., Hughes, N. C., Paulsen, T. S., Heim, N. A., & Parcha, S. K. (2006). Cambrian depositional history of the Zanskar Valley region of Indian Himalaya: Tectonic implications. *Journal of Sedimentary Research*, *76*, 364–381.

Nábělek, J., Hetenyi, G., Vergne, J., Sapkota, S., Kafle, B., Jiang, M., et al. (2009). Underplating in the Himalaya-Tibet collision zone revealed by the Hi-CLIMB experiment. *Science*, *325*, 1371–1374. https://doi.org/10.1126/science.1167719

Najman, Y., Garzanti, E., Pringle, M., Bickle, M., Stix, J., & Khan, I. (2003). Early-middle Miocene paleodrainage and tectonics in the Pakistan Himalaya. *Geological Society of America Bulletin*, *115*, 1265–1277.

Najman, Y., Jenks, D. et al. (2017). The Tethyan Himalayan detrital record shows that India-Asia terminal collision occurred by 54 Ma in the Western Himalaya. *Earth and Planetary Science Letters*, *459*, 301–310.

Najman, Y., Pringle, M., Godin, L., & Oliver, G. (2001). Dating of the oldest continental sediments from the Himalayan foreland basin. *Nature*, *410*, 194– 197.

Najman, Y., Pringle, M., Godin, L., & Oliver, G. (2002). A reinterpretation of the Balakot formation: implications for the tectonics of the NW Himalaya, Pakistan. *Tectonics, 21* (Art. No. 1045).

Nelson, K. D., Zhao, W., Brown, L. D., & others (1996). Partially molten crust beneath Southern Tibet: Synthesis of project INDEPTH results. *Science, 274*, 1684–1688.

Ni, J., & Barazangi, M. (1984). Seismotectonics of the Himalayan collision zone; geometry of the underthrusting Indian Plate beneath the Himalaya. *Journal of Geophysical Research, 89*, 1147–1163.

Oldham, R. D. (1883). The geology of Jaunsar and the Lower Himalayas. *Records of the Geological Survey of India, 16*, 193–198.

Pandey, M. R., Lavé, J., & Massot, J. P. (1995). Interseismic strain accumulation on the Himalayan Crustal Ramp (Nepal). *Geophysical Research Letters, 22*, 751–754.

Parrish, R. R., & Hodges, K. V. (1996). Isotopic constraints on the age and provenance of the Lesser and Greater Himalaya sequences, Nepalese Himalaya. *Geological Society of America Bulletin, 108*, 904–911.

Parsons, A. J., Hosseini, K., Palin, R. M., & Sigloch, K (2020). Geological, geophysical and plate kinematic constraints for models of the India-Asia collision and the post-Triassic central Tethys oceans. *Earth-Science Reviews, 208*, 103084. https://doi.org/10.1016/j.earscirev.2020.103084

Parsons, A. J., Law, R. D., Lloyd, G. E., Phillips, R. J., & Searle, M. P. (2016). Thermo-kinematic evolution of the Annapurna-Dhaulagiri Himalaya, central Nepal: The composite orogenic system. *Geochemistry, Geophysics, Geosystems, 17*, 1511–1539. https://doi.org/10.1002/2015GC006184

Patel, R. C., Singh, S., Asokan, A., Manickavasagam, R. M., & Jain, A. K. (1993). Extensional tectonics in the Himalayan orogen, Zanskar, NW India. In P. J. Trealor & M. P. Searle (Eds.), *Himalayan tectonics* (pp. 445–459). Geological Society Special Publication, 74.

Patra, A., & Saha, D. (2019). Stress regime changes in the Main Boundary Thrust zone, Eastern Himalaya, decoded from fault-slip analysis. *Journal of Structural Geology, 120*, 29–47. https://doi.org/10.1016/j.jsg.2018.12.010

Patriat, P., & Achache, J. (1984). India-Eurasia collision chronology has implications for crustal shortening and driving mechanism of plates. *Nature, 311*, 615– 621.

Pearson, O. N., & DeCelles, P. G. (2005). Structural geology and regional tectonic significance of the Ramgarh thrust, Himalayan fold-thrust belt of Nepal, *Tectonics, 24*, TC4008. https://doi.org/10.1029/2003TC001617

Pêcher, A. (1989). The metamorphism in the central Himalaya. *Journal of Metamorphic Geology, 7*, 31–41.

Pêcher, A. (1991). The contact between the Higher Himalayan crystallines and the Tibetan sedimentary series: Miocene large-scale dextral shearing. *Tectonics, 10*, 587–598.

Pêcher, A., & Le Fort, P. (1986). The metamorphism in central Himalaya, its relations with the thrust tectonic. In P. Le Fort et al. (Eds.), *Évolution des domains orogéniques d'Asie méridionale (de la Turquie à l'Indoneasie)* (pp. 285–309). Science de la Terre, 47.

Phukon, P., Sen, K., Singh, P. C., Sen, A., Srivastava, H. B., & Singhal, S. (2019). Characterizing anatexis in the Greater Himalayan Sequence (Kumaun, NW India) in terms of pressure, temperature, time and deformation. *Lithos, 344–345*, 22–50. https://doi.org/10.1016/j.lithos.2019.04.018

Pilgrim, G. E., & West, W. D. (1928). The structure and correlation of the Simla Rocks. *Memoirs of the Geological Survey of India, 53*, 1–138.

Pognante, U., & Benna, P. (1993). Metamorphic zonation, migmitization and leucogranites along the Everest transect of Eastern Nepal and Tibet: Record of an exhumation history. In P. J. Trealor & M. P. Searle (Eds.), *Himalayan tectonics* (pp. 323–340). Geological Society Special Publication, 74.

Powers, P. M., Lillie, R. J., & Yeats, R. S. (1998). Structure and shortening of the Kangra and Dehra Dun reentrants, Sub-Himalaya, India. *Geological Society of America Bulletin, 110*, 1010–1027.

Quade, J., Cater, J. M. L., Ohja, T. P., Adam, J., & Harrison, T. M. (1995). Late Miocene environmental change in Nepal and the northern Indian subcontinent: stable isotopic evidence from paleosols. *Geological Society of America Bulletin, 107*, 1381–1397

Quade, J., Cerling, T., & Bowman, J. (1989). Development of Asian monsoon revealed by marked ecological shift during the latest Miocene in northern Pakistan. *Nature, 342*, 163–166. https://doi.org/10.1038/342163a0

Quidelleur, X., Grove, M., Lovera, O. M., Harrison, T. M., Yin, A., & Ryerson, F. J. (1997). The thermal evolution and slip history of the Renbu Zedong Thrust, southeastern Tibet. *Journal of Geophysical Research, 102*, 2659–2679.

Raiverman, V., Kunte, S. V., & Mukherjea, A. (1983). Basin geometry, Cenozoic sedimentation, and hydrocarbon prospects in northwestern Himalaya and Indo-Gangetic plains. *Petroleum Asia Journal, 6*, 67–92.

Rajendran, K., Parameswaran, R. M., & Rajendran, C. (2017). Seismotectonic perspectives on the Himalayan arc and contiguous areas: Inferences from past and recent earthquakes. *Earth-Science Reviews, 173*, 1–30. https://doi.org/10.1016/j.earscire2017.08.003

Rapa, G., Mosca, P., Groppo, C., & Rolfo, F. (2018). Detection of tectonometamorphic discontinuities within the Himalayan orogen: Structural and petrological constraints from the Rasuwa district, central Nepal Himalaya. *Journal of Asian Earth Sciences, 158*, 266–286. https://doi.org/10.1016/j.jseaes.2018.02.021

Ray, S. (1947). Zonal metamorphism in the eastern Himalayas and some aspects of local geology. *The Quarterly journal of the Geological, Mining, and Metallurgical Society of India, 19*, 117–140.

Reuber, I. (1986). Two peridotite units superposed by intra-oceanic thrusting in the Spongtang Klippe (Ladakh-Himalaya). *Sciences Geologiques [Bulletin], 39*, 391–402.

Roberts, A. G., Weinberg, R. F., Hunter, N. J. R., & Ganade, C. E. (2020). Large-scale rotational motion within the Main Central Thrust Zone in the Darjeeling-Sikkim Himalaya, India. *Tectonics, 39*, e2019TC005949. https://doi.org/10.1029/2019TC005949

Robinson, D. M., DeCelles, P. G., & Copeland, P. (2006). Tectonic evolution of the Himalayan thrust belt in western Nepal: Implications for channel flow models, *Geological Society of America Bulletin, 118*, 865–885. https://doi.org/10.1130/B25911.1

Robinson, D. M., DeCelles, P. G., Garzione, C. N., Pearson, O. N., Harrison, T. M., & Catlos, E. J. (2003). Kinematic model for the Main Central Thrust in Nepal. *Geology, 31*, 359–362.

Robinson, D. M., DeCelles, P. G., Patchett, P. J., & Garzione, C. N. (2001). The kinematic evolution of the Nepalese Himalaya interpreted from Nd isotopes. *Earth and Planetary Science Letters, 192*, 507–521.

Robyr, M., & Lanari, P. (2020). Kinematic, metamorphic, and age constraints on the Miyar Thrust Zone: Implications for the Eohimalayan history of the High Himalayan Crystalline of NW India. *Tectonics, 39*, e2020TC006379. https://doi.org/10.1029/2020TC006379

Rolfo, F., Groppo, C., & Mosca, P. (2014). Petrological constraints of the "Channel Flow" model in eastern Nepal. In S. Mukherjee et al. (Eds), *Tectonics of the Himalaya*. Geological Society, London, Special Publications, 412. https://doi.org/10.1144/SP412.4

Rolland, Y., Picard, C., Pecher, A., Lapierre, H., Bosch, D., & Keller, F. (2002). The cretaceous Ladakh arc of NW Himalaya: Slab melting and melt-mantle interaction during fast northward drift of Indian Plate. *Chemical Geology, 182*(2–4), 139–178. https://doi.org/10.1016/S0009-2541(01)00286-8

Rowley, D. B. (1996). Age of collision between India and Asia: A review of the stratigraphic data. *Earth and Planetary Science Letters, 145*, 1–13.

Sachan, H. K., Kohn, M. J., Saxena, A., & Corrie, S. L. (2010). The Malari leucogranite, Garhwal Himalaya, northern India: Chemistry, age, and tectonic implications. *Geological Society of America Bulletin, 122*, 1865–1876. https://doi.org/10.1130/B30153.1

Sakai, H., Iwano, H., Danhara, T., Takigami, Y., Rai, S. M., Upreti, B. N. et al. (2013). Rift-related origin of Kuncha Formation. *Island Arc, 22*, 338–360. https://doi.org/10.1111/iar.12021

Sanyal, P., & Sinha, R. (2010). Evolution of the Indian summer monsoon: Synthesis of continental records. *Geological Society, London, Special Publications, 342*, 153–183. https://doi.org/10.1144/SP342.11

Saxena, M. N. (1971). The crystalline axis of the Himalaya: The Indian shield and continental drift. *Tectonophysics, 12*(6), 433–447. https://doi.org/10.1016/0040-1951(71)90044-8

Schelling, D. (1992). The tectonostratigraphy and structure of the Eastern Nepal Himalaya. *Tectonics, 11*, 925–943.

Schelling, D., & Arita, K. (1991). Thrust tectonics, crustal shortening, and the structure of the far-eastern Nepal Himalayas. *Tectonics, 10*, 851–862.

Schmalholz, S. M., & Podladchikov, Y. Y. (2013). Tectonic overpressure in weak crustal-scale shear zones and implications for the exhumation of high-pressure rocks. *Geophysical Research Letters, 40*, 1984–1988. https://doi.org/10.1002/grl.50417

Schneider, C., & Masch, L. (1993). The metamorphism of the Tibet Series from the Manang area, Marsyandi Valley, central Nepal. In P. J. Trealor & M. P. Searle (Eds.), *Himalayan tectonics* (pp. 357–374). Geological Society Special Publication, 74.

Scotese, C. R., Gahagan, L. M., & Larson, R. L. (1988). Plate tectonic reconstructions of the Cretaceous and Cenozoic ocean basins. *Tectonophysics, 155*, 27–48.

Searle, M. P. (1999). Extensional and compressional faults in the Everest-Lhotse massif, Khumbu Himalaya, *Nepal. Journal of the Geological Society London, 156*, 227–240.

Searle. M. P. (2010). Low-angle normal faults in the compressional Himalayan orogen; Evidence from the Annapurna–Dhaulagiri Himalaya, Nepal. *Geosphere, 6*(4), 296–315. https://doi.org/10.1130/GES00549.1

Searle, M. P., & Rex, A. J. (1989). Thermal model for the Zanskar Himalaya. *Journal of Metamorphic Geology, 7*, 127–134.

Searle, M., Avouac, J., Elliott, J., & Dyck, B. (2017). Ductile shearing to brittle thrusting along the Nepal Himalaya: Linking Miocene channel flow and critical wedge tectonics to April 25 2015 Gorkha earthquake. *Tectonophysics, 714–715*, 117–124. https://doi.org/10.1016/j.tecto.2016.08.003

Searle, M. P., Law, R. D., & Jessup, M. J. (2006). Crustal structure, restoration and evolution of the Greater Himalaya in Nepal-South Tibet: Implications for channel flow and ductile extrusion of the middle crust. *Geological Society, London, Special Publications, 268*, 355–378. https://doi.org/10.1144/GSL.SP.2006.268.01.17

Searle, M. P., Law, R. D., Godin, L., Larson, L. P., Streule, M. J., Cottle, J. M., et al. (2008). Defining the Himalayan Main Central Thrust in Nepal. *Journal of the Geological Society, 165*(2), 523–534.

Searle, M. P., Metcalfe, R. P., Rex, A. J., & Norry, M. J. (1993). Field relations, petrogenesis and emplacement of the Bhagirathi leucogranite, Garhwal Himalayas. In P. J. Trealor & M. P. Searle (Eds.), *Himalayan tectonics* (pp. 429–444). Geological Society Special Publication, 74.

Şengör, A. M. C., & Atayman, S. (2009). The Permian extinction and the Tethys; an exercise in global geology. *Special Paper - Geological Society of America, 448*. https://doi.org/10.1130/2009.2448

Siddiqui, R. H., Jan, M. Q., & Khan, M. A. (2012). Petrogenesis of Late Cretaceous lava flows from a Ceno-Tethyan Island Arc: The Raskoh Arc, Balochistan, Pakistan. *Journal of Asian Earth Sciences, 59*, 24–28.

Siddiqui, R. H., Jan, M. Q., Khan, M. A., Kakar, M. I., & Foden, J. D. (2017). Petrogenesis of the Late Cretaceous Tholeiitic volcanism and ocean island arc affinity of the Chagai arc, western Pakistan. *Acta Geologica Sinica, 91*, 1248–1263.

Simpson, R. L., Parrish, R. R., Searle, M. P., & Waters, D. J. (2000). Two episodes of monazite crystallization during metamorphism and crustal melting in the Everest region of the Nepalese Himalaya. *Geology, 28*, 403–406.

Sinha-Roy, S. (1982). Interactions of Tethyan blocks and the evolution of Asian fold belts. *Tectonophysics, 82*, 271–297.

Spear, F. S. (1986). P-T PATH:A FORTRAN program to calculate pressure-temperature paths from zoned metamorphic garnets. *Computers in Geoscience, 12*, 247–266.

Spear, F. S. (1993). *Metamorphic phase equilibria and pressure-temperature-time paths*. Washington, D.C.: Mineralogical Society of America.

Spear, F. S., & Daniel, C. G. (1998). Three-dimensional imaging of garnet porphyroblast sizes and chemical zoning: Nucleation and growth history in the garnet zone. *Geological Materials Research*, *1*, 44.

Spear, F. S., & Peacock, S. M. (1989). Metamorphic pressure-temperature-time paths. *American Geophysical Union Short Course in Geology*, *7*, 102.

Spear, F. S., & Rumble, D., III (1986). Pressure, temperature and structural evolution of the Orfordville Belt, west-central New Hampshire. *Journal of Petrology*, *27*, 1071–1093.

Spear, F. S., & Selverstone, J. (1983). Quantitative P-T paths from zoned minerals: Theory and tectonic applications. *Contributions to Mineralogy and Petrology*, *83*, 348–357.

Spear, F. S., Kohn, M. J., & Paetzold, S. (1995). Petrology of the regional sillimanite zone, west-central New Hampshire, USA, with implications for the development of inverted isograds. *American Mineralogist*, *80*, 361–376.

Spear, F. S., Selverstone, J., Hickmott, D., Crowley, P. and Hodges, K. V. (1984). P-T paths from garnet zoning: A new technique for deciphering tectonic processes in crystalline terranes. *Geology*, *12*, 87–90.

Srivastava, V., Mukul, M., Barnes, J. B., & Mukul, M. (2018). Geometry and kinematics of Main Frontal thrust-related fault propagation folding in the Mohand Range, northwest Himalaya. *Journal of Structural Geology*, *115*, 1–18. https://doi.org/10.1016/j.jsg.2018.06.022

Stäubli, A. (1989). Polyphase metamorphism and the development of the Main Central thrust: *Journal of Metamorphic Geology*, *7*, 73–93.

Stephenson, B. J., Waters, D. J., & Searle, M. P. (2000). Inverted metamorphism and the Main Central Thrust: Field relations and thermobarometric constraints from the Kishtwar Window, NW Indian Himalaya. *Journal of Metamorphic Geology*, *18*(5), 571–590. https://doi.org/10.1046/j.1525-1314.2000.00277.x

Stickroth, S. F., Carrapa, B., DeCelles, P. G., Gehrels, G. E., & Thomson, S. N. (2019). Tracking the growth of the Himalayan fold-and-thrust belt from lower Miocene foreland basin strata: Dumri Formation, western Nepal. *Tectonics*, *38*, 3765–3793. https://doi.org/10.1029/2018TC005390

Stöcklin, J. (1980). Geology of Nepal and its regional frame. *Journal of the Geological Society*, *137*, 1–34.

Subedi, S., Hetényi, G., Vergne, J., Bollinger, L., Lyon-Caen, H., Farra, V., et al. (2018). Imaging the Moho and the Main Himalayan Thrust in Western Nepal with receiver functions. *Geophysical Research Letters*, *45*, 13,222–13,230. https://doi.org/10.1029/2018GL080911

Thakur, C., & Kumar, S. (1994). Seismotectonics of the October 20 1991 Uttarkashi earthquake in Garhwal, Himalaya, North India. *Terra Nova*, *6*, 90–94. https://doi.org/10.1111/j.1365-3121.1994.tb00637.x

Thakur, S. S., Patel, S. C., & Singh, A. K., (2015). A P-T pseudosection modelling approach to understand metamorphic evolution of the Main Central Thrust Zone in the Alaknanda valley, NW Himalaya. *Contributions to Mineralogy and Petrology*, *170*, 2. https://doi.org/10.1007/s00410-015-1159-y

Thakur, V. C., Jayangondaperumal, R., & Malik, M. A. (2010). Redefining Medlicott-Wadia's main boundary fault from Jhelum to Yamuna: An active fault strand of the main boundary thrust in northwest Himalaya. *Tectonophysics*, *489* (1–4), 29–42. https://doi.org/10.1016/j.tecto.2010.03.014

Tobgay, T., McQuarrie, N., Long, S., Kohn, M. J., & Corrie, S. L. (2012). The age and rate of displacement along the Main Central Thrust in the western Bhutan Himalaya. *Earth and Planetary Science Letters*, *319–320*, 146–158. https://doi.org/10.1016/j.epsl.2011.12.005

Tripathi, C., & Singh, G. (1987). Gondwana and associated rocks of the Himalaya and their significance. In G. D. McKenzie (Ed.), *Gondwana six; Stratigraphy, sedimentology, and paleontology* (pp. 195–205). AGU Geophysical Monograph, 41.

Trivedi, J. R., Gopalan, K., & Valdiya, K. S. (1984). Rb-Sr ages of granitic rocks within the Lesser Himalaya nappes, Kumaun, India. *Journal of the Geological Society of India*, *25*, 641–654.

Upreti, B. N. (1999). An overview of the stratigraphy and tectonics of the Nepal Himalaya. *Journal of Asian Earth Sciences*, *17*, 741–753.

Upreti, B. N., & Le Fort, P. (1999). Lesser Himalayan crystalline nappes of Nepal: Problems of their origin. In A. Macfarlane et al. (Eds.), *Himalaya and Tibet: Mountain roots to mountain tops* (pp. 225–238). Geological Society of America Special Paper, 328.

Upreti, B. N., & Yoshida, M. (2005). *Basement history and provenance of the Tethys sediments of the Himalaya: an appraisal based on recent geochronologic and tectonic data.* Abstract. The 1st International Conference on the Geology of Tethys, 2005, 12–14 November 2005, Cairo.

Valdiya, K. S. (1980). *Geology of the Kumaon Lesser Himalaya.* Wadia Institute of Himalaya, Dehra Dun, India.

Valdiya, K. S. (1988). Tectonics and the evolution of the central sector of the Himalaya. *Philosophical Transactions of the Royal Society of London, A.*, *326*, 151–175.

Valdiya, K. S. (1992). The Main Boundary Thrust Zone of the Himalaya, India. *Annales Tectonicae*, *6, Suppl.*, 54–84.

van Hinsbergen, D. J. J., Lippert, P. C.; Dupont-Nivet, G., McQuarrie, N., Doubrovine, P. V., Spakman, W. et al. (2012). Greater India Basin hypothesis and two-stage Cenozoic collision between India and Asia. *Proceedings of the National Academy of Sciences of the United States of America*, *109*, 7659–7664.

Vannay, J.-C., & Grasemann, B. (1998). Inverted metamorphism in the High Himalaya of Himachal Pradesh (NW India): phase equilibria versus thermobarometry. *Schweizerische Mineralogische und Petrographische Mitteilungen*, *78*, 107–132.

Vannay, J.-C., & Hodges, K. V. (1996). Tectonometamorphic evolution of the Himalayan metamorphic core between Annapurna and Dhaulagiri, central Nepal. *Journal of Metamorphic Geology*, *14*, 635–656.

Vannay, J.-C., & Steck, A. (1995). Tectonic evolution of the High Himalaya in upper Lahul (NW Himalaya, India). *Tectonics*, *14*, 253–263.

Vannay, J.-C., Grasemann, B., Rahn, M., Frank, W., Carter, A., Baudraz, V., et al. (2004). Miocene to Holocene exhumation of metamorphic crustal wedges in the NW Himalaya: Evidence for tectonic extrusion coupled to fluvial erosion. *Tectonics*, *23*, TC1014.

Von Loczy, L. (1907). Beobachtungen im östlichen Himalaya (vom 8. his 28. Febr.. 1878). *Foldtani Kozlony*, *35*(9), 1–24

Wakita, K., & Metcalfe, I. (2005). Ocean plate stratigraphy in East and Southeast Asia. *Journal of Asian Earth Sciences*, *24*, 679–702.

Wang, J. M., Zhang, J. J., & Wang, X. X. (2013). Structural kinematics, metamorphic P-T profiles and zircon geochronology across the Greater Himalayan Crystalline Complex in south-central Tibet: Implication for a revised channel flow. *Journal of Metamorphic Geology*, *31*, 607–628. https://doi.org/10.1111/jmg.12036

Wang, J. M., Zhang, J. J., Liu, K., Zhang, B., Wang, X.-X., Rai, S., et al. (2016). Spatial and temporal evolution of tectonometamorphic discontinuities in the central Himalaya: Constraints from P-T paths and geochronology. *Tectonophysics*, *679*, 41–60. https://doi.org/10.1016/j.tecto.2016.04.035

Wang, X., Wei, S., & Wu, W. (2017). Double-ramp on the Main Himalayan Thrust revealed by broadband waveform modeling of the 2015 Gorkha earthquake sequence. *Earth and Planetary Science Letters*, *473*, 83–93. https://doi.org/10.1016/j.epsl.2017.05.032

Waters, D. J. (2019). Metamorphic constraints on the tectonic evolution of the High Himalaya in Nepal: The art of the possible. In J. Treloar & M. Searle (Eds.), *Himalayan tectonics: A modern synthesis*. Geological Society, London, Special Publications, 483. https://doi.org/10.1144/SP483-2018-187

Webb, A. G., Schmitt, A. K., He, D., & Weigand, E. L. (2011). Structural and geochronological evidence for the leading edge of the Greater Himalayan Crystalline Complex in the central Nepal Himalaya. *Earth and Planetary Science Letters*, *304*, 483–495. https://doi.org/10.1016/j.epsl.2011.02.024

Webb, A. G., Yin, A., Harrison, T. M., Célérier, J., & Burgess, W. P. (2007). The leading edge of the Greater Himalayan Crystalline complex revealed in the NW Indian Himalaya: Implications for the evolution of the Himalayan orogen. *Geology*, *35* (10), 955–958. https://doi.org/10.1130/G23931A.1

Whipple, K. X., Shirzaei, M., Hodges, K., & Arrowsmith, J. R. (2016). Active shortening within the Himalayan orogenic wedge implied by the 2015 Gorkha earthquake. *Nature Geoscience*, *9*, 711–716.

White, R. W., Powell, R., Holland, T. J. B., Johnson, T. E., & Green, E. C. R. (2014). New mineral activity-composition relations for thermodynamic calculations in metapelitic systems. *Journal of Metamorphic Geology*, *32*, 261– 286.

Whitney, D. L., & Evans, B. D. W. (2010). Abbreviations for names of rock-forming minerals. *American Mineralogist*, *95*, 185–187.

Wiesmayr, G., & Grasemann, B. (2002). Eohimalayan fold and thrust belt: Implications for the geodynamic evolution of the NW Himalaya (India). *Tectonics*, *2*, 1058.

Willems, H., Zhou, Z., Zhang, B., & Gräfe, K-U. (1996). Stratigraphy of the upper cretaceous and lower tertiary strata in the Tethyan Himalayas of Tibet (Tingri area, China). *Geologische Rundschau*, *85*, 723. https://doi.org/10.1007/BF02440107

Wu, C., Nelson, K. D., Wortman, G., Samson, S., Yue, Y., Li, J., et al. (1998). Yadong cross structure and South Tibetan Detachment in the east central Himalaya (89–90E). *Tectonics*, *17*, 28–45.

Wu, F-Y., Liu, X-C., Liu, Z-C., Wang, R-C., Xie, L., Wang, J-M., et al. (2020). Highly fractionated Himalayan leucogranites and associated rare-metal mineralization. *Lithos*, *352–353*, 105319. https://doi.org/10.1016/j.lithos.2019.105319

Yang, T., Ma, Y., Bian, W., Jin, J., Zhang, S., Wu, H., et al. (2015). Paleomagnetic results from the Early Cretaceous Lakang Formation lavas: Constraints on the paleolatitude of the Tethyan Himalaya and the India-Asia collision. *Earth and Planetary Science Letters*, *428*, 120–133. https://doi.org/10.1016/j.epsl.2015.07.040

Ye, H., Zhang, W., Yu, Z., & Xia, G. (1981). The seismicity and regional crustal movement in the Himalaya region. *Geological and Ecological Studies of the Qinghai-Xizang Plateau*, *1*, 65–80.

Yeats, R. S., Nakata, T., Farah, A., Fort, M., Mirza, M. A., Pandey, M. R., et al. (1992). The Himalayan Frontal Fault System. *Annales Tectonicae*, *6, Suppl.*, 85–98.

Yin, A. (2006). Cenozoic tectonic evolution of the Himalayan orogen as constrained by along-strike variation of structural geometry, exhumation history, and foreland sedimentation. *Earth-Science Reviews*, *76*, 1 –131.

Yin, A., & Harrison, T. M. (2000). Geologic evolution of the Himalayan-Tibet orogen. *Annual Reviews in Earth and Planetary Science*, *28*, 211–280.

Yin, A., Harrison, T. M., Murphy, M. A., Grove, M., Nie, S., Ryerson, F. J., et al. (1999). Tertiary deformation history of southeastern and southwestern Tibet during the Indo-Asian collision. *Geological Society of America Bulletin*, *111*, 1644–1664.

Yin, A., Harrison, T. M., Ryerson, F. J., Wenji, C., Kidd, W. S. F., & Copeland, P. (1994). Tertiary structural evolution of the Gangdese thrust system, southeastern Tibet. *Journal of Geophysical Research*, *99*, 18175–18201.

Yoshida, M., & Upreti, B. N. (2006). Neoproterozoic India within East Gondwana: Constraints from recent geochronologic data from Himalaya. *Gondwana Research*, *10*(3–4), 349–356. https://doi.org/10.1016/j.gr.2006.04.011

Yuan, J., Yang, Z., Deng, C., Krijgsman, W., Hu, X., Li, S., et al. (2020). Rapid drift of the Tethyan Himalaya terrane before two-stage India-Asia collision. *National Science Review*, *0*, 1–13. https://doi.org/10.1093/nsr/nwaa173

Zhang, H., Harris, N., Parrish, R., Zhang, L., & Zhao, Z. (2004). U-Pb ages of Kude and Sajia leucogranites in Sajia dome from North Himalaya and their geological implications. *Chinese Science Bulletin*, *49*, 2087. https://doi.org/10.1360/04wd0198

Zhao, W., Nelson K. D., & Project INDEPTH (1993). Deep seismic reflection evidence for continental underthrusting beneath southern Tibet. *Nature*, *366*, 557–559.

Zhou, Z., Kusky, T. M., & Tang, C-C. (2019). Coulomb stress change pattern and aftershock distributions associated with a blind low-angle megathrust fault, Nepalese Himalaya. *Tectonophysics*, *767*, 228161. https://doi.org/10.1016/j.tecto.2019.228161

Zhu, B., Kidd, W. S. F., Rowley, D. B., Currie, B. S., & Shafique, N. (2005). Age of initiation of the India-Asia collision in the east-central Himalaya. *Journal of Geology*, *113*, 265–285.

Zhu, D. C., Zhao, Z. D., Niu, Y. L., Dilek, Y., Hou, Z. Q., & Mo, X. X. (2013). The origin and pre-Cenozoic evolution of the Tibetan Plateau. *Gondwana Research*, *23*, 1429–1454.

Zyabrev, S. V., Kojima, S., & Ahmad, T. (2008). Radiolarian biostratigraphic constraints on the generation of the Nidar ophiolite and the onset of Dras arc volcanism: Tracing the evolution of the closing Tethys along the Indus–Yarlung-Tsangpo suture. *Stratigraphy*, *5*(1), 99–112.

7
Tectonics of the Southeast Anatolian Orogenic Belt

Yücel Yılmaz[1], Erdinç Yiğitbaş[2], and İbrahim Çemen[3]

ABSTRACT

The tectonic development of the Southeast Anatolian Orogenic Belt (SAOB) is closely related to the demise of the NeoTethys Ocean that existed between the Arabian and Eurasian plates from the Late Cretaceous to Late Miocene. This ocean contained several continental slivers and intraoceanic magmatic arcs. The continental slivers represent narrow tectonic belts rifted off and drifted away from the Arabian Plate while the NeoTethyan Ocean and the backarc basins were opened. These slivers later collided with each another during which the branches of the oceans were eliminated and the continental slivers were integrated in the subduction zone and turned into metamorphic massifs. During the Late Cretaceous, the first collision occurred when an accretionary complex was thrust over the Arabian Plate's leading edge. Despite the collision, the ocean survived in the north and its northward subduction generated a new intraoceanic arc, which collided later with the northerly located continental slivers. During the Middle Eocene, the metamorphic massifs and the intraoceanic arc front migrated to the south. The new magmatic arc collided with the southerly transported nappe package during the Late Eocene. The amalgamated nappe pile eventually obducted onto the Arabian Plate during the Late Miocene. The collision produced escape structures during the Neotectonic period.

7.1. INTRODUCTION

The Anatolian Orogen is a tectonic mosaic formed during the collisions of the continental slivers rifted off from the African-Arabian Plate and accreted to the Eurasian Plate. During the collisions, two oceans were closed: (1) the PaleoTethys during the Late Paleozoic–Early Mesozoic and (2) the NeoTethys during the Mesozoic-Cenozoic (Şengör & Yılmaz, 1981). The tectonic events in association with the closure of the Paleo-Tethyan ocean are recorded mainly in the Pontide Range (see also Yılmaz et al., Chapter 8 in this volume).

[1] *Department of Geology, Istanbul Technical University, Istanbul, Turkey*

[2] *Department of Geology, Çanakkale Onsekiz Mart Üniversity, Çanakkale, Turkey*

[3] *Department of Geological Sciences, The University of Alabama, Tuscaloosa, Alabama, USA*

The Southeast Anatolian Orogenic Belt (SAOB) is the southernmost component of the Anatolian Orogen extending eastward along with the Zagros Mountains of Iran to the Oman-Makran subduction system (Fig. 7.1). Despite the collision of the surrounding continents in Anatolia and Iran, on both ends, the Indian Ocean and the eastern Mediterranean are still open as remaining parts of this ocean (Fig. 7.1, inset). Therefore, the geology of the SAOB provides data to decipher the tectonic development of the Tethyan system. There is a rich literature about the eastern Mediterranean ophiolites, significantly increased after the pioneering papers of Gass (1968) and Moores and Vine (1971) on the Troodos Ophiolite of Cyprus.

Within the NeoTethys ophiolites, the northern and southern branches were differentiated in the Anatolian Orogen (Şengör & Yılmaz, 1981). They were developed because of the consecutive separation of the continental slivers from the Afro-Arabian Plate during the Early

Compressional Tectonics: Plate Convergence to Mountain Building, Geophysical Monograph 277, First Edition.
Edited by Elizabeth J. Catlos and İbrahim Çemen.

DOI:10.1002/9781119773856.ch07

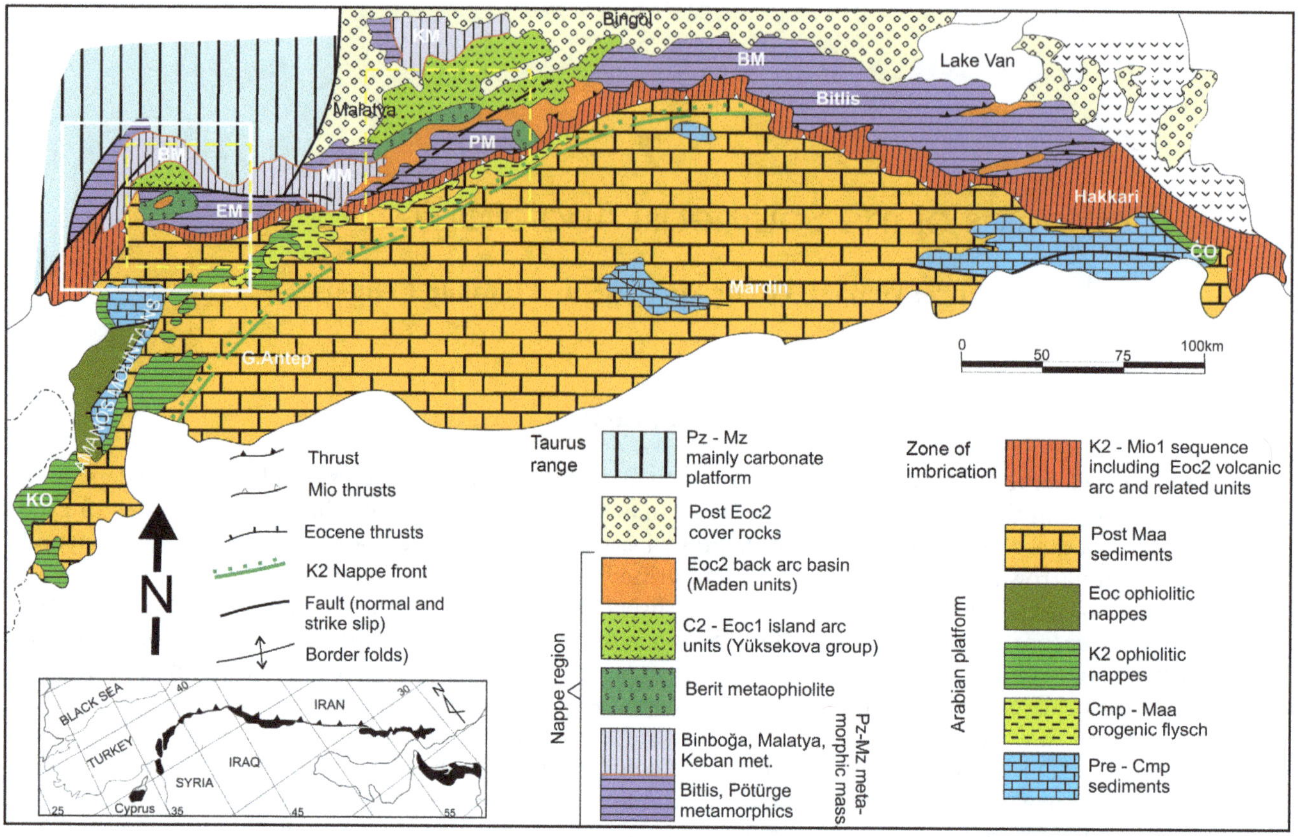

Figure 7.1 Geological map of Southeast Anatolian region. The white and yellow squares show the location of the maps in Figs. 7.4a and 7.5a,b. (Inset) Location map of the Bitlis Suture Mountains within the Bitlis-Zagros orogenic belt showing the NeoTethyan suture (the black strip) along the northern margin of the Arabian Plate.

Mesozoic. The Southern NeoTethys, generating the SAOB, evolved between the Afro-Arabian Plate and the Bitlis Massif–Tauride Platform (Şengör & Yılmaz, 1981). The latter belt was possibly connected with the narrow strip of the Pelagonian continent and Adria in the west, separating the southern NeoTethys from the Alpine Tethys (Dilek & Furnes, 2019). To the east, the southern NeoTethys widened toward the present Indian Ocean.

The SAOB represents the northwestern part of the Bitlis-Zagros Orogen (Fig. 7.1, inset). It is a composite tectonic entity that consists of nappes of the metamorphic massifs and ophiolites. The SAOB is subdivided into three east-west trending zones. They are from south to north, the Arabian Platform, the imbricated zone, and the nappes. The zones are separated by major thrusts (Yılmaz, 1993; Fig. 7.1).

Since the late 1970s, numerous researchers have worked in this region and published many papers concerning different aspects of the SAOB, providing valuable information on the local and regional scales. Among the papers are Şengör et al., 1979; Şengör & Yılmaz, 1981; Dilek & Moores, 1990; Yılmaz & Yiğitbaş, 1991; Aktaş & Robertson, 1991; Beyarslan & Bingöl, 2000; Parlak et al., 2004, 2009, 2012; Rızaoğlu et al., 2009; Dilek & Sandvol, 2009; Silja et al., 2009; Oberhänsli et al., 2010; Rolland et al., 2012; Yeşilova & Helvacı, 2013; Karaoğlan et al., 2016; Robertson, 2002; Robertson et al., 2006, 2007, 2012, 2016a, 2016b; Pourteau et al., 2013; Akıncı et al., 2016; Seyitoğlu et al., 2017; Yeşilova et al., 2018; Dilek & Furnes, 2019, Van Hinsbergen et al., 2020; Yılmaz, 2017, 2019, 2021). Most of the post-2005 works are related to petrological problems of the metamorphic and ophiolitic associations based mainly on the isotope and geochemical data (Rızaoğlu et al., 2006; Bağcı et al., 2008; Parlak et al., 2010; Karaoğlan et al., 2013a,b,c,d; Oberhänsli et al., 2010, 2012, 2014; Candan et al., 2012; Yıldırım, 2015; Parlak, 2016; Nurlu et al., 2016; Awalt & Whitney, 2018; Beyarslan & Bingöl, 2018; Bingöl et al., 2018;). These geochemical data contributed significantly to the available field evidence (Yılmaz, 1993, 2019) to reassess critical tectonic problems associated with the development of the SAOB.

The Southeast Anatolia has also been affected by the post-Miocene indentation tectonics, which formed the major strike-slip faults of the region (Çemen et al., 1993; Korucu & Çemen, 1998; Yılmaz, 2017, 2020, 2021).

The main purpose of this chapter is to review the tectonic evolution of the orogenic belt based on the analytical data and our geologic data from the region.

7.2. GEOLOGICAL OUTLINES OF THE SOUTHEAST ANATOLIAN OROGENIC BELT

The SAOB is subdivided into approximately three east-west trending zones. From south to north, they are the Arabian Platform, the Imbricated zone, and the Nappes. The zones are separated by major thrusts (Yılmaz et al., 1993; Fig. 7.1).

7.2.1. The Arabian Platform

The Arabian Platform represents the northwestern part of the Arabian Plate, where a thick sedimentary succession was deposited, mostly in the marine environment from Cambrian to present (Fig. 7.2; Tuna, 1973; Perinçek, 1979; Yiğitbaş, 1989; Yılmaz, 1984, 1990; Yılmaz et al., 1988; Siyako et al., 2013; Robertson et al., 2012a,b, 2016b). The succession contains several regional unconformities. However, regional unconformities correspond to the three nappe emplacement stages (Figs. 7.2a,b; Yılmaz, 1993). The sequence is therefore, divided into three autochthonous successions with respect to the nappes (the allochthonous units) (Fig. 7.2b; Yılmaz, 2021). The first period of sediment deposition ended when the first ophiolite nappe package (the lower ophiolite nappe, LN) was tectonically emplaced onto the Arabian Platform during the Late Campanian–Early Maastrichtian period (Fig. 7.2a,b: Yılmaz, 1993). The ophiolites of this period display the supra subduction zone (SSZO) affinities (Pearce, 1975; Dilek & Thy, 1992; Parlak et al., 2004). They were developed above the northerly subducting Tethyan oceanic lithosphere (Robertson, 2012; Dilek & Furnes, 2019; Yılmaz 2019, 2021).

Overlying unconformably, the nappes is a marine transgressive sequence of Maastrichtian to Middle Miocene age (Fig. 7.2; Tuna, 1973; Yılmaz, 1984, 1993; Robertson et al., 2012a,b; Siyako et al., 2013). The basal clastic rocks of the transgressive unit transit to neritic limestones (Fig. 7.2a), which pass upward to shales (the Germav Formation) of Late Maastrichtian–Paleocene age (Fig. 7.2a). A thick neritic limestone succession of Eocene age (the Midyat Group) grades into carbonate flysch of Upper Eocene–Oligocene age (the Fırat Formation) (Fig. 7.2a; Tuna, 1973; Yilmaz et al., 1987).

A new ophiolite nappe was tectonically emplaced above the Arabian Platform during the Middle Eocene (the middle ophiolite nappe, MN) (Yılmaz, 1984, 2019). Above the ophiolite slab is an epiophiolitic pelagic chalk-radiolarite sequence of Upper Cretaceous–Lower Eocene age range (the Cona Group of Yılmaz, 1993). Basal sandstones of Late Eocene transgression rest on the middle nappes and transit to a neritic limestone succession (the Midyat Group, Fig. 7.2).

A regional unconformity of the Late Eocene–Oligocene age separates the Eocene marine sediments from the overlying Upper Eocene–Oligocene terrestrial and shallow marine clastics rocks (Yeşilova & Helvacı, 2017). They grade laterally into the Lower Miocene flysch unit (the Lice Fm; Fig. 7.2b). A regressive sequence of late Early Miocene–Middle Miocene age follows the flysch beginning with thick (> 500 m) olistostromes (Azgıt Fm in Fig. 7.2b).

A giant nappe pile (UN) was thrust over the Lower-Middle Miocene clastic units during the Middle-Late Miocene (Fig. 7.2b; Yılmaz, 1993, 2019, 2021). The south-vergent compressional stress severely deformed the units of the imbricated zone, which were compressed between the Arabian Plate and the southerly transporting nappes.

7.2.2. The Zone of Imbrication

This east-west trending belt (Fig. 7.4a; Yıldırım & Yılmaz, 1991; Yılmaz, 1993, 2019) is 500 m to 2 km wide and consists of south-vergent thrust sheets (Fig. 7.3). The successions are complimentary within the imbricated zone, revealing a continuous sequence before the imbrication (Yılmaz, 1993). The sequence is Upper Cretaceous to Early Miocene in age (Fig. 7.3).

There is a deep-sea sedimentary succession of the Upper Cretaceous to Lower-Middle Eocene in age that conformably overlies the ophiolite at the top of the imbricated zone. The pelagic sediments consist of limestone (chalk), chert, clayey limestone, marl, and calciturbidites. The pelagic sedimentary sequence is identical to the succession overlying the middle nappe (the Cona Group, Yılmaz, 1984). Intermediate and felsic volcanic rocks (the Helete volcanics) alternate with the pelagic sediments (Aktaş & Robertson, 1985, 1991; Yılmaz, 1993; Bağcı, 2013). Above an unconformity surface, Upper Eocene–Oligocene olistostromes overlie the deep-sea sequence. A nappe pile (< 8 km thick) tectonically overlies the imbricated zone consisting of metamorphic and ophiolitic thrust sheets (Figs. 7.4 and 7.5).

7.2.3. The Nappes

Five metamorphic and ophiolite thrust sheets are recognized in the nappe zones based on their lithological and tectonostratigraphic features and the stratigraphic order. Fig. 7.4 displays the main components of the SAOB. From the bottom to the top, they are (1) the lower ophiolite nappes (LO), (2) the middle (MO) ophiolite nappes, (3) the lower (southern) metamorphic massifs

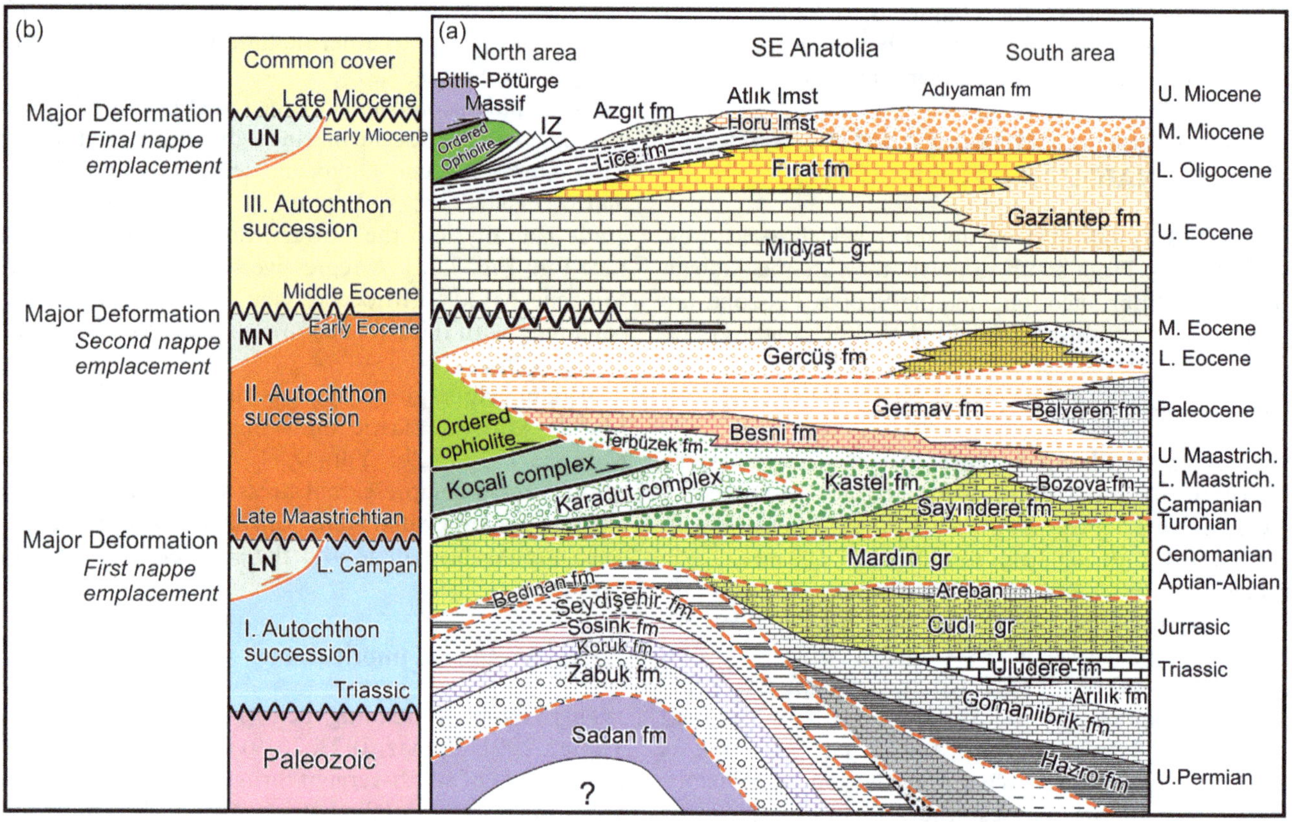

Figure 7.2 (a) Generalized stratigraphic section of the Arabian Platform in southeastern Anatolia, from the suture mountains to the north of the Arabian Platform. The lithology and age of the rock units shown in the here are as follows: Bitlis-Poturge Massif represents the nappe of the metamorphic massifs of the southeast Anatolia. Ordered Ophiolite represents the ophiolite nappe. IZ = the imbricated zone. Azgıt Formation (coarse clastic rocks; Middle-Lower Miocene). Horu and Atlık limestones (reefal limestone; Middle Miocene); Adıyaman Formation (fluvial and lacustrine sedimentary rocks; Middle-Upper Miocene); Lice flysch (Lower Miocene); Gaziantep Formation (pelagic limestone; Upper Eocene-Lower Miocene); Fırat Formation (reefal limestone, Oligocene-Lower Miocene); Midyat Formation (platform carbonate succession, Middie-Upper Eocene); Gergus Formation (basal conglomerate and sandstone, Lower-Middle Eocene); Belveren Formation (pelagic limestone, Paleocene-Lower Eocene); Germav Formation (shale, Lower Maastrichtian-Paleocene); Besni Formation (reefal limestone, Upper Maastrichtian); Terbuzek Formation (basal sandstone-conglomerate; Upper Maastrichtian); ordered ophiolite sequence (Upper Cretaceous); Koçali complex (ophiolitic mélange association, Upper Cretaceous); Karadut complex (wild flysch-flysch, Upper Triassic-Upper Cretaceous); Kastel Formation (flysch and olistostrome, Upper Campanian-Lower Maastrichtian); Bozova Formation (limestone-marl alternations, Campanian-Lower Maastrichtian); Sayındere Formation (clayey limestone; Campanian); Mardin Group (platform carbonate succession, Aptian-Cenomanian); Areban Formation (basal sandstone, limestone, Aptian-Albian); Cudi Group (platform carbonate succession, Triassic-Upper Jurassic); Uludere Formation (siltstone-marl-limestone alternations, Triassic); Atlık Formation (quartzite, Lower Triassic); Gomaniibrik Formation (limestone, Djulfian); Hazro Formation (sandstone, siltstone, Upper Permian); Bedinan Formation (shale and clastic rocks, Upper Ordovician); Seydisehir Formation (shale, sandstone, Upper Cambrian–Lower Ordovician); Sosink Formation (shale-sandstone alternations, Upper Cambrian); Koruk Dolomite (Middle Cambrian); Zabuk Formation (arkosic sandstone, Lower–Middle Cambrian?); Sadan Formation (shale-slate, Precambrian? Lower Cambrian?); Telbesmi Formation (metamorphosed tuff and felsic lava, Precambrian?). Names of the rock stratigraphic units were adopted from the Turkish Petroleum Company (revised from Yılmaz 1993). (b) Columnar section showing three nappe emplacement stages and the consequent subdivisions of the Arabian Platform sequence into allochthonous and autochthonous successions. Note: LN = lower nappe, ON = middle nappe, UP = upper nappe.

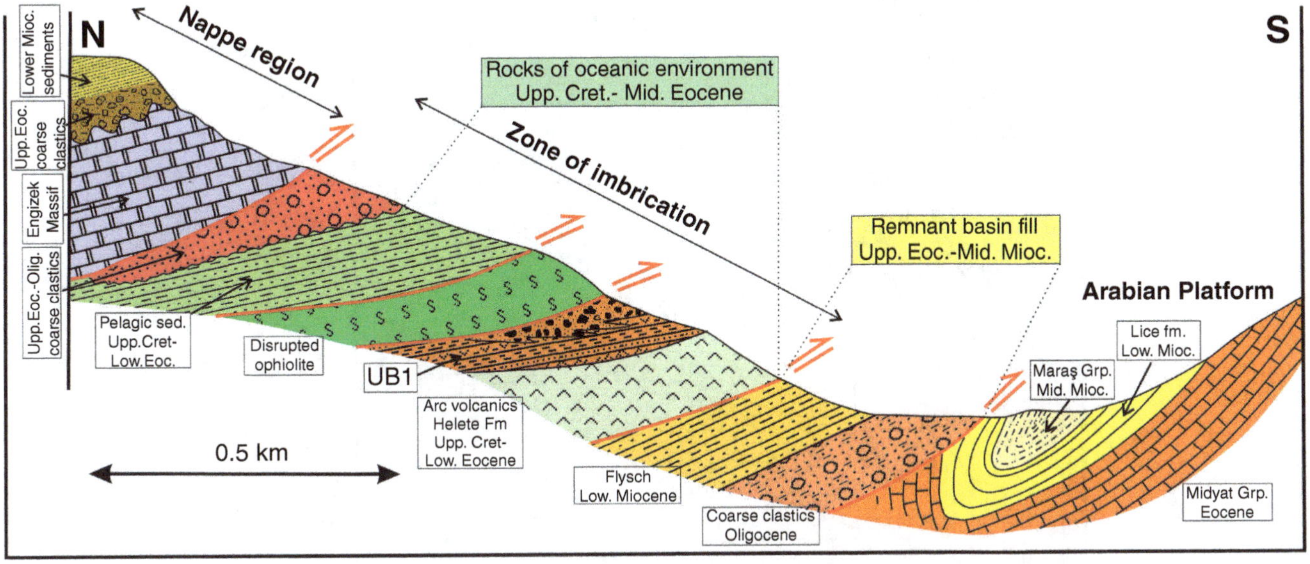

Figure 7.3 Major tectonostratigraphic units of the zone of imbrication. Lithologies and ages of the tectonostratigraphic units within the imbricated zone are as follows: the overturned syncline at the top of the Arabian Platform sequence is the Lower Miocene Lice Flysch and the regressive Middle Miocene sandstones. The coarse clastics–Oligocene are wild flysch grading into the Lower Miocene flysch. The flysch in the imbricated zone is the distal equivalent of the Lice Flysch. The Helete Formation represents the volcanic arc consisting mainly of andesitic lavas and pyroclastic rocks of the Middle Eocene age. UB1= the cover sediments of the arc sequence formed during the late stage of the arc development in the Middle-Late Eocene. Pelagic sediments of Upper Cretaceous-Middle Eocene ages above the ophiolite represent epiophiolitic deep-sea sediments; the Cona Grp. Upp. Eoc-Olig. coarse clastics are the postnappe cover sediments that sealed the amalgamated nappe pile. The Engizek Massif represents the lower metamorphic nappe (LN).

(LM), (4) the upper ophiolite nappe (UO), and (5) the upper (northern) metamorphic massifs (UM). During the first two nappe emplacement stages, ophiolite slabs were thrust over the Arabian Platform (Fig. 7.2a, b). The present orogenic belt was developed in the last phase when a nappe pile consisting of the ophiolite nappes and the overlying metamorphic massifs were tectonically emplaced onto the Arabian Plate in Miocene (Figs. 7.2, 7.4, and 7.5; Yılmaz, 2019, 2021).

Within the SAOB, two approximately east-west trending metamorphic belts may be differentiated as the northern (the Binboğa-Malatya-Keban metamorphic massifs) and the southern (the Engizek-Pötürge-Bitlis metamorphic massifs) metamorphic belts (Fig. 7.1). The former is tectonically above the latter (Figs. 7.4b and 7.5). Both metamorphic massifs consist of two major lithostratigraphic components: a Paleozoic core and a Mesozoic cover (Yılmaz, 1975; Yılmaz et al., 1993, Yılmaz, 2019). Within the cover rocks, the metamorphic grade decreases steadily upward in the sequence, where the primary sedimentary features may be identified in the thick recrystallized limestones (Hall, 1974; Yılmaz et al., 1993). The age of the cover rocks ranges from Triassic to Upper Cretaceous–Paleocene (?) (Hall, 1974; Perinçek & Kozlu, 1984; Hempton, 1985; Yılmaz, 1993; Yılmaz et al., 1993; Yiğitbaş & Yılmaz, 1996a,b; Robertson et al., 2016b). The ages and lithological characteristics of the core and the cover units correlate closely with the pre-Mesozoic basement and the overlying Mesozoic carbonate platform succession of the Taurus Range (Şengör & Yılmaz, 1981; Göncüoğlu & Turhan, 1984; Yılmaz, 2019) leading to the interpretation that both have a common origin, and they were rifted off and drifted away from the Arabian Plate during the Triassic (Şengör & Yılmaz, 1981). The metamorphic massifs underwent the major phase of metamorphism after the development of the complete sedimentary succession. In that sense, the metamorphic massifs do not fit the classical term of a massif in an orogenic belt representing tectonically elevated or protruded bodies of basement rocks consolidated during earlier orogeneses.

Genetically associated with the nappe emplacements, two belts of basins are also differentiated as the lower

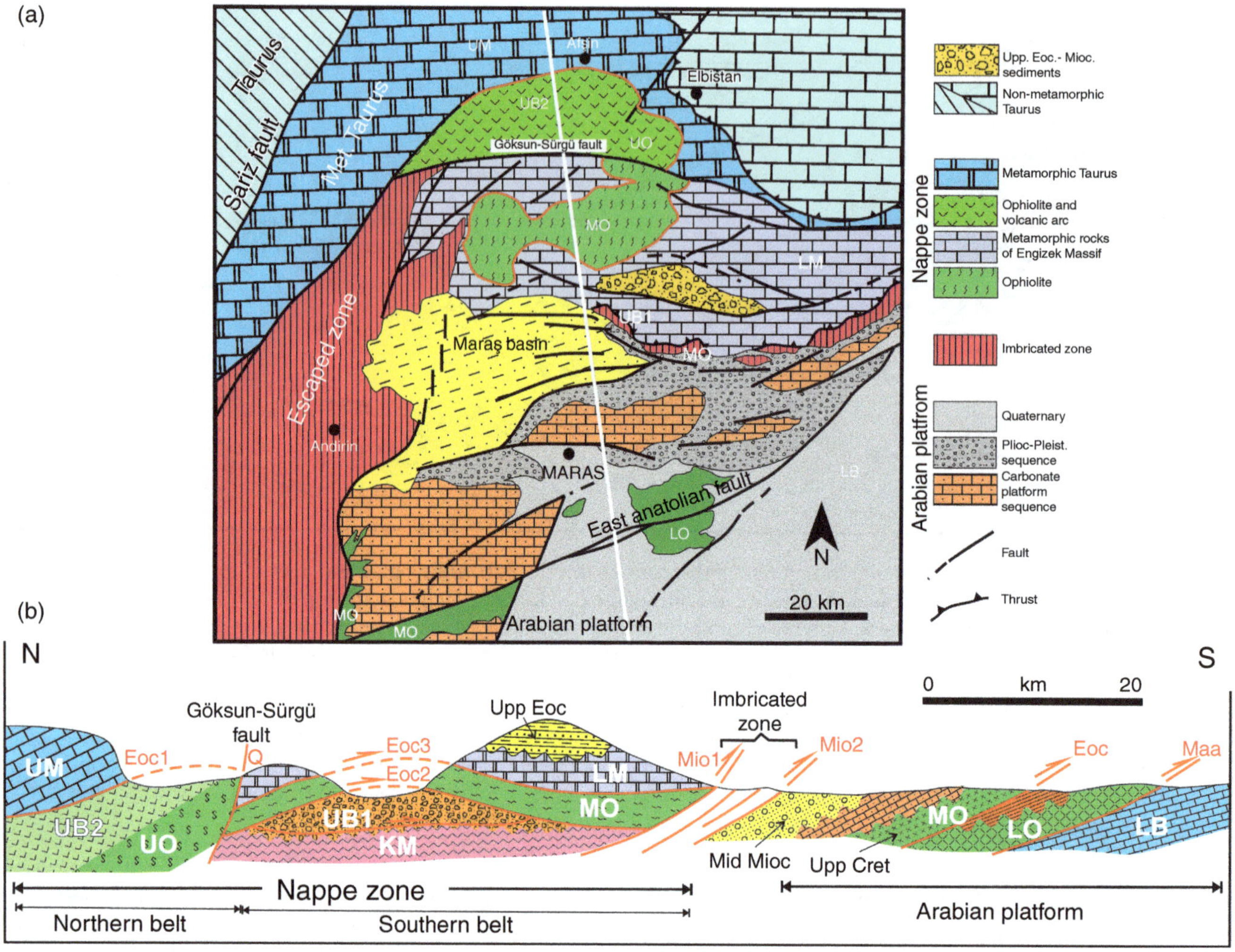

Figure 7.4 (a) Geological map of the western part of the Southwest Anatolian Orogenic Belt showing major tectonic zones and structural elements of the region (modified after Yılmaz, 1993). The white line is the cross-section direction displayed in Fig. 7.4b. The Surgu Fault (The Goksun-Surgu strike-slip fault) is one of the prominent faults of the orogenic belt, which separates the nonmetamorphic Goksun Ophiolite (Upper Ophiolite nappe; UO) and the Elbistan arc (UB2) from the metamorphic nappes (MO and LM). Note: UM=Upper Metamorphic massifs (the Binboğa Massif); UB2=the upper basin (upper part of the Elbistan arc succession); UO=the Upper Ophiolite nappe (Goksun Ophiolite); MO=the Middle Ophiolite nappe (Berit meta-ophiolite); LM=The Lower metamorphic massif (Engizek Massif); LO=Lower ophiolite nappe; LB=Lower basin, Maden=Maden Basin; UB1=The Kızılkaya, the weakly metamorphic upper Cretaceous-Lower Eocene volcanic arc unit. Small red letters accompanying the trusts indicate thrusting order of the nappes, Eoc1= late Early Eocene; Eoc2=late Early Eocene–Middle Eocene; Eo3=late Middle Eocene–early Late Eocene; Mio1=Early Miocene; Mio2=Middle-Late Miocene; Eoc=late Middle Eocene; Maa=Late Campanian–Early Maastrichtian; Q=Quaternary. (b) Geologic cross section across the Southeastern Anatolian Orogenic Belt. Abbreviations are the same as in (a). Arrows and numbers indicate the stacking order of the nappes; main thrusting stages = 1: late Early Eocene; 2: late Middle Eocene; 3: Early Miocene; 4: Middle-Late Miocene.

basin (LB) and the upper (UB) basin. Concerning their tectonic connections with the nappes and the distance from the Arabian Platform, the upper basins may also be subdivided into two belts as inner (UB1) and outer (UB2) upper basins (Fig. 7.4a,b). Geological characteristics of the nappes and the basins are outlined in the following section.

The Lower Ophiolite Nappe and the Inner Basin

Development of the lower basin (LB) is spatially, temporally, and genetically related to the lower ophiolite (LO) nappe's emplacement. The LO is a nappe stack consisting of two major groups of related rocks. Large outcrops of a thick (< 4,000 m) ordered ophiolite slab are exposed at the Cilo and Kızıldağ mountains

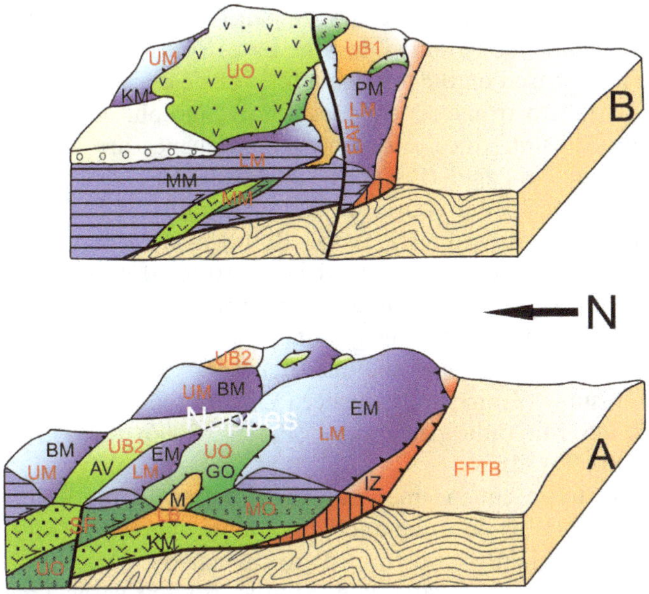

Figure 7.5 Block diagrams from the western and central part of the nappe regions of the SAOB showing order of the nappe piles. Locations of A and B are shown in Figure 7.1a. Abbreviations: black letters and red letters indicate names and tectonic orders of the nappes. In A, red letters: LM= the Lower Metamorphic Massifs; MO=the Middle Ophiolite Nappe; UO= the Upper Ophiolite Nappe; UM= the Upper Ophiolite Nappe; LB= the Lower Basin; UB1=the Upper Internal Basin; IZ= the Zone of imbrication; EAF=The east Anatolian Transform Fault; SF= the Srugu Fault; FFTB= Foreland fold and thrust belt. The black letters: PM=Poturge Metamorphic Massif; MM=Malatya Metamorphic Massif; KM=Keban Metamorphic Massif. In B, the red letters: EM=Engizek Metamorphic Massif; GO=Goksun Ophiolite; BM=Binboga Metamorphic Massif; AV=Elbistan arc volcanics; M= the Maden Basin; KM=Kizilkaya Metamorphics. The red letters are the same as in A.

(Fig. 7.1; Yılmaz, 1994; Dilek & Delaloye, 1992). Their ages vary between 92 and 80 Ma (Bağcı et al., 2005, 2008; Parlak et al., 2010; Karaoğlan et al., 2013a). In the Cilo Mountains, an upper basaltic lava layer is seen above the ophiolite, followed upward by an intermediate volcanic suite. Felsic plutonic rocks cut the entire volcanic association. Collectively, they form an intrusive-extrusive complex (Yılmaz, 1985).

The geochemical and isotope studies on the southeast Anatolian, Tauride and Cyprus ophiolites are consistent with a suprasubduction zone origin (Pearce, 1975; Dilek et al., 1990; Dilek & Eddy, 1992; Dilek & Thy, 1998; Dilek et al., 2007; Parlak et al., 2009; Parlak, 2016; Dilek & Furnes; 2019; Robertson et al., 2012a; Karaoğlan et al., 2013a, 2013d) indicating that the older NeoTethyan oceanic lithosphere was eliminated and its demise by intraoceanic subduction generated a younger SSZ ophiolite during the Late Cretaceous (92–73 Ma) (Karaoğlan et al., 2013a, 2013b, 2013c, 2013d; Dilek & Furnes, 2019, and the references therein).

Dragged under the LO, there are two distinctly different subophiolitic thrust sheets, separated by thrust faults. An ophiolitic mélange of Upper Cretaceous age, the Koçali Complex, is underlain tectonically by a flysch-wild flysch succession, the Karadut Complex (Fig. 7.2a; Yılmaz, 1984). The Karadut Complex is a severely sheared and internally commonly chaotic sedimentary unit whose age ranges from Late Triassic to Campanian. The lower part of the succession consists of a hemipelagic limestone and calcareous turbidite unit followed by a flysch and an overlying pelagic limestone, red chert, radiolarite, radiolarian mudstone unit (the Şebker Fm). The Karadut Complex represents outer-shelf and continental-slope environments. Towards the top, the complex contains multiple debris-flow deposits of Upper Campanian age containing limestone fragments and calciturbidites derived from the Arabian carbonate platform.

The Koçali Complex (Fig. 7.2a) is an ophiolitic mélange composed of blocks of ophiolite and pelagic sedimentary rocks. Its matrix comprises sheared serpentinite and multicolored radiolarian mudstone, shale, and splitized basaltic lavas. Basal clastics of Upper Maastrichtian transgressive succession unconformably overlie the LO.

The Metamorphic Massifs

The metamorphic massifs of southeastern Anatolia (LM and UM) display polyphase metamorphism (Yılmaz, 1975; Okay et al., 1985; Yılmaz et al., 1992; Parlak et al., 2012; Oberhänsli et al., 2012, 2014; Awalt & Whitney, 2018). Initially, they underwent HP metamorphism followed by an HT metamorphism (Yılmaz, 2019, and the references therein). Oberhänsli et al., (2012, 2014) described blueschist facies metamorphic rocks from the Bitlis Massif cover units and estimated the peak conditions about ca. 480°C–540°C/1.9–2.4 Gpa. They calculated the age of blueschist as 79–71 Ma. The Bitlis-Pötürge-Engizek massifs were later experienced a retrograde greenschist facies metamorphism, possibly during the Paleocene (Yılmaz, 2019).

The Middle (MO) and Upper Ophiolite (UO) Nappes and the Upper Basins (UB1 and UB2)

Away from the thrust front to the north, the middle ophiolite nappe is exposed in a tectonic window (Fig. 7.4a). The MO consists of three thrust sheets (Fig. 7.4b), which contain the following rock units.

1. The lower thrust sheet is represented by a low-grade metamorphic basaltic lava and its metasedimentary cover (the Kızılkaya Metamorphics, KM, in Figs. 7.4b and 7.5a; Yılmaz et al., 1987, 1993). The metamorphic grade does not extend beyond the lower limit of the greenschist facies.

2. The middle thrust sheet is a nonmetamorphic volcano-sedimentary sequence of Middle Eocene age known as the Maden Complex or Maden Group (Maden in Fig. 7.4b and M in Fig. 7.5a; Aktaş & Robertson, 1985; Yılmaz et al., 1993; Yiğitbaş & Yılmaz, 1996a,b), which developed a fragmenting nappe package during the Middle Eocene (Yılmaz et al., 1993; Yiğitbaş & Yılmaz, 1996a,b).

3. The upper thrust sheet is the major component of the lower nappe pile represented by a thick (< 3 km) metamorphic ophiolite slab (the Berit metaophiolite; MO in Figs. 7.4a,b and 7.5a) (Genç et al., 1993; Yılmaz et al., 1993; Yiğitbaş & Yılmaz, 1996a,b; Robertson et al., 2006; Karaoğlan et al., 2013d; Kozlu et al., 2014; Awalt & Whitney, 2018) consisting of several thrust slices. Each one of these thrust sheets exhibits an apparent ophiolite stratigraphy representing the mantle and crustal layers of an ordered ophiolite (Genç et al., 1993; Yılmaz et al., 1993). However, the ophiolite stratigraphy was reversed across the thrust sheets (Genç et al., 1993).

The Berit Ophiolite displays polyphase metamorphism, the initial granulite-eclogite facies followed by the amphibolite facies. A retrograde greenschist facies metamorphism superimposed on the earlier metamorphic phases (Yiğitbaş, 1989; Genç et al., 1993; Candan et al., 2012; Oberhänsli et al., 2012, 2014; Awalt & Whitney, 2018).

In the northern part of the SAOB, across the Göksun-Sürgü Fault (Fig. 7.4a), an ordered nonmetamorphic ophiolite slab is exposed (UO in Fig. 7.5b). The ophiolite is referred to as Göksun Ophiolite (GO in Fig. 7.5a) and its epiophiolitic cover, the Elbistan volcano-sedimentary sequence of Cretaceous-Paleocene age (AV and UB2 in Fig. 7.4a). In the epiophiolitic sequence, the pelagic sedimentary rocks are overlain by andesite-dacite lavas of island arc affinity (Yılmaz et al.,1993; Yiğitbaş & Yılmaz, 1996a; Parlak et al., 2004, 2020; Karaoğlan et al., 2013c). The lavas are followed upward by the–Lower Eocene flysch (Yılmaz et al., 1987; Yılmaz, 1993; Robertson et al., 2006). The Binboğa-Keban-Malatya metamorphic massif (UM) tectonically overlies the UO and AV (Figs. 7.4a,b and 7.5).

7.3. TIME CONSTRAINTS ON THE AMALGAMATION OF THE NAPPES

Figure 7.4b shows field relations of the time and order of piling of the nappes. Thrusting of the Binboğa-Malatya metamorphic massif (UM) above the Göksun Ophiolite-Elbistan volcanic arc pair (UO and UB2) may be tightly constrained to the Late Ypresian-Lutetian. This is based on the following lines of field evidence.

1. Olistostrome deposits resting stratigraphically above the Elbistan volcano-sedimentary sequence are Middle Eocene in age (Perinçek & Kozlu, 1984; Yılmaz et al., 1987). The source of the rapid influx of the internally chaotic sediment is the Binboğa-Malatya massif. The olistostromes may thus be interpreted as the precursor of the approaching metamorphic nappe.

2. The post-thrusting granites that intruded into the Binboğa Massif and the underlying Göksun Ophiolite–Elbistan volcanic arc sequence (Yılmaz et al., 1987) yield 51–45 Ma radiometric ages (Parlak, 2006; Karaoğlan et al., 2013d).

Tectonic emplacement of the Engizek-Pötürge-Bitlis Massif (LM) onto the middle ophiolite nappe pile (MO) is also tightly constrained to a narrow time span corresponding to the late Middle Eocene–early Late Eocene because (1) the UB1 (the Maden Complex) composes the Middle Eocene units (Yiğitbaş & Yılmaz, 1996a,b; Escartin et al., 2017), and (2) the oldest marine sediments deposited above the LM are Upper Eocene sandstones (Yılmaz, 1978; Perinçek & Kozlu, 1984; Yiğitbaş, 1989; Yılmaz, 1993; Yılmaz et al., 1981, 1987, 1993; Yılmaz & Yıldırım, 1996).

The amalgamation of the northern and the southern nappe piles corresponds to the period between the end of the Middle Eocene and the beginning of the Late Eocene because the youngest rocks under the nappe package are the Middle Eocene volcano-sedimentary unit, and the first cover sediments that seals the nappe package is the Upper Eocene sandstones (Fig. 7.4b). From the Late Eocene onward, the nappe pile moved as a coherent body (Fig. 7.5a,b).

7.4. DISCUSSION ON THE MAJOR TECTONIC EVENT LEADING TO THE DEVELOPMENT OF THE SAOB

In this section, we discuss a tectonic evolution model based on our many years of fieldwork in Southeast Anatolia (Figs. 7.6 and 7.7) that provide lines of evidence leading to this model. The suprasubduction ophiolite origin of the lower ophiolite (LO) (Fig. 7.4) suggests that an older oceanic lithosphere was consumed by northward subduction of the Arabian Plate and generated a suprasubduction ophiolite in the upper plate during the Cenomanian (95–73 Ma; Karaoğlan et al., 2013b,c; Fig. 7.6a).

The stratigraphy in the thrust slices (Fig. 7.2a) indicates that an ophiolitic slab was detached from its root, dragged underneath the ophiolitic mélange assemblage, began to move southward toward the Arabian continent (Fig. 7.6b). The nappe hit the leading edge of the Arabian

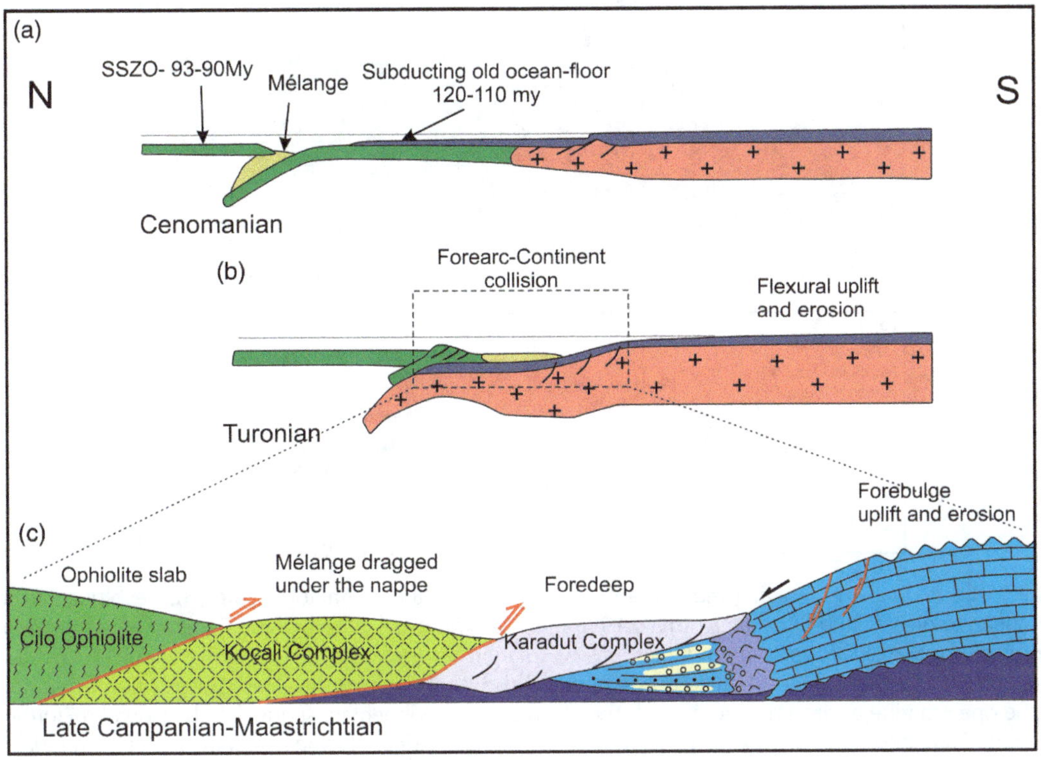

Figure 7.6 Cartoons showing tectonic evolution of the Southeastern Anatolia during the Late Cretaceous. (a) Cenomanian-A north-facing passive continental margin was developed on the Arabian Platform. The northward intraoceanic subduction generated a young SSZO. Following the total consumption of the old ocean, the young oceanic lithosphere reached the leading edge of the Arabian Plate. (b)Turonian. An ophiolitic slab detached from its root. The ophiolite and the mélange dragged underneath were thrust over the Arabian Plate's leading edge. This tectonic event may be interpreted as a forearc (accretionary complex) continent (the Arabian Plate) collision. In front of the nappe pile, a foredeep and an accompanying forebulge formed. The rectangle defines the region detailed in (c) (inspired from Casey, 1980). (c) Late Campanian–Maastrichtian. The foredeep subsided beneath the CCD. Blocks and olistostromes derived from the continental slope, and the outer-shelf areas were deposited rapidly into this foredeep. The adjacent forebulge was eroded (the Turonian unconformity in Fig. 7.2). The nappes' continuing advance rapidly lowered the formerly elevated and eroded platform areas beneath sea level and formed a deep basin (the Sayindere basin in Fig 7.2). The thick nappe pile gradually reached above sea level and formed a structural high along the continental platform's outer margin. Debris flows and blocks, derived mostly from this high, were deposited rapidly into this basin lying in front of the nappes. Therefore, the basin where the pelagic limestone was formerly deposited turned gradually into an environment of clastic deposition (Kastel basin in Fig 7.2a) (inspired from Robertson, 1987, Fig. 14).

continent during Turonian and then synchronously obducted onto the northwestern margin of the Arabian Plate during the Late Campanian–Early Maastrichtian period (Robertson, 1987; Yılmaz, 1993, 2019) when the Tethyan subduction system from west to east possibly extended for more than 8,000 km. The consumption of the oceanic crust along this subduction zone caused the generation of the wide accretionary prism. The remnant of this accretionary prism is the present-day Makran Accretionary prism and the Late Cretaceous SSZ ophiolites, observed along with the northeastern periphery of the Arabian Plate. The Cilo Ophiolite in southeast Turkey (Yılmaz, 1994), the Neyriz Ophiolite in northwest Iran (Moghadam et al., 2014), and the Semail Ophiolite in the Oman Mountains (Searle & Cox, 1999) represent part of this ophiolitic belt. The tectonic interaction between the ophiolite nappe and the Arabian Plate is considered as forearc-continent collision (Fig. 7.6b,c). The initial stage of the collision is recorded in a regional unconformity caused by uplift of the Arabian Platform (Figs. 7.2 and 7.6c; Yılmaz, 1993, 2019).

The stratigraphic data from the Arabian Platform together with the rapid facies changes from the northern to the southern parts of the region (Fig. 7.2a; Robertson, 1987) may be interpreted as follows: a coeval foredeep and a forebulge were developed in front of the

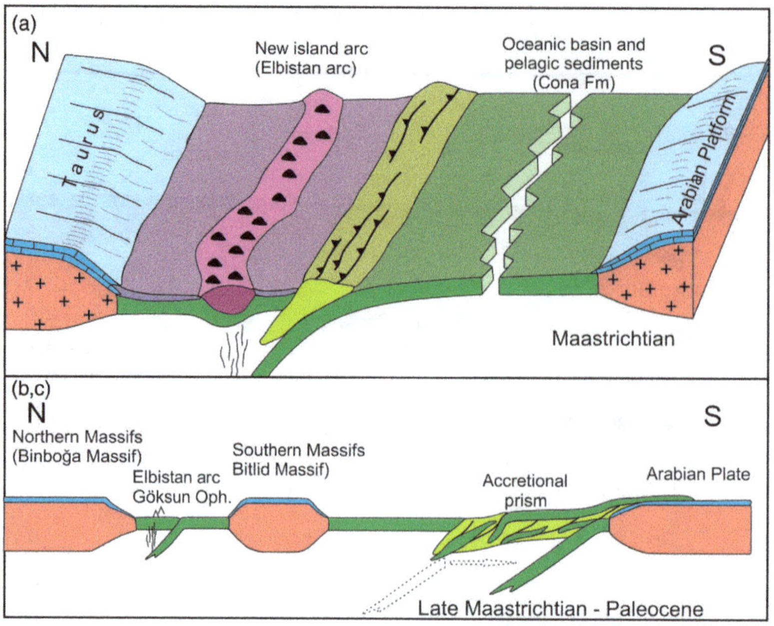

Figure 7.7 Block diagrams showing the subsequent stages of southeast Anatolian orogenic evolution from Late Maastrichtian to present (modified after Yilmaz, 1993). (a) Maastrichtian. After the Late Campanian ophiolite obduction onto the Arabian Platform, a north-facing passive margin formed once again during Late Maastrichtian and continued uninterruptedly to the Middle Eocene epoch. This was the marine invasion's resumption from the north, where the open marine environment remained. The abyssal-plain sedimentary sequence (the Cona Fm) formed during this period are presently seen among the tectonic slices of the imbricated zone and at the top of the middle ophiolite nappe (the MO; Fig 7.2b). The ocean separating the Arabian continent from the northern continental fragment (the metamorphic massifs) began to be consumed by northward subduction, which generated a younger ensimatic island arc. (b) Later periods of Maastrichtian. Due to the retreat of the subducting slab, the northerly located continent was split into two continental slivers. They were later incorporated in the orogeny and turned into the southern and northern metamorphic massifs. A younger SSZO (the Goksun Ophiolite, GO) and an ensimatic magmatic arc (the Elbistan arc, EV) was developed between them. (c) Late Maastrichtian–Paleocene. Retreat of the subducting ocean lithosphere continued, which caused southward migration of the arc front. Volcanic activity in the southern arc continued until the Late Eocene (the Helete volcanics). (d) Paleocene-Early Eocene. The southerly located continental sliver attached to the oceanic slab involved in the subduction zone. They underwent HP and HT metamorphisms. Partly simultaneously, the subducting oceanic slab retreated (rollback). Hot asthenosphere wedged into the space generated by the rollback. The asthenospheric inflow contributed unusually high heat, which caused HT metamorphism, which superimposed on the previous HP metamorphism. The rollback also promoted the exhumation. The oceanic and continental fragments, when exhumed, formed the Bitlis Massif and the Berit metaophiolite. The northerly located continental sliver hit and moved onto the Goksun ophiolite (GO) and the overlying Elbistan arc (EV) during the Late Eocene (the continent-arc collision). The 51–45 my posttectonic granites (Gr) intruded into the nappe package. (e) Middle Eocene. Volcanoes of the magmatic arc rose above sea level, and fringing carbonate reefs formed. A short-lived backarc/ interarc basin, the Maden Basin, opened fragmenting the nappe package. (f) Late Eocene–Oligocene. As a result of the continuing southward transport, the nappes moved over and destroyed the Maden Basin to the end of the Middle Eocene. Different tectonostratigraphic units; the northerly located metamorphic massifs, the ophiolite nappes (MO, UO), the Elbistan and Helete volcanic arcs, and UB2 were tectonically amalgamated. This event may be considered as the magmatic arc-continent collision. The oceanic basin was totally consumed. The development of the subduction mélange and the deep-sea sediment deposition ended before the Late Eocene. Above the elevated nappe pile formed a rugged topography, which supplied olistostrome deposits and coarse clastics into the adjacent lowlands. The Upper Eocene–Oligocene sediments deposited above the nappes as a first common cover. From this time onward, the nappe pile began to move as a coherent body. (g) Early Miocene. The remnant sea left after the oceanic lithosphere consumption was initially filled with coarse-grained sediment accumulation from the adjacent topographic highs. They were gradually replaced by more orderly flysch deposition during the Early Miocene. A transition from the shallow sea to a linear flysch basin (the Lice Flysch) may be observed from the Miocene sections across the mountain range (Yılmaz et al., 1987, 1988). (h) Middle–Late Miocene. The flysch basin was severely deformed under the southerly transported nappes, which also caused the imbrication of the belts squeezed between the nappes and the Arabian Plate (the imbricated zone). The nappes were then trust onto the Arabian Plate (the latest phase of the continent-continent collision). (i) Late Miocene–present. Further convergence due to the continuing southward advance of the nappes and northward movement of the Arabian Plate caused elevation of the suture mountains. Consequently, the sea retreated from the Arabian Platform toward the Mediterranean. The continental foredeep (the Maras Basin in Fig. 7.4a) began to be filled with terrestrial deposits.

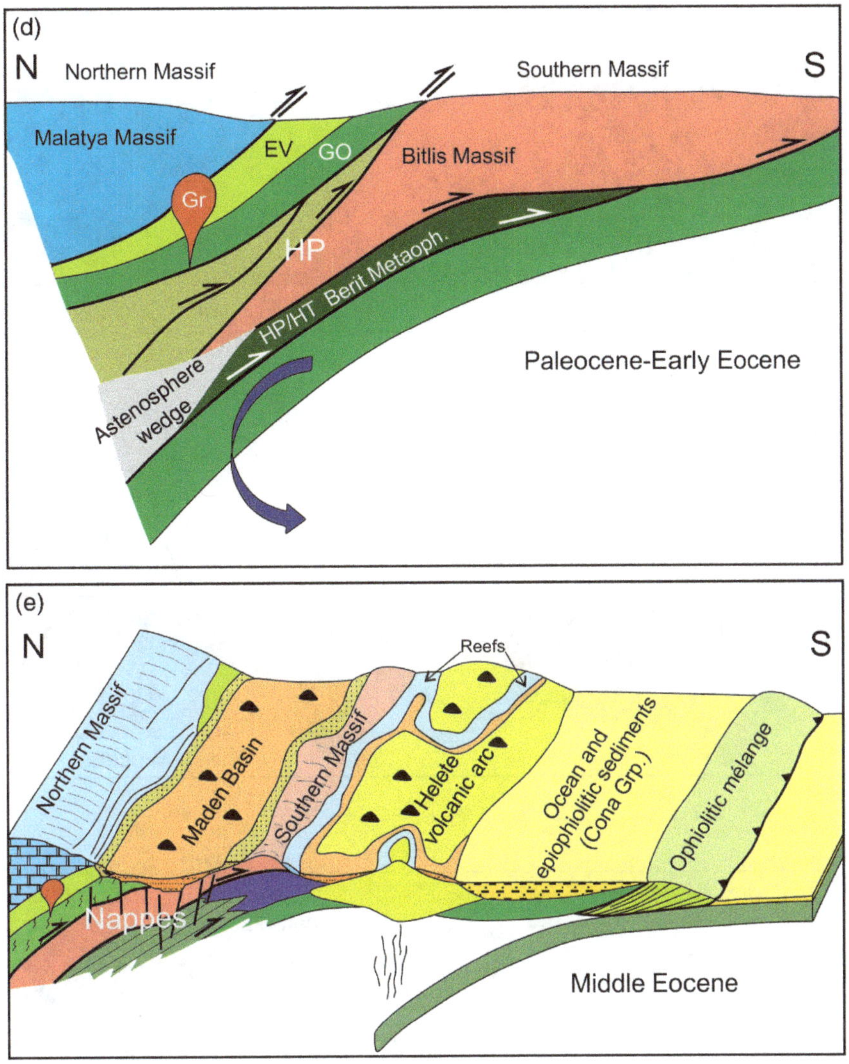

Figure 7.7 (Continued)

nappe pile (Fig. 7.6b,c). Under the nappe pile's heavy load, the foredeep subsided below the CCD (the Şebker Fm). The blocks and olistostromes derived from the steep slope, and the outer-shelf areas were deposited rapidly into the foredeep (Fig. 7.6b). The elevated land supplied olistoliths and olistostrome deposits into the foredeep basin (Figs. 7.2a and 7.6c).

Generations of the two internally chaotic assemblages defined as mélanges in the previous studies owe their origins to sedimentary (the Karadut Complex) and tectonic processes (the Koçali Complex). The former was developed throughout the Mesozoic on the continental slope and then slid into the foredeep developed in front of the ophiolite nappe. The latter is an ophiolitic mélange generated during the demise of the ocean along the subduction zone. The forearc-continent collision and the following events, the thickening of continental crust, and the consequent elevation of the topography (Yeşilova et al., 2018) are synchronously developed all along the Arabian Plate margin from the SAOB to the Oman Mountains.

The nappes' continuing advance rapidly lowered the formerly elevated and eroded platform areas beneath the sea level. This formed a progressively deepening foreland basin (the Sayındere Fm in Fig. 7.2). The thick nappe pile rising above the sea-level formed a structural barrier along the continental platform's outer margin.

Basal sandstones of new transgression were deposited above the LO during the Late Maastrichtian (Fig. 7.2), indicating that the thickened continental crust collapsed rapidly (Fig. 7.6c), and the sea transgressed onto the Arabian Plate once again (Fig. 7.2 a). The overlying neritic limestone, which grades into the Upper Maastrichtian-Paleocene pelagic limestone and shale interbedded sequence (Germav Fm in Fig. 7.2), reveals that the north-facing passive continental margin reestablished during the Late Maastrichtian (Fig. 7.6c).

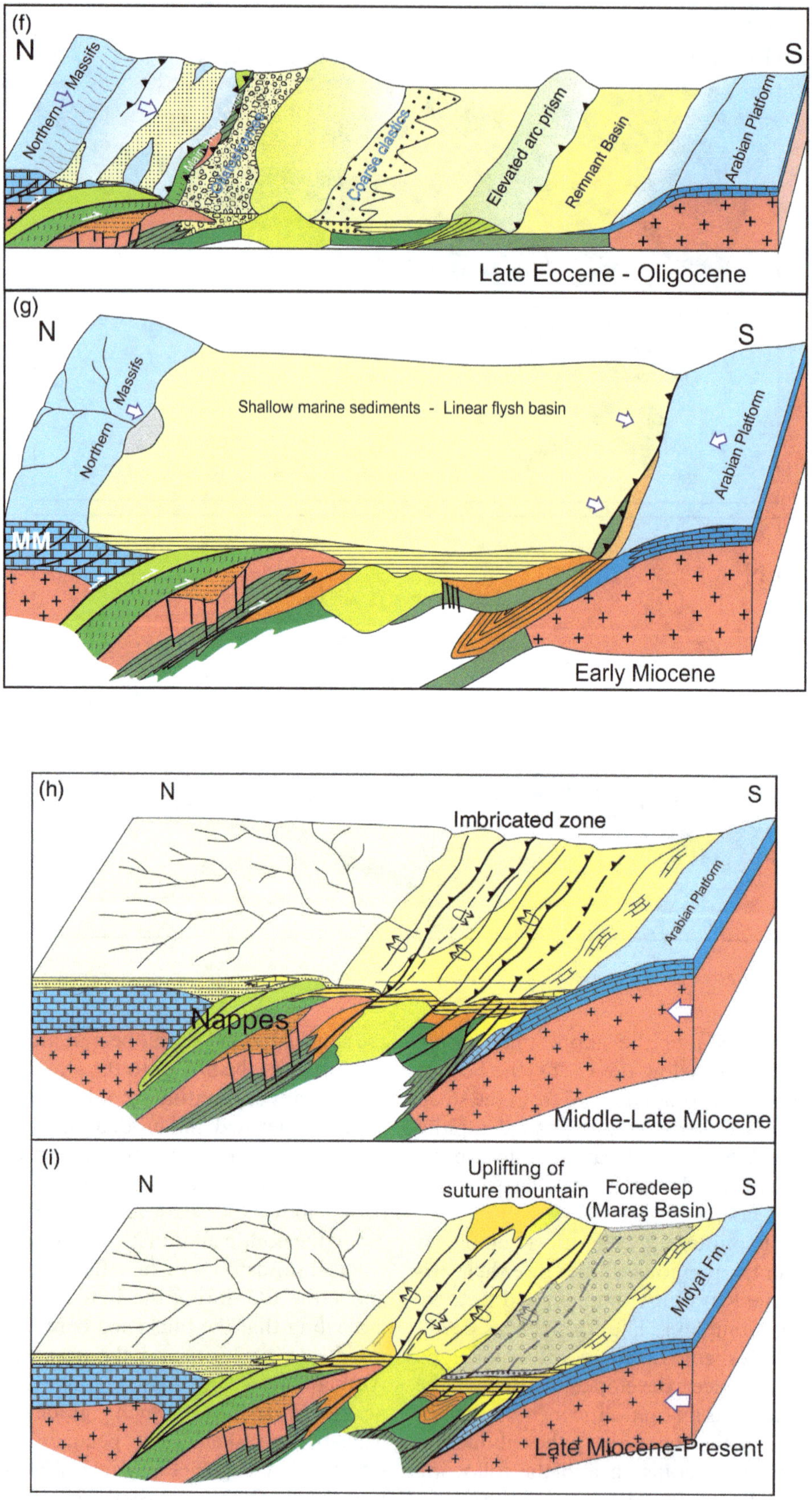

Figure 7.7 (Continued)

Despite the emplacement of the LO onto the Arabian continent, the oceanic environment continued to exist in the northern regions of the SAOB throughout the Late Cretaceous (Figs. 7.7a to c; Yılmaz, 2019). This is supported by the presence of uninterrupted Upper Cretaceous–Lower Eocene epiophiolitic deep-sea sedimentary sequence transported above the MO during the late Middle Eocene (Yılmaz, 2019, 2021).

The geochemical and isotope data on the sooutheast Anatolian-Tauride and Cyprus ophiolites consistent with a suprasubduction zone origin (Pearce, 1975; Dilek & Moores, 1990; Dilek & Eddy, 1992; Dilek & Thy, 1998; Robertson et al., 2012a,b; Karaoğlan et al., 2013a–d; Dilek & Furnes, 2019). The isotope ages support further that the older NeoTethyan oceanic lithosphere was eliminated, and younger SSZ ophiolites were continually generated in an intraoceanic environment toward the end of the Late Cretaceous (Fig. 7.6a) (i.e., 83–73 Ma; Bağcı et al., 2008; Parlak et al., 2010; Karaoğlan et al., 2013a–d; Dilek & Furnes, 2019). Ages of the fragments from the ophiolitic mélange show that the demise of the oceanic lithosphere by the subduction processes continued until the end of the Middle Eocene (Figs. 7.6c and 7.7a; Yılmaz, 2019).

Following the rifting from the Arabian Plate, the continental slivers that were located between the Taurus Plate and the Arabian Plate (Şengör & Yılmaz, 1981) underwent metamorphism and formed the metamorphic massifs during the progression of the orogen between the Late Cretaceous–Early Cenozoic (Fig 7.7b,c,d; Şengör & Yılmaz, 1981; Yılmaz, 2019, and the references therein). The Berit metaophiolite (HP/HT) and the Bitlis Massif (HP) underwent penecontemporaneous, synkinematik metamorphisms (Fig. 7.7d; Yılmaz, 1975; Okay et al., 1985; Pourteau et al., 2013; Oberhänsli et al., 2014; Yılmaz, 2019). For the eclogite and blueschist metamorphic facies, Oberhänsli et al. (2014) inferred a burial of 65 km and 35 km (Fig. 7.7d) and calculated the peak conditions of the blueschist metamorphism around 79–74 Ma. The P/T path (Oberhänsli et al., 2014) indicates that the continental slab was attached to the subducting oceanic lithosphere and deeply buried along a subduction zone (Fig. 7.7d; Yılmaz, 2019). The Amphibolite facies minerals developed on the HP metamorphic rocks require an unexpectedly high temperature, possibly added by the asthenospheric wedge injection into the space created due to the subducting plate's rollback (Fig. 7.7d; Dilek & Flower, 2003). The seismic images of the southern Tethyan oceanic slab under the eastern Anatolia display the rollback and associated retreat (Piromallo & Regard, 2006: Şengör et al., 2008; Özaçar et al., 2010).

The Göksun-Sürgü Fault (Figs. 7.1 and 7.4) presently separates the subduction involved HP lower plate associations (the LM and MO) from the nonmetamorphic ophiolitic association of the upper plate units (the UO and EV; Fig. 7.7b,c). Therefore, this fault may be viewed as a large-scale detachment fault, part of which was taken up later by a strike-slip fault during the Neotectonic period in Pleistocene-Holocene (Yılmaz 2017, 2019). The structural fabrics of ductile to brittle deformation recorded within the metamorphosed mafic-ultramafic rocks were developed during lithospheric-scale detachment faulting associated with upper mantle exhumation. These events may be compared closely with the oceanic core complex formation documented from the modern and ancient oceanic lithosphere such as the western and southern Alpine ophiolites (Miranda & Dilek, 2010; Festa et al., 2015; Escartin et al., 2017; Pohl et al., 2018, and references therein).

Following the metamorphism that occurred during the Late Cretaceous–Early Eocene period, several kilometers thick rock columns were exhumed as evidenced by the field and thermochronological data (Cavazza et al., 2018; Yılmaz, 2019, and the references therein). The Engizek-Bitlis Massif and the Berit metaophiolite reached the surface in a relatively short period by the end of the Early Eocene. For this, the following data may be given: the metamorphic nappes were thrust above the Middle Eocene Maden Basin units (Fig. 7.4b), and Upper Eocene marine sandstones were deposited above the nappe package (Fig. 7.4b).

The thrusting of the northern metamorphic belt over the Göksun ophiolite-magmatic arc pair may be evaluated as an arc-continent collision (Fig. 7.7b,c,d), which occurred between the end of the Early Eocene and early Middle Eocene. The post-tectonic granites, which intruded into this nappe pile (Fig. 7.7b,d; Yılmaz et al., 1987; Rızaoğlu et al., 2006) were dated 51–45 Ma (Karaoğlan et al., 2013d). From this time onward, the nappe package moved as a coherent body (Figs. 7.7d,e). The data summarized above refute the previous claims that the metamorphic massifs were old and collided with the Arabian Plate during the Late Cretaceous (Yazgan, 1984). The field data also disfavor the view that the Taurus belt (the northern metamorphic massifs; the upper nappes) collided with the southern metamorphic massifs during the Late Cretaceous (Bingöl et al., 2018). Recently, Ertürk et al. (2022) documented Late Cretaceous isotope ages from plutonic rocks of the Keban regions, and based on the data they claimed that the final collision along the southeast Anatolian orogen occurred during this period. However, the data are not adequate enough in regional scale to restrict the time of the collision to the Late Cretaceous with the reason outlined in the preceding paragraphs.

The geological record indicates that a remnant oceanic basin survived in the south of the nappe pile until the end of the Middle Miocene (Figs. 7.7d and 7.8a; Yiğitbaş, 1989; Yiğitbaş & Yılmaz, 1996a; Escartin et al., 2017; Yılmaz, 1993, 2019). During this period, the arc front migrated to the south due possibly to the subducting slab's rollback (Fig. 7.7c; Dilek & Furnes, 2009). The arc volcanic succession continued to grow in the southern region until the Late Eocene (Fig. 7.7e,f). A thick calc-alkaline andesitic-dacitic volcanic sequence was built above an ophiolite foundation (Fig. 7.7b,c,e; Yılmaz, 1993, 2019). The volcanic rocks were gradually replaced upward by shallowing marine sandstone-siltstone and reefal limestones (Fig. 7.7e; Yılmaz, 1993, 2019; Yiğitbaş & Yılmaz, 1996a and b; Kuşcu et al., 2010). At higher layers, olistostromes fluxed into the sandstones of the Upper Eocene-Oligocene age (Fig. 7.7f). The stratigraphic order of the units in the volcanic arc sequence may be interpreted that the volcanic activity waned, and the oceanic lithosphere was possibly totally obliterated before the Late Eocene (Fig. 7.7f; Yılmaz, 1993; Şengör et al., 2003; Rolland et al., 2012; Karaoğlan et al., 2016.

The nappes thrusted over the Helete volcanic arc sequence during the Late Eocene–Oligocene when the nappe pile collided with the southerly migrated younger magmatic arc (Fig. 7.7f). The nappes elevated above the sea level and began supplying coarse clastics into the basin in front of the southerly advancing nappes (Fig. 7.7f). Southward, the coarse clastics graded into sandstones and flysch of the Lower Miocene Lice Formation (Fig. 7.7f,g).

The platform carbonates deposited above the Arabian Plate throughout Eocene (the Midyat Group in Fig. 7.2a) are replaced upward by shallow marine sandstone-conglomerates, debris flow deposits, and continental red beds of the Late Eocene–Oligocene age (Figs. 7.2a and 7.7g). Basal clastics of a new transgressive sequence rest above the Oligocene sediments over a marked unconformity (Yeşilova & Helvacı, 2017). They pass rapidly to a flysch sequence of Early Miocene age (the Lice Fm) (Figs. 7.3 and 7.7g; Derman & Atalık, 1993; Siyako et al., 2013; Özdoğan et al., 2011). The rapid transition from the time-regressive to the time-transgressive successions from the north to south indicates a flexural foredeep (the linear flysch basin): The Lice Formation developed in front of the southerly transporting nappe pile (Fig. 7.7g). As the nappe pile transportation continued, the flysch and the underlying units were severely deformed, tightly folded, and imbricated by the south-vergent compressional stress (Fig. 7.7h). These are the initial phases of the continent-continent collision between the nappes and the Arabian Plate (Yılmaz, 1993, 2017, 2019; Dilek, 2006; Hüsing et al., 2009; Silja et al., 2009).

After development of the imbricated zone (Figs. 7.1 and 7.7h), the nappes were thrust over the Arabian Plate's leading edge during the Middle-Late Miocene period. This event is the collision of the nappe pile with the Arabian Plate. During the continuing north-south convergence, the suture zone began to rise and formed the Southeast Anatolian suture mountains (Fig. 7.7h,i), which started to develop during the Late Miocene and is continuing today (Yılmaz, 1993; Yılmaz et al., 1987, 1988; Akıncı et al., 2016; Yılmaz, 2017). Consequently, the sea retreated from the orogenic belt toward the Mediterranean (Özdoğan et al., 2011; Siyako et al., 2013; Yılmaz, 2017, 2019), which remains as the surviving part of the ocean that extended to the Indian Ocean before the development of the Bitlis-Zagros Orogenic Belt (inset in Fig. 7.1).

The northward advance of the Arabian Plate continued after the collision. The resulting compression has been initially accommodated along the Arabian Platform's northern boundary with the development of a wide fold and thrust belt (Fig. 7.8; Yılmaz, 2017). Later, when the compressional deformation reached an excessive stage, the shortening deformation was replaced by escape tectonics (Perinçek & Çemen, 1990; Elmas & Yılmaz, 2003; Boulton and Robertson 2008; Yilmaz 2017; 2020). Several E-W trending left-lateral strike-slip faults cut and displaced the fold and thrust belt (Fig. 7.8) and began to transfer the stress to the SW direction (Yılmaz, 2020).

7.5. CONCLUDING SUMMARY

The SAOB was developed due to the collisional events that followed the demise of the NeoTethyan Ocean and its dependencies such as backarc/interarc and remnant basins. The following successive major events are differentiated in the tectonic development of the SAOB.

1. Collision of the forearc (the Koçali and Karadut complexes and the overlying ophiolite slab) with the Arabian continent during the Late Cretaceous. Similar coeval events were recorded along the northern boundary of the Arabian Plate from the Amanos Mountains of southern Turkey (Fig. 7.1; Yılmaz, 1984) to the Oman Mountains (Searle & Cox, 1999; Goodenough et al., 2014)

2. (a) Development of new northward subduction in the surviving ocean; (b) Involvement of a continental crust into the subduction zone, which formed the southern metamorphic belt (the Engizek- Pötürge-Bitlis massifs) during the Maastrichtian–Early Eocene period; (c) Fragmentation of south Taurus platelet (development of the Göksun ophiolite and Elbistan volcanic arc) during the Maastrichtian–Middle Eocene

3. Collision of the Elbistan intraoceanic magmatic arc with the northerly located continent (the northern metamorphic belt; the Malatya-Pötürge massifs) during the Middle Eocene

4. Collision of the southern and the northern nappe piles (the continent-continent collision) during the late Middle Eocene

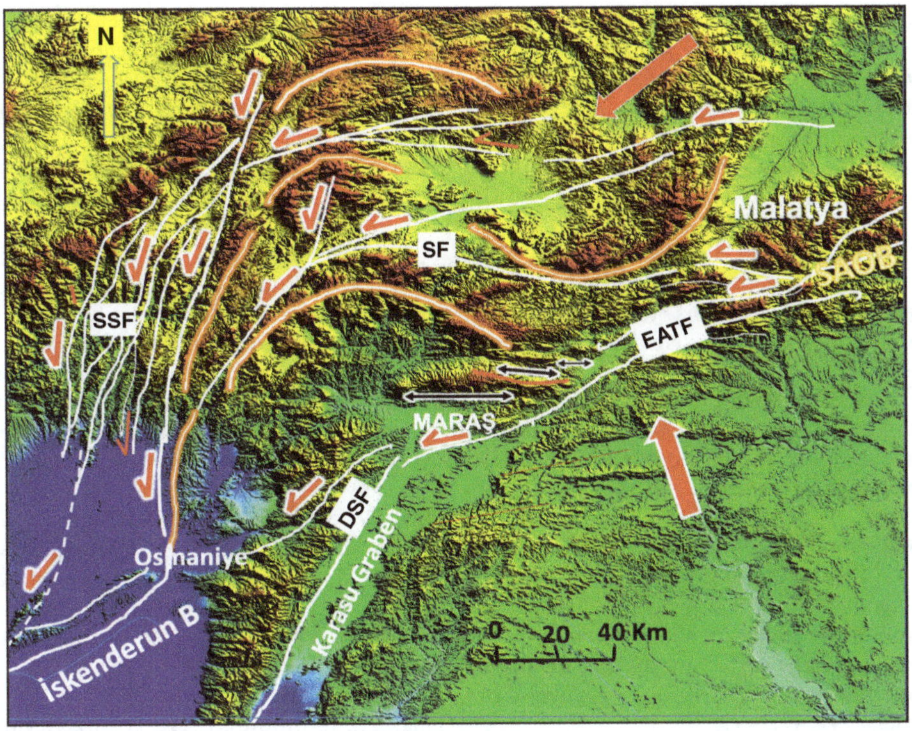

Figure 7.8 The physiographic map of western regions of the Southeastern Anatolian Orogenic Belt (SAOB) and the adjacent areas. The red arrows indicate the motion directions of the Arabian and Anatolian plates. The brown curvilinear lines show the trend lines of the mountain ranges, which correspond to the axes of the regional folds formed due to the compressional forces exerted by the escape regime, which also generated strike-slip faults (the white lines). The double-headed black arrows indicate prominent foreland folds displaced by left-lateral strike-slip faults. Note: SSF= the fault bundle in the Sarız-Saimbeyli mega shear zone comprises several fault-bound blocks or a tectonic wedge transferring the compressional stress to the south; SF=the Surgu Fault, which connects the East Anatolian Transform Fault Zone (EATF) to the Mediterranean Region; DSF=the Dead Sea Fault.

5. The southward advance of the amalgamated nappe pile and the destruction of the remnant basin during the Late Eocene
6. Emplacement of the nappes onto the Arabian Plate during the early Late Miocene, the final stage of the collisional development of the SAOB
7. Development of a wide fold and thrust belt along the nappe front of the Arabian Plate during Plio-Pleistocene
8. Replacement of the orthogonal shortening by the escape tectonics and formation of strike-slip faults that transfer the north-south shortening deformation to west and southwest during Pleistocene to present

ACKNOWLEDGMENTS

We thank our colleagues from the universities in Turkey, Europe, and North America and TPAO with whom we discussed many important aspects of the geology and tectonic setting of the Southeast Anatolian throughout many years. We extend our sincerest gratitude to TPAO, which supported the field work of Yücel Yılmaz, Erdinç Yiğitbaş, and İbrahim Çemen. Special thanks to Fevzi Gürer for drawing figures.

REFERENCES

Akıncı, A. C., Robertson, A. H. F., & Ünlügenç, U. C. (2016). Late Cretaceous-Cenozoic subduction-collision history of the Southern NeoTethys: New evidence from the Çağlayancerit-Gölbaşı area, SE Turkey. *International Journal of Earth Sciences*, *105*, 315–337.

Aktaş, G., & Robertson, A. H. F. (1985). The Maden complex SE Turkey: Evolution of a NeoTethyan active margin. In J. H. Dixon & A. H. F. Robertson (Eds.), The geological evolution of the eastern Mediterranean (pp. 375–402). *Geological Society of London Special Publication*, 17.

Aktaş, G., & Robertson, A. H. F. (1991). Tectonic evolution of the Tethys suture zone in SE Turkey: evidence from the petrology and geochemistry of Late Cretaceous and Middle Eocene extrusives. In J. Malpas et al. (Eds.), *Ophiolites oceanic crustal analogues* (pp. 311–328). Proceedings of the Symposium Troodos 1987. Geological Survey Department, Ministry of Agriculture and Natural Resources, Nicosia, Cyprus.

Awalt, M. B., & Whitney, D. (2018). Petrogenesis of kyanite- and corundum-bearing mafic granulite in a meta-ophiolite, SE Turkey. *Journal of Metamorphic Geology*, *36*(7), 881–904. https://doi.org/10.1111/jmg.12317

Bağcı, U. (2013). The geochemistry and petrology of the ophiolitic rocks from the Kahraman Maraş region, southern Turkey. *Turkish Journal of Earth Sciences*, *22*, 536–562. https//doi.org/10.3906/yer.1203.1

Bağcı, U., Parlak, O., & Höck, V. (2005). Whole rock and mineral chemistry of cumulates from the Kızıldağ (Hatay) ophiolite (Turkey); clues for multiple magma generation during crustal accretion in the southern Neotethyan ocean. *Mineralogical Magazine*, *69*, 53–76.

Bağcı, U., Parlak, O., & Höck, V. (2008). Geochemistry and tectonic environment of diverse magma generations forming the crustal units of the Kızıldağ (Hatay) ophiolite, Southern Turkey. *Turkish Journal of Earth Sciences*, *17*, 43–71.

Beyarslan, M., & Bingöl, A. F. (2000). Petrology of a supra-subduction zone ophiolite (Elazığ, Turkey). *Canadian Journal of Earth Sciences*, *47*, 1411–1424.

Beyarslan, M., & Bingöl, A. F. (2018). Zircon U-Pb age and geochemical constraints on the origin and tectonic implications of late cretaceous intra-oceanic arc magmatics in the Southeast Anatolian Orogenic Belt (SE-Turkey). *Journal of African Earth Sciences*, *147*, 477–497.

Bingöl, A. F., Beyarslan, M., Lin, Y., et al. (2018). Geochronological and geochemical constraints on the origin of the Southeast Anatolian ophiolites, *Turkey. Arabian Journal of Geosciences*, *11*(18), 569. https://doi.org/10.1007/s12517-018-3880-0

Boulton, S. J., & Robertson, A. H. F. (2008).The Neogene and recent Hatay Graben area, south central Turkey: Graben formation in setting of oblique extension (transtension) related to post collisional tectonic escape. *Geological Magazine*, *145*, 800–821.

Candan, O., Çetinkaplan, M., Topuz, G., Koralay, E., Oberhansli, R., Yiğitbaş, et al. (2012). *Eclogites in the Berit area (Kahraman Maraş, Turkey) and their tectonic implications*. International Earth Science Colloquium on the Aegean Region (IESCA-2012), 1–5 October 2012, Abstract 54.

Casey, J. F. (1980). *The geology of the southern part of the North Arm Mountain Massif, Bay of Islands Ophiolite Complex, western Newfoundland with application to ophiolite obduction and the genesis of the plutonic portions of oceanic crust and upper mantle*. Ph.D. dissertation. State University of New York, Albany.

Cavazza, W., Cattò, S., Zattin, M., Okay, A. I., & Reiners, P. (2018). Thermochronology of the Miocene Arabia-Eurasia collision zone of southeastern Turkey. *Geosphere*, *14*(5), 2277–2293. https://doi.org/10.1130/GES01637.1

Çemen, I., Göncüoğlu, M. C., Erler, A., Kozlu, H., & Perinçek, D. (1993). Indentation tectonics and associated lateral extrusion in east, southeast and central Anatolia. *GSA Programs and Abstracts*, A116–A117.

Derman, A. S., & Atalık, E. (1993). Sequence stratigraphic analysis of Miocene sediments in Maraş Miocene basin and effect of tectonism in the development of sequences. *Special Publications Sequence Stratigraphy, Sedimentology Study Group*, *1*, 43–52.

Dilek, Y. (2006). Collision tectonics of the Mediterranean region: Causes and consequences. In Y. Dilek & Y. Pavlides (Eds.), *Post collisional tectonics and magmatism in the Mediterranean region and Asia* (pp. 1–13). Geological Society of America Special Paper, 409.

Dilek, Y., & Delaloye, M. (1992). Structure of the Kızıldağ ophiolite, a slow spread Cretaceous ridge segment north of the Arabian promontory. *Geology*, *2*, 19–22. https://doi.org/10.1130/00917613 (1992) 020<0019: SOTKOA>2.3.CO;2

Dilek, Y., & Eddy, C. A. (1992). The Troodos (Cyprus) and Kızıldag (S Turkey) ophiolites as structural models for slow-spreadıng ridge segments. *Journal of Geology*, *100*(3), 305–322.

Dilek, Y., & Flower, M. F. J. (2003). Arc-trench rollback and fore arc accretion: 2. A model template for ophiolites in Albania, Cyprus, and Oman. In Y. Dilek & P. T. Robinson (Eds.), *Ophiolites in Earth history* (pp. 43–68). Geological Society of London, 218.

Dilek, Y., & Furnes, H. (2009). Structure and geochemistry of Tethyan ophiolites and their petrogenesis in subduction rollback systems. *Lithos*, *13*, 1–20. https://doi.org/10.1016/j.lithos.2009.04.022

Dilek Y., & Furnes, H. (2019). Tethyan ophiolites and Tethyan seaways. *Journal of the Geological Society*, *176*, 899–912. https://doi.org/10.1144/jgs2019-129

Dilek, Y., & Moores, E. M. (1990). Regional tectonics of the eastern Mediterranean ophiolites. In J. Malpas et al. (Eds.), *Ophiolites, oceanic crustal analogues* (pp. 295–309). Proceedings of the Symposium Troodos 1987. Geological Survey Department, Nicosia.

Dilek, Y., & Sandvol, E. (2009). Seismic structure, crustal architecture and tectonic evolution of the Anatolian-African plate boundary and Cenozoic orogenic belts in the eastern Mediterranean region. In J. R. Murphy et al. (Eds.), *Ancient orogens and modern analogs* (pp. 127–160). Geological Society of London, Special Publication, 327. https://doi.org/10.1144/SP327. 7.& 0305-8719/09/&15 00

Dilek, Y., & Thy, P. (1998). Structure, petrology, and seafloor spreading tectonics of the Kızıldağ Ophiolite (Turkey). In R. Mills & K. Hawkins (Eds), *Modern ocean floor processes and the geological record* (pp. 43–69). Geological Society London, Special Publication, 148.

Dilek, Y., Furnes, H., & Shallo, M (2007). Suprasubduction zone ophiolite formation along the periphery of Mesozoic Gondwana. *Gondwana Research*, *11*, 453–475. https://doi.org/10.1016/j.gr.2007.01.005

Dilek, Y., Thy, P., Moores, E. M., & Ramsden, T. W. (1990). Tectonic evolution of The Troodos Ophiolite within the Tethyan framework. *Tectonics*, *9*(4), 811–823.

Elmas, A., & Yılmaz, Y. (2003). Development of an oblique subduction zone: Tectonic evolution of the tethys suture zone in southeast Turkey. *International Geology Review*, *45*(9), 827–840.

Ertürk, M. A., Beyarslan, M., Chungbc, S-L., & Lin, Te-H. (2017). Eocene magmatism (Maden Complex) in the Southeast Anatolian Orogenic Belt: Magma genesis and tectonic implications. *Geoscience Frontiers*, 1–19.

Ertürk, M. A., Sar, A., & Rizeli, M. E. (2022). Petrology, zircon U-Pb geochronology and tectonic implications of the

A1-type intrusions, Keban region, eastern Turkey. *Geochemistry*, *82*(3), 125882. https://doi.org/10.1016/j.chemer.2022.125882

Escartín, J., Mével, C., et al. (2017). Tectonic structure, evolution, and the nature of oceanic core complexes and their detachment fault zones (13°20'N and 13°30'N, Mid Atlantic Ridge). *Geochemistry, Geophysics, Geosystems*, *18*, 1451–1482. https://doi.org/10.1002/2016GC006775

Festa, A., Balestro, G., Dilek, Y., & Tartarotti, P. (2015). A Jurassic oceanic core complex in the high-pressure Monviso ophiolite (western Alps, NW Italy). *Lithosphere*, *7*, 646–652,

Genç, Ş. C., Yiğitbaş, E., & Yılmaz, Y. (1993). Geology of the Berit metaophiolite. In Suat Erk Jeoloji Simpozyumu Bildirileri, Ankara Üniversitesi Fen Fakültesi Jeoloji Bölümü (pp. 37–52).

Göncüoğlu, M. C., & Turhan, N. (1984). Geology of the Bitlis metamorphic belt. In O. Tekeli & M. C. Göncüoğlu (Eds.), *Geology of the Taurus belt* (pp. 237–244). Proceedings of the International Symposium on the Geology of the Taurus Belt, 26–29, September 1983. Ankara: Mineral Research and Exploration Institute of Turkey (MTA).

Goodenough, K. M., Thomas, R.J., Styles, M. T., Schofield, D. J., & MacLeod, C. J. (2014). Records of ocean growth and destruction in the Oman-UAE ophiolites. *Elements*, *10*, 105110. https://doi.org/10.2113/gselements.10.2.109

Hall, R. (1974). *The structure and petrology of an Ophiolitic Mélange near Mutki, Bitlis Province, Turkey*. Unpublished Ph.D. thesis, University of London.175p.

Hempton, M. R. (1985). Structure and deformation history of the Bitlis suture near Lake Hazar, southeastern Turkey. *Geological Society of America Bulletin*, *96*, 33–243.

Hüsing S. K., Zachariasse W. J., van Hinsbergen W., Krijgsman M., İnceöz M., et al. (2009). Oligocene-Miocene basin evolution in SE Anatolia, Turkey; constrains on the closure of the eastern Tethys gateway. *Geological Society Special Publications*, *311*. 107–132. https://doi.org/10.1144/SP3114

Karaoğlan, F., Parlak, O., Hejl, E., Neubauer, F., & Klötzli, U. (2016). The temporal evolution of the active margin along the Southeast Anatolian Orogenic belt (SE Turkey): Evidence from U-Pb, Ar-Ar and fission track chronology. *Gondwana Research*, *33*, 190–208.

Karaoğlan, F., Parlak, O., Klötzli, S. U., Rızaoğlu, T., & Koller, F. (2013c). Age and duration of intraoceanic arc volcanism built on a suprasubduction zone type oceanic crust in southern NeoTethys, *SE Anatolia. Geoscience Frontiers*, *4* (4), 399–408. https://doi.org/10.1016/j.gsf.2012.11.011

Karaoğlan, F., Parlak, O., Klötzli, U., & Thöni, M. (2013b). U/Pb and Sm/Nd geochronology of the ophiolites from the SE Turkey: Implications for the Neotethyan evolution. *Geodinamica Acta*, *26*, 1–16, 201.

Karaoğlan, F., Parlak, O., Klötzli, U., Thöni, M., & Koller, F. (2013a). U-Pb and Sm-Nd geochronology of the Kızıldağ (Hatay, Turkey) ophiolite: Implications for the timing and duration of suprasubduction zone type oceanic crust formation in southern NeoTethys. *Geological Magazine*, *150*(2), 283–299. https://doi.org/10.1017/S0016756812000477

Karaoğlan, F., Parlak, O., Robertson, A., Thöni, M., Klötzli, U., Koller, F., et al. (2013d). Evidence of Eocene high temperature/ high pressure metamorphism of ophiolitic rocks and granitoid intrusion related to Neotethyan subduction processes (Doğanşehir area, SE Anatolia). In A. H. F. Robertson et al. (Eds.), *Geological development of Anatolia and the easternmost Mediterranean region* (pp. 249–272). Geological Society, London, Special Publications, 372.

Karaoğlan, F., Yıldırım, N., Yıldırım, E., & Topak, Y. (2021). The geology of Gölbaşı (Adıyaman) Region: The Upper Cretaceous–Eocene evolution of the Southeast Anatolian Orogenic Belt. *Proceedings 73th Geological Congress of Turkey, Ankara*, 453–545.

Korucu, M., & Çemen, I. (1998). *Seismic expression of structural traps in frontal imbricate zones and foreland structures in the western part of southeast Anatolia fold and thrust belt*, Turkey. American Association of Petroleum Geologists. Annual Convention Program, 19.

Kozlu, H., Prichard, H., Melcher, F., Fisher, P., & Broug, C. (2014). Platinum-group (PGE) mineralization and chromite geochemistry in the Berit Ophiolite (Elbistan/Kahramanmaraş) SE Turkey. *Ore Geology Reviews*, *60*, 97–111.

Kuşcu, L., Gençalioğlu-Kuşcu, G., Ulrich, I. D., & Friedman, R. (2010). Magmatism in southeastern Anatolian orogenic belt: Transition from arc to post-collisional setting in an evolving orogen. In M. Sosson et al. (Eds.), Sedimentary basin tectonics from the Black Sea and Caucasus to the Arabian platform. *Geological Society of London*, *340*, 437–460.

Miranda, E. A., & Dilek, Y. (2010). Oceanic core complex development in modern and ancient oceanic lithosphere: Gabbro-localized versus peridotite-localized detachment models. *Journal of Geology*, *118*, 95–110.

Moghadam, H. S., Khedr, M. Z., Chiaradia, M., Stern, R. J., et al. (2014). Supra-subduction zone magmatism of the Neyriz ophiolite, Iran: Constraints from geochemistry and Sr-Nd-Pb isotopes. *International Geology Review*, *56*(11), 1395–1412. https://doi.org/10.1080/00206814.2014.942391

MTA (2002). *Explanatory text of the geological map of Turkey on the scale of 1/500.000; The Sıvas and Hatay sheets*. General Directorate of Mineral Research and Exploration, Ankara, Turkey.

Nurlu, N., Parlak, O., Robertson, A. H. F., et al. (2016). Implications of late Cretaceous U-Pb zircon ages of granitoid intrusions cutting ophiolitic and volcanogenic rocks for the assembly of the Tauride allochthon in SE Anatolia (Helete area, Kahramanmaraş Region, Turkey). *International Journal of Earth Science*, *105*, 283–314.

Oberhänsli, R., Bousquet, R., Candan, O., & Okay, A. I. (2012). Dating subduction events in East Anatolia. *Turkish Journal of Earth Sciences*, *21*, 1–18. https://doi.org/10.3906/yer-1006-26

Oberhänsli, R., Candan, O., Bousquet, R., Rimmele, G., Okay, A., & Goff, J. (2010). Alpine HP evolution of the eastern Bitlis complex, SE Turkey. In M. Sosson et al. (Eds.), *Sedimentary basins, tectonics from Black Sea and Caucasus to the Arabian platform* (pp. 461–483). Geological Society, London, Special Publications, 340. https://doi.org/10.1144/SP340.20

Oberhänsli, R. E., Koralay, O., Candan, A., Pourteau, R., & Bousquet, R. (2014). Late Cretaceous eclogitic high-pressure relics in the Bitlis Massif. *Geodinamica Acta*, *26*, 175–190. https://doi.org/10.1080/09853111.2013.858951

Okay, A., Arman, M. B., & Göncüoğlu, M. C. (1985). Petrology and phase relations of the kyanite-eclogites from Eastern Turkey. *Contributions to Mineralogy and Petrology*, *91*, 196–204.

Özaçar, A. A. G., Zandt, H., Gilbert, S., & Beck, S. L. (2010). Seismic images of crustal variations beneath the East Anatolian Plateau (Turkey) from teleseismic receiver functions. In M. Sosson et al. (Eds.), *Sedimentary basin tectonics from the Black Sea and Caucasus to Arabian platform* (pp. 485–496). Geological Society of London, Special Publication, 340.

Özdoğan, T. O., Kaya, I., Acıkbaş, D., Bahtiyar I., & Siyako, M. (2011). *The Miocene Lice basin of southeastern Turkey: An example of a shallow to non-marine foreland basin. Achaean to Anthropocene*. Geological Society of America Annual Meeting and Exposition Paper, 146–148.

Parlak, O. (2006). Geodynamic significance of granitoid magmatism in the Southeast Anatolian Orogen: Geochemical and geochronological evidence from Göksun-Afşin (KahramanMaraş, Turkey) region. *International Journal of Earth Sciences*, *95*, 609–627.

Parlak, O. (2016). The Tauride ophiolites of Anatolia (Turkey): A review. *Journal of Earth Science*, *27*, 901–934. https://doi.org/10.1007/s12583-016-0679-3

Parlak, O., Bağcı, U., Rızaoğlu, T., et al. (2020). Petrology of ultramafic to mafic cumulate rocks from the Göksun (Kahraman Maraş) ophiolite, southeast Turkey. *Geoscience Frontiers*, *11*(1), 109–128.

Parlak, O., Höck, V., Kozlu, H., & Delaloye, M. (2004). Oceanic crust generation in an island arc tectonic setting, SE Anatolian Orogenic Belt (Turkey). *Geological Magazine*, *141*, 583–603.

Parlak, O., Karaoğlan, F., Klötzli, U., Koller, F., & Rızaoğlu, T. (2010). Geochronology of ophiolites in Turkey: Implications for Neotethyan geodynamics in eastern Mediterranean. *Acta Mineralogica-Petrographica*, *6*, 585.

Parlak, O., Karaoğlan, F., Thöni, M., Robertson, A. H. F., Okay, A., & Koller, F. (2012). *Geochemistry, geochronology, and tectonic significance of high- temperature meta-ophiolitic rocks: Possible relation to Eocene south-Neotethyan arc magmatism (Malatya area, SE Anatolia)*. EGU General Assembly 2012, 22–27 April 2012, Vienna, Austria, 1251.

Parlak, O., Rızaoğlu, T, Bağcı, U., Karaoğlan, F., & Höck, V. (2009). Tectonic significance of the geochemistry and petrology of ophiolites in southeast Anatolia, Turkey. *Tectonophysics*, *473* (1–2), 173–187.

Pearce, J. A. (1975). Basalt geochemistry to investigate past tectonic environments in Cyprus. *Tectonophysics*, *23*, 41–67.

Perinçek, D., & Çemen, I. (1990). The structural relationship between the East Anatolian and Dead Sea fault zones in southeastern Turkey. *Tectonophysics*, *172* (3–4), 331–340.

Perinçek, D., & Kozlu, H. (1984). Stratigraphy and structural relation of the units in the Afşin-Elbistan-Doğanşehir Region. In O. Tekeli & C. Göncüoğlu (Eds.), *International symposium on the geology of the Taurus belt, 1983* (pp. 181–198). Mineral Research and Exploration, Ankara.

Perinçek, D. (1979). *Guidebook for excursion «B», Interrelations of the Arab and Anatolian plates*. First Geological Congress on Middle East, Ankara, Turkey, 34.

Piromallo, C. V., & Regard, V. (2006). Slab detachment beneath eastern Anatolia: A possible cause for the formation of the North Anatolian Fault. *Earth and Planetary Science Letters*, *214*, 85–97.

Pohl, F., Froitzheim, N., et al. (2018). Kinematics and age of syn-intrusive detachment faulting in the Southern Alps: Evidence for Early Permian crustal extension and implications for the Pangea A versus B controversy. *Tectonics*, *37*, 3668–3689. https://doi.org/10.1029/2018TC004974

Pourteau, A., Sudo, M., Candan, O., Lanari, P., Vidal, O., & Oberhansli, R. (2013). NeoTethys closure history of Anatolia: Insight from 40 Ar/39Ar geochronology and P-T estimation in high-pressure metasediments. *Journal of Metamorphic Geology*, *31*, 585–606. https://doi.org/10.1111/jmg.12034

Rızaoğlu, T., Parlak, O., Höck, V., & Isler, F. (2006). Nature and significance of Late Cretaceous ophiolitic rocks and its relation to the Baskil granitoid in Elazığ region, SE Turkey. In A. H. F. Robertson & D. Mountrakis (Eds), *Tectonic development of the Eastern Mediterranean* (pp. 327–350). Geological Society London, Special Publication, 260.

Rızaoğlu, T., Parlak, O., Höck, V., Koller, F., Hames, W. E., & Billor, Z. (2009). Andean-type active margin formation in the eastern Taurides: Geochemical and geochronogical evidence from the Baskil granitoid (ElazIg, SE Turkey). *Tectonophysics*, *473*(1–2), 188–207.

Robertson, A. H. F. (1987). The transition from a passive margin to an Upper Cretaceous foreland basin related to ophiolite emplacement in the Oman Mountains. *GSA Bulletin*, *99*(5), 633–653.

Robertson, A. H. F. (2002). Overview of the genesis and emplacement of Mesozoic ophiolites in the Eastern Mediterranean Tethyan region. *Lithos*, *65*(1–2), 1–67. https://doi.org/10.1016/S0024-4937(02)00160-3

Robertson, A. H. F. (2012). Late Palaeozoic-Cenozoic tectonic development of Greece and Albania in the context of alternative reconstructions of Tethys in the Eastern Mediterranean region. *International Geology Review*, *54* (4), 373–454. https://doi.org/10.1080/00206814.2010.543791

Robertson, A. H. F., Parlak, O., & Ustaömer, T. (2012a). Overview of the Palaeozoic-Neogene evolution of Neotethys in the Eastern Mediterranean region (southern Turkey, Cyprus, Syria). *Petroleum Geoscience*, *18*(4), 381–404.

Robertson A. H. F., Parlak, O., & Ustaömer, T. (2016b). Permian-Recent palaeogeographical and tectonic development of Anatolia: Some recent contributions. *International Journal of Earth Sciences*, *105*, 1–5.

Robertson, A. H. F., Parlak, O., Rızaoğlu, T., Ünlügenç, U. C., Inan, N., Taşlı, K., et al. (2007). Tectonic evolution of the South Tethyan ocean: Evidence from the Eastern Taurus Mountains (Elazığ region, SE Turkey). In A. C. Ries et al. (Eds.), *Deformation of the continental crust: The legacy of Mike Coward* (pp. 231–270). Geological Society London, Special Publications, 272.

Robertson A. H. F., Parlak, O., Yıldırım, N., Dumitrica, P., & Taşlı, K. (2016a). Late Triassic rifting and Jurassic?

Cretaceous passive margin development of the Southern NeoTethys: Evidence from the Adıyaman area, SE Turkey. *International Journal of Earth Sciences*, *105*, 167–201.

Robertson, A. H. F., Tasli, K., & Inan, N. (2012b). Evidence from the Kyrenia Range, Cyprus, of the northerly active margin of the Southern Neotethys during Late Cretaceous-Early Cenozoic time. *Geological Magazine*, *149* (2), 264–290. https://doi.org/10.1017/S0016756811000677

Robertson, A. H. F, Ustaömer, T., Parlak, O., Ünlügenç, U. C., Taşlı, K., & Inan, N. (2006). The Berit transect of the Tauride thrust belt, S. Turkey: Late Cretaceous–Early Cenozoic accretionary/ collisional processes related to closure of the southern NeoTethys. *Journal of Asian Earth Science*, *27*, 108–145.

Rolland, Y., Perinçek, D., Kaymakçı, N., Sosson, S., Barrier, E., & Avagyan, A. (2012). Evidence for ~80–75 Ma subduction jump during Anatolide-Tauride-Armenian block accretion and ~48 Ma Arabia-Eurasia collision in Lesser Caucasus-East Anatolia. *Journal of Geodynamics*, *56–57*, 76–85.

Searle, M. P., & Cox, J. S. (1999). Tectonic setting, origin and obduction of the Oman ophiolite. *Geological Society of America Bulletin*, *111*, 104–122.

Şengör, A. M. C., & Yilmaz, Y. (1981). Tethyan evolution of Turkey; a plate tectonic approach. *Tectonophysics*, *75*, 181–241.

Şengör, A. M. C., Özeren, M. S., Keskin, M., Sakınc, M., Obakır, A. D., & Kayan, I. (2008). Eastern Turkish high plateau as a small Turkic-type orogen: Implications for post collisional crust forming processes in Turkic-type orogen. *Earth-Science Reviews*, *90*, 1–48. https://doi.org/10.1016/j.earscirev.2008.05.002

Şengör, A. M. C., Özeren, S., Genc, T., & Zor, E. (2003). East Anatolian high plateau as a mantle-supported, north-south shortened domal structure. *Geophysical Research Letters*, *30*(24), 1–3. https://doi.org/10.1029/2003GL017858

Şengör, A. M. C., White, G., & Dewey, J. F. (1979). Tectonic evolution of the Bitlis suture, southeastern Turkey: Implications for the tectonics of the Eastern Mediterranean. *Rapp. Comm. Int. Mer. Medit.*, *25*/26, 95–97.

Seyitoğlu, G., Esat, K., & Kaypak, B. (2017). The neotectonics of southeast Turkey, northern Syria, and Iraq: The internal structure of the Southeast Anatolian Wedge and its relationship with recent earthquakes. *Turkish Journal of Earth Sciences*, *26*(2), 105–126.

Silja, K., Hüsing, W., et al. (2009). Oligocene-Miocene basin evolution in SE Anatolia, Turkey: Constraints on the closure of the eastern gateway. In D. J. J. van Hinsbergen et al. (Eds.), *Collision and collapse at the Africa-Arabia-Eurasia subduction zone* (pp. 107–132). Society of London, Special Publication, 311.

Siyako, M., Bahtiyar, I., Özdoğan, I., Acıkbaş, D., & Kaya, O. (2013). Batman cevresinde mostra veren birimlerin stratigrafisi. *TPAO. Arama Grubu Arşivi Teknik Rapor*, *5463*, 131.

Tuna, D. (1973). Bölge litostratigrafi adlamasının acıklayıcı raporu. *Turkiye Petrolleri Anonim Ortaklığı Rapor*, *813*, 131.

Van Hinsbergen, D. J. J., Torsvik, T. H., Schmid, S. M., Majenc, L. C., Maffione Visser, M. R. et al. (2020). Orogenic architecture of the Mediterranean region and Kinematic reconstruction of since the Triassic. *Gondwana Research*, *81*, 79–229.

Yazgan, E. (1984). Tauric-subduction (Malatya-Elazığ provinces) and its bearing on tectonics of the Tethyan realms in Turkey. In J. E. Dixon & A. H. F. Robertson (Eds.), *The geological evolution of the eastern Mediterranean* (pp. 361–373). Geological Society of London Special Publication, 17.

Yeşilova, Ç., & Helvacı, C. (2013). Batman Siirt Kuzeyi Sedimantolojisi. *Türkiye Petrol Jeologları Derneği Bülteni*, *23*, 7–43.

Yeşilova, Ç., Helvacı, C., & Carrillo, E. (2018). Evaporitic sedimentation in the Southeastern Anatolian Foreland Basin: New insights on the Neotethys closure. *Sedimentary Geology*, *369*, 13–27. https://doi.org/10.1016/j.sedgeo.2018.03.012

Yeşilova, P. G., & Helvacı, C. (2017). Petrographic study and geochemical investigation of the evaporites associated with the Germik Formation (Siirt Basin, Turkey). *Carbonates and Evaporites*, *32* (2), 177–194. https://doi.org/10.1007/s13146-015-0285-y

Yiğitbaş, E. (1989). *Engizek Dağı (Kahraman Maraş) dolayındaki tektonik birliklerin petrolojik incelenmesi (Petrological studies of the tectonic units in the Engizek Mountain, Kahraman Maras)*. Doctoral thesis. Istanbul Üniversitesi, Fen Fakültesi.

Yiğitbaş, E., & Yılmaz, Y. (1996a). New evidence and solution to the Maden Complex controversy of the Southeast Anatolian orogenic belt (Turkey). *Geol. Rundschau.*, *85*, 250–263.

Yiğitbaş, E., & Yılmaz, Y. (1996b). Post-late Cretaceous strike -slip tectonics and its implication on the southeast Anatolian Orogen, Turkey. *International Geology Review*, *38*, 818–831.

Yıldırım, E. (2015). Geochemistry, petrography, and tectonic significance of the ophiolitic rocks, felsic intrusions, and Eocene volcanic rocks of an imbrication zone (Helete area, Southeast Turkey). *Journal of African Earth Sciences*, *7*, 89–107.

Yıldırım, M., & Yılmaz, Y. (1991). Güneydoğu Anadolunun ekaylı zonu (Imbricated zone of the southeast Anatolian orogenic belt). *Bulletin of the Turkish Association of the Petroleum Geologists*, *3*(1), 57–73.

Yılmaz, O. (1975). Petrographic and stratigraphic study of the rocks of the Cacas region (Bitlis Massif). *Türkiye Jeoloji Kurumu Bülteni*, *18*, 33–40.

Yılmaz, Y. (1978). Bitlis massif and ophiolite relationship around Gevaş. Van: 4th Petroleum Congress of Turkey. *Proceedings, Turkish Association of Petroleum Geologists*, *23*, 83–93.

Yılmaz, Y. (1984). Amanos Dağlarının Jeolojisi (1–4). *Türkiye Petrolleri Anonim Ortaklığı Rapor*, *1920*.

Yılmaz, Y. (1985). Geology of the Cilo Ophiolite: An ancient ensimatic island arc fragment on the Arabian Platform, *SE Turkey. Ofioliti*, *10* (2/3), 457–484.

Yılmaz, Y. (1990). Allochthonous terranes in the Tethyan Middle East: Anatolia and the surrounding regions. *Royal Society of London Philosophical Transactions*, *A331*, 611–625.

Yılmaz, Y. (1993). New evidence and model on the evolution of southeast Anatolian orogen. *Geological Society of America Bulletin*, *105*, 251–271.

Yılmaz, Y. (1994). Geology of the Cilo Ophiolite and the surrounding region, southeast Turkey; comparison with Oman. *Bulletin of Tech. Univ. Istanbul*, *47*, 509–533.

Yılmaz, Y. (2017). Morphotectonic development of Anatolia and surrounding regions. In I. Çemen & Y. Yılmaz (Eds.), *NeoTectonics and earthquake potential of the Eastern Mediterranean region* (pp. 11–92). AGU Geophysical Monograph 225.

Yılmaz, Y. (2019). Southeast Anatolian Orogenic Belt revisited. *Canadian Journal of Earth Sciences*, *1–18*. https://doi.org/10.1139/cjes-1170

Yılmaz, Y. (2020). Morphotectonic development of the Adana plain and the surrounding mountains, South Turkey. *Mediterranean Geoscience Reviews*, *2*, 341–358. https://doi.org/10.1007/s42990-020-00043-4

Yılmaz, Y. (2021). Geological correlation between Northern Cyprus and Southern Anatolia. *Canadian Journal of Earth Science*, *58*(7), 640–657.

Yılmaz, Y., & Yiğitbaş, E. (1991). *The different ophiolitic-metamorphic assemblages of S. E. Anatolia and their significance in the geological evolution of the region*. 8th Petroleum Congress of Turkey, Geology Proceedings, Ankara, Turkey, Turkish Association of Petroleum Geologists, 128–140.

Yılmaz, Y., & Yıldırım, M. (1996). Geology of the Nappe region of the southeast Anatolian orogenic belt with emphasis on the metamorphic massifs (Güneydoğu Anadolu orojenik kuşağında nap alanının metamorfik masiflerin jeolojisi ve evrimi). *Turkish Journal of Earth Sciences*, *38*, 21–38.

Yılmaz, Y., Dilek, Y., & Işik, H. (1981). Gevaş (Van) ofiyolitinin jeolojisi ve sinkinematik bir makaslama zonu. *Türkiye Jeoloji Kurumu Bülteni. 24*(1), 37–45.

Yılmaz, Y., Gürpınar, O., & Yiğitbaş, E. (1988). Amanos Dağları ve dolaylarında Miyosen havzalarının tektonik evrimi (Tectonic evolution of the Miocene basins at the Amanos mountains and the Maraş Region). *Türkiye Petrol Jeologları Derneği Bülteni*, *Ill*, 52–72.

Yılmaz, Y., Gürpınar, O., Kozlu, H., Gül, M. A., Yiğitbaş, E., Yıldırım, M., et al. (1987). Maraş Kuzeyinin Jeolojisi (Andırın-Berit-Engizek-Nurhak-Binboğa Dağları). *Türkiye Petrolleri Anonim Ortaklığı Rapor,* 2028 (Cilt 1,2,3).

Yılmaz, Y., Yiğitbaş, E., & Genc, C. (1993). Ophiolitic and metamorphic assemblages of southeast Anatolia and their significance in the geological evolution of the orogenic Belt. *Tectonics*, *12*, 1280–1297.

Yılmaz Y., Yiğitbaş, E., Yıldırım, M., & Genc, C. (1992). *Origin of the southeast Anatolian metamorphic massifs*. 9th Petroleum Congress and Exhibition of Turkey. Abstracts, Turkish Association of Petroleum Geologists, 170–180.

8

Tectonics of Eastern Anatolian Plateau: Final Stages of Collisional Orogeny in Anatolia

Yücel Yılmaz[1], İbrahim Çemen[2], and Erdinç Yiğitbaş[3]

ABSTRACT

The East Anatolian High Plateau, which is a part of the Alpine-Himalayan orogen, is a 200 km wide, approximately east-west trending belt surrounded by two peripheral mountains of the Anatolian Peninsula. The plateau is covered by thick, interbedded Neogene volcanic and sedimentary rocks. Outcrops of the underlying rocks are rare and, therefore, contrasting views were proposed on the nature of the basement rocks. New geological and geophysical data suggest the presence of an ophiolitic mélange-accretionary complex under cover rocks of Eastern Anatolia. The Neogene cover units began to be deposited during the closure of the NeoTethyan Ocean that was located between the Pontide arc to the north, and the continental slivers drifted away from the Arabian Plate to the south. The bordering orogenic belts, the Pontides in the north, and the Bitlis-Zagros Mountains in the south have undergone entirely different evolution. The Eastern Anatolian orogen was formed during the later stages of the development of the surrounding orogenic belts. In this period, the mélange-accretionary prism that occupied a large terrain behaved like a wide and thick cushion, which did not allow a head-on collision of the bordering continents. The NeoTethyan oceanic lithosphere was eliminated from the entire eastern Anatolia by northward subduction that lasted till the Late Eocene. The Eastern Anatolia began to rise when the northern advance of the Arabian Plate continued after the total demise of the oceanic lithosphere. The present stage of the elevation of the East Anatolian Plateau as a coherent block started during the Late Miocene.

8.1. INTRODUCTION

The Eastern Anatolian region is part of the Alpine-Himalayan belt. It is usually referred to as an East Anatolian High Plateau because it is on average 2,000 m in elevation. The region is a 200 km wide belt between the Pontide Mountains to the north and the Bitlis-Zagros suture mountains to the south (Fig. 8.1).

The Pontides were formed during consecutive collisions between the Andean-type volcanic arcs and continental blocks of Gondwanan origin (Yılmaz et al., 1997). The Bitlis-Zagros suture mountains were formed as a result of the continent-continent collision (Yılmaz, 2019, and the references therein; see chapter 7 in this volume for the accompanying chapter on the Southeast Anatolian Orogenic Belt). The Eastern Anatolian orogen was formed during the later stages of development of the surrounding orogenic belts.

The most significant structural features of Eastern Anatolia are the North Anatolian Transform Fault (NATF) and the East Anatolian Transform Fault (EATF) (Figs. 8.1, 8.2). The two faults converge in the Karlıova Junction (KJ in Fig. 8.1) and define the Anatolian Plate.

[1] *Department of Geology, Istanbul Technical University, Istanbul, Turkey*

[2] *Department of Geological Sciences, The University of Alabama, Tuscaloosa, Alabama, USA*

[3] *Department of Geology, Çanakkale Onsekiz Mart Üniversity, Çanakkale, Turkey*

Compressional Tectonics: Plate Convergence to Mountain Building, Geophysical Monograph 277, First Edition.
Edited by Elizabeth J. Catlos and İbrahim Çemen.

DOI:10.1002/9781119773856.ch08

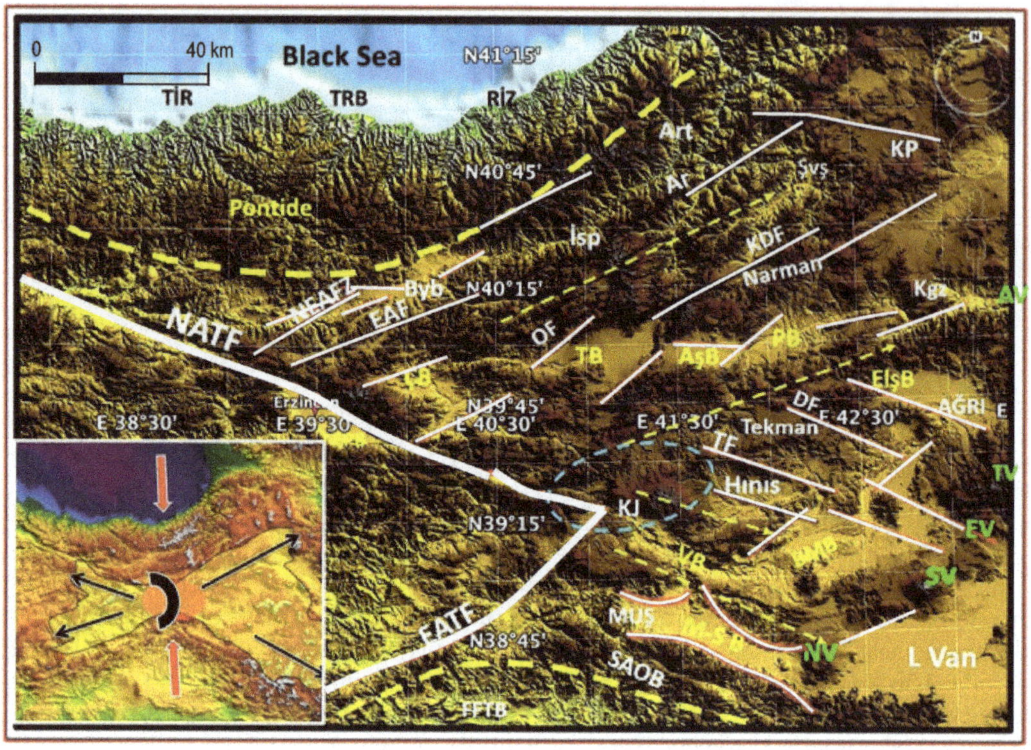

Figure 8.1 Morphotectonic map of Eastern Anatolia showing major faults (straight lines) and trend lines of the mountain ranges (broken lines). Thick, broken, curvilinear lines represent trend lines of the peripheral orogenic belts, the Pontide, and the Southeastern Anatolian Orogenic Belt (SAOB). The white lines with the red glove are reverse faults. The inset map shows the central high resembling sheaved wheat and the dispersing major morphological features. Note: NATF = North Anatolian Transform Fault; EATF = East Anatolian Transform Fault; EAF = East Anatolian fault zone; NEAFZ = Northeast Anatolian fault zone; OF = Olur Fault; DF = Doğu Beyazıt Fault; TF = Tutak Fault, the ellipse represents the center of the virgation; KJ = The Karlıova Junction; FFTB = Foreland fault and thrust belt of the Southeastern Anatolian Orogen. Basins: ÇB = Çayırlı basin; TB = Tercan basin; AşB = Aşkale basin; PB = Pasinler basin; VB = Varto basin; BMB = Bulanık-Malazgirt basin; M-SB = Muş-Solhan basin. Volcanoes; NV = Nemrut; SV = Süphan; EV = Etrüsk; TV = Tendürek; AV = Ağrı (Ararat). Towns and cities (black letters along the coastal zone): TİR = Tirebolu; TRB = Trabzon; RİZ = Rize. White letters inland: Art = Artvin, Ar = Ardanuç; Byb = Bayburt, İsp = İspir; Şvş = Şavşat; KP = Karst Plateau; LVan = Lake Van.

The transform faults are long recognized as the major manifestation of the escape tectonics and associated lateral extrusion of the Anatolian Plate (e.g., Şengör, 1979; Şengör & Yilmaz, 1981; Çemen et al., 1993; Yılmaz, 2017).

The Eastern Anatolia is covered by a thick, interbedded Neogene volcanic and sedimentary rocks and contains many conical peaks and east-northeast and west-northwest trending hills (Figs. 8.1, 8.2). The individual peaks correspond to volcanic cones (Fig. 8.1) (Yılmaz et al., 1987, 1998; Pearce et al., 1990; Yılmaz, 2017). The volcanoes produced a wide range of edifices from plateau basalts to ignimbrite deposits (Yılmaz et al., 1998; Kaygusuz et al., 2018).

The Neogene sedimentary cover rocks of Eastern Anatolia extend mainly along with two separate stripes of depressions adjacent to the peripheral mountains (Figs. 8.1, 8.2; Yılmaz, 2017). Rates of uplift in the bordering mountains are greater (0.2–0.3 mm/y; Keskin et al., 2011) than the plateau's uplift (0.1–0.2 mm/y; McNab et al., 2018). Therefore, headword erosion across the peripheral mountains cannot keep pace with the elevation increase in Eastern Anatolia. Consequently, major rivers in the plateau flow generally in east-west directions (Fig. 8.1).

The thick Neogene cover sequence is commonly flat but is locally tightly folded and faulted. The morphological pattern of Eastern Turkey resembles a sheaf of wheat tied at the center (see inset in Fig. 8.1), reflecting strict structural control of the ongoing tectonics. The peripheral mountains on both sides curve around a central dome, which determines the regional structures and the present drainage network (Şaroğlu & Güner, 1981; Maggi & Priestley, 2005; Yılmaz, 2017; Fig. 8.1 and inset). The hills, depressions, and rivers fan out from the central high (Fig. 8.1).

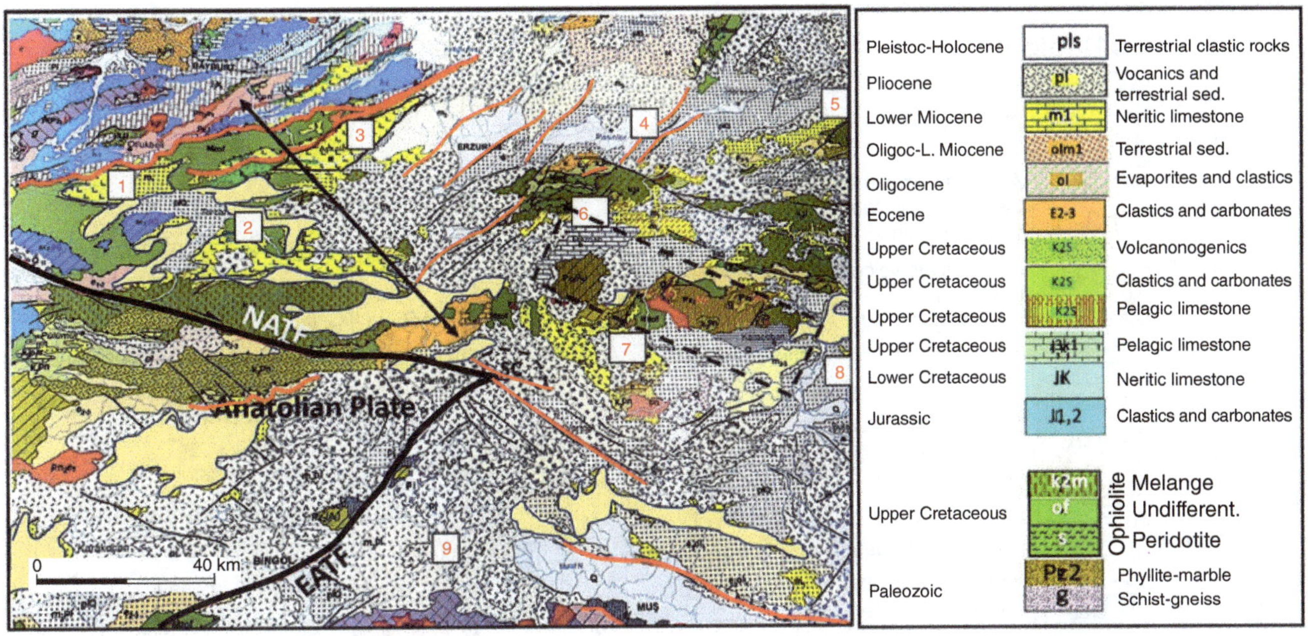

Figure 8.2 Geology map of the Eastern Anatolia (modified after MTA 1/500 000 scale geology map of Turkey covering regions from the Erzurum, Van, Diyarbakır, and Trabzon sheets). Numbers 1 to 9 show approximate locations of the young continental basins: 1 = Çayırlı, 2 = Tercan, 3 = Aşkale, 4 = Pasinler, 5 = Kağızman, 6 = Tekman, 7 = Hınıs, 8 = Bulanık-Malazgirt, 9 = Muş-Bingöl. Straight lines are major strike-slip faults. Curvilinear lines along the northern and southern edges of Eastern Anatolia are the major thrusts separating Eastern Anatolia from the neighboring orogenic belts. The rectangle defined by black broken lines shows the location of the map in Figure 8.8a. The black line with arrows at both ends indicates the direction of the cross section in Figure 8.5c. Note: SC = the Solhan volcano's caldera; Broken black half-circle defines the caldera's northern half.

Outcrops of the basement rocks below the thick Neogene volcano-sedimentary layer are rare. As a result, contrasting views were proposed on the nature of the basement rocks, which made the orogenic evolution of the belt controversial. This paper aims to document new data leading to clarify the nature of basement rocks in Eastern Anatolia and discuss the orogenic development based on the new data.

8.2. GEOLOGIC OVERVIEW

In this section, we will summarize stratigraphic, structural, and igneous features of Eastern Anatolia.

8.2.1. Stratigraphy

Stratigraphic columnar sections covering the entire Eastern Anatolian region are displayed in Figure 8.3. The sections summarize the data gathered mainly from our fieldwork together with the TPAO reports and the previous studies (Kurtman & Akkuş, 1971; Özdemir, 1981; Şenel et al., 1984; Koçyiğit et al., 1985; Şaroğlu & Yılmaz, 1984, 1986, 1987, 1991; Uysal, 1986; Gedik, 1986; Yılmaz et al., 1987a,b; Yılmaz et al., 1988; Tarhan, 1997a,b, 1998a,b; Akay et al., 1989; Bozkuş, 1990; Temiz et al., 2002; MTA, 2002; Konak & Hakyemez, 2008; Yılmaz, 2107; Bedi & Yusufoğlu, 2018; Yılmaz & Yılmaz, 2019; Üner, 2021) enabling correlations and comparisons along and across the East Anatolian Plateau possible.

The generalized stratigraphic columnar section of Eastern Anatolia (GSS in Fig. 8.3a) shows the presence of an ophiolitic mélange below the Neogene cover in most outcrops (MTA, 2002; Özdemir, 1981; Önal & Kaya, 2007; Şenel et al., 1984; Konak & Hakyemez, 2008; Elitok & Dolmaz, 2008; Yılmaz & Yılmaz, 2019; Üner, 2021; TPAO field reports and drilling data, and our field observations). The overlying Neogene cover is represented commonly by terrestrial sedimentary rocks. However, the stratigraphy in the northeastern part of East Anatolia (i.e., north of the Kars province; Fig. 8.1) is entirely different (the Karst region in Fig. 8.3a). It consists of two major components, an old metamorphic basement and overlying thick Paleozoic, Mesozoic, and Cenozoic successions. This is shown in the stratigraphic column of the Kars region in Fig. 8.3a, where the succession is identical to that of the eastern Pontide region. The western and northwestern parts of the Kars province are thus considered easterly continuation of the eastern Pontide.

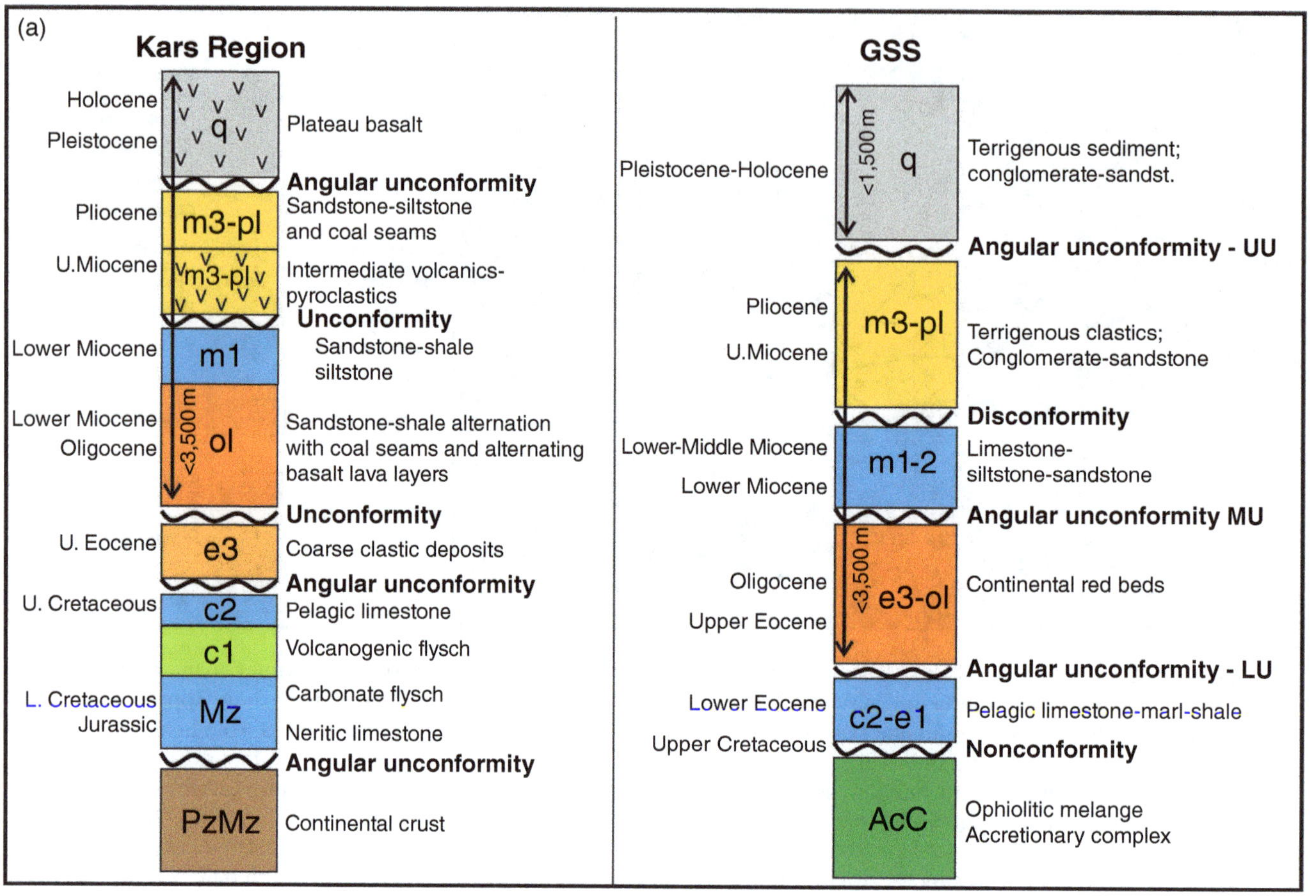

Figure 8.3 (a) GSS, Generalized stratigraphic section representing the eastern Anatolian basins. Note: LU = Lower unconformity; MU = Middle unconformity; UU = Upper unconformity. Kars region; Stratigraphic section representing the northeastern part of Eastern Anatolia, the Kars region, and surroundings. (b, c) Stratigraphic columnar sections of the major basins of Eastern Anatolia. Locations of the basins are shown in Figure 8.2.

There is no detailed study on the Eastern Anatolian ophiolitic mélange-accretionary complex. The previous works generally outline its major constituents (Altınlı, 1966a,b; Yılmaz et al., 1988; Tarhan, 1989, 1997a,b, 1998 a,b; Bedi & Yusufoğlu, 2018; Yılmaz & Yılmaz, 2019). A pelagic sedimentary succession consisting of limestone (chalk), radiolarite, and siltstone of the Upper Cretaceous to Lower Eocene age range is observed in association with the mélange. They are either blocks incorporated into the mélange or deep-sea sedimentary sequences deposited above the mélange foundation. Altınlı (1966a,b), Şenel et al., 1984; Tarhan (1989, 1997a,b, 1998a,b); Kaya (2009), and Bedi and Yusufoğlu (2018) documented Upper Cretaceous, Paleocene, and Eocene fossil lists from the pelagic sediments. In some exposures, Upper Cretaceous–Eocene shallow sea sediments are also observed lying stratigraphically over the ophiolitic mélange (Bedi & Yusufoğlu, 2018; Yılmaz & Yılmaz, 2019).

The Neogene cover succession consists of four major rock groups separated by angular unconformities corresponding to (1) Late Eocene-Oligocene (the lower unconformity); (2) Late Oligocene–Early Miocene (the middle unconformity); and (3) Late Pliocene (the upper unconformity). The lower unconformity separates the ophiolitic mélange-accretionary complex from the Upper Eocene–Oligocene terrestrial units consisting mainly of coarse clastic deposits (1 to 9 in Fig. 8.3-b,c). The middle unconformity separates the terrestrial sediments from the overlying Upper Oligocene–Lower Miocene transgressive sequence. The marine sediments begin with fine-grained sandstone-siltstones-marl alternations followed by reefal limestone and Lower–Middle Miocene neritic limestone (the Adilcevaz limestone; Şaroğlu & Yılmaz, 1986). The limestone unit is observed in the entire eastern Anatolian region (Şaroğlu & Yılmaz, 1986, 1987; Tarhan et al., 1992; Tarhan, 1998; Kaya, 2009; Yılmaz, 2017; Bedi & Yusufoğlu, 2018; Yılmaz & Yılmaz, 2019).

The shallow marine sediments grade upward into evaporites (Akkuş, 1970; Yılmaz, 2017; Bedi & Yusufoğlu, 2018; Yılmaz & Yılmaz, 2019; Helvacı, 2021) and lacustrine limestone, shale-sandstone alternations of Upper

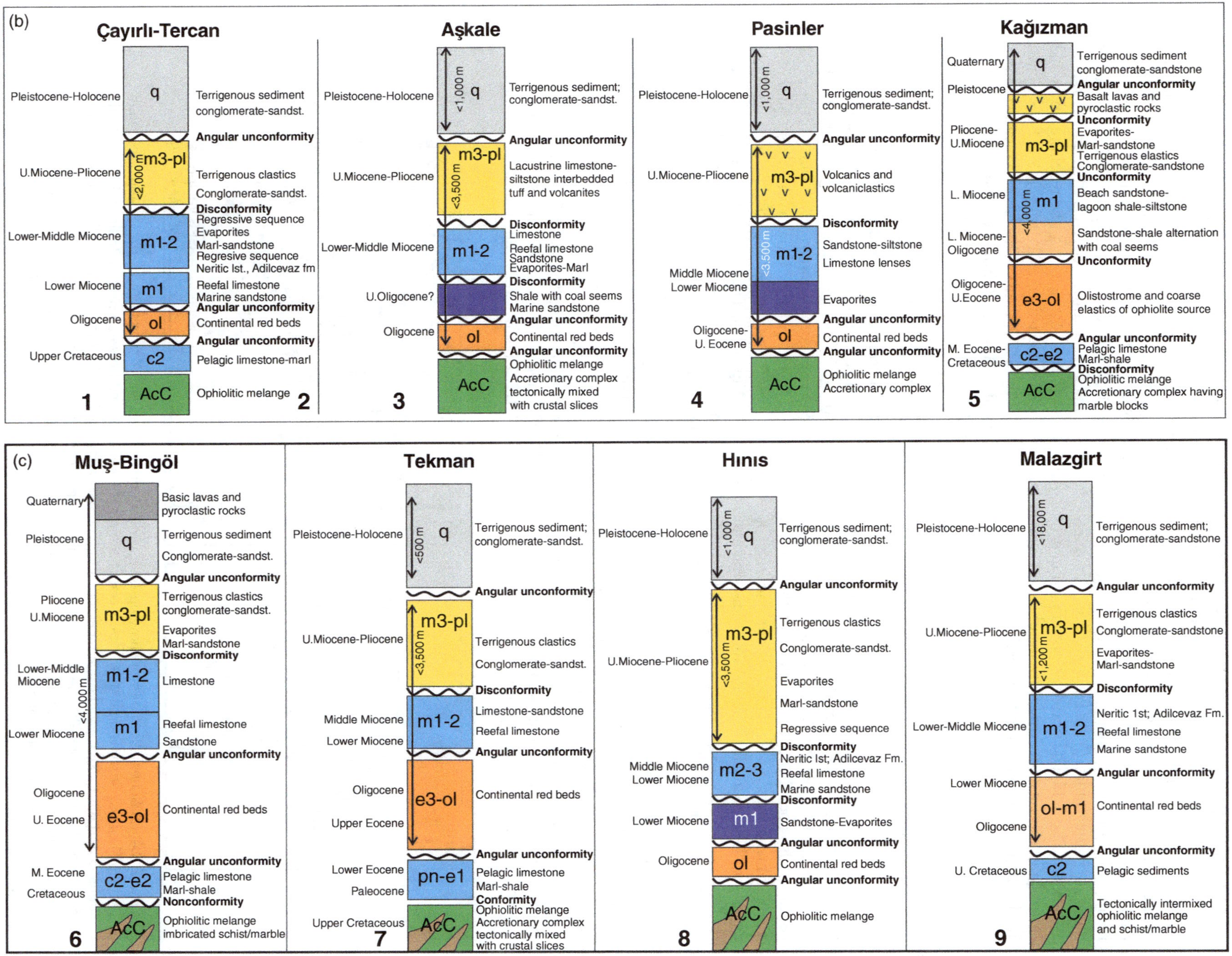

Figure 8.3 (Continued)

Miocene–Lower Pliocene age (Akkuş, 1970, 19071; Tarhan, 1998; Yılmaz, 2017; Bedi & Yusufoğlu, 2018; Yılmaz & Yılmaz, 2019). Thick Pleistocene-Holocene lacustrine-fluvial conglomerates and sandstones unconformably overlie them.

The upper unconformity marks a critical change in the region's morphotectonics (Yılmaz, 2017, and the references therein). The field data obtained from regional geological mapping in Eastern Anatolia (unpublished maps of Turkish Petroleum Association) show that the Miocene and Pliocene sediments are regionally distributed. However, the post-Pliocene deposits are confined to the local depressions and display abrupt vertical-lateral facies changes within the basins (Şaroğlu & Yılmaz, 1987; Tarhan, 1989, 1997a, 1998a; Yılmaz, 2017; Bedi & Yusufoğlu, 2018).

The stratigraphic data outlined above reveal that the final stage of the uplift in Eastern Anatolia began during the Late Miocene (Yılmaz, 2017; Bedi & Yusufoğlu, 2018). This event is also coeval with the elevation of the Central Anatolia, which leads us to assume that the rise of Eastern Anatolia also accelerated the uplift and magmatism in Central Anatolia (Schildgen et al., 2014; Bartol & Govers, 2014; Govers & Fichtner, 2016; McNab et al., 2018).

Stratigraphic and isotopic age data reveal that volcanic activity in the Eastern Anatolian Plateau began sporadically during the Late Miocene, possibly about 13–11 my ago (Şaroğlu & Yılmaz, 1984; Yılmaz et al., 1987a,b, 1998; Yılmaz, 2017). The volcanic edifices increased in the northern regions around 7–8 my ago and then migrated to the central areas around 5–4 my ago (Pearce et al., 1990; Yılmaz et al., 1998; Keskin, 2003, 2007). However, the volcanic activity intensified in the central and the southern regions about 3 my ago and has continued almost uninterruptedly to the present (Yılmaz et al., 1987a,b, 1998; Yılmaz, 1990; Pearce et al., 1990; Keskin, 2003, 2007; Keskin et al., 2006, 2012). As a result, a thick volcanic blanket covered the entire Eastern Anatolian Plateau. A thick lava pile reaching up to 2 km in thickness was measured and drilled in the Kars Plateau (unpublished TPAO data). The major volcanic centers, the Nemrut, Süphan, Tendürek, and Ağrı volcanoes, were built during the Quaternary (Yılmaz et al., 1988).

8.2.2. Structural Geology

Eastern Anatolia displays most of the geological features of a collisional orogen, which is similar in many respects to the Tibetan Plateau (Şengör & Kidd, 1979; Şengör, 1979; Şengör & Yılmaz, 1981; Dewey et al., 1986; Şaroğlu & Yılmaz, 1984; Barazangi, 1989; Şengör & Natalin, 1996; Şengör et al., 2003, 2008; Seyitoğlu et al., 2018; Fig. 8.4a). The GPS measurements of crustal displacements (Fig. 8.4b; Reilenger et al., 2006; Şengör et al., 2008), focal mechanism of the earthquakes, and the distribution of active faults (Figs. 8.1, 8.2, 8.4a; Şengör et al., 2003; Bozkurt, 2001; Yılmaz, 2017) confirm that Eastern Turkey experiences an ongoing north-directed compressional stress (Fig. 8.4a; Yılmaz, 2017, and the references therein).

The present-day morphology in Eastern Anatolia was developed under a significant structural control (Yılmaz, 2017; Figs. 8.1, 8.2). The major morphotectonic

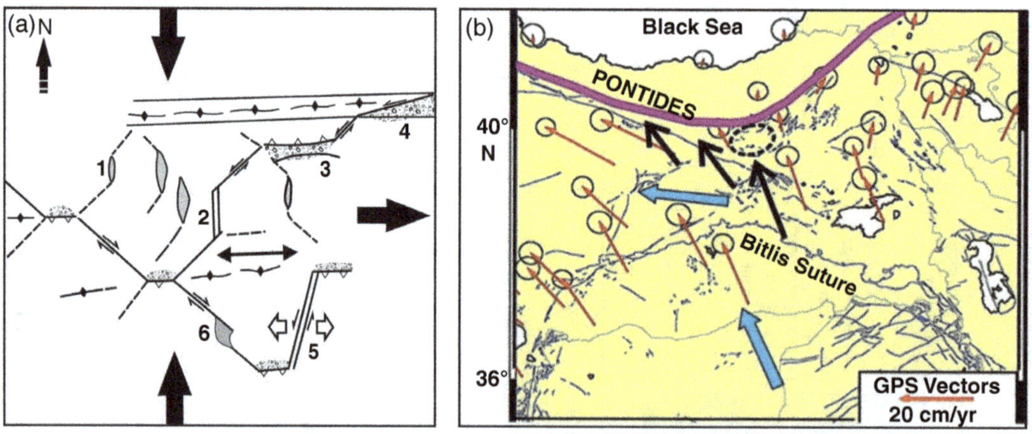

Figure 8.4 (a) Schematic diagram showing types and trends of the major structures of the Eastern Anatolian Plateau: 1 and 2 are tensional openings; 3 and 4 are ramp basins formed between reverse fault and oblique-slip faults; 5 and 6 are extensional opening associated with pull-apart basins. The large arrows indicate compression and extension directions. (b) The GPS vectors measured in Eastern Anatolia (modified after Şengör et al., 2008). The thick line displays the crest line of the Pontide Range. The wide blue arrows show the motion directions of the Arabian and Anatolian plates. The black arrows show the major stress direction deforming southern regions of the Pontides.

features, northeast, and northwest-trending hills, and depressions (Fig. 8.1) correspond to anticlines and synclines, respectively (Fig. 8.1; Yılmaz, 2017). There is a centrally located structural dome, which may be regarded as the center of virgation (Fig. 8.1), the maximum indentation location. The peripheral mountains on both sides make curves around the virgation (inset, Fig. 8.1). The strike-slip faults disperse away from this high (Figs. 8.1, 8.2). The arrangement and interactions of the structures demonstrate that Eeastern Anatolia has undergone a complex tectonic evolution (Şengör & Kidd, 1979; Şengör & Yılmaz, 1981; Şaroğlu & Yılmaz, 1986; Copley & Jackson, 2006; Yılmaz, 2017; Seyitoğlu et al., 2018) from the time of collision along with the Southeastern Anatolian suture zone in the Late Eocene. This event corresponds to a wholesale elevation reflected in Eastern Anatolian stratigraphy as a marked angular unconformity (Fig. 8.3; Yılmaz, 2019; see also chapter 7, in this volume, by Yılmaz et al.).

In the East Anatolian High Plateau, the following major groups of structures are readily observed (Fig. 8.4a).

1. Northeast and northwest striking strike-slip faults (Figs. 8.1, 8.4a) forming conjugated pairs (Şengör & Kidd, 1979; Şaroğlu & Yılmaz, 1984; Bozkurt, 2001; Seyitoğlu et al., 2017). The northeast striking faults are commonly longer and more prominent (Fig. 8.1; Bozkurt, 2001; Seyitoğlu et al., 2017). Most of these faults are young and active that generate frequent earthquakes (i.e., the Elazığ earthquake on 24 January 2020, Mw = 6.7; the Iran-Turkey border earthquake on 23 February 2020, Mw = 6.0; and the Malatya earthquake on 24 January 2020, Mw = 6.7).
2. Approximately east-west or east-southeast–west-northwest- striking reverse faults (Fig. 8.4a)
3. East-west trending open and tight folds (Şaroğlu et al., 1980; Şaroğlu & Yılmaz, 1984, 1986, 1987; Koçyiğit et al., 2001; Yılmaz, 2017).
4. North-south trending extensional structures (Fig. 8.4a; Şengör & Kidd, 1979; Şaroğlu & Yılmaz, 1984, 1987; Yılmaz, 2017).

The trends of all these structures are compatible with the north-south compressional stress field (Fig. 8.4a) generated from the northward advance of the Arabian Plate (Fig. 8.4b), which also caused the peripheral mountains to thrust over the Eastern Anatolian Plateau (Figs. 8.2, 8.5a,c). Along the thrust boundary, the basement metamorphic rocks are exposed on the hanging walls (Figs. 8.2, 8.5c). As a result, the crust along the overthrust zones is thicker than the average, reaching up to 50 km (48 km beneath the Bitlis-Pötürge massifs) in the south and 50 km under the Pontide Range (Barazangi et al., 2006; Şengör et al., 2003, 2008; Tezel et al., 2013; Pamukçu et al., 2007; Medved et al., 2021).

Two narrow, fault-bounded chains of depression were formed between the central high and the peripheral mountains during Pleistocene (Figs. 8.1, 8.2, 8.5a,b). The boundary faults give the young basins their distinct rhombohedral or parallelogram geometrical patterns (Fig. 8.5b). The transpressional faults with the major strike slip, coupled with reverse slip components, strike parallel with, and determine the basins' long axes (Fig. 8.5b). Due to the reverse slip displacements along the boundary faults, the depressions may be viewed as ramp basins or intermontane basins (Fig. 8.5a,c) as exemplified from the Muş, Bingöl, Murat, Elazığ, Malatya basins to the south (Figs. 8.1 and 8.5b) and the Tercan-Aşkale, Pasinler, Kağızman basins to the north (Figs. 8.1 and 8.5b). The Eastern Anatolian High Plateau may also be regarded as a giant ramp basin between the thrust elevated peripheral mountains (Fig. 8.5a).

Within the rhombohedral depressions where normal faults define the releasing bends, the local basins were alternatively described as pull-apart basins (Keskin et al., 1998). Some volcanoes were built above the extensional openings (Fig. 8.1; Dewey et al., 1986; Pearce et al., 1990; Keskin et al., 1998; Yılmaz, 2017).

The two transform faults, the North Anatolian Transform Fault and the East Anatolian Transform Fault define the Anatolian Plate (Figs. 8.1, 8.2), which protrudes away from the area of convergence in the Karlıova Junction (KJ in Fig. 8.1; McKenzie, 1972; Şengör & Yılmaz, 1981; McKenzie & Yılmaz, 1991; Şaroğlu & Yılmaz, 1991; Çemen et al., 1993; Özeren & Holt, 2010; Aktuğ et al., 2012; Karaoğlu et al., 2017; Figs. 8.1, 8.2, 8.5b). The area where the transform faults intersect is a wedge-shaped depression (Fig. 8.5b), which widens progressively as the Anatolian Plate moves away from the point of convergence (Şaroğlu & Yılmaz, 1991; Fig. 8.5b). The oldest lateral fan deposits sourced from the basin boundary faults (Fig. 8.5b) is Upper Pliocene (?)–Pleistocene in age (Şaroğlu & Yılmaz, 1991), which sets a lower limit to the time of the westerly escape of the Anatolian Plate. The escape tectonics partitions the north-south compressional stress (Yılmaz, 2017).

Associated with the development of the two transform faults and the consequent escape tectonics of Anatolia, a new tectonic regime, the NeoTectonics, began and has drastically changed the morphotectonic character of the Anatolia and surrounding regions (e.g., Şengör, 1979; Şengör & Yılmaz, 1981; Çemen et al., 1993; Yılmaz, 2017). The GPS vectors in Fig. 8.4b show that the eastward motion of Eastern Anatolia is slower than its westward movement. Field and earthquake data indicate that the north-south compressional stress is accommodated in this region along with several small-size strike-slip faults with right-lateral displacements (Fig. 8.2).

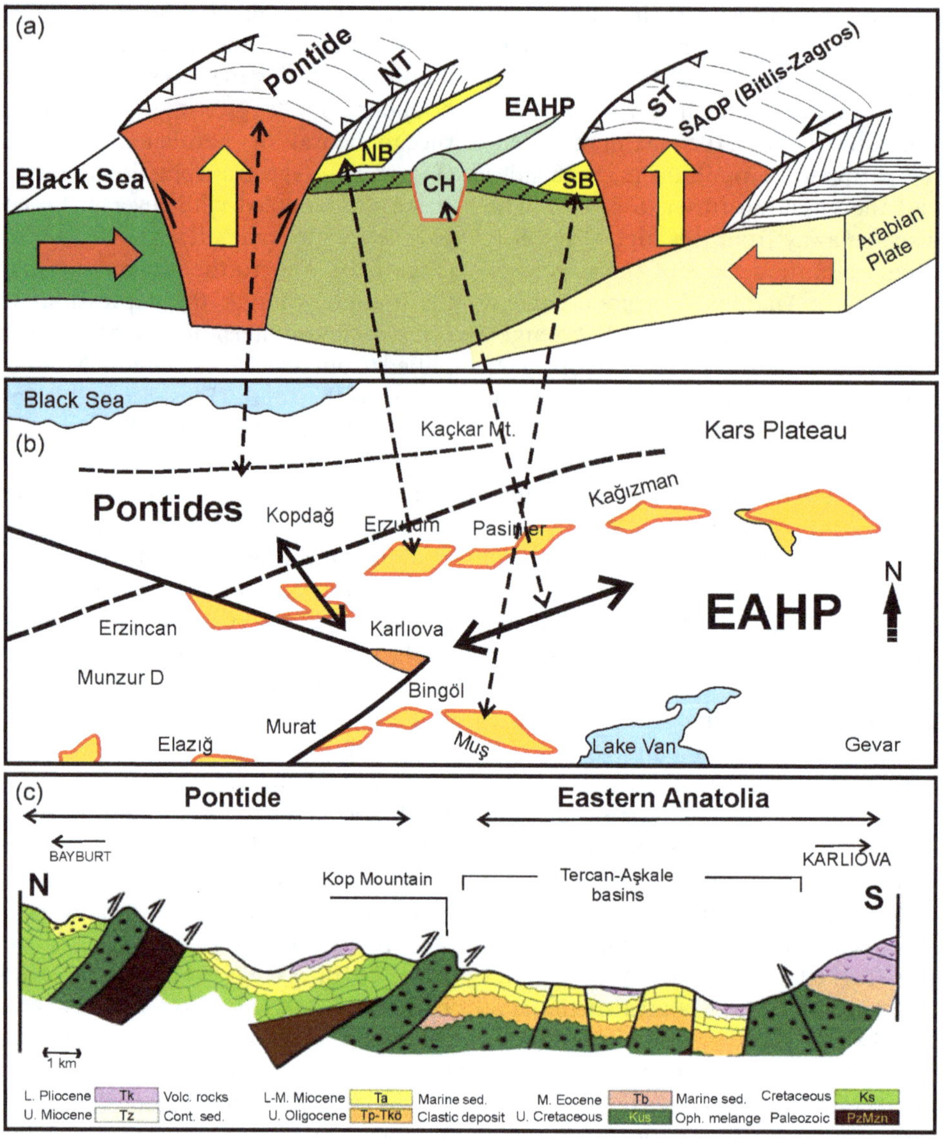

Figure 8.5 (a) Schematic block diagram displaying major tectonic belts and morphotectonic features of Eastern Turkey. Eastern Anatolia is squeezed (the dark arrows) between the northward advance of the Arabian Plate, and the resisting old oceanic lithosphere under the Black Sea is shortened, thickened, and elevated. The Pontide Range and the Southeastern Anatolian Orogenic Belt (SAOB; the Bitlis-Zagros Mountains) underlain by old and rigid continental crust were thrust over Eastern Anatolia and elevated with a higher rate (the pale arrows) because the ophiolitic mélange-accretionary complex underlying Eastern Anatolia absorbs the bulk of north-south compression. Along the trust fronts, two narrow chains of fault-bound basins (NB and SB) were formed. The oblique faults (major strike-slip coupled with reverse slip displacements) give the basins their distinct parallelogram shapes and ramp basin characters in (b). The center of Eastern Anatolia responded to the north-south compressional stress by protruding upward to form a central high (CH). Note: EAHP = the East Anatolian High Plateau; SAOB = Southeast Anatolian Orogenic Belt, the Bitlis-Zagros Mountains; SB, NB = southern and northern basins; ST and NT = southern and northern thrusts along with the peripheral mountains were thrust over the young basins. (b) Schematic structural map of Eastern Anatolia showing fault-bound chains of basins and centrally located structural high (the black line with arrows at both ends. The broken lines correspond to the trend lines of the mountain ranges. The short arrow between Kopdağ and Karlıova shows the direction of the geological cross section in (c). The thin broken vertical lines connect the structural elements between the map and the block diagram in (a) and (b). Note: NATF = North Anatolian Transform Fault; EATF = East Anatolian Transform Fault; Munzur D = The Munzur Mountains; Mt = mountains. (c) Geological cross section from Eastern Anatolia to the Pontide Mountains along the cross-section direction shown in Figures 8.2 and 8.5b (modified from the cross section provided by Ö. Şahintürk).

8.2.3. Thickness of Crust and Lithosphere

Geophysical data suggest that the Eastern Anatolian crust is thick (Çınar & Alkan, 2015; Pamukçu et al., 2007; Cırmık, 2018; Medved et al., 2021), but the uppermost mantle below the crust is very thin because it strongly attenuates Sn waves (Barazangi et al., 2006). The Pn velocities are also low (−7.6 km/Sn; Türkelli et al., 2003; Sandvol et al., 2003a,b; Zor et al., 2003; Gök et al., 2003, 2007; Skobeltsyn et al., 2014; Angus et al., 2015; Barazangi et al., 2006; Özaçar et al., 2008; Göğüş & Pysklywec, 2008; Biryol et al., 2011; Zor, 2008; Elitok & Dolmaz, 2008; Bartol et al., 2012; Warren et al., 2013; Tezel et al., 2013; Cırmık, 2018). The heat flow is high (Tezcan, 1987; Tezcan & Turgay, 1989; Dolmaz et al., 2005; Ateş et al., 2005; İlkışık, 1995; Bektaş et al., 2007; Bektaş, 2013; Pamukçu et al., 2014) and the Curie point depth is around 12–13 km below surface (Ateş et al., 2005; Bektaş, 2013; Pamukçu et al., 2014; Cırmık, 2018).

Based on data from the East Anatolian seismic experiment, Barazangi et al. (2006) proposed that the East Anatolian crust floats on a partly molten asthenosphere. Almost the whole thickness of the mantle lithosphere was removed from the overlying crust (Piromallo & Regard, 2006; Barazangi et al., 2006; Gök et al., 2007; Lei & Zhao, 2007; Medved et al., 2021). The following mechanisms were proposed to explain the obliteration of the lithospheric upper mantle.

1. The delamination of upper mantle (Al-Lazki et al., 2003; Göğüş & Pysklywec, 2008; Biryol et al., 2011; Bartol et al., 2012)
2. The steepening and break-off of the subducting slab (Davies & Blackenburg, 1995; Piromallo & Morelli, 2003; Piromallo & Regard, 2006; Facenna et al., 2006; Lei & Zhao, 2007; Gans et al., 2009; Dilek & Sandvol, 2009; Govers & Fithcher, 2016)
3. Combination of both mechanisms that are listed above (Mahatsente et al., 2018)

The space created by the removal of the mantle lithosphere was filled with a hot, upwelling asthenosphere or asthenospheric wedge (Şengör et al., 2003; Piromallo & Regard 2006), which caused adiabatic decompression. It is the hot mantle considered responsible for the high heat flow, the consequent regional uplift, and coeval volcanism (Pearce et al., 1990; Keskin, 2003; Piromallo & Regard, 2006; Lei & Zhao, 2007; Şengör et al., 2008. Yılmaz, 2017, 2019),

8.3. DISCUSSION

In the following paragraphs, the two contrasting views proposed on the nature of the Eastern Anatolian crust are discussed based on geological, geophysical, and geochemical data:

1. An ophiolitic mélange-accretionary complex forms the basement, which was developed during the demise of the NeoTethyan Ocean. Later, it was trapped between the approaching continents (Fig. 8.6a) (Şengör & Kidd 1979; Şengör et al., 2008; Yılmaz, 2021).
2. The entire Eastern Anatolia is underlain by an old and thick continental crust (Fig. 8.6 b) (Topuz et al., 2017).

Both views listed above were based on inconclusive evidence. The following critical data set provides verifications for the two hypotheses.

1. Detailed geological data to examine the applicability of the view on the regional scale, including comparative data to correlate the Eastern Anatolian geology with the neighboring tectonic belts (i.e., with the Southeastern Anatolian Orogenic belt to the south and the Pontide belt to the north to provide evidence of how each hypothesis fits with the geology and tectonic evolution of the neighboring regions in a time-space reference.
2. Geophysical data, particularly seismic data, to enlighten the nature of the basement

The hypothesis proposed by Topuz et al. (2017) is based on the observations basically from two metamorphic rock inliers in Eastern Anatolia (Fig. 8.2) and the following three statements:

1. *Metagranite obtained from the metamorphic rock outcrops have a Late Ordovician–Early Silurian protolith and were therefore connected with the Menderes Massif of western Anatolia, which displays a similar protolith.*

The old metamorphic minerals obtained from Eastern Anatolia's metamorphic inliers are expected to be similar to those obtained from the Menderes Massif because the metamorphic massifs of Western, Central, and Southeast Anatolian orogenic belts share the Pan African–Gondwanan origin (Şengör & Yılmaz, 1981).

2. *Multiple continental fragments separated by oceanic accretionary complexes are absent in Eastern Anatolia.*

This interpretation dismisses the essential characteristics of the Anatolian orogen, including Eastern Turkey where the orogen was developed by accretions of small continental fragments following the closure of the separating oceans (Şengör & Yılmaz, 1981; Yılmaz & Şengör, 1985; Dercourt et al., 1986; Dilek, 2006; Robertson et al., 2012; Barrier et al., 2018; Yılmaz, 2021; McNab et al., 2018). Regional heating generated during the plate reorganization is stated to have caused major obduction of these ophiolites (Hassig et al., 2013, 2016a,b; Rolland et al., 2020).

3. *The metamorphic mineral assemblages that crop out in the metamorphic rocks of the region display medium to high T metamorphisms, but not high-pressure metamorphism.*

This interpretation disregards the critical geological data obtained from the neighboring orogenic belts. Particularly, the recent studies on the Southeast Anatolian metamorphic massifs indicate that the ophiolitic rocks and the Bitlis Massif underwent an initial HP metamorphism,

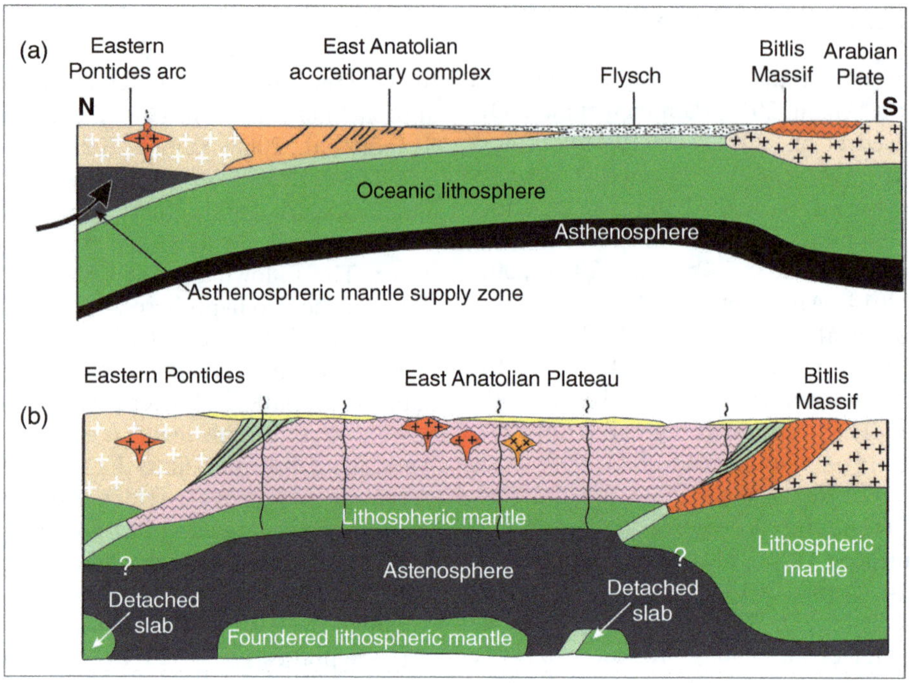

Figure 8.6 Two alternative geological models proposed for the nature of Eastern Anatolia's basement (after Topuz et al., 2017): (a) An ophiolitic mélange-accretionary complex forms the basement. (b) An old metamorphic basement underlies eastern Anatolia.

followed by an HT metamorphic phase during the Late Cretaceous–Early Eocene period (Rolland et al., 2012; Oberhansli et al., 2012, 2014; Pourteau et al., 2013; Awalt & Whitney, 2018; Yılmaz, 2019). The Bitlis Massif, a fragment of continental crust, was involved in northward subduction, together with the oceanic lithosphere (the Berit metaophiolite; Yılmaz, 2019), under the Eastern Anatolia where they were buried down to 35–65 km at depths (Oberhansli et al., 2014). Their exhumation occurred mainly during the Early Eocene period (Rolland et al., 2012; Oberhansli et al., 2014; Yılmaz, 2019). A transgressive sequence was deposited above the elevated metamorphic massifs during the Middle Eocene (Yiğitbaş, 1989; Yiğitbaş & Yılmaz, 1996; Yılmaz, 1993; Yılmaz, 2019). The metamorphic massifs were then thrust over the ophiolitic mélange during the Late Eocene before they were tectonically emplaced as a giant nappe package onto the Arabian Plate in the Late Miocene (Yılmaz, 2019). Consequently, the Southeast Anatolian metamorphic massifs represent allochthonous belts (Fig. 8.7b) resting tectonically on the Late Cretaceous–Middle Eocene metamorphic ophiolites (the Kızılkaya metamorphics and the Berit metaophiolite), and nonmetamorphic ophiolites (the Göksun Ophiolite) and an accretionary complex (Fig. 8.7b) extending from the Eastern Anatolia (Fig. 8.7a). Tens of kilometers of the nappe transport of the metamorphic massifs may be estimated from the Muş region in the north to the Sason region in the south (Fig. 8.7a,b; Yılmaz, 2019, and the references therein).

8.3.1. Geological Data

Based on limited geological data, Topuz et al. (2017) state that the metamorphic rocks and the overlying ophiolitic layer of Eastern Anatolia were exhumed during the Late Cretaceous, above which the sediments of Maastrichtian transgression were deposited. This view necessitates two highly improbable consequences.

1. The continental crust must have remained buried under the thick, dense ophiolitic layer for more than 60–70 Ma, which seems physically highly unlikely and does not fit with tectonic behavior of the Alpine-Himalayan Mountain ranges.
2. The stratigraphic data documented in Fig. 8.3 display that the wholesale elevation of the Eastern Anatolia above the sea level occurred during the Late Eocene–Oligocene. The bulk of Eastern Anatolia became emergent in Neogene (Şengör & Yılmaz, 1981; Bedi & Yusufoğlu, 2018; Okay et al., 2020; McNab et al., 2018).

Moreover, geological data from the neighboring orogenic belts, the Pontide, and the southeastern orogens reveal that the oceanic environment survived in these regions until the Middle Eocene (Dilek & Sandvol, 2009; Hassig et al., 2013; Nikishin et al., 2015; Sosson et al., 2017; Meijers et al., 2017; Yılmaz, 2019).

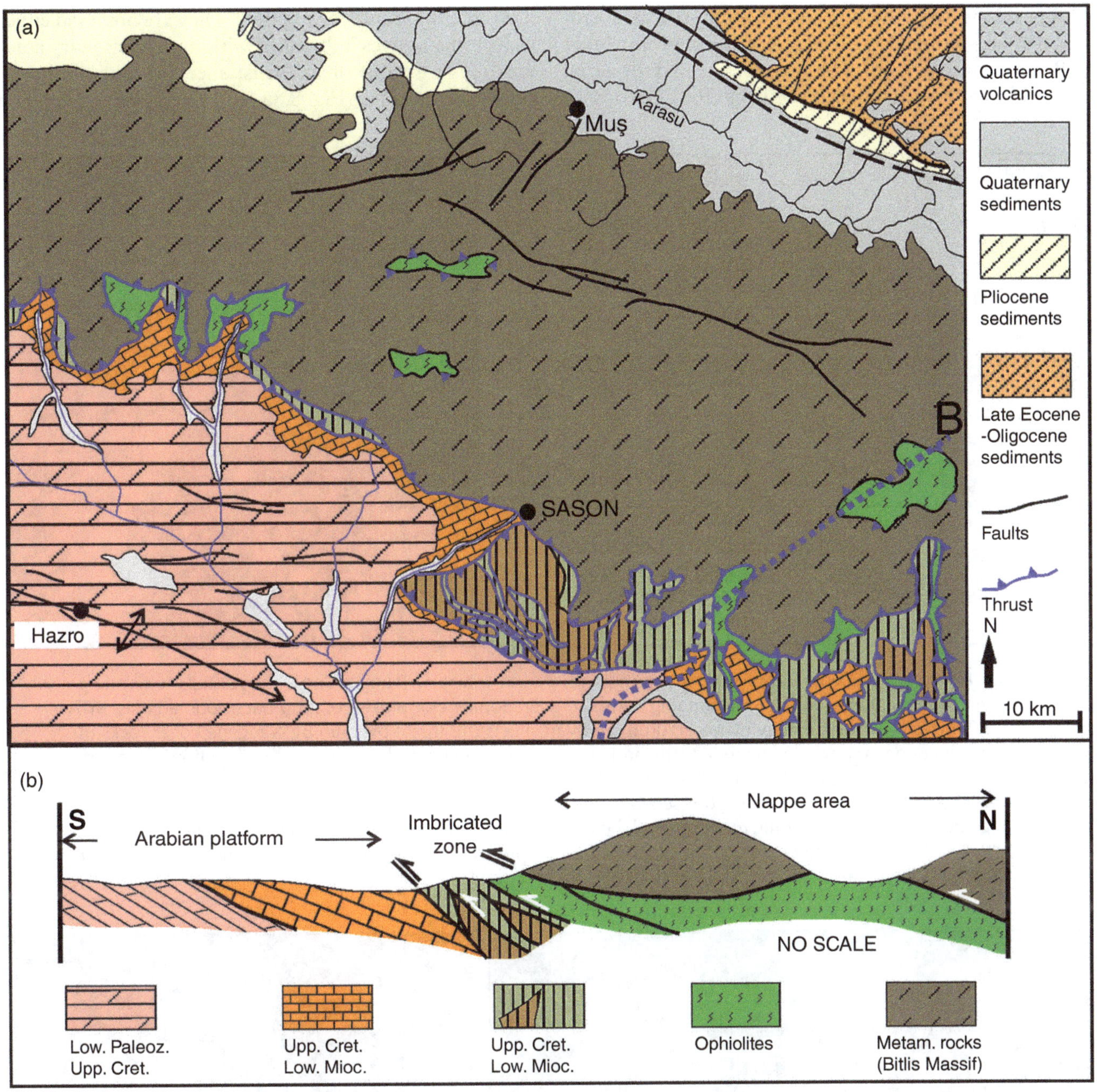

Figure 8.7 (a) The geology map of the Southeast Anatolian Orogenic Belt (SAOB; modified after the MTA 1/500 000 scale geology map of the Erzurum sheet). (b) Geological cross section across the SAOB along the direction indicated by the A-B broken line in the map. The cross section shows metamorphic massifs as thrust sheets above the ophiolitic mélanges (modified after Yılmaz, 1993).

Elevation of fragments of the accretionary complex, reaching somewhere above the sea during the Late Cretaceous–Early Eocene period, does not rule out the ocean's existence in the region, similar to many growing accretionary prisms of the world today, such as the Aegean–Eastern Mediterranean region (Robertson et al., 1996).

The hypothesis by Topuz et al. (2017) is based essentially on the presence of the metamorphic inliers surrounded by the accretionary complex located between the towns of Tekman, Hınıs, Karayazı, and Aktuzla (Figs. 8.2, 8.8a). They argue that these metamorphic assemblages represent fragments of the elevated old metamorphic basement. The following evidence questions validity of this hypothesis. Figure 8.8b illustrates a geological cross section along the largest metamorphic inlier (A,B section in Fig. 8.8b). The cross section and the aerial photos in Figs. 8.8c and 8.9a,b show that the outcrops do not represent intact rock masses. They are observed either as tectonic wedges imbricated with the accretionary complex (Figs. 8.9a,b) or as thrust sheets resting above the ophiolitic mélange (Fig. 8.8a,b,c). Furthermore, the data

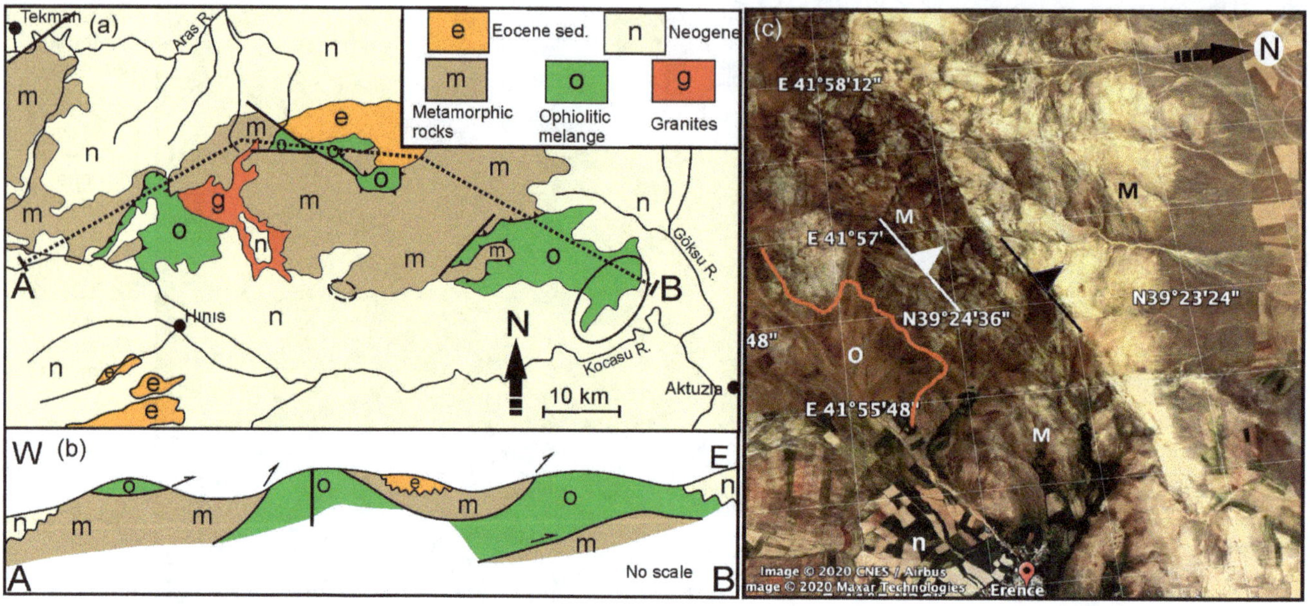

Figure 8.8 (a) Simplified geology map of the region between towns of Tekman-Varto and Aktuzla where large metamorphic inliers (m) and the surrounding ophiolite-accretionary complexes (o) crop out (modified after MTA 1/500 000 scale Turkish geology maps, Erzurum and Van sheets). Both are overlain by Neogene cover rocks (n). The map location is shown in Figure 8.2. The broken line connecting A and B indicates the direction of the cross section in (b). The small and big ellipses show the location of the aerial photos in Figures 8.8c and 8.9a, respectively. (b) A schematic geological cross section along the direction of A and B in (a) showing the structural arrangement of the metamorphic rocks (m) and the accretionary complex (o). (c) Aerial photo showing the metamorphic rocks tectonically overlying the accretionary complex; north of the Erence village, northwest of Varto. The small ellipse in (a) marks the location of (b).

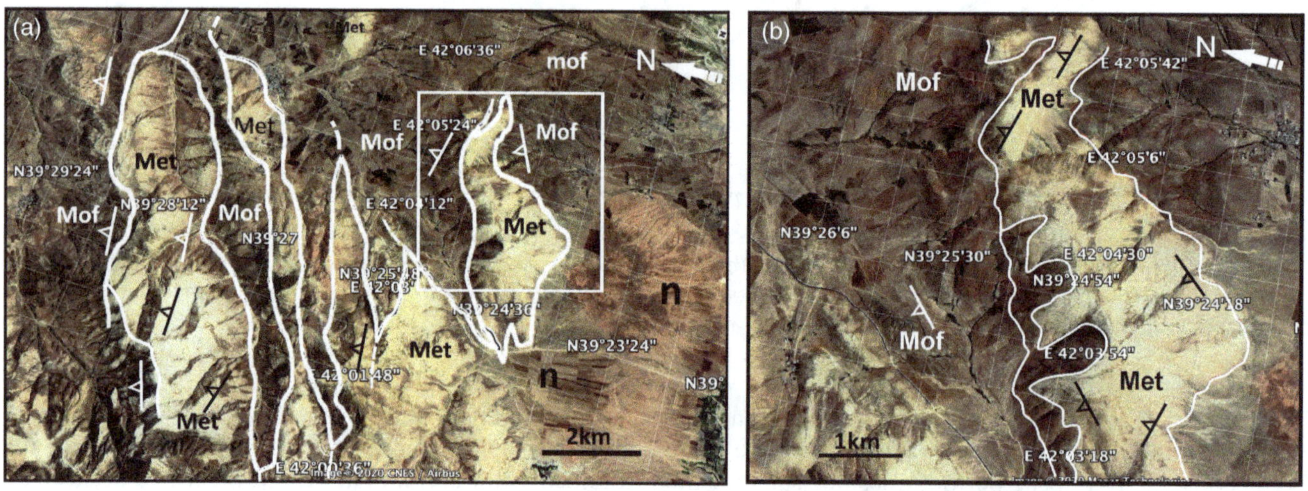

Figure 8.9 (a) Aerial photo of the area marked with the big ellipse in Figure 8.8a, the northwest of Aktuzla shows tectonically intermixed metamorphic rocks (Met) and the ophiolitic mélange (Mof). The white rectangle indicates the location of the aerial photo in (b); n = the Neogene cover rocks. (b) Aerial photo showing a slice of metamorphic rocks (Met) tectonically sandwiched between the ophiolitic mélange (Mof) units.

summarized below do not favor the presence of an old metamorphic basement under the ophiolitic mélange.

1. The field evidence from the northern (Çayırlı-Tercan-Aşkale-Pasinler basins), central (Tekman-Hınıs-Malazgirt-Tekman basins), and southern (Muş-Bingöl-Gevaş basins) (Fig. 8.2) regions of the East Anatolia, particularly from the basin boundaries where the base rocks are exposed, reveal that there is almost always an ophiolitic mélange under the Neogene cover (Figs. 8.3 and 8.5c).

2. All the wells drilled by the oil and gas industry in the region cutting through the Neogene succession penetrated the ophiolitic mélange (unpublished TPAO data).

8.3.2. Geophysical Data

The distributions of seismic b-values from the entire Eastern Anatolia and the major high-wave number magnetic anomaly values favor high-density rocks under the Neogene successions matching the density of an ophiolite association (Bektaş et al., 2007; Bektaş, 2013; Mahatsente et al., 2018). The seismic waves down to 5–6 km at depths display a relatively uniform pattern (Fig. 8.10a,b) without abrupt vertical changes to infer a lower density layer under the ophiolitic rocks.

Maden and Öztürk (2015), examining rich data derived from the entire Eastern Anatolia and the surrounding areas, concluded that the Eastern Anatolian crust is thick, where the large negative gravity anomalies and low b-values are observed.

Fowler (2004) calculated a Bouguer gravity anomaly of −300 mgal in the case of 100% isostatic equilibrium (roc = 2.85 gr/cm^3, rum = 3.3 gr/cm^3) in a region with a topographic height of 1 km. The Eastern Anatolia region has an average topography of 2 km. Therefore, it would be expected to yield gravity values lower than −300 mgal. However, the values observed in Eastern Anatolia range between −100 and −200 mgal. This may be explained by the rapid uplift of the Eastern Anatolian region due to delamination of the mantle lithosphere (see the discussion in the following paragraphs), insufficient immersion of the root of the lithosphere in the mantle, and the consequent isostatic disequilibrium.

8.3.3. Geochemical Data From the Neogene Volcanic Rocks

The Eastern Anatolian lavas' modal and normative compositions show a broad spectrum from nepheline

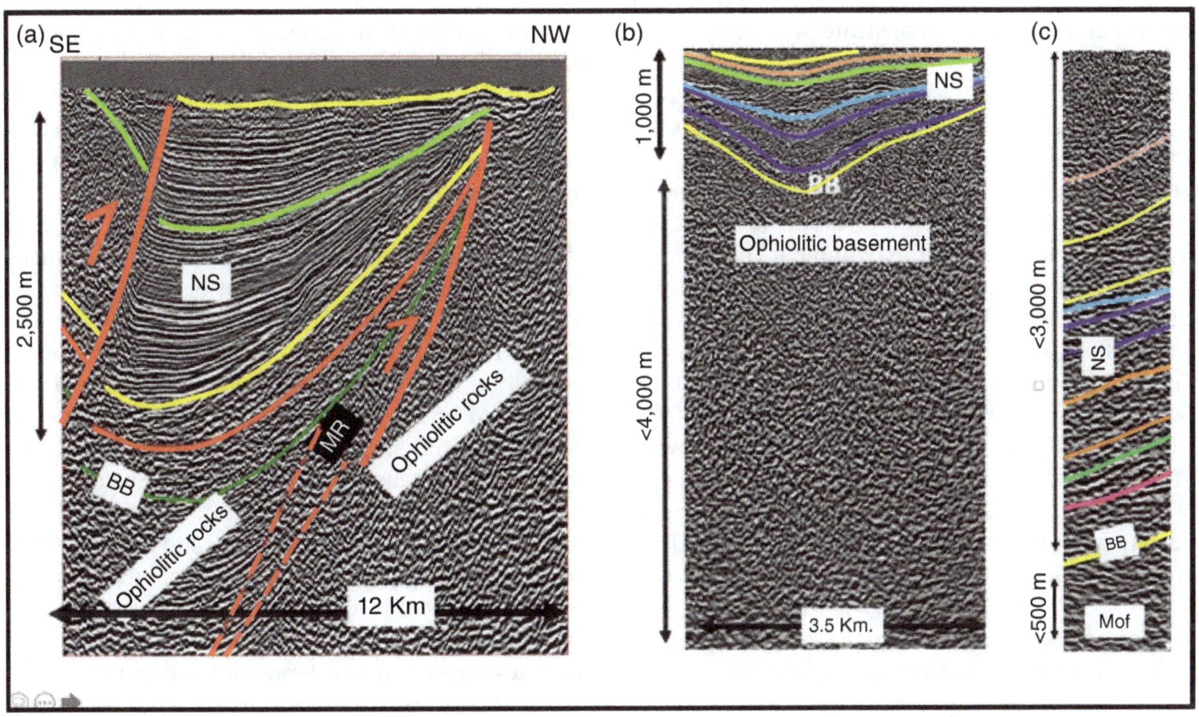

Figure 8.10 Seismic profiles across the young basins of Eastern Anatolia from northern, southern, and central regions showing the ophiolitic rocks under the Neogene succession (seismic sections provided by Ö. Şahintürk). (a) From the Hınıs Basin, central Eastern Anatolia, located between 7 and 8 in Figure 8.2. (b) From the southern central region of Eastern Anatolia to the west of no. 8 in Figure 8.2. (c) From the Pasinler basin, the northern region of Eastern Anatolia, 4 in Figure 8.2. Note: NS = Neogene sequence; BB = Basement boundary; MR = Metamorphic rocks; Mof = Ophiolitic rocks.

normative basalts to quartz normative felsic lavas (Yılmaz et al., 1987a.b, 1998; Pearce et al., 1990; Keskin, 2003). The southern volcanic centers produced large quantities of basic and intermediate lavas. The northern volcanoes extruded voluminous intermediate volcanic edifices in which pyroclastic rocks dominate, particularly in the northeast of the Erzurum city (Fig. 8.2).

Based on the geochemistry of the Eastern Anatolian Neogene lavas, the previous studies (e.g., Yılmaz et al., 1987a,b, 1988; Yılmaz, 1990; Pearce et al., 1990; Notsu et al., 1995; Keskin et al., 1998, 2006, 2012; Lustrino et al., 2012; Özdemir & Güleç, 2014; Oyan et al., 2016a, 2016b; Lebedev et al., 2016a,b; Kaygusuz et al., 2018; Açlan et al., 2020; Üner, 2021) reached the following conclusions on the magma compositions.

1. The magmas were derived from the heterogeneous mantle source and retained their compositions during the transit with minor modifications.

2. Magma chemistry of the alkaline, mildly alkaline, and calc-alkaline lavas displays distinct enrichment by the subduction components (enriched in LILE and LREE and depleted in HFSE; e.g., Pearce et al., 1990; Jean et al., 2010; Lebedev et al., 2018a,b; Kaygusuz et al., 2018; Üner, 2021, and the references therein) and also boninitic enrichment (Üner, 2021). In accordance with this conclusion, the seismic images obtained by various analytical methods demonstrate a remnant subducted oceanic lithosphere that is broken into pieces underneath Eastern Anatolia (Piromallo & Morelli, 2003; Piromallo & Regard, 2006; Facenna et al., 2006; Gans et al., 2009; Warren et al., 2013). (For the discussion on the time, place, and role of the subducting events in the development of the surrounding orogenic belts, see chapter 7 by Yılmaz et al. in this volume.) The enriched mantle source generated a wide compositional range in the consecutive magma batches.

3. The Sr 86/ Sr 87 ratios of lavas of the Eastern Anatolian volcanic provinces are on an average of 0.704, which varies in a narrow range commonly between 0.703 and 0.705. The Nd 143 Nd/144 Nd ratio ranges between 0.5127 and 0.5128. The isotope values plot in the mantle array (Kaygusuz et al., 2018). The isotope ratios together with the magma composition compare favorably with the OIB lavas (Üner, 2021; Yang et al., 2019; McNabb et al., 2018).

4. The petrochemical characteristics indicate that the Eastern Anatolian magmas were derived from an ophiolitic host and underwent significant fractionation in the magma chamber and during the transit.

5. Upper crustal contribution into the magmas is negligible (Kaygusuz et al., 2018; Üner, 2021).The petrochemical data derived from the Eastern Anatolian magmas may be used to test the validity of the hypothesis that assumes the presence of a thick continental crust under Eastern Anatolia. Substantial chemical modifications would be expected in the magma composition if they passed through the thick and hot continental crust. These may be listed as follow: (a) higher Sr and Nd isotope values, (b) increasing amount of continental crustal components, (c) large volumes of magmas of continental crust origin (granitic magmas), (d) mixing and mingling of magmas of diverse compositions.

The petrochemical properties of the lavas from the entire Eastern Anatolian region show boninitic or metasomatic enrichment but do not confirm these alternatives and thus do not support presence of a thick continental crust.

8.4. CONCLUDING SUMMARY

Geological field observations and the geophysical data suggest the presence of an ophiolitic mélange-accretionary complex under the cover sequence of Eastern Anatolia (A in Fig. 8.11-I). The Eastern Anatolian volcanic and sedimentary cover units were piled up during the closure of the NeoTethyan Ocean that was located between the Pontide arc to the north, and the continental slivers drifted away from the Arabian Plate to the south (e.g., Şengör & Yılmaz, 1981). The Upper Cretaceous–Lower-Middle Eocene deep-sea sedimentary rocks, associated genetically with the ophiolitic mélange indicate that the NeoTethyan oceanic lithosphere survived during this period and was finally eliminated from entire Eastern Turkey by the Late Eocene. Continental fragments of various sizes were tectonically incorporated into the mélange prism, possibly during the growth of the accretionary complex.

The northward advance of the Arabian Plate continued after the elimination of the oceanic environment. The mélange-accretionary prism that occupied a large terrain behaved like a wide and thick buffer unit, which did not allow a head-on collision of the bordering continents (cf., the Turkic type orogen of Şengör & Natalin, 1996). The northward advance of the Arabian Plate that continued after the initial stage of the collision caused shortening deformation. It began squeezing the Eastern Anatolian accretionary complex. As a result, Eastern Anatolia was elevated above the sea during the Late Eocene–Oligocene. An irregular topography was developed on the elevated land as indicated by coarse-grained thick Upper Eocene–Oligocene terrestrial conglomerates deposited in irregularly developed narrow depressions; B in Fig. 8.11-I). The rough topography was smoothed, and the region subsided steadily during the Late Oligocene–Early Miocene when an epeiric sea invaded the region again (Yılmaz, 2017, and the references therein). This is evidenced by the low-energy marine sediments laid down on the Upper Eocene–Oligocene terrestrial sedimentary rocks. The smooth topography survived during the Early-Middle Miocene.

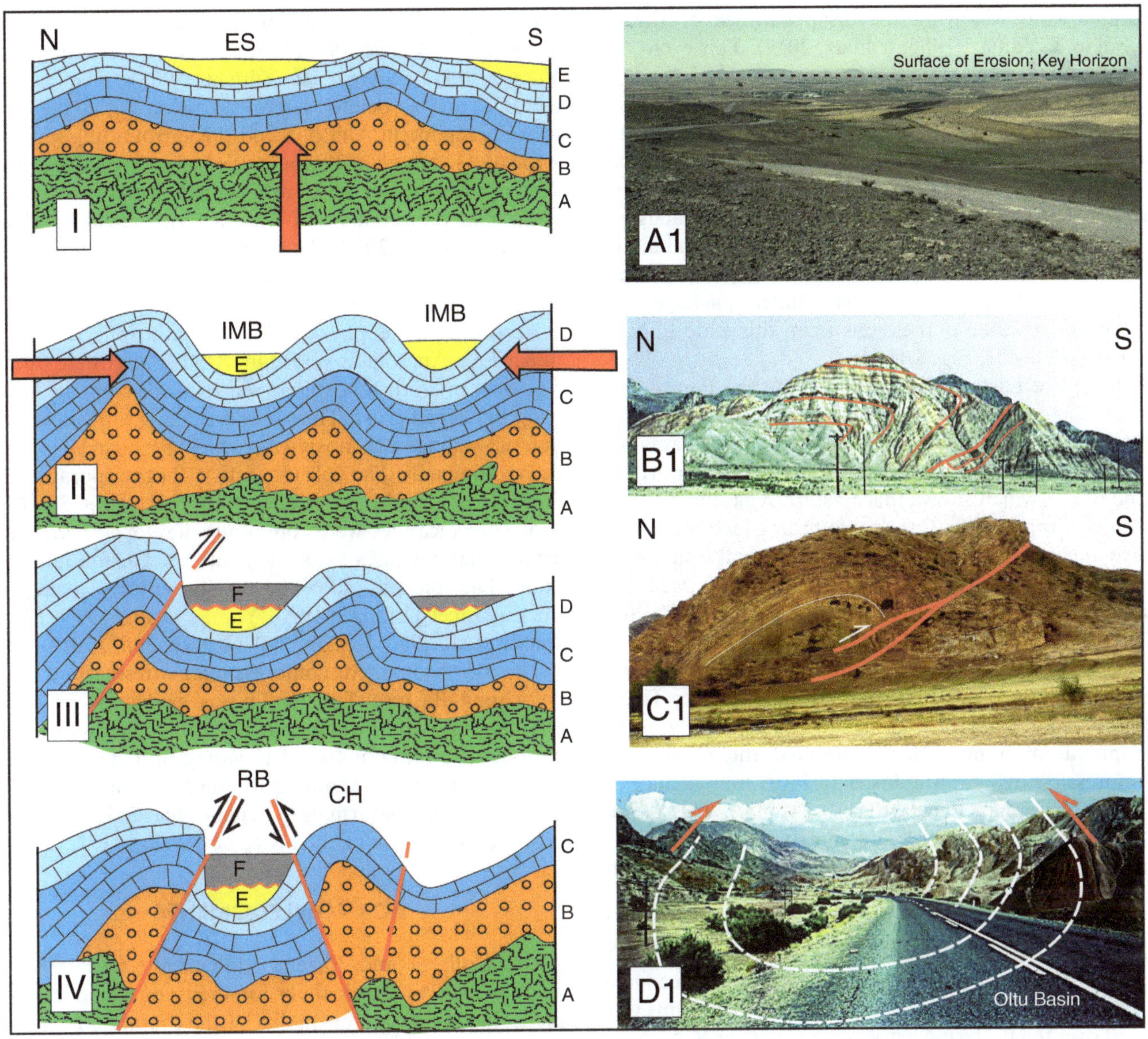

Figure 8.11 Schematic geological cross sections illustrating consecutive stages of tectonic development of Eastern Anatolia. Legend for I to IV: A, Ophiolitic mélange-accretionary complex; B, Upper Eocene–Oligocene terrestrial coarse-grained sedimentary rocks; C, Lower-Middle Miocene shallow marine limestone; D, Upper Miocene–Lower Pliocene lacustrine limestone; E (in II, III, and IV), Upper Pliocene terrestrial sediments; F (in III and IV), Pleistocene-Holocene fluvial conglomerates. I shows Eastern Anatolia during the Late Oligocene–Early Miocene. Following the denudation phase, a regionwide flat-lying erosional surface (ES) was developed (photo A1). II shows the continuing north-south compression that caused large-scale folds (photo B1). III shows tight folds and thrusts that were developed in the following periods (photo C1). IV shows big-scale tight folds accompanied by thrusts from the opposite directions that caused the development of ramp basins (photo D1) during the more advanced stage of the compressional stress. Note: ES = Erosional surface; IMB = Intermountain basins; AB = Asymmetrical basins; RB = Ramp basin; CH = Central High. Photos A1 to C1 are folds and thrusts from different Eastern Anatolian regions showing the deformed Neogene sequence under the north-south compressional stress. A1 = A northerly view from the southwest of Erzurum; B1 and C1= Kağızman-Tuzluca areas; D1= Oltu-Olur city road.

Shallow marine limestones (the Adilcevaz Limestone) were deposited above the fine-grained marine sediments (C in Fig. 8.11-I) (Şaroğlu & Yılmaz, 1986; Bedi & Yusufoğlu, 2018). The limestones graded upward into evaporates and lacustrine limestones during the Late Miocene (Fig. 8.3 and D in Fig. 8.11-I). The gradual transition observed in the entire Eastern Anatolian region reveals that interconnected lakes were developed over the elevated land following the disappearance of the sea (Şaroğlu & Yılmaz, 1986; Yılmaz, 2017). This event may also be interpreted that the Eastern Anatolia began to rise as a coherent block (en mass) during the Late Miocene (Phase I, Fig. 8.11-I). Following the disappearance of the interconnected lakes, the elevated land underwent a severe denudation phase, which formed a flat-lying erosional surface above the Upper Miocene lacustrine limestones (ES in Fig. 8.11-I and the accompanying photo A1; Yılmaz, 2017). The smooth topography disappeared after this period. Various sediment packages were formed in separate depressions from this Late Pliocene onward (Phase II, Fig. 8.11-II).

The entire Eastern Turkey, including the Pontide and the Arabian Platform, has behaved as an interconnected tectonic entity since their tectonic amalgamation. Paleomagnetic studies and the GPS data support this conclusion, which shows that Eastern Anatolia has been deformed together with the surrounding tectonic entities since the Late Miocene following the collision of the Arabian plate with the Anatolian blocks (Reilenger et al., 2006; Şengör et al., 2008; Çinku et al., 2014, 2016; Gürer et al., 2017; Bakkal et al., 2019).

The continuing north-south compressional stress caused a complex pattern of structures in Eastern Turkey (Fig. 8.4a). The rigid continental crust underlying the peripheral mountains accommodated the north-south compression by elevating faster (0.2–0.3 mm/y) than the Eastern Anatolian Plateau (0.1–0.2 mm/y). This is possibly because the blocks and matrix of the ophiolitic mélange underlying Eastern Anatolia partly absorbed the compression.

Starting from the Late Pliocene–Pleistocene, big scale folds and thrusts began to form in the Eastern Anatolia (phases III and IV; Fig. 8.11-III, IV) and the accompanying photos C1 and D1). The peripheral mountains were thrust over the Eastern Anatolian Plateau (Fig. 8.2; 8.5a,c). Two narrow, fault-bound chains of east-west trending depressions were formed along the thrust fronts as intermountain or ramp basins (Figs. 8.1 and 8.5a,b,c). The boundary faults give the young basins their distinct rhombohedral or parallelogram geometrical patterns (Fig. 8.5b).

When the north-south compression and associated shortening reached an excessive stage, which could no longer be accommodated within the volume of Eastern Anatolia, the stress permutation occurred. This led to the development of two transform faults, the North Anatolian Transform Fault (NATF) and the East Anatolian Transform Fault (EATF) (Figs. 8.1, 8.2; Şengör, 1979; Şengör & Kidd, 1979; Şengör & Yılmaz, 1981; Çemen et al., 1993; Yılmaz, 2017). They defined an independent tectonic entity, the Anatolian Plate, which began escaping away from the area of convergence to transfer part of the north-south compressional stress to the west (McKenzie, 1972, 1978; Şengör, 1979; Şaroğlu & Yılmaz, 1991; Şengör & Yılmaz, 1981). The escape tectonics and associated lateral extrusion initiated a new tectonic regime in Anatolia and the surrounding regions known as the NeoTectonics, which determined the development of the major morphotectonic entities in the peripheral mountains and the Eastern Anatolian High Plateau (Yılmaz, 2017). This event also caused anticlockwise rotations of the semi-independent fault-bounded blocks of Central Anatolia (Yılmaz, 2017).

According to geophysical data, the lithospheric mantle under East Anatolia is very thin. Almost the whole thickness of the mantle lithosphere was removed from the overlying thickened crust. The space created was filled with a hot, upwelling asthenosphere, which produced mantle-derived magmas. The volcanic activity began sporadically during the Late Miocene and intensified about 5–6 Ma ago. The volcanoes were commonly developed above the extensional openings associated with the basin boundary faults. The volcanic edifices covered the entire plateau as a thick blanket (Yılmaz, 1990; Yılmaz et al., 1987, 1998; Pearce et al., 1990; Keskin, 2007; Keskin et al., 2012).

The north-directed compressional stress is actively deforming Eastern Turkey. This is evidenced by GPS measurements indicating that the high plateau and the peripheral mountains are still elevating, and the Anatolian Plate's westward escape is continuing at about 20 mm/y. This continuing deformation may be regarded as the late-post-tectonic phase of the orogenic development.

ACKNOWLEDGMENTS

Yan Rolland and Rezene Mahatsente reviewed the manuscript. We thank them for their constructive criticisms that improved the quality of the chapter. Many colleagues were most helpful in allowing us to use some of the seismic data and geological maps. Among those, we particularly thank Ö. Şahintürk for his close collaboration. Onur Tunç helped enormously during the preparation of the figures and manuscript, and we greatly appreciate his cooperation.

REFERENCES

Açlan, M., Oyan, V., & Köse, O (2020). Petrogenesis and evolution of Pliocene Timar basalts in the east of Lake Van, Eastern Anatolia, Turkey; A consequence of a metasomatized spinel-rich lithospheric mantle source. *Journal of African Earth Science*, *168*, 103844.

Akay, E. Erkan, E., & Ünay, E. (1989). Stratigraphy of the Muş Tertiary Basin. *Bulletin of the Mineral Research and Exploration*, *109*, 59–76.

Akkuş, M. F. (1970). Darende Balaban Havzasındaki (Malatya, ESE Anadolu) stratigrafik birimler ve jipsli formasyonların yaşı hakkında yeni bilgiler. *M.T.A. Dergisi*, *75*, 1–15.

Akkuş, M. F. (1971). Darende-Balaban Havzasındaki (Malatya, ESE Anadolu) jeolojik ve stratigrafik incelenmesi. *M.T.A. Dergisi*, *76*, 1–61.

Aktuğ, B., Dikmen, Ü., Doğru, A., & Özener, H. (2012). Seismicity and strain accumulation around Karlıova Triple Junction (Turkey). *Journal of Geodynamics*, *67*. https://doi.org/10.1016/j.jog.2012.04.008

Al-Lazki, A. D., Seber, E., Sandvol, N., Türkelli, R., Mohamad, & Barazangi, M. (2003). Pn velocity and anisotropy structure beneath the Anatolian plateau (eastern Turkey) and the surrounding regions. *Geophysical Research Letters*, *30*(24), 8043. https://doi.org/10.1029/2003GL017391

Altınlı, E. (1966a). Geology of Eastern and Southeastern Anatolia. Part I. *M.T.A. Bulletin*, *66*, 35–76.

Altınlı, E. (1966b). Geology of Eastern and Southeastern Anatolia, Part II. *M.T.A. Bulletin*, *67*, 1–23.

Angus, D. A., Wilson, E., Sandvol, N., & Ni, J. F. (2015). Lithospheric structure of the Arabian and Eurasian collision zone in eastern Turkey from S-wave receiver functions. *Geophysical Journal International*, *166*, 1335–1346. https://doi.org/10.1111/j.1365-246X. 2006. 03070.x

Ateş, A., Bilim, F., & Büyüksaraç, A. (2005). Curie point depth investigation of Central Anatolia, Turkey. *Pure and Applied Geophysics*, *162*, 357–371.

Awalt, M. B., & Whitney, D. (2018). Petrogenesis of kyanite- and corundum-bearing mafic granulite in a meta-ophiolite, SE Turkey. *Journal of Metamorphic Geology*, *36*(7), 881–904. https://doi.org/10.1111/jmg.12317

Bakkal, B., Çinku, M. C., & Heller, F. (2019). Paleomagnetic results along the Bitlis-Zagros suture zone in SE Anatolia, Turkey: Implications for activation of the Dead Sea fault Zone. *Journal of Asian Earth Science*, *172*, 14–29. https://doi.org/10.1016/j.jseaes.2018.08.026

Barazangi, M. (1989). Continental collision zones: Seismotectonics and crustal structure. In D. James (Ed.), *Encyclopedia of solid Earth geophysics* (pp. 58–75). New York: Van Nostrand Reinhold Company.

Barazangi, M., Sandvol, E., & Seber, D. (2006). Structure and tectonic evolution of the Anatolian plateau in eastern Turkey. In Y. Dilek & S. Pavlides (Eds.), *Post collisional tectonics and magmatism in the Mediterranean region and Asia* (pp. 463–473). Geological Society of America Special Papers, 409.

Barrier, E., Vrielynck, B., Brouillet, J. F., Brunet, M. F., Angiolini, L., Kaveh, F., et al. (2018). *Paleotectonic reconstruction of the Central Tethyan Realm*. Tectono-Sedimentary-Palinspastic maps from Late Permian to Pliocene. CCGM/CGMW, Paris, atlas of 20 maps (scale: 1/15 000 000). http://www.ccgm.org

Bartol, J., & Govers, R. A. (2014). Single cause for uplift of the Central and Eastern Anatolian plateau? *Tectonophysics*, *637*, 116–136. https://doi.org/10.1016/j.tecto.2014.10.002

Bartol, J., Govers, R., & Wortel, M. J. R. (2012). Mantle delamination as the cause for the Miocene-Recent evolution of the Central and Eastern Anatolian Plateau. *Geophysical Research Abstracts*, *14*, 2012–11778. EGU General Assembly 2012.

Bedi, Y., & Yusufoğlu, H. (2018). 1/1000 000 ölçekli Türkiye Jeoloji Haritaları Serisi Malatya L40 paftası no 261. Maden Tetkik ve Arama Genel Müdürlüğü Jeoloji Etütleri Dairesi Ankara.

Bedi, Y., Yusufoğlu, H., Usta, D., Özkan, M. K., Beyazpirinç, M., Baran, C., et al. (2017). Doğu Torosların Jeodinamik Evrimi (Elbistan-Malatya-Dolayı) Maden Tetkik ve Arama Genel Müdürlüğü Cilt I, Cilt II (554 s), Ankara.

Bektaş, O. (2013). Thermal structure of the crust in inner east Anatolia from aeromagnetic and gravity data. *Physics of the Earth and Planetary Interiors*, *221*, 27–37.

Bektaş, O., Ravat, D., Büyüksaraç, A., Bilim, F., & Ateş, A. (2007). Regional geothermal characterization of East Anatolia from aeromagnetic, heat flow and gravity data. *Pure and Applied Geophysics*,*164*, 975–998.

Biryol, C. B., Beck, S. L., Zandt, G., & Özaçar. A. A. (2011). Segmented African lithosphere beneath the Anatolian region inferred from teleseismic P-wave tomography. *Geophysical Journal International*, *184*(3), 1037–1057. doi:10.1111/j.1365-246X.2010.04910.x

Bozkurt, E. (2001). NeoTectonic of Turkey: A synthesis. *Geodinamica Acta*, *31*, 3–30.

Bozkuş, C. (1990). Stratigraphy (coal) of the north-eastern part of Oltu-Narman Tertiary basin. *Bulletin of the Geological Society of Turkey*, *33*, 47–56.

Çemen, I., Göncüoğlu, M. C., Erler, A., Kozlu, H., & Perinçek, D. (1993). Indentation tectonics and associated lateral extrusion in east, southeast and central Anatolia. *GSA Programs and Abstracts*, A116–A117.

Çınar, H. H., & Alkan, H. (2015). Crustal structure of Eastern Anatolia from single-station Rayleigh Wave Group Velocities. *Eastern Anatolian Journal of Science*, *I*(2), 57–69.

Çinku, M., Hisarlı, M., Yılmaz, Y., Özbey, Z., et al. (2016). The tectonic history of the Niğde-Kırşehir Massif and the Taurides since the late Mesozoic: Paleomagnetic evidence for two-phase orogenic curvature in Central Anatolia. *Tectonics*, *35*(3), 772–811. https://doi.org/10.1002/2015TC003956. https://doi.org/org/10.1002/2015TC003956

Çinku, M. C., Hisarlı, M., Keskin, M., Ustaömer, T., & Orbay, N. (2014). *Paleomagnetic evidence of the Neogene tectonic block rotations in Eastern Anatolia*. EUG General Assembly 16, 27 April 2014, Abs. Book, 0–16.

Cırmık, A. (2018). Examining the crustal structures of eastern Anatoliausing thermal gradient, heat flow, radiogenic heat production and seismic velocities (Vp and Vs) derived from Curie Point depth. Examining the crustal structures of eastern Anatolia. *Bulletin of Geophysics and Oceanography*, *59*, 117–134. https://doi.org/10.4430/bgta0230

Copley, A., & Jackson, J. (2006). Active tectonics of the Turkish-Iranian Plateau. *Tectonics*, *25*, 6006.

Davies, J. H., & von Blackenburg, F. (1995). Slab break off: A model of lithospheric detachment and its test in the magmatism and deformation of collisional orogens. *Earth and Planetary Science Letters*,*129*, 85–102.

Dercourt, J., Zonenshaın, L. P., Ricou, L. E., Kazmin, V. G., Lepichon, X., et al. (1986). Geologıcal evolutıon of the Tethys belt from the Atlantıc to the Pamirs since the Lias. *Tectonophysics*, *123*(1–4), 41–315.

Dewey J. F., Hempton, M. R., Kidd, W. S. F., Şaroğlu, F., & Şengör, A. M. C. (1986). Shortening of continental lithosphere: the neotectonics of Eastern Anatolia: A young collision zone. In M. P. Coward & A. C. Ries, *Collision tectonics* (pp. 3–36). Geological Society Special Publications 19.

Dilek, Y. (2006). Collision tectonics of the Mediterranean region: Causes and consequences. In Y. Dilek & Y. Pavlides (Eds.), *Post collisional tectonics and magmatism in the Mediterranean region and Asia* (pp. 1–13). Geological Society of America Special Paper, 409.

Dilek, Y., & Sandvol, E. (2009). Seismic structure, crustal architecture and tectonic evolution of the Anatolian-African plate boundary and Cenozoic orogenic belts in the eastern Mediterranean region. In J. R. Murphy et al. (Eds.), *Ancient orogens and modern analogs* (pp. 127–160). Geological Society of London, Special Publication, 327. https://doi.org/10.1144/SP327. 7.& 0305-8719/09/&15 00

Dolmaz, M. N., Hisarli, Z. M., Ustaömer, T., & Orbay, N. (2005). Curie-point depth variations to infer thermal structure of the crust at the African-Eurasian convergence zone, SW Turkey. *Earth Planets and Space*, *57*, 373–383.

Elitok, Ö., & Dolmaz, M. N. (2008). Mantle flow-induced crustal thinning in the area between the easternmost part of the Anatolian plate and the Arabian foreland (E. Turkey) deduced from the geological and geophysical data. *Gondwana Research*, *13*(3), 302–318.

Facenna, C., Bellier, O., Martinod, J., Piromallo, C., & Regard, V. (2006). Slab detachment beneath eastern Anatolia: A possible cause for the formation of the North Anatolian fault. *Earth and Planetary Science Letters*, *242*, 85–97.

Fowler, C. M. R. (2004). *The solid Earth: An introduction to global geophysics*. Cambridge: Cambridge University Press.

Gans, C. R., Beck, S. L., Zandt, G., Biryol, C. B., & Özaçar, A. A. (2009). Detecting the limit of slab break-off in central Turkey: New high-resolution Pn tomography results. *Geophysical Journal International*, *179*, 1566–1577.

Gedik, A. (1986). Tekman (Erzurum) havzasının jeolojisi ve petrol olanakları. *MTA Dergisi*, *103*/104, 1–24.

Göğüş, O. H., & Pysklywec, R. N. (2008). Mantle lithosphere delamination driving plateau uplift and synconvergent extension in eastern Anatolia. *Geology*, *36*(9), 723–726. doi:10.1130/G24982A.1

Gök, R., Pasyanos, M., & Zor, E. (2007). Lithospheric structure of the continent-continent collision zone: Eastern Turkey. *Geophysical Journal International*, *169*, 1079–1088.

Gök, R., Sandvol, E., Türkelli, N., Seber, D., & Barazangi, M. (2003). Sn attenuation in the Anatolian and Iranian plateau and surrounding regions. *Geophysical Research Letters*, *30*(24), 8042. doi:10.1029/2003GL018020

Govers, R., & Fichtner, A. (2016). Signature of slab fragmentation beneath Anatolia from full-waveform tomography. *Earth and Planetary Science Letters*, *450*, 10–19.

Gürer, D., van Hinsbergen, J. J. G. D., Özkaptan, M., Creton, I., et al. (2017). Paleomagnetic constraints on the timing and distribution of Cenozoic rotations in Central and Eastern Anatolia. *Solid Earth*, *9*, 295–322. https://doi.org/pangea.de/10.1594/se-9-295

Hässig, M., Duretz, T., Rolland, Y., & Sosson, M. (2016b). Obduction of old oceanic lithosphere due to reheating and plate reorganization: insights from numerical modeling and the NE Anatolia–Lesser Caucasus case example. *Journal of Geodynamics*, *96*, 35–49.

Hässig, M., Rolland, Y., Duretz, T., & Sosson, M. (2016a). Obduction triggered by regional heating during plate reorganization. *Terra Nova*, *28*(1), 76–82.

Hässig, M., Rolland, Y., Sosson, M., Galoyan, G., Sahakyan, L., Topuz, G., et al. (2013). Linking the NE Anatolian and Lesser Caucasus ophiolites: Evidence for large-scale obduction of oceanic crust and implications for the formation of the Lesser Caucasus-Pontides Arc. *Geodinamica Acta*, *26*(3–4), 311–330. https://doi.org/10.1080/09853111.2013.877236.

Helvacı, C. (2021). Geochemistry of Miocene evaporites from the Aşkale (Erzurum, eastern Turkey) area. *Bulletin MTA*, *164*, 1–45.

Hüsing, S. K., Zachariasse, W. J., van Hinsbergen, W., Krijgsman, M., İnceöz, M., et al. (2009). Oligocene-Miocene basin evolution in SE Anatolia, Turkey: Constrains on the closure of the eastern Tethys gateway. *Geological Society of London Special Publications*, *311*, 107–132. https://doi.org/10.1144/SP3114

İlkışık, O. M. (1995). Regional heat flow in Western Anatolia: Silica temperature estimates from thermal springs. *Tectonophysics*, *244*, 175–184.

Jean, M. M., Shervais, J. W., Choi, S. H., & Mukasa, S. B. (2010). Melt extraction and melt refertilizing in mantle peridotite of the Coast range ophiolite: An LA-ICP-MS study. *Contributions to Mineralogy and Petrology*, *159*, 113–136.

Karaoğlu, Ö., Selçuk, A. S., & Gudmundsson, A. (2017). Tectonic controls on the Karlıova triple junction (Turkey): Implications for tectonic inversion and the initiation of volcanism. *Tectonophysics*, *694*, 368–384.

Kaya, M. C. (2009). Benthic foraminiferal biostratigraphy of the tertiary sediments from the Elazig and Malatya Basins, Eastern Turkey. *Journal of the Geological Society of India*, *74*, 209–222.

Kaygusuz, A., Aslan, Z, Aydınçakır, E., Yücel, C., Güçer, A., & Şen, C. (2018). Geochemical and Sr-Nd-Pb isotope characteristics of the Miocene to Pliocene volcanic rocks from the Kandilli (Erzurum) area, Eastern Anatolia: Implications for magma evolution in extension-related origin. *Lithos*, *296*, 332–351.

Keskin, M. (2003). Magma generation by slab steepening and breakoff beneath a subduction accretion complex: An alternative model for collision-related volcanism in Eastern Anatolia, *Turkey. Geophysical Research Letters*, *30*(24), 8046. https://doi.org/10.1029/2003GL018019

Keskin, M. (2007). Eastern Anatolia: A hotspot in a collisional zone without a mantle plume. In G. R. Foulger & D. M. Jurdy (Eds.), *Plates, plumes, and planetary processes* (pp. 693–722). Geological Society of America Special Papers, 430.

Keskin, M., Chugaev, A. V., Lebedev, A., Sharkov, V., Oyan, V., & Kavak, O. (2012). The geochronology and origin of mantle sources for late Cenozoic intraplate volcanism in the frontal part of the Arabian Plate in the Karacadağ Neovolcanic area of Turkey. Part I, The result of isotope-geochronological studies. *Journal of Volcanology and Seismology*, *6*, 352–360.

Keskin, M., Pearce, J. A., & Mitchell, J. G. (1998). Volcano-stratigraphy and geochemistry of collision-related volcanism on the Erzurum-Kars Plateau, Northeastern Turkey. *Journal of Volcanology and Geothermal Research*, *85*(1–4), 355–404.

Keskin, M., Pearce, J. A., Kempton, P. D., & Greenwood, P. (2006). Magma-crust interactions and magma plumbing in a collision setting: Geochemical evidence from the Erzurum-Kars Plateau, Eastern Turkey. *Geological Society of America Special Paper*, *409*, 475–505. https://doi.org/10.1140/2006.2409(23)

Keskin, S., Pedoja, K., & Bektaş O. (2011). Coastal uplift along the eastern Black Sea coast: New marine terrace data from eastern Pontides, Trabzon (Turkey) and a review. *The Journal of Coastal Research*, *27*, 63–73.

Koçyiğit, A., Öztürk, A., İnan, S., & Gürsoy, H. (1985). Karasu havzasının (Erzurum) tektonomorfolojisi ve mekanik yorumu. *Cumhuriyet University Journal of Engineering Faculty Series A Earth Sciences*, *2*, 3–15.

Koçyiğit, A., Yılmaz, A., Adamia, S., & Kuloshvili, S. (2001). Neotectonics of East Anatolian Plateau (Turkey) and Lesser Caucasus: Implication for transition from thrusting to strike slip faulting. *Geodinamica Acta*, *14*, 177–195.

Konak, N., & Hakyemez, Y. (2008). *Geological map of Turkey in scale 1:100.000, Tortum H47 sheet (in Turkish)*. Ankara, Turkey: General +Directorate of Mineral Research and Exploration.

Kurtman, F., & Akkuş, M. F. (1971). Doğu Anadolu'daki ara basenler ve bunların petrol olanakları. *Bulletin of Mineral Research and Exploration*, *77*, 1–9.

Lebedev, V. A., Chernyshev, I. V., Sagatelyan, A. K., et al. (2018b). Miocene-Pliocene volcanism of Central Armenia: Geochronology and the role of AFC processes in magma petrogenesis. *Journal of Volcanology and Seismology*, *12*, 310–331. https://doi.org/10.1134/S0742046318050056

Lebedev, V. A., Chugaev, A. V., Ünal, E., Sharkov, E. V., & Keskin, M. (2016b). Late Pleistocene Tendürek volcano (Eastern Anatolia, Turkey). II. Geochemistry and petrogenesis of the rocks. *Petrology*, *24*(3), 234–270.

Lebedev, V. A., Parfenov, A. V., & Yakushev, A. I. (2018a). Neogene-Quaternary magmatism of the Çaldıran Plain and its vicinity (Eastern Turkey): An example of post-collisional transition from subduction to intraplate type. *Petrology*, *26*, 469–491. https://doi.org/10.1134/S0869591118050053

Lebedev, V. A., Sharkov, E. V., Ünal, E., & Keskin, M. (2016a). Late Pleistocene Tendürek volcano (Eastern Anatolia, Turkey). I. Geochronology and petrographic characteristics of igneous rocks. *Petrology*, *24*(2), 127–152.

Lei, J., & Zhao, D. (2007). Teleseismic evidence for a break-off subducting slab under Eastern Turkey. *Earth and Planetary Science Letters*, *257*, 14–28.

Lustrino, M., Keskin, M., Mattioli, M., & Kavan, O. (2012). Heterogeneous mantle sources feeding the volcanic activity of Mt. Karacadag. *Journal of Asian Earth Science*, *46*, 120–139.

Maden, N., & Öztürk, S. (2015). Seismic b-values, Bouguer gravity and heat flow data beneath Eastern Anatolia, Turkey: Tectonic implications. *Surveys in Geophysics*. https://doi.org/10.1007/s10712-015-9327

Maggi, A., & Priestley, K. (2005). Surface waveform tomography of the Turkish-Iranian plateau. *Geophysical Journal International*, *160* (3), 1068–1080.

Mahatsente, R., Önal, G., & Çemen, I. (2018). Lithospheric structure and the isostatic state of Eastern Anatolia: Insight from gravity data modeling. *Lithosphere*, *685*, 1. https://doi.org/10.1130/L685.1; IF:3.195

McKenzie, D. (1972). Active tectonic of the Mediterranean region, *Geophysical Journal of the Royal Astronomical Society*, *30*, 109–185. https://doi.org/10.1111/j.1365-246X.1972.tb02351

McKenzie, D. (1978). Active tectonics of the Alpine-Himalayan belt: the Aegean Sea and surrounding regions. *Geophysical Journal of the Royal Astronomical Society*, *55*, 217– 254.

McKenzie, D., & Yılmaz, Y. (1991). Deformation and volcanism in western Turkey and the Aegean. *Bulletin of the Technical University of Istanbul, Special Issue on Tectonics*, *44*, 345–373.

McNab, F., Ball, P. W., Hoggard, M. J., & White, N. J. (2018). Neogene uplift and magmatism of Anatolia: Insights from drainage analysis and basaltic geochemistry. *Geochemistry, Geophysics, Geosystems*, *19*, 175–213. https://doi.org/10.1002/2017GC007251

Medved, I., Polat, A., & Koulakov, G. (2021). Crustal structure of the Eastern Anatolia region (Turkey) based on seismic tomography. *Journal of Geosciences*, *11*(2). https://doi.org/10.3390./geosciences11020091

Meijers, M. J. M., Smith, B., Pastor-Galán, D., Degenaar, R., et al. (2017). Progressive orocline formation in the Eastern Pontides–Lesser Caucasus. In *Tectonic evolution of the Eastern Black Sea and Caucasus* (pp. 117–143). Geological Society London Special Publication 428.

MTA (2002). *Geological map of Turkey, at 1/500.000 scale*. Ankara, Turkey, General Directorate of Mineral Research and Exploration.

Nikishin, A. M., Okay, A., Tüysüz, O., Demirer, A., Wannier, M., Amelin, N., et al. (2015). The Black Sea basins structure and history: New model based on new deep penetration regional seismic data. Part 2: Tectonic history and paleogeography. *Marine and Petroleum Geology*, *59*, 656–670. https://doi.org/10.1016/j.marpetgeo.2014.08.018

Notsu, K., Fujitoni, T., Ui, T., Matsuda, J., & Ercan, T. (1995). Geochemical features of collision related volcanic rocks in central and Eastern Anatolia, Turkey. *Journal of Volcanology and Geothermal Research*, *64*, 171–192.

Oberhänsli, R., Bousquet, R., Candan, O., & Okay, A. I. (2012). Dating subduction events in East Anatolia. *Turkish Journal of Earth Sciences*, *21*, 1–18. https://doi.org/10.3906/yer-1006-26

Oberhänsli, R. E., Koralay, O., Candan, A., Pourteau, R., & Bousquet, R. (2014). Late Cretaceous eclogitic high-pressure relics in the Bitlis Massif. *Geodinamica Acta*, *26*, 175–190. https://doi.org/10.1080/09853111.2013.858951

Okay, A. I., Zattini, M., Özcan. E., & Sunal, G. (2020). Uplift of Anatolia. *Turkish Journal of Earth Science*, *29*, 696–713. https://doi.org/10.3906/yer-2003-10

Önal, M., & Kaya, M. (2007). Stratigraphy and tectono-sedimentary evolution of Upper Cretaceous-Tertiary sequence in the southern part of the Malatya basin, East Anatolia, Turkey. *Journal of Asian Earth Science*, *29*, 878–890.

Oyan, V., Keskin, M., Lebedev, V. A., Chugaev, A. V., & Sharkov, E. V. (2016b). Magic evolution of the Early Pliocene Etrüsk strato volcano, Eastern Anatolian Collision Zone, Turkey. *Lithos*, *256–257*, 88–108.

Oyan, V., Özdemir, Y., Keskin, M., & Güleç, N. (2016a). *Geochemical and isotopic evolution of Pliocene basaltic volcanism in the Eastern Anatolia, Turkey*. World Multidisciplinary Earth Sciences Symposium 2016, Prague, Çek Cumhuriyeti, 5-09 Eylül 2016, ss.180.

Özaçar, A. A., Gilbert, H., & Zandt, G. (2008). Upper mantle discontinuity structure beneath East Anatolian Plateau (Turkey) from receiver functions. *Earth and Planetary Science Letters*, *269*, 426–434.

Özdemir, İ. (1981). *Oltu-Balkaya (Erzurum) kÖmürlü Neojen havzasının ekonomik jeolojisi.* MSc, Ankara University, Ankara, Turkey.

Özdemir, Y., & Güleç, N. (2014). Geological and geochemical evolution of the Quaternary Süphan stratovolcano, Eastern Anatolia, Turkey: Evidence for the lithosphere- asthenosphere interaction in post-collision volcanism. *Journal of Petrology*, *55*, 37–52.

Özeren, M. S., & Holt, W. E. (2010). The dynamics of the eastern Mediterranean and eastern Turkey. *Geophysical Journal International*, *183*(3), 1165–1184. https://doi.org/10.1111/j.1365-246X2010.04819. x

Pamukçu, O., Akçığ, Z., Hisarlı, M., & Tosun, S. (2014). Curie point depths and heat flow of eastern Anatolia (Turkey). *Energy Sources Part A*, *36*, 2699–2706.

Pamukçu, O. A., Akçığ, Z., Demirbaş, S., & Zor, E. (2007). Investigation of crustal thickness in eastern Anatolia using gravity, magnetic and topographic data. *Pure Applied Geophysics*, *164*, 2345–2358.

Pearce, J. A., Bender, J. F., De Long, S. E., Kidd, P. J., Low, Y., et al. (1990). Genesis of collision volcanism in Eastern Anatolia, Turkey. *Journal of Volcanology and Geothermal Research*, *44*, 189–229.

Piromallo, C., & Morelli, A. (2003). P wave tomography of the mantle under Alpine-Mediterranean area. *Journal of Geophysical Research*, *108*(B2), 2065.

Piromallo, C. V., & Regard, V. (2006). Slab detachment beneath eastern Anatolia: A possible cause for the formation of the North Anatolian Fault. *Earth and Planetary Science Letters*, *214*, 85–97.

Pourteau, A., Sudo, M., Candan, O., Lanari, P., Vidal, O., & Oberhansli, R. (2013). NeoTethys closure history of Anatolia: Insight from 40Ar/39Ar geochronology and P-T estimation in high-pressure metasediments. *Journal of Metamorphic Geology*, *31*, 585–606. https://doi.org/10.1111/jmg.12034

Reilenger, R., McClusky, S., Vernant, P., Lawrence, S., Ergintav, S., Çakmak, R., et al. (2006). GPS constraints on continental deformation in the Africa-Arabia-Eurasia continental collision zone and implications for the dynamics of plate interactions. *Journal of Geophysical Research: Solid Earth*, *111*(5), 1–26. https://doi.org/10.1029/2005JB004051

Robertson and Shipboard Scientific Party (1996). Tectonic introduction: Drill hole reports in Hellenic trench. *Proceedings of the Ocean Drilling Program Initial Report*, *160*, 1–18.

Robertson, A. H. F., Parlak, O., & Ustaömer, T. (2012). Overview of the Palaeozoic-Neogene evolution of Neotethys in the Eastern Mediterranean region (southern Turkey, Cyprus, Syria). *Petroleum Geoscience*, *18*(4), 381–404.

Rolland, Y., Hässig, M., Bosch, D., Bruguier, O., Melis, R., Galoyan, G., et al. (2020). The East Anatolia–Lesser Caucasus ophiolite: An exceptional case of large-scale obduction, synthesis of data and numerical modelling. *Geoscience Frontiers*, *11*(1), 83–108.

Rolland, Y., Perinçek, D., Kaymakçı, N., Sosson, S., Barrier, E., & Avagyan, A. (2012). Evidence for ~80–75 Ma subduction jump during Anatolide-Tauride-Armenian block accretion and ~48 Ma Arabia-Eurasia collision in Lesser Caucasus-East Anatolia. *Journal of Geodynamics*, *56–57*, 76–85.

Sandvol, E., Türkelli, N., & Barazangi, M. (2003a). The Eastern Turkey seismic experiment: The study of a young continent collision. *Geophysical Research Letters*, *24*, 8038–8041.

Sandvol, E., Türkelli, N., Zor, E., Gök, R., Bekler, T., Gürbüz, C., et al. (2003b). Shear wave splitting in a young continent-continent collision: An example from Eastern Turkey. *Geophysical Research Letters*, *30*(24), 8041. https://doi.org/10.1029/2003GL017390

Şaroğlu, F., & Güner, Y. (1981). Doğu Anadolu 'nun jeomorfolojik gelişimine etki eden öğeler; jeomorfoloji, tektonik, volkanizma ilişkileri. *Türkiye Jeoloji, Kur. Bült.*, *24*(2), 39–50.

Şaroğlu, F., & Yılmaz, Y. (1984). Doğu Anadolu 'nun neotektoniği ve ilgili magmatizması. Türkiye Jeol. Kur. Ketin Sempozyumu Bildiriler Kitabı, 149–162.

Şaroğlu, F., & Yılmaz, Y. (1986). Doğu Anadolu'da Neotektonik dÖnemdeki jeolojik evrim ve havza modelleri. *Bulletin of the Mineral Research and Exploration*, *107*, 73–94.

Şaroğlu, F., & Yılmaz, Y. (1987). Geological evolution and basin models during neotectonic episode in eastern Anatolia. *Bulletin of the Mineral Research and Exploration*, *107*, 61–83.

Şaroğlu, F., & Yılmaz, Y. (1991). Geology of the Karlıova region: Intersection of the North Anatolian and the East Anatolian transform faults. *Bulletin of the Technical University of Istanbul, Special Issue on Tectonics*, *44*(1), 475–493.

Şaroğlu, F., Güner, Y., Kidd, W. S. F., & Şengör, A. M. C. (1980). Neotectonic of Eastern Turkey: New evidence for crustal shortening and thickening in a collision zone. *EOS, Transactions, American Geophysical Union*, *61* (17), 360.

Schildgen, T. F., Yıldırım, C., Cosentino, D., & Strecker, M. R. (2014), Linking slab break-off, Hellenic trench retreat, and uplift of the central and eastern Anatolian plateaus, *Earth-Science Review*, *128*, 147–168.

Şenel, M., Acarlar, M., Çakmakoğlu, A., et al. (1984). *Özalp-İran sınırı arasındaki bölgenin Jeolojisi (Geology of the area between Özalp (city of Van) and the Iranian border)*. MTA

(Mineral Research and Exploration Institute of Turkey) Report No. 663.

Şengör, A. M. C. (1979). *Türkiye'nin Neotektonik Esasları (Principles of the Neotectonics of Turkey) (Vol. 2)*. Ankara Türkiye Jeoloji Kurumu Yayınları Serisi.

Şengör, A. M. C., & Kidd, W. S. F. (1979). Post-collisional tectonics of the Turkish-Iranian Plateau and a comparison with Tibet. *Tectonophysics, 55*, 361–376.

Şengör, A. M. C., & Natalin, B. (1996). Turkic-type orogeny and its role in the making of the continental crust. *Annual Review of Earth and Planetary Sciences, 24*, 263–337.

Şengör, A. M. C., & Yilmaz, Y. (1981). Tethyan evolution of Turkey: A plate tectonic approach. *Tectonophysics, 75*, 181–241.

Şengör, A. M. C., Özeren. S., Zor, E., & Genç, T. (2003). East Anatolian high plateau as a mantle-supported, north-south shortened domal structure. *Geophysical Research Letters, 30*(24), 8045. doi:10.1029/2003GL017858

Şengör, A. M. C., Özeren, M. S., Keskin, M., Sakınc, M., Obakır, A. D., & Kayan, I. (2008). Eastern Turkish high plateau as a small Turkic-type orogen: Implications for post collisional crust forming processes in Turkic-type orogen. *Earth-Science Reviews, 90*, 1–48. https://doi.org/10.1016/j.earscirev.2008.05.002

Seyitoğlu, G., Esat, K., & Kaypak, B. (2017). The neotectonics of Southeast Turkey, Northern Syria, and Iraq: The internal structure of the Southeast Anatolian wedge and its relationship with recent earthquakes. *Turkish Journal of Earth Sciences, 26*(2), 105–126. https://doi.org/10.3906/yer-1605-21

Seyitoğlu, G., Esat, K., Kaypak, B., Tooric, M., & Aktuğ, B. (2018). The neotectonics of Eastern Turkey, Northwest Iran, Armenia, Nahçivan and Southern Azerbaijan: The rhomboidal cell model in the internal deformation of Turkish-Iranian Plateau. In *71st Geological Congress of Turkey Proceedings* (pp. 661–664).

Skobeltsyn, G. R., Mellors, R., GÖk, N., Türkelli, G., et al. (2014). Upper mantle S wave velocity structure of the East Anatolian-Caucasus region. *Tectonics, 33*(3), 207–221. https://doi.org/10.1002/2013TC003334

Sosson, M., Stephenson, R., & Adamia, S. (2017). Tectonic evolution of the Eastern Black Sea and Caucasus: An introduction. *Geological Society, London, Special Publications, 428*(1), 19. https://doi.org/10.1144/SP428.16

Tarhan, N. (1989). Hınıs-Varto dolayının Jeolojisi ve Petrolojisi. Ph. D Thesis. Istanbul University Institute of Science.

Tarhan, N., Deveciler, E., Karabalık, N. N., Akdoğan, E., Çolak, T., & Kar, H. (1992). Aşkale-Çat (Erzurum) dolayının jeolojisi. MTA Rapor, 9447.

Tarhan, N. (1997a). 1/ 100 000 ölçekli Açınsama nitelikli Türkiye Jeoloji Haritaları no 52. Erzurum-G 31 Paftası. Jeoloji Etütleri Dairesi Ankara MTA Genel Müdürlüğü.

Tarhan, N. (1997b). 1/ 100 000 ölçekli Açınsama nitelikli Türkiye Jeoloji Haritaları no 53. Erzurum-G 32 Paftası. Jeoloji Etütleri Dairesi Ankara MTA Genel Müdürlüğü.

Tarhan, N. (1998a). 1/ 100 000 ölçekli Açınsama nitelikli Türkiye Jeoloji Haritaları no 52. Erzurum-F 31 Paftası. Jeoloji Etütleri Dairesi Ankara MTA Genel Müdürlüğü.

Tarhan, N. (1998b). 1/ 100 000 ölçekli Açınsama nitelikli Türkiye Jeoloji Haritaları no 56. Erzurum-G 31 Paftası. Jeoloji Etütleri Dairesi Ankara MTA Genel Müdürlüğü.

Temiz, H., Guezou, J. C., Tatar, O., Ünlügenç, C., & Poisson, A. (2002). Tectonostratigraphy of the Tercan-Çayirli Basin: Implications for the Neogene-Quaternary tectonic deformation of the Northeast Anatolian Block, Turkey. *International Geological Review, 44*, 243–253.

Tezcan, A. K. (1987). Geothermal studies, their present status and contribution to heat flow contouring in Turkey. In V. Cermak & L. Rybach (Eds.), *Terrestrial heat flow in Europe* (pp. 283–291). Berlin: Springer Verlag.

Tezcan, A. K., & Turgay, M. I. (1989). Türkiye ısı akısı haritası. General Directorate of Mineral Research and Exploration (MTA), Ankara.

Tezel, T., Shibutani, T., & Kaypak, B. (2013). Crustal thickness of Turkey determined by receiver function. *Journal of Asian Earth Science, 75*, 36–45.

Topuz, G., Candan, O., Zack, T., & Yılmaz, A. (2017). East Anatolian plateau constructed over a continental basement: no evidence for the East Anatolian accretionary complex. *Geology, 45*, 791–794.

Türkelli, N., Sandvol, E., Zor, R. G. K. T., Bekler, A., Al-Lazki, H., Karabulut, D., et al. (2003). Seismogenic zones in eastern Turkey. *Geophysical Research Letters, 30*(24), 8039.

Üner, T. (2021) Supra-subduction zone mantle peridotites in the Tethyan ocean (East Anatolian Accretionary Complex-Eastern Turkey); petrological evidence for melting-rock interaction. *Mineralogy and Petrology*. https://doi.org/10.1007/s00710-021-00760-0

Uysal, S. (1986). MusTersiyer havzasının stratigrafisi ve Üst Lütesiyen'in varlığı. *M.T.A Dergisi, 105*/106, 69–74.

Warren, L., Beck, S. L., Biryol, C. B., Zandt, A. A., Özaçar, A. A., & Yang, G. (2013). Crustal velocity structure of central and eastern Turkey from ambient noise tomography. *Geophysical Journal International, 194*, 1941–1954.

Yang, G., Li, T., Tong, L., Duan, F., Xu, Q., & Li, H. (2019). An overview of oceanic island basalts in accretionary complexes and seamount accretion in the western Central Asian Orogenic Belt. *Journal of Asian Earth Science, 179*, 385–398.

Yiğitbaş, E. (1989). *Engizek Dağı (Kahraman Maraş) dolayındaki tektonik birliklerin petrolojik incelenmesi (Petrological studies of the tectonic units in the Engizek Mountain, Kahraman Maras)*. Doctoral thesis. Istanbul Üniversitesi, Fen Fakültesi.

Yiğitbaş, E., & Yılmaz, Y. (1996a). New evidence and solution to the Maden Complex controversy of the Southeast Anatolian orogenic belt (Turkey). *Geol. Rundschau., 85*, 250–263.

Yıldırım, C., Melnick, D., Ballato, P., Schildgen, T. F., Echtler, H., Erginal, A. E., et al. (2013). Differential uplift along the northern margin of the Central Anatolian Plateau: Inferences from marine terraces. *Quaternary Science Reviews, 81*, 12–28.

Yılmaz, A., & Yılmaz, H. (2019). Structural evolution of the Eastern Anatolian Basins: an example from collisional to post collisional tectonic processes, Turkey. *Turkish Journal of*

Earth Science, *28*, 329–350. https://doi.org/10.3906/yer-1805-1820

Yılmaz, A., Terlemez, I., & Uysal, S. (1988). Some stratigraphic and tectonic characteristics of the area around Hınıs (southeast of Erzurum). *Bulletin of the Mineral Research and Exploration*, *108*, 1–21.

Yılmaz, Y. (1990). Comparison of young volcanic associations of western and eastern Anatolia formed under a compressional regime: A review. *Journal of Volcanology and Geothermal Research*, *44*, 69–87.

Yılmaz, Y. (1993). New evidence and model on the evolution of the Southeast Anatolian orogen. *Geological Society of America Bulletin*, *105*(2), 251–271.

Yılmaz, Y. (2017). Morphotectonic development of Anatolia and surrounding regions. In I. Çemen & Y. Yılmaz (Eds.), *NeoTectonics and earthquake potential of the Eastern Mediterranean region* (pp. 11–92). AGU Geophysical Monograph 225.

Yılmaz, Y. (2019). Southeast Anatolian Orogenic Belt revisited. *Canadian Journal of Earth Sciences*, *1–18*. https://doi.org/10.1139/cjes-1170

Yılmaz, Y. (2021). Geological correlation between Northern Cyprus and Southern Central Anatolia. *Canadian Journal of Earth Science.* https://doi.org/10.1139/cjes-2020-0129.

Yılmaz, Y., & Şengör, A. M. C. (1985). Palaeo-Tethyan ophiolites in northern Turkey; petrology and Tectonic setting. *Ophioliti*, *10* (2/3), 485–504.

Yılmaz, Y., Güner, Y., & Şaroğlu, F. (1998). Geology of the Quaternary volcanic centers of the east Anatolia. *Journal of Volcanology and Geothermal Research*, *85*, 173–210.

Yılmaz, Y., Şaroğlu, F., & Güner, Y. (1987a). Doğu Anadolu'da Solhan (Muş) volkanitlerinin petrojenetik incelenmesi. Yerbilimleri 14, 133–167.

Yılmaz, Y., Şaroğlu, F., & Güner, Y. (1987b). Initiation of the neomagmatism in East Anatolia. *Tectonophysics*, *134*, 177–199.

Yılmaz, Y., Tüysüz, O., Yiğitbaş, E., Genç, Ş. C., & Şengör, A. M. C. (1997). Geology and tectonic evolution of the Pontides. In A. G. Robinson (Ed.), *Regional and petroleum geology of the Black Sea and surrounding region* (pp. 183–226). American Association of Petrolium Geologists Memoir 68.

Zor, E. (2008). Tomographic evidence of slab detachment beneath eastern Turkey and the Caucasus, *Geophysical Journal International*, *175*, 1273–1282.

Zor, E., Sandvol, E., Gürbüz, C., Türkelli, N., Seber, D., & Barazangi, M. (2003). The crustal structure of the East Anatolian Plateau (Turkey) from receiver functions. *Geophysical Research Letters*, *30*(24), 8044.

9

When and Why the NeoTethyan Subduction Initiated Along the Eurasian Margin: A Case Study From a Jurassic Eclogite in Southern Iran

Bo Wan[1], Yang Chu[1], Ling Chen[1], Zhiyong Zhang[1], Songjian Ao[1], and Morteza Talebian[2]

ABSTRACT

Tethyan evolution is characterized by cyclical continent-transfer from Gondwana to the continents in the Northern Hemisphere, similar to a "one-way" train. Subduction has been viewed as the primary driver of transference. Therefore, it is important to understand the tectonic evolution of all past subduction zones that occurred along Eurasia's southern margin. We studied the earliest known eclogite located at the NeoTethyan suture in the Iranian segment. A prograde-E-MORB-like eclogite reached a peak metamorphic condition of 2.2 GPa and 560°C, at 190 ± 11 Ma (1σ rutile U-Pb ages), which constrains the youngest age for subduction initiation of the NeoTethyan slab. Combined with regional magmatic and structural data, the oldest age for NeoTethys subduction initiation is 210–192 Ma, which is younger than the Paleo-Tethyan closure time of 228–209 Ma. These data, used with previous numerical modeling, support collision-induced subduction initiation. The collision-induced force, together with the Paleo-Tethyan subduction-driven mantle flow, is likely to have exploited weak inherited structures from earlier NeoTethyan rifting, resulting in a northward directed subduction zone along the southern margin of the Central Iran Block.

9.1. INTRODUCTION

Currently, with the exception of Antarctica, most continents are connected and encircle the North Pole (Fig. 9.1). Almost all continents, including Gondwana, were located in the Central to Southern Hemisphere to cap the South Pole at about 500 Ma (Lawver et al., 2015). The one-way megatransferring of continents from south to north was a key event of Earth, that occurred during the last 500 Myr, with the Australian continent moving north at a rate of 70 mm/yr as a representative active example (DeMets et al., 2010). During this megatransferring of continents, the sea-land paleogeography changed, which influenced the Earth's surface temperature swinging between an icy and warmer world (Merdith et al., 2019; Bergman et al., 2021; Scotese et al., 2021). However, the driving mechanism for this one-way/single-directed transfer is still debated. Two competing theories are (1) whole mantle convection, which involves plume upwelling and subduction (e.g., Becker & Faccenna, 2011; Jolivet et al., 2016; Faccenna et al., 2021), and (2) northward oceanic subduction (Wan et al., 2019; Wu et al., 2020). After 50 Ma, the Indo-Australian oceanic plate began subducting beneath the southern Eurasian continent (Sunda Shelf)

[1] State Key Laboratory of Lithospheric Evolution, Institute of Geology and Geophysics, Chinese Academy of Sciences, Beijing, China

[2] Research Institute for Earth Sciences, Geological Survey of Iran, Tehran, Iran

Compressional Tectonics: Plate Convergence to Mountain Building, Geophysical Monograph 277, First Edition.
Edited by Elizabeth J. Catlos and İbrahim Çemen.

DOI:10.1002/9781119773856.ch09

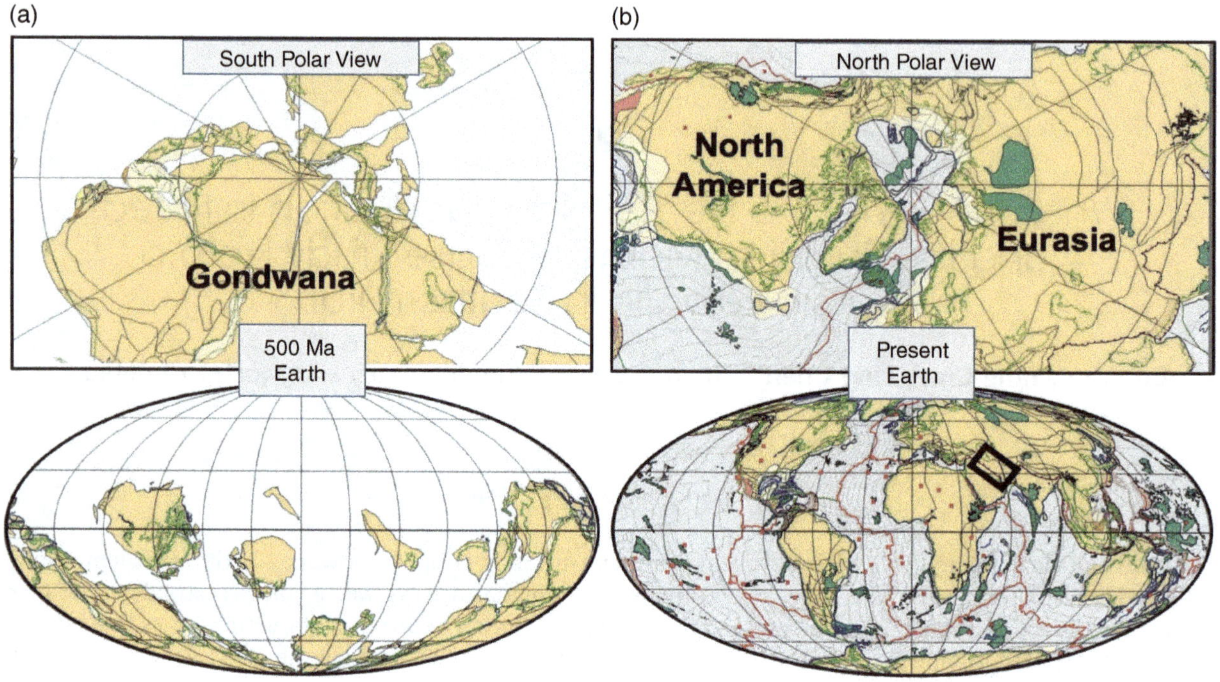

Figure 9.1 Views of Earth: (a) the South Polar view of the Earth and world map (at 500 Ma); (b) the North Polar view of the Earth and world map (present day) (adapted from Lawver et al., 2015). Light grey = continent; dark grey = large igneous provinces. The study area is denoted by a black square.

margin along the Java trench (Hall, 2017). The major eruption of the Kerguelen large igneous province (90–120 Ma) has been proposed to have fragmented the Antarctic-Australia plate. However, this eruption was earlier than the abrupt velocity change of the Australia plate at 45 Ma (Whittaker et al., 2013; Williams et al., 2019). Instead, the Australian plate acceleration is temporally closer with that of the 50 Ma Java trench subduction zone activity. This supports the idea that subduction is the driving force of the northward migration of the Australian plate (Forsyth & Uyeda, 1975).

The cycle of Tethyan oceans involved the Proto-Tethys (440–420 Ma closure), the Paleo-Tethys (330–220 Ma closure), and the NeoTethys (65–15 Ma closure). This cycle merged many Gondwana-derived continents with continents to the north. All surviving Tethyan sutures are presently located in the Northern Hemisphere (Stampfli et al., 2013; Torsvik & Cocks, 2017; Wu et al., 2020). It is unclear if the closure of the Tethyan oceans was caused by the same tectonic forces that are controlling the current evolution of the Indian Ocean and the northward migration of the Australian plate. To compare these two scenarios, it is necessary to first constrain the subduction initiation for the Tethyan oceans, in particular the youngest NeoTethyan Ocean. However, subduction initiation is a challenging and poorly understood topic (Stern & Gerya, 2018), especially given the discrepancy between numerical modeling (Gerya et al., 2015; Leng & Gurnis, 2015; Zhong & Li, 2020; Zhou et al., 2020; Zhong & Li, 2022), and geological observations (Whattam & Stern, 2011; Guilmette et al., 2018; Van Hinsbergen et al., 2021).

The NeoTethyan Iranian segment preserves many NeoTethyan ophiolites (Moghadam & Stern, 2015; Ao et al., 2016), subduction-related magmatic episodes (Omrani et al., 2008; Chiu et al., 2013; Chiu et al., 2017; Zhang et al., 2018; Moghadam et al., 2022), and metamorphic events (Agard et al., 2011; Davoudian et al., 2016; Bonnet et al., 2020). This makes the Iranian segment an ideal location to study the NeoTethyan subduction initiation problem. There is a consensus about NeoTethyan subduction initiation along the southern Iranian continent's margin (Hassanzadeh & Wernicke, 2016; Stern et al., 2021). However, the timing of NeoTethyan subduction initiation in Iran is debated, and occurred either in the Triassic-Jurassic (Arvin et al., 2007) or the Cretaceous (Moghadam & Stern, 2015). In addition, the driving mechanism has been explained through either collision-induced (Wan et al., 2019) or spontaneous nucleation (Moghadam & Stern, 2011). Additionally, it is crucial to understand why the subduction of the Indian, NeoTethyan, and Paleo-Tethyan ocean basins consistently initiate along the southern Eurasian margin with a north-dipping slab, as opposed to how subduction polarity frequently reverses in the Southern Hemisphere in the Pacific realm (Brown & Ryan, 2011).

In this study, an eclogite sample along the NeoTethyan suture zone in southern Iran was analyzed to find out its photolith nature, in situ metamorphic age, and corresponding metamorphic condition based on geochemical-petrological studies. The new in situ U-Pb rutile age is the earliest subduction-related high-pressure metamorphic event observed in the NeoTethyan suture zone. This event defines the latest (youngest) timing of subduction initiation. In addition, the mechanism for how and why subduction occurred is discussed with the goal to better understand other Tethyan regions and subduction zones globally.

9.2. GEOLOGICAL BACKGROUND

Three major tectonic domains dominate the Tethyan evolution in Iran. Along the Talesh-Alborz-Binalud mountain range in northern Iran, the Paleo-Tethyan suture divides the Kopeh-Dagh domain to the north from the Central Iran domain to the south. The NeoTethyan closure sutured the Central Iran domain and Arabia domain in southern Iran. The Kopeh-Dagh is dominated by Jurassic to Cenozoic shallow marine to continental sedimentary rocks, such as limestone, sandstone, shale, and conglomerate. The basement of the Kopeh-Dagh domain is a Paleozoic–Early Mesozoic volcanic arc located on the Baltic or Siberian Precambrian continents (Natal'in & Şengör, 2005; Zanchetta et al., 2013; Chu et al., 2021).

According to sedimentary records, the entire Central Iran domain was formerly part of the Arabian continent but separated as a ribbon continent (Cimmeria) in the late Permian (Koop et al., 1982). During the rifting and drifting process, Central Iran was surrounded by passive margins, as indicated by sedimentary facies analyses (Leven & Gorgij, 2011), which share a similar scenario of Indian drifting northward during the Cretaceous. Recent research in northern Iran, along the Paleo-Tethyan suture near Mashhad, indicates that the first arrival of Eurasian material to the passive margin of Central Iran occurred at 209–228 Ma (Chu et al., 2021). Previous stratigraphic studies suggest that the collision must have occurred between the late Triassic (228 Ma) and the mid Jurassic (174 Ma), and possibly initiated at the Carnian-Norian boundary (Fürsich et al., 2009). Additionally, the extensive time span might be restricted by a stitching pluton at 217 ± 1.7 Ma (2σ) that crosscuts through Triassic compressional structures in the Alborz (Zanchetta et al., 2013). Thus, the new detrital zircon results (209–228 Ma; Chu et al., 2021) corroborate prior stratigraphic and structural geological studies (217–228 Ma), implying a Triassic collision event.

Magmatism is extensive in Central Iran, with two magmatic flare-ups occurring in the Jurassic along southern Iran (Fig. 9.2), and in the Cenozoic throughout Central Iran (Verdel et al., 2011; Chiu et al., 2013; Zhang et al., 2018; Moghadam et al., 2022). The Jurassic plutons are mostly granite, granodiorite, quartz diorite, and gabbro (Hassanzadeh & Wernicke, 2016), while the Jurassic volcanic rocks are mostly basaltic, andesitic lava, and volcanoclastics (Emami & Khalili, 2008). Geochemical studies on the Jurassic igneous rocks found $\varepsilon Hf_{(T)}$ values ranging from +13 to –3 (Chiu et al., 2017; Zhang et al., 2018), indicating that they originated from mixed juvenile and reworked crustal sources. The Hf isotopic ratios of Jurassic igneous rocks are in the range of Cenozoic igneous rocks $\varepsilon Hf_{(T)}$ (+14 to –7) along southern Iran (Chiu et al., 2017; Moghadam et al., 2022). The similar petrological and geochemical features between Jurassic and Cenozoic magmatic rocks imply a similar subduction-related tectonic environment, while some researchers think that the Jurassic rocks have inherited the geochemical signature from a Mesozoic continental rifting episode (Azizi & Stern, 2019).

Along southern Iran, Jurassic high-pressure low-temperature (HP/LT) metamorphic rocks have been reported (Davoudian et al., 2016; Jamali Ashtiani et al., 2020). The HP/LT rock is an eclogite with a peak metamorphic condition of 2.35–2.5 GPa and 520°C–600°C, and metamorphic ages of 172–184 Ma by Ar-Ar dating of phengite (Davoudian et al., 2016). The eclogites outcrop along a regional northwest-striking shear zone (Fig. 9.3), and the field relationship has been described in detail by Davoudian et al. (2016). Eclogite bodies and marbles occur as lenses in the schist, and the lenses are mostly meters scale. The marble lenses contain dark eclogite, indicating the marble has also undergone HP metamorphism. The schist is at amphibolite grade, with greenschist in the surrounding region. The ages of the schists are still unknown. Recent studies show that the gneisses are from Cadomian basement, with an age of 552 Ma, which also contain a younger intrusion dated as 176 ± 3.3 Ma (2σ) (Jamali Ashtiani et al., 2020). Permian limestone, Jurassic volcanic-sedimentary rocks are thrust over the metamorphic complex with various grades of metamorphic rocks such as schist, gneiss, and amphibolite. The Jurassic volcanic-sedimentary rocks have experienced low-grade metamorphism, at prehnite–pumpellyite facies. This type of rock assemblage, with different origin with contrasting metamorphic grades, is very similar to the accretionary complex associated with subduction zones observed globally (Wakabayashi, 2011). Additionally, understanding the origin of the HP/LT rocks will aid in comprehending the subduction environment at the time.

Many early Cretaceous ophiolites occur in the south of Central Iran (Moghadam & Stern, 2011, 2015), and extensive Cretaceous to Cenozoic magmatic rocks are viewed as the subduction products of the NeoTethyan slab (Omrani et al., 2008; Moghadam et al., 2022).

Figure 9.2 Tectonic sketch map of Iran, highlighting tectonic domains, NeoTethyan suture, Jurassic magmatism, and high-pressure metamorphism. Note: Mashhad (Chu et al., 2021); Malayer earliest arc magmatism at 187 Ma (Ahadnejad et al., 2011); Shahrekord earliest eclogite (Fig. 9.3). Based on the map from the Geological-Survey-of-Iran (2009) (strata observation positions from Leven & Gorgij, 2011).

The final closure of the NeoTethys merged the Central Iran domain with the Zagros domain along the Zagros suture during the Oligocene or the Miocene (McQuarrie & Van Hinsbergen, 2013; Zhang et al., 2017).

9.3. SAMPLE, ANALYTICAL METHODS, AND RESULTS

In the Shahrekord region, we sampled an eclogite (TM107, N32°45′47″ E50°39′37″) near Chadegan (Fig. 9.3). The eclogite is composed of garnet, omphacite, muscovite, chlorite, barroisite, calcic-amphibole, epidote, quartz, rutile, and ilmenite. The metamorphic mineral assemblage at its peak is composed of amphibole, muscovite, garnet, omphacite, quartz, and rutile.

The sample was ground into 200-mesh powder and the bulk sample's major and trace elements geochemistry was analyzed to determine the rock's origin. To understand the metamorphic processes and peak metamorphic conditions, we chose a representative garnet for major oxide measurements and P-T pseudosection modeling. Additionally, rutile was separated and the U-Pb isotopes were analyzed

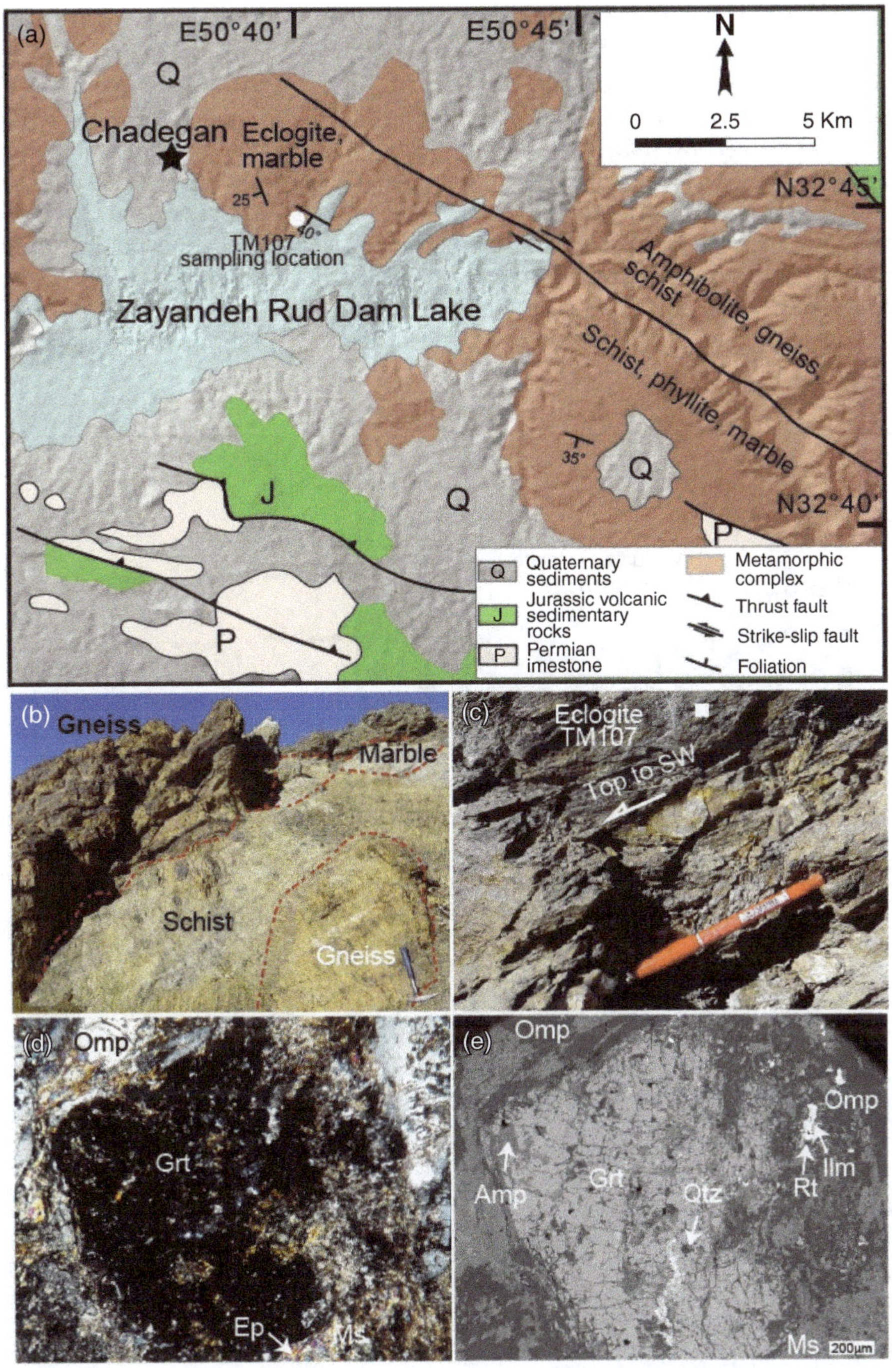

Figure 9.3 (a) Simplified geological map showing the juxtaposition of various rock types and metamorphic grades of rocks, based on the Shahrekord sheet of the Geological Survey of Iran (2009). (b) Marble lenses in schist and gneiss; (c) Top-to-the-SW fabric in eclogite (white square marks the thin section site). Two thin section photos, (d) cross polarized light, (e) backscattered electron, showing the major mineral associations. Note: Grt = garnet, Omp = omphacite, Rt = rutile, Amp = amphobile, Qtz = quartz, Ilm = ilimnite, ep = epidote, and Ms = mica.

to determine the absolute age of peak metamorphism. Rutile was used as it formed at a peak metamorphic stage, is in equilibrium with garnet, and has an appropriate U-Pb system closure temperature of ~600°C with a proper crystal size (Cherniak, 2000). The analyses were undertaken by various facilities at the Institute of Geology and Geophysics, Chinese Academy of Sciences. The details are described in the appendix at the end of this chapter.

The eclogite sample (TM107) has SiO_2 of (47.02 wt. %), Fe_2O_{3T} (13.92 wt. %), TiO2 (1.84 wt. %), MgO (7.34 wt. %),

Na_2O (2.55 wt. %), CaO (7.94 wt. %), K_2O (0.65 wt. %), and negligible P_2O_5. The eclogite is characterized by a slightly enriched light-rare-earth element pattern (La/Yb = 2.68). It shares a typical composition with that of enriched middle-ocean ridge basalt (E-MORB) (Sun & McDonough, 1989; Fig. 9.4a; Table 9.1). The garnet is relatively homogenous with Fe and Al, belonging to the almandine end member, and shows a decrease in Mn from the core to the rim along a profile across the mineral (Fig. 9.4b). The compositions of minerals and bulk rock are presented in detail in Table 9.1 and Appendix Table 9.A1. The garnet-derived P-T conditions constrain a prograde path with a maximum P-T condition of 2.2 GPa and 560°C and a geothermal gradient of 7.7°C/km (Fig. 9.4c). Exhumation of the eclogite resulted in the production of epidote and amphibole following peak metamorphism. The modeling result agrees with the thin section observations, indicating that rutile formed at the garnet rim during peak metamorphism. A total of 32 rutile grains (each with a 100 μm crystal size) was measured for U-Pb isotopes, with 24 grains providing valid data (Table 9.2). The U concentration is modest (1.8–0.3 ppm), while the

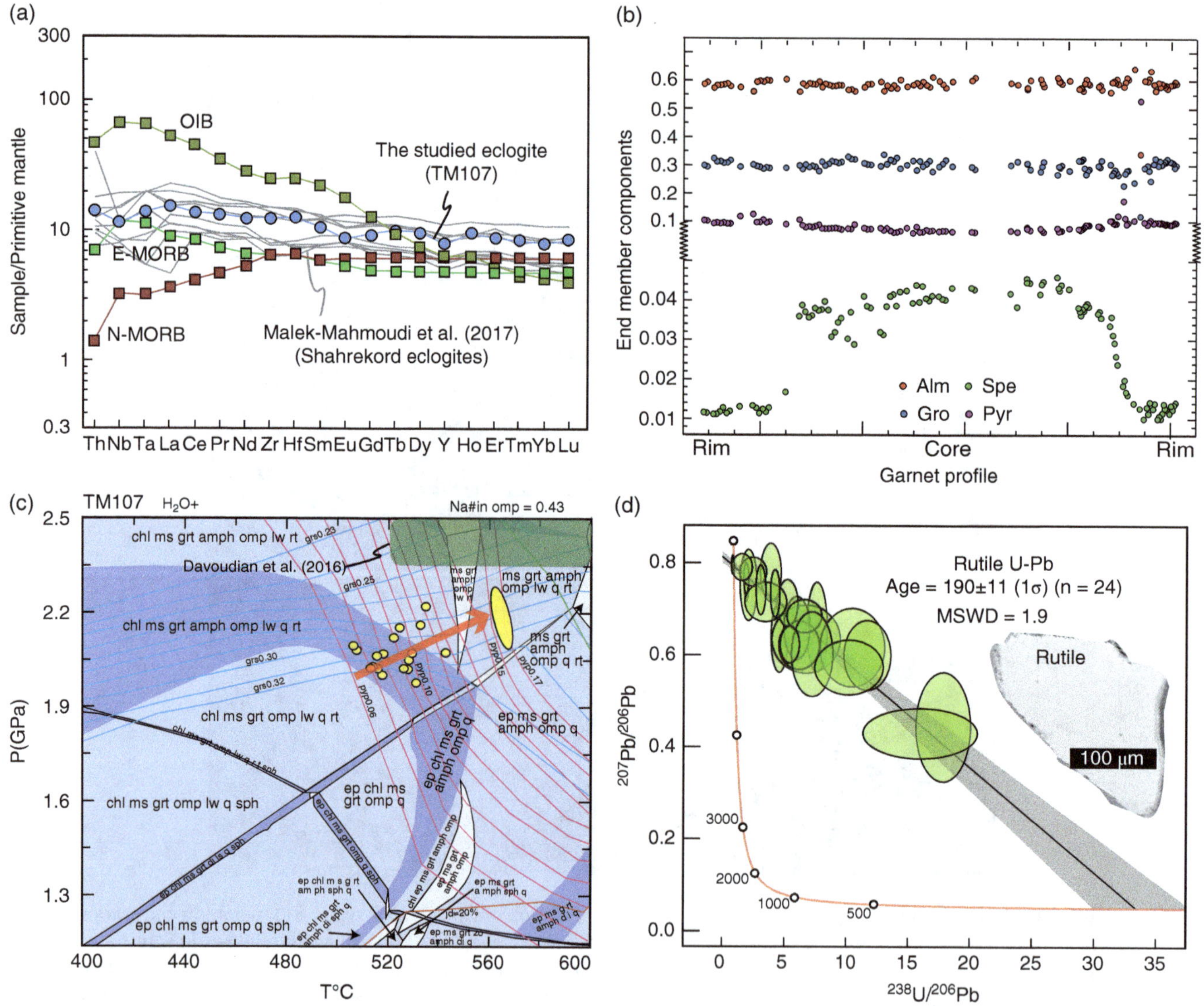

Figure 9.4 Analytical results from bulk sample or minerals from studied eclogite sample TM107. (a) Primitive mantle-normalized trace-element pattern, normalizing data from Sun and McDonough (1989). (b) EMPA profiles showing homogenous components of Mg, Ca, Al, and decreasing Mn garnet end member components from core to rim in garnet grain; (c). P-T pseudosection (SiO_2 48.23, TiO_2 1.89, Al_2O_3 14.51, FeO 12.86, MgO 7.54, CaO 11.18, Na_2O 2.62, MnO 0.24, K_2O 0.65 wt.%, which is corrected from Table 9.1). Ellipse region is calculated from EPMA data of garnet (Appendix, Table 9.A1), and rectangle region from Davoudian et al. (2016). (d) Rutile SIMS U-Pb age.

Table 9.1 Composition of whole rock and garnet of Shahrekord eclogite (TM107)

Whole rock		Garnet	(Rim)	(Core)
SiO_2	47.02	SiO_2	37.57	37.04
TiO_2	1.84	TiO_2	0.13	0.16
Al_2O_3	14.13	Al_2O_3	20.94	20.06
TFe_2O_3	13.92			
		FeO	26.18	27.04
MnO	0.23	MnO	00.59	1.88
MgO	7.34	MgO	2.25	1.54
CaO	7.94	CaO	10.88	9.91
		Cr_2O_3	.01	0.01
Na_2O	2.55			
K_2O	0.65			
P_2O_5	0.16			
LOI	4.26			
TOTAL	100.04			
Trace element		**a.p.f.u. & calculated end member**		
Nb	8,20	Si	2.993	2.983
Ta	0.56	Al	1.967	1.95
La	10.6			
Ce	24.	Ti	0.008	0.009
Pr	3.67	Cr	0.001	0.0
Nd	16.8	Fe^{3+}	0.041	0.07
Zr	139	Fe^{2+}	1.745	1.791
Hf	3.90	Mn	0.04	0.129
Sm	4.77	Mg	0.267	0.189
Eu	1.46	Ca	0.929	0.877
Gd	5.61			
Tb	1.08			
Dy	7.12	Pyr	0.09	0.06
Y	36.	And	0.00	0.00
Ho	1.57	Spe	0.01	0.04
Er	4.23	Gro	0.31	0.29
Tm	0.62	Alm	0.59	0.60
Yb	3.96			
Lu	0.62			

Th/U ratio varies between 0.01 and 0.33. The valid 24 analyses yield a lower intercept age of 190 ± 11 Ma (1σ) with an MSWD of 1.9., as shown on the Terra-Wasserburg diagram (Fig. 9.4d).

9.4. DISCUSSION

9.4.1. NeoTethyan Subduction Initiation Time

Most researchers agree that NeoTethyan subduction began at the southern margin of Central Iran but disagree on the timing of initiation between the Triassic-Jurassic (Arvin et al., 2007; Ahadnejad et al., 2011; Chiu et al., 2013; Davoudian et al., 2016; Zhang et al., 2018) and the late Cretaceous (Moghadam & Stern, 2015). The widespread magmatism in the early Jurassic in Central Iran clearly shows arc signatures of depleted Hf isotopes (Chiu et al., 2017; Zhang et al., 2018), which are regarded as an upper constraint (earliest age) of subduction initiation. Some researchers attribute a late Triassic Siah-Kuh granite (200 ± 30 Ma (2σ) Sm-Nd isochron) as the earliest evidence (Arvin et al., 2007). However, the updated LA-ICPMS zircon U-Pb dating gave ages of 175 ± 1.8 Ma (2σ) from the Siah-Kuh granite (Chiu et al., 2013), and now the earliest and most reliable arc magmatism in southern Iran is a 187 ± 6 Ma (2σ) granodiorite (Ahadnejad et al., 2011; Fig. 9.2). Stern (2004) suggested that the age of the SSZ-type ophiolite could represent subduction initiation. The earliest SSZ-type ophiolites in southern Central Iran along the Turkish-Zagros suture formed 90 ± 10 Ma (Whattam & Stern, 2011; Moghadam & Stern, 2015), which is significantly younger than the Jurassic subduction arc magmatism (Zhang et al., 2018).

In contrast, the oldest HP rocks associated with subduction in the accretionary complex may provide better evidence for subduction initiation. Using the phengite Ar-Ar technique, the earliest eclogite was dated as being 184 ± 1 Ma (1σ) (Davoudian et al., 2016). The subarc depth is known to be 100 km, which corresponds to the location of the majority of arc magma production (Syracuse & Abers, 2006). HP metamorphism in a subduction zone mostly occurs at a forearc depth that is shallower than the subarc depth (Agard et al., 2009). As a result, the earliest HP metamorphic age in the subduction zone should be older than the earliest magmatic arc rock formed following subduction initiation (only if these rocks have been preserved and discovered). The eclogite contains E-MORB geochemistry that is consistent with a previous report by Malek-Mahmoudi et al. (2017), indicating an oceanic slab subduction environment (Fig. 9.4). The in situ rutile U-Pb dating method yielded an age of 190 ± 11 Ma (1σ) with an MSWD of 1.9. This new in situ rutile age is slightly older than the oldest arc magmatism (187 Ma) in Central Iran. The Mn concentration of garnet from the eclogite is enriched in the core and is depleted in the rim, indicating that the eclogite was under prograde metamorphism (Fig. 9.4d).

This study calculates that the eclogites experienced a geothermal gradient of 7°C–7.7 °C/km, which replicates the results from Davoudian et al. (2016). These temperatures are within typical subduction zone geothermal gradients of 5°C–10°C/km (Wang et al., 2021). However, numerous geological and numerical investigations revealed a hot subduction environment during subduction initiation that is characterized by high-temperature rocks such as boninite and a metamorphic sole (Stern, 2004; Esna-Ashari et al., 2016; Maunder et al., 2020; Coulthard et al., 2021). While the new rutile

Table 9.2 SIMS U-Pb data of rutile from Shahrekord eclogite (TM107)

			TW concordia columns					
Name	U	Th/U	$^{238}U/^{206}Pb$	1σ (%)	$^{207}Pb/^{206}Pb$	1σ (%)	age (Ma)	1σ
TM107-01@01	0.9	0.07	0.795947	5.811998	0.795785	1.071056	400	107
TM107-01@02	1.1	0.18	1.886874	3.803730	0.745883	1.660098	381	66
TM107-01@03	0.4	0.25	4.562589	6.089317	0.637769	4.815065	348	74
TM107-01@05	0.3	0.06	7.595185	4.445783	0.660784	6.042157	186	50
TM107-01@06	0.3	0.27	2.068980	13.609286	0.751823	4.198377	325	164
TM107-01@08	0.5	0.07	3.865549	8.885439	0.779239	3.327236	118	65
TM107-01@10	0.6	0.12	6.724610	16.781395	0.634139	4.741298	242	76
TM107-01@11	0.8	0.01	5.959918	7.198755	0.659189	4.216732	239	54
TM107-01@12	0.5	0.10	1.876448	4.438371	0.766228	1.783636	297	71
TM107-01@14	0.7	0.06	6.147637	10.920344	0.621137	3.422747	281	58
TM107-01@16	0.8	0.05	2.440012	18.177573	0.780591	1.661653	182	75
TM107-01@17	0.5	0.02	5.211729	6.554152	0.712620	2.928143	191	44
TM107-01@18	0.7	0.00	17.839482	5.094676	0.444147	10.959140	178	31
TM107-01@19	0.5	0.01	6.568532	8.785013	0.684605	2.165168	186	34
TM107-01@20	0.5	0.02	10.458270	10.316299	0.616103	5.649829	170	44
TM107-01@21	0.4	0.05	7.203603	9.476491	0.601567	6.237684	262	66
TM107-01@23	1.4	0.04	15.979384	11.841301	0.429663	5.180421	205	35
TM107-01@24	0.5	0.00	11.684900	6.670011	0.599235	5.169264	164	32
TM107-01@25	0.6	0.04	5.394691	5.845419	0.622866	2.601378	317	42
TM107-01@26	0.7	0.01	10.195017	11.150205	0.573205	4.234457	208	42
TM107-01@27	1.0	0.23	1.537830	22.030820	0.790401	1.524525	237	114
TM107-01@28	1.8	0.33	3.450675	19.975521	0.715067	2.434675	281	95
TM107-01@30	0.6	0.03	4.213443	3.646025	0.733375	2.896049	196	47
TM107-01@32	0.4	0.11	3.060977	5.397827	0.741265	2.977421	248	71

U-Pb age from the eclogite sample constrains the oldest subduction-related timing in the Iranian NeoTethyan region to date, it represents the closest time lagged after the subduction initiation with respect to previous studies (Ahadnejad et al., 2011; Davoudian et al., 2016). Unlike the records in Iran, Early Jurassic ophiolites and accretionary complexes are well documented in Turkey, to the west of Iran, along the strike (Topuz et al., 2013; Okay et al., 2020). The geology in Turkey would help the correlation of the Jurassic subduction zone along the southern Eurasian margin, in southern Iran with the nearby region.

Nikolaeva et al. (2010) modeling experiments suggest that the transition from a stable margin to subduction initiation at a passive margin is controlled by the ductile strength of the lower continental crust, subcontinental lithospheric mantle, and density contrast with the suboceanic lithospheric mantle. The modeled transition takes 1 Myr to ~45 Myr to initiate subduction. However, new numerical modeling examined the time of subduction initiation at passive margins, and determined that subduction could occur only in the presence of a weak zone at the passive margin (Zhong & Li, 2020). If a horizontal convergence force (larger than 3.0×10^{12} N/m) is exerted on the passive margin, the transition duration from passive to subduction is between ~2 Myr to ~20 Myr, but for most cases is less than 10 Myr (Zhong & Li, 2020). As previously estimated, the subduction plate at 700 km depth would exert 4.9×10^{13} N/m force on the trench plate (Wan et al., 2021). Previous modeling shows that the net slab force to pull the trailing plate is 10% of the slab pull force (Schellart, 2004), ~ 4.9×10^{12} N/m for the Iranian Tethys case. After initial collision, and before slab break off, the net slab pull force together with the ridge push force from NeoTethyan midocean should match the numerical modeling requirements of Zhong and Li (2020). The new rutile age of 190 + 11 Ma (1σ) from the geological observation does not conflict with the results from numerical tests in Zhong and Li (2020; Fig. 9.5). The earliest subduction initiation at a passive margin should occur between 192 and 210 Ma.

9.4.2. Mechanism of NeoTethyan Subduction Initiation From the Eurasian Margin

The boundary between the NeoTethyan oceanic plate and Central Iran continental crust is the location where the NeoTethyan subduction zone initiated. This is because no earlier oceanic island arc (older than 100–120 Ma) has been reported in the southern Central Iran (Moghadam & Stern, 2011; Moghadam & Stern, 2015). Central Iran has

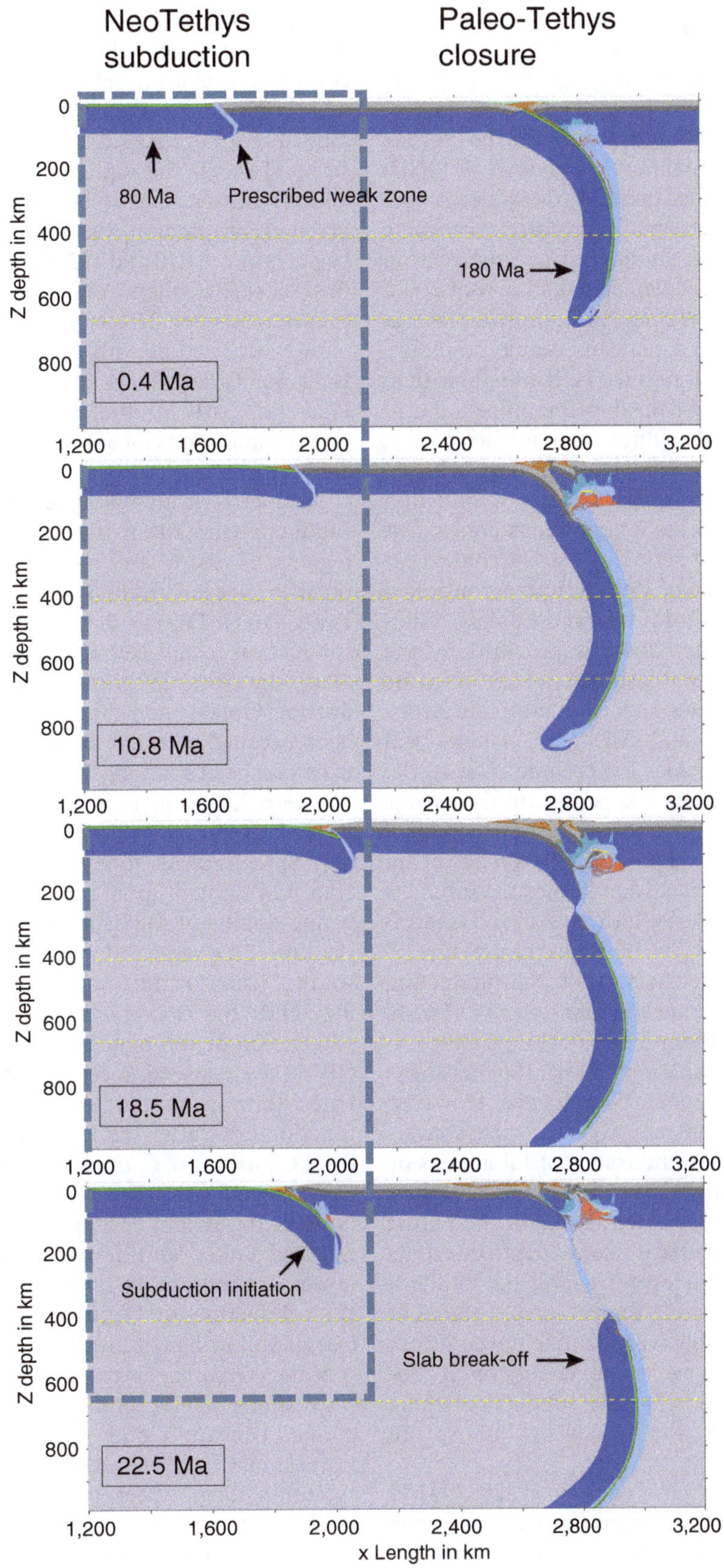

Figure 9.5 Convergent boundary force of 3.0 × 10^{12} N/m, with a prescribed weak zone at the passive margin model showing the subduction initiation taking 22.5 Myr after collision. Dashed boxes mark the passive margin (after Zhong & Li, 2020).

a current size of approximately 1 million km^2. Considering the shortening rate of Central Iran during the Arabian-Eurasian collision, including the older Cimmerian orogeny, its original size must be larger than its current size and could be comparable with the size of the Ontong Java oceanic plateau of 1.5 million km^2. Niu et al. (2003) showed that a plume-modified oceanic lithosphere is ~1% less dense than a normal oceanic lithosphere. According to new numerical modeling studies, an oceanic plateau with a strong rheological and depleted mantle root could assist the subduction zone by transferring from a plateau-continent collision zone to a plateau-oceanic boundary (Yan et al., 2021). Central Iran must be even lighter than an oceanic plateau because of the lighter continental lithosphere and thicker and lighter sedimentary cover. Therefore, its resistance to subduction during collision should be even more likely. To induce a new subduction zone, it is necessary to provide a continuous convergent force to overcome the lithospheric rigidity. The convergent force at the southern margin of Central Iran is controlled by the interplay between Central Iran and Eurasia colliding. The northward collision force is provided by the Paleo-Tethyan slab's northward subduction prior to its separation from Central Iran and its lateral continents. According to Van Hunen and Allen (2011) numerical modeling experiments, the older the oceanic crust is, the stronger it is, and the longer the period of time (over 20 Myr) it requires to break off. Because the Paleo-Tethyan opening occurred earlier than the Devonian (Chu et al., 2021), the late Triassic collision formed the subducting boundary between the Paleo-Tethyan slab and Central Iran. Such a prolonged time span indicates the presence of a strong oceanic lithosphere. As a result, the subducting Paleo-Tethyan slab may give at least another 20 Myr of convergence following collision. The docking/amalgamation and growth of mountains in northern Iran may have absorbed part of the horizontal convergence. However, kilometer-scale uplift cannot absorb all of the convergence from hundreds or even thousands of kilometers of continental collision. Following the ultimate break off of the Paleo-Tethyan slab, the low-density continental crust in the mantle will be exhumed to a shallow depth due to its buoyancy, providing an extra force to supply the southern Central Iran margin with the convergent force required to initiate subduction. Thus, following continental collision, several geological processes may generate horizontal convergence stresses sufficient to initiate subduction along preexisting weak zones in the Gondwana lithosphere (Fig. 9.5d).

The newly constrained timing of collision at 228–209 Ma indicates that a continuous convergent force can continue to 208–189 Ma before slab break off (Van Hunen & Allen, 2011; Zhong & Li, 2020; Chu et al., 2021). This coincides with our new predicted NeoTethyan subduction initiation time range of 210–192 Ma. This study's limitation is that observation of geological events, such as continental collision, subduction initiation, and accompanying slab break off, cannot be as precise as numerical modeling, and geological events may be diachronous along strike. However, the sequence between continental collision and subduction initiation remains constant. According to updated information, the Iranian NeoTethys subduction began between 210 and 192 Ma, which is younger than the final closure of the Paleo-Tethys in northern Iran, which occurred between 228 and 209 Ma (Fig. 9.6a). Similar scenarios occurred in other Tethyan regions, such as in Bulgaria, Turkey, Tibet, and southeast Asian regions (see Wan et al., 2019, and references therein), which may have broad implications for subduction zone transfer in Tethyan collision events.

The NeoTethyan region's continental collision-induced subduction initiation may be distinct from the Pacific region, as there are numerous cases of subduction polarity reversal in the south to west Pacific (Brown & Ryan, 2011). During the Tethyan development, the new subduction in the NeoTethys and Indian oceans began along the Eurasian continental margin and moved the drifting Gondwanan blocks toward the Eurasian continent because of northward oceanic subduction. There may have been a unique feature during the creation of the southern Eurasian passive margin. The rifting process created several faults along the continental margin, which may serve as weak zones for future subduction. Central Iran was located upon a mantle convection cell defined by the sinking of the Paleo-Tethyan slab in the north and the upwelling of mantle in the NeoTethyan ridge in the south, during the interval between the NeoTethys opening and the Paleo-Tethyan closing. According to subduction-driven plate tectonics of Forsyth and Uyeda (1975), the convection cell was driven by the downgoing Paleo-Tethyan oceanic slab and should continue to function even after the Paleo-Tethyan oceanic slab split from Central Iran (Conrad & Lithgow-Bertelloni, 2002). Taken together, the normal convergent force ($3–5 \times 10^{12}$ N/m) at the abrupt continent-oceanic boundary via preexisting faults and a continuing northward mantle movement beneath the continent are likely the reasons for the subduction to begin along the southern Eurasian margin following continental collision (Fig. 9.6b).

Many researchers proposed that the whole-mantle convection that is bounded by upwelling of mantle plumes beneath Gondwana and downwelling beneath Eurasia may have played important roles for the Tethyan evolution (e.g., Becker & Faccenna, 2011; Jolivet et al., 2016; Faccenna et al., 2021). In our studies, we proposed that the subduction beneath Eurasia could cause continental rifting from Gondwana (Wan et al., 2021), drive continental northward drifting, and

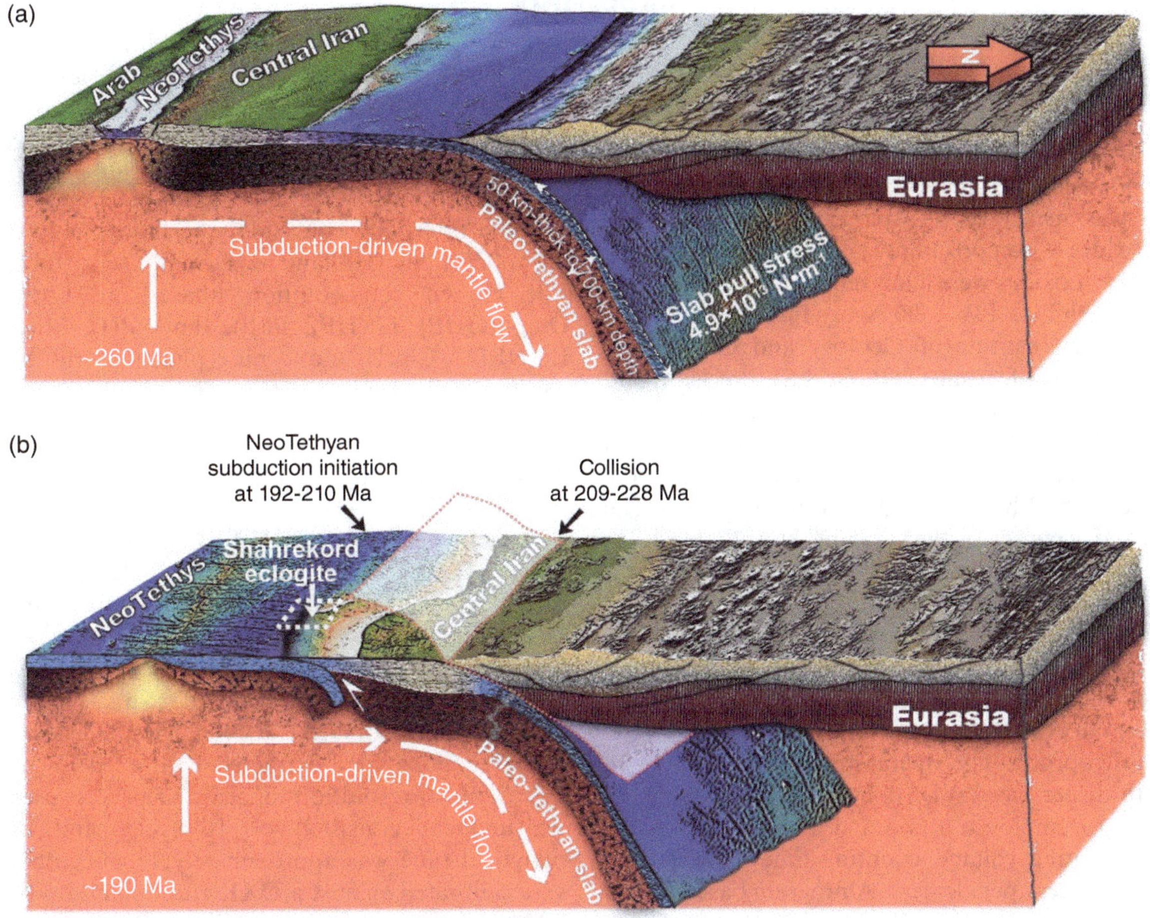

Figure 9.6 (a) Paleo-Tethyan subduction leads rifting to open NeoTethys in Iran (revised from Wan et al., 2021); (b) Collision induced NeoTethyan subduction initiation along the passive margin of Central Iran.

trigger subduction initiation along the southern Eurasian margin. The major difference between the whole-mantle convention model and the subduction-driven model is that the upwelling of mantle plumes plays an active role in the long-lived single-directed Tethyan evolution or it is only a passive feedback of the slab subduction toward Eurasia (Chen et al., 2020). Gaining deep insight into this issue demands more whole-mantle scale studies and more geological cases in the earlier history than Paleo-Tethys (Coltice et al., 2019; Robert et al., 2020; Wu et al., 2020).

9.5. CONCLUSIONS

Consider the following conclusions:

1. At 190 ± 11 Ma (1σ) from rutile U-Pb dates, a prograde-E-MORB-like eclogite along the NeoTethyan suture reached peak metamorphic condition of 2.2 GPa and 560°C. This age represents the youngest (latest) age for subduction initiation of NeoTethys.

2. Based on regional geological events and numerical modeling results, the oldest NeoTethyan subduction initiation date is 192–210 Ma, which is slightly younger than the Paleo-Tethyan closure timing of 209–228 Ma, implying collision-induced subduction initiation.

3. During the rifting of NeoTethys, the southern margin of Central Iran inherited preexisting faults. At such a boundary, the collision-induced force, combined with subduction-driving mantle flow, may aid in the initiation of northward subduction.

ACKNOWLEDGMENTS

We thank Elizabeth J. Catlos for the invitation for this contribution. We appreciate the logistical help from colleagues at GSI during the fieldwork in Iran. We thank J. M. Wang and Y. Chen for the P-T reconstructions. Discussions with W. Lin, F. Y. Wu, and W. J. Xiao helped to improve our interpretations. We thank R. Hansman for improving our writing. Comments from three anonymous reviewers and Jamshid Hassanzadeh clarified our original idea. This study was supported by the NSFC grants (91855207, 41888101).

APPENDIX

METHODS

Major and Trace Elements Analysis

The major and trace elemental contents of an eclogite (TM107) was measured as well as the representative major oxides compositions from the sample. Major-element oxides were analyzed on fused glass disks employing a Phillips PW 1500 X-ray fluorescence spectrometer at the Institute of Geology and Geophysics, Chinese Academy of Sciences (IGGCAS). Precision and accuracy of the major-element data as determined with Chinese whole-rock basalt standard GSR-3 are ≤ 5% and ca. 5%, respectively. Trace-element abundances were determined by ICP-MS (VG-PQII) at the IGGCAS. The precision and accuracy of the data are better than 5 %, as determined on GSR-3.

Mineral Analysis

Mineral chemistry of garnet, omphacite, and mica were determined at IGGCAS using a JEOL JXA-8100 electron microprobe. The operating conditions were in wavelength-dispersive (WDS) mode. Analyses were undertaken to employ a 1–2 μm beam spot size, with 15 kV accelerating voltage, counting time of 20 s, and 20 nA beam current per element. A program based on the ZAF procedure was used for data correction.

P-T Pseudosection Modeling

P-T pseudosections for the eclogite were calculated in the model system MnNCKFMASHT(O) [MnO-Na2O-CaO-K2O-FeO-MgO-Al2O3-SiO2-H2O-TiO2(–O)] considering the actual mineral assemblage. The fluid phase was assumed to be pure H2O. Chemical compositions of bulk rock and garnet were used in pseudosection calculations that were performed using the Perple_X software package (Connolly, 1990, 2009; version 6.7.8) and the internally consistent thermodynamic database of hp02ver.dat (Holland & Powell, 2011; Table 9.A1). The chosen solid-solution phases are cAmph(DP), Omph(GHP), Gt(HP), Pl(h), Ilm(WPH), Mica(CHA), Ep(HP11), Chl(HP), and pure phases of rutile, sphene, zoisite and quartz.

Rutile U-Pb Dating

Measurements of rutile U-Pb isotopes were performed using a Cameca IMS 1280HR ion microprobe (SIMS) (IGGCAS). Detailed analytical procedures are given in Li et al. (2011) for rutile; only brief descriptions are provided here. An O_2^- primary ion beam was accelerated at 13 kV, with an intensity of ca. 10–15 nA. The ellipsoidal spot was approximately 20 × 30 μm in size. Positive secondary ions were extracted with a 10-kV potential, and a mass resolution of ca. 7000 was used. Each measurement consisted of 10 cycles and the total analytical time was approximately 15 min. Pb/U ratios were calibrated against a DXK rutile standard (1782.6 ± 2.8 Ma) and were monitored using a JDX rutile standard (518 ± 4 Ma) (Li et al., 2013). Common-lead uncorrected data were used to construct a Tera-Wasserburg plot to calculate the lower intercept and U-Pb age by IsopoltR (Vermeesch, 2018). The ^{207}Pb-based common-lead correction method was used for single spot age calculations (Li et al., 2011).

Table 9.A1 Representative composition along the profile across garnet of Shahrekord eclogite (TM107)

Sample	L12-3	L12-4	L12-5	L13-34	L13-35	L13-36	L13-82	L13-83	L13-84
	Rim	Rim	Rim	Mantle	Mantle	Mantle	Rim	Rim	Rim
SiO_2	37.33	37.57	39.39	36.65	37.01	37.22	37.31	37.44	37.36
TiO_2	0.14	0.13	0.14	0.21	0.16	0.19	0.11	0.10	0.07
Al_2O_3	20.97	20.94	21.19	20.49	20.63	20.64	20.82	21.00	21.00
Cr_2O_3	0.01	0.01	0.02	0.02	0.01	0.02	0.01	0.04	0.01
Fe_2O_3	0.48	0.68	0	1.56	1.08	0.74	1.17	1.05	0.85
FeO	26.13	26.18	25.87	25.56	26.15	26.03	26.49	26.82	26.55
MnO	0.55	0.59	0.50	1.72	1.92	1.74	0.55	0.50	0.56
MgO	2.30	2.25	2.23	1.56	2.03	1.49	2.64	2.43	2.41
CaO	10.73	10.88	10.87	10.81	9.77	11.00	9.96	10.27	10.33
Na_2O	0.05	0.06	0.37	0.01	0.03	0.03	0.03	0.01	0.02
K_2O	0	0.01	0.01	0.01	0.01	0	0.02	0	0
Totals	98.69	99.31	100.6	98.61	98.8	99.09	99.11	99.66	99.16
Oxygens	12	12	12	12	12	12	12	12	12
Si	2.990	2.993	3.072	2.964	2.981	2.990	2.981	2.978	2.983
Ti	0.008	0.008	0.008	0.013	0.010	0.011	0.007	0.006	0.004
Al	1.980	1.967	1.948	1.953	1.959	1.955	1.961	1.969	1.977
Cr	0.001	0.001	0.001	0.001	0.001	0.001	0.001	0.003	0.001
Fe^3	0.029	0.041	0	0.095	0.065	0.045	0.070	0.063	0.051
Fe^2	1.750	1.745	1.687	1.729	1.761	1.749	1.770	1.784	1.773
Mn	0.037	0.040	0.033	0.118	0.131	0.118	0.037	0.034	0.038
Mg	0.275	0.267	0.259	0.188	0.244	0.178	0.314	0.288	0.287
Ca	0.921	0.929	0.908	0.937	0.843	0.947	0.853	0.875	0.884
Na	0.008	0.009	0.056	0.002	0.005	0.005	0.005	0.002	0.003
K	0	0.001	0.001	0.001	0.001	0	0.002	0	0
Sum	8	8	8	8	8	8	8	8	8
pyr	0.0922	0.0896	0.0897	0.0633	0.0819	0.0595	0.1056	0.0966	0.0962
spe	0.0124	0.0134	0.0114	0.0397	0.0440	0.0394	0.0124	0.0114	0.0127
gro	0.3087	0.3116	0.3145	0.3153	0.2830	0.3165	0.2868	0.2935	0.2964
alm	0.5867	0.5854	0.5843	0.5818	0.5911	0.5846	0.5952	0.5985	0.5946

REFERENCES

Agard, P., Omrani, J., Jolivet, L., Whitechurch, H., Vrielynck, B., Spakman, W., et al. (2011). Zagros orogeny: a subduction-dominated process, *Geological Magazine*, *148*(5–6), 692–725. https://doi.org/10.1017/S001675681100046x.

Agard, P., Yamato, P., Jolivet, L., & Burov, E. (2009). Exhumation of oceanic blueschists and eclogites in subduction zones: Timing and mechanisms. *Earth-Science Reviews*, *92*(1–2), 53–79. https://doi.org/10.1016/j.earscirev.2008.11.002

Ahadnejad, V., Valizadeh, M.-V., Deevsalar, R., & Rezaei-Kahkhaei, M. (2011). Age and geotectonic position of the Malayer granitoids: Implication for plutonism in the Sanandaj-Sirjan Zone, *W Iran, Neues Jahrbuch für Geologie und Paläontologie-Abhandlungen*, *261*(1), 61–75.

Ao, S. J., Xiao, W. J., Jafari, M. K., Talebian, M., Chen, L., Wan, B., et al. (2016). U-Pb zircon ages, field geology and geochemistry of the Kermanshah ophiolite (Iran): From continental rifting at 79 Ma to oceanic core complex at ca. 36 Ma in the southern Neo-Tethys. *Gondwana Research*, *31*, 305–318. https://doi.org/10.1016/j.gr.2015.01.014

Arvin, M., Pan, Y., Dargahi, S., Malekizadeh, A., & Babaei, A. (2007). Petrochemistry of the Siah-Kuh granitoid stock southwest of Kerman, Iran: Implications for initiation of Neotethys subduction. *Journal of Asian Earth Sciences*, *30* (3–4), 474–489.

Azizi, H., & Stern, R. J. (2019). Jurassic igneous rocks of the central Sanandaj–Sirjan zone (Iran) mark a propagating continental rift, not a magmatic arc. *Terra Nova*, *31*(5), 415–423. https://doi.org/10.1111/ter.12404

Becker, T. W., & Faccenna, C. (2011). Mantle conveyor beneath the Tethyan collisional belt. *Earth and Planetary Science Letters*, *310*(3–4), 453–461. https://doi.org/10.1016/j.epsl.2011.08.021

Bergman, S. C., Eldrett, J. S., & Minisini, D. (2021). Phanerozoic Large Igneous Province, Petroleum System, and Source Rock Links. In *Large Igneous Provinces* (pp. 191–228). https://doi.org/10.1002/9781119507444.ch9

Bonnet, G., Agard, P., Whitechurch, H., Fournier, M., Angiboust, S., Caron, B., et al. (2020). Fossil seamount in southeast Zagros records intraoceanic arc to back-arc transition: New constraints for the evolution of the Neotethys. *Gondwana Research*, *81*, 423–444. https://doi.org/10.1016/j.gr.2019.10.019

Brown, D., & Ryan, P. D. (2011). *Arc-continent collision*. London, New York: Springer Science & Business Media, Springer Heidelberg Dordrecht.

Chen, L., Wang, X., Liang, X., Wan, B., & Liu, L. (2020). Subduction tectonics vs. plume tectonics: Discussion on driving forces for plate motion. *Science China Earth Sciences*, *63*(3), 315–328. https://doi.org/10.1007/s11430-019-9538-2

Cherniak, D. J. (2000). Pb diffusion in rutile, *Contributions to Mineralogy and Petrology*, *139*(2), 198–207. https://doi.org/10.1007/PL00007671

Chiu, H.-Y., Chung, S.-L., Zarrinkoub, M. H., Melkonyan, R., Pang, K.-N., Lee, H.-Y., et al. (2017). Zircon Hf isotopic constraints on magmatic and tectonic evolution in Iran: Implications for crustal growth in the Tethyan orogenic belt. *Journal of Asian Earth Sciences*, *145*, 652–669. https://doi.org/10.1016/j.jseaes.2017.06.011

Chiu, H.-Y., Chung, S.-L., Zarrinkoub, M. H., Mohammadi, S. S., Khatib, M. M., & Iizuka, Y. (2013). Zircon U-Pb age constraints from Iran on the magmatic evolution related to Neotethyan subduction and Zagros orogeny. *Lithos*, *162–163*, 70–87. https://doi.org/10.1016/j.lithos.2013.01.006

Chu, Y., Wan, B., Allen, M. B., Chen, L., Lin, W., Talebian, M., et al. (2021). Detrital zircon age constraints on the evolution of Paleo-Tethys in NE Iran: Implications for subduction and collision tectonics. *Tectonics*, *40*, e2020TC006680. https://doi.org/10.1029/2020TC006680

Coltice, N., Husson, L., Faccenna, C., & Arnould, M. (2019). What drives tectonic plates? *Science Advances*, *5*(10), eaax4295. https://doi.org/10.1126/sciadv.aax4295

Conrad, C. P., & Lithgow-Bertelloni, C. (2002). How mantle slabs drive plate tectonics. *Science*, *298*(5591), 207–209. https://doi.org/10.1126/science.1074161

Connolly, J. A. D. (1990). Multivariable phase diagrams: An algorithm based on generalized thermodynamics. *American Journal of Science*, *290*(6), 666–718. doi:10.2475/ajs.290.6.666

Connolly, J. A. D. (2009). The geodynamic equation of state: What and how. *Geochemistry, Geophysics, Geosystems*, *10*, 10.1029/2009GC002540.https://doi.org/10.1029/2009GC002540

Coulthard, D. A., Jr., Reagan, M. K., Shimizu, K., Bindeman, I. N., Brounce, M., Almeev, R. R., et al. (2021). Magma Source Evolution Following Subduction Initiation: Evidence From the Element Concentrations, Stable Isotope Ratios, and Water Contents of Volcanic Glasses From the Bonin Forearc (IODP Expedition 352). *Geochemistry, Geophysics, Geosystems*, *22*(1). https://doi.org/10.1029/2020gc009054

Davoudian, A. R., Genser, J., Neubauer, F., & Shabanian, N. (2016). 40Ar/39Ar mineral ages of eclogites from North Shahrekord in the Sanandaj-Sirjan Zone, Iran: Implications for the tectonic evolution of Zagros orogen. *Gondwana Research*, *37*, 216–240. https://doi.org/10.1016/j.gr.2016.05.013.

DeMets, C., Gordon, R. G., & Argus, D. F. (2010). Geologically current plate motions. *Geophysical Journal International*, *181*(1), 1–80. https://doi.org/10.1111/j.1365-246X.2009.04491.x

Emami, N., & Khalili, M. (2008). Mineralogical and geochemical constraints of Jurassic fossil hydrothermal alteration associated with an calc-alkaline volcano-sedimentary complex in Sanandaj-Sirjan Zone, Southwest of Iran. *Journal of Applied Sciences*, *8*(9), 1600–1611.

Esna-Ashari, A., Tiepolo, M., & Hassanzadeh, J. (2016). On the occurrence and implications of Jurassic primary continental boninite-like melts in the Zagros orogen. *Lithos*, *258–259*, 37–57. https://doi.org/10.1016/j.lithos.2016.04.017

Faccenna, C., Becker, T. W., Holt, A. F., & Brun, J. P. (2021). Mountain building, mantle convection, and supercontinents: revisited. *Earth and Planetary Science Letters*, *564*. https://doi.org/10.1016/j.epsl.2021.116905

Forsyth, D., & Uyeda, S. (1975). On the relative importance of the driving forces of plate motion. *Geophysical Journal International*, *43*(1), 163–200.

Fürsich, F. T., Wilmsen, M., Seyed-Emami, K., Majidifard, M. R., Brunet, M.-F., Wilmsen, M., et al. (2009). Lithostratigraphy of the Upper Triassic–Middle Jurassic Shemshak Group of Northern Iran. In *South Caspian to Central Iran basins* (pp. 129–160). Geological Society of London. https://doi.org/10.1144/sp312.6

Geological Survey of Iran (2009). *1:5,000,000 International geological map of the Middle East*. Geological Survey of Iran, Tehran.

Gerya, T. V., Stern, R. J., Baes, M., Sobolev, S. V., & Whattam, S. A. (2015). Plate tectonics on the Earth triggered by plume-induced subduction initiation. *Nature*, *527*(7577), 221–225. https://doi.org/10.1038/nature15752

Guilmette, C., Smit, M. A., Van Hinsbergen, D. J. J., Gürer, D., Corfu, F., Charette, B., et al. (2018). Forced subduction initiation recorded in the sole and crust of the Semail Ophiolite of Oman. *Nature Geoscience*, *11*(9), 688–695. https://doi.org/10.1038/s41561-018-0209-2

Hall, R. (2017). Southeast Asia: New views of the geology of the Malay archipelago. *Annual Review of Earth and Planetary Sciences*, *45*(1), 331–358. https://doi.org/10.1146/annurev-earth-063016-020633

Hassanzadeh, J., & Wernicke, B. P. (2016). The Neotethyan Sanandaj-Sirjan zone of Iran as an archetype for passive margin-arc transitions. *Tectonics*, *35*(3), 586–621. https://doi.org/10.1002/2015tc003926

Holland, T. J. B., & Powell, R. (2011). An improved and extended internally consistent thermodynamic dataset for phases of petrological interest, involving a new equation of state for solids. *Journal of Metamorphic Geology*, *29*(3), 333–383.

Jamali Ashtiani, R., Hassanzadeh, J., Schmitt, A. K., Sudo, M., Timmerman, M., Günter, C., et al. (2020). Geochronology and geochemistry of subducted Cadomian continental basement in central Iran: Decompressional anatexis along the Jurassic Neotethys margin. *Gondwana Research*, *82*, 354–366. https://doi.org/10.1016/j.gr.2020.01.005

Jolivet, L., Faccenna, C., Agard, P., Frizon de Lamotte, D., Menant, A., Sternai, P., et al. (2016). Neo-Tethys geodynamics and mantle convection: from extension to compression in Africa and a conceptual model for obduction. *Canadian Journal of Earth Sciences*, *53*(11), 1190–1204. https://doi.org/10.1139/cjes-2015-0118

Koop, W., Stoneley, R., Ridd, M., Murphy, R., Osmaston, M., & Kholief, M. (1982). Subsidence history of the middle east Zagros Basin, Permian to recent. *Philosophical Transactions of The Royal Society A: Mathematical, Physical and Engineering Sciences*, *305*, 149–167. https://doi.org/10.1098/rsta.1982.0031

Lawver, L. A., Dalziel, I. W., Norton, I. O., Gahagan, L., & Davis, J. (2015). *The PLATES 2014 Atlas of Plate Reconstructions (550 Ma to Present Day)*. PLATES Progress Report No. 374–0215, University of Texas Institute for Geophysics Technical Reports, 220.

Leng, W., & Gurnis, M. (2015). Subduction initiation at relic arcs. *Geophysical Research Letters*, *42*(17), 7014–7021.

Leven, E. J., & Gorgij, M. N. (2011). Fusulinids and stratigraphy of the Carboniferous and Permian in Iran. *Stratigraphy and Geological Correlation*, *19*(7), 687–776. https://doi.org/10.1134/s0869593811070021

Li, Q., Lin, W., Su, W., Li, X.-h, Shi, Y.-h., Liu, Y., et al. (2011). SIMS U-Pb rutile age of low-temperature eclogites from southwestern Chinese Tianshan, *NW China. Lithos*, *122* (1–2), 76–86.

Li, Q., Yang, Y., Shi, Y.-h., & Lin, W. (2013), Eclogite rutile U-Pb dating: constrant for formation and evolution of continental collisional orogen. *Chinese Science Bulletin*, *58*(23), 2279–2284. https://doi.org/10.1360/972013-590

Malek-Mahmoudi, F., Reza Davoudian, A., Shabanian, N., Azizi, H., Asahara, Y., Neubauer, F., et al. (2017). Geochemistry of metabasites from the North Shahrekord metamorphic complex, Sanandaj-Sirjan Zone: Geodynamic implications for the Pan-African basement in Iran. *Precambrian Research*, *293*, 56–72. https://doi.org/10.1016/j.precamres.2017.03.003

Maunder, B., Prytulak, J., Goes, S., & Reagan, M. (2020). Rapid subduction initiation and magmatism in the Western Pacific driven by internal vertical forces. *Nature Communications*, *11*(1), 1874. https://doi.org/10.1038/s41467-020-15737-4

McQuarrie, N., & Van Hinsbergen, D. J. J. (2013). Retrodeforming the Arabia-Eurasia collision zone: Age of collision versus magnitude of continental subduction. *Geology*, *41*(3), 315–318. https://doi.org/10.1130/G33591.1

Merdith, A. S., Williams, S. E., Brune, S., Collins, A. S., & Müller, R. D. (2019). Rift and plate boundary evolution across two supercontinent cycles. *Global and Planetary Change*, *173*, 1–14. https://doi.org/10.1016/j.gloplacha.2018.11.006

Moghadam, H. S., & Stern, R. J. (2011). Geodynamic evolution of Upper Cretaceous Zagros ophiolites: Formation of oceanic lithosphere above a nascent subduction zone. *Geological Magazine*, *148*(5–6), 762–801.

Moghadam, H. S., & Stern, R. J. (2015). Ophiolites of Iran: Keys to understanding the tectonic evolution of SW Asia: (II) Mesozoic ophiolites. *Journal of Asian Earth Sciences*, *100*(0), 31–59. https://doi.org/10.1016/j.jseaes.2014.12.016

Moghadam, H. S., Li, Q.-L., Griffin, W. L., Stern, R. J., Santos, J. F., Ducea, M. N., et al. (2022). Temporal changes in subduction- to collision-related magmatism in the Neotethyan orogen: The Southeast Iran example. *Earth-Science Reviews*, *226*, 103930. https://doi.org/10.1016/j.earscirev.2022.103930

Natal'in, B. A., & Şengör, A. M. C. (2005). Late Palaeozoic to Triassic evolution of the Turan and Scythian platforms: The pre-history of the Palaeo-Tethyan closure. *Tectonophysics*, *404*(3), 175–202. https://doi.org/10.1016/j.tecto.2005.04.011

Nikolaeva, K., Gerya, T. V., & Marques, F. O. (2010). Subduction initiation at passive margins: Numerical modeling. *Journal of Geophysical Research*, *115*(B3). https://doi.org/10.1029/2009jb006549

Niu, Y., O'Hara, M. J., & Pearce, J. A. (2003). Initiation of subduction zones as a consequence of lateral compositional buoyancy contrast within the lithosphere: A petrological perspective. *Journal of Petrology*, *44*(5), 851–866. https://doi.org/10.1093/petrology/44.5.851

Okay, A. I., Sunal, G., Sherlock, S., Kylander-Clark, A. R. C., & Özcan, E. (2020). İzmir-Ankara Suture as a Triassic to Cretaceous Plate Boundary: Data From Central Anatolia. *Tectonics*, *39*(5). https://doi.org/10.1029/2019tc005849

Omrani, J., Agard, P., Whitechurch, H., Benoit, M., Prouteau, G., & Jolivet, L. (2008). Arc-magmatism and subduction history beneath the Zagros Mountains, Iran: A new report of adakites and geodynamic consequences. *Lithos*, *106*(3–4), 380–398. https://doi.org/10.1016/j.lithos.2008.09.008

Robert, B., Domeier, M., & Jakob, J. (2020). Iapetan Oceans: An analog of Tethys? *Geology*, *48*(9), 929–933. https://doi.org/10.1130/g47513.1

Schellart, W. (2004). Quantifying the net slab pull force as a driving mechanism for plate tectonics. *Geophysical Research Letters*, *31*(7). https://doi.org/10.1029/2004gl019528

Scotese, C. R., Song, H., Mills, B. J. W., & van der Meer, D. G. (2021). Phanerozoic paleotemperatures: The earth's changing climate during the last 540 million years. *Earth-Science Reviews*, *215*. https://doi.org/10.1016/j.earscirev.2021.103503

Stampfli, G. M., Hochard, C., Vérard, C., Wilhem, C., & VonRaumer, J. (2013). The formation of Pangea. *Tectonophysics*, *593*, 1–19. https://doi.org/10.1016/j.tecto.2013.02.037

Stern, R. J. (2004). Subduction initiation: spontaneous and induced. *Earth and Planetary Science Letters*, *226*(3–4), 275–292. https://doi.org/10.1016/j.epsl.2004.08.007

Stern, R. J., & Gerya, T. (2018). Subduction initiation in nature and models: A review. *Tectonophysics*, *746*, 173–198. https://doi.org/10.1016/j.tecto.2017.10.014

Stern, R. J., Moghadam, H. S., Pirouz, M., & Mooney, W. (2021). The Geodynamic Evolution of Iran. *Annual Review of Earth and Planetary Sciences*, *49*(1), 9–36. https://doi.org/10.1146/annurev-earth-071620-052109

Sun, S. S., & McDonough, W. F. (1989). Chemical and isotopic systematics of ocean basins: implications for mantle composition and processes. *Geological Society of London, Special Publications*, *42*, 313–345.

Syracuse, E. M., & Abers, G. A. (2006). Global compilation of variations in slab depth beneath arc volcanoes and implications. *Geochemistry, Geophysics, Geosystems*, *7*(5). https://doi.org/10.1029/2005GC001045

Topuz, G. l., Göçmengil, G., Rolland, Y., Çáelik, Ö. F., Zack, T., & Schmitt, A. K. (2013). Jurassic accretionary complex and ophiolite from northeast Turkey: No evidence for the Cimmerian continental ribbon. *Geology*, *41*(2), 255–258. https://doi.org/10.1130/g33577.1

Torsvik, T. H., & Cocks, L. R. M. (2017). *Earth history and palaeogeography*. Cambridge University Press, United Kingdom. https://doi.org/10.1017/9781316225523

Van Hinsbergen, D. J. J., et al. (2021). A record of plume–induced plate rotation triggering subduction initiation. *Nature Geoscience*, *14*(8), 626–630. https://doi.org/10.1038/s41561-021-00780-7

Van Hunen, J., & Allen, M. B. (2011). Continental collision and slab break-off: A comparison of 3-D numerical models with observations. *Earth and Planetary Science Letters*, *302*(1), 27–37.

Verdel, C., Wernicke, B. P., Hassanzadeh, J., & Guest, B. (2011). A Paleogene extensional arc flare-up in Iran. *Tectonics*, *30*(3). https://doi.org/10.1029/2010tc002809

Vermeesch, P. (2018). IsoplotR: A free and open toolbox for geochronology. *Geoscience Frontiers*, *9*(5), 1479–1493. https://doi.org/10.1016/j.gsf.2018.04.001

Wakabayashi, J. (2011). Mélanges of the Franciscan Complex, California: Diverse structural settings, evidence for sedimentary mixing, and their connection to subduction processes. *Geological Society of America Bulletin Special Papers*, *480*, 117–141. https://doi.org/10.1130/2011.248005

Wan, B., Chu, Y., Chen, L., Liang, X., Zhang, Z., Ao, S., et al. (2021). Paleo-Tethys subduction induced slab-drag opening the Neo-Tethys: Evidence from an Iranian segment of Gondwana. *Earth-Science Reviews*, *103788*. https://doi.org/10.1016/j.earscirev.2021.103788

Wan, B., Wu, F., Chen, L., Zhao, L., Liang, X., Xiao, W., et al. (2019). Cyclical one-way continental rupture-drift in the Tethyan evolution: Subduction-driven plate tectonics. *Science China Earth Sciences*, *62*, 2005–2016. https://doi.org/10.1007/s11430-019-9393-4

Wang, J.-M., Lanari, P., Wu, F.-Y., Zhang, J.-J., Khanal, G. P., & Yang, L. (2021). First evidence of eclogites overprinted by ultrahigh temperature metamorphism in Everest East, Himalaya: Implications for collisional tectonics on early Earth. *Earth and Planetary Science Letters*, *558*. https://doi.org/10.1016/j.epsl.2021.116760

Whattam, S. A., & Stern, R. J. (2011). The "subduction initiation rule": A key for linking ophiolites, intra-oceanic forearcs, and subduction initiation. *Contributions to Mineralogy and Petrology*, *162*(5), 1031–1045. https://doi.org/10.1007/s00410-011-0638-z

Whittaker, J. M., Williams, S. E., & Müller, R. D. (2013). Revised tectonic evolution of the Eastern Indian Ocean. *Geochemistry, Geophysics, Geosystems*, *14*(6), 1891–1909. https://doi.org/10.1002/ggge.20120

Williams, S. E., Whittaker, J. M., Halpin, J. A., & Müller, R. D. (2019). Australian-Antarctic breakup and seafloor spreading: Balancing geological and geophysical constraints. *Earth-Science Reviews*, *188*, 41–58. https://doi.org/10.1016/j.earscirev.2018.10.011

Wu, F. Y., Wan, B., Zhao, L., Xiao, W. J., & Zhu, R. X. (2020). Tethyan geodynamics. *Acta Petrologica Sinica*, *36*(6), 1627–1674 (in Chinese with English abstract). https://doi.org/10.18654/1000-0569/2020.06.01

Yan, Z., Chen, L., Xiong, X., Wan, B., & Xu, H. (2021). Oceanic plateau and subduction zone jump: two-dimensional thermomechanical modeling. *Journal of Geophysical Research: Solid Earth*, *126*(7). https://doi.org/10.1029/2021jb021855

Zanchetta, S., Berra, F., Zanchi, A., Bergomi, M., Caridroit, M., Nicora, A., et al. (2013). The record of the Late Palaeozoic active margin of the Palaeotethys in NE Iran: Constraints on the Cimmerian orogeny. *Gondwana Research*, *24*(3), 1237–1266. https://doi.org/10.1016/j.gr.2013.02.013

Zhang, Z., et al. (2018). Geochemistry, zircon U-Pb and Hf isotope for granitoids, NW Sanandaj-Sirjan zone, Iran: Implications for Mesozoic-Cenozoic episodic magmatism during Neo-Tethyan lithospheric subduction. *Gondwana Research*, *62*, 227–245. https://doi.org/10.1016/j.gr.2018.04.002

Zhang, Z., Xiao, W., Majidifard, M. R., Zhu, R., Wan, B., Ao, S., et al. (2017). Detrital zircon provenance analysis in the Zagros Orogen, SW Iran: implications for the amalgamation history of the Neo-Tethys. *International Journal of Earth Sciences*, *106*(4), 1223–1238. https://doi.org/10.1007/s00531-016-1314-3

Zhong, X., & Li, Z.-H. (2020). Subduction initiation during collision-induced subduction transference: numerical modeling and implications for the Tethyan evolution. *Journal of Geophysical Research: Solid Earth*, *125*(2), e2019JB019288. https://doi.org/10.1029/2019jb019288

Zhong, X., & Li, Z.-H. (2022). Wedge-Shaped Southern Indian Continental Margin Without Proper Weakness Hinders Subduction Initiation, *Geochemistry, Geophysics, Geosystems*, *23*(2), e2021GC009998. https://doi.org/10.1029/2021GC009998

Zhou, X., Li, Z. H., Gerya, T. V., & Stern, R. J. (2020). Lateral propagation-induced subduction initiation at passive continental margins controlled by preexisting lithospheric weakness. *Sci Adv*, *6*(10), eaaz1048. https://doi.org/10.1126/sciadv.aaz1048

Part III

North America Mountain Building

10

Stratigraphic and Thermal Maturity Evidence for a Break-Back Thrust Sequence in the Southern Appalachian Thrust Belt, Alabama, USA

Jack C. Pashin

ABSTRACT

The timing and sequence of deformation of the southern Appalachian thrust belt in Alabama are analyzed by integrating structural, stratigraphic, and thermal maturity data. The analyses indicate that thrusting in the distal foreland began during the Late Mississippian with structures becoming progressively younger toward the proximal foreland. Structural and stratigraphic evidence constrain the thrusting age of the distal foreland and, in contrast, thermal maturity data was essential for determining the thrusting time in the proximal foreland. This approach provides new insight into the spatial and temporal dynamics of the southern Appalachian thrust belt. The application of similar methods to other thrust belts may help constrain the age and sequence of deformation.

10.1. INTRODUCTION

Determining the sequence of thrusting in the Appalachian foreland is complicated by a lack of synsedimentary growth structures in much of the orogen. Accordingly, most researchers have relied on cross-cutting relationships among faults and fault-related folds to recognize in-sequence (break-forward) and out-of-sequence (break-back) thrust elements (e.g., Tull, 1984; Thomas & Bayona, 2005; Solis, 2010). Where Appalachian thrust belt structures intersect the greater Black Warrior Basin, large-scale folds and faults of compressional and extensional origin are developed within the synorogenic sedimentary cover, which helps provide insight into the age of the foreland thrust belt as well as the sequence of thrusting (Pashin, 1994; Groshong et al., 2010; Ettensohn et al., 2019). This chapter synthesizes research on the southern Appalachian foreland to provide a new perspective on the age and relative timing of thrusting and folding in the greater Black Warrior Basin.

The first evidence for Pennsylvanian synsedimentary deformation in the southern Appalachian thrust belt became apparent during the analysis of the Oak Grove coalbed methane field, where the Sequatchie anticline, which is a major first-order Appalachian thrust-related fold, plunges into and terminates in the synorogenic sedimentary cover (Pashin, 1994, 1998). Since that time, a range of geologic evidence has been assembled from compressional and extensional structures that helps resolve the timing of deformation in the southern Appalachian thrust belt of Alabama (e.g., Pashin et al., 1995; Groshong et al., 2010; Ettensohn et al., 2019), and this chapter presents the first synthesis of this evidence.

The hypotheses of this research are (1) that deformation of the frontal Appalachian structures in this region began near the Mississippian-Pennsylvanian boundary and (2) that this deformation left a record within the synorogenic sedimentary cover that reveals the sequence of deformation. These hypotheses have been tested using a

Boone Pickens School of Geology, Oklahoma State University, Stillwater, Oklahoma, USA

Compressional Tectonics: Plate Convergence to Mountain Building, Geophysical Monograph 277, First Edition.
Edited by Elizabeth J. Catlos and İbrahim Çemen.

DOI:10.1002/9781119773856.ch10

variety of subsurface stratigraphic, sedimentologic, and structural techniques. Analysis of thermal maturity data from the numerous coal beds preserved in the sedimentary cover provides further evidence that helps resolve the timing and sequence of deformation.

10.1.1. Geologic Setting

The Black Warrior Basin is a late Paleozoic foreland basin that formed adjacent to the Appalachian-Ouachita orogenic juncture on the Alabama continental promontory (Fig. 10.1). The greater Black Warrior Basin includes the main part of the Black Warrior Basin, which is northwest of the Birmingham anticlinorium, as well as the Birmingham anticlinorium, Cahaba synclinorium and Coosa synclinorium, which contain a thick succession of sedimentary rocks that were originally connected to those in the main part of the Black Warrior Basin (Thomas, 1997; Fig. 10.2). Appalachian folds and associated thrust faults are developed in the southeastern part of the greater Black Warrior Basin, and comprise the Wiley dome, the Sequatchie anticline, the Blue Creek anticline, the Birmingham anticlinorium, the Cahaba synclinorium, and the Coosa synclinorium, which are the focus of this study (Fig. 10.2).

10.1.2. Stratigraphic Framework

Strata in the greater Black Warrior Basin range from Cambrian through Pennsylvanian in age (Fig. 10.3). Crystalline basement is typically overlain by shale of the Cambrian Rome and Conasauga formations, which are up to 1.8 km thick in the thrust belt. This shale section is overlain by a Cambrian-Ordovician carbonate succession that is about 1.5–4.0 km thick and includes the upper part of the Conasauga Formation, the Ketona Dolomite,

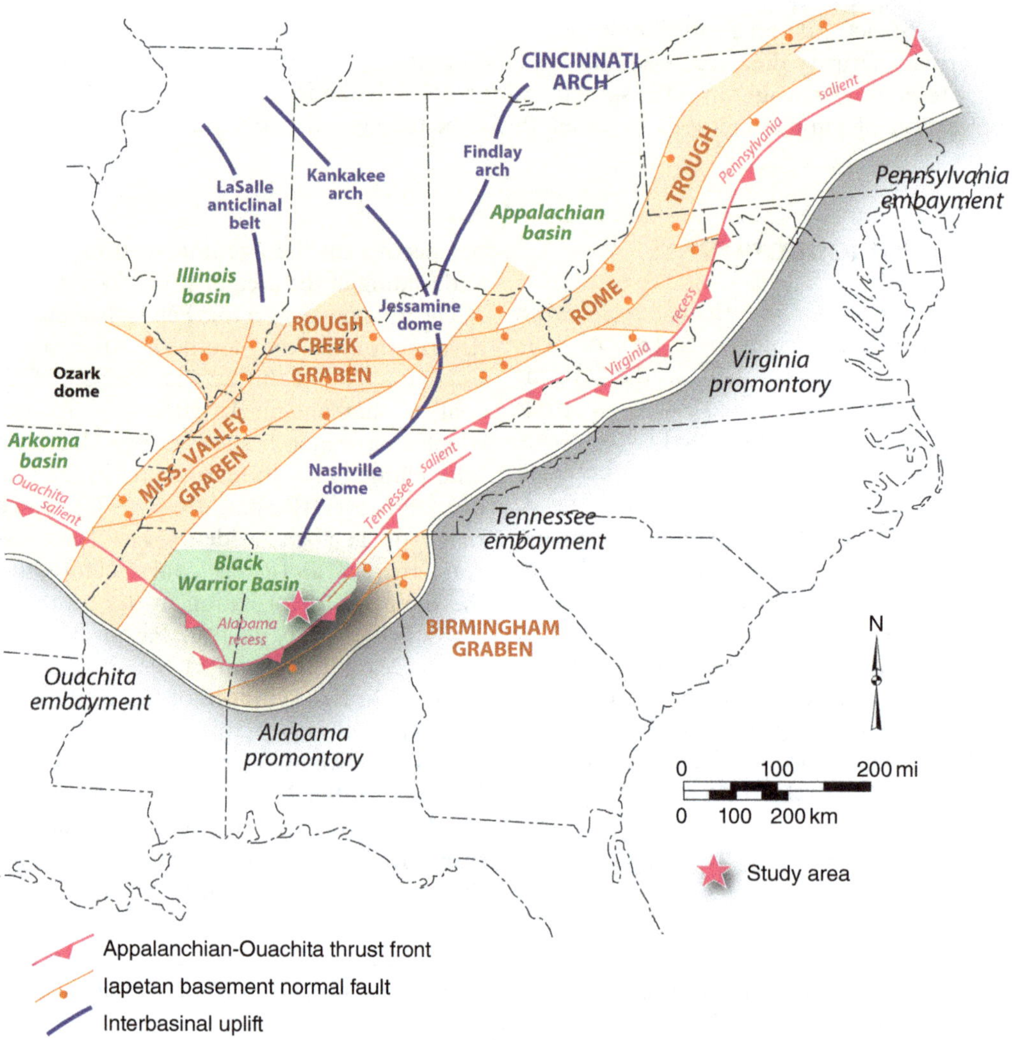

Figure 10.1 Map showing tectonic setting of the greater Black Warrior Basin (after Thomas, 1991).

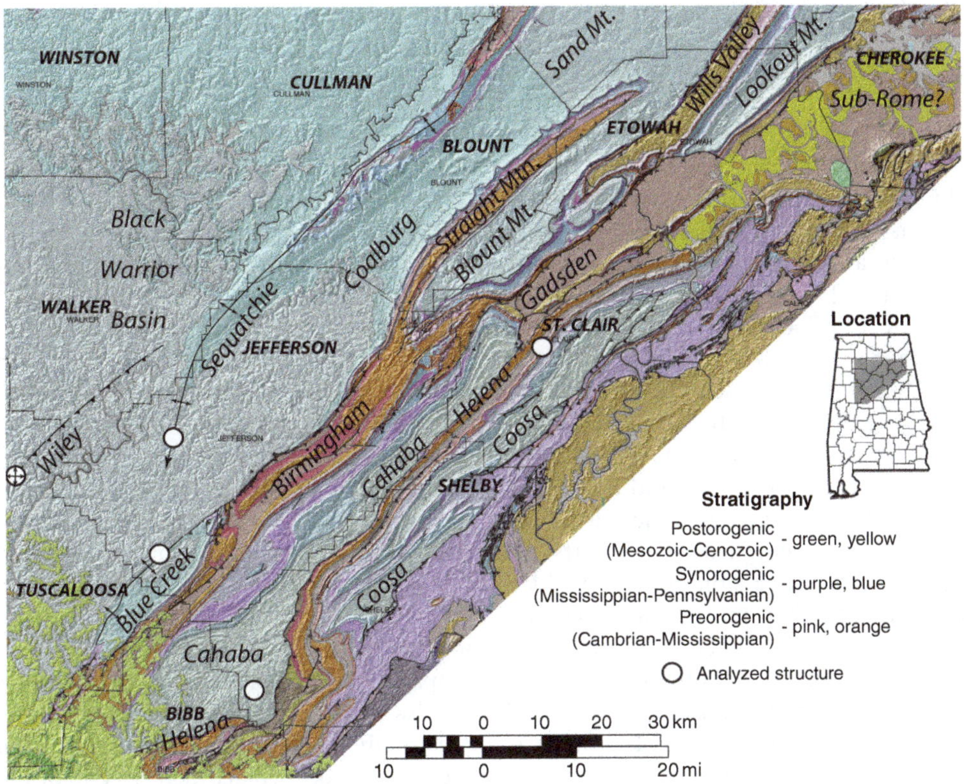

Figure 10.2 Geologic map showing geologic framework and locations of major structures studied in the greater Black Warrior Basin, which includes the Black Warrior Basin, the Birmingham anticlinorium, the Cahaba synclinorium, and the Coosa synclinorium (digital geologic map from Geological Survey of Alabama).

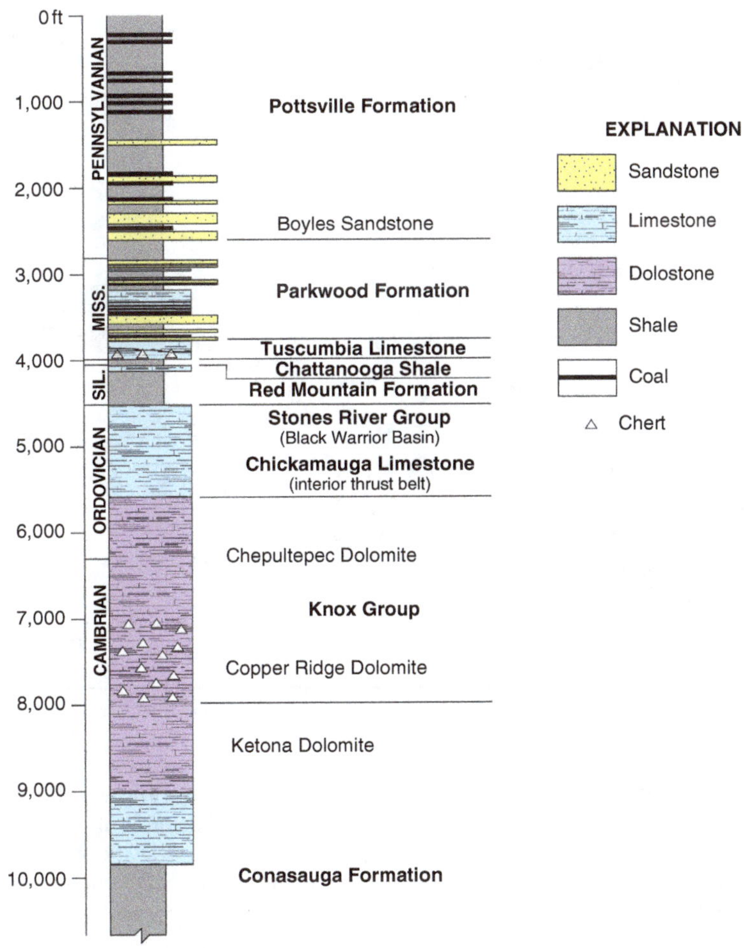

Figure 10.3 Generalized stratigraphic section of the Black Warrior Basin showing major stratigraphic units (after Clark et al., 2013).

the Knox Group, and the Chickamauga Limestone–Stones River Group. The Silurian-Devonian section is typically 0.1–0.7 km thick and includes mixed carbonate-siliciclastic strata assigned to the Red Mountain Formation, Frog Mountain Sandstone, and the organic-rich Chattanooga Shale. The Mississippian section is composed of carbonate and siliciclastic strata and ranges in thickness from about 0.5 to 0.8 km. Carbonate units, including the Fort Payne Chert, Tuscumbia Limestone, and Bangor Limestone dominate the section in the northeastern part of the basin, whereas siliciclastic-dominated units, such as the Floyd Shale and Parkwood Formation, form the bulk of the section in the southwest. The Pennsylvanian section is assigned to the Pottsville Formation and contains a succession of interbedded shale, sandstone, and coal that reaches a maximum thickness of about 2.6 km in the Cahaba and Coosa synclinoria. The Pottsville is generally considered to be of Lower Pennsylvanian age (late Bashkirian), but detrital geochronology reveals that muscovite and biotite near the top of the Pottsville in Alabama and Mississippi is probably Middle Pennsylvanian (late Moscovian; 308 ± 1.2 Ma; Moore, 2012; Uddin et al., 2016). Paleozoic strata of the greater Black Warrior Basin are overlain with angular unconformity by an onlapping wedge of Cretaceous-Tertiary strata that are part of the Gulf of Mexico Basin.

The Rome-Conasauga section is thought to record the late stages of Iapetan rifting, whereas the overlying carbonate section records the transition to a passive shelf (e.g., Thomas, 1988, 1991; Thibodeaux, 2015; Fig. 10.4). Subsidence of the greater Black Warrior Basin was modest from Middle Ordovician through Early Mississippian time, and the effects of the Taconic and Acadian orogenies were limited in most of the region. Subsidence began accelerating during Mississippian time, signaling the start of Appalachian-Ouachita orogenesis and foreland basin development on the southwestern margin of the Alabama Promontory (Thomas, 1972; Whiting & Thomas, 1994). Subsidence continued accelerating during the latest Mississippian and the Pennsylvanian as Appalachian-Ouachita orogenesis proceeded along the southwest (Ouachita) and southeast (Appalachian) margins of the Alabama Promontory with maximum burial most likely taking place near the end of the Pennsylvanian or during the Early Permian (Thibodeaux, 2015).

Following this was a prolonged episode of postorogenic unroofing in which about 2–4 km of the foreland

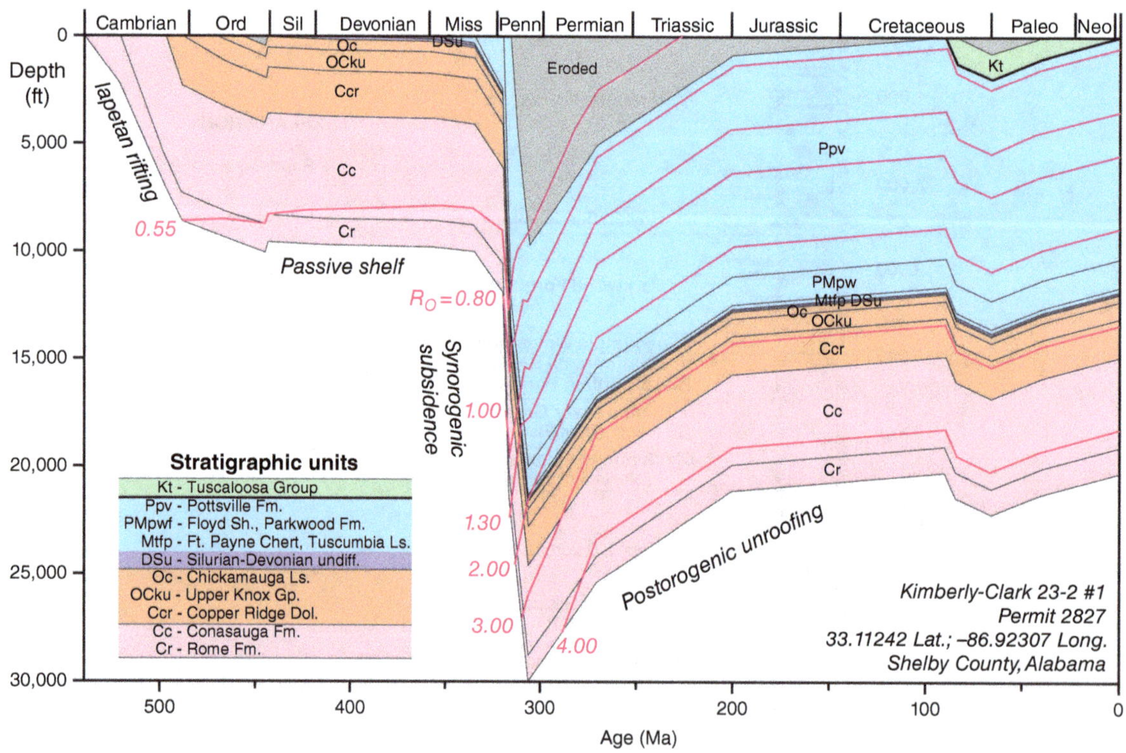

Figure 10.4 Burial and thermal maturation history of the Paleozoic section in the Cahaba synclinorium of the Appalachian thrust belt (after Thibodeaux, 2015). The plot was developed using Schlumberger PetroMod software showing initial thermal maturation during Paleozoic rifting and passive shelf development and major maturation coincident with late Paleozoic synorogenic subsidence and the early stages of unroofing. Maturation was effectively complete by the end of Permian time.

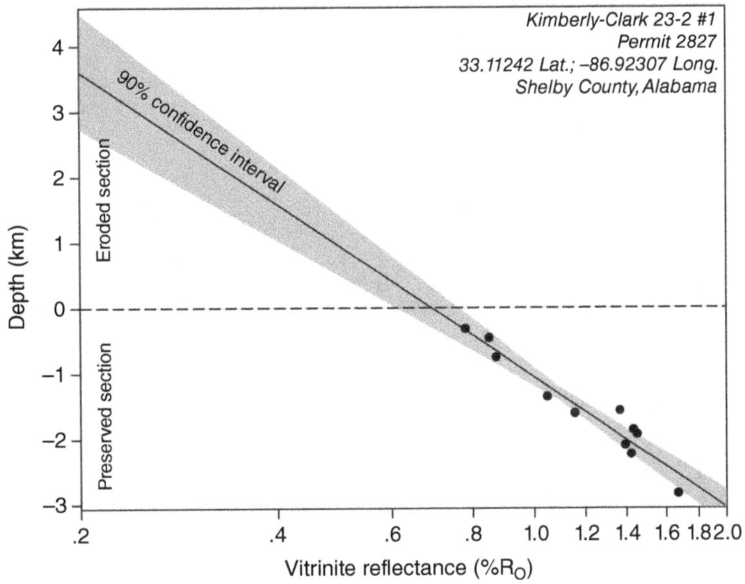

Figure 10.5 Vitrinite reflectance profile for the Kimberly Clark 23-2 #1 well, Cahaba synclinorium, Shelby County, Alabama. Projection of the regression line for the profile suggests about 3.6 km of sediment was eroded from the top of the Pottsville Formation during postorogenic unroofing.

basin fill was eroded in the greater Black Warrior Basin (Telle et al., 1987; Carroll et al., 1995; Thibodeaux, 2015; Fig. 10.5). The region was buried below Cretaceous-Tertiary strata of the Gulf of Mexico Basin, which have subsequently been eroded from the northeastern part of the region.

Structural Framework

Strata in the Black Warrior Basin generally dip homoclinally southwest at about 0.8° and are offset by numerous normal faults that form horst-and-graben structures (Fig. 10.6). Fault displacement increases southwest toward the Ouachita orogen; faults penetrate basement in the southwestern part of the basin, having vertical separation in places that exceeds 400 m and trace lengths on the order of 10 to >100 km. The history of thick-skinned normal faulting is complex, with some faults terminating near the top of the Rome-Conasauga section, indicating an origin related to Iapetan rifting, and many offsetting the complete preserved Paleozoic section (Groshong et al., 2010).

Growth strata related to folding and faulting indicate that different structures were active at a variety of times during the Paleozoic, and some lack growth strata and thus appear to postdate the preserved Paleozoic stratigraphy. All of the faults in the Paleozoic section terminate at or below the angular unconformity marking the base of the homoclinal Cretaceous section, which lacks significant faults in the study area. Locally, some normal faults appear to be detached in Cambrian shale, marking a transition zone between thick-skinned and thin-skinned faulting. In the northeastern part of the basin, normal faults typically have vertical separation of less than 70 m, and geologic mapping and area balancing indicates that these faults formed above a widespread detachment near the base of the Pottsville Formation (Pashin & Groshong, 1998; Groshong et al., 2010).

A large part of the thrust belt is developed above the Birmingham Graben, which is an Iapetan rift (Figs. 10.1, 10.7). Indeed, Iapetan basement faults appear to have played an important role in localizing large-scale frontal thrust ramps in the southern Appalachians (Thomas, 1985, 2001, 2007; Thomas & Bayona, 2005). Within the greater Black Warrior Basin, the thrust belt is detached in Cambrian shale, which is overlain by a thick succession of brittle Cambrian-Ordovician carbonate. Mississippian-Pennsylvanian synorogenic strata have been translated toward the foreland above the brittle carbonate, and only localized detachments associated with second-order structures are developed above the Cambrian shale (e.g., Pashin & Groshong, 1998). Southeast of the Coosa synclinorium, however, detachment geometry becomes much more complex, and a major upper level detachment is developed near the base of the Mississippian section (Thomas & Drahovzal, 1974; Thomas, 2007; Thomas & Bayona, 2005).

Structural geometry varies considerably in the Appalachian thrust belt within the greater Black Warrior Basin. Some frontal structures have been characterized as various types of detachment folds and fault-propagation folds with steep forelimbs and gently dipping backlimbs (e.g., Pashin & Groshong, 1998; Gates,

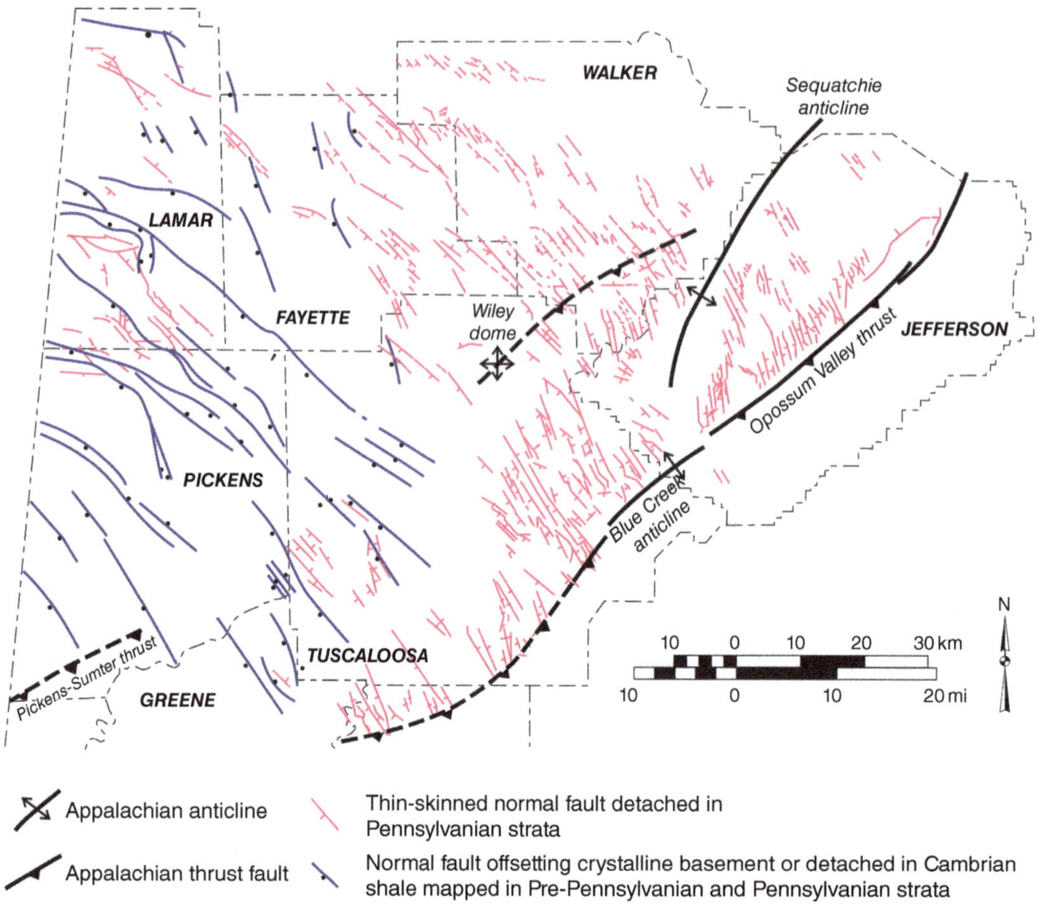

Figure 10.6 Structural trace map showing locations of major geologic structures in the Black Warrior Basin (after Pashin et al., 1991; Groshong et al., 2010).

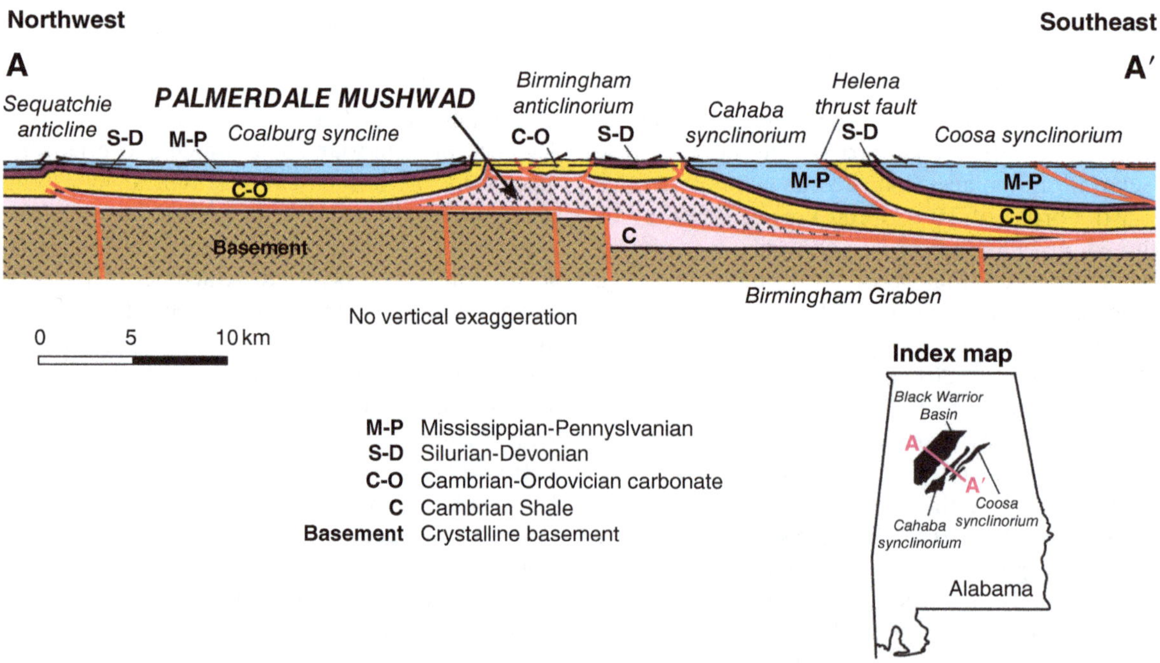

Figure 10.7 Structural cross section showing regional structural style and relationship of the Appalachian thrust belt to the Birmingham Graben (after Thomas & Bayona, 2005).

2006; Groshong et al., 2010). The southwest tip of the Sequatchie anticline has been characterized as a detachment fold, and en-echelon normal faults at the terminus and in the backlimb of the structure demonstrate left-lateral shear and a kinematic linkage between the thin-skinned extensional faults in the Black Warrior Basin and distal foreland compression (Pashin, 1994; Groshong et al., 2010). The Birmingham anticlinorium is the largest of the anticlinal structures and is cored by giant duplexes in Conasauga shale that are locally thicker than 4 km and overlie the northwest margin of the Birmingham Graben (e.g., Thomas, 2001; Thomas & Pashin, 2011; Pashin et al., 2012). The Cahaba synclinorium includes the backlimb of the Birmingham anticlinorium and contains Pennsylvanian strata that are roughly equivalent to those in the main part of the Black Warrior Basin (Pashin et al., 1995). The southeastern margin of the Cahaba synclinorium contains a large-scale footwall syncline adjacent to the Helena fault, which is a long-trace thrust fault that defines the southeastern boundary of the synclinorium. Translation along the Helena fault was great enough that hanging-wall strata dip with the frontal thrust ramp, and no hanging-wall bed cutoffs have been preserved. The Coosa synclinorium was translated toward the foreland along the Helena thrust and is the most interior structure of the southern Appalachians containing a thick Pottsville section. Here, the thickness of the Pottsville section is similar to that in the Cahaba synclinorium, but the distribution of marine strata in the fill suggests that the youngest preserved strata correlate with the middle part of the Cahaba Pottsville (Pashin, 1997; Pashin & Gastaldo, 2009).

The Cretaceous-Tertiary strata that onlap the greater Black Warrior Basin lack significant folds and faults (Thomas, 1988; Pashin et al., 1991). These strata form a simple southwest-dipping homocline that postdates folding and faulting in the greater Black Warrior Basin. Accordingly, the Mesozoic-Cenozoic section records a simple southwest tilting of the region during formation of the Gulf of Mexico Basin.

10.2. METHODS

Analysis of structures in the greater Black Warrior Basin utilized a variety of stratigraphic, structural, and thermal maturity data. Geophysical well logs and sample logs from the State Oil and Gas Board of Alabama were used to identify rock types based on spontaneous potential, gamma-ray, resistivity, and density log signatures. Wells are identified by State Oil and Gas Board of Alabama permit number. Stratigraphic units were identified and correlated, and stratigraphic cross sections, structural cross sections, and structural contour maps were constructed. A subsurface geologic map was made of an unconformity surface in the Wiley dome. Near the terminus of the Sequatchie anticline, which has been drilled extensively at 40 acre spacing, coal and sandstone thickness were mapped to identify structurally controlled patterns of sedimentation. Dip was calculated for key stratigraphic markers in the forelimb and backlimb of the structure to test for growth strata. Stratigraphic patterns, specifically the frequency of marine strata, were analyzed to compare the Pennsylvanian fill in the Black Warrior Basin with that in the Cahaba synclinorium. Surface geologic maps were used to determine cross-cutting relationships between the Helena thrust fault and adjacent structures in Cahaba synclinorium.

Thermal maturity data based on coal rank parameters are abundant in the Pottsville Formation in the study area and include data from proximate analyses and vitrinite reflectance measurements on file at the Geological Survey of Alabama (e.g., Winston, 1990a,b; Carroll et al., 1995; Pashin et al., 1995). Volatile matter content was calculated on a dry, ash-free basis, and used as the primary indicator of thermal maturity. Vitrinite reflectance was correlated with volatile matter based on the coal rank chart in Taylor et al. (1997). Volatile matter content was mapped and superimposed on structural cross sections to determine the relationship between thermal maturity and folding using the classic method of Teichmüller and Teichmüller (1966).

Analytical results were synthesized to determine the relative age of each structure studied. Synsedimentary growth strata and facies relationships provide the most precise constraint on the age of structural movement, whereas cross-cutting structural relationships provides a more general constraint that provides insight into the relative ages of different structures. The results were then synthesized into a structural chronology designed to reveal the timing and sequence of deformation in the southern Appalachian thrust belt.

10.3. RESULTS

In this section, structures are described from the distal foreland of the Black Warrior Basin into the interior of the Appalachian thrust belt. The focus is on the structural geometry and indicators of the age of the structure. For each structure, a different mix of stratigraphic, structural, and thermal maturity data helps characterize the age of deformation.

10.3.1. Wiley Dome

The Wiley dome is a domal uplift that was first discovered in Pennsylvanian strata of the Pottsville Formation (Semmes, 1929). In the Pottsville, the dome is a subtle, subcircular, four-way structural closure with about 120 m of structural relief (Semmes, 1929; Pashin et al., 2009). Between 1974 and 1977, four wells were drilled that demonstrate that the uplift is cored by strata as old as the

Ordovician Stones River Group, which are overlain disconformably by the Mississippian-Pennsylvanian Parkwood Formation. Natural gas was discovered in Ordovician and Mississippian strata, and 24 conventional wells penetrating sub-Pottsville have now been drilled into the structure and penetrate strata as deep as the Copper Ridge Dolomite of the Knox Group (Figs. 10.8, 10.9). In addition, 55 coalbed methane wells in the Blue Creek Field have been drilled into and produce from the Pottsville Formation within the structural closure. Although a simple domal structure is apparent in the Pottsville Formation, deep drilling indicates that a complex, asymmetrical structure is preserved below the Pottsville.

Well control indicates that Cambrian-Devonian strata near the southeast end of the cross section (Fig. 10.8) dip about 4° southeast, and a dipmeter log indicates that strata in the backlimb near the crest of the asymmetrical structure at depth dip between 25 and 29° southeast. The oldest strata drilled in the forelimb of this structure are Devonian-Mississippian strata of the Chattanooga Shale, Fort Payne Chert, and Tuscumbia Limestone. Geophysical logs from well 13311 demonstrate that the Chattanooga-Tuscumbia section has been duplicated by thrust faulting (Fig. 10.10). In well 13905 (not shown), the Chattanooga Shale and overlying strata of the Fort Payne Chert are repeated five times. Similar duplication of the Chattanooga-Tuscumbia section is recorded in the backlimb of the structure (well 9987) (Figs. 10.9, 10.10). Duplication of these strata is associated with major overthickening of the Chattanooga-Tuscumbia section and thus provides primary evidence that the Wiley dome includes compressional structure. Repetition of the Chattanooga Shale and overlying strata in the forelimb and backlimb suggests development of an

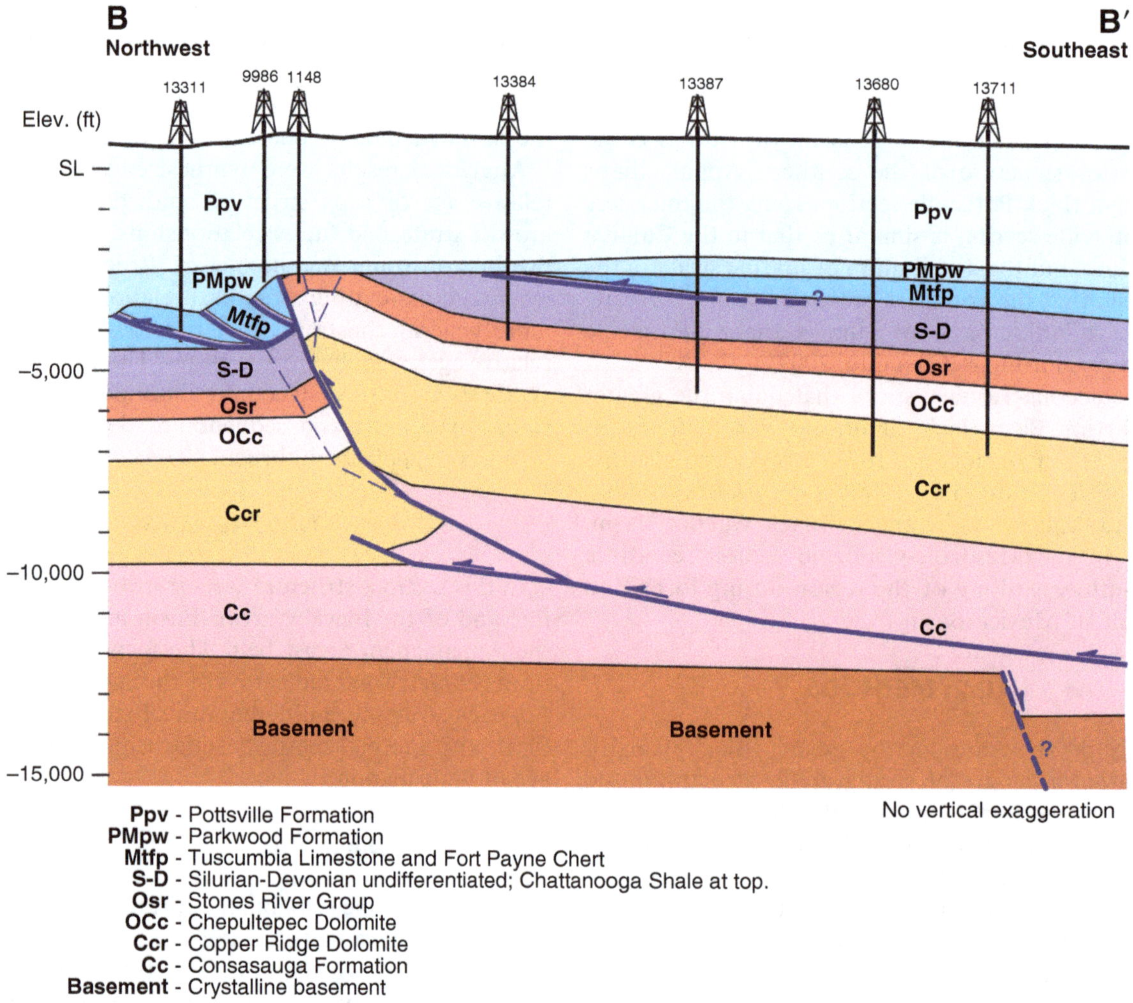

Figure 10.8 Structural cross section showing geometry of the Wiley dome and angular unconformity near the Mississippian-Pennsylvanian boundary (see Fig. 10.9 for location).

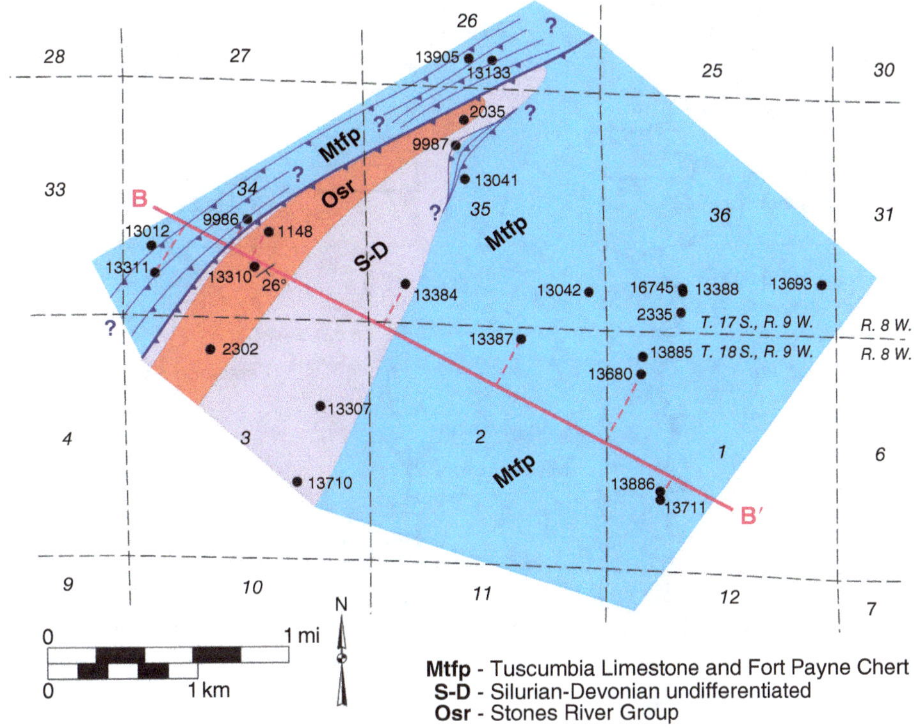

Figure 10.9 Sub-Parkwood geologic map of Wiley dome based on well control.

upper-level detachment at the base of the Chattanooga and associated thin-skinned duplexes in the Chattanooga-Tuscumbia section.

Only one seismic profile traverses the structure and shows a broad basement anticline below the backlimb and a thick Conasauga shale section between basement and the Cambrian-Ordovician carbonate section. Accordingly, interpretation of the structure is based mainly on well control. Well control indicates that the core of the Wiley dome is strongly asymmetrical, with a gently dipping backlimb and a structurally complex forelimb containing Chattanooga-Tuscumbia duplexes. Above most of the structure, the Paleozoic section is truncated by a marked angular unconformity that places an anomalously thin section of the Parkwood Formation above older Paleozoic strata (Fig. 10.8). Ordovician strata of the Stones River Group sit below the unconformity in the crestal region of the structure, and only the uppermost part of the Parkwood section is preserved above the unconformity, suggesting that the main structure was exposed and was ultimately onlapped by sediment near the Mississippian-Pennsylvanian boundary (Figs. 10.8, 10.10). Along the northwest margin of the structure, however, the complete Parkwood section is preserved. A sub-Parkwood geologic map showing the formations underlying the unconformity indicates that the dome marks part of a structural culmination that abuts the Chattanooga-Tuscumbia duplex (Fig. 10.9). This culmination appears part of a much larger structure that continues blindly northeastward along strike below the Pottsville Formation and may ultimately intersect the Sequatchie anticline (Ed Hollingsworth, personal communication).

Sparse well control and a lack of seismic data in the crestal region of the structure hampers detailed interpretation, but it appears that uplift in the hanging wall truncated the Chattanooga-Tuscumbia duplexes. Local preservation of a similar duplex in the backlimb of the structure (Fig. 10.9) may further indicate that uplift and exposure of Stones River strata in the crestal region postdate duplex formation.

Based on the available evidence, the Wiley dome is interpreted as a structure resembling a breakthrough fault-propagation fold (e.g., Suppe & Medwedeff, 1990), and a similar structural style has been documented in the Wills Valley anticline in the southern Appalachian thrust belt (Gates, 2006). The southern part of the Wills Valley anticline includes a listric thrust (Groshong, 2006), and a similar thrust geometry may underlie the Wiley dome as well as thrust belt structures in other regions (Suppe et al., 2004; Groshong & Porter, 2019). Formation of the structural closure in the Pottsville Formation indicates that minor deformation continued at least until the end of Bashkirian time, and the broad, gentle nature of the dome as originally

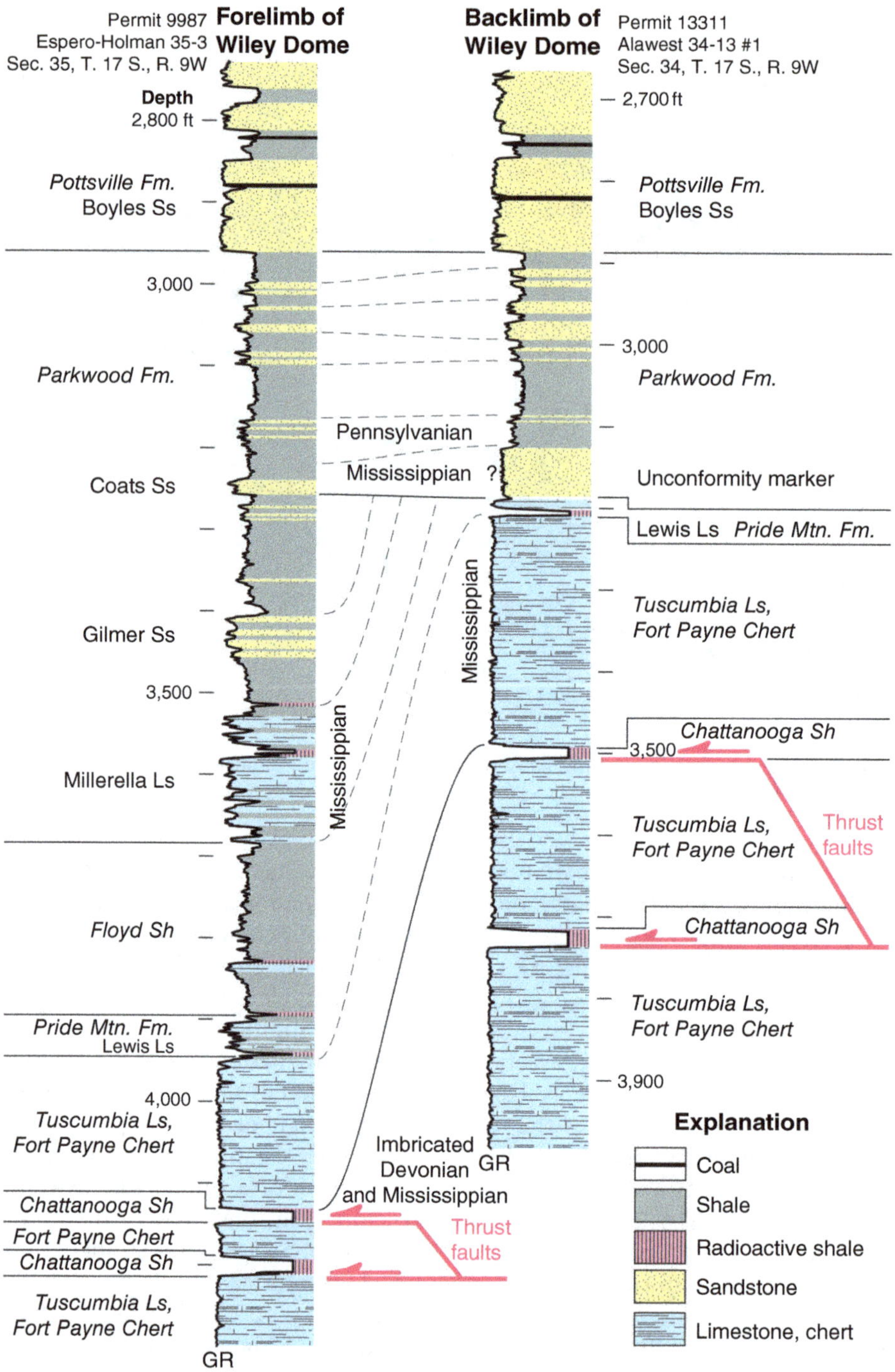

Figure 10.10 Interpreted gamma-ray logs showing thrust duplication of the Chattanooga-Tuscumbia section and angular unconformity at the base of the Parkwood Formation in the backlimb of the structure (see Fig. 10.8 for well locations).

observed by Semmes (1929) at the surface apparently formed after the asymmetrical thrust-related structure.

Sequatchie Anticline

The Sequatchie anticline plunges southwestward into the Pennysylvanian section in the Black Warrior Basin. The anticline has been drilled intensively in the Oak Grove Field which is the world's first coalbed methane field (Fig. 10.11a). The anticline in the Oak Grove Field forms a three-way a structural closure with about 150 m (500 ft) of structural relief. The anticline terminates at a series of en echelon normal faults, indicating that

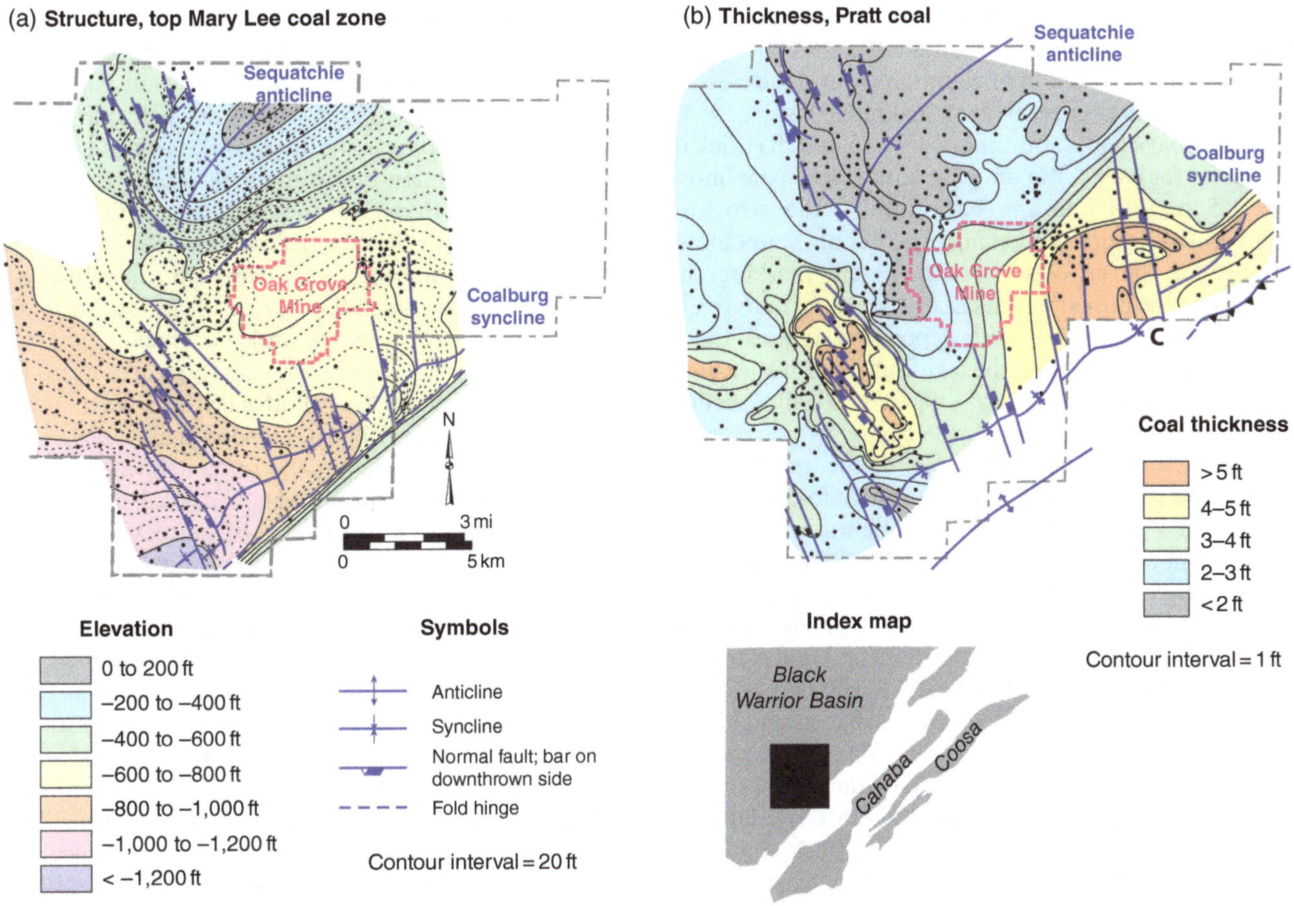

Figure 10.11 Maps showing structure and coal thickness at the southwest terminus of the Sequatchie anticline (after Pashin et al., 1991): (a) Structural contour map, top Mary Lee coal zone; (b) coal thickness map, Pratt coal seam.

displacement was accommodated by left-lateral shear. Numerous en echelon normal faults also are mapped in the backlimb of the structure, suggesting that shearing helped accommodate increasing displacement as the anticline gained structural relief toward the northeast.

In the Oak Grove Field, the anticline is a very broad, open structure with an interlimb angle of 178.2°. Within the Pottsville Formation, the forelimb of the structure dips about 0.6°, and the backlimb dips about 1.1°. Wells penetrate the economic coal-bearing strata of the upper Pottsville Formation, specifically the Black Creek through Pratt coal zones, which are of Early Pennsylvanian (late Bashkirian) age. Marine flooding surfaces separating coal zones within the upper Pottsville form more reliable structural markers than the coal seams, which can be discontinuous and are commonly associated with significant depositional topography (Pashin, 1994; Pashin & Groshong, 1998). Gamma-density well logs have accuracy of about 2 ft, and thus enable precise calculation of dip. Plotting limb dip for each successive coal zone indicates a general upward decrease in dip (Fig. 10.12).

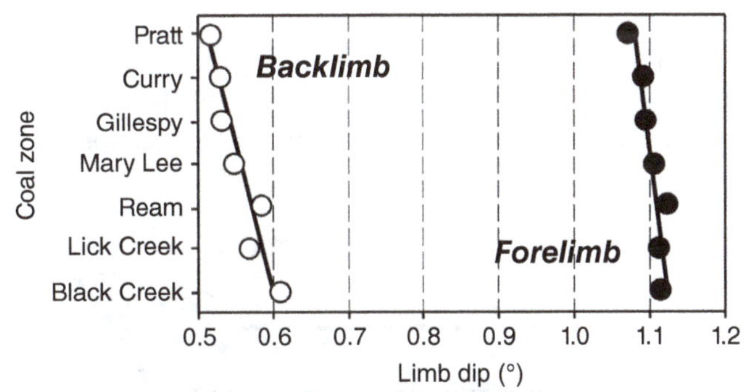

Figure 10.12 Plot of limb dip versus coal zone in the Sequatchie anticline. Dip generally decreases upward within the forelimb and backlimb, indicating synsedimentary growth.

Dip of the forelimb decreases by 0.05°, whereas that in the backlimb decreases by 0.1°. Decreasing dip appears to reflect synsedimentary growth of the anticline during the late Bashkirian and accounts for 8.3% of the total structural relief of the anticline.

Synsedimentary growth has been interpreted to have affected depositional patterns around the anticline. These patterns include diversion of channels around the crest of the structure and changes of coal thickness (Pashin, 1994). The Pratt coal seam, for example, is < 0.6 m (2 ft) thick in the crestal region of the anticline and thickens to more than 1.5 m (5 ft) in the backlimb of the structure (Fig. 10.11b). Major thickening of the coal occurs along the edge of the en echelon fault system marking the terminus of the anticline, indicating that some faults were already active during Pottsville deposition. Indeed, facies changes across the edge of the tear fault system have been documented throughout the upper Pottsville section (Pashin, 1994; Groshong et al., 2010).

The extremely broad, open nature of the anticline suggests that the structure may be a detachment fold in the Oak Grove Field. A structure of similar scale was documented seismically in the southern part of the Black Warrior Basin by Groshong et al. (2010), where a broad detachment fold with greater structural relief (~500 m; 1,600 ft) is imaged in the footwall of the Pickens-Sumter anticline. Here, a detachment is developed in the Parkwood Formation and has greatest structural relief in lower Pottsville strata. Upper Pottsville strata from the Black Creek through Pratt coal zones thin and lap onto the forelimb of the structure, and the Cobb coal zone, which overlies the Pratt, overlaps the structure, indicating growth during the late Bashkirian. The Sequatchie anticline in Oak Grove Field may have a similar detachment geometry, but was clearly active at least into Middle Pennsylvanian (Moscovian) time because the complete upper Pottsville section is deformed (Black Creek through Brookwood coal zones).

Blue Creek Anticline

The Blue Creek anticline is along the southeastern margin of the main Black Warrior Basin near the Birmingham anticlinorium (Fig. 10.2). The anticline has a steep forelimb and gently dipping backlimb that leads into the Blue Creek syncline. The Blue Creek anticline is developed largely in the Pottsville Formation, and upper Pottsville strata including the late Bashkirian Mary Lee coal zone are preserved along the axis of the Blue Creek syncline. A small structural culmination brings the Parkwood Formation to the surface along the axis of the anticline, and the structure appears to be detached primarily in the Cambrian-Ordovician section and secondarily in the Parkwood Formation (Cates et al., 2004; Groshong et al., 2010). The Pennsylvanian section in the anticline-syncline pair contains no clear evidence of synsedimentary structural growth, although thickening of lower Pottsville strata is apparent in the forelimb of the anticline (Groshong et al., 2010).

Thermal maturity data (volatile matter and vitrinite reflectance) from surface and subsurface coal are abundant in the Black Warrior Basin and the Blue Creek syncline and have been mapped in detail by Winston (1990a,b) and Pashin et al. (2009). These data help provide constraint on the relationship between thermal maturity and folding. Where isovols (lines of equal maturity) are folded with strata, thermal maturation is prekinematic, whereas where isovols cut squarely across structure, the maturation is postkinematic (Fig. 10.13). Where isovols are folded, but not as tight as the strata, by contrast, thermal maturation is synkinematic. Along the southeastern margin of the Black Warrior Basin, a major metallurgical coal province contains medium and low volatile bituminous coal (Fig. 10.14). In the Blue Creek syncline, coal at the surface is of high volatile A bituminous rank. Superimposing coal rank on a structural cross section shows that isovols cut squarely across the forelimb of the Blue Creek anticline, indicating that thermal maturation is postkinematic. Hence, a lack of synsedimentary growth strata and a postkinematic maturation pattern indicates that the Blue Creek anticline is significantly younger than Wiley dome and the southwestern terminus of the Sequatchie anticline.

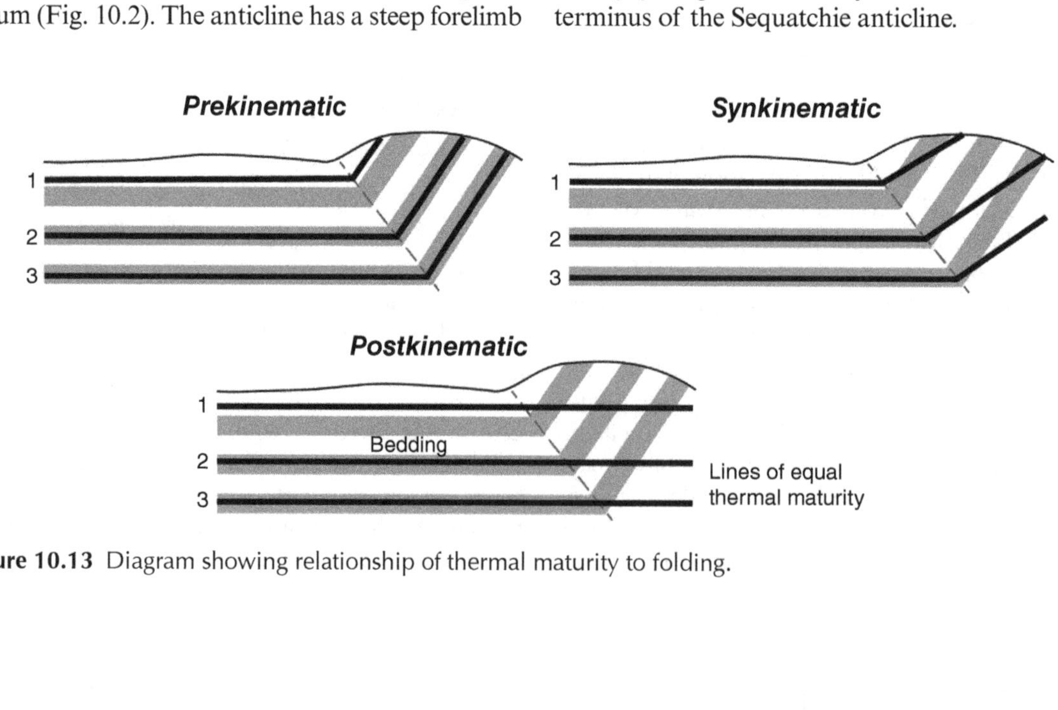

Figure 10.13 Diagram showing relationship of thermal maturity to folding.

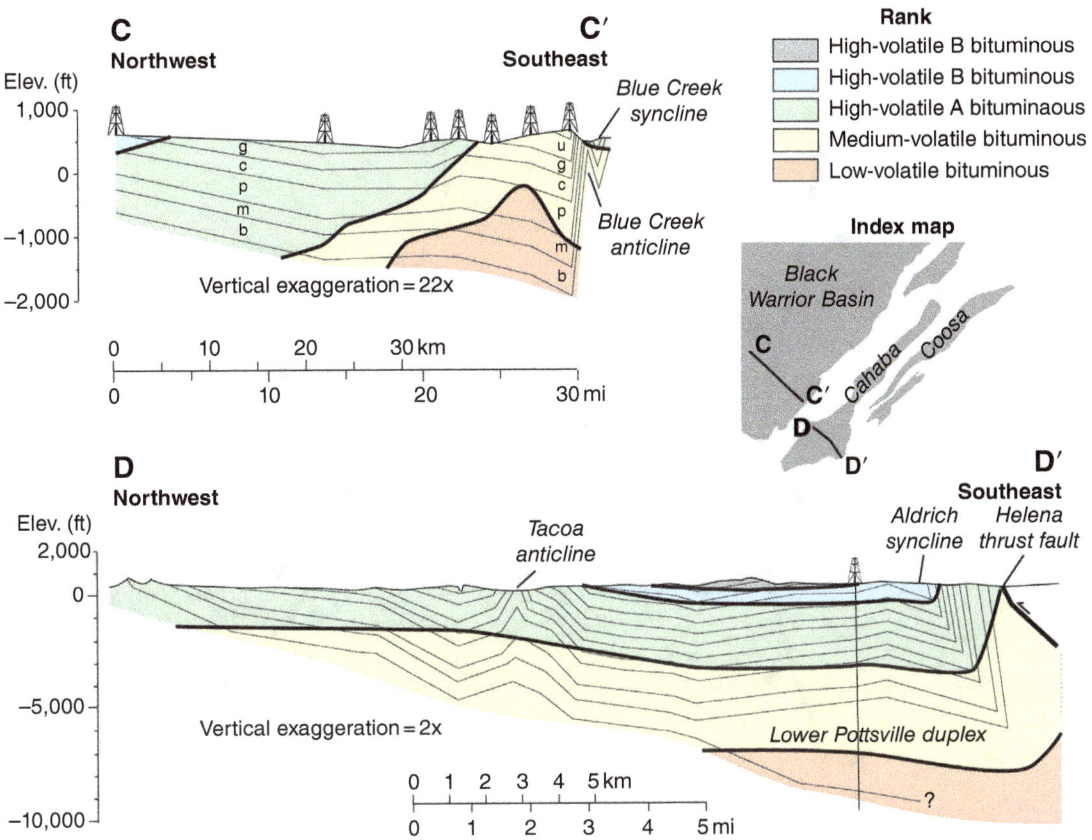

Figure 10.14 Structural cross sections with thermal maturity (coal rank) patterns superimposed. Lines of equal maturity (isovols) cut across the Blue Creek anticline, indicating postkinematic thermal maturation. In a footwall syncline in the Cahaba synclinorium adjacent to the Helena thrust, isovols are folded, but less tightly than bedding, indicating synkinematic thermal maturation.

Birmingham Anticlinorium and Cahaba Synclinorium

The Birmingham anticlinorium has a steep forelimb, a broad crest containing Cambrian-Ordovician carbonate that overlies a ductile duplex in Cambrian shale. The anticlinorium has a broad backlimb that leads into the Cahaba synclinorium, which contains a thick Mississippian-Pennsylvanian section (Fig. 10.7). Growth strata have not been documented in the forelimb of the anticlinorium, and thermal maturity patterns are effectively the same as those in the forelimb of the Blue Creek anticline. Pashin et al. (2012) and Thibodeaux (2015) interpreted emplacement of the shale duplexes to be largely synchronous with synorogenic subsidence, which made the space to accommodate the duplexes. They found that major thermal maturation occurred during the late stages of burial and the early stages of postorogenic unroofing, resulting in what appears to be a postkinematic maturation pattern that was influenced by advective heat flow within the Cambrian shale masses. Accordingly, structural and thermal maturity data indicate that major deformation and thermal maturation largely postdates the preserved Paleozoic section (Fig. 10.14). Burial history modeling based on a vitrinite reflectance profile in the Cahaba synclinorium indicates that about 3.6 km of synorogenic sediment has been eroded since maximum burial (Thibodeaux, 2015; Fig. 10.5).

Structural cross sections of the Pottsville Formation in the Cahaba synclinorium (Pashin et al., 1995) show no obvious evidence of structural growth beyond a general southeastward thickening that can be attributed to flexural subsidence and sediment loading in the greater Black Warrior Basin. However, comparison of the stratigraphic section in the Cahaba synclinorium with that in the Black Warrior Basin proper shows some critical differences (Fig. 10.15).

In the Black Warrior Basin, the percentage of marine strata (i.e., strata containing marine body fossils and trace fossils) per 152 m (500 ft) of section shows no clear systematic vertical change, varying inconsistently between 48% and 88%. Intervening strata interpreted as nonmarine contain abundant terrestrial flora, including root systems. The percentage of sandstone and conglomerate per unit section, however, increases from about

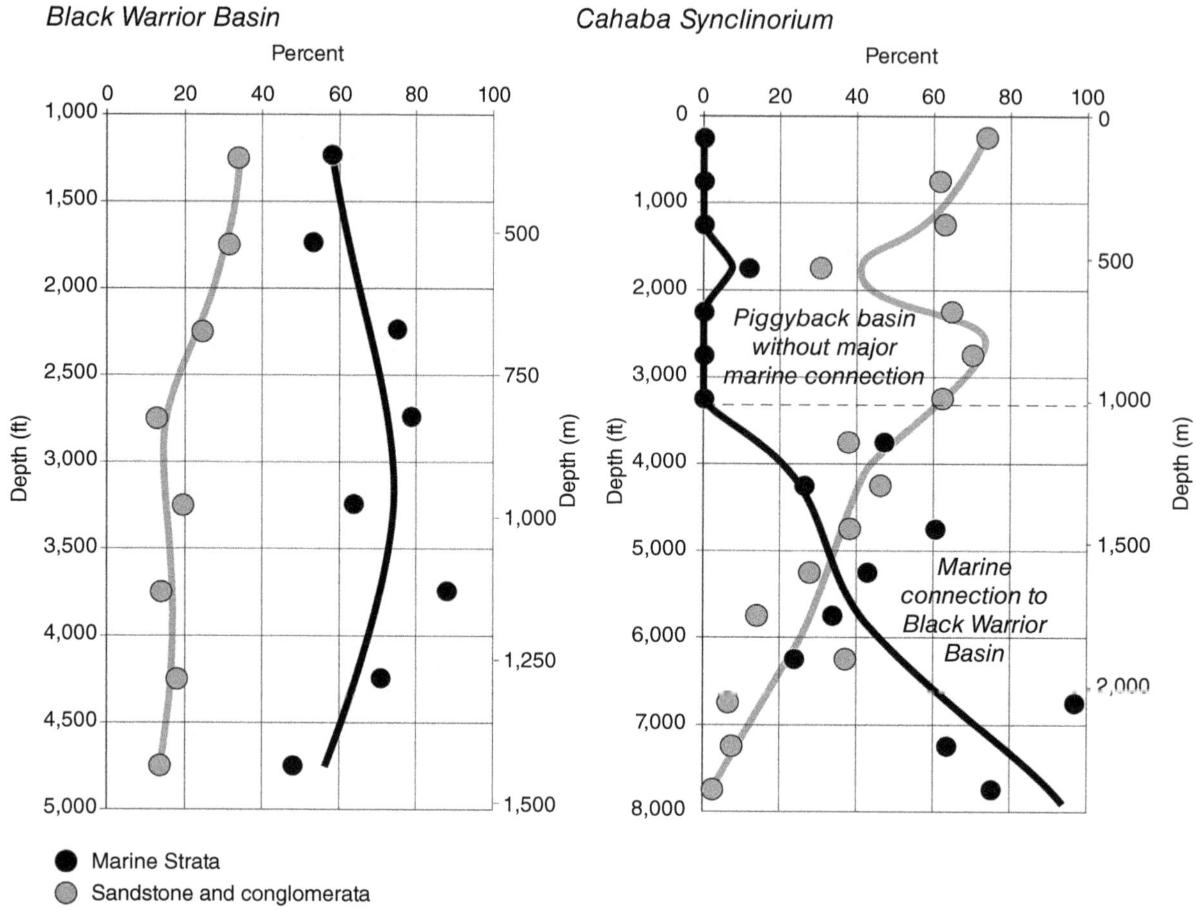

Figure 10.15 Plots showing vertical change in the percentage of sandstone/conglomerate and marine deposits in the Black Warrior Basin and the Cahaba synclinorium. Only one marine zone is in the upper kilometer of the Cahaba section, and the absence of marine deposits corresponds with deposition of numerous extraformational conglomerate beds.

15% to 20% in the lower half of the section to 34% at the top. In the Cahaba synclinorium, by contrast, the percentage of sandstone and conglomerate increases upward in section from 3% to 74%. The percentage of marine strata, interpreted mainly as deltaic and tidal flat facies (Pashin et al., 1995), is very high near the base of the Pottsville Formation in the Cahaba synclinorium (63–97%) and decreases to zero in the upper part of the section, which is dominated by interpreted fluvial, interfluvial, and paludal strata containing abundant terrestrial flora. Indeed, only one marine zone has been identified in the upper kilometer of the Pottsville in this area (Pashin et al., 1995). This change is associated with the deposition of numerous widespread extraformational conglomerate units containing pebbles and cobbles of chert, igneous rocks, metamorphic rocks, and volcanics that have been interpreted as braidplain facies (Pashin et al., 1995; Uddin et al., 2016).

These stratigraphic differences indicate that the Black Warrior Basin was repeatedly flooded by the sea throughout Pottsville deposition, whereas the Cahaba synclinorium was largely terrestrialized sometime near the Bashkirian-Moscovian boundary (315 Ma). The increasing abundance of sandstone and conglomerate in the Black Warrior Basin reflects progressively greater input of coarse-grained sediment from the Appalachian-Ouachita orogen. The abundance of syntectonic conglomerate and nonmarine strata in the upper kilometer of the Pottsville in the Cahaba synclinorium points toward a major episode of thrusting that led to a stratigraphic decoupling from the main Black Warrior Basin. During highstands of sea level, the Black Warrior Basin was filled with progradational sediment dominated by mudstone, and once the basin was filled, sand and extraformational lithoclasts could in places be transported across the Birmingham anticlinorium into the main part of the Black Warrior Basin. Accordingly, the favored interpretation is that the Birmingham anticlinorium had gained enough relief around the Lower-Middle Pennsylvanian (Bashkirian-Moscovian) boundary to isolate the Cahaba synclinorium from most marine transgressions, and, at this point, the Cahaba synclinorium had become a piggyback basin.

Helena Thrust and Coosa Synclinorium

The Helena thrust fault defines the southeast edge of the Cahaba synclinorium and truncates a large footwall syncline that is developed along the southeastern margin of the synclinorium in the Pottsville Formation. The Helena thrust clearly crosscuts faults along the lateral ramp at the northeastern end of the synclinorium (Fig. 10.2), and thus has been recognized as a break-back thrust on the basis of structural relationships (e.g., Thomas & Bayona, 2005; Solis, 2010). The Coosa synclinorium is developed in the hanging wall of the Helena thrust and contains a Pottsville section that is similar in thickness to that in the Cahaba synclinorium, but, based on the distribution of economic coal seams and marine strata, appears to be stratigraphically equivalent to about the lower half of the Cahaba Pottsville (Pashin, 1997). A lack of deep well control combined with internal structural complexity precludes testing for synsedimentary growth strata in the Coosa synclinorium, but examination of the footwall syncline in the Cahaba synclinorium provides some insight.

In the southwestern part of the Cahaba synclinorium, the Helena thrust wraps around a complete overturned Pottsville section that is preserved in the footwall. Volatile matter data are abundant at the surface from coal exploration, and deep control on thermal maturity is provided by vitrinite reflectance data from the Kimberly-Clark 23-2 #1 well (permit 2827) (Figs. 10.5, 10.14). Based on the available data, the isovols are clearly folded, but not as tightly as the strata, indicating synkinematic thermal maturation (Figs. 10.12, 10.15). This relationship stands in stark contrast to the postkinematic signature in the Blue Creek anticline and the northwest margin of the Birmingham anticlinorium, indicating that folding in the footwall of the Helena thrust is younger than that associated with the forelimb of the anticlinorium.

10.4. DISCUSSION

Analysis of structural timing based on structural, stratigraphic, and thermal maturity evidence indicates that the parts of the southern Appalachian thrust belt preserving the synorogenic sedimentary fill formed out of sequence (Fig. 10.16). Deformation began during Late Mississippian (Serpukhovian) time in the Wiley dome, which consists of a large fault-related fold resembling a fault-propagation fold detached in Cambrian shale and a duplex in Devonian-Mississippian strata that is developed above an upper-level detachment in the

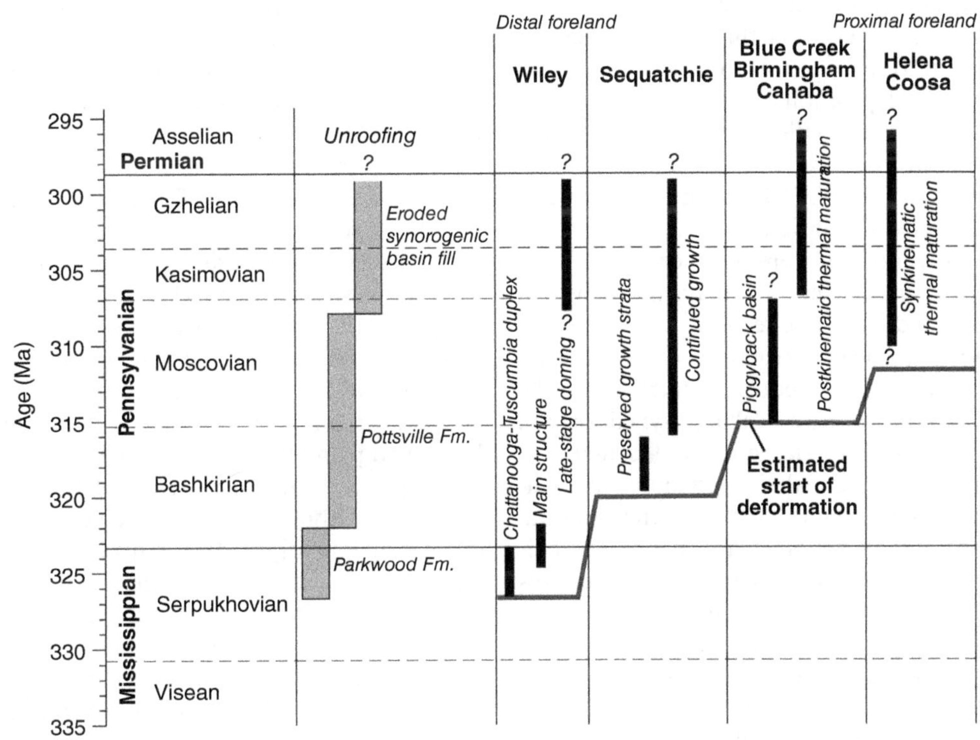

Figure 10.16 Chart showing estimated timing of structural events in the southern Appalachian thrust belt of the greater Black Warrior Basin. Thrust-related structures become progressively younger toward the proximal foreland, indicating a break-back thrust sequence.

Chattanooga shale (Figs. 10.7–10.9). This event is apparently synchronous with thrusting of volcanic arc rocks of the Hillabee Greenstone onto the easternmost part of the continental shelf in Alabama and Georgia (Tull et al., 2007), which may have provided the initial driving force for distal foreland deformation.

Crosscutting relationships suggest that formation of the duplex largely preceded uplift of the main structure. An angular unconformity is developed atop the structure in the Mississippian-Pennsylvanian Parkwood Formation, and gentle, late-stage folding of the synorogenic Parkwood-Pottsville section resulted in a broad domal structure with four-way closure.

The southern terminus of the Sequatchie anticline appears to be a broad, plunging detachment fold that is developed within the synorogenic stratigraphic section. Growth strata indicate that deformation had begun by the Early Pennsylvanian (late Bashkirian; Fig. 10.16). Although deformation was extremely subtle, it had a significant impact on facies distribution in the Pottsville Formation (Figs. 10.10b, 10.11). Documented growth strata account for only 8.3% of the structural relief, and so it appears that the anticline had a prolonged history of uplift that may have continued into the Late Pennsylvanian or Permian.

The Cahaba synclinorium appears to have become a piggyback basin near the Early-Middle Pennsylvanian boundary (Bashkirian-Moscovian boundary; Fig. 10.16). The principal lines of evidence for this are terrestrialization of the synclinorium relative to the main Black Warrior Basin and the deposition of widespread extraformational conglomerate units (Fig. 10.15). Postkinematic thermal maturity patterns in the Blue Creek anticline and Birmingham anticlinorium (Fig. 10.14) indicate that these folds and the associated Cambrian shale duplex forming the core of the anticlinorium were in place by the time of deep burial and major thermal maturation, which postdates the preserved Paleozoic section.

The Helena thrust truncates structures in its footwall, indicating that it is out of sequence relative to the Birmingham anticlinorium and Cahaba synclinorium. Thermal maturity data from a large footwall syncline at the southeast margin of the Cahaba synclinorium indicate that isovols are folded, but less tightly than the strata. Accordingly, movement of the Helena thrust and associated footwall structures was synkinematic and thus was deformed later than the major structures associated with the Birmingham anticlinorium and Cahaba synclinorium. Therefore, the Helena thrust probably began propagating during Moscovian time, and deformation may have continued until maximum burial was reached in the Late Pennsylvanian or Early Permian.

The results of this study indicate that break-back deformation prevailed in the southern Appalachian foreland of Alabama in a belt spanning more than 60 km from the distal Appalachian foreland to the interior of the sedimentary thrust belt. Break-back deformation also has been recognized in the frontal hinterland thrusts, which crosscut metamorphic isograds (Tull, 1984), and thus may indicate very widespread out-of-sequence deformation in the southern Appalachians. One contributing factor is that Cambrian-Ordovician carbonate rocks form a stiff layer that was apparently capable of transmitting horizontal stress across large distances (Thomas, 2001; Thomas & Bayona, 2005). Association of frontal thrust ramps with Iapetan basement faults further suggests that inherited crustal weaknesses, especially those manifested in the mechanically stiff carbonate section, influenced where thrust faults and related folds would form, but this does not explain why the structures would break out of sequence across such a large area. And considering that palinspastic restorations indicate that the thrust belt was shortened by about 30%, with the amount of shortening increasing southeastward (e.g., Thomas & Bayona, 2005), explaining why structures would form first in the distal foreland remains elusive and should be a subject for further research.

10.5. CONCLUSIONS

The key hypotheses tested in this research are that deformation in the southern Appalachian thrust belt in the greater Black Warrior Basin began near the Mississippian-Pennsylvanian boundary and that structural, stratigraphic, and thermal maturity evidence from the synorogenic basin fill can reveal the timing and sequence of deformation. Each structure analyzed required a different mix of evidence to determine timing and thrust sequence. In the Wiley dome, which is the most distal foreland structure, an angular unconformity near the Mississippian-Pennsylvanian boundary confirms early growth, and preservation of a complete Parkwood section above a Chattanooga-Tuscumbia duplex in the forelimb of the structure indicates that deformation actually began during Serpukhovian time. Analysis of bedding dip and facies patterns helps constrain the age of the Sequatchie anticline where the structure plunges to a terminus within the synorogenic stratigraphy. Here, decreasing limb dip and structurally controlled facies patterns can be demonstrated in the Early Pennsylvanian section. Synsedimentary growth strata were not apparent closer to the Appalachian hinterland, but the distribution of marine strata and syntectonic conglomerate suggests that thrusting was initiated in the Birmingham anticlinorium and Cahaba synclinorium sometime near the Early-Middle Pennsylvanian boundary.

Thermal maturity data proved useful for constraining the timing of deformation relative to the burial history.

This approach proved particularly useful in the Blue Creek anticline and the forelimb of the Birmingham anticlinorium, where a postkinematic maturation pattern indicates that fold limbs were emplaced prior to deep burial and thermal maturation. In the footwall of the Helena thrust, crosscutting relationships support break-back thrusting, and a synkinematic thermal maturity pattern in a large footwall syncline indicates that the thrust and footwall syncline were active during deep burial and thermal maturation. And so, movement of the Helena thrust apparently postdates major deformation in the Birmingham anticlinorium and Cahaba synclinorium and was perhaps initiated during the Middle Pennsylvanian.

The processes that caused break-back thrusting in a wide area remain unclear. One contributing factor appears to include the ability of a mechanically stiff Cambrian-Ordovician carbonate section to transmit stress deep into the craton and localization of stress by Iapetan basement structures. Results of this study indicate that a unified approach integrating structural, stratigraphic, and thermal maturity data can provide insight into the timing and sequence of deformation, and similar approaches may be applicable to other regions where the structural history can be resolved with increased detail.

ACKNOWLEDGMENTS

This chapter includes research funded by the U.S. Department of Energy and the U.S. Geological Survey under various agreements. This chapter was reviewed by Rick Groshong, Paul Hackley, Jim Tull, and one anonymous reviewer. Their comments contributed greatly to the quality of this contribution and are greatly appreciated.

REFERENCES

Carroll, R. E., Pashin, J. C., & Kugler, R. L. (1995). Burial history and source-rock characteristics of Upper Devonian through Pennsylvanian strata, Black Warrior basin, Alabama. *Alabama Geological Survey Circular 187*.

Cates, L. M., Mcintyre, M. R., Hawkins, W. B., & Groshong, R. H., Jr. (2004). *Structure and oil and gas production in the Black Warrior Basin*. Tuscaloosa, University of Alabama, College of Continuing Studies, 2004 International Coalbed Methane Symposium Proceedings, Paper 0440.

Clark, P. E., Pashin, J. C., Carlson, E. S., Goodliffe, A. M., McIntyre-Redden, M. R., Guohai, J., et al. (2013). *Site characterization for CO2 storage from coal-fired power facilities in the Black Warrior Basin of Alabama: Final Report*. U.S. Department of Energy Award DE-FE0001910.

Ettensohn, F. R., Pashin, J. C., & Gilliam, W. (2019). The Appalachian and Black Warrior foreland basins in the eastern United States. In A. D. Miall (Ed.), *Sedimentary basins of United States and Canada (2d ed.)* (pp. 129–237). Amsterdam: Elsevier.

Gates, M. G. (2006). *Structure of the Wills Valley anticline in the vicinity of Mentone, Alabama*. M. S. thesis. University of Alabama, Tuscaloosa.

Groshong, R. H., Jr. (2006). *3D structural geology*. Heidelberg: Springer-Verlag.

Groshong, R. H., Jr., & Porter, R. (2019). Predictive models for the deep geometry of a thick-skinned thrust matched to crustal structure: Wind River Range, western USA. *Lithosphere*, *11*, 448–464.

Groshong, R. H., Jr., Hawkins, W. B., Jr., Pashin, J. C., & Harry, D. L. (2010). Extensional structures of the Alabama promontory and Black Warrior foreland basin: Styles and relationship to the Appalachian fold-thrust belt. *Geological Society of America Memoir 206*, 579–605.

Moore, M. F. (2012). *$^{40}Ar/^{39}Ar$ dating of detrital muscovite and sediment composition analysis from the southern carboniferous Appalachian basin in Alabama: Implications for tectonics and sedimentation*. M. S. thesis. Auburn University, Auburn.

Pashin, J. C. (1994). Coal-body geometry and synsedimentary detachment folding in Oak Grove coalbed-methane field, Black Warrior basin, Alabama. *American Association of Petroleum Geologists Bulletin*, *78*, 960–980.

Pashin, J. C. (1997). Stratigraphy and depositional environments of the Pottsville Formation (Lower Pennsylvanian) in the Coosa coalfield. In D. N. Bearce et al. (Eds.), *Geology of the Coosa coalfield* (pp. 19–28). Alabama Geological Society Guidebook.

Pashin, J. C. (1998). Stratigraphy and structure of coalbed methane reservoirs in the United States: An overview. *International Journal of Coal Geology*, *35*, 207–238.

Pashin, J. C., & Gastaldo, R. A. (2009). Carboniferous of the Black Warrior basin. In S. Greb & Chesnut, D. (Eds.), *Carboniferous geology and biostratigraphy of the Appalachian basin* (pp. 10–21). Kentucky Geological Survey Special Publication 10.

Pashin, J. C., & Groshong, R. H., Jr. (1998). Structural control of coalbed methane production in Alabama. *International Journal of Coal Geology*, *38*, 89–113.

Pashin, J. C., Carroll, R. E., Barnett, R. L., & Beg, M. A. (1995). Geology and coal resources of the Cahaba coal field. *Alabama Geological Survey Bulletin 163*.

Pashin, J. C., Kopaska-Merkel, D. C., Arnold, A. C., McIntyre, M. R., & Thomas, W. A. (2012). Gigantic, gaseous mushwads in Cambrian shale: Conasauga Formation, southern Appalachians, USA. *International Journal of Coal Geology*, *103*, 70–91.

Pashin, J. C., McIntyre, M. R., Carroll, R. E., Groshong, R. H., Jr., & Bustin, R. M. (2009). Carbon sequestration and enhanced recovery potential of mature coalbed methane reservoirs in the Black Warrior basin. *American Association of Petroleum Geologists Studies in Geology*, *59*, 125–147.

Pashin, J. C., Ward, W. E., II, Winston, R. B., Chandler, R. V., Bolin, D. E., Richter, K. E., et al. (1991). *Regional analysis of the Black Creek-Cobb coalbed-methane target interval, Black Warrior basin*, Alabama. Alabama Geological Survey Bulletin 145.

Semmes, D. R. (1929). *Oil and gas in Alabama. Alabama Geological Survey Special Report*, *15*.

Solis, M. P. (2010). *Tectonic controls on Lower Devonian sandstone distribution, Alabama*. M.S. thesis. University of Kentucky, Lexington.

Suppe, J., & Medwedeff, D. A. (1990). Geometry and kinematics of fault-propagation folding. *Eclogae Geologicae Helvetiae, 83*, 409–454.

Suppe, J., Connors, C. S., & Zhang, Y. (2004). Shear fault-bend folding. *American Association of Petroleum Geologists Memoir, 82*, 303–323.

Taylor, G. H., Teichmüller, M., Davis, A., Diessel, C. F. K., Littke, R., & Robert, P. (1997). *Organic petrology*. Berlin: Gebrüder Borntraeger.

Teichmüller, M., & Teichmüller, R. (1966). *Geological causes of coalification*. Washington, D.C., *Advances in Chemistry Series 55*, 133–155.

Telle, W. R., Thompson, D. A., Lottman L. K., & Malone, P. G. (1987). Preliminary burial-thermal history investigations of the Black Warrior basin: Implications for coalbed methane and conventional hydrocarbon development. *1987 Coalbed Methane Symposium Proceedings*, 37–50.

Thibodeaux, B. J. (2015). *Emplacement, burial history, and thermal maturation of giant deformed shale masses in the Conasauga Formation, southern Appalachian thrust belt*. Master's thesis. Stillwater, Oklahoma State University'.

Thomas, W. A. (1972). *Mississippian stratigraphy of Alabama*. Alabama Geological Survey Monograph 12.

Thomas, W. A. (1985). Northern Alabama sections. In N. B. Woodward (Ed.), Valley and Ridge thrust belt: Balanced structural sections, Pennsylvania to Alabama. *University of Tennessee Department of Geological Sciences Studies in Geology, 12*, 54–60.

Thomas, W. A. (1988). The Black Warrior basin. In L. L. Sloss (Ed.), *Sedimentary cover: North American craton* (pp. 471–492). Geological Society of America, The Geology of North America, D-2.

Thomas, W. A. (1991). The Appalachian-Ouachita rifted margin of southeastern North America. *Geological Society of America Bulletin, 103*, 415–431.

Thomas, W. A. (1995). Diachronous thrust loading and fault partitioning of the Black Warrior foreland basin within the Alabama recess of the late Paleozoic Appalachian-Ouachita thrust belt. *Society of Economic Paleontologists and Mineralogists Special Publication, 52*, 111–126.

Thomas, W. A. (1997). Nd isotopic constraints on sediment sources of the Ouachita-Marathon fold belt: alternative interpretation and reply: Alternative interpretation. *Geological Society of America Bulletin, 109*, 779–785.

Thomas, W. A. (2001). Mushwad: Ductile duplex in the Appalachian thrust belt in Alabama. *American Association of Petroleum Geologists Bulletin, 85*, 1847–1869.

Thomas, W. A. (2007). Balancing tectonic shortening in contrasting deformation styles through a mechanically heterogeneous stratigraphic succession. In J. W. Sears et al. (Eds.), *Whence the mountains? Inquiries into the evolution of orogenic systems: A volume in honor of Raymond A. Price* (pp. 277–290). Geological Society of America Special Paper, 433.

Thomas, W. A., & Bayona, G. (2005). *The Appalachian thrust belt in Alabama and Georgia: thrust-belt structure, basement structure, and palinspastic reconstruction*. Alabama Geological Survey Monograph 16.

Thomas, W. A., & Drahovzal, J. A. (1974). *Geology of the Coosa deformed belt*. Tuscaloosa, Alabama Geological Society, 45–75.

Thomas, W. A., & Pashin, J. C. (Eds.) (2011). *Of mushwads and mayhem: Disharmonically deformed shale in the southern Appalachian thrust belt*. Alabama Geological Society 48th Annual Field Trip Guidebook.

Tull, J. F. (1984). Polyphase late Paleozoic deformation in the southeastern foreland and northwestern Piedmont of the Alabama Appalachians. *Journal of Structural Geology, 6*, 223–234.

Tull, J. F., Barineau, C. I., Mueller, P. A., & Wooden, J. L. (2007). Volcanic arc emplacement onto the southernmost Appalachian Laurentian Shelf: Characteristics and constraints. *Geological Society of America Bulletin, 119*, 261–274.

Uddin, A., Hames, W. E., Peavy, T., & Pashin, J. C. (2016). Detrital history of the Lower Pennsylvanian Pottsville Formation in the Cahaba synclinorium of Alabama. *Journal of Sedimentary Research, 86*, 1287–1297.

Whiting, B. M., & Thomas, W. A. (1994). Three-dimensional controls on subsidence of a foreland basin associated with a thrust-belt recess, Black Warrior basin, Alabama and Mississippi. *Geology, 22*, 727–730.

Winston, R. B. (1990a). Vitrinite reflectance of Alabama's bituminous coal. *Alabama Geological Survey Circular 139*.

Winston, R. B. (1990b). Preliminary report on coal quality trends in upper Pottsville Formation coal groups and their relationships to coal resource development, coalbed methane occurrence, and geologic history in the Warrior coal basin, Alabama. *Alabama Geological Survey Circular 152*.

11

Strain Partitioning in Foreland Basins: An Example From the Ouachita Fold-Thrust Belt Arkoma Basin Transition Zone in Southeastern Oklahoma and Western Arkansas

İbrahim Çemen[1] and Donald J. Yezerski[2]

ABSTRACT

We examined strain partitioning between the Ouachita fold-thrust belt and the Arkoma Foreland basin in southeastern Oklahoma and western Arkansas. The Choctaw Fault and Ross Creek Fault are the leading-edge thrusts of the fold-thrust belt and form the southern boundary of the Arkoma basin in Oklahoma and Arkansas, respectively. Strain portioning is accommodated by the Wilburton Triangle Zone in the footwall of the Choctaw Fault zone in Oklahoma. The triangle zone continues east towards Arkansas where the Choctaw Fault loses its separation and the Ross Creek Fault becomes the leading-edge thrust. Our structural interpretation of depth-converted 2-D seismic profiles and well log data in north-central Scott County, western Arkansas suggest the presence of a triangle zone containing the surficial tip-out of the Choctaw Fault in the footwall of the Ross Creek Fault. This triangle zone is named here the Waldron Triangle Zone and is composed of three stacked wedges that share a north-dipping roof thrust, which meets a south-dipping floor thrust at the tip line below the Poteau Syncline. Our interpretation of surface and subsurface geology suggests that the Waldron Triangle Zone continues along strike westward and coincides with its lateral equivalent, the Wilburton Triangle Zone.

11.1. INTRODUCTION

The Late Paleozoic Ouachita Orogeny resulted from a continental collision between the southern margin of North America and Gondwana (e.g., Poole et al., 2005). Orogenic structures crop out for 400 kms along strike across eastern Oklahoma and western Arkansas and are unconformably overlain by Cretaceous units to the south. The Arkoma basin is a foreland basin (Fig. 11.1) of the orogeny formed during Pennsylvanian. The Ouachita-Arkoma system may be continuous with the Marathon fold-thrust belt of Texas to the south (Poole et al., 2005), and may be continuous with the Appalachian fold-thrust belt to the east (e.g., Thomas, 1989 and Robinson et al., 2012).

Strain partitioning between the frontal portion of fold-thrust belts and associated foreland basins in many regions is expressed by a tectonically thickened wedge known as a *triangle zone*, a term that first entered the literature as a description of the triangular zone of poor seismic imaging between the foreland- and hinterland-dipping reflectors in the foothills region of the eastern Cordillera of Alberta, Canada (e.g., Gordy et al., 1977; Jones, 1982; DeCelles & Giles, 1996). Subsequent study of the Canadian Cordillera and other orogenic belts containing triangle zones demonstrated that this type of frontal structure exhibits similar geometries worldwide (Vann et al., 1986; Çemen et al., 2001a, 2001b).

[1] *Department of Geological Sciences, The University of Alabama, Tuscaloosa, Alabama, USA*
[2] *Chevron Corporation, Houston, Texas, USA*

Compressional Tectonics: Plate Convergence to Mountain Building, Geophysical Monograph 277, First Edition.
Edited by Elizabeth J. Catlos and İbrahim Çemen.

DOI:10.1002/9781119773856.ch11

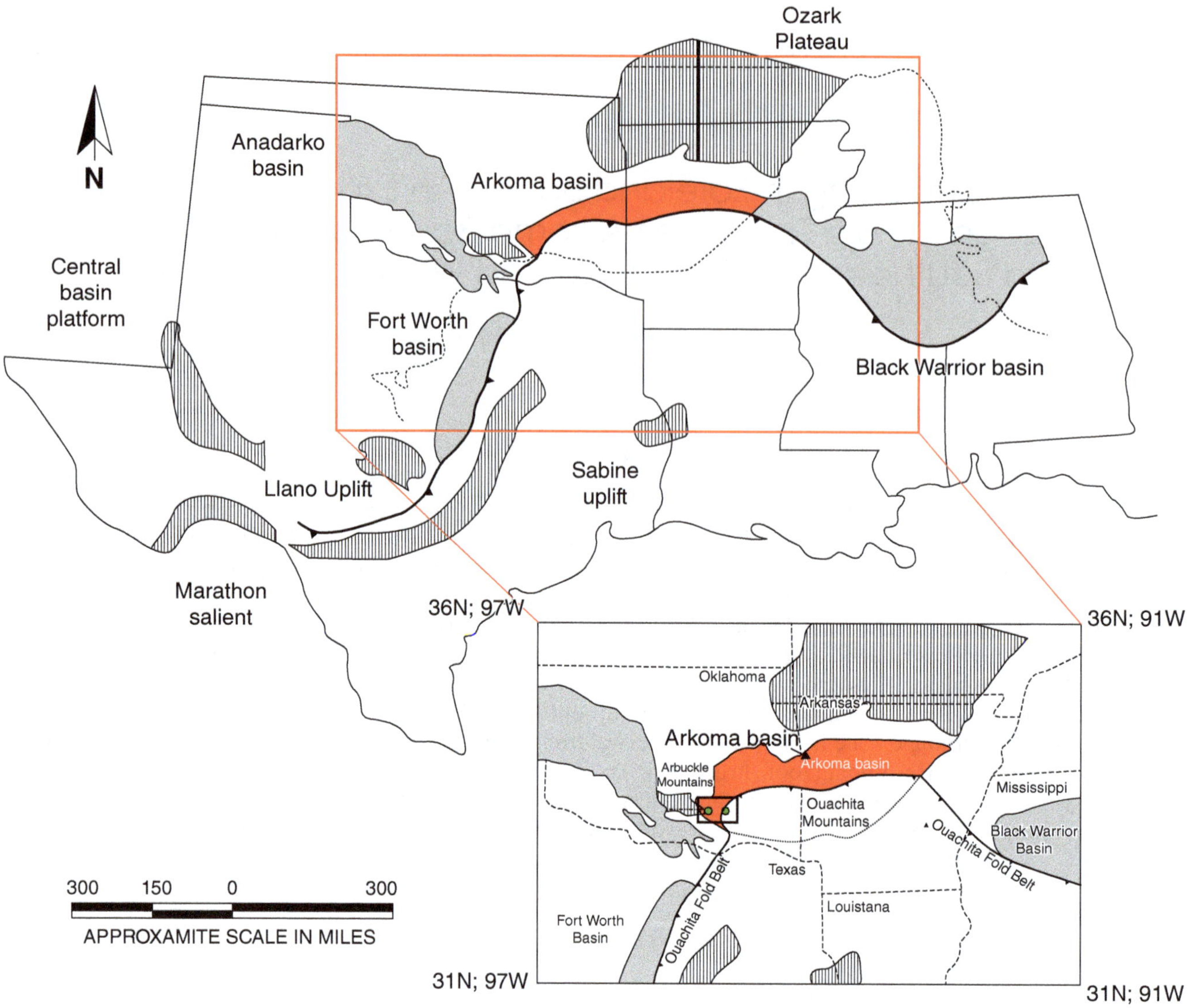

Figure 11.1 Location of the Arkoma Foreland Basin with respect to the major structural features of the midcontinent region, the inset on the lower left shows the location of the basins with respect to major structural features in Oklahoma, Northern Texas, Mississippi, Tennessee, and Alabama.

Çemen et al. (2001a, 2001b) proposed that strain partitioning along the frontal Ouachitas to Arkoma basin transition zone in eastern Oklahoma is accomplished via a triangle zone in the footwall of the Choctaw Fault (CHF). They named this structure the Wilburton Triangle Zone (WTZ) after nearby Wilburton, Oklahoma (Fig. 11.1). The triangle zone continues to eastern Arkansas where it is named the Waldron Triangle Zone (WLTZ).

The main purpose of this chapter is to describe how strain partitioning occurs (a) from the highly deformed Ouachita fold-thrust to mildly deformed Arkoma basin and (b) along the strike of the Arkoma Basin from the Wilburton area in the western part of the Arkoma basin to the Waldron area in the eastern part of the basin in Arkansas. Our interpretation of the strain portioning is based on available surface and subsurface data. Describing the structures that accommodate strain partitioning between the Ouachita fold-thrust belt and the Arkoma basin will aid in understanding of the structural evolution of the Ouachita system, in addition to providing insight about the lateral continuity with the WTZ in Oklahoma.

11.2. TECTONIC OVERVIEW

The Arkoma basin is located south of the Ozark uplift and north of the Ouachita fold-thrust belt (Fig. 11.1). Sedimentation began in the Late Precambrian to Early Cambrian during breakup and rifting of the Rodinian

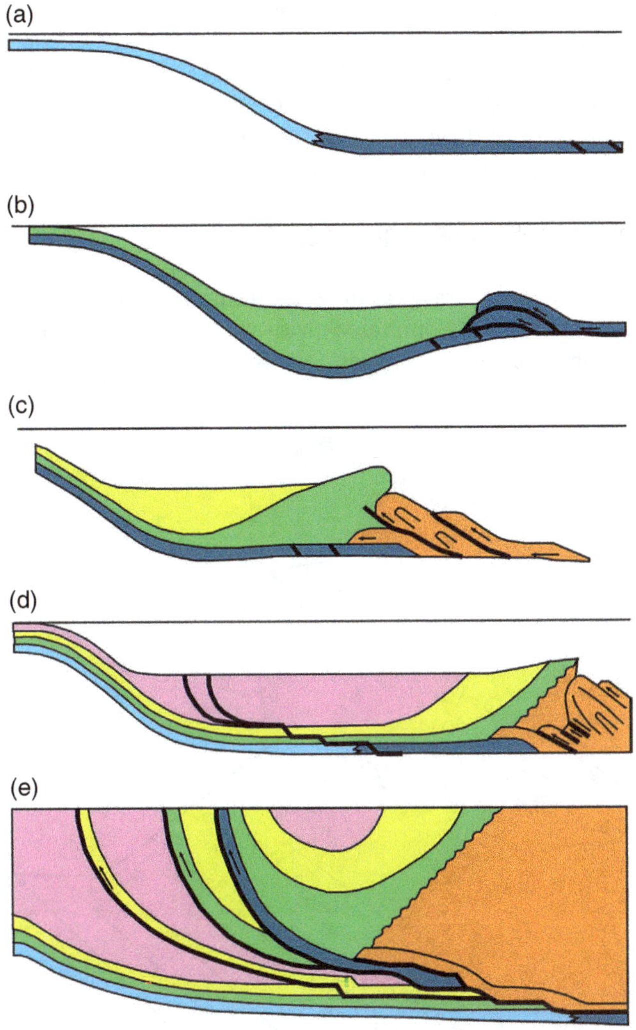

Figure 11.2 Tectonic evolution of the Arkoma Basin: (a) Stable shelf from Cambrian to Devonian; (b) subsidence begins in early Mississippian in response to thrusting and tectonic loading; (c) subsidence continues through Pennsylvanian as thrusting and tectonic loading progresses; (d) continued thrusting into middle Pennsylvanian shifts axis of deposition north; (e) middle Pennsylvanian, Ouachita tectonic belt is uplifted ending basin formation (modified from Suneson, 2012).

supercontinent, which resulted in the development of the Laurentian platform (Fig. 11.2). Divergence eventually resulted in the formation of oceanic crust outboard of the southern continental margin, which was a passive margin through Early Mississippian time (Houseknect, 1986; Poole et al., 2005; Suneson, 2012). The southward-thickening passive margin wedge consists of carbonate-dominated shallow shelf facies to the west, with increasing clastic input to the east (Sutherland, 1988). South of the Laurentian platform, a sediment-starved deep water basin, known as the Ouachita embayment, hosted a succession of dark shale-containing minor carbonate and chert that characterize the "Ouachita facies," exposed in the thrust sheets of the fold-thrust belt (Houseknecht, 1986).

The onset of the Ouachita orogeny is marked by deposition of a thick flysch sequence above deep water sediments beginning in Mississippian time (Figs. 11.2 and 11.3) due to the northward advance of a Gondwanan volcanic-plutonic arc, above south-directed subduction of Laurentian oceanic crust (Viele, 1979; Houseknecht, 1986; Suneson, 2012). Continued convergence between southern Laurentia and the Yucatan block consumed the oceanic crust of the Ouachita oceanic basin resulting in a growing accretionary wedge (Houseknecht, 1986). As the basin closed, the accretionary wedge was uplifted to near the sea surface and was a major source of sediment that began filling the Arkoma foreland basin (Houseknecht & Kacena, 1983; Suneson, 2012).

Flexure and sediment loading of the continental platform resulted in major tectonic subsidence of the foreland basin, along with movement of south-dipping, east- to northeast- striking syndepositional normal faults that propagated northward into the basin. These extensional faults resulted from flexural extension of the lower plate during subduction (Van Arsdale & Schweig, 1990; Bradley & Kidd, 1991). Flexure-induced extension involved the passive margin sedimentary sequence and caused extreme sediment thickening above the downthrown block, especially during deposition of the middle Atoka Formation in Pennsylvanian (Sutherland, 1988). The greatest syndepositional movement occurred in the southwestern portion of the Arkoma basin in eastern Oklahoma (Houseknecht, 1986; Suneson, 2012). Continuity of the base Pennsylvanian disconformity eastward into Arkansas limits the onset of major basement-involved down-to-south faulting to earlier than 318 Ma (Arbenz, 1989; Van Arsdale & Schweig, 1990; Thomas, 2010) in Mississipian. During Pennsylvanian, northward movement of the southern continent and associated thrusting of the Ouachita orogeny provided crustal loading and deep basins formed in front of the northward moving thrust sheets. Figure 11.3 shows the northward movement of thrust and down to the south movement along the normal faults creating a deep basin for the large sediment load associated with the northeast to southwest flowing fluvial system of Appalachian Mountains during the Pennsylvanian.

Folding of synorogenic sediments in the southern Arkoma basin indicates that compression began during Late Pennsylvanian time after the deposition of the Atoka Formation (Houseknecht & Kacena, 1983; Suneson, 2012). Gentle folding of the post-Atokan section suggests minor compressional deformation continued through the end of the Pennsylvanian, controlling the deposition of shallow marine through alluvial systems along the axis of the gently contracting basin before complete tectonic quiescence beginning in Permian time (Houseknecht & Kacena, 1983; Suneson, 2012).

11.3. PALEOZOIC STRATIGRAPHY AND STRUCTURAL GEOLOGY

Paleozoic stratigraphy of the Ouachita Mountains–Arkoma foreland basin is traditionally divided into two distinctive facies with a deepening north-to-south transition: (1) North American continental platform section and (2) Ouachita deep-water facies. In the Ouachita Mountains and Arkoma basin, the platform section contains the Arkoma basin section of the Arbuckle facies and foredeep section of the deep water facies (Figs. 11.3 and 11.4). The Paleozoic sedimentary rock successions in the frontal Ouachita fold-thrust belt and the Arkoma basin is divided here into preorogenic and synorogenic stratigraphy.

11.3.1. Preorogenic Stratigraphy

Preorogenic stratigraphy is Laurentian passive margin sediment that accumulated prior to tectonic subsidence during the early phases of the Ouachita orogeny

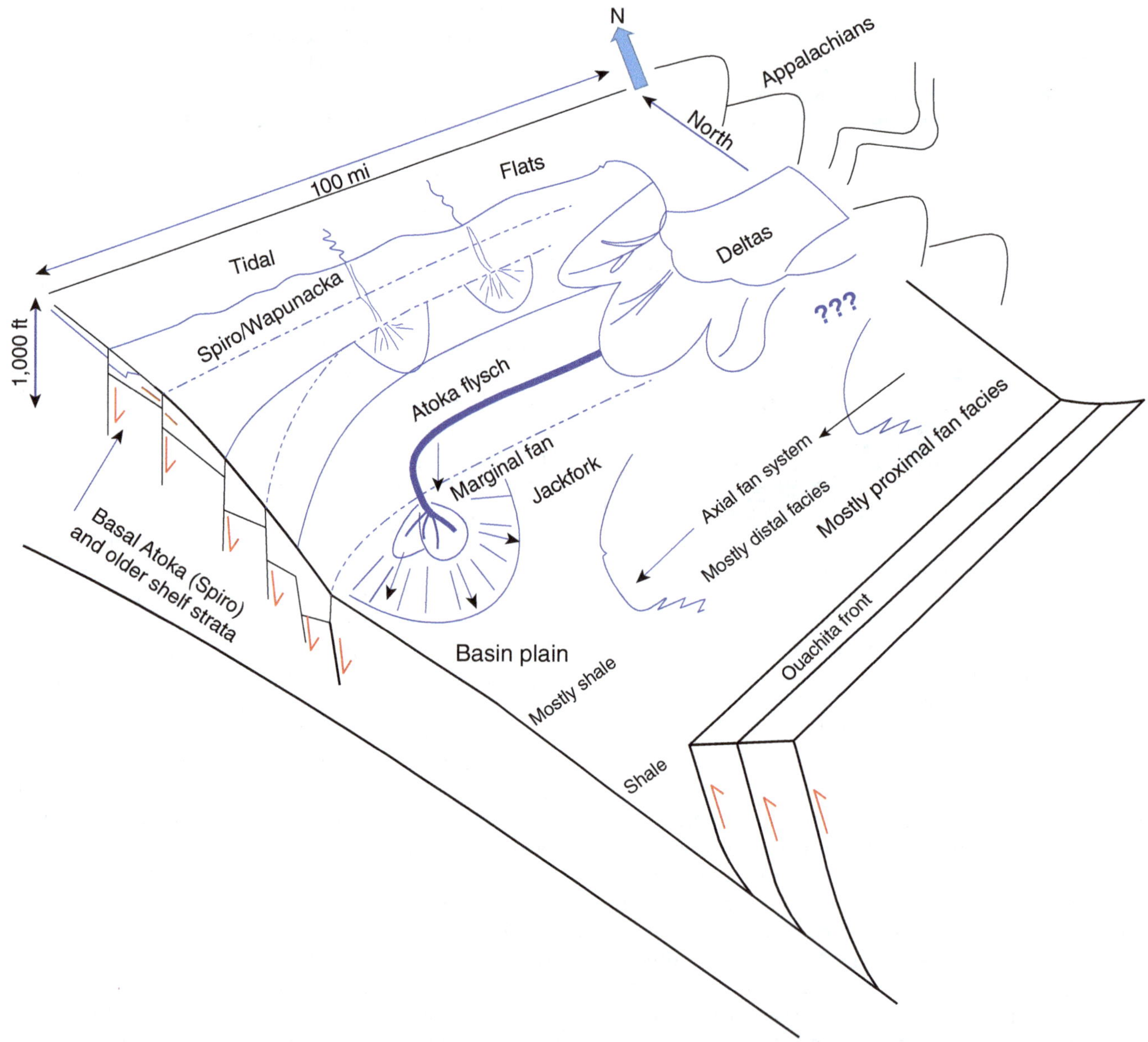

Figure 11.3 Schematic diagram showing sedimentary depositional environments of the Arkoma Basin during the lower to middle Pennsylvanian flysch sediments. The thrust faults of Ouachita orogeny to the south and normal faults along the North American continental margin to the north provided deep basin for the sediments driven from the rising Appalachian mountains to the east.

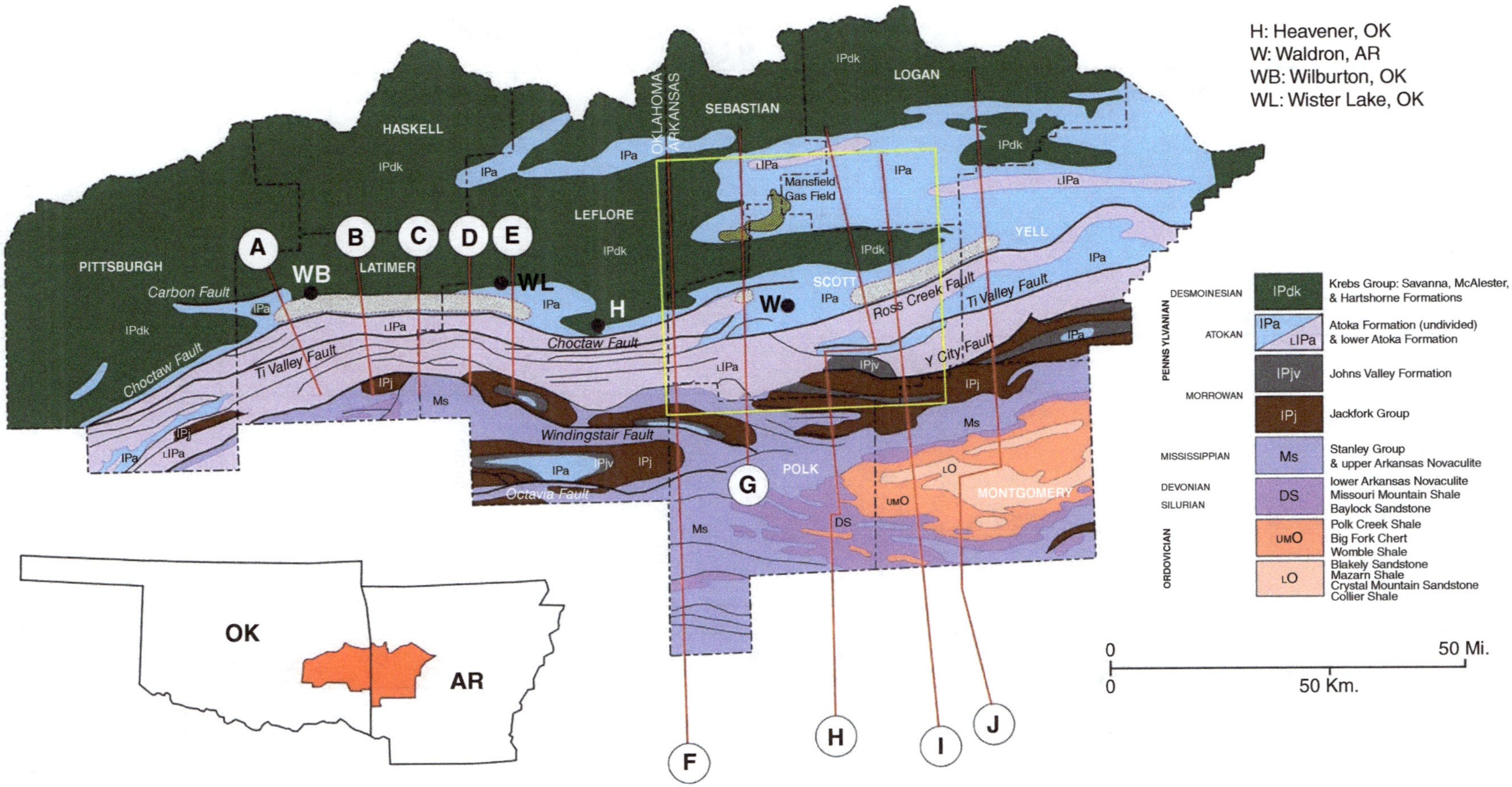

Figure 11.4 Simplified geologic map of the frontal Ouachitas and southern Arkoma basin. Lettered lines show the location of previous cross-sectional studies. Solid black lines indicate major south-dipping thrusts. Red-and-white diagonal hash pattern indicates areas where a triangle zone is described. Yellow box indicates area covered by the topographic map shown in Figure 11.5. Solid lines show the position of previous workers' cross sections: A = Çemen et al. (2001a); B = Evans (1997); C = Medhi (1997); D = Ronck (1997); E = McPhail (2001); F = Arbenz (2004); G = Roberts (2004); H = Blythe et al. (1988); I = Arbenz (2004); J = Blythe et al. (1988). Base map modified from Bennison (1986), Geological highway map of the mid-continent region.

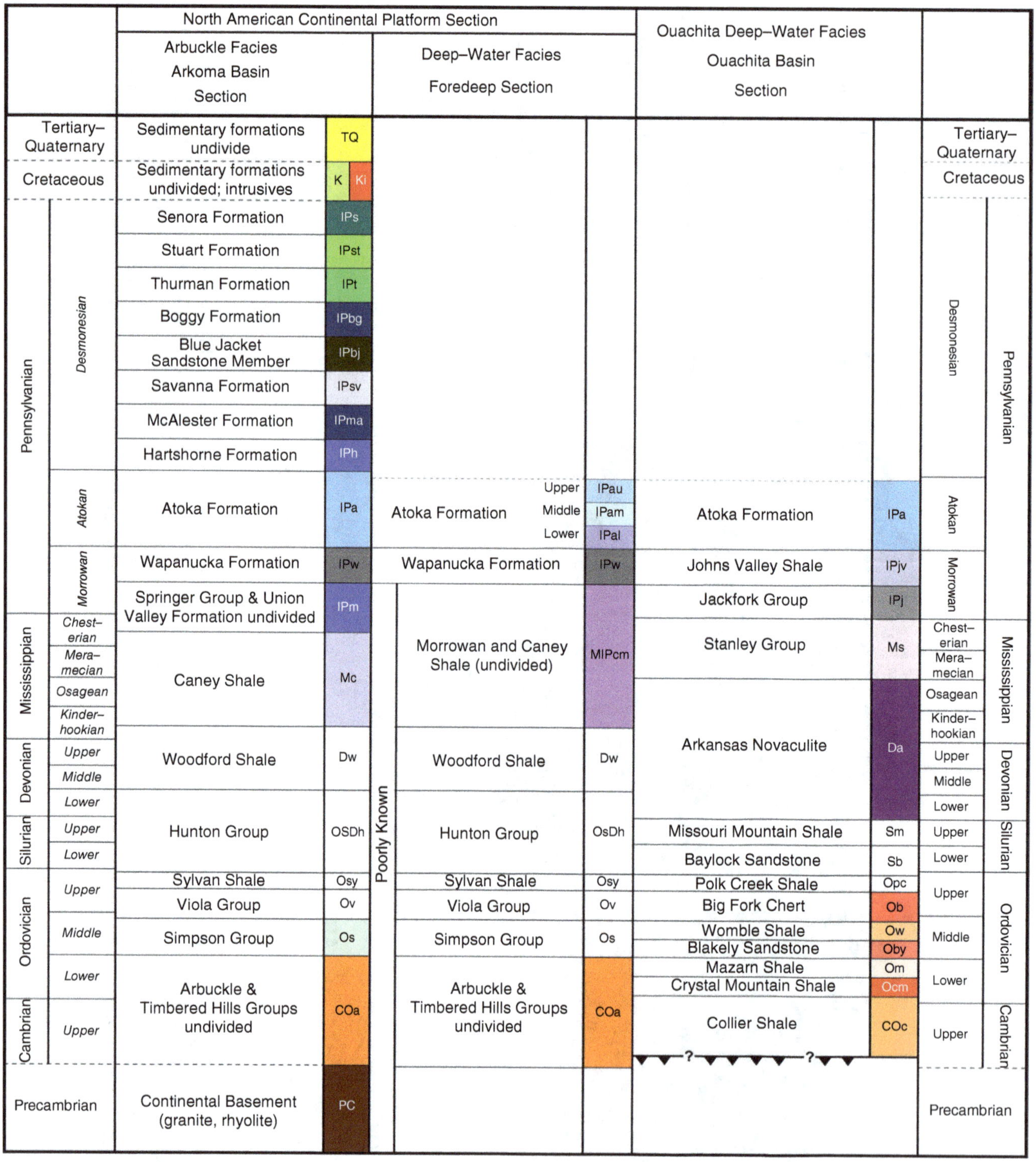

Figure 11.5 Stratigraphic column and correlative facies of the Ouachita-Arkoma system (adapted from Arbenz, 2004, and Suneson, 2012).

(Figs. 11.2, 11.4). The preorogenic Late Cambrian to Early Mississippian stratigraphy is not exposed in the frontal portion of the Ouachita fold-thrust belt and has not been penetrated by local wells; therefore, descriptions presented here are drawn from measured sections in the central Ouachita Mountains (Lowe, 1989).

The oldest sedimentary unit of the Ouachita basin section is the Late Cambrian to Early Ordovician Collier

Formation. Although its basal contact with underlying crystalline rocks is not exposed, more than 1,000 ft (305 m) of section is exposed that contains gray to black shale, containing minor amounts of chert, sandstone, and conglomerate, as well as limestone beds up to 15 m thick in the upper portion of the formation (McFarland, 2004). In Arkansas, the upper limestone beds contain a greater abundance of sand and are often lenticular (Lowe, 1989). Lying above Collier Formation is the 150 to 250 m thick Early Ordovician Crystal Mountain Formation, which consists of massive, coarse-grained, and carbonate-cemented gray quartzose sandstone interbedded with thin shale (McFarland, 2004). Few large boulders of meta-arkose up to 15 m in diameter have been reported in conglomerate beds in eastern Arkansas (Lowe, 1989). The overlying Early Ordovician Marzan Formation ranges from 300 to 750 m thick gray to black shale that contains minor siltstone, sandstone, conglomerate, and chert (McFarland, 2004). The overlying Middle Ordovician Blakely Formation contains up to 200 m thick black to green shale interbedded with few coarse-grained quartzose sandstone beds less than 1 m thick (Lowe, 1989). The overlying Middle Ordovician Womble Formation is a 150 to 350 m thick section of predominately black and green shale (McFarland, 2004). Quartzose sandstone beds less than 30 m thick are present in the lower portion of the formation, while thin limestone and chert beds are present in the upper portion (Lowe, 1989). The overlying Middle to Late Ordovician Bigfork Formation consists of up to 150 m thick, highly fractured chert and is interbedded with thin shale and limestone (McFarland, 2004). The overlying Late Ordovician Polk Creek Formation consists of up to 70 m thick black carbonaceous shale containing interbedded black chert near the base of the formation and a characteristic conglomerate at the top (Lowe, 1989).

The Early Silurian period is recorded by about 350 m thick Baylock Formation, which consists of fine- to medium-grained quartzose gray and tan sandstone, interbedded with black shale (Lowe, 1989). The Late Silurian period is recorded by the Missouri Mountain Formation, a 90 m thick dark shale succession interbedded with thin quartzose sands and containing about 1 m thick conglomerate at its base and thin bedded novaculite near the top of the formation (McFarland, 2004).

The overlying Devonian through Early Mississippian section is recorded in the Arkansas novaculite, a white and black chert unit that is up to 300 m thick in the Ouachita Mountains but thins quickly to the north (McFarland, 2004). Throughout the formation, the novaculite is interbedded with thin shale and contains sparse conglomerate beds (McFarland, 2004). The dominantly clastic preorogenic interval recorded slow deposition rates of pelagic and hemipelagic deep marine clays between periods of sand deposition by turbidity currents and sediment-gravity flows (Lowe, 1989). Paleocurrent measurements indicate that sediment dispersal was northeast-southwest with periodic quartzose influx from the south (Fig. 11.4).

11.3.2. Synorogenic Stratigraphy

The Late Mississippian Stanley Group overlies the Arkansas novaculite as the first synorogenic unit (Figs. 11.3, 11.4) and is 1,000 to 3,000 m thick, thinning to the north (McFarland, 2004). Although mostly conformable with the underlying novaculite, basal chert conglomerate indicates some submarine erosion (McFarland, 2004). The Stanley Group consists dominantly of dark gray shale; however, beds of laterally discontinuous sandstone are common as well as olistostromes that contain exotic sandstone and limestone blocks (Morris, 1989). The occurrence of tuff and the paleocurrent indicators contained therein demonstrate dispersal from a southern volcanic arc (Morris, 1989).

The 1,000 to 1,800 m thick Early Pennsylvanian Jackfork Group conformably overlies the Stanley Group and consists dominantly of brown to tan, thin- to thick-bedded, quartzose sandstone (Morris, 1989) that fines to the north and supports the elevated topography south of the transition zone in western Arkansas (Suneson, 2012). Interbedded gray to black shale increases in abundance to the north (McFarland, 2004).

The Early Pennsylvanian Johns Valley Formation overlies the Jackfork Group and consists of dark shale with minor amounts of sandstone. Despite the high degree of internal deformation, the original thickness is estimated at 450 m (McFarland, 2004). The Johns Valley Formation also contains a characteristic Cambrian to Pennsylvanian passive margin facies olistostromal blocks (see Fig. 11.7; Morris, 1989). The blocks may have originated from a combination of slumping of the depositional shelf to the north and erosion from the footwall of south-dipping syndepositional normal faults (Sutherland & Manger, 1979).

The overlying 20 m thick Morrowan Cromwell Sandstone of the Union Valley Formation is a discontinuous calcareous sandstone interbedded with thin shale and overlain by less than 20 ft (6 m) of arenaceous limestone (Sutherland, 1988). Although local well tops indicate that the Cromwell Sandstone overlies the Johns Valley Formation in the transition zone, Morris (1989) interprets the Cromwell Sandstones equivalent to the Johns Valley Formation in Oklahoma, and Sutherland (1988) shows the Cromwell as equivalent to basin shale in the frontal Ouachita Mountains immediately to the south and presumably in the hanging wall of the Ross Creek Fault. The overlying Early Pennsylvanian Bloyd

Formation consists of interbedded dark shale, limestone, calcareous sandstone, and minor amounts of coal (McFarland, 2004). The Bloyd Formation is primarily restricted to the Ozark facies of northern Arkansas where it is up to 60 m thick (McFarland, 2004); however, the shelfal Kessler Limestone member likely extends into the transition zone area as interpreted in local wells. The presence of deep-water facies at the Morrowan-Atokan boundary in the hanging wall of the Ross Creek Fault (RCF) suggests that the underlying Late Morrowan stratigraphy is carbonate-poor basinal equivalents (Sutherland, 1988).

Carbonate influence along the Early Pennsylvanian shelf ceased at the beginning of Atokan time with rapid deposition of the Atoka Formation (Figs. 11.3, 11.4), an asymmetric synorogenic clastic wedge with an estimated thickness greater than 6,000 m at the southern margin of the Arkoma basin where the lithology is dominantly shale (Zachry, 1983). The Atoka Formation is subdivided into three members, the lower, middle, and upper (Sutherland, 1988). The base of the lower member is recognized by the incision of 10–60 m thick Spiro Sandstone into shale of the Bloyd Formation (Zachry, 1983) and marks the end of northerly sourced passive margin deposition before rapid subsidence accommodated deposition of a thick succession of turbidites preserved in the hanging wall of the RCF in Arkansas that grade toward neritic strata to the west in Oklahoma above the Choctaw Fault (CHF) (Figs. 11.1, 11.5; Sutherland, 1988; Arbenz, 2004).

Syndepositional normal faulting during Middle Atokan time controlled deep seafloor bathymetry and resulted in axial channelization of easterly sand-rich sediment-gravity flows that rapidly thicken above the downthrown hanging walls that define the middle portion of the Atoka Formation in Oklahoma (Fig. 11.3). Accurate quantification of the extreme thicknesses reached by the middle Atoka Formation is precluded by erosion of the upper intraformational contact, but the thickest measurable section is located within the study area between the overlap of the traces of the CHF and RCF (Figs. 11.1, 11.4). Restored thickness estimates by Haley (1988) indicate that the formation thickens southward at 150 m per mile. Due to the limited lateral extent of the submarine axial channels and fans, sandstone of the middle Atoka Formation are discontinuous and commonly change or lose character over distances of about half a kilometer and cannot be correlated over significant distances (Zachry & Sutherland, 1984). Cessation of down-to-south normal faulting marked the end of the deep marine facies of the middle Atoka Formation. The upper Atoka Formation is not cut by these faults and consists of fluvial and deltaic systems that prograded southward in Late Atokan time (Sutherland, 1988).

Conformably overlying the Atoka Formation is the Desmoinesian Krebs Group. Exposures are limited to the lower formations in the Poteau Syncline at the northern edge of the Waldron quadrangle. The basal unit of the Krebs Group is the tidal-deltaic Hartshorne Formation (Fig. 11.4), from a few to 100 m thick, brown to light gray, massive sandstone section (McFarland, 2004). Conformably overlying the Hartshorne Formation is the McAlister Formation, from 150 to 700 m thick, which consists dominantly of shale, isolated sandstone beds, and an increasing abundance of coal toward the top of the formation. Conformably overlying the McAlister Formation is the Savanna Formation, about 500 m thick, which consists of dark gray shale, siltstone, fine-grained quartzose sandstone, and several coal beds (McFarland, 2004).

11.3.3. Structural Geology of the Ouachitas-Arkoma Basin Transition Zone

Structural mapping of the southern Arkoma basin north of the Choctaw Fault (CHF) in eastern Oklahoma (Fig. 11.1) shows a series of broad synclines separated by tight, north-verging anticlines (Houseknecht, 1987). Availability of subsurface data increased in the late 1980s and 1990s with the discovery of natural gas in the Wilburton area of Oklahoma. Interpretations of well log and 2D seismic data led Çemen et al. (2001a,b) to propose the presence of a triangle zone, the Wilburton Triangle Zone (WTZ), in the footwall of the CHF to explain strain partitioning from the highly deformed frontal Ouachita fold-thrust belt in the south to the mildly deformed Arkoma basin in the north (Fig. 11.6, line A).

Here, the intercutaneous wedge of the WTZ is composed of middle Atoka Formation that is detached above the lower Atokan Detachment (LAD), a horizontal to gentle (6°) south-dipping floor thrust that merges with the Carbon Fault (CF) at a tip line. The north-dipping CF, while mostly eroded, serves as the roof thrust that bounds the passively uplifted and transported foreland syncline (San Bois Syncline). The WTZ is underlain by a duplex that imbricates the basal Atokan Spiro Sandstone, which utilizes the LAD as a roof thrust and is detached above a floor thrust that ramps from the Late Devonian to Early Mississippian Woodford Shale through the Late Mississippian to Early Pennsylvanian Springer Shale. Elements of the triangle zone extend eastward near Wister Lake in Leflore County, Oklahoma (Fig. 11.6, line B; McPhail, 2001). Although the geometry is similar to the WTZ, the triangle zone incorporates both lower and middle Atoka Formation in the intercutaneous wedge.

In 2010 we obtained three 2D seismic reflection profiles that trend approximately perpendicular to the structures

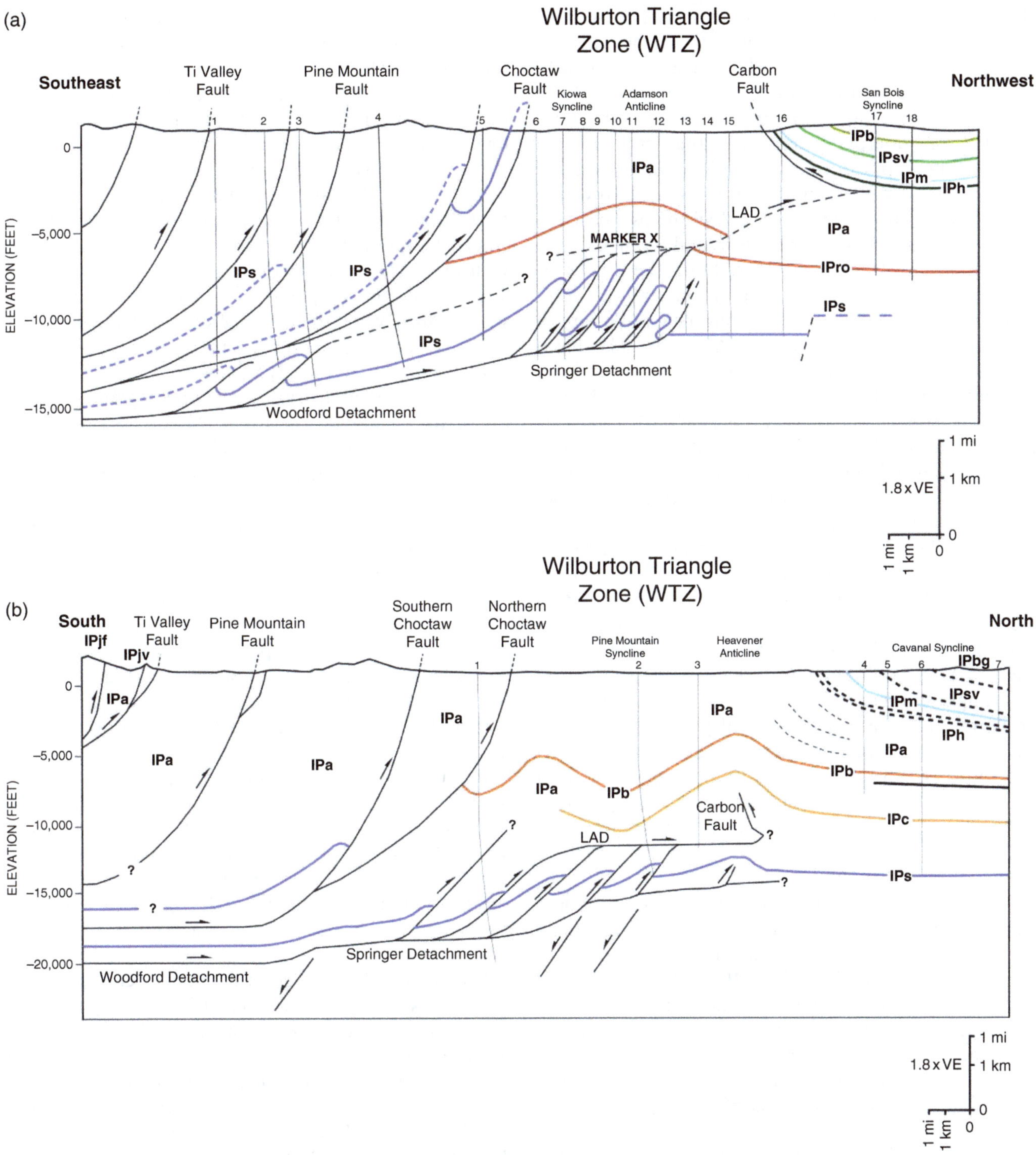

Figure 11.6 Cross sections of the Wilburton Triangle Zone (WTZ) in Oklahoma: (a) Modified from Çemen et al. (2001a); (b) modified from McPhail (2001). LAD = lower Atokan Detachment; IPb/IPbg = Boggy Formation; IPsv = Savanna Formation; IPm = McAlister Formation; IPh = Hartshorne Formation; IPa = Atoka Formation; IPro = Red Oak Sandstone; IPc = Cecil Sandstone; IPs = Spiro Sandstone; IPjv = Johns Valley Formation; IPjf = Jackfork Group.

within the Waldron and Boles quadrangles in western Arkansas. The seismic lines provide the highest quality available images of the subsurface. They were depth-converted based on the velocity model developed from seismic line A-A′ (Fig. 11.7). Figure 11.8 shows the uninterpreted and interpreted profile in the depth domain from the topographic surface to –9,150 m with no vertical exaggeration (H = V). Formation tops interpreted within local wells and correlation from geologic maps are shown as seismic horizons to illustrate structural geometry.

Along the cross section in Figure 11.6b, the LAD similarly merges with the CF at a tip line beneath the foreland syncline (Cavanal Syncline), but the CF dies out in the subsurface of the Heavener Anticline. The basal Spiro Sandstone-involved duplex is likewise present and does not change character from its lateral equivalent to the west. Geologic mapping by Suneson (2012) suggests that the Carbon Fault (CF) terminates in the Wister Lake area west of the Oklahoma-Arkansas state border while the CHF continues into western Arkansas before tipping out at the surface in central Scott County (Fig. 11.3). The cross section plane for the seismic reflection profile along the line A-A′ strikes 002° for 26 km within the Waldron quadrangle and crosses the Ross Creek Fault (RCF) in the northern Boles quadrangle imaging the hanging wall south to Buffalo Mountain (Fig. 11.7). Due to the plunge of the folds in the Waldron quadrangle and the strike of the RCF, a minor apparent dip is introduced along the central portion of the line while a true structural dip is preserved along the northern and southern portions of the profile. The structural geometry suggests a dominantly forward-propagating series of thrusts with two detachment levels toward the foreland that form a triangle zone, composed of three wedges (Figs. 11.8, 11.9, 11.10).

At the southern edge of the profile, the south-dipping RCF intersects the topographic surface as a high angle (80°) bedding parallel thrust immediately north of High Point Mountain (Fig. 11.8). The RCF exhibits the listric characteristic of decreasing dip angle with depth to a bedding-parallel flat detached within the upper Jackfork Group at –4,575 m. The hanging wall shows moderately south-dipping reflectors of the lower Atoka Formation near the surface that similarly decrease in dip with depth. Morrowan stratigraphy below the basal Atokan Spiro Sandstone and above the upper Jackfork Group detachment surface is interpreted at the base of the hanging wall block and extends to the shallow subsurface. The RCF splays into two segments near the topographic surface and ramps up-section, offsetting the south-dipping limb of a minor hanging wall anticline of sub-Spiro Sandstone section.

The upper wedge interpreted along the profile is a triangular region of chaotic reflectors in the immediate footwall of the RCF and a north-dipping backthrust within the lower Atoka Formation. This backthrust continues north where it occupies the same structural position as the LAD in profile 1; therefore, it is interpreted as the southern continuation of the LAD where it cuts shallowly up section before emerging south of Ross Mountain (Fig. 11.8). This poorly imaged region is floored by a bedding-parallel thrust within the Union Valley Formation above the Cromwell Sandstone from where it enters the profile from the south to a ramp that merges with the LAD at the northern edge of the chaotic zone and is probably composed entirely of ductily deformed shale of the lower Atoka Formation. The hanging wall of the LAD consists of north-dipping middle and upper Atoka Formation that is passively uplifted in an in-sequence series of thrusting. The unconstrained depositional thickening of the lower Atoka Formation and poor imaging prohibit reasonable seismic interpretation and estimation of the magnitude of shortening accommodated within the upper wedge. Therefore, the interpreted position of the Spiro Sandstone near the base of the structure is schematic and represents a minimum (undeformed) bed length.

The middle wedge is located below the Waldron Syncline and consists of a thrust sheet within the lower Atoka Formation and upper Union Valley Formation, and contains the basal Spiro Sandstone (Fig. 11.9). The thrust sheet is bound by sigmoidal south-dipping thrusts D1 and D2 that splay from the up-dip limit of the decollement within the Union Valley Formation (floor thrust) and sole into the LAD (roof thrust) (see Fig. 11.12). Thrust D1 separates the upper wedge from the Spiro Sandstone-involved thrust sheet of the middle wedge. Thrust D2 is the basal thrust of the middle wedge and soles into the LAD and its up-dip limit. Within the thrust sheet, a minor south-dipping thrust offsets the Spiro Sandstone before displacement dies out into the overlying section.

The lower wedge consists of the Stanley Group through the lower Atoka Formation folded in a hanging wall anticline detached above the SD, which passively transports the overlying upper and middle wedges northward in sequence. The frontal edge of the lower wedge (Fig. 11.10) is the lateral continuation of the structure interpreted in profile 1 as the SD ramps above the normal fault, N1. The SD cores the fault propagation fold before merging with the LAD at the tip line underlying the Poteau Syncline. The interpretation does not show a stratigraphically and structurally higher tip line within the middle Atoka Formation, requiring the CF and associated north-dipping listric thrust to die out west of the profile. North of normal fault N1, a second south-dipping normal fault N2 is interpreted to offset the pre-Morrowan stratigraphy.

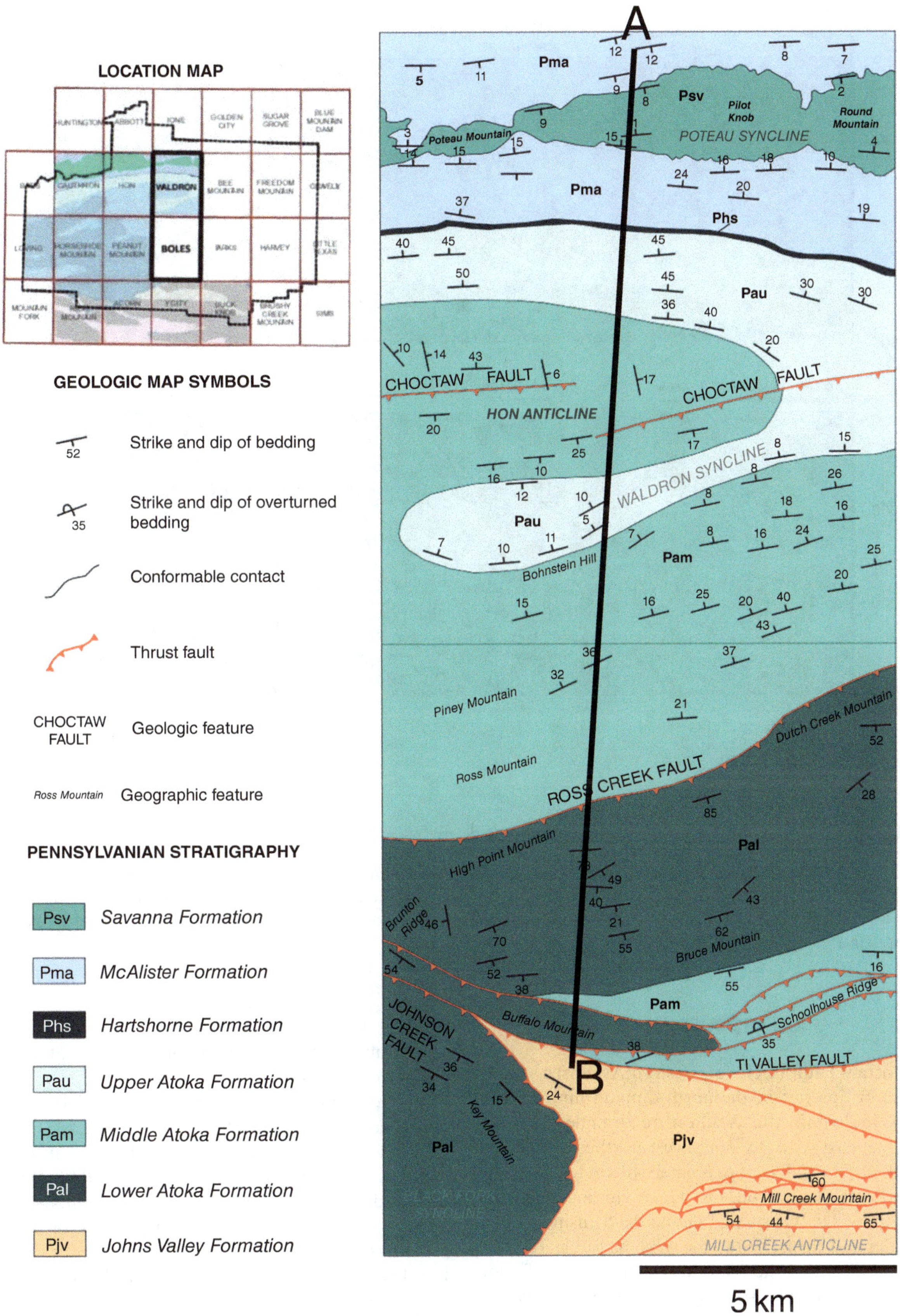

Figure 11.7 Geologic map of the Waldron and Boles quadrangles in Arkansas. The black line shows the location of depth-converted available seismic profile used for structural interpretation.

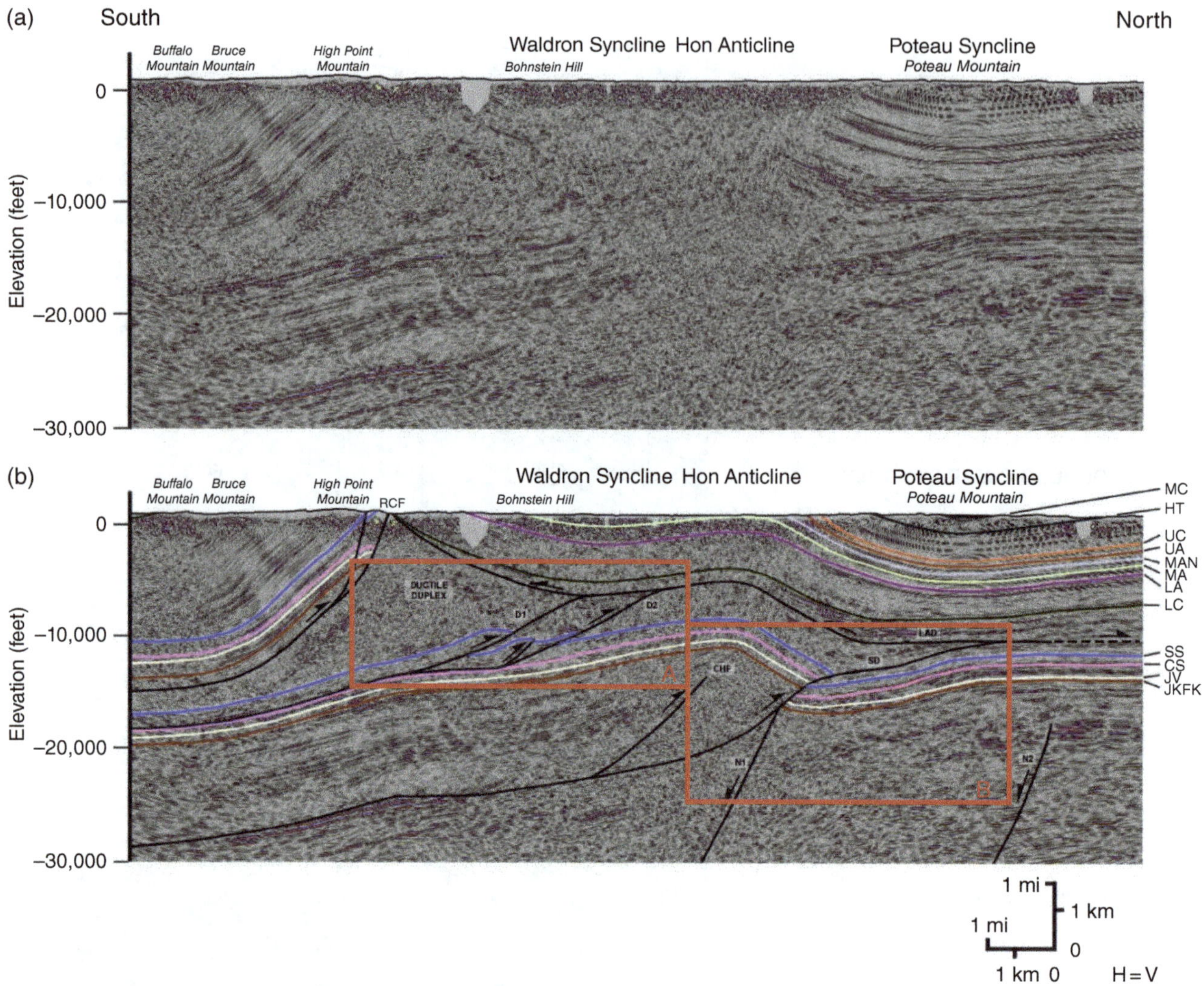

Figure 11.8 (a) Uninterpreted and (b) interpreted seismic 2D depth-converted reflection profile along line A-B, Figure 11.7: MC = McAlister Formation; HT = Hartshorne Sandstone; UC = upper Carpenter Sandstone; UA = upper Alma Sandstone; MAN = Mansfield Sandstone; MA = middle Alma Sandstone; LA = lower Alma Sandstone; LC = lower Carpenter Sandstone; SS = Spiro Sandstone; CS = Cromwell Sandstone; JV = Johns Valley Formation; JKFK = Jackfork Group. The rectangles A and B mark the areas of the section enlarged in Figures 11.9 and 11.10, respectively.

11.4. DISCUSSION

The Choctaw Fault (CHF) in western Arkansas strikes east-west and tips out in the north-dipping limb of the Hon Anticline within the Waldron quadrangle, where it overlaps the east-west to southwest-northeast striking Ross Creek Fault (RCF) approximately 10 km to the south within the Boles quadrangle (Fig. 11.7). This arrangement of the leading-edge thrusts that define the transition from the Ouachita fold-thrust belt to the Arkoma foreland basin is consistent with the surface expression of an accommodation zone, where slip is transferred between separate faults as a decrease in slip along one fault segment coincides with an increase in displacement along a parallel fault segment sharing a common decollement surface at depth (Fig. 11.11).

Displacement along the northern edge of the Ouachita fold-thrust belt (Fig. 11.1) is accommodated by the CHF in Oklahoma (Figs. 11.4, 11.6) and the RCF in Arkansas (Figs. 11.4, 11.6, 11.8). Structural interpretation of depth-converted 2D seismic reflection profiles demonstrate that the RCF is the dominant leading-edge thrust of the frontal Ouachitas in north-central Scott County, western Arkansas (Fig. 11.7). The CHF is interpreted as a blind thrust in the footwall of the RCF that merges with the south-dipping Stanley Decollement (SD) (Figs. 11.8, 11.9, 11.10).

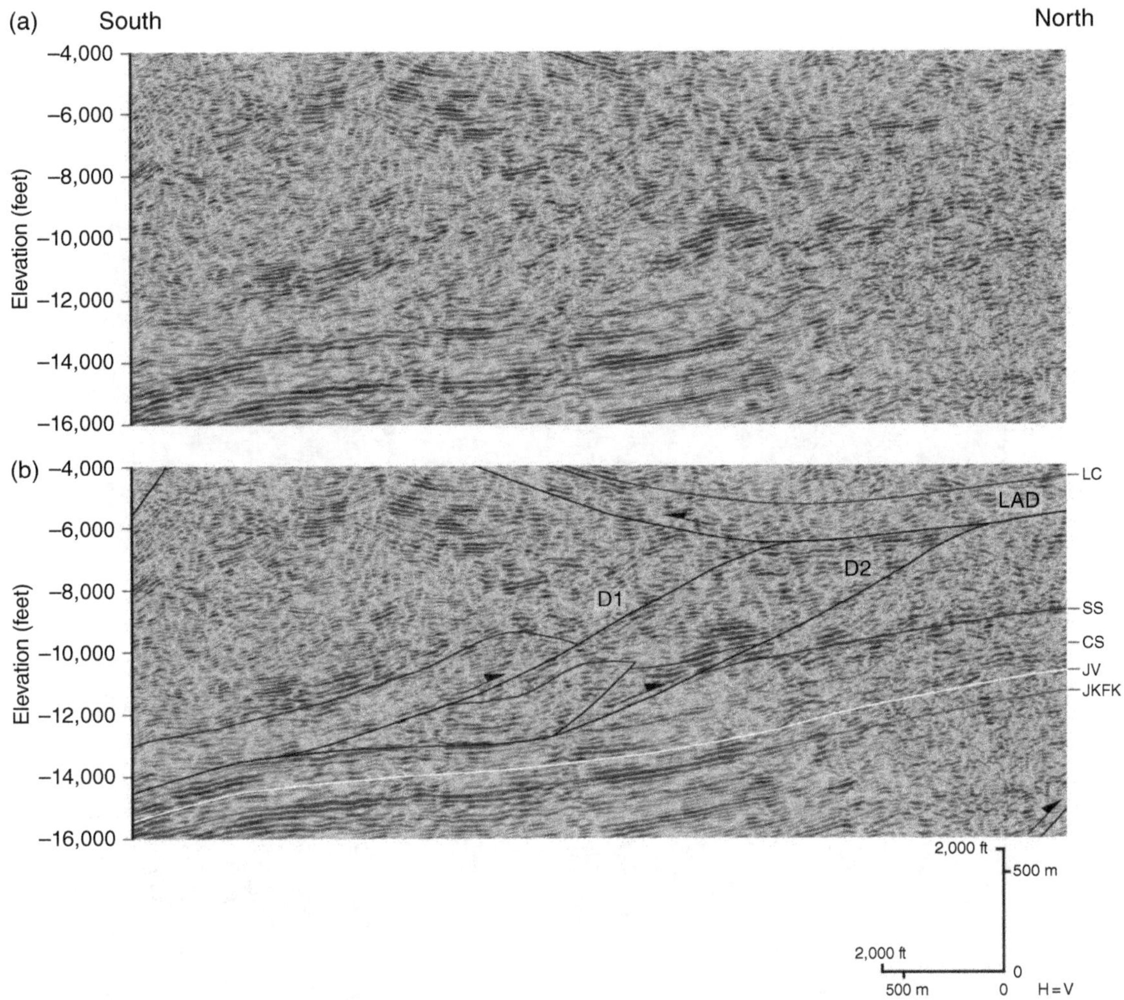

Figure 11.9 Detailed (a) uninterpreted and (b) interpreted images of the middle wedge (rectangle A, Fig. 11.8) of seismic profile along line A-B, Figure 11.7. LC = Lower Carpenter Sandstone; SS = Spiro Sandstone; CS = Cromwell Sandstone; JV = Johns Valley Formation; JKFK = Jackfork Group.

The RCF, LAD, and SD thrusts bound a triangular region internally composed of three stacked wedges. The CHF is interpreted within the lower wedge, however, as displacement increases to the west, the Wilburton Triangle Zone (WTZ) develops within the footwall (Fig. 11.9). The geometry of an accommodation zone predicts that displacement along the RCF decreases to the west with increasing displacement along the CHF, although the nature of the westward termination of the RCF is not well known (Arbenz, 2004). The triangle zone in the footwall of the RCF also may die out to the west with the increasing structural dominance of the CHF and WTZ in Oklahoma, although further study is required to document the structure to the west along strike.

Continued compression resulted in the propagation of the decollement within the Union Valley Formation and establishment of thrust D2 north of D1, forcing the LAD and tip line into the foreland, further uplifting the passive roof above the LAD, and completing the development of the thrust sheet belonging to the middle wedge (Fig. 11.9). Slip was then transferred from the upper and middle wedges to the basal SD, which ramps from the base of the Stanley Group into the lower Atoka Formation due to the buttress effect of the preorogenic carbonates in the footwall of the south-dipping N1 (Fig. 11.12b). Ramping of the SD into shale of the lower Atoka Formation synchronously establishes a new tip line with the LAD north of the middle wedge. The final stage of Late Pennsylvanian thrusting forming the current triangle zone geometry in western Arkansas is the out-of-sequence blind thrust of the CHF, which splays from the SD south of and in response to the buttressing effect of the N1 footwall (Fig. 11.12). Formation of the CHF in the subsurface increases the structural elevation of the hanging wall anticline composing the lower wedge, during which the LAD is passively folded.

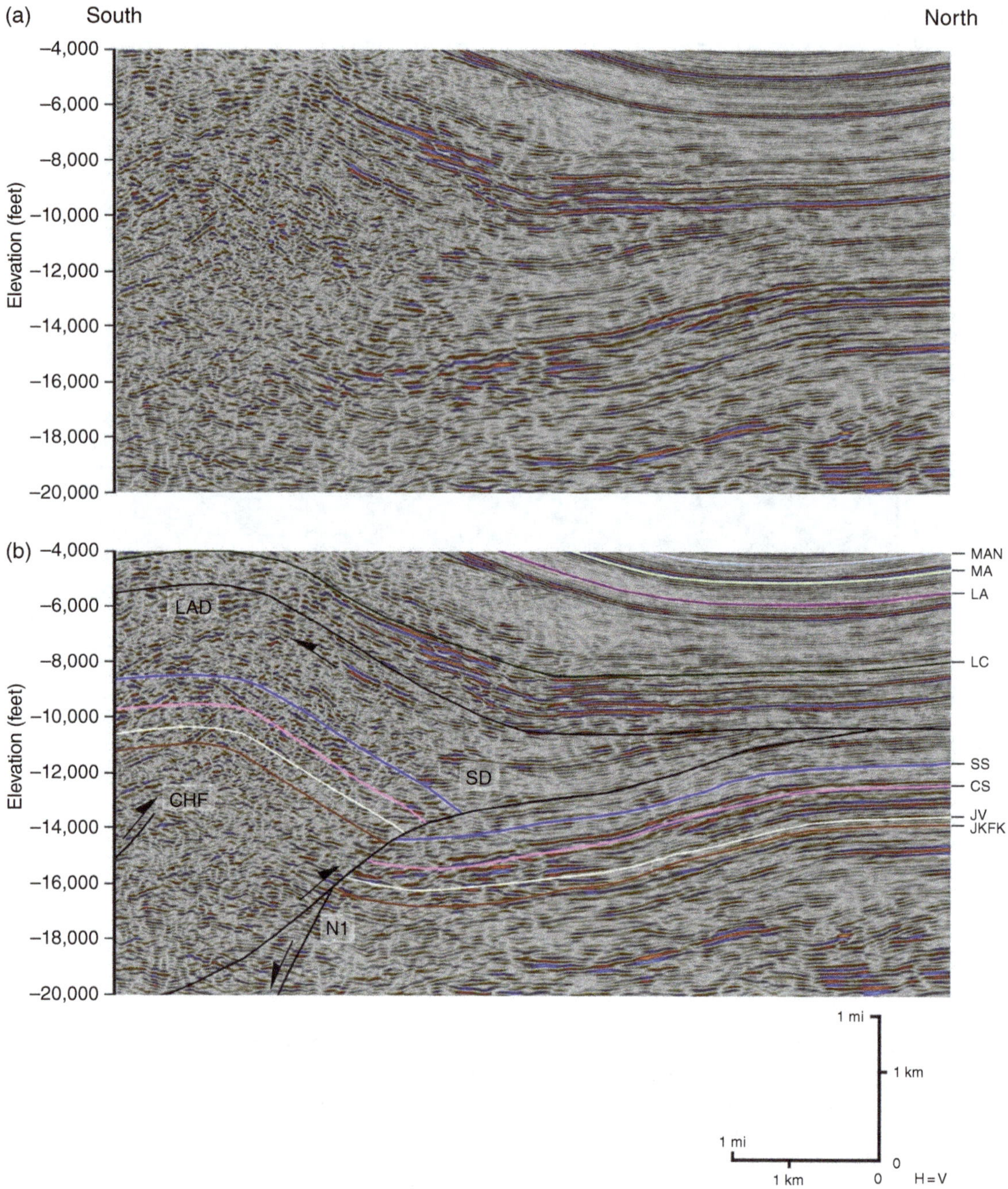

Figure 11.10 Detailed (a) uninterpreted and (b) interpreted images of the lower wedge (rectangle B, Fig. 11.8) of seismic profile along line A-B, Figure 11.7. UC = upper Carpenter Sandstone; UA = upper Alma Sandstone; MAN = Mansfield Sandstone; MA = middle Alma Sandstone; LA = lower Alma Sandstone; LC = lower Carpenter Sandstone; SS = Spiro Sandstone; CS = Cromwell Sandstone; JV = Johns Valley Formation; JKFK = Jackfork Group.

Although the maximum age of thrusting within the transition zone is constrained by the presence of deformed Desmoinesian stratigraphy, it is not possible to quantify the duration of compression and the age of each stage with the available data. Regional study by Houseknecht and Kacena (1983) suggest that compression ceased by the end of Pennsylvanian at 298.9 Ma ago.

11.4.1. Comparison of the Wilburton and Waldron Triangle Zones

In a triangle zone, foreland thrusting is commonly an intercutaneous wedge bound by a floor thrust and a roof thrust. Couzens and Wiltschko (1996) propose an end-member model for the internal geometry of the wedge

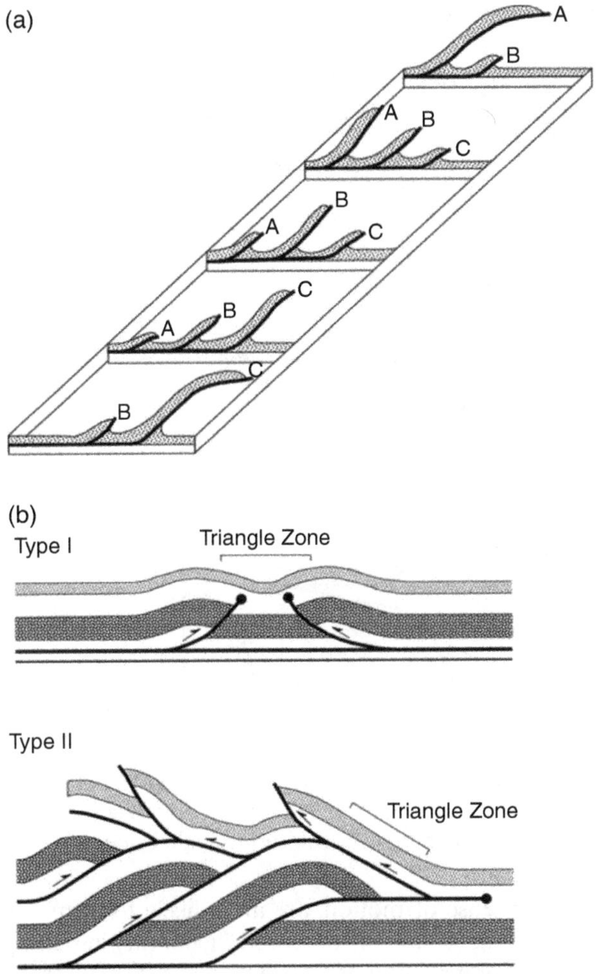

Figure 11.11 (a) Schematic cross sections demonstrating the three dimensional geometry of an accommodation zone along strike of the transition zone from the highly deformed fold-thrust belt and mildly deformed foreland basin. From the upper right to lower left, slip along thrust fault A decreases to zero while slip along thrust fault B reaches a maximum in the center of the accommodation zone and thrust fault C increases from zero (modified from Dahlstrom,1969b). (b) Type I and II of triangle zone geometry proposed for strain partitioning from the highly deformed fold-thrust belt to mildly deformed foreland basin structures (modified from Couzens & Wiltschko, 1996).

(Fig. 11.11b). Type I triangle zones contain a wedge that remains relatively undeformed between opposing thrusts that emanate from a master detachment. Type II triangle zones contain internally duplexed wedges that possess two or more detachment levels. In the case of both end-member geometries, the roof and floor thrusts merge at a tip line or point of zero displacement, passively uplifting a frontal syncline as transport of the wedge forces the tip line to propagate into the foreland (Jones, 1982, 1996).

Triangle zones yield information about the processes of deformation during the formation of fold-thrust belts, although the specific role that the triangle zone occupies in the spatial and temporal evolution of a thrust belt is not well understood and could vary among orogenic belts (Jones, 1996). Finite-element models demonstrate the viability of a propagating syncline or continuous generation model of triangle zone development if the roof strata are of sufficiently low friction (Jamison, 1996). The triangle zone of western Arkansas exhibits geometries consistent with frontal structures of both the North American Cordillera and the southern Appalachians; however, the most significant and relevant comparison is with the Wilburton Triangle Zone (WTZ) of eastern Oklahoma (Çemen et al., 2001a,b), 60 km to the west along strike and occupying the same structural position at the southern boundary of the Arkoma basin (Fig. 11.4; Arbenz, 2004).

The primary difference in structural geometry between the two triangle zones is that the upper wedge of the Waldron Triangle Zone (WLTZ) is composed of a ductile duplex of lower Atoka Formation, while the intracutaneous wedge of the Wilburton Triangle Zone (WTZ) involves the younger middle Atoka Formation, which is broadly folded (Figs. 6, 12a). This example illustrates the transient appearance of ductile duplexes. Additionally, the LAD in western Arkansas shifts to a north-dipping backthrust from a south-dipping in-sequence thrust within the WTZ.

The geometry of the Spiro Sandstone-involved thrust sheet of the middle wedge presented here is similar in many respects to the duplex at the base of the WTZ (Figs. 11.6, 11.12a; Çemen et al., 2001a,b) as each imbricate the Spiro Sandstone between a floor thrust and roof thrust in age-equivalent formations. Within the WTZ, the Woodford and Springer detachments mark the base of thrusting (Figs. 11.6, 11.12a). In western Arkansas, the base of thrusting also lies within Mississippian shale of the Stanley Group. Although no wells penetrate Early Mississippian strata in the frontal Ouachitas or southernmost Arkoma basin, this deepest detachment is inferred at the base of the Stanley Group (Caney Shale equivalent), based on well imaged stratigraphy down to the presumed top of the Arkansas novaculite at −9,150 m at the southern end of the seismic profile (Fig. 11.8), increasing in structural elevation to −3,050 m where it tips out into the LAD. The presence of a major decollement in the Stanley Group has been previously proposed by Arbenz (2004); its continuation into the transition zone area is inferred from outcrop relationships by Suneson et al. (2005) who concluded that the SD is the eastern continuation of the Woodford Detachment of the WTZ (Figs. 11.6, 11.12). Previous work in the Wilburton area of Oklahoma also suggests that south-dipping normal faults analogous to N1 and N2 underlie and likely control the placement of ramps in the Woodford and Springer detachments (Figs. 11.6, 11.12; Çemen et al, 2001a,b; Ronck, 1997; Medhi, 1997; McPhail, 2001).

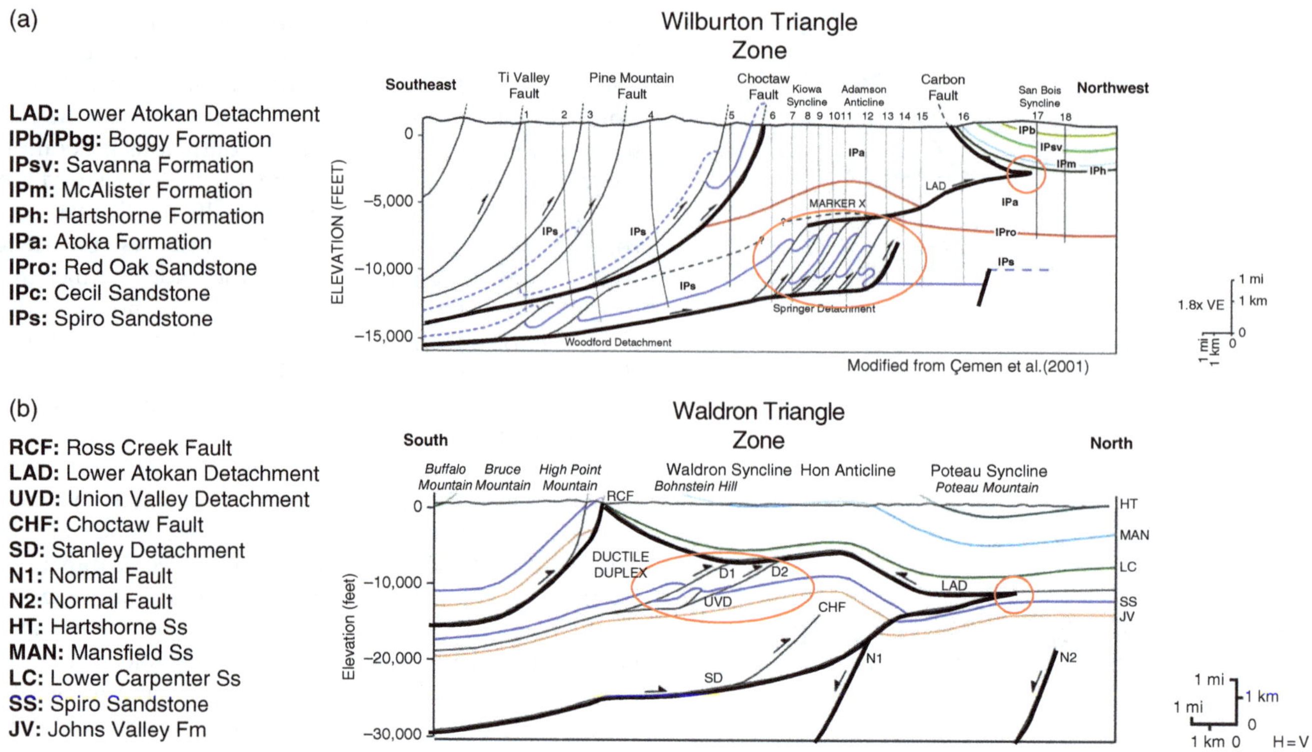

Figure 11.12 Diagrams showing the general structural geometry of (a) the Wilburton and (b) the Waldron triangle zones. The Woodford Detachment of Wilburton Triangle Zone becomes Stanley Detachment, and Choctaw Fault becomes Ross Creek fault and is intersected by the Lower Atokan Detachment in the Waldron Triangle Zone.

Because of the similarity in structural position and geometry, the triangle zone of western Arkansas is the eastern continuation of the WTZ of eastern Oklahoma (Çemen, et al., 2001a,b); however, due to the changing frontal thrusts and facies changes to deep-water strata, the triangle zone in western Arkansas is the named Waldron Triangle Zone, after the nearby town of Waldron, Arkansas. Further study is required to document the geometry of the Ouachita fold-thrust belt to Arkoma basin transition zone between the WTZ in the Wister Lake area, Oklahoma and the WLTZ area of Arkansas (Figs. 11.4, 11.6, 11.12).

11.4.2. Worldwide Structural Analogues

Triangle zones are structures commonly recognized at the boundary between fold-thrust belts and associated foreland basin transition zones where strain partitioning occurs (DeCelles & Giles, 1996), however, triangle zones are not recognized in several orogenic systems (Table 11.1). The factors controlling the development of triangle zones are not fully understood but may be related to the mechanical stratigraphy of the deforming section and the preexisting structure prior to compression that forms the foreland basin (Couzens & Wiltschko, 1996). Empirical evidence compiled from known triangle zones, or transition zones containing a backthrust sense of motion, are more likely to form when the deforming section is composed of contrasting competent and incompetent units, allowing a wedge of competent strata to detach above (floor thrust) and below (roof thrust) an incompetent horizon (Couzens & Wiltschko, 1996). The triangle zones are less likely to form when a deforming section is composed of mostly competent units (Couzens & Wiltschko, 1996). It is unknown whether triangle zone formation is a continuous process at work during orogenesis or is restricted to late stage evolution of a fold-thrust belt (Jones, 1996), however, incompetent shale-rich horizons provided by burial of late synorogenic deposits host backthrusts in numerous triangle zones worldwide (Couzens & Wiltschko, 1996; Couzens-Schultz & Wiltschko, 2000). Finite-element and sandbox models also suggest that the presence of backthrusts promotes triangle zone formation (Jamison, 1996). It is noted in the Arkoma basin that the triangle zones form as the deformation required to form a backthrust is reduced toward the undeformed foreland basin sedimentary units (Çemen et al., 2001a,b). The geometry of the Waldron Triangle Zone is consistent with these requirements, having developed within thick synorogenic deposits of contrasting mechanical behavior (Fig. 11.12b) above a large south-dipping normal fault that exposes a competent footwall block as a barrier to northward thrusting. Although transition zones lacking a triangle zone

Table 11.1 Chart showing the location of orogenic belts with dominant backthrust or forethrust sense of motion at the foreland limit of thrusting

Orogenic belt	Frontal thrust type	Geometry
Southern Candian Rockies	Backthrust	II
Central Canadian Rockies	Backthrust	II
West MacKenzie Mountains	Backthrust	II
Brooks Range, Alaska	Backthrust	II
Wyoming thrust belt	Backthrust	II
Ouachitias, E. OK and W. AR	Backthrust	II
Great Valley, California	Backthrust	II
Southern Coast Ranges, CA	Backthrust	II
Barbados	Backthrust	II
Ellesmerian Front, Greenland	Backthrust	II
N. Argentina Precordillera	Backthrust	II
S. Argentina Precordillera	Backthrust	II
Venetian Alps	Backthrust	II
Northern Urals	Backthrust	II
Dagestan Belt, Caucasus	Backthrust	II
Kohat Plateu, Pakistan	Backthrust	II
Kirthar and Sulaiman ranges	Backthrust	II
Potwar Plateau, Pakistan	Backthrust	I
Central Pyreness	Backthrust	I
Central Appalachians	Backthrust	I
Southern Urals	Backthrust	I
Southern Alps, Italy	Forethrust	N/A
Central Appalachians (PA/VA/WV)	Forethrust	N/A
Zagros, Far Region	Forethrust	N/A
French Alps	Forethrust	N/A
Eastern Pyrenees	Forethrust	N/A
Jura Mountains	Forethrust	N/A

Source: Modified from Couzens and Wiltschko (1996).
Note: The orogenic belts with a backthrust sense of motion contain triangle zones while the orogenic belts lacking a significant backthrust do not contain a triangle zone.

(Table 11.1) may be the result of bias due to erosion of all or part of the structure that contained a triangle zone, a deforming section composed dominantly of competent units without layered incompetent horizons inhibits backthrusting and are instead expressed by forethrusts with decreasing displacement toward undeformed section in the foreland rock units (Couzens & Wiltschko, 1996; Çemen et al., 2001a,b).

11.5. CONCLUSIONS

Our surface and subsurface interpretations in southeastern Oklahoma suggest that strain partitioning between the highly deformed Ouachita fold-thrust belt and the mildly deformed Arkoma basin is accomplished via the Wilburton Triangle Zone (Fig. 11.6). In north-central Scott County, Arkansas, an accommodation zone exists between the leading-edge thrusts of the Ouachita fold-thrust belt. Slip along the Choctaw Fault dies out in the subsurface north of the Ross Creek Fault (Figs. 11.7, 11.8), which continues as the leading-edge thrust fault of the fold-thrust belt to the east. The accommodation zone between the Choctaw Fault and Ross Creek Fault serves as the boundary between the highly deformed frontal Ouachitas and the mildly deformed Arkoma basin sedimentary successions. Strain partitioning between the fold-thrust belt and the basin is accommodated by a triangle zone consisting of three wedges in the footwall of the Ross Creek Fault (Figs. 11.8, 11.12). The present structural geometry evolved through both in-sequence and out-of-sequence thrusting during the Ouachita orogeny in Late Pennsylvanian time. This triangle zone is the lateral continuation of the Wilburton Triangle Zone of southeastern Oklahoma (Fig. 11.6). However, frontal thrust geometry and associated mechanical stratigraphy are different in Arkansas. Therefore, the triangle zone of western Arkansas exhibits a different geometry (Fig. 11.8) and is named here the Waldron Triangle Zone.

The upper wedge of the Waldron Triangle Zone consists of a region of ductily deformed shale of the lower Atokan Formation bounded by two oppositely dipping thrust faults, the Ross Creek Fault and the lower Atokan Decollement, similar to the ductile duplex or MUSHWAD recognized in the foreland basins of the southern Appalachians (Thomas, 2001). The middle wedge consists of a thrust sheet of lower Atoka Formation that contains the regionally extensive Spiro Sandstone, however, this thrust sheet is only observed along the seismic profile in Arkansas (Fig. 11.8). The thrust sheet dies out eastward through the southern Waldron quadrangle. The north-dipping lower Atokan Decollement serves as the roof thrust of both the ductile duplex and the Spiro Sandstone-involved thrust sheet, which soles into and accommodates slip transferred from the south-dipping Stanley Decollement, or floor thrust of the lower wedge, via a tip line below the Poteau Syncline (Fig. 11.8). The location and mechanics of the Stanley Decollement ramp down-dip of the tip line is controlled by strain localization above a major south-dipping pre-Atokan normal fault, N1.

The geometry and subsurface formations indicate that normal fault N1 is likely the source of the large exotic blocks of early Paleozoic shelf strata observed within the Stanley Group and Johns Valley Formation to the south. The Waldron Triangle Zone is likely the site of the Late Mississippian and Early Pennsylvanian shelf, consistent with the fault-controlled depositional model for the flysch units (Stanley Group, Jackfork Group, and Johns Valley Formation) of the Ouachita basin section.

The presence of the Waldron Triangle Zone in the footwall of the Ross Creek Fault is consistent with previous studies that find deforming strata of variable mechanical strength at the foreland limit of thrusting in an orogenic belt typically host triangle zones. Furthermore, the presence of the Waldron Triangle Zone in the footwall of the Ross Creek Fault demonstrates that the triangle zone structure at the transition from the Ouachita fold-thrust belt to the Arkoma basin is not an ephemeral feature of the orogenic belt and persists across the accommodation zone eastward into central Arkansas. The surface tip-out and decreasing subsurface displacement along the Wilburton Triangle Zone and Choctaw Fault into western Arkansas also shows that the footwall structures follow the same accommodation zone pattern as the leading-edge thrusts.

ACKNOWLEDGMENTS

We thank Southwestern Energy corporation for providing three 2D seismic reflection profiles for Donald Yezerski's M.S. thesis project in eastern Arkansas. We also thank many of our colleagues who provided discussions in the field and during the constructions of our cross sections. We specifically thank Neil Suneson, Roger Slat, and William Coffey. İbrahim Çemen also owes thanks to the late Zuhair Al-Shaeib and the graduate students who completed their M. S. theses with his supervision at Oklahoma State University from 1995 to 2003, namely, Ata Sagnak, Justin Evans, Jeffrey Ronck, Syed Mehdi, Marline Straight, and Kristopher McPhail.

REFERENCES

Arbenz, J. K. (1989). Ouachita thrust belt and Arkoma basin. In R.D. Hatcher Jr., W.A. Thomas, & G.W. Viele (Eds.), *The Appalachian-Ouachita orogen in the United States* (pp. 621–634). Geological Society of America, Geology of North America, F-2.

Arbenz, J. K. (2004). Structural framework of the Ouachita Mountains. In N. H. Suneson (Ed.), *Stratigraphic and structural evolution of the Ouachita Mountains and Arkoma Basin, southeastern Oklahoma and west-central Arkansas* (pp. 4–40). Applications to Petroleum Exploration: 2004 Field Symposium, 2008 symposium, Oklahoma Geological Survey Circular 112A.

Bennison, A. P. (1986). *Geological highway map of the mid-continent region.* 1:1,900,800. Tulsa, Oklahoma: The American Association of Petroleum Geologists.

Blythe, A. E., Sugar, A., & Phipps, S. P. (1988). Structural profiles of the Ouachita Mountains, western Arkansas. *The American Association of Petroleum Geologists Bulletin, 72*(7), 810–819.

Bradley, D. C., & Kidd, W. S. F. (1991). Flexural extension of the upper continental crust in collisional foredeeps. *Geological Society of America Bulletin, 103*, 1416–1438.

Çemen, İ., Evans, J., & Sagnak, A. (2001b). Eastern continuation of the Wilburton triangle zone in the Red Oak gas-field area, frontal Ouachitas-Arkoma basin transition zone, southeastern Oklahoma. In K. S. Johnson & D. F. Merriam (Eds.), *Petroleum systems of sedimentary basins in the southern mid-continent 2000 symposium* (pp. 81–95). Oklahoma Geological Survey Circular 106.

Çemen, İ., Sagnak, A., & Akthar, S. (2001a). Geometry of the triangle zone and duplex structure in the Wilburton gas field area of the Arkoma basin, southeastern Oklahoma. In K. S. Johnson (Ed.), *Pennsylvanian and Permian geology and petroleum in the southern midcontinent 1998 symposium* (pp. 87–98). Oklahoma Geological Survey Circular 104.

Couzens, B. A., & Wiltschko, D. V. (1996). The control of mechanical stratigraphy on the formation of triangle zones. *Bulletin of Canadian Petroleum Geology, 44*(2), 165–179.

Couzens-Schultz, B. A., & Wiltschko, D. V. (2000). The control of the smectite-illite transition on passive-roof duplex formation: Canadian Rockies Foothills, Alberta. *Journal of Structural Geology, 22*, 207–230.

Dahlstrom, C. D. A. (1969). Balanced cross sections. *Canadian Journal of Earth Sciences, 6*, 743–757.

DeCelles, P. G., & Giles, K. A. (1996). Foreland basin systems. *Basin Research, 8*(2), 105–123.

Evans, J. (1997). *Structural geometry and thrust faulting in the Baker Mountain and Panola quadrangles, southeastern Oklahoma*. M.S. thesis. Oklahoma State University, Stillwater.

Gordy, P. L., Frey, F. R., & Norris, D. K. (1977). *Geological guide for the CSPG 1977 Waterton-Glacier Park field conference*. Canadian Society of Petroleum Geologists, Calgary, Alberta, Canada.

Haley, B. R. (1988). *Restored thickness map of Atoka Formation, West-Central Arkansas*. Arkansas Geological Commission Oil and Gas Chart 7.

Houseknecht, D. W. (1986). Evolution from passive margin to foreland basin: The Atoka Formation of the Arkoma basin, south-central U.S.A. In P. A. Allen & P. Homewood (Eds.), *Foreland basins* (pp. 327–345). Boston, Blackwell Scientific Publications, International Association of Sedimentologists, Special Publication 8.

Houseknecht, D. W. (1987). *The Atoka Formation of the Arkoma basin: Tectonics, sedimentology, thermal maturity, sandstone petrology*. Tulsa Geological Society, Short Course Notes.

Houseknecht, D. W., & Kacena, J. A. (1983). *Tectonic-sedimentary evolution of the Arkoma Basin. SEPM, Mid-Continent Section, 1*, 3–33.

Jamison, W. R. (1996). Mechanical models of triangle zone evolution. *Bulletin of Canadian Petroleum Geology, 44*(2), 180–194.

Jones, P. B. (1982). Oil and gas beneath east-dipping underthrust faults in the Alberta foothills. In R. B. Powers (Ed.), *Studies of the Cordilleran thrust belt* (pp. 6–74). Rocky Mountain Association of Geologists.

Jones, P. B. (1996). Triangle zone geometry, terminology, and kinematics. *Bulletin of Canadian Petroleum Geology, 44*(2), 139–152.

Lowe, D. R. (1989). Stratigraphy, sedimentology, and depositional setting of pre-orogenic rocks of the Ouachita Mountains, Arkansas and Oklahoma. In R. D. Hatcher Jr. et al. (Eds.), *The Appalachian-Ouachita orogen in the United*

States (pp. 621–634). Geological Society of America, Geology of North America, F-2.

McFarland, J. D. (2004). *Stratigraphic summary of Arkansas. Arkansas Geological CommissionInformation Circular 36*.

McPhail, K. R. (2001). *Structural geometry of thrust faulting in the Summerfield and Leflore SE 7.5-minute quadrangles, Leflore County, SE Oklahoma*. M.S. Thesis. Oklahoma State University, Stillwater.

Mehdi, S. Y. (1997). Structural geometry of thrust systems in the Red Oak and Talihina Quadrangles, Latimer and Leflore counties, Arkoma basin, southeastern Oklahoma. M.S. Thesis. Oklahoma State University, Stillwater.

Morris, R.C. (1989). Stratigraphy and sedimentary history of pre-Permian Paleozoic rocks of the Marathon uplift. In R. D. Hatcher Jr. et al. (Eds.), *The Appalachian-Ouachita orogen in the United States* (pp. 621–634). Geological Society of America, Geology of North America, F-2.

Poole, F. G., Perry, W. J., Madrid, R. J., & Amaya-Martínez, R. (2005). *Tectonic synthesis of the Ouachita-Marathon-Sonora orogenic margin of southern Laurentia: Stratigraphic and structural implications for timing of deformational events and plate-tectonic model*. Geological Society of America Special Paper *393*, 543–596.

Roberts, M. T. (2004). Geologic profiles derived from seismic data west and east of the Arkansas-Oklahoma state line, Arkoma Basin and frontal Ouachita Mountains. In N. H. Suneson et al. (Eds.), *Stratigraphic and structural evolution of the Ouachita Mountains and Arkoma Basin, southeastern Oklahoma and west-central Arkansas, applications to petroleum exploration* (pp. 17–24). 2008 Field Symposium, Oklahoma Geological Survey Circular 112B.

Robinson, D. M., Bailey, R. M., & Goodliffe, A. M. (2012). Structure of the Appalachian thrust belt under the Gulf Coastal Plain sediments. *GCAGS Journal*, *1*, 44–54.

Ronck, J.L. (1997). Structural geology of thrust faulting in the Wister Lake area of the frontal Ouachita Mountains, Arkoma basin, southeastern Oklahoma. M.S. Thesis. Oklahoma State University, Stillwater.

Suneson, N. H. (2012). Arkoma Basin petroleum, past, present, future: Shale shaker. *The Journal of the Oklahoma City Geological Society*, *63*(1), 38–70.

Suneson, N. H., Çemen, İ, Kerr, D. R., Roberts, M. T., Slatt, R. M., & Stone, C. G. (Eds.) (2005). *Stratigraphic and structural evolution of the Ouachita Mountains and Arkoma Basin, southeastern Oklahoma and west-central Arkansas: Applications to petroleum exploration*. Oklahoma Geological Survey Guidebook *34*.

Sutherland, P. K. (1988). Late Mississippian and Pennsylvanian depositional history in the Arkoma Basin area, Oklahoma and Arkansas. *Geological Society of America Bulletin*, *100* (11), 1787–1802.

Sutherland, P. K., & Manger, W. L. (Eds.) (1979). *Mississippian-Pennsylvanian shelf-to basin transition, Ozark and Ouachita regions, Oklahoma and Arkansas*. Oklahoma Geological Survey Guidebook 19.

Thomas, W. A. (1989). The Appalachian-Ouachita orogen beneath the Gulf Coastal Plain between the outcrops in the Appalachian and Ouachita Mountains. In R. D. Hatcher Jr. et al. (Eds.), *The Appalachian-Ouachita orogen in the United States* (pp. 537–553). Geological Society of America, The Geology of North America, F-2.

Thomas, W. A. (2001). Mushwad: Ductile duplex in the Appalachian thrust belt in Alabama. *The American Association of Petroleum Geologists Bulletin*, *85*(10), 1847–1869.

Thomas, W. A. (2010). Interactions between the southern Appalachian-Ouachita orogenic belt and basement faults in the orogenic footwall and foreland. In R. P. Tollo et al. (Eds.), *From Rodinia to Pangea:* The lithotectonic region of the Appalachian region (pp. 897–916). Geological Society of America Memoir 206.

Van Arsdale, R. B., & Schweig, E. S., III (1990). Subsurface structure of the Eastern Arkoma Basin. *The American Association of Petroleum Geologists Bulletin*, *74*(7), 1030–1037.

Vann, I. R., Graham, R. H., & Hayward, A. B. (1986). The structure of mountain fronts. *Journal of Structural Geology*, *8*, 215–228.

Viele, G.W. (1979). Geologic map and cross section, eastern Ouachita Mountains, Arkansas: map summary: Geological Society of America Bulletin, v. *90*, p. 1096–1099.

Zachry, D. L. (1983). Sedimentologic framework of the Atoka Formation, Arkoma basin, Arkansas. In D. W. Houseknecht (Ed.), *Tectonic-sedimentary evolution of the Arkoma basin, v. 1* (pp. 34–52). SEPM Midcontinent Section.

Zachry, D. L., & Sutherland, P. K. (1984). Stratigraphy and depositional framework of the Atoka Formation (Pennsylvanian) Arkoma Basin of Arkansas and Oklahoma. *Oklahoma Geological Survey Bulletin*, *136*, 9–18.

12

Extensional Collapse of Orogens: A Review and Example From the Southern Appalachian Orogen

David A. Foster[1], Chong Ma[2], Ben D. Goscombe[1,3], and Paul A. Mueller[1]

ABSTRACT

The extensional collapse of an orogenic belt occurs when the gravitational potential energy in thickened crust is released by spreading laterally. Collapse may occur during shortening and mountain building (synorogenic collapse) or after a change in plate dynamics terminates mountain building and results in divergence (postorogenic collapse). In collisional orogens like the Cenozoic Himalaya–Tibet or Paleozoic Appalachian systems, collapse typically starts tens of millions of years after crustal thickening. The lag between thickening and collapse is related to the time it takes for radiogenic heat to accumulate within thickened crust. Heating causes partial melting and high metamorphic temperatures at mid-crustal levels, which enhances ductile flow. The Southern Appalachian Orogen underwent extensional collapse near the end of the continent-continent collisions that formed Pangea. Collapse occurred during the Carboniferous-Permian and was accompanied by high-temperature–low-pressure metamorphism, partial melting of midcrustal rocks, and magma emplacement. Extension was accommodated by a southeast dipping detachment fault system that developed near the location of the Laurentia-Gondwana suture with the Triassic South Georgia rift in the hanging wall.

12.1. INTRODUCTION

Synorogenic, or postorogenic, collapse (extension) of an orogenic belt occurs when the topographic load of the orogen reaches a point of gravitational instability, the maximum principal stress on the orogen is vertical, and the strength of the crustal or lithospheric profile is weak enough for deep seated rocks to flow (e.g., Dewey, 1988; Rey et al., 2001; Vanderhaeghe & Teyssier, 2001; Jamieson et al., 2011). This situation typically occurs after significant crustal thickening at a collisional plate boundary, at an accretionary plate boundary, or through magmatic loading in the overriding plate of a subduction zone. Spreading may occur through extension and thinning of the entire lithosphere or through thinning of the most ductile layers, which commonly are in the middle or lower crust.

In collisional orogens like the Cenozoic-Recent Himalaya-Tibet Orogen or the Paleozoic Appalachian–Caledonian Orogen, collapse typically starts tens of millions of years after large-scale crustal thickening. The lag between thickening and orogenic collapse is related to the time it takes for radiogenic heat to build up in the thickened crust, which increases metamorphic temperatures and expands the thickness of the ductile zone (e.g., England &Thompson, 1984; Bird, 1991; Royden, 1996; Vanderaeghe et al., 2003). The buildup of radiogenic heat may also cause partial melting in the lower and middle crust. Magma migration into the middle crust enhances

[1] *Department of Geological Sciences, University of Florida, Gainesville, Florida, USA*

[2] *Mineral Exploration Research Centre, Harquail School of Earth Sciences, Laurentian University, Sudbury, Ontario, Canada*

[3] *Integrated Terrane Analysis Research, Adelaide, Australia*

Compressional Tectonics: Plate Convergence to Mountain Building, Geophysical Monograph 277, First Edition.
Edited by Elizabeth J. Catlos and İbrahim Çemen.

DOI:10.1002/9781119773856.ch12

flow by further decreasing the strength of the middle crust through the introduction of fluids and advection of heat (Vanderhaeghe &Teyssier, 2001). Progressive attenuation enhances weakening of the lower and middle crust by instigating decompression melting, and may be accompanied by lithospheric drips or delamination (tectonothermal recycling of the lithospheric mantle into the convective mantle), which promotes upwelling of the asthenosphere and additional heat advection, and facilitates further collapse (Houseman et al., 1981). Collapse may occur during shortening and mountain building (synorogenic collapse) or may occur after a change in plate boundary conditions that results in an end to mountain building and even divergence (postorogenic collapse).

Depending on plate motions, postorogenic extension may lead to eventual rifting of the continent near the site of maximum crustal thickening of a former mountain belt (e.g., Ma et al., 2019; Wiest et al., 2020; Osmundsen et al., 2020). Many rifted passive margins around the world today formed along the axis of Precambrian and Phanerozoic orogenic belts, and in many respects the collapse of an orogen is an expected consequence of the "Wilson Cycle" (Dewey, 1988).

The first part of this chapter is an overview of orogenic collapse in the context of this volume. This section is followed by an overview of an example of the extensional collapse of an orogenic plateau (one of many possible examples): the Southern Appalachian Orogen in late Carboniferous-Permian times.

12.2. EXTENSIONAL COLLAPSE OF OROGENS

The gravitational force that drives orogenic collapse arises due to contrasts in mass, while the rate at which collapse occurs is controlled, in part, by the strength profile of the lithosphere (Molnar &Lyon-Caen, 1988; Rey et al., 2001). In lithosphere with layers of different viscosity and density, gravitational collapse may be concentrated in weak layers, or channels, such as the lower crust or a melt-infused middle crust (e.g., Block & Royden, 1990; McKenzie et al., 2000; Beaumont et al., 2001; Vanderhaeghe et al., 2003; Searle, et al., 2006; Jamieson et al., 2011).

The initiation of gravitational collapse is the result of changes in the forces acting on the lithosphere, including (1) plate boundary conditions that reduce horizontal compressive stresses; (2) reduction in basal shear stress of the lithospheric plate above a subduction zone; (3) subduction slab rollback resulting in extension in the upper plate; (4) convective thinning of the lithosphere (removal of mantle lithosphere) leading to a relative increase in potential energy; (5) reduction in strength of the orogenic pile or adjacent regions due to radioactive, conductive, or convective heating of the crust and/or lithospheric mantle (e.g. Dewey, 1988; England & Houseman, 1988; Rey et al., 2001). Rey et al. (2001) described two alternative geodynamic regimes of gravitational collapse: (1) when excess gravitational potential energy results from convergent deformation and crustal thickening, orogenic collapse occurs in a direction away from the region of convergent deformation; (2) when a deficit in gravitational potential energy results from divergence or lack of convergence, orogenic collapse occurs in the direction of thinner crust.

12.2.1. Modes of Extensional Collapse

The two endmember modes of collapse of a thickened orogenic belt are illustrated in Figure 12.1 from Rey et al. (2001). Fixed boundary collapse involves flow from the thickened orogenic crust to adjacent areas of thinner crust (Mode 1), whereas free boundary collapse involves flow and thinning of the orogen because of divergence of the bounding plate (Mode 2) (Rey et al., 2001).

Fixed boundary collapse could be accomplished by thinning of the upper crust via extensional detachments within the thickened crust that may connect to thrusts over adjacent regions, thinning of the lower crust by ductile flow into adjacent thinner crust without upper crustal deformation, and a combination of these two processes (Fig. 12.1a,b,c). Free boundary collapse results in thinning of the entire orogen via normal faulting in the brittle upper crust and by ductile flow in the lower crust. Upper and lower crustal thinning are typically separated by low-angle detachments. Expansion of the Tibetan plateau is an excellent example of fixed boundary collapse with extension of the upper crust synchronous with boundary thrusting (Fig. 12.2). In this case, the excess gravitational potential energy results from extensive crustal thickening combined with changes in the strength profile due to heating of the thickened crust by radioactive decay, partial melting, and convective thinning of the subcrustal lithosphere. Extension of the North American Basin and Range Province or the Aegean Sea due to changes in relative plate motion, subducted plate rollback, subduction retreat, and/or oceanic ridge subduction are examples of free boundary collapse (Fig. 12.2).

Exhumation of middle (and lower) crustal rocks results from thinning and/or removal of the upper crust by extensional detachment faults combined with erosion in Modes 1 and 2. In the case of lower-middle crustal flow, the deeper level rocks contain mineral assemblages and compositions that indicate decompression and document the thinning of the crust. Exhumation may be tectonic in the sense of unroofing beneath a detachment if extensional deformation transitions to the upper crust, or

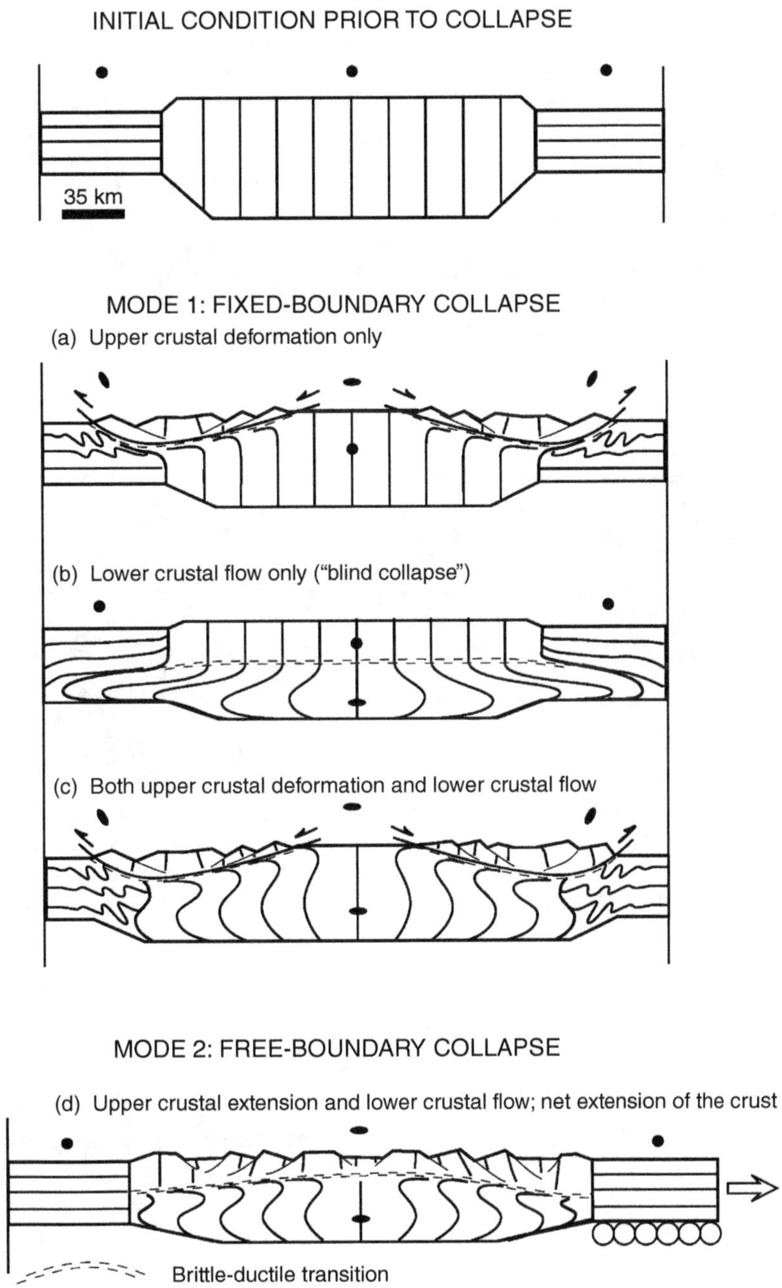

Figure 12.1 Modes of gravitational collapse affecting thick continental crust depicted in cross sections of the crust from Rey et al. (2001).

erosional over time if thinning is restricted to the lower crustal layers. In most cases, a combination of these processes is likely to have occurred.

12.2.2. Crustal Strength, Partial Melting, and Orogenic Collapse

Nelson et al. (1996) presented geophysical data that suggest that the middle crust of the Himalaya-Tibet Orogen is currently partially molten (Fig. 12.3). More recent magnetotelluric data support the interpretation that areas of Tibet contain high melt fractions at middle crustal depths (e.g., Xie et al., 2021). The recognition that a significant fraction of partial melt is present in the middle crust was subsequently tied to evidence for extensive partial melting (migmatites) and high temperature mineral assemblages in rocks that were at middle crustal depths less than 10 Mya, which are now exhumed along parts of the Himalaya orogenic front (e.g., Goscombe et al., 2018, and references therein; Ding et al., 2021; Fig. 12.4).

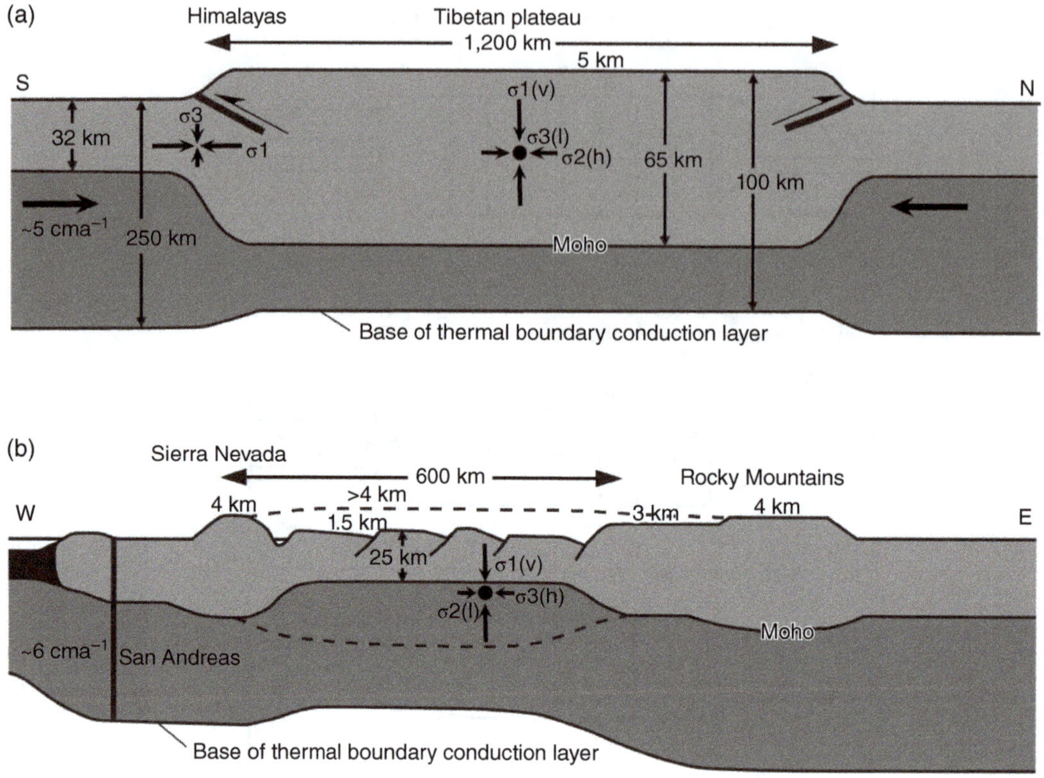

Figure 12.2 Cross sections showing two examples of regions undergoing extensional collapse modified after Dewey (1988). (a) The Tibetan Plateau of the Himalaya-Tibet Orogen, and (b) the Basin and Range Province of the North American Cordilleran Orogen. Stress indicated by fine black arrows as maximum (σ1) to minimum (σ3) directions. Light gray represents continental crust, dark gray is lithosphere, and black is oceanic crust.

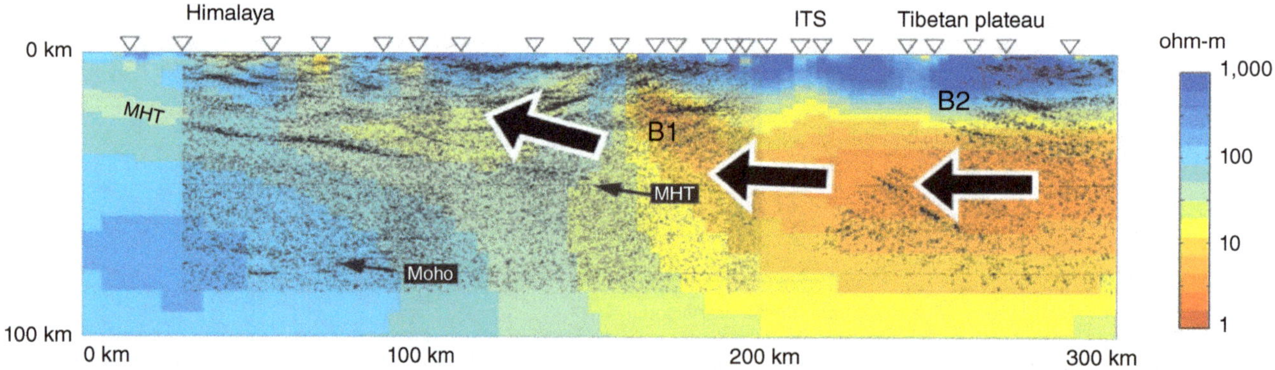

Figure 12.3 Geophysical data from the INDEPTH profile across the Himalaya and Tibet reproduced from Jamieson et al.'s (2011) reproduction from Nelson et al. (1996). The background image shows seismic reflection data, and the contours show electrical resistivity values from magnetotelluric data (upside down triangles are data recording points). The highlighted areas have resistivity values that are anomalously low indicating the presence of a fluid (partial melt). The seismic reflectors labeled B1 and B2 are interpreted to be the top of the fluid-bearing region. MHT is the Main Himalayan Thrust with underthrust Indian plate crust in the footwall. Black arrows show possible direction of crustal flow.

Geochronologic data (primarily U-Pb zircon and monazite) show that rocks within the middle crust of thick orogens contained partial melt for up to tens of millions of years while these rocks were at middle crustal depths (e.g., Cavalcante et al., 2018; Ding et al., 2021; Guergouz et al., 2018; Turlin et al., 2018; Goscombe et al., 2022). Migration of partial melt into the middle crust and long-term radiogenic heating leads to very high midcrustal

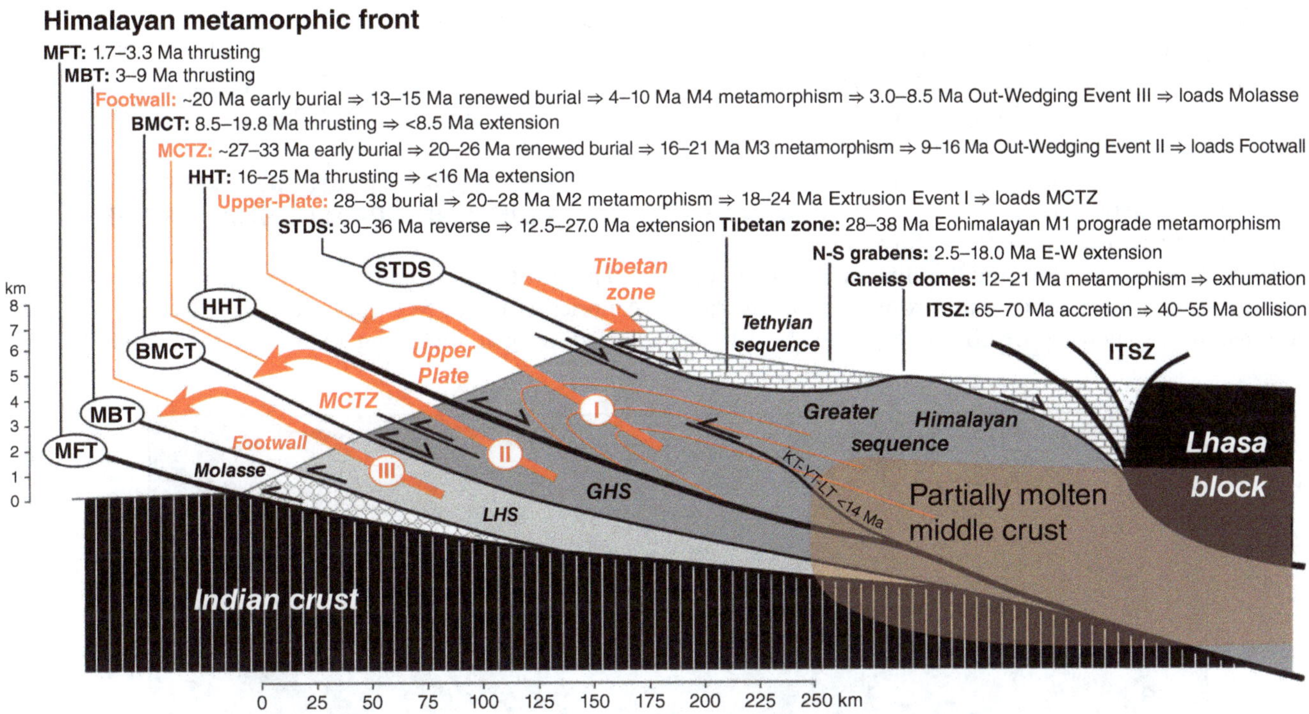

Figure 12.4 Diagrammatic cross section through the Himalaya-Tibet Orogen. Partial melt in the middle crust reduces the strength of the orogen and contributes to outwedging of the middle and lower crust along the Himalayan metamorphic front. The cross section also shows the time sequence of thrusting and collapse of the orogenic plateau (modified from Goscombe et al., 2018). Each successive out-wedging stage and progression of the orogenic front to the south over the Indian crust is the result of continued shortening and extensional collapse of the plateau as it reaches critical thickness. Note: MFT = Main Frontal Thrust; MBT = Main Boundary Thrust; BMCT = Basal Main Central Thrust; MCTZ = Main Central Thrust Zone; HHT = High Himal Thrust; STDS = South Tibet Detachment System; LHS = Lesser Himalayan Sequence; GHS = Greater Himalayan Sequence. I, II, and III represent the progressively extruded or outwedged packages of middle crust at the Himalayan front.

temperatures (>700°C), high geothermal gradients, and a reduction in strength of the crustal profile (Cavalcante et al., 2018; Chen et al., 2019). High middle crustal temperatures develop during protracted growth of a thick orogenic belt and are revealed by mineral assemblages that indicate sustained temperatures >700°C (at mid-crustal depths of ~12–25 km). The structures associated with magmatic/migmatitic rocks at middle crustal levels in high-temperature–low-pressure (high-T–low-P) domains typically show evidence for large-scale ductile flow (e.g., Vanderhaeghe & Teyssier, 2001).

The strength profile of the crust in an orogenic pile and the lithospheric rheology are major controls on the intensity and style of collapse (e.g., Rey et al., 2009, 2010; Jamieson et al., 2011). Within an orogenic system, these parameters may change over time due to progressive heating of the thickened crust by decay of radioactive elements, shear heating, partial melting of the lower and middle crust and subsequent upward migration of these magmas (e.g., Laurent et al., 2017; Jamtveit et al., 2021). Intrusion of magma from the mantle, and lithospheric thinning due to convective removal or thermally activated and density driven drips may also contribute. These processes are certainly not mutually exclusive and typically occur concurrently or progressively within orogens (Chen et al., 2019). The rheology of the middle crust in high-T–low-P domains is dependent on the percentage of partial melt and if the melt is interconnected rather than simply accumulating in isolated pockets (Vanderhaeghe, 2009; Brown et al., 2011). Once there is an interconnected network of melt between crystals, rocks experience a significant drop in strength, which occurs with as little as about 7% melt (Rosenberg & Handy, 2005).

Ancient collisional orogens that expose middle and lower crustal rocks that record high geothermal gradients, contain extensive migmatites, and histories of protracted magma intrusion are relatively common in deeply eroded Precambrian and Paleozoic terranes (Clark et al., 2011; Brown et al., 2011; Rivers, 2012; Goscombe et al., 2017, 2022; Holder et al., 2019; Regis et al., 2021; Spencer et al., 2021). The pressure-temperature-time histories of these high-T–low-P metamorphic belts show that high geothermal gradients were maintained over tens of millions of years at midcrustal levels during which time the rocks were slowly decompressing (Clark et al., 2011). It is common for the midcrustal rocks to record a subsequent phase of rapid decompression to upper crustal depths (Clark et al., 2011). Large- and small-scale structures and mineral fabrics typically show evidence for extensive flow and thinning, including nappe structures and shear zones that strongly attenuate rock sections; these structures are often overprinted by discrete brittle-ductile detachment faults (Dewey, 1988; Vanderhaeghe & Teyssier, 2001; Yin, 2004; Foster et al., 2007; Toe et al., 2013; Whitney et al., 2013; Goscombe et al., 2017, 2018, 2022).

12.3. ALLEGHANIAN COLLAPSE OF THE SOUTHERN APPALACHIAN OROGEN

The Southern Appalachian Orogen in eastern North America (Fig. 12.5) exhibits classical features of a collapsed orogen including exhumation of high-T–low-P

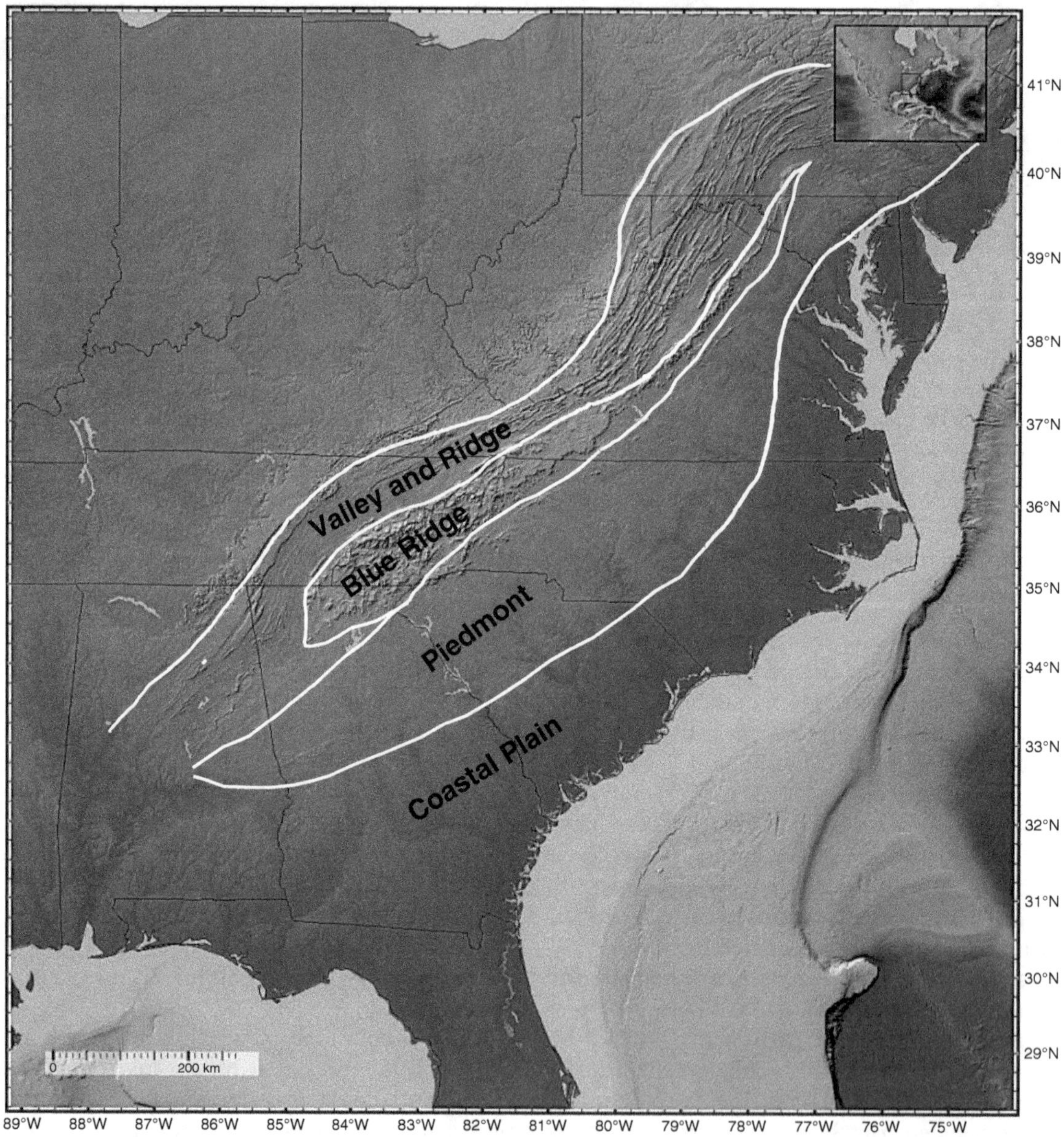

Figure 12.5 Physiographic provinces of the Southern Appalachian Orogen of eastern North America showing the current outcrop area of the orogenic belt. Extensive areas of the Southern Appalachian Orogen exist in the subsurface of the Coastal Plain south and east of the Piedmont Province. Global Multi Resolution Topography (GMRT) base map is from GeoMapApp (http://www.geomapapp.org).

middle crust that developed during continent-continent collisions leading to formation of the supercontinent Pangea. Extension of the collapsed Appalachian Orogen as a whole eventually led to rifting and seafloor spreading in the Atlantic Ocean basin (Dewey, 1988; Ma et al., 2023).

12.3.1. Appalachian Orogen

The Appalachian Orogen records multiple orogenic events that occurred during the Rodinia to Pangea supercontinent cycle, culminating with the late Mississippian–Permian Alleghanian orogeny (e.g., Hatcher, 2010). The Alleghanian orogeny was most intense in the southern Appalachians where it is characterized by an extensive foreland fold-thrust belt and composite, crystalline thrust sheets in the hinterland (Hatcher et al., 2007). The hinterland thrust sheets largely consist of polydeformed intermediate-to-high-grade metamorphic rocks that host late syn- to post-Alleghanian, epi- and mesozonal granitic plutons across the core of the orogen, including in the Blue Ridge, Inner Piedmont, and Carolina (super) terrane (Fig. 12.6; e.g., Speer et al., 1994; Mueller et al., 2014; Lin, 2015).

During the late Paleozoic Alleghanian orogeny, the southern part of the Appalachian Orogen developed at the nexus of the Laurentia-Gondwana collision that formed Pangea (e.g., Hatcher, 2010; Fig. 12.7). The southern Appalachian crust and lithosphere reached maximum thickness during the Alleghanian orogeny and then underwent extension that progressed from orogenic collapse in the core of the orogen, to continental rifting, and eventually to the opening of the Atlantic Ocean and the Gulf of Mexico (Dewey, 1988). Orogenic collapse occurred via ductile flow and thinning of the magmatically infused middle crust linked to regional scale, core complex-style extensional detachment faults, and oblique transtension on regional-scale strike-slip faults (Ma et al., 2019).

The crust in the southern Appalachians was thickened and then presumably thinned during the Taconian, Acadian, and Neoacadian orogenic events associated with collisions of Gondwana-derived terranes with the Laurentian margin (Hatcher, 2010). In the orogen core, thickening associated with the Neoacadian in the early Carboniferous transitioned into the beginning of the Alleghanian orogeny (late Carboniferous) such that the southern Appalachian orogenic plateau was progressively constructed and metamorphosed over a long interval from Mississippian to early Permian times. The trace element composition of late Paleozoic granitic plutons in

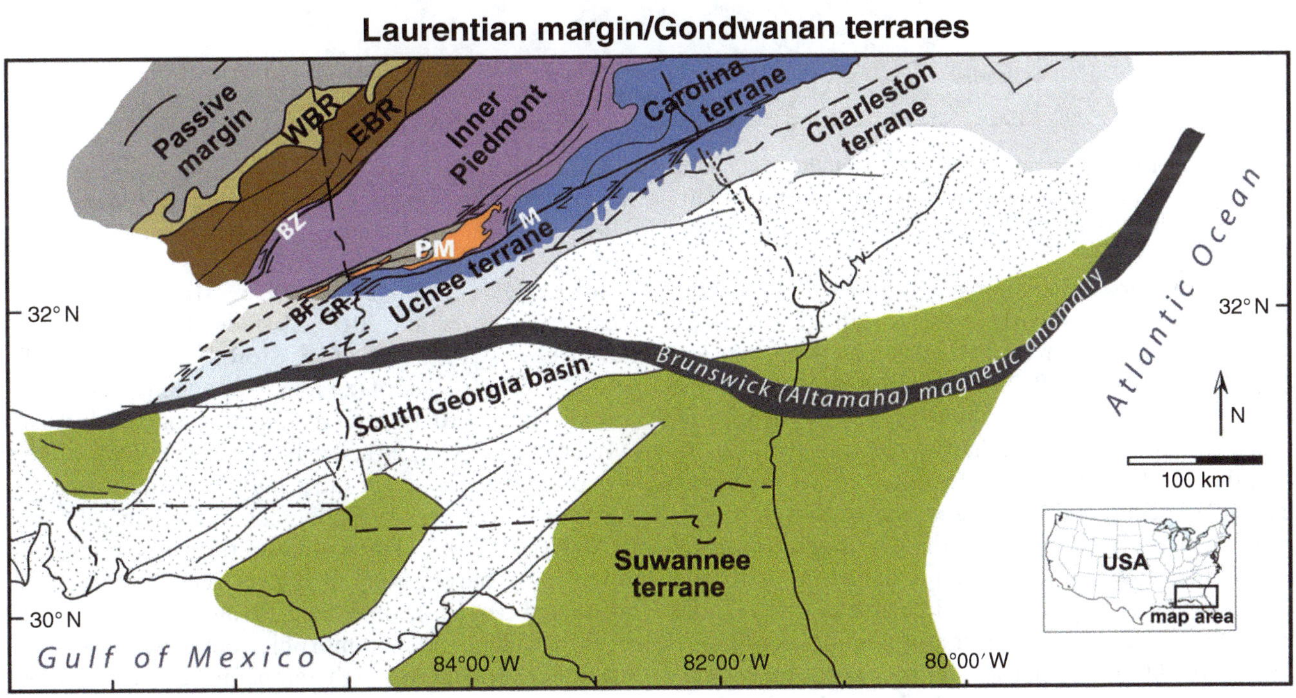

Figure 12.6 Schematic depiction of the major terranes and structures within and adjacent to the Suwannee suture zone (Laurentia-Gondwana suture) in the southern Appalachians (modified from Mueller et al., 2014). Also shown are the Brunswick (Altamaha) magnetic anomaly (a Proterozoic or Mesozoic feature that does not mark the suture (Duff & Kellogg, 2019)) and the Mesozoic South Georgia basin, which is composed of several subbasins. Note: PM = Pine Mountain Inlier; EBR = Eastern Blue Ridge; WBR = Western Blue Ridge; BZ = Brevard zone; BF = Bartlett's Ferry shear zone; GR = Goat Rock shear zone; M = Modoc shear zone.

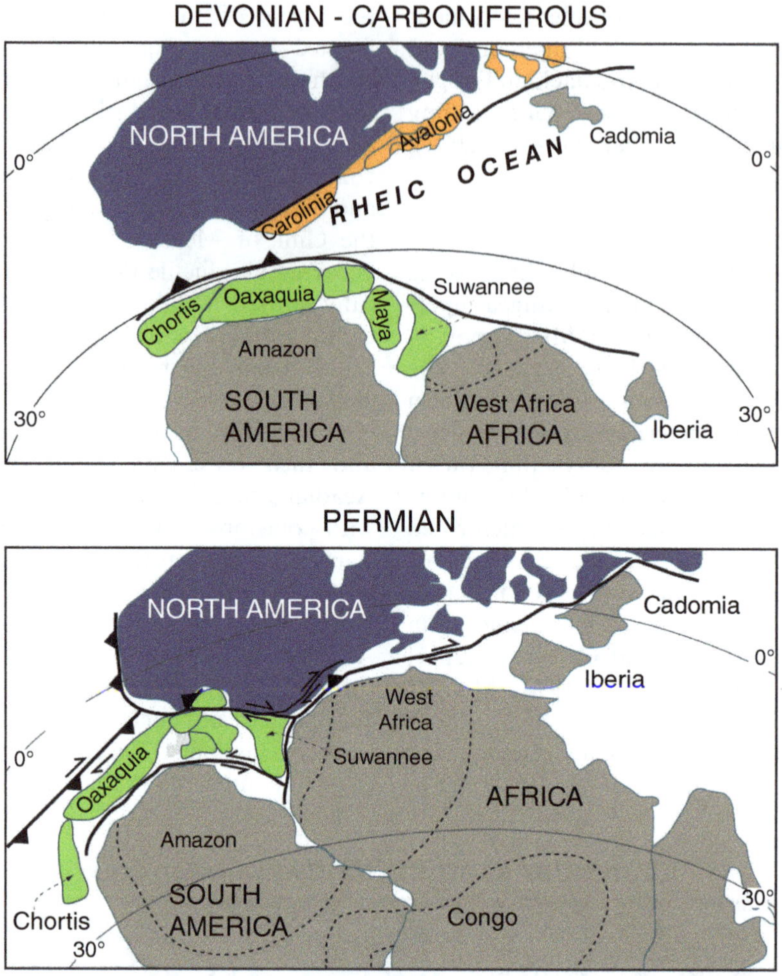

Figure 12.7 Schematic depiction of relative plate motions associated with the sequential arrival of the Suwannee and other peri-Gondwanan terranes to Laurentia from Devonian to Permian times leading to formation of Pangea (modified from Mueller et al., 2014, and Keppie et al., 2008).

the southern Appalachian Orogen indicates that the crust dramatically thickened early in the Alleghanian orogeny (Stowell et al., 2019). Areas where the crust is still >30 km thick (Wagner et al., 2012) expose rocks at the surface that were intruded and metamorphosed at depths of about 20 km during the Alleghanian (Stowell et al., 2019). These data and scale of the continent-continent collision between Laurentia and Gondwana suggest that the thickness of the crust at the height of the Alleghanian orogeny was on the order of 55–70 km, analogous with the Himalaya-Tibet Orogen. The middle crustal rocks of the core of the orogen were at upper amphibolite to granulite facies with widespread partial melting. These midcrustal rocks are now exposed in the Piedmont Province and sampled via drilling through the Coastal Plain sediments. Widespread syn- to postorogenic Alleghanian plutons (named plutons shown on Fig. 12.8) give U-Pb zircon ages that span a range from about 340 to 290 Ma for magmas intruded at similar midcrustal levels (Speer et al., 1994; Samson et al., 1995; Miller et al., 2006; Heatherington et al., 2010; Mueller et al., 2011, 2014; Foster et al., 2012; Lin, 2015; Stowell et al., 2019). The youngest U-Pb ages of zircons from each pluton most commonly record final crystallization of the intrusion (Miller et al., 2006; Mueller et al., 2011). The range in ages of the antecrysts (or older magmatic zircons) in each intrusion, however, typically cover a span of about 10 Myr (Lin, 2015). These indicate the (near) continuous presence of partial melt in the orogenic system over a protracted period. Monazite U-Pb ages and Sm-Nd ages of garnet from the metamorphic rocks record a similar history of protracted high temperatures in the middle crust during the Alleghanian (Mueller et al., 2011; Stowell et al., 2019). These data show that Alleghanian metamorphism in the Southern Appalachian Orogen took place progressively and continuously for some 30–50 Myr as thermal, barometric, and fluid conditions evolved within the orogenic pile from about 345 to 295 Ma.

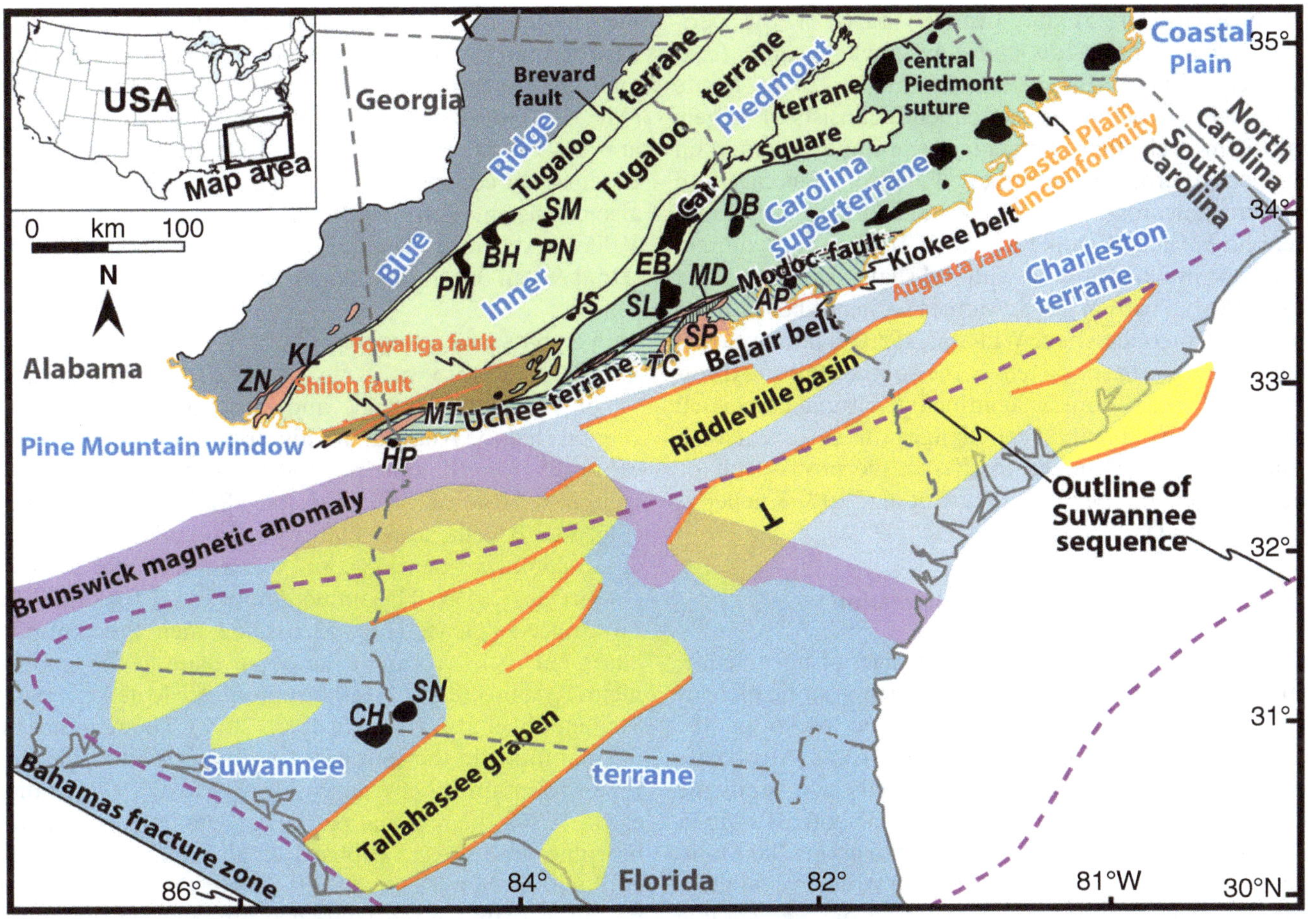

Figure 12.8 Geological map of the Southern Appalachian Orogen from Ma et al. (2019) showing the Laurentian and Gondwanan tectonic terranes, location of the South Georgia basin, and the Mississippian-Permian Alleghanian plutons (also shown are the Ordovician Zana (ZN) and Kowaliga (KL) plutons; Barineau et al., 2022). Pluton names: AP = Appling, BH = Ben Hill, CH = Calhoun, DB = Danburg, EB = Elberton, HP = Hospilika, IS = Indian Springs, MD = Modoc, MT = Motts, PM = Palmetto-Tyrone, PN = Panola, SL = Siloam, SM = Stone Mountain, SN = Seminole, SP = Sparta, TC = Town Creek.

12.3.2. Synorogenic to Postorogenic Collapse

High-grade (high-T–low-P) metamorphism in the hinterland was synchronous with thrusting in the Appalachian foreland along Alleghanian-age thrusts, such that the orogen was spreading or growing laterally during collision. The initial stages of collapse were, therefore, likely by ductile thinning of the partially melted middle and lower crust, perhaps linked to detachments that extended out into the foreland as thrusts. In this sense, collapse occurred without a fixed boundary like the Himalayan example (Dewey, 1988; Rey et al., 2001). The record of any hypothetical early detachments is not preserved at the present level of exposure in the Piedmont or orogenic core. After the final welding of Pangea, the supercontinent began to break up, presumably because of the thermal blanketing of the mantle and the evolution of new subduction zones at the margins of the supercontinent. Early break up of Pangea in the southern Appalachians was concentrated within the weak orogenic pile, which had partial melt within the middle crust until at least 290 Ma (Heatherington et al., 2010).

The dominant contractional structures formed during the Alleghanian collision in the Southern Appalachian Orogen were locally extended and reactivated by northeast-striking normal faults and oblique transtensional motion on regional-scale strike-slip faults. These faults include the down-to-the-southeast Augusta fault between the high-grade Kiokee belt and the low-grade Belair belt (Fig. 12.8; Maher et al., 1994), and top-down-to-the-southeast Shiloh fault in the Pine Mountain window (Fig, 12.8; Steltenpohl et al., 2010). To the southeast of these normal faults, a thick (4–6 km) Upper Cambrian to Middle Devonian siliciclastic succession, known as the Suwannee sequence (Fig. 12.8), is present beneath the Coastal Plain south and east of the Piedmont

Province (Fig. 12.5) and overlaps parts of the peri-Gondwanan Charleston and Suwannee terranes (e.g., Mueller et al. 2014; Boote & Knapp, 2016; Boote et al., 2018). The Suwannee sequence consists of largely undeformed clastic sedimentary strata and is underlain by Neoproterozoic to earliest Paleozoic igneous and metamorphic rocks (Mueller et al., 2014). The South Georgia basin, which is filled with Triassic and younger arkosic strata interlayered with red shales, siltstones, and basalt flows, is above the Suwannee sequence beneath the Coastal Plain (Chowns & Williams, 1983). The composite South Georgia basin system is the southernmost expression of Triassic-Jurassic continental rifting associated with the early breakup of Pangea in eastern North America, and was linked to rifts in the proto–Gulf of Mexico (Fig. 12.8; e.g., Withjack et al., 2013; Kneller & Johnson, 2011).

12.3.3. Evidence for Tectonic Exhumation

Ma et al. (2019) reviewed $^{40}Ar/^{39}Ar$ ages of hornblende, muscovite, and biotite from Alleghanian granitic plutons and other plutonic and metamorphic rocks in the Southern Appalachian Orogen (Fig. 12.9). Muscovite cooling ages show consistent regional-scale gradients including a decrease of cooling age from about 310 Ma in the northwest along the Brevard fault to about 280 Ma in the southeast near the Pine Mountain structural window (Inlier) that exposes Mesoproterozoic crystalline basement. East of the Pine Mountain Inlier, a locally inverted gradient that decreases from about 280 Ma near the window to 250 Ma to the northwestern limit of crystalline basement (Fig. 12.9). These data, along with the U-Pb crystallization ages, indicate protracted high temperatures in the middle crust and provide important evidence for the extensional collapse of the Alleghanian orogenic plateau.

The $^{40}Ar/^{39}Ar$ ages of muscovite from granitic plutons reflect regional cooling of the middle crust, because (1) the muscovite mineral ages are 17–150 Myr younger than the zircon U-Pb ages of the same plutons, and (2) there are no abrupt changes of cooling ages between the plutons and proximal country rocks. The interpreted smooth, systematic trends in cooling ages (Fig. 12.9) could reflect diachronous cooling of the Blue Ridge–Piedmont megathrust sheet during its northwestward displacement over a southeast-dipping basement ramp and subsequent erosion (e.g., Dallmeyer, 1989; Hatcher et al., 1989), or indicate progressive extension along a regional normal fault system that produced a systematic decrease in cooling ages in the slip direction in the footwall rocks (e.g., Foster & John, 1999; Stockli, 2005; Foster et al., 2010). If the thrusting hypothesis outlined above is accurate, the Alleghanian hinterland would still have been under contraction at ~260–240 Ma in the area to the east of the Pine Mountain window. However, structural and geochronologic data along the Modoc fault (Fig. 12.8) show that extension in the orogen core started at ~300–280 Ma (Snoke & Frost, 1990; Sacks et al., 1993). Extensional collapse of the orogen core after 280 Ma is also consistent with isotopic evidence for thrusting in the Valley and Ridge foreland as young as about 270 Ma (Hnat & van der Pluijm, 2014).

12.3.4. Postorogenic Extension

The thermochronologic and structural data strongly suggest that the Alleghanian hinterland was progressively exhumed from the northwest at the Brevard fault to the southeast toward the Coastal Plain unconformity (Fig. 12.10) as the lower plate of a regional metamorphic core complex beneath an extensional detachment (Ma et al., 2019). The unmetamorphosed rocks of the Suwannee sequence (Fig. 12.10) and their Precambrian basement rocks represent the upper plate to the detachment system (e.g., Heatherington & Mueller, 1997; Steltenpohl et al., 2008; Ma et al., 2019). The presence of normal faults in the southern Carolina superterrane and Pine Mountain Inlier (e.g., Snoke & Frost, 1990; Maher et al., 1994), the regional pattern of late Paleozoic metamorphic grade (e.g., Hibbard et al., 2002; Steltenpohl et al., 2008; Mueller et al., 2014), and the structure and timing of rifting in the Triassic South Georgia rift basins (e.g., Clendenin, 2013; Heffner, 2013) all indicate that extensional collapse of the orogen started in the Permian and continued through the Triassic along a major southeast dipping low-angle detachment system that cut through the upper crust and penetrated the lower crust.

Other fault systems in the southern Appalachians are consistent with crustal extension and collapse of the orogen. The Augusta fault (Fig. 12.8) records significant crustal extension following Alleghanian shortening (Maher et al., 1994). Seismic reflection profiles show that the Augusta fault projects southeastward at a low angle >80 km from its surface expression (Cook & Vasudevan, 2006) and that the high-angle border fault of the Riddleville subbasin of the South Georgia basin soles into the Augusta fault (Petersen et al., 1984). This suggests that the Augusta normal fault is listric and at least partially accommodated the Triassic rifting of the South Georgia basin. In the Uchee terrane (Fig. 12.8), a southeast-dipping low-angle (~25°) normal shear zone and synthetic high-angle (~50–65°) normal faults in the hanging wall, crosscut the Alleghanian gneissic fabrics (Ma et al., 2019) and were active at about 280 Ma (Ma et al., 2023). This ductile-brittle structure along with the kinematically compatible Augusta and Shiloh faults, likely represent normal faulting in the uppermost footwall

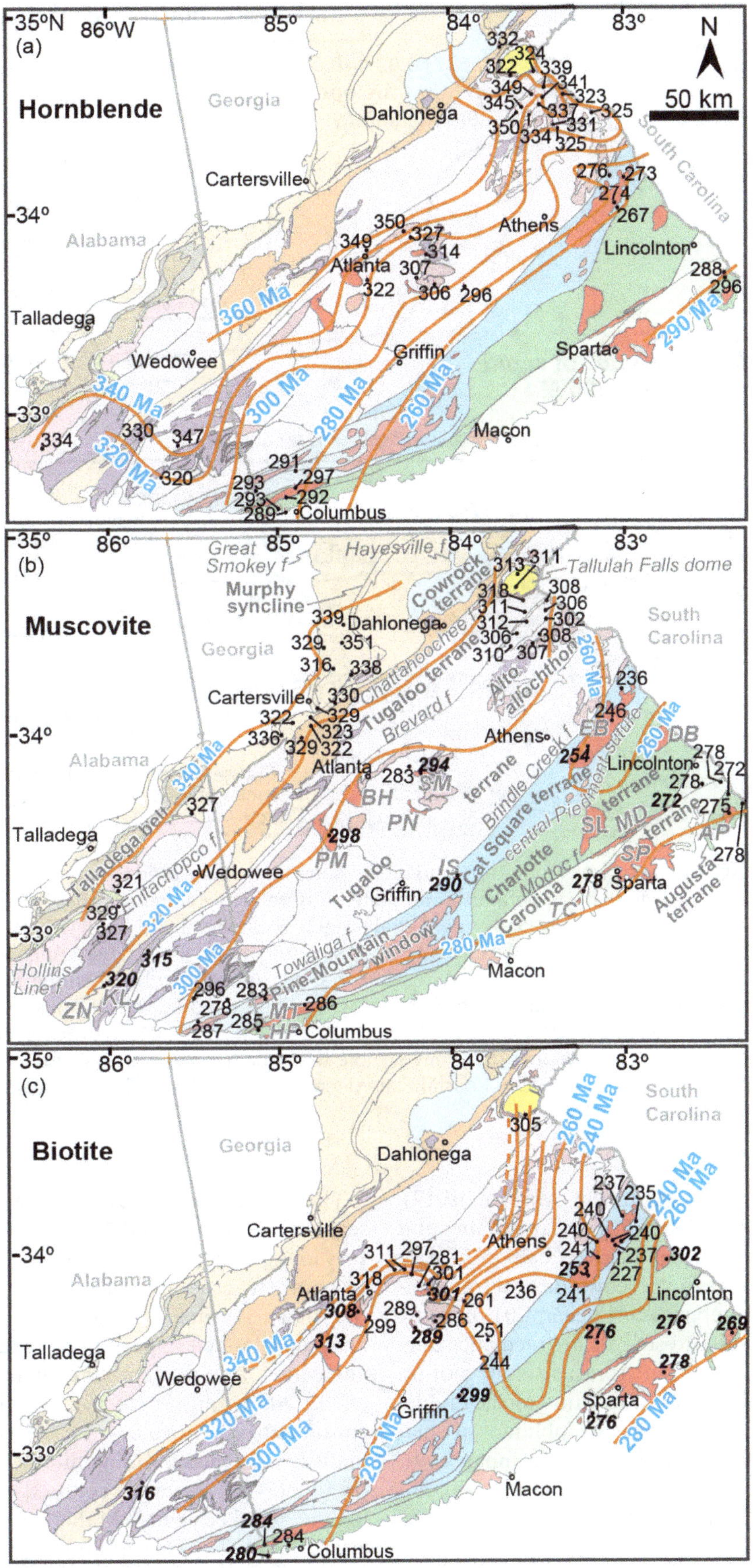

Figure 12.9 Maps of (a) hornblende, (b) muscovite, and (c) biotite $^{40}Ar/^{39}Ar$ and K-Ar data from the Southern Appalachian Orogen (from Ma et al., 2019).

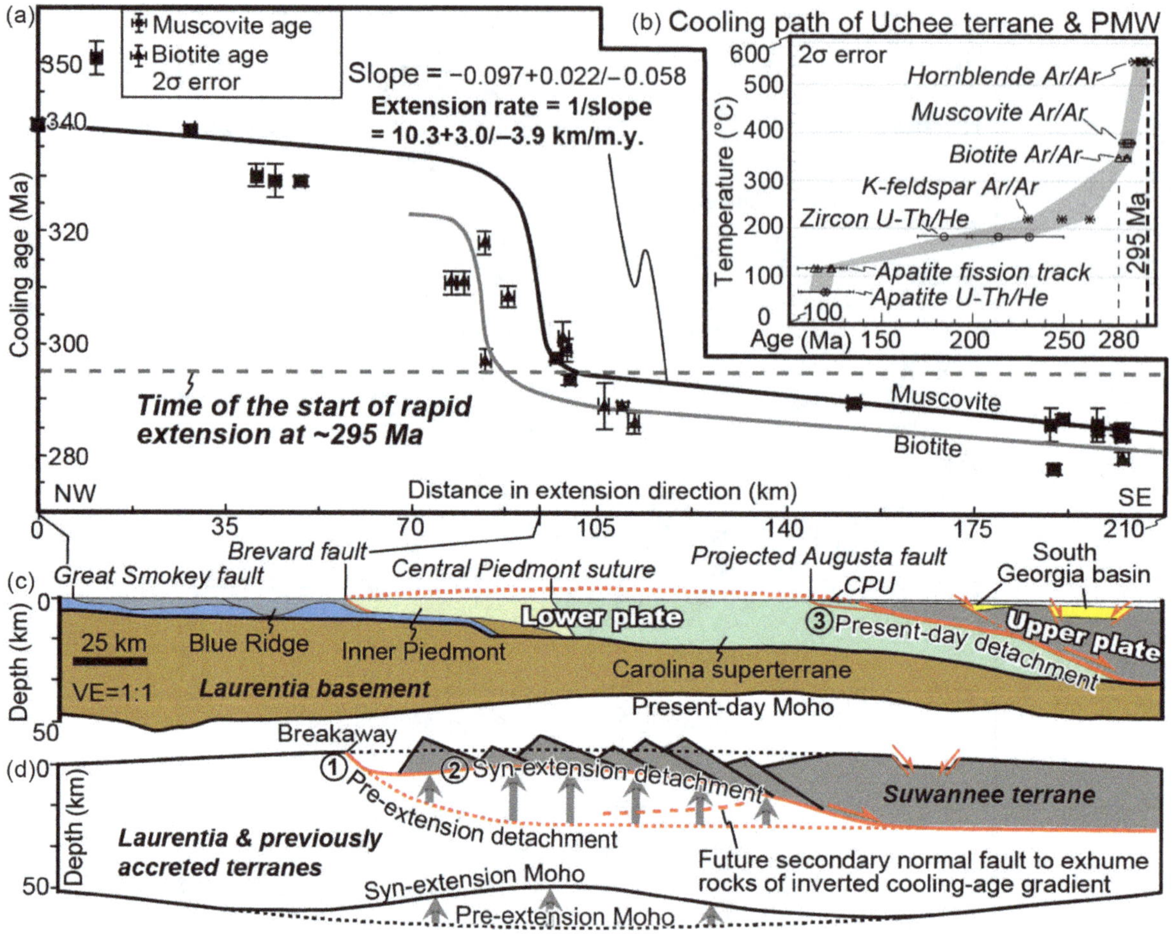

Figure 12.10 (a) Age-distance plot of muscovite and biotite dates that can be used to constrain the time and rate of extension. (b) Cooling path of the uppermost lower plate. Closure temperatures are after Reiners and Brandon (2006). Data sources: Anderson (1981), Steltenpohl et al. (1992), Steltenpohl and Kunk (1993), Layfield (2009), and Ma et al. (2019). (c) Cross section of the current crustal structure across the southern Appalachians showing the inferred extensional detachment fault system developed at the suture zone between the Suwannee terrane and the previously accreted Carolina superterrane (locally Uchee where exposed) (after Cook & Vasudevan, 2006; Hatcher et al., 2007; Hopper et al., 2017; Duff & Kellogg, 2019; Marzen et al., 2019). (d) Schematic cross section showing the evolution of the extensional detachment; note the dynamic geometries of the detachment (from listric to antilistric) and Moho. Note: CPU = Coastal Plain unconformity; PMW = Pine Mountain window (figure modified from Ma et al., 2019).

of the presumed extensional detachment (Ma et al., 2019), and are consistent with the late Alleghanian arching and extensional exhumation of the Kiokee belt (Snoke & Frost, 1990). The Towaliga fault records dextral transtension in the Triassic (Huebner & Hatcher, 2013); many other major faults also record strike-slip kinematics during the Alleghanian. Concurrent transtensional motion on faults like the Towaliga and exhumation of the middle crust suggests linkages between an oblique normal detachment system and strike-slip faults (compare with McQuarrie & Wernicke, 2005, for opening the Basin and Range Province; Foster et al., 2009, for exhumation in the Cambrian Damara Orogen).

The sedimentary rocks of the Paleozoic Suwannee sequence are nearly undeformed; they underwent at most very low-grade metamorphism (e.g., Mueller et al., 2014). In contrast, crust exposed directly to the northwest in the Uchee terrane is strongly deformed at uppermost amphibolite-facies, partially melted, and intruded by anatectic Alleghanian magmas (e.g., Steltenpohl et al., 2008; Lin, 2015; Fig. 12.11). High-grade rocks continue to the northeast of the Uchee terrane as migmatitic paragneiss, sillimanite-bearing schist, granitic orthogneiss, and metamorphosed mafic/ultramafic rocks in the Kiokee belt (Maher et al., 1994; Hibbard et al., 2002). The clear difference in metamorphic grade

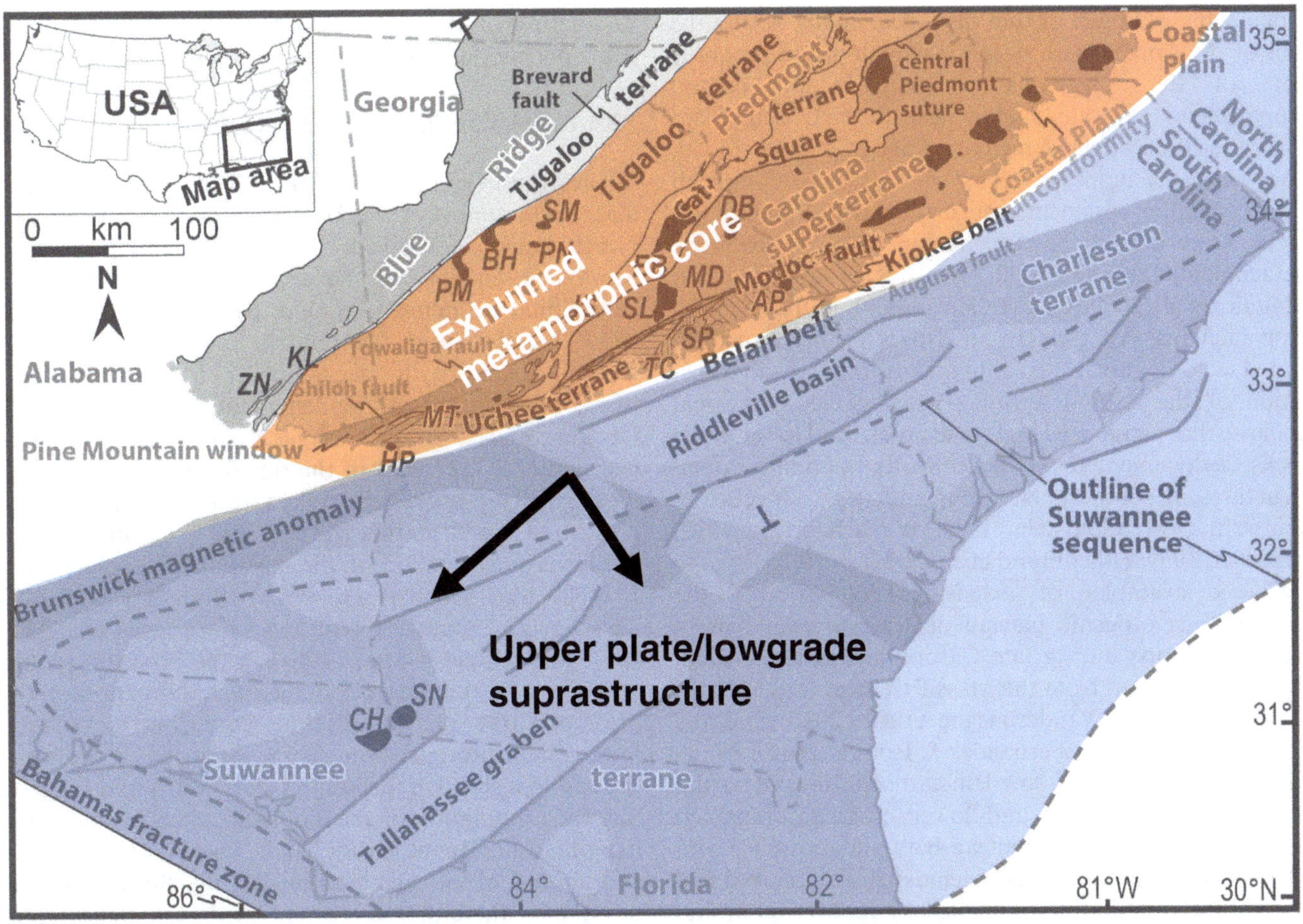

Figure 12.11 Sketch map depicting the relationship between exhumed middle crustal rocks of the Inner Piedmont and Uchee terranes and rocks of the Suwannee terrane that remained in the upper crust during the Alleghanian orogeny. The arrows show the range of possible relative displacements of the upper plate based on southeast dipping low-angle faults and southwest oriented strike-slip faults along the boundary. The lack of outcrop along the boundary between the Suwannee and Uchee terranes makes direct observations of the kinematics difficult.

between the Suwannee sequence and its upper plate basement compared with lower plate Appalachian crystalline rocks between the Brevard fault and the Coastal Plain unconformity is consistent with the low-grade overlying high-grade relationship typical of extensional detachment faults (e.g., Whitney et al., 2013). In Permian time, rocks now exposed in the southernmost Appalachians were at >10–15 km depth and the rocks within the Suwannee terrane were at <5 km depth. These middle crustal rocks (Uchee) and upper crustal rocks (Suwannee) are now separated by only a few kilometers across an enigmatic, partially buried normal fault (Goat Rock-Bartlett's Ferry; Mueller et al., 2014) or regional scale detachment system (or perhaps megadetachment, e.g., Wernicke et al., 2008).

12.3.5. Other Parts of the Alleghanian-Variscan Orogen

The Southern Appalachian Orogen preserves more clear evidence for large-scale Alleghanian extensional collapse than other parts of the Appalachian system in North America. The southern Appalachians contain the largest area of exposed crust that underwent significant thickening during the Alleghanian orogeny. On the basis of plate reconstructions and amount of overthrusting, this area was located at the nexus of the Permian continental collisions involving Gondwanan (Africa and Amazonia) and peri-Gondwanan terranes with Laurentia. For these reasons, it is likely that the crust of the orogenic plateau was as thick or thicker than other areas of the exposed

Appalachian Orogen during the Alleghanian. Along the axis of the Appalachian system, the high-grade, prograde core of the Alleghanian orogeny in the central Appalachians is located mainly beneath the Coastal Plain or in the rifted offshore region, or underlie large, folded, Taconic thrust nappes in Pennsylvania, which, along with their crystalline basement, were displaced at the multikilometer scale during the Alleghanian, prior to extensive detachment faulting (Faill, 1998; Krol et al., 1999; Maguire et al., 2009; Wise & Ganis, 2009). In the northern Appalachians, Alleghanian structures record transpressional tectonics associated with underthrusting and translation of the Avalon terrain. Postorogenic extensional collapse that juxtaposed high-grade rocks and lower grade rocks is documented for detachments bounding domal structures in southern New England (e.g., Moecher & Wintsch, 1994; Goldstein, 1989; Wintsch et al., 1992; Moecher, 1999; Hillenbrand et al., 2021).

Classic examples of extensional collapse of the Carbonifous orogenic plateau of the Variscan belt in western Europe during late Carboniforous to Permian time are described from the Massif Central (e.g. Ménard & Molnar, 1988; Vanderhaeghe et al., 2020) and the Iberian Massif (Diez Fernandez & Pereira, 2016). In the Massif Central, high-T–low-P metamorphism and partial melting of the middle and lower crust accompanied building of an orogenic plateau between about 345 and 310 Ma. Formation of the orogenic plateau occurred in a transpressive setting accompanied by motion on strike-slip shear zones and perhaps the transition to orogen-parallel extension (Vanderhaeghe et al., 2020), like the tectonics of the southern Appalachian Orogen. The growth of the Variscan orogenic plateau was coeval with the foreland propagation of the deformation front, again like the Alleghanian orogenic plateau. Collapse and regional-scale extension of the Variscan belt, and the Massif Central in particular, was marked by the activation of low-angle detachments, which led to the exhumation of migmatite cored domes between about 305 and 295 Ma. In the Iberian Massif, detachment faulting and exhumation, and intrusion of granite into the high-grade orogen core, accompanied by transpressional overprints on strike-slip shear zones, occurred between about 321 and 300 Ma (Gutierrez-Alanso et al., 2011; Rubio Pascual et al., 2013; Diez Fernandez & Pereira, 2016). U-Pb data from the exhumed middle and lower crustal rocks of the Massif Central indicate that partial melt was present from about 340 to 295 Ma, which is strikingly similar in duration to the data from the Alleghanian in the southern Appalachians. The development of detachments and progressive exhumation of the middle crustal core of the Massif Central part of the Variscan orogen is reflected in cooling-age gradients ($^{40}Ar/^{39}Ar$, K-Ar, etc.). These data have been attributed to gradual delamination and foundering of the lithosphere beneath the mature orogenic plateau (e.g., Vanderhaeghe et al., 2020).

12.3.6. Relationship to Triassic Rifts

The South Georgia basins (Figs. 12.6, 12.8), the oldest and southernmost of the Triassic-Jurassic rifts in eastern North America (Withjack et al., 2013), developed on the Suwannee sequence of the upper plate of the regional extensional detachment system. Seismic reflection data show that the South Georgia rift basin is a horst-and-graben system with steep border faults that strike northeast (Fig. 12.8; Clendenin, 2013; Heffner, 2013). Opening of the South Georgia basin is constrained as Carnian (237–227 Ma) in the timescale of Cohen et al. (2013) by fossil spores and pollen from shale in the upper portion of the basin fill (Traverse, 1987). Clastic rocks in the lowermost portion of the basin are thus older than 237–227 Ma, which places initial rifting prior to 237–227 Ma. The youngest biotite $^{40}Ar/^{39}Ar$ ages (ca. 240–235 Ma; Dallmeyer et al., 1981) in the Elberton pluton area (Fig. 12.9) indicate cooling and exhumation of the presumed lower plate at this time. Cooling of rocks in the lower plate through about 300°C occurred at the same time as deposition in the South Georgia rifts and suggests that the exhumation of the footwall and extension of the hanging wall were at least partially coeval. The structure and timing of normal faulting of the South Georgia basin are, therefore, consistent with the detachment hypothesis.

Trend lines defined by muscovite and biotite $^{40}Ar/^{39}Ar$ cooling ages from footwall rocks below the proposed detachment plotted against distance along the extension direction show breaks in slope followed by linear segments (Fig. 12.10), which is typical for extensional detachments (Foster & John, 1999). The ~295 Ma break in slope is interpreted to indicate the start of extension, the rate of which is estimated by the inverse of the slope of the muscovite-age gradient, yielding a rate of ~10 km/Myr. This relatively rapid extension rate likely declined after ~280 Ma as the lower plate underwent slower cooling. Syn-extensional thrusting apparently extended until ~280–276 Ma in the foreland (Hnat & van der Pluijm, 2014), such that extensional detachments may have bound the upper part of a ductile midcrustal channel with the thrusts bounding the lower level. The inverted cooling-age gradient in the area east of the Pine Mountain window (Fig. 12.9) suggests local exhumation of the footwall rocks (likely in the form of arching; Snoke & Frost, 1990) from 280 Ma to 240 Ma beneath a secondary northwest-dipping normal fault splay from the main detachment (Fig. 12.10d).

Identifying remnants of the Permian-Triassic extensional detachment fault system in the southern

Appalachians remains a challenge because of the extensive Coastal Plain sediment cover. The master detachment fault is likely buried beneath the corridor between the Coastal Plain unconformity (Fig. 12.8) and the northwestern margins of the Suwannee sequence and South Georgia basin. Hopper et al. (2017) documented a southeast-dipping, low-angle crustal discontinuity under this corridor based on common conversion point stacking of S-to-P waves of EarthScope seismic data and interpreted it as the Alleghanian suture. Marzen et al. (2019) also identified a major transition in modeled crustal and upper mantle seismic velocity in the Suwannee suture and Georgia rift basin refraction seismic experiment data at the Suwannee suture and interpreted these as a shallow dipping detachment. The crustal seismic velocity discontinuity identified by both studies is structurally compatible with an extensional detachment fault system formed by reactivation of the Alleghanian suture.

Extensional detachments facilitated the breakup of Pangea and diachronous opening of nascent ocean basins from Triassic to Cretaceous time along the Atlantic margins (e.g., Lister et al., 1991). Detachments associated with the Devonian collapse of the Caledonian orogen are shown to have been reactivated by brittle normal faults bounding Permian-Triassic rifts of the North Atlantic–North Sea margin (e.g., Fossen, 2010; Wiest et al., 2020; Osmundsen et al., 2020). Similarly, it is possible that the detachment system that we infer to have developed between the Alleghanian metamorphic core and the Suwannee terrane played a role in the opening of the South Georgia basins and in the early stages of continental-scale rifting of the Central Atlantic margin. The Suwannee detachment system did not lead to full continental breakup and seafloor spreading in the South Georgia basin, which would have linked the Atlantic to incipient spreading in the Gulf of Mexico (e.g., Kneller & Johnson, 2011). The South Georgia basin developed into a failed arm of a triple junction when transform faults formed along the eastern side of the Blake plateau and magmatic crust of the Bahamas, and central Atlantic rifting jumped to the east in a more favorable orientation for the separation of Africa from North America (Kneller & Johnson, 2011; Kneller et al., 2012). Had the Suwannee detachment remained active to the stage of breakup, presumably the exhumed Alleghanian orogen core and the South Georgia basin would have developed into a lower plate style margin while the Suwannee terrane and either Africa or South America would have developed into an upper plate margin.

12.4. CONCLUSIONS

The extensional collapse of collisional orogenic belts is a result of processes within the orogenic pile that increase heat and decrease the strength of the middle or lower crust, and/or the result of extra-orogenic processes including changes in the dynamics of plate motion and plate boundary interactions. The Alleghanian orogeny was caused by the final collision of North America with Gondwana in late Paleozoic time. The Alleghanian orogenic plateau collapsed after the crust was thickened and experienced high-grade metamorphism and partial melting at midcrustal depths. The relationship of the exhumed orogenic core to the upper crustal rocks of the Suwannee terrane along with other data including thermochronology suggest that middle crustal rocks of the Inner Piedmont and Uchee terranes were once under the Suwannee terrane before development of a regional-scale southeast or south dipping detachment system. These relationships allow us to infer that the progressive collapse of the Southern Appalachian Orogen occurred in a similar way to areas of the Alleghanian-Variscan system and other collapsed orogenic belts worldwide like the Damara Orogen and Grenville Orogen that have better outcrop exposures of detachment fault systems.

ACKNOWLEDGMENTS

We thank volume editors Elizabeth Catlos and İbrahim Çemen for the invitation to contribute this chapter. The manuscript was improved by review comments of Olivier Vanderhaeghe, Alex Zagorevski, and anonymous individuals.

REFERENCES

Anderson, T. M. (1981). *Application of fission track studies to the dynamo-thermal history of a metamorphic terrane, South Georgia Piedmont*. M.S. thesis. Florida State University, Tallahassee.

Barineau, C., Sagul, A., & Mueller, P.(2022). Ordovician back-arc silicic magmatism in the southern Appalachians. *Geological Society of America Bulletin, 134* (9–10), 2321–2334.

Beaumont, C., Jamieson, R. A., Nguyen, M. H., & Lee, B. (2001). Himalayan tectonics explained by extrusion of a low-velocity crustal channel coupled to focused surface denudation. *Nature, 414*, 738–742.

Bird, P. (1991). Lateral extrusion of lower crust from under high topography in the isostatic limit. *Journal of Geophysical Research, 96*, 10, 275–10, 286.

Block, L., & Royden, L. H. (1990). Core complex geometries and regional scale flow in the lower crust. *Tectonics, 7*, 557–567.

Boote, S. K., & Knapp, J. H. (2016). Offshore extent of Gondwanan Paleozoic strata in the southeastern United States: The Suwannee suture zone revisited. *Gondwana Research, 40*, 199–210.

Boote, S. K., Knapp, J. H., & Mueller, P. A. (2018). Preserved Neoproterozoic continental collision in southeastern North

America: The Brunswick suture zone and Osceola continental margin arc. *Tectonics*, *37*, 305–321.

Brown, M., Korhonen, F. J., & Siddoway, C. S. (2011). Organizing melt flow through the crust. *Elements*, *7*, 261–266.

Cavalcante, C., Hollanda, M. H., Vauchez, A., & Kawata, M. (2018). How long can the middle crust remain partially molten during orogeny? *Geology*, *46*(10), 839–842.

Chen, L., Song, X., Gerya, T. V., Xu, T., & Chen, Y. (2019). Crustal melting beneath orogenic plateaus: Insights from 3-D termomechanical modeling. *Tectonophysics*, *761*, 1–5.

Chowns, T., & Williams, C. (1983). *Pre-Cretaceous rocks beneath the Georgia coastal plain*. U.S. Geological Survey Professional Paper 1313-L, 1–42.

Clark, C., Fitzsimons, E. C. W., Healy, D., & Harley, S. L. (2011). How does the continental crust get really hot? *Elements*, *7*, 235–240.

Clendenin, C. (2013). Insights into the mode of the South Georgia rift extension in eastern Georgia, USA. *Tectonophysics*, *608*, 613–621.

Cohen, K. M., Finney, S. C., Gibbard, P. L., & Fan, J. X. (2013). The ICS International Chronostratigraphic Chart. *Episodes*, *36*, 199–204.

Cook, F. A., & Vasudevan, K. (2006). Reprocessing and enhanced interpretation of the initial COCORP Southern Appalachians traverse. *Tectonophysics*, *420*, 161–174.

Dallmeyer, R., Hess, J., & Whitney, J. (1981). Post-magmatic cooling of the Elberton Granite: Bearing on the late Paleozoic tectonothermal history of the Georgia Inner Piedmont. *The Journal of Geology*, *89*, 585–600.

Dallmeyer, R. D. (1989). Late Paleozoic thermal evolution of crystalline terranes within portions of the U.S. Appalachian orogen. In R. D. Hatcher et al. (Eds.), The Appalachian-Ouachita Orogen in the United States: Boulder, Colorado (pp. 417–444). Geological Society of America, Geology of North America, F-2.

Dewey, J. F. (1988). Extensional collapse of orogens. *Tectonics*, *7*, 1123–1139.

Diez Fernandez, R., & Pereira, M. F. (2016). Extensional orogenic collapse captured by strike-slip tectonics: constraints from structural geology and U-Pb geochronology of the Pinhel shear zone (Variscan orogen, Iberian Massif). *Tectonophysics*, *691*, 290–310.

Ding, H., Kohn, M. J., & Zhang, Z. (2021). Long-lived (ca. 22–24 Myr) partial melts in the eastern Himalaya: Petrochronologic constraints and tectonic implications. *Earth and Planetary Science Letters*, *558*, 116764.

Duff, P. D., & Kellogg, J. N. (2019). The Brunswick magnetic anomaly: Geophysical signature and geologic source. *Geology*, *47*(4), 355–358.

England, P. C., & Houseman, G. A. (1988). The mechanics of the Tibetan Plateau. *Philosophical Transactions of the Royal Society of London, Series A, Mathematical and Physical Sciences*, *326*(1589), 301–320.

England, P. C., & Thompson, A. B. (1984). Pressure-temperature-time paths of regional metamorphism I. Heat transfer during the evolution of regions of thickened continental crust. *Journal of Petrology*, *25*, 894–928.

Faill, R. T. (1998). A geologic history of the north-central Appalachians; Part 3, The Alleghany Orogeny. *American Journal of Science*, *298*(2), 131–179.

Fossen, H. (2010). Extensional tectonics in the North Atlantic Caledonides: A regional view. *Geological Society London, Special Publications*, *335*, 767–793.

Foster, D. A., & John, B. E. (1999). Quantifying tectonic exhumation in an extensional orogen with thermochronology: examples from the southern Basin and Range Province. *Geological Society, London, Special Publications*, *154*(1), 343–364.

Foster, D. A., Doughty, P. T., Kalakay, T. J., Fanning, C. M., Coyner, S., Grice, W. C., et al. (2007). Kinematics and timing of exhumation of metamorphic core complexes along the Lewis and Clark fault zone, northern Rocky Mountains, USA. In A. B. Till et al. (Eds.), *Exhumation associated with continental strike-slip fault systems* (pp. 207–232). Geological Society of America Special Paper 434.

Foster, D. A., Goscombe, B. D. & Gray, D. R, (2009). Rapid exhumation of deep crust in an obliquely convergent orogen: The Kaoko Belt of the Damara Orogen. *Tectonics*, *28*, TC4002. https://doi.org/10.1029/2008TC002317

Foster, D. A., Grice, W. C., & Kalakay, T. J. (2010). Extension of the Anaconda metamorphic core complex: $^{40}Ar/^{39}Ar$ thermochronology and implications for Eocene tectonics of the northern Rocky Mountains and the Boulder batholith. *Lithosphere*, *2*, 232–246.

Foster, D. A., Mueller, P. A., Heatherington, A., Wooden, J. L., Daitch, P., & Chong, M. (2012). Alleghanian magmatism in the southern Appalachians: Implications for Pangean tectonic models. *Geological Society of America Abstracts with Programs*, *44*(4), 23.

Goldstein, A. G. (1989). Tectonic significance of multiple motions on terrane-bounding faults in the northern Appalachians. *Geological Society of America Bulletin*, *101*(7), 927–938.

Goscombe, B., Foster, D., Gray, D., & Wade, B. (2017). Metamorphic response and crustal architecture in a classic collisional orogen: Damara Belt, Namibia. *Gondwana Research*, *52*, 80–124.

Goscombe, B., Foster, D. A., Kelsey, D., Wade, B., Jiang, P., Gray, D., et al. (2022). Episodic intra-continental reactivation during collapse of a collisional orogen: The Damara Belt, Namibia. *Gondwana Research*, *109*, 285–375.

Goscombe, B., Gray, D., & Foster, D. A. (2018). Metamorphic response to collision in the Central Himalayan Orogen. *Gondwana Research*, *57*, 191–265.

Guergouz, C., Martin, L., Vanderhaeghe, O., Thebaud, N., & Fiorentini, M. (2018). Zircon and monazite petrochronologic record of prolonged amphibolite to granulite facies metamorphism in the Ivrea-Verbana and Strona-Ceneri Zones, NW Italy. *Lithos*, *308–309*, 1–18.

Gutierrez-Alonso, G., Murphy, J. B., Fernandez-Suarez, J., Weil, A. B., Franco, P., & Gonzalo, J. C. (2011). Lithospheric delamination in the core of Pangea: Sm-Nd insights from the Iberian mantle. *Geology*, *39*, 155–158.

Hatcher, R. D. (2010). The Appalachian orogen: A brief summary. In R. P. Tollo et al. (Eds.), *From Rodinia to Pangea: The lithotectonic record of the Appalachian region* (pp. 1–19). Geological Society of America Memoir 206.

Hatcher, R. D., Lemiszki, P. J., & Whisner, J. B. (2007). Character of rigid boundaries and internal deformation of the southern Appalachian foreland fold-thrust belt. In J. W. Sears et al. (Eds.), *Whence the mountains? Inquiries into the evolution of orogenic systems* (pp. 243–276). Geological Society of America Special Paper 433.

Hatcher, R. D., Thomas, W. A., Geiser, P. A., Snoke, A. W., Mosher, S., & Wiltschko, D. V. (1989). Alleghanian orogen. In R.D. Hatcher et al. (Eds.), The Appalachian-Ouachita orogen in the United States (Geology of North America, F-2) (pp. 233–318). Boulder, Colorado, Geological Society of America.

Heatherington, A. L., & Mueller, P. A. (1997). Geochemistry and origin of Florida crustal basement terranes. In A. F. Randazzo & D. S. Jones (Eds.), *The geology of Florida* (pp. 27–37). Gainesville: University Press of Florida.

Heatherington, A. L., Mueller, P. A., & Wooden, J. L. (2010). Alleghanian plutonism in the Suwannee terrane, USA: Implications for late Paleozoic tectonic models. In R. P Tollo et al. (Eds.), From Rodinia to Pangea: The lithotectonic record of the Appalachian region (pp. 607–620). Geological Society of America Memoir 206.

Heffner, D. M. (2013). *Tectonics of the South Georgia Rift*. Ph.D. Thesis. University of South Carolina, Columbia.

Hibbard, J., Stoddard, E., Secor, D., & Dennis, A. (2002). The Carolina Zone: Overview of Neoproterozoic to Early Paleozoic peri-Gondwanan terranes along the eastern flank of the southern Appalachians. *Earth-Science Reviews*, *57*, 299–339.

Hillenbrand, I. W., Williams, M. L., Li, C., & Gao, H. (2021). Rise and fall of the Acadian altiplano: Evidence for a Paleozoic orogenic plateau in New England. *Earth and Planetary Science Letters*, *560*, 116797.

Hnat, J. S., & van der Pluijm, B. A. (2014). Fault gouge dating in the Southern Appalachians, USA. *Geological Society of America Bulletin*, *126*, 639–651.

Holder, R. M., Viete, D. R., Brown, M., & Johnson, T. E. (2019). Metamorphism and the evolution of plate tectonics. *Nature*, *572*, 378–381.

Hopper, E., Fischer, K. M., Wagner, L. S., & Hawman, R. B. (2017). Reconstructing the end of the Appalachian orogeny. *Geology*, *45*, 15–18.

Houseman, G. A., McKenzie, D. P., & Molnar, P. (1981). Convective instability of a thickened boundary layer and its relevance for the thermal evolution of continental convergent belts. *Journal of Geophysical Research*, *86*, 6115–61321.

Jamieson, R. A., Unsworth, M. J., Harris, N. B., Rosenberg, C. L., & Schulmann, K. (2011). Crustal melting and the flow of mountains. *Elements*, *7*(4), 253–260.

Jamtveit, B., Dunkel, K. G., Petley-Ragan, A., Austrheim, H., Corfu, F. & Schmid, D. W. (2021). Rapid fluid-driven transformation of lower continental crust associated with thrust-induced shear heating. *Lithos*, *396–397*. https://doi.org/10.1016/j.lithos.2021.106216

Keppie, J. D., Dostal, J., Murphy, J. B., & Nance, R. D. (2008). Synthesis and tectonic interpretation of the westernmost Paleozoic Variscan orogen in southern Mexico: From rifted Rheic margin to active Pacific margin. *Tectonophysics*, *461*, 277–290.

Kneller, E. A., & Johnson, G. A. (2011). Plate kinematics of the Gulf of Mexico based on integrated observations from the Central and South Atlantic. *Gulf Coast Association of Geological Society Transactions*, *61*, 283–299.

Kneller, E. A., Johnson, C. A., Karner, G. D., Einhorn, J., & Queffelec, T. A. (2012). Inverse methods for modeling non-rigid plate kinematics: Application to Meoszoic plate reconstructions of the Central Atlantic. *Computers & Geosciences*, *49*, 217–230.

Krol, M. A., Muller, P. D., & Idleman, B. D. (1999). *Late Paleozoic deformation within the Pleasant Grove shear zone, Maryland: results from $^{40}Ar/^{39}Ar$ dating of white mica* (pp. 93–111). *Geological Society of America Special Paper 330*.

Laurent, O., Couzinié, S., Zeh, A., Vanderhaeghe, O., Moyen, J. F., Villaros, A., Gardien, V., & Chelle-Michou, C. (2017). Protracted, coeval crust and mantle melting during Variscan late-orogenic evolution: U-Pb dating in the eastern French Massif Central. *International Journal of Earth Sciences*, *106*(2), 421–451.

Layfield, N. T. (2009). *Permian to Cretaceous evolution of the Piedmont along the Alabama-Georgia coastal plain unconformity*. M.S. Thesis. Auburn University, Auburn, Alabama.

Lin, Q. (2015). Spatial-temporal distribution of the Alleghanian magmatism in the southern Appalachians: Implication of the assembly of supercontinent Pangaea. M.S. Thesis. University of Florida, Gainesville.

Lister, G. S., Etheridge, M. A., & Symonds, P. A. (1991). Detachment models for the formation of passive continental margins. *Tectonics*, *10*, 1038–1064.

Ma, C., Hames, W. E., Foster, D. A., Xiao, W., Mueller, P. A., & Steltenpohl, M. (2023). Transformation of eastern North America from compression to extension in the Permian-Triassic. In Whitmeyer, S. J., Williams, H. L., Kellett, D. A., and Tikoff, B. (Eds.), *Laurentia: Turning points in the evolution of a continent*. Geological Society of America Memoir 220, p. 1–16, doi.org/10.1130/2022.1220.

Ma, C., Foster, D. A., Hames, W. E., Mueller, P. A., & Steltenpohl, M. G. (2019). From the Alleghanian to the Atlantic: Extensional collapse of the southernmost Appalachian orogen. *Geology*, *47*(4), 367–370.

Maguire, T. J., Volkert, R. A., Swisher, C., III, & Sheridan, R. E. (2009). Evidence for an Alleghanian (Early Carboniferous to Late Permian) tectonothermal event in the New Jersey Coastal Plain basement from $^{40}Ar/^{39}Ar$ biotite data, geochemistry and gravity modeling. *Journal of Geodynamics*, *48*(1), 23–36.

Maher, H. D., Dallmeyer, R. D., Secor, D. T., & Sacks, P. E. (1994). $^{40}Ar/^{39}Ar$ constraints on chronology of Augusta fault zone movement and late Alleghanian extension, Southern Appalachian Piedmont, South Carolina and Georgia. *American Journal of Science*, *294*, 428–448.

Ménard, G., & Molnar, P. (1988). Collapse of a Hercynian Tibetan plateau into a late Palaeozoic European basin and range province. *Nature*, *334*, 235.

Marzen, R. E., Shillington, D. J., Lizarralde, D., & Harder, S. H. (2019). Constraints on Appalachian orogenesis and continental rifting in the southeastern United States from wide-angle seismic data. *Journal of Geophysical Research Solid Earth*, *124*(7), pp. 6625–6652.

McKenzie, D., Nimmo, F., Jackson, J. A., Gans, P. B., & Miller E. L. (2000). Characteristics and consequences of flow in the lower crust. *Journal of Geophysical Research, 105*, 11029–11046.

McQuarrie, N., & Wernicke, B. P. (2005). An animated tectonic reconstruction of southwestern North America since 36 Ma. *Geosphere, 1*, 147–172, https://doi.org/10.1130/GES00016.1

Miller, B. V., Fetter, A. H., & Stewart, K. G. (2006). Plutonism in three orogenic pulses, eastern Blue Ridge Province, southern Appalachians. *Geological Society of America Bulletin, 118*(1–2), 171–184.

Moecher, D. P. (1999). The distribution, style, and intensity of Alleghanian metamorphism in south-central New England: petrologic evidence from the Pelham and Willimantic domes. *The Journal of geology, 107*(4), 449–471.

Moecher, D. P., & Wintsch, R. P. (1994). Deformation-induced reconstitution and local resetting of mineral equilibria in polymetamorphic gneisses: tectonic and metamorphic implications. *Journal of Metamorphic Geology, 12*(4), 523–538.

Molnar, P., & Lyon-Caen, H. (1988). *Some simple physical aspects of the support, structure, and evolution of mountain belts*. Geological Society of America Special Paper 218, 179–207.

Mueller, P., Heatherington, A., Foster, D., &Wooden, J. (2011). Alleghanian granites of the southern Appalachian orogen: Keys to Pangean reconstructions. *Georgia Geological Society Guidebook, 31*, 29–48.

Mueller, P. A., Heatherington, A. L., Foster, D. A., Thomas, W. A., & Wooden, J. L. (2014). The Suwannee suture: Significance for Gondwana-Laurentia terrane transfer and formation of Pangaea. *Gondwana Research, 26*, 365–373.

Nelson, K. D., Zhao, W., Brown, L. D., Kuo, J., Che, J., Liu, X., et al. (1996). Partially molten middle crust beneath southern Tibet: Synthesis of project INDEPTH results. *Science, 274*(5293), 1684–1688.

Osmundsen, P. T., Péron-Pinvidic, G., & Bunkholt, H. (2020). Rifting of collapsed orogens: successive incision of continental crust in the proximal margin offshore Norway. *Tectonics*, e2020TC006283. https://doi.org/10.1029/2020TC006283

Petersen, T. A., Brown, L. D., Cook, F. A., Kaufman, S., & Oliver, J. E. (1984). Structure of the Riddleville basin from COCORP seismic data and implications for reactivation tectonics. *The Journal of Geology, 92*, 261–271.

Regis, D., Pehrsson, S., Martel, E., Thiessen, E., Peterson, T., & Kellett, D. (2021). Post-1.9 Ga evolution of the south Rae craton (Northwest Territories, Canada): a Paleoproterozoic orogenic collapse system. *Precambrian Research, 355*, 106105.

Reiners, P. W., & Brandon, M. T. (2006). Using thermochronology to understand orogenic erosion. *Annual Review of Earth and Planetary Science, 34*, 419–466.

Rey, P., Vanderhaeghe, O., & Teyssier, C. (2001). Gravitational collapse of the continental crust: Definition, regimes and modes. *Tectonophysics, 342*(3–4), 435–449.

Rey, P. F., Teyssier, C., & Whitney, D. L. (2009). Extension rates, crustal melting, and core complex dynamics. *Geology, 37*, 391–394. https://doi.org/10.1130/G25460A.1

Rey, P. F., Teyssier, C., & Whitney, D. L. (2010). Limit of channel flow in orogenic plateaux. *Lithosphere, 2*, 328–332. https://doi.org/10.1130/L114.1

Rivers, T. (2012). Upper-crustal orogenic lid and mid-crustal core complexes: signature of a collapsed orogenic plateau in the hinterland of the Grenville Province. *Canadian Journal of Earth Science, 49*, 1–42.

Rosenberg, C. L., & Handy, M. R. (2005). Experimental deformation of partially melted granite revisited: implications for the continental crust. *Journal of Metamorphic Geology, 23*, 19–28.

Royden, L. (1996). Coupling and decoupling of crust and mantle in convergent orogens: Implications for strain partitioning in the crust. *Journal of Geophysical Research, 101*, 17, 679–17, 705.

Rubio Pascual, F. J., Arenas, R., Martinez Catalan, J. R., Rodrigues Fernandez, L. R., & Wijbrans, J. R. (2013). Thickening and exhumation of the Variscan roots in the Iberian Central System: Tectonothermal processes and $^{40}Ar/^{39}Ar$ ages. *Tectonophysics, 587*, 207–221.

Sacks, P. E., Secor, D. T., Maher, H. D., Wright, J., & Dallmeyer, R. D. (1993). Early Alleghanian oblique dextral extension and magmatism along the Modoc fault zone, eastern Appalachian Piedmont, SC-GA. *Geological Society of America Abstracts with Programs, 44*(7), 286.

Samson, S. D., Coler, D. G., & Speer, J. A. (1995). Geochemical and Nd-Sr-Pb isotopic composition of Alleghanian granites of the southern Appalachians: Origin, tectonic setting, and source characterization. *Earth and Planetary Science Letters, 134*(3–4), 359–376.

Searle, M. P., Law, R. D., & Jessup, M. J. (2006). Crustal structure, restoration and evolution of the Greater Himalaya in Nepal-South Tibet: Implications for channel flow and ductile extrusion of the middle crust. In R. D. Law, et al. (Eds.), Channel Flow, ductile extrusion and exhumation in continental collision zones (pp. 355–378). Geological Society, London, Special Publication, 268.

Snoke, A. W., & Frost, B. R. (1990). Exhumation of high pressure pelitic schist, Lake Murray spillway, South Carolina: Evidence for crustal extension during Alleghanian strike-slip faulting. *American Journal of Science, 290*, 853–881.

Speer, J. A., McSween, H. Y., Jr., & Gates, A. E. (1994). Generation, segregation, ascent, and emplacement of Alleghanian plutons. *Journal of Geology, 102*, 249–267.

Spencer, B. M., Thigpen, J. R., Gallen, S. F., Dortch, J. M., Hodges, K. V., Law, R. D., et al. (2021). An evaluation of erosional-geodynamic thresholds for rapid orogenic denudation. *Journal of Geophysical Research Solid Earth, 126*, https://doi.org/10.1029/202JB022353

Steltenpohl, M. G., & Kunk, M. J. (1993). $^{40}Ar/^{39}Ar$ thermochronology and Alleghanian development of the southernmost Appalachian Piedmont, Alabama and southwest Georgia. *Geological Society of America Bulletin, 105*, 819–833.

Steltenpohl, M. G., Goldberg, S. A., Hanley, T. B., & Kunk, M. J. (1992). Alleghanian development of the Goat Rock fault zone, southernmost Appalachians: Temporal compatibility with the master decollement. *Geology, 20*, 845–848.

Steltenpohl, M. G., Hatcher, R. D., Mueller, P. A., Heatherington, A. L., & Wooden, J. L. (2010). Geologic history of the Pine Mountain window, Alabama and Georgia: Insights from a new geologic map and U-Pb isotopic dates. In R. P. Tollo et al. (Eds.), From Rodinia to Pangea: The lithotectonic record of the Appalachian region (pp. 837–857). Geological Society of America Memoir 206.

Steltenpohl, M. G., Mueller, P. M., Heatherington, A. L., Hanley, T. B., & Wooden, J. L. (2008). Gondwanan/peri-Gondwanan origin for the Uchee terrane, Alabama and Georgia: Carolina zone or Suwannee terrane (?) and its suture with Grenvillian basement of the Pine Mountain window. *Geosphere*, *4*, 131–144.

Stockli, D. F. (2005). Application of low-temperature thermochronometry to extensional tectonic settings. *Reviews in Mineralogy and Geochemistry*, *58*, 411–448.

Stowell, H. H., Schwartz, J. J., Ingram, S. B., III, Madden, J., Jernigan, C., Steltenpohl, M., et al. (2019). Linking metamorphism, magma generation, and synorogenic sedimentation to crustal thickening during Southern Appalachian mountain building, USA. *Lithosphere*, *11*(5), 722–749.

Toe, W., Vanderhaeghe, O., Andre-Mayer, A. S., Feybesse, J. L., & Milesi, J. P. (2013). From migmatites to granites in the Pan-African orogenic belt, Namibia. *Journal of African Earth Sciences*, *85*, 62–74.

Traverse, A. (1987). Pollen and spores date origin of rift basins from Texas to Nova Scotia as early Late Triassic. *Science*, *236*, 1469–1472.

Turlin, F., Deruy, C., Eglinger, A., Vanderhaeghe, O., Andre-Mayer, S., Poujol, M., et al. (2018). A 70 Ma record of supra-solidus conditions in the large, hot, long-deration Granville Orogen. *Terra Nova*, *30*, 233–243.

Vanderhaeghe, O. (2009). Migmatites, granites and orogeny: Flow modes of partially-molten rocks and magmas associated with melt/solid segregation in orogenic belts. *Tectonophysics*, *477*, 119–134.

Vanderhaeghe, O., & Teyssier, C. (2001). Crustal-scale rheological transitions during late-orogenic collapse. *Tectonophysics*, *335*, 211–228.

Vanderhaeghe, O., Laurent, O., Gardien, V., Moyen, J-F., Gebelin, A., Chelle-Michou, C., et al. (2020). Flow of partially molten crust controlling construction, growth and collapse of the Variscan orogenic belt: The geological record of the French Massif Central. *BSGF Earth Science Bulletin*, *191*, 1–56. https://doi.org/10.1051/bsgf/2020013

Vanderhaeghe, O., Medvedev, S., Fullsack, P. Beaumont, C., & Jamieson, R. A. (2003). Evolution of orogenic wedges and continental plateau: insights from crustal thermal-mechanical models overlying subducting mantle lithosphere. *Geophysical Journal International*, *153*, 27–51.

Wagner, L. S., Stewart, K., & Metcalf, K. (2012). Crustal-scale shortening structures beneath the Blue Ridge Mountains, North Carolina, USA. *Lithosphere*, *4*, 242–256.

Wernicke, B., Davis, J., Niemi, N. A., Luffi, P., & Bisnath, S. (2008). Active megadetachment beneath the western United States. *Journal of Geophysical Research*, *113*. https://doi.org/10.1029/2007/JB005375

Whitney, D. L., Teyssier, C., Rey, P., & Buck, W. R. (2013). Continental and oceanic core complexes. *Geological Society of America Bulletin*, *125*(3–4), 273–298.

Wiest, J. D., Wrona, T., Bauck, M. S., Fossen, H., Gawthorpe, R. L., Osmundsen, P. T., et al. (2020). From Caledonian collapse to North Sea rift: The extended history of a metamorphic core complex. *Tectonics*, *39*(10), e2020TC006178.

Wintsch, R. P., Sutter, J. F., Kunk, M. J., Aleinikoff, J. N., & Dorais, M. J. (1992). Contrasting P-T-t paths: Thermochronologic evidence for a Late Paleozoic final assembly of the Avalon Composite Terrane in the New England Appalachians. *Tectonics*, *11*(3), 672–689.

Wise, D. U., & Ganis, G. R. (2009). Taconic orogeny in Pennsylvania: A~ 15–20 my Apennine-style Ordovician event viewed from its Martic hinterland. *Journal of Structural Geology*, *31*(9), 887–899.

Withjack, M., Schlische, R., Malinconico, M., & Olsen, P. (2013). Rift-basin development: Lessons from the Triassic-Jurassic Newark Basin of eastern North America. *Geological Society, London, Special Publications*, *369*, 301–321.

Xie, C., Jin, S., Wei, W., Ye, G., Jing, J., Zhang, L., et al. (2021). Middle crustal partial melting triggered since the mid-Miocene in Southern Tibet: insights from magnetotelluric data. *Journal of Geophysical Research, Solid Earth*, *126*. https://doi.org/10.1029/2021JB022435

Yin, A. (2004). Gneiss domes and gneiss dome systems. In D. L. Whitney et al. (Eds.), Gneiss domes in orogeny (pp. 1–14). Boulder, Colorado, Geological Society of America Special Paper 380.

INDEX

Tables are indicated by **bold** page references. Figures are indicated by *italicized* page references.

Compressional Tectonics: Plate Convergence to Mountain Building, Geophysical Monograph 277, First Edition.
Edited by Elizabeth J. Catlos and İbrahim Çemen.

DOI:10.1002/9781119773856.index

www.ingramcontent.com/pod-product-compliance
Lightning Source LLC
LaVergne TN
LVHW082115200926
848484LV00008B/1956